Cover Illustration: The cover illustrates some of the aspects covered in this workshop. The interests of the participants ranged from presolar composition (illustrated by the famous "Hubble pillars") through solar composition (SOHO/EIT image), stellar evolution (SN1987A), to abundances in the galaxy (M100 background image). The workshop was held in the beautiful city of Bern, the silhouette of which defines the lower border of the cover. Credits: All individual images are not copyrighted. The "Hubble pillars", the image of M100, and of SN 1987A are by courtesy of STSCI. The image of the Sun is by courtesy of the SOHO/EIT consortium. SOHO is a project of international cooperation between ESA and NASA. The silhouette of the medieval city of Bern is by courtesy of Bern Tourism.

Related Titles from the AIP Conference Proceedings Subseries on Astronomy and Astrophysics

556 Explosive Phenomena in Astrophysical Compact Objects: First KIAS Astrophysics Workshop
Edited by Heon-Young Chang, Chang-Hwan Lee, Mannque Rho, and Insu Yi, March 2001, 1-56396-987-4

537 Waves in Dusty, Solar, and Space Plasmas
Edited by F. Verheest, M. Goossens, M. A. Hellberg, and R. Bharuthram, October 2000, 1-56396-962-9

528 Acceleration and Transport of Energetic Particles Observed in the Heliosphere: ACE 2000 Symposium
Edited by Richard A. Mewaldt, J. R. Jokipii, Martin A. Lee, Eberhard Möbius, and Thomas H. Zurbuchen, July 2000, 1-56396-951-3

522 Cosmic Explosions: Tenth Astrophysics Conference
Edited by Stephen S. Holt and William W. Zhang, June 2000, 1-56396-943-2

499 Small Missions for Energetic Astrophysics: Ultraviolet to Gamma-Ray
Edited by Steven P. Brumby, December 1999, 1-56396-912-2

471 Solar Wind Nine: Proceedings of the Ninth International Solar Wind Conference
Edited by Shadia Rifai Habbal, Ruth Esser, Joseph V. Hollweg, Philip A. Isenberg, May 1999, 1-56396-865-7

402 Astrophysical Implications of the Laboratory Study of Presolar Materials
Edited by Thomas J. Bernatowicz and Ernst Zinner, 1997, 1-56396-664-6

To learn more about these titles, or the AIP Conference Proceedings Series, please visit the webpage **http://proceedings.aip.org**

SOLAR AND GALACTIC COMPOSITION

A Joint SOHO/ACE Workshop

Bern, Switzerland *6-9 March 2001*

EDITOR
Robert F. Wimmer-Schweingruber
University of Bern, Switzerland

CD-ROM INCLUDED

Melville, New York, 2001
AIP CONFERENCE PROCEEDINGS ■ VOLUME 598

Editor:

Robert F. Wimmer-Schweingruber
Physikalisches Institut
Universität Bern
Sidlerstrasse 5
CH-3012 Bern
SWITZERLAND

E-mail: robert.wimmer@phim.unibe.ch

L.C. Catalog Card No. 2001096654
ISBN 0-7354-0042-3
ISSN 0094-243X

Printed in the United States of America

CONTENTS

OVERVIEW

SOLAR AND SOLAR SYSTEM ABUNDANCES

CORONAL AND SOLAR WIND COMPOSITION

ENERGETIC PARTICLE COMPOSITION

GALACTIC COMPOSITION

FRACTIONATION, ACCELERATION, AND TRANSPORT PROCESSES

APPLICATIONS AND CONSTRAINTS ON COSMOCHEMISTRY

Preface and Thanks

Somebody somewhere and sometime in 1998 had the idea to hold a workshop that would join the various communities concerned with the composition of the Sun and its surroundings. It should unite workers from two very active space missions, the SOlar and Heliospheric Observatory (SOHO)[1] and the Advanced Composition Explorer (ACE)[2]. It should not be limited to scientists analyzing and interpreting data gathered with instruments on these two spacecraft, but be open to the wider community. The idea was to have input into our work from scientists not related to SOHO or ACE - people who would tell us what the "outside" (scientific) world expected from us. We chose the workshop title accordingly - "Solar and Galactic Composition" is attractive to workers in a wide range of fields.
We began preparing the workshop in late 1998, a first meeting with a preliminary Scientific Organizing Committee (SOC) took place in May 1999. A final SOC was formed soon thereafter, with the following members:

P. Bochsler (chair), University of Bern, Switzerland
A. C. Cummings, Caltech, USA
H. Kunow, University of Kiel, Germany
G. M. Mason, University of Maryland, USA
H. Mason, University of Cambridge, UK
J.-C. Vial, IAS, Université de Paris XI, France
M. E. Wiedenbeck, JPL/Caltech, USA
K. Wilhelm, Max-Planck-Institut für Aeronomie, Lindau, Germany
R. F. Wimmer-Schweingruber, University of Bern, Switzerland
T. H. Zurbuchen, University of Michigan, USA

A steadily growing Local Organizing Committee soon became active at the Physikalisches Institut of the University of Bern. We are indebted to the following members for their work:

Frédéric Allegrini
Frédéric Buclin
Erwin Flückiger
Edith Hertig
Markus Hohl
Karine Issautier
James Weygand
Louise Wilson
Robert F. Wimmer-Schweingruber (chair)
Sylvia Wirth
Peter Wurz

[1]SOHO is a project of international collaboration between the European Space Agency (ESA) and its American counterpart, NASA. It was launched on December 2, 1995

[2]ACE is a NASA mission, launched on August 27, 1997.

The workshop was generously sponsored by the following institutions, without their help we could not have held the workshop:

European Space Agency
Schweizerische Akademie der Naturwissenschaften (SANW)
Schweizerischer Nationalfonds zur Förderung der Wissenschaft
Oerlikon Contraves Space SA
Physikalisches Institut, University of Bern

The workshop was hosted by the Physikalisches Institut of the University of Bern in the building "Exakte Wissenschaften". We thank its director, H. Balsiger for the hospitality and M. Niederhäuser, head of house services for his help.
All papers in this volume were refereed by one or two referees, many of them not participants of the workshop. We thank the following referees for their services in evaluating the papers submitted for these proceedings: (in parenthesis the number of evaluated papers)

R. Binns	F. Bühler (2)	C. M. S. Cohen (2)
N. Crooker	W. Curdt	H. Debrunner
L. Desorgher	P. Eberhardt	D. Ellison
O. Eugster	B. Fleck	G. Gloeckler
P. K. F. Grieder	H. Grünwaldt	B. Heber (2)
V. Heber	M. Hilchenbach	G. Ho (2)
M. Hofer	H. Holweger	K. Issautier
R. Kallenbach	E. Kallio	B. Klecker
Y.-K. Ko	H. Kucharek	M. Laming
D. Lario	M. A. Lee	R. A. Leske (2)
Y. Litvinenko	S. Lorenzetti	G. M. Mason (2)
H. Mason	F. Matteucci	R. A. Mewaldt
E. Möbius	R. Müller-Mellin	C. Ng
K. Ogilvie	U. Ott	B. P. Pagel
J. Paquette (2)	R. Pepin	M. Popecki
K. Pretzl	J. Raymond	D. Reames (2)
J. A. le Roux	N. A. Schwadron	P. Slocum
S. Suess	F.-K Thielemann	C. Tranquille
S. Turcotte	A. Tylka	M. E. Wiedenbeck
R. Wieler	R. Wiens	M. Wieser
K. Wilhelm	R. F. Wimmer-Schweingruber (5)	J. Withby
P. Wurz (2)	N. Yanasak	P. Young

Finally, thanks go to Charles Doering and Sabine Kessler of the publishing house, AIP, for their patience with all of us.

Robert F. Wimmer-Schweingruber   Peter Bochsler

Group Photograph

and

Identification

1. R. von Steiger
2. M. Popecki
3. L. Wilson
4. R.C. Wiens
5. J.E. Mazur
6. Y.E. Litvinenko
7. R. F. Wimmer-Schweingruber
8. A. Reinard
9. U. Sofia
10. F. Buclin
11. C.M.S. Cohen
12. R. A . Leske
13. F.M. Ipavich
14. D. Spadoro
15. E. Antonucci
16. A.J. Lazarus
17. M. Laming
18. H. Kimura
19. E.R. Christian
20. N.E. Yanasak
21. K.W. Ogilvie
22. J.C. Vial
23. B. Fleck
24. O.K. Manuel
25. J.S. George
26. W.R. Binns
27. P. K. Wenzel
28. S. Yusainee
29. M.E. Wiedenbeck
30. B. Klecker
31. D. Reisenfeld
32. G. Del Zanna
33. P. Bochsler
34. H. Kunow
35. J.W. Keller
36. S. Giordano
37. J.C. Raymond
38. A.C. Cummings
39. S. Suess
40. E. Landi
41. D.V. Reames
42. G. Poletto
43. L. Zangrilli
44. E. Möbius
45. R.A. Mewaldt
46. L. Teriaca
47. J.A. Paquette
48. K. Wilhelm
49. M. Den
50. C. Chiappini
51. S. Parenti
52. P. Hoppe
53. M.Y. Hofer
54. J. Rodriguez-Pacheco
55. C. Tranquille
56. H. Shimazu
57. P. Wurz
58. R. Wieler
59. G.C. Ho
60. J. Sequeiros-Ugarte
61. M. Desai
62. G.M. Mason
63. F.C. Jones
64. M. Chaussidon
65. L. Abbo
66. V. Heber
67. P.L. Slocum
68. Y.-K. Ko
69. H. Busemann
70. E. Salerno
71. J.R. Jokipii
72. T.T. von Rosenvinge
73. M. Hohl
74. S. Wirth
75. F.-K. Thielemann
76. K. Issautier
77. H. Holweger

Not on picture:

F. Allegrini, K. Altwegg, K. Bamert,
V. Bothmer, F. Bühler, P. Eberhardt,
E.O. Flückiger, G. Gloeckler, P. Grieder,
E. Hertig, R. Kallenbach, D. Kirilova,
L. Kocharov, O. Kryakunova, V.E. Timofeev,
I.S. Veselovsky, T.H. Zurbuchen

1
2
3
4
5
6
7
8
9
10
11
12
13
14
15
16
17
18
19
20
21
22
23
24
25
26
27
28
29
30
31
32
33
34
35
36
37
38
39
40
41
42
43
44
45
46
47
48
49
50
51
52
53
54
55
56
57
58
59
60
61
62
63
64
65
66
67
68
69
70
71
72
73
74
75
76
77

Overview

Solar and Galactic Composition

Robert F. Wimmer-Schweingruber

Physikalisches Institut, Universität Bern, Sidlerstrasse 5, CH-3012 Bern, Switzerland

Abstract. This is a personal overview of our joint SOHO/ACE workshop "Solar and Galactic Composition" held in Bern, Switzerland, between the 6th to the 9th of March, 2001. The overview is not complete, although I have tried to capture the spirit of the workshop. We were all confronted with results from fields we were unfamiliar with. This broadened our horizons and showed us that our work is important to other - sometimes unexpected - fields. Linking solar and galactic composition was an exciting experience.

INTRODUCTION

How representative of the cosmic abundances are solar-system abundances? What are the cosmic abundances? And how well do we know solar and solar-system abundances? How do we measure abundances, and what processes can alter them? What can we learn from composition measurements? A set of questions I am certain all of us have pondered. The Joint SOHO/ACE workshop "Solar and Galactic Composition" addressed these questions in a fashion that each of us could not do on his own. We managed to unite workers in as diverse fields as solar remote-sensing and in-situ studies, meteoritics, stellar, interstellar, and cosmic-ray physics, as well as galactic chemical evolution and Big Bang nucleosynthesis. This workshop was an attempt to put the work of all of us into a broader context and to help us understand what workers in other fields expect to learn from us.

I will not attempt to give an objective summary of this workshop. I freely admit that I can't. While I had many questions at the beginning of the workshop, I believe we have uncovered many more. Relating solar and galactic composition was like poking into an ants nest.

We all knew that composition studies are important. Some of us used abundance measurements to trace physical processes, others used them to study our origins. Both subjects are interesting, fascinating, and could occupy us for a substantial fraction of our lives. But how relevant are they? Does a better understanding of fractionation effects in gradual solar energetic particle events tell us any more than just that? Do measurements of the composition of the local interstellar cloud tell us more than just that? After all, it is extremely local, viewed on a galactic scale! Nevertheless, we found that, yes, solar composition is important, even on a "galactic scale"

Let me illustrate some of the implications of solar abundance determinations. The solar system formed from the interstellar medium as it existed at that time, $(4.57 \pm 0.02) \times 10^9$ years ago. It formed in a fragment of a giant molecular cloud, maybe similar to one of the Hubble "pillars". We know from certain isotopic anomalies preserved in presolar grains (see Hoppe [1], this volume) that various nucleosynthetic sources such as AGB stars, Wolf-Rayet stars (see [2]), or supernovae contributed material to the "stuff we're made of". As we gradually begin to understand more and more about how our solar system evolved, we see more and more clearly that this contribution may not be accidental.

It is well possible that we would not exist if the massive, short-lived stars that formed in the same galactic neighborhood as the solar system had not halted the accretion of the protoplanetary disk by the Sun in time for the Earth (and the other planets) to survive the solar "cannibalism". Observations of a multitude of young stars has shown that their surrounding protoplanetary disks are often subject to intense UV irradiation of neighboring young, massive stars [3, 4]. Gas drag on the forming planetesimals and protoplanets makes them gradually spiral nearer to the central star. Depending on the time of gas removal, planets can survive in that stellar system, or they are engulfed by the star they orbited. At least for the outer parts of protoplanetary disks, the evaporation times by photoionization are short compared to their viscous lifetimes [4]. This picture has consequences for the surface composition of stars. Butler *et al.* [5] found that planet-bearing stars had a higher metallicity than the Sun, see Figure 1. So far, only stars with at least Saturn-sized companions have been detected [5]. Smaller planets cannot presently be detected because their influence on the central star is too small. Because giant gas planets only form beyond the ice line, their detection

CP598, *Solar and Galactic Composition*, edited by R. F. Wimmer-Schweingruber

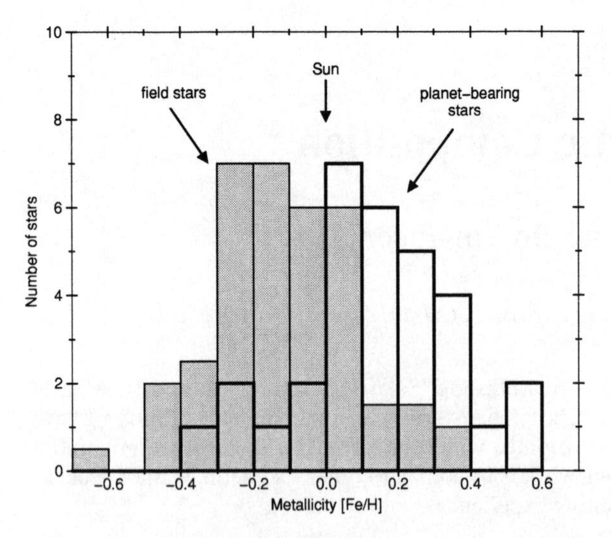

FIGURE 1. Comparison of the metallicity of planet-bearing and field stars. Data are from Butler *et al.* [5]. Plotted are the number of field stars (shaded, thin line) and planet-bearing stars (empty, thick line) in a given metallicity bin. Metallicities are normalized to solar and given on a logarithmic scale (dex).

closer to their star means that they have migrated inward. And any Earth-like, rocky planet must have migrated too far... Planet-bearing stars also show a striking surface enhancement of Li [6], an element that is largely destroyed by nucleosynthetic processes in the early phases of the evolution of Sun-like stars. That Li is depleted in the solar photosphere would then imply that the Sun did not "eat up" a substantial amount of planetary matter after the formation of an outer convective zone.

Let us now consider the solar metallicity. We find that it lies on the high side of field stars (0.2 dex, i. e. 58%, higher than the median), and near the median of planet-bearing stars. Did the Sun "eat up" some small planets in its youth? Or is the Sun just simply metal-rich - the well known "old Sun" problem? The latter could indeed be the case. Low-metallicity clouds probably have less dust in them than high-metallicity clouds, simply because the metals condense more easily than H and He. Dust is very likely an important ingredient in forming planetary systems [see e. g. 7, and references therein].

This workshop brought a new twist to the "old Sun" problem. The presently most detailed considerations of the solar abundances of some key elements were reported by Holweger [8]. These new determinations of solar abundances that take into account non-equilibrium and granulation effects seem to indicate that the Sun is quite typical of the neighboring stars at this galactocentric distance. For instance, the newly recommended oxygen abundance is nearly 0.2 dex lower that that given in the classical work of Anders and Grevesse [9]. Iron is 0.22 dex less abundant in the photosphere than previously thought. So the Sun lies right at the median of field stars and below the median of planet-bearing stars - solving the "old Sun" problem[1]. This should serve to illustrate that solar abundances have implications beyond just finding some reference abundances against which to compare others. Understanding our history is quite intimately linked to solar (and galactic) composition.

SOLAR ABUNDANCES

Let us continue to investigate solar abundances. Most of the mass in the solar system is concentrated in the Sun. As such, it is *the* reference against which solar-system abundances must be measured. Consequently, many different methods have been used to determine solar composition. Photospheric solar abundance determinations aren't easy. Holweger [8] (this volume) gives an account of some of the pitfalls such as non-equilibrium and granulation effects. For chromospheric or coronal abundance determinations, in addition to instrumental uncertainties, the charge-state composition of the solar plasma must be taken into account, moreover, the atomic properties governing line emission and absorption are not known with sufficient precision (See e. g. Del Zanna *et al.*, [13], Raymond *et al.* [14], von Steiger *et al.* [15], all in this volume). All of these uncertainties are on the order of 20% - 30% or more. Adding quadratically we arrive at uncertainties of about 0.2 dex - the difference between field and planet-bearing stars.

As Busemann *et al.* [16] show, meteoritic and solar abundances agree remarkably well - well within the discussed photospheric abundance uncertainties. This is illustrated by Figure 2. On the right-hand side it compares photospheric and (CI) meteoritic abundances. The logarithm of the ratio of their abundances (in the dex notation) is plotted versus 50% condensation temperature. While some of the uncertainties are appreciable, the overall agreement is good. The left-hand side shows a histogram of the right-hand panel. It is well fit by a Gaussian with standard deviation of 7% (dex). How good are the meteoritic abundances as a proxy for solar composition? Elemental abundances in meteorites are different in the various mineral separates in different meteorites. Possibly they have been altered by aqueous processes or, in the differentiated meteorites, due to processes accompanying differentiation of their parent bodies. The most primitive meteorites, the CI chondrites have retained many of their volatile elements and are the most volatile-rich bodies in the solar system short of the Sun

[1] Note however, that it still exists for isotopes. For most elements, the Sun is isotopically lighter than the interstellar medium[Compare e. g. 10, 11, 12].

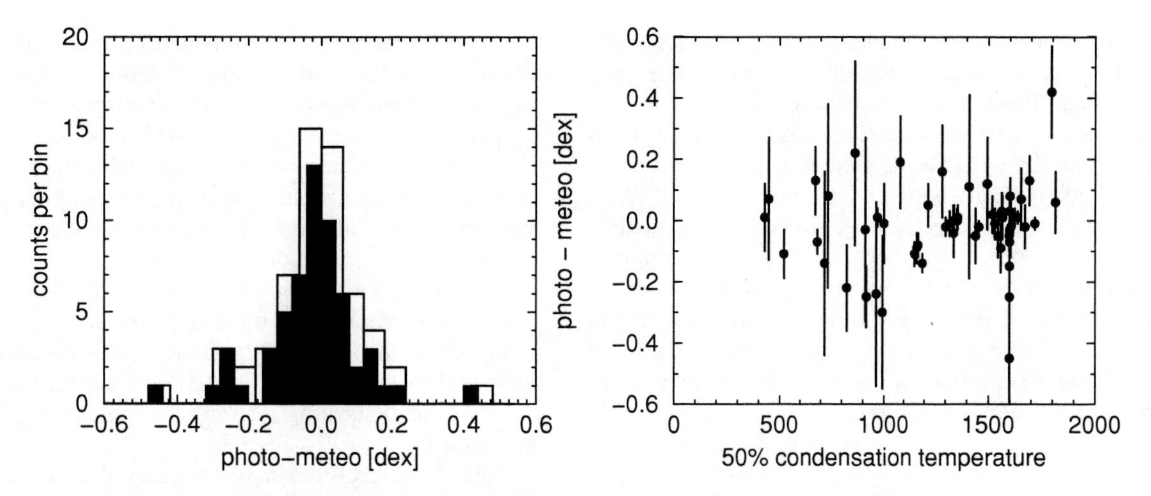

FIGURE 2. Comparison of photospheric and (CI) meteoritic element abundances. The right-hand panel shows the logarithm of the ratio between photospheric and meteoritic element abundances plotted versus 50% condensation temperature. The agreement between the two different determinations of solar abundances is good. This is illustrated by the histogram in the left panel. It is well fitted by a Gaussian with $\sigma_{dex} = 7\%$.

and comets. Therefore, we presume that they have preserved a faithful sample of the composition of the presolar nebula. However, as all classes of meteorites, they have lost their primordial noble gas inventory; moreover, because they have retained a large amount of volatiles such as water, they are especially susceptible to aqueous alteration. Furthermore, CI chondrites are an extremely rare class of meteorites[2]. When we speak of element abundances based on CI chondrite*s* (in the plural), we tend to ignore the fact that these abundances are largely based on the analysis of one single CI chondrite, Orgueil. Only few others have fallen onto Earth (Alais, Ivuna, Revelstoke, and Tonk) and they often were very small or little has been preserved[3]. Hence they have not been analysed as extensively as Orgueil. To put it bluntly, we base our solar system abundances on the analysis of one single rock. The remarkable agreement between solar (photospheric and coronal and solar wind (after some corrections)) abundances and meteoritic abundances shows that this rock was wisely chosen. Nevertheless, with the enormous progress being made in the determinations of abundances from remote-sensing and in-situ observations, we will soon have to begin investigating possible differences.

The composition of the Sun can also be measured more directly, by acquiring a sample of it. The Sun emits a constant stream of particles in the form of the solar wind (at energies of about 1 keV/nuc). Their abundances can be used to measure solar isotope and element composition after correction for fractionation effects. These affect composition mostly by first ionization potential or time, but possibly also by other atomic properties. Lunar soils contain an archive of the ancient solar wind (see e. g. Heber *et al.*, [18], this volume). Solar energetic particles are accelerated to much higher energies out of the solar atmosphere, corona, and even the solar wind. Their composition can be measured with high accuracy, but again needs to be corrected for fractionation effects (see e. g. Reames, [19] this volume).

Helioseismology has considerably improved the knowledge about the internal structure of the Sun. From observations of solar oscillations it is possible to derive detailed knowledge of the solar interior [see, e.g., 20]. Standard solar models today include the effects of element diffusion [see, e.g., 21] which leads to elemental and isotopic segregation. Models including elemental segregation exhibit the best agreement with parameters derived from helioseismological data. Element and isotopic segregation are due to a competition between two processes. Minor species in the solar interior are influenced by various external forces, for example, gravity, radiation pressure, etc. The particles also collide with each other and scatter randomly. This competition between external forces and random scattering leads to elemental segregation [22]. Consistent solar evolution models which include diffusion and radiative acceleration effects (apart from gravitational settling) predict a depletion of the heavy elements at the present day solar surface of the order of 10% or 0.04 dex [23]. This is less than the 0.07 dex seen in Figure 2 and we are not yet in a position to verify this theoretical prediction by

[2] Only 5 CI out of 22507 authenticated meteorites listed in the catalogue by Grady [17].

[3] While more than 10 kg of Orgueil have been preserved, onlyless than about 90 g have been preserved of the originally recovered 6 kg of Alais. 704.5 g of Ivuna were recovered, 1 g of Revelstoke, and 7.7 g of Tonk.[17]

direct abundance measurements. Future mission, such as Genesis, launched on the 8th of August, 2001, may yield data with sufficient precision to settle this question. However, the 0.04 dex depletion of the heavy elements is also about the precision of meteoritic abundance determinations (typically between 0.03 and 0.05 dex). While these are not limited by badly known atomic or possibly instrumental properties, there is a limit due to possible selection biases. Obviously, in spite of the proximity of the Sun, we do not yet know its composition with an accuracy that is sufficient to confidently relate it to surrounding stars with a precision that is relevant to answer some of our questions..

THE SUN AND THE INTERSTELLAR MEDIUM

As the Sun moves through the galactic interstellar medium, it encounters different environments. Neutral particles enter the heliosphere where they can be measured (Salerno *et al.*, [24] this volume). Thanks to the large collecting area of the COLLISA experiment [24] these workers obtain a rather precise value for the cosmologically important ratio ^{3}He/^{4}He. On its way through the heliosphere, the interstellar neutral gas is ionized and can then be measured as pick-up ions (Gloeckler and Geiss, [25] this volume) or, after a further stage of acceleration (see Jones *et al.*, [26]), as anomalous cosmic rays. Lunar soils may have preserved an archive of the varying galactic environment the solar system passed through during its history (Wimmer-Schweingruber and Bochsler, [27] this volume).

The Sun and the interstellar medium have evolved since the time the solar system formed. Nucleosynthesis (see Thielemann *et al.*, [28] this volume) in stars and supernovae contributes freshly synthesized material to the interstellar medium. The composition of the galaxy is slowly changed by stellar nucleosynthesis. It's evolution can be modeled (Chiappini and Matteucci, [29] this volume) and measured (see e. g. Sofia, [30] this volume) with new instrumentation. Galactic cosmic rays form another sample of this medium (see e. g. Wiedenbeck *et al.*, [31] and the review by Klecker *et al.*, [32] in this volume).

Remarkably, the local interstellar medium if anything is less metallic than the Sun, as reported by Gloeckler and Geiss [25]. This is also seen in galactic cosmic rays [31] and in most elements in nearby interstellar clouds [30]. Given this observational agreement, it does make sense to speak of galactic abundances. However, we expect that Big Bang nucleosynthesis, the Sun, and the interstellar medium form a temporal sequence composition wise [33]. We expect BBN to give us the primordial composition, the Sun to be representative of the composition of the interstellar medium some 4.6 billion years ago, and the local interstellar medium to be representative of the present-day interstellar medium [33]. So the reported observations make us wonder wonder what happened to galactic chemical evolution in the past 4.6 billion years. Was there essentially no evolution, or is the Sun currently in a less evolved region of the galaxy?

This puzzle is augmented by the observations of Binns *et al.* [2] who determined the isotope abundance ratio of ^{22}Ne/^{20}Ne in the source for galactic cosmic rays (GCR). They find a value which is five times higher the the solar (wind) value, indicating a strong contribution by Wolf-Rayet stars.

Other work reported here indicates that the source of the GCR is interstellar dust grains. GCRs also show fractionation by first ionization potential or some related property. According to George *et al.* [34], this controlling parameter is volatility, implying that interstellar dust is the origin of the GCR, as has been proposed by Meyer *et al.* and Ellison *et al.* [35, 36].

COMPOSITION AS A TRACER

One of the many uses of composition as a tracer is illustrated by Figure 1 in Reinard *et al.* [37] in these proceedings. Solar wind composition is a good indicator of the coronal origin of the solar wind. In this figure, anomalously high iron charge states indicate the passage material ejected in a coronal mass ejection. In the solar wind in the ecliptic plane coronal mass ejections often have elemental composition very similar to the slow solar wind, with the exception of He/H, which is sometimes anomalously high.

Charge-state and elemental composition in CMEs are determined by very different mechanisms on quite different time scales. Elements are fractionated by some process which fractionates according to their first ionization potential and is probably intimately linked to the loop-like structure of the plasma prior to onset of the CME. Hence elements have a long time, on the order of the life time of the loop, to be fractionated. Observations by Feldman *et al.* [38] show that the FIP bias in loops increases linearly with time, ranging from no FIP bias in emerging loops up to a factor of about 10. This allows us to determine the average life time of the loops which turns out to be on the order of a few days. The observation of Neukomm [39] that CMEs in the high-speed solar wind sampled out of the ecliptic plane by Ulysses have composition virtually indistinguishable of that of the fast solar wind tells us that there the loop life times must be shorter, on the order of a day or less. A similar story seems to hold for the mass fractionation observed in a special class of CMEs [40, 41].

Ko and coworkers [42] compare solar wind abundances with measurements in the solar corona using 3-dimensional MHD models. They find a likely correlation of electron temperatures and elemental abundances in the observed coronal region with those in the solar wind. This may imply that the slow wind originates in regions with mixing of closed field lines which are penetrated by open field lines. As the open field line moves through the close region, the resulting reconnection would then liberate the confined plasma, feeding the slow solar wind [e. g. 43].

The work by Antonucci and Giordano [44] as well as of Parenti *et al.* [45] confirm the picture that the slow wind originates in the legs of streamers [46, 47]. Streamers have long been known to be the source of the slow solar wind [48]. They are often associated with crossings of the heliospheric current sheet [49]. Solar wind measurements show a strong depletion of He/H at the sector boundaries [49] accompanied by a depletion of He/O [46, 47, 50]. Together, the optical and in-situ observations suggest that the solar wind flows out from the "legs" of the streamer, and does not originate at the top of streamers. On it's way to interplanetary space, this type of slow wind is additionally fractionated by inefficient Coulomb drag [51, 52].

The work of Morris *et al.* [53] is another example of the use of composition as a tracer. They observed that at the medium energies observed with ACE/SEPICA the ratio of singly-charged to doubly-charge He increases from the beginning to the end of corotating events. They interpret this as due to the changing magnetic connection of the spacecraft and the corotating interaction region. The later on in the event, the farther away lies the connection point and the larger is the contribution of accelerated interstellar pick-up helium ions.

Let me end this section on the use of composition as a tracer with the word of caution that we heard from Don Reames at this workshop. His work [19] shows that variations in the elemental abundances of energetic particles do not always imply new processes or new sources. Often, they can be explained by one single (but not necessarily simple) process, in this case wave-particle interaction between the heavy ions and the proton-generated waves generated in the vicinity of interplanetary shocks. That the same properties are not seen at higher energies [see the paper by von Rosenvinge *et al.*, 54] is simply a consequence of the conservation of the number of particles.

CONCLUSIONS

A key result of this workshop is that solar abundances are important to workers in a wide range of fields. A change in solar abundances can strongly influence interpretations in vastly different questions than what they were originally determined for. We should remember this point when we consider audiences for future publications..

The workshop lasted for four long days, full of hard work, as many of you will agree. Of course, this is a short time given the background of medieval city of Bern (founded 1191 A. D.). Yet even this time span pales against the time scales governing the evolution of the Sun, solar system, and the galaxy.

ACKNOWLEDGMENTS

I thank the participants of the workshop for a very stimulating time and many interesting discussions. This work was supported by the Schweizerischer Nationalfonds and the Kanton Bern.

REFERENCES

1. Hoppe, P., "Elemental and Isotopic Abundances in Meteorites", in *Solar and Galactic Composition*, edited by R. F. Wimmer-Schweingruber, AIP conference proceedings, Woodbury, NY, 2001, this volume.
2. Binns, W. R., Wiedenbeck, M. E., Christian, E. R., Cummings, A. C., George, J. S., Israel, M. H., Leske, R. A., Mewaldt, R. A., Stone, E. C., von Rosenvinge, T. T., and Yanasak, N. E., "GCR Neon Isotopic Abundances: Comparison with Wolf-Rayet Star Models and Meteoritic Abundances", in *Solar and Galactic Composition*, edited by R. F. Wimmer-Schweingruber, AIP conference proceedings, Woodbury, NY, 2001, this volume.
3. O'Dell, C. R., and Beckwith, S. V. W., *Science*, **276**, 1355 – 1359 (1997).
4. Hollenbach, D. J., Yorke, H. W., and Johnstone, D., "Disk Dispersal Around Young Stars", in *Protospars and Plantes IV*, edited by V. Mannings, A. P. Boss, and S. S. Russell, University of Arizona Press, Tucson, Arizona, 2000, pp. 401 – 428.
5. Butler, R. P., Vogt, S. S., Marcy, G. W., Fischer, D. A., Henry, G. W., and Apps, K., *Astrophys. J.*, **545**, 5–4 – 511 (2000).
6. Israelian, G., Santos, N. C., Mayor, M., and Rebolo, R., *Nature*, **411**, 163 – 166 (2001).
7. Lissauer, J. J., *Icarus*, **114**, 217 – 236 (1995).
8. Holweger, H., "Photospheric Abundances: Problems, Updates, Implications", in *Solar and Galactic Composition*, edited by R. F. Wimmer-Schweingruber, AIP conference proceedings, Woodbury, NY, 2001, in press.
9. Anders, E., and Grevesse, N., *Geochim. Cosmochim. Acta*, **53**, 197 – 214 (1989).

10. Wimmer-Schweingruber, R. F., Bochsler, P., and Wurz, P., "Isotopes in the solar wind: New results from ACE, SOHO, and WIND", in *Solar Wind Nine*, edited by S. R. Habbal, R. Esser, J. V. Hollweg, and P. A. Isenberg, American Institute of Physics, 1999, pp. 147 – 152.
11. Kallenbach, R., "Isotopic Composition Measured In-Situ in Different Solar Wind Regimes by CELIAS/MTOF on board SOHO", in *Solar and Galactic Composition*, edited by R. F. Wimmer-Schweingruber, AIP conference proceedings, Woodbury, NY, 2001, this volume.
12. Wilson, T. L., and Rood, R. T., *Annu. Rev. Astron. Astrophys.*, **32**, 191 – 226 (1994).
13. Del Zanna, G., Bromage, B. J. I., and Mason, H. E., "Elemental abundances of the low corona as derived from SOHO/CDS observations", in *Solar and Galactic Composition*, edited by R. F. Wimmer-Schweingruber, AIP conference proceedings, Woodbury, NY, 2001, this volume.
14. Raymond, J. C., Mazur, J. E., Abbo, L., Allegrini, F., Antonucci, E., Zanna, G. D., Den, M. H., Desai, M. I., Giordano, S., Ho, G., Ko, Y.-K., Landi, E., Lazarus, A., Mason, G., Ogilvie, K. W., Parenti, S., Poletto, G., Reinard, A., Rodriguez-Pacheco, J., Spadaro, D., Teriaca, L., Wurz, P., Zangrilli, L., and Zuccarello, F., "Coronal Abundances", in *Solar and Galactic Composition*, edited by R. F. Wimmer-Schweingruber, AIP conference proceedings, Woodbury, NY, 2001, this volume.
15. von Steiger, R., Vial, J.-C., Chaussidon, M., Cohen, C. M. S., Fleck, B., Heber, V. S., Holweger, H., Issautier, K., Lazarus, A. J., Ogilvie, K. W., Paquette, J. A., Reisenfeld, D. B., Teriaca, L., Wilhelm, K., Yusainee, S., Laming, J. M., and Wiens, R. C., "Measuring Solar Abundances", in *Solar and Galactic Composition*, edited by R. F. Wimmer-Schweingruber, AIP conference proceedings, Woodbury, NY, 2001, this volume.
16. Busemann, H., Altwegg, K., Binns, W. R., Chiappini, C., Gloeckler, G., Hoppe, P., Kirilova, D., Leske, R. A., Manuel, O. K., Mewaldt, R. A., Möbius, E., Suess, S. T., Wieler, R., Wiens, R. C., Wimmer-Schweingruber, R. F., and Yanasak, N. E., "Applications of abundance data and requirements for cosmochemical modeling", in *Solar and Galactic Composition*, edited by R. F. Wimmer-Schweingruber, AIP conference proceedings, Woodbury, NY, 2001, this volume.
17. Grady, M. M., *Catalogue of Meteorites*, Cmbridge University Press, 2000.
18. Heber, V. S., Baur, H., and Wieler, R., "Solar Krypton and Xenon in gas-rich Meteorites: New Insights into a Unique Archive of the Solar Wind", in *Solar and Galactic Composition*, edited by R. F. Wimmer-Schweingruber, AIP conference proceedings, Woodbury, NY, 2001, in press.
19. Reames, D. V., "Energetic Particle Composition", in *Solar and Galactic Composition*, edited by R. F. Wimmer-Schweingruber, AIP conference proceedings, Woodbury, NY, 2001, this volume.
20. Gough, D. O., Leibacher, J. W., Scherrer, P. H., and Toomre, J., *Science*, **272**, 1281 – 1282 (1996).
21. Christensen-Dalsgaard, J., Däppen, W., Ajukov, S. V., Anderson, E. R., Antia, H. M., Basu, S., Baturin, V. A., Berthomieu, G., Chitre, S. M., Cox, A. N., Demarque, P., Donatowicz, J., Dziembowski, W. A., Gabriel, M., Gough, D. O., Guenther, D. B., Guzik, J. A., Harvey, J. W., Hill, F., Houdeck, G., Iglesias, C. A., Kosovichev, A. G., Leibacher, J. W., Morel, P., Proffitt, C. R., Provost, J., Reiter, J., Rhodes Jr., E. J., Rogers, F. J., Roxburgh, I. W., Thompson, M. J., and Ulrich, R. K., *Science*, **272**, 1286 – 1292 (1996).
22. Richard, O., Vauclair, S., Charbonnel, C., and Dziembowski, W. A., *Astron. Astrophys.*, **312**, 1000 – 1011 (1996).
23. Turcotte, S., Richer, J., Michaud, G., Iglesias, C. A., and Rogers, F. J., *Astrophys. J.*, **504**, 539 – 558 (1998).
24. Salerno, E., Bühler, F., Bochsler, P., Busemann, H., Eugster, O., Zastenker, G. N., Aganof, Y. N., and Eismont, N. A., "Direct measurement of $^3He/^4He$ in the LISM with the COLLISA experiment", in *Solar and Galactic Composition*, edited by R. F. Wimmer-Schweingruber, AIP conference proceedings, Woodbury, NY, 2001, this volume.
25. Gloeckler, G., and Geiss, J., "Composition of the local interstellar cloud from observations of interstellar pickup ions", in *Solar and Galactic Composition*, edited by R. F. Wimmer-Schweingruber, AIP conference proceedings, Woodbury, NY, 2001, this volume.
26. Jones, F. C., Baring, M. G., and Ellison, D. C., "The Effect of Self-Consistent Stochastic Preacceleration of Pickup Ions on the Composition of Anomalous Cosmic Rays", in *Solar and Galactic Composition*, edited by R. F. Wimmer-Schweingruber, AIP conference proceedings, Woodbury, NY, 2001, this volume.
27. Wimmer-Schweingruber, R. F., and Bochsler, P., "Lunar Soils: A Long-Term Archive for the Galactic Environment of the Heliosphere?", in *Solar and Galactic Composition*, edited by R. F. Wimmer-Schweingruber, AIP conference proceedings, Woodbury, NY, 2001, this volume.
28. Thielemann, F.-K., Argast, D., Brachwitz, F., and Martinez-Pinedo, G., "Stellar Nucleosynthetic and Galactic Abundances", in *Solar and Galactic Composition*, edited by R. F. Wimmer-Schweingruber, AIP conference proceedings, Woodbury, NY, 2001, this volume.
29. Chiappini, C., and Matteucci, F., "Galactic Chemical Evolution", in *Solar and Galactic Composition*, edited by R. F. Wimmer-Schweingruber, AIP conference proceedings, Woodbury, NY, 2001, this volume.
30. Sofia, U., "Limits to Galactic Abundances based on Gas-Phase Measurements in the Interstellar Medium", in *Solar and Galactic Composition*, edited by R. F. Wimmer-Schweingruber, AIP conference proceedings, Woodbury, NY, 2001, this volume.
31. Wiedenbeck, M. E., Binns, W. R., Christian, E. R., Cummings, A. C., Davis, A. J., George, J. S., Hink, P. L., Israel, M. H., Leske, R. A., Mealdt, R. A., Stone, E. C., von Rosenvinge, T. T., and Yanasak, N. E., "Constraints on the Nucleosynthesis of Refractory Nuclides in Galactic Cosmic Rays", in *Solar and Galactic Composition*, edited by R. F. Wimmer-Schweingruber, AIP conference proceedings, Woodbury, NY, 2001, this volume.
32. Klecker, B., Bothmer, V., Cummings, A. C., George, J. S., Keller, J., Salerno, E., Sofia, U. J., Stone, E. C., Thielemann, F.-K., Wiedenbeck, M. E., Buclin, F., Christian, E. R., Flückiger, E. O., Hofer, M. Y., Jones, F. C., Kirilova, D., Kunow, H., Laming, M., and abd K.-P. Wenzel, C. T., "Galactic Abundances: Report of Working Group 3", in *Solar and Galactic Composition*,

edited by R. F. Wimmer-Schweingruber, AIP conference proceedings, Woodbury, NY, 2001, this volume.
33. Bochsler, P., Wimmer-Schweingruber, R. F., and Wurz, P., "Sun, solar wind, meteorites, and interstellar medium: What are the compositional relations?", in *Solar and Galactic Composition*, edited by R. F. Wimmer-Schweingruber, AIP conference proceedings, Woodbury, NY, 2001, in press.
34. George, J. S., Wiedenbeck, M. E., Binns, W. R., Christian, E. R., Cummings, A. C., Hink, P. L., Leske, R. A., Mewaldt, R. A., Stone, E. C., von Rosenvinge, T. T., and Yanasak, N. E., "The Phosphorus/Sulfur Abundance Ratio as a Test of Galactic Cosmic-Ray Source Models", in *Solar and Galactic Composition*, edited by R. F. Wimmer-Schweingruber, AIP conference proceedings, Woodbury, NY, 2001, this volume.
35. Meyer, J., Drury, L. O., and Ellison, D. C., *Astrophys. J.*, **487**, 182 – 196 (1997).
36. Ellison, D. C., Drury, L. O., and Meyer, J., *Astrophys. J.*, **487**, 197 – 217 (1997).
37. Reinard, A. A., Zurbuchen, T. H., Fisk, L. A., Lepri, S. T., Skoug, R. M., and Gloeckler, G., "Average composition signatures of CMEs in the solar wind", in *Solar and Galactic Composition*, edited by R. F. Wimmer-Schweingruber, AIP conference proceedings, Woodbury, NY, 2001, in press.
38. Widing, K. G., and Feldman, U., *Astrophys. J.*, **555**, 426 – 434 (2001).
39. Neukomm, R. O., *Composition of CMEs*, Ph.D. thesis, University of Bern, Switzerland (1998).
40. of magnetic cloud plasmas during 1997, C., and 1998, "P. Wurz and R. F. Wimmer-Schweingruber and K. Issautier and P. Bochsler and A. b. Galvin and J. A. Paquette and F. M. Ipavich", in *Solar and Galactic Composition*, edited by R. F. Wimmer-Schweingruber, AIP conference proceedings, Woodbury, NY, 2001, this volume.
41. Wurz, P., Bochsler, P., and Lee, M. A., *J. Geophys. Res.*, **105**, 27239 – 27249 (2000).
42. Ko, Y.-K., Zurbuchen, T., Strachan, L., Riley, P., and Raymond, J. C., "A Solar Wind Coronal Origin Study from SOHO/UVCS and ACE/SWICS Joint Analysis", in *Solar and Galactic Composition*, edited by R. F. Wimmer-Schweingruber, AIP conference proceedings, Woodbury, NY, 2001, in press.
43. Zurbuchen, T. H., Fisk, L. A., Gloeckler, G., and Schwadron, N. A., *Space Sci. Rev.*, **85**, 215 – 226 (1998).
44. Antonucci, E., and Giordano, S., "Oxygen Abundance in the Extended Corona at Solar Minimum", in *Solar and Galactic Composition*, edited by R. F. Wimmer-Schweingruber, AIP conference proceedings, Woodbury, NY, 2001, this volume.
45. Parenti, S., Poletto, G., Bromage, B. J. I., Suess, S. T., Raymond, J. C., Noci, G., and Bromage, G. E., "Preliminary results from coordinated SOHO-Ulysses observations", in *Solar and Galactic Composition*, edited by R. F. Wimmer-Schweingruber, AIP conference proceedings, Woodbury, NY, 2001, this volume.
46. Wimmer-Schweingruber, R. F., *Oxygen, Helium, and Hydrogen in the Solar Wind: SWICS/ULYSSES results*, Ph.D. thesis, Univ. of Bern, Bern, Switzerland (1994).
47. von Steiger, R., Wimmer-Schweingruber, R. F., Geiss, J., and Gloeckler, G., *Adv. Space Res.*, **15**, 3 – 12 (1995).
48. Gosling, J. T., Borrini, G., Asbridge, J. R., Bame, S. J., Feldman, W. C., and Hansen, R. F., *J. Geophys. Res.*, **86**, 5438 – 5448 (1981).
49. Borrini, G., Wilcox, J. M., Gosling, J. T., Bame, S. J., and Feldman, W. C., *J. Geophys. Res.*, **86**, 4565 – 4573 (1981).
50. Wimmer-Schweingruber, R. F., *Adv. Space Sci.* (2001), invited review for COSPAR 2000, in press.
51. Geiss, J., Hirt, P., and Leutwyler, H., *Sol. Phys.*, **12**, 458 – 483 (1970).
52. Bodmer, R., and Bochsler, P., *J. Geophys. Res.*, **105**, 47 – 60 (2000).
53. Morris, D., Möbius, E., Lee, M. A., Popecki, M. A., Klecker, B., Kistler, L. M., and Galvin, A. B., "Implications for Source Populations of Energetic Ions in Co-Rotating Interaction Regions from Ionic Charge States", in *Solar and Galactic Composition*, edited by R. F. Wimmer-Schweingruber, AIP conference proceedings, Woodbury, NY, 2001, in press.
54. von Rosenvinge, T. T., Cohen, C. M. S., Christian, E. R., Cummings, A. C., Leske, R. A., Mewaldt, R. A., Slocum, P. L., Cyr, O. C. S., Stone, E. C., and Wiedenbeck, M. E., "Time Variations in Elemental Abundances in Solar Energetic Particle Events", in *Solar and Galactic Composition*, edited by R. F. Wimmer-Schweingruber, AIP conference proceedings, Woodbury, NY, 2001, this volume.

Solar and Solar System Abundances

Measuring Solar Abundances

R. von Steiger[*], J.-C. Vial[†], P. Bochsler[**], M. Chaussidon[‡], C. M. S. Cohen[§], B. Fleck[¶], V. S. Heber[||], H. Holweger[††], K. Issautier[**], A. J. Lazarus[‡‡], K. W. Ogilvie[§§], J. A. Paquette[¶¶], D. B. Reisenfeld[***], L. Teriaca[†††], K. Wilhelm[‡‡‡], S. Yusainee[§§§], J. M. Laming[¶¶¶] and R. C. Wiens[***]

[*]*International Space Science Institute, Hallerstrasse 6, CH-3012 Bern, Switzerland*
[†]*Institut d'Astrophysique Spatiale, Université Paris XI - C.N.R.S., Bâtiment 121, F-91405 ORSAY Cedex, France*
[**]*Physikalisches Institut, University of Bern, Sidlerstr. 5, CH-3012 Bern, Switzerland*
[‡]*CRPG-CNRS, Vandoeuvre-lès-Nancy, 54501, France*
[§]*California Institute of Technology, Pasadena, CA 91125, USA*
[¶]*ESA Space Science Department, NASA/GSFC Mailcode 682.3, Greenbelt, MD 20015, USA*
[||]*Institute for Isotope Geology and Mineral Resources, ETH Zürich, 8092 Zürich, Switzerland*
[††]*Inst. für Theoret. Physik und Astrophysik, Universität Kiel, Kiel, 24098, Germany*
[‡‡]*MIT, Room 37-687, 77 Mass. Ave., Cambridge, MA 02139, USA*
[§§]*NASA/GSFC, Code 692, Greenbelt, MD 20771, USA*
[¶¶]*Department of Physics, University of Maryland, College Park, MD 20742, USA*
[***]*Los Alamos National Laboratory, Mail Stop D466, Los Alamos, NM 87545, USA*
[†††]*Osservatorio Astrofisico di Arcetri, Largo Enrico Fermi 5, Firenze 50125, Italy*
[‡‡‡]*Max-Planck-Institut für Aeronomie, 37191 Katlenburg-Lindau, Germany*
[§§§]*Univerisiti Teknologi Maralecturer Building, Bandar Jengka, Pahang, 26400, Malaysia*
[¶¶¶]*Naval Research Laboratory, Naval Research Laboratory, Code 7674, Washington, DC 20375, USA*

Abstract. This is the rapporteur paper of Working Group 2 on Measuring Solar Abundances. The working group presented and discussed the different observations and methods for obtaining the elemental and isotopic composition of the Sun, and critically reviewed their results and the accuracies thereof. Furthermore, a few important yet unanswered questions were identified, and the potential of future missions to provide answers was assessed.

INTRODUCTION

This paper is an attempt at summarising the deliberations of Working Group 2 in the Joint SOHO-ACE Workshop on Solar and Galactic Composition. The tasks of this working group were defined by the Scientific Organising Committee, and somewhat extended by the group, to be the following:

- How do we derive solar abundances from remote-sensing data?
- What are the major agreements and disagreements?
- What are the advantages and disadvantages of the different methods?
- What are the currently best solar abundances, and how accurate are they?
- What future observations and experiments do we need to improve the situation?

The tasks were addressed in three group meetings during the workshop. First, the group addressed some basic questions, to which the answers caused no controversy. It then went on to present and discuss the various methods with which abundances can be measured in different reservoirs, what elements are accessible to the different methods, how the solar abundances can be inferred from the measurements, and how accurate the resulting values are. Several elemental abundances and isotopic ratios are still poorly known even today, leaving some important questions unanswered. The group identified a few such questions, and finally discussed how future missions will improve our knowledge of solar abundances. The structure of this rapporteur paper follows this outline quite closely.

CP598, *Solar and Galactic Composition,* edited by R. F. Wimmer-Schweingruber

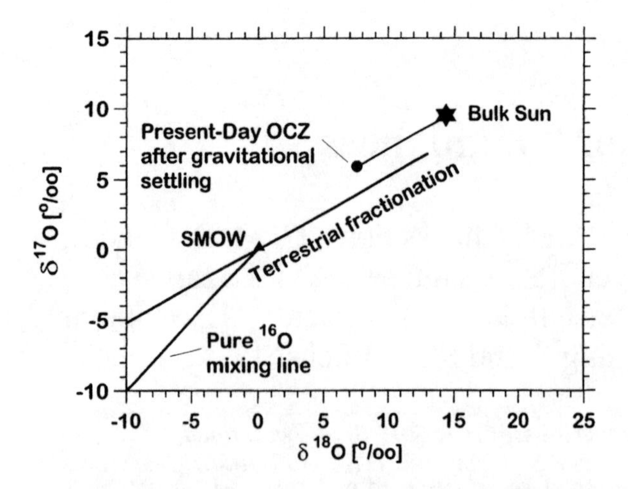

FIGURE 1. Oxygen isotope plot indicating that the heavy isotopes may have been depleted by a few tenths of a percent per mass unit over the life of the Sun [8], adapted from [50]. The axes are labelled in the δ notation, which indicates the variations of the isotope ratios ($^{17}O/^{16}O$ *vs.* $^{18}O/^{16}O$) relative to a reference value (standard mean ocean water, SMOW) in permil.

What solar abundances?

The first question is: What *are* solar abundances, anyway? What is the reservoir that contains them?

Solar abundances are contained in the outer convective zone (OCZ), as represented by the photosphere. Indeed, it is (nearly) pristine, or representative of the protosolar nebula, because it is so large and well-mixed, and because it is sufficiently shallow so as not to reach and dredge up the products of nuclear burning in the core [9]. Nevertheless, a few exceptions have to be kept in mind:

- Deuterium was burnt to form ^{3}He already in the pre-main sequence phase of the nascent Sun.
- Because of partial mixing below the OCZ, large Li depletion and slight ^{3}He increase can occur.
- ^{4}He has gravitationally settled out of the OCZ into the top of the radiative zone, an effect of the order of $\approx$10 %.
- The same gravitational settling also affects the abundances of heavier elements such as CNO by a few percent, but this change is below the threshold of detectability with today's instrumentation.
- This is not true for the relative abundances of isotopes, though, which are measured with higher accuracy than the elements. Modelling indicates that isotopic abundances can be affected by up to 5 % in the extreme case of ^{3}He/^{4}He [40], and a few permil per mass unit in the case of heavier elements such as oxygen. Figure 1 (from [8], adapted from [50]) illustrates this for O, which may have shifted from a protosolar value (indicated by a star) along a line with slope 1/2 (as expected for mass-dependent fractionation) to an isotopically lighter value (indicated by a circle).

Why solar abundances?

The next question is: Why do we need solar abundances? And how accurately do we need to know them?

Solar abundances represent the baseline, or ground truth, for all regimes in the solar system. Only a solid baseline, in particular of the volatiles, allows inferences on how the different regimes were formed and how they evolved over the past 4.6 Gy.

Obviously the baseline must be known more accurately than the natural variability within and between the reservoirs. This variability can be very large, e. g., between the inner and the outer planets, or very small, e. g., in the case of the O isotopes between Earth, Mars, and asteroids, which is of the order of 10^{-4} [50]. Moreover, in the case of natural variability within a reservoir (e. g., the solar wind), the relevant processes causing the fractionations should be understood in sufficient detail so they can be accounted for when deducing the underlying solar abundances.

MEASURING SOLAR ABUNDANCES

It is not possible, for obvious reasons, to fly a sample-return mission to the Sun. Consequently, all observations and measurements of solar abundances are indirect. The working group discussed a number of methods for abundance determination and their advantages and disadvantages when deriving solar abundances from them:

- Helioseismology
- optical remote-sensing observations in the solar photosphere (and corona)
- observations of the solar wind, both with in-situ instrumentation and by trapped materials in lunar samples (and meteorites)
- in-situ observations of solar energetic particles

These methods are now discussed in sequence.

Helioseismology

This method has the advantage to directly probe the OCZ with high accuracy. The determination of solar interior abundances has been recently improved with the help of helioseismology measurements with SOHO. The basic tool is the sound speed variation in the interior where the mean molecular weight depends on the composition of the medium. In the OCZ the reference value for helium is $Y = 0.249 \pm 0.003$, given by Basu and Antia [6], who used opacities calculated with the OPAL code [36]. Brun *et al.* [12] concentrated on the discrepancies between "observations" and models at the tachocline (where the turbulent outer region, the convective zone, meets the orderly interior, or radiative zone), and adopting the above-mentioned Y value, and including macroscopic diffusion, they could derive a better determination of the CNO abundances (found to be underestimated by 2 to 3 %). According to Turck-Chièze [39], the abundances of heavy elements are determined with 1-σ uncertainties of 12 % (C), 17 % (N), 17 % (O), 15 % (Ne), and 9.5 % (Fe). Overall, the metallicity value quoted in table 3 of [12] is (2.45 ± 0.2) %.

As far as isotopes are concerned, the $^3He/^4He$ "surface" ratio is increased by 10 % at most. As for the ^7Li, it seems that its present depletion (by a factor larger than 100) could be accounted for by a small mixing [40].

The results from helioseismology generally give values for the He abundance close to that accepted for standard big bang nucleosynthesis, around $Y = 0.25$ [38] or $He/H = Y/4(1-Y) = 8.3\%$. The value of Y usually assumed for the protosolar nebula is a few times 0.01 higher than this, which is also consistent with observations of the He abundance in the gas planets. Uncertainties in helioseismology inversions stem from the approximations made in deriving the equation of state and the opacities used for the solar envelope, but seem unlikely to seriously affect the results (see section 7 and table 14 of [4] for a survey of how the inferred He abundance varies with changes in the solar model, and see [35] for the influence of diffusion mixing). This lower-than-expected He abundance in the OCZ thus has been taken as evidence of gravitational settling of He.

Optical remote sensing

This technique has been in use since the 1920s, when Payne-Gaposchkin and Russell first analysed the spectrum of the solar photosphere and extracted the abundances of 56 elements therefrom. Since it is the oldest method it is often taken as the one giving the "standard" solar composition. Yet, like all methods, it has its advantages and disadvantages. The primary advantage is that it directly probes the photosphere, which is representative for the OCZ without any correction. It yields abundances with good accuracy of most elements and even of some isotopes, provided the oscillator strengths are known sufficiently well. On the downside it has to be noted that noble gas abundances, which are particularly important for solar system taxonomy, are inaccessible to the method because these elements have no optical transitions at photospheric temperatures. Moreover, each line stems from one particular ionisation state of an element, so the abundance has to be inferred by dividing by the ionisation fraction of that state at the temperature where it is formed. This is usually done by assuming local thermal equilibrium, which is quite likely a good assumption in the photosphere (but probably less good in the corona – see the report of WG 1 on coronal abundances). Most element abundances are derived from several lines stemming from different charge states, thus reducing the importance of this effect, but they may still be affected as different charge states rarely are observed at the same site in the atmosphere, and as non-Maxwellian distribution functions (e. g., suprathermal tails) may affect the ionisation equilibria.

Photospheric abundances are discussed in detail by Holweger in this volume [24]. Moreover, Patsourakos *et al.* in this volume [33] show a comparison between observed ion effective temperatures in polar coronal holes (PCH) and model calculations. The overall good agreement between the observations and the model that is based on the ion-cyclotron resonance mechanism of solar wind heating and acceleration, provides new piece of evidence in favour of this mechanism. The above authors also presented some initial theoretical results concerning the iron and oxygen abundances in PCH within the framework of the ion-cyclotron mechanism. They showed a depletion of iron with respect to oxygen in PCH, a result still to be confirmed.

Optical remote sensing can also be used to pinpoint the sources of the solar wind, such as fast streams emanating from coronal holes: Bright plumes are striking features that, within coronal holes, can extend to more than 30 solar radii from the Sun [e. g., 14]. However, recent analyses of SUMER and UVCS data have shown that the width of UV lines is larger in interplume than in plume regions, hinting to interplumes as the site where energy is preferentially deposited and, possibly, fast wind emanates [e. g., 19, 5, 51].

An important factor that may differentiate plume and interplume regions is their elemental abundance. Ulysses observations have identified a difference in the elemental abundances of the fast and slow wind: low FIP elements are more enriched, with respect to their photospheric abundance, in the slow wind than in the high speed wind. Hence, if interplume regions are really the sources of the fast wind, we may possibly find a different elemental composition between plumes and interplume plasma at coronal levels. Recent measurements of the O VI 1032/1037 Å line intensity ratio in the interplume lanes are presented by Teriaca *et al.* in this volume together with an attempt to model the O VI line intensities as a function of height. Their analysis shows that negligible outflow velocities are present below 1.3 solar radii, supporting the idea that the fast wind may start being accelerated only above that altitude. They were also able to find a lower limit of 8.5 for the interplume oxygen abundance (on the log scale with H≡12). This value is lower than what is measured in the fast solar wind [45] as it could be expected if the oxygen and hydrogen ions move at different speeds at 1.2 solar radii and at about the same speed at 1 AU.

In situ solar wind

The elemental and charge state composition of the solar wind can be observed in situ by space-borne mass spectrometers such as SWICS on Ulysses or CELIAS on SOHO, or using the foil collection technique as on the Apollo missions; the results are summarised in several review papers [43, 8]. These in situ observations are complementary to the optical remote sensing observations in several ways:

- They comprise a true abundance measurement as all charge states of an element are observed at the same time, so no assumption is needed about thermal equilibrium. The statistical accuracy of the observations can be very high – unless high time resolution is required – but the results are often limited by systematic, instrumental uncertainties, which are very difficult to squeeze below 20 %.
- Unlike optical observations in the photosphere, they give good results for the noble gases He, Ne, and Ar (and see the next subsection regarding Kr and Xe), in particular the foil collection technique [17].
- The foil collection technique and the newest generation of space mass spectrometers such as CELIAS-MTOF or ACE-SWIMS can even resolve isotopes of most elements that can be observed, as reviewed in [52].

However, these features come at a price: Due to the relatively small geometric factors of space-borne sensors, or the relatively short exposure time of lunar foils, the observations are still limited to the 10–15 most abundant elements (H, He, C, N, O, Ne, Mg, Si, S, Ar, and Fe [reviewed in 43], with recent additions of the less abundant Na [25], Al [10], Ca [26, 46], and Cr [32] from SOHO-CELIAS). More importantly, solar wind abundances are not a genuine, unbiased sample of solar abundances, but they are fractionated. One such fractionation depends on the first ionisation potential (FIP): When comparing solar wind to solar abundances, elements with low FIP (<10 eV) are enriched by a significant factor, the FIP bias, over those with a high FIP. To complicate matters further, the solar wind comes in two quasi-stationary varieties (plus transients such as coronal mass ejections), which differ in strength of the FIP bias: It is about 1.5–2 and relatively constant in fast streams from coronal holes, while in the slow solar wind from the streamer belt region it is about 2.5–3 on average but highly variable [44] (see also [41] for a brief review of ideas that have been put forward to explain the FIP fractionation). It is tempting to take the fast wind abundances with a modest correction as a good sample of the solar abundances, but care must be taken: If we don't understand the FIP effect and in particular its variability we don't understand solar abundances as derived from the solar wind. Another fractionation process affects mainly helium, causing its abundance in the SW to be only about half of the solar abundance [see, e. g., 1]. It most likely due to insufficient Coulomb drag between protons and alpha particles in the accelerating solar wind.

Trapped solar wind

Much of the current knowledge on the noble gases in the solar wind is based on the analysis of lunar samples and gas-rich meteorites. The solar wind noble gas data obtained from these "targets" are in good agreement with data obtained from in-situ measurements and the Apollo foils (e. g., for He, Ne and Ar [7, 34, 30]), which gives us confidence to make predictions also for Kr and Xe. Because of the long exposure time of the meteoritic and lunar samples, their analysis is so far the only method to obtain solar wind Kr and Xe abundances as well as the Kr and Xe isotopic composition. Precisions on the order of 1 % for elemental ratios and a few permil for isotopic ratios are possible. The solar wind composition deduced from these extraterrestrial dust samples is very important to investigate the solar history, because some of them collected their noble gases several billion years ago.

However, the noble gas record in lunar and meteoritic regolith samples is not straightforward to read. Diffusive loss or redistribution of light noble gases within the grains may influence the original solar wind composition. Experiments on mineral separates have shown that the most retentive Fe-Ni grains conserve the true relative abundances of solar He, Ne, Ar [34, 30], but Fe-Ni is very rare in lunar samples. Ilmenite, another retentive mineral phase, which is abundant in lunar samples, has lost some of its solar He and Ne, but apparently without altering the isotopic composition of these gases [7]. A further difficulty arises when one compares samples that contain solar wind of different antiquities. Alteration processes on the Moon and meteoritic parent bodies may remove part of the solar wind containing grain surfaces probably in a time-dependent manner. Furthermore, other noble gas components can compromise the solar wind signal, e. g., cosmogenic noble gases due to long exposure to galactic cosmic rays, or primordial noble gases (especially Kr and Xe) in chondrites. With the help of the in-vacuo-etch-technique (closed system stepped etching in a device directly connected to a mass spectrometer [37]) these components can at least partly be separated from each other. This is done by dissolving only the uppermost solar wind bearing grain layers, thus avoiding e. g., cosmogenic contributions from deeper layers. Furthermore, this release technique avoids noble gas fractionation in the laboratory, because the gases are released at room temperature by slowly etching the carrier minerals and not by diffusion as is the case for stepwise heating.

Now two examples show how solar wind noble gases in extraterrestrial dust samples are important to investigate the solar history. The ^{3}He/^{4}He ratio in the solar wind and its long time evolution are of great interest, because this ratio is a very sensitive indicator for admixing of material of interior layers into the outer convective zone [9]. New data, obtained by very high-resolution in-vacuo etching of lunar grains of different solar wind antiquity show at face value an enrichment of ^{3}He of around 5 %/Gy. However, even this small apparent increase can probably be explained by secondary processes on the Moon [23].

Lunar soil samples appear to be very well suited to study the isotopic composition of solar wind Kr and Xe because the moon is extremely depleted in indigenous Kr and Xe. Wieler *et al.* [47, 48, 49] showed variable Xe and Kr enrichments in the solar wind, compared to bulk solar composition [3]. The enrichment factor for Xe is similar to that observed for low-FIP elements in the slow solar wind [43], whereas Kr is less enriched. At least some regolithic meteorites probably collected solar wind from the early Sun. Analysis of these samples therefore enlarges the data base. Unfortunately there is so far no method to determine the solar wind antiquity in meteorites as is possible for lunar samples. But the data available so far suggest that the FIP-related enrichment of Xe and Kr has been even more variable in the past than it was known from lunar samples and – if at least some of the meteorites were irradiated very early – the FIP effect was already active in the early Sun, as further discussed by Heber *et al.* in this volume.

Solar energetic particles

Solar energetic particle (SEP) events are transient occurrences which are related to coronal mass ejections (CMEs) and flares on the Sun. The particles are typically accelerated out of the corona and solar wind by CME-driven shocks. They can also be accelerated at the site of flares and propagate outward to be observed at 1 AU. Current instrumentation for measuring SEPs is very mature and enables low fluxes of particles to be measured. The mass resolution of current measurements is good enough to resolve many isotopes of heavy ions between C and Ni. Because of such excellent instrumentation, the relative abundances of elements and isotopes in SEP events as observed at 1 AU can be quite accurately determined, not only averaged over a single SEP event but often as a function of time within an event.

It has been observed that the composition of SEP events can vary substantially from event to event [28]. In comparing these variations to an average composition, they can often be organised by the charge to mass ratios of the elements [11]. Such fractionation is probably an acceleration and/or transport effect. An additional fractionation, with respect to the photospheric composition (as determined from optical measurements), governed by the first ionisation potential (FIP) of an element is also apparent in the SEP composition [13]. This is the same type of fractionation that is so often observed in the solar wind and like the solar wind, the degree of fractionation varies. Only by correcting for these fractionations a measurement of the photospheric abundances can be obtained using SEP measurements for both elements and isotopes. The abundances of SEPs are therefore better suited to accurately probing acceleration processes and plasma effects rather than to inferring the bulk solar composition.

Summary

So, what are the best solar abundances that are currently available? Of course, that depends on the element, and as we have seen above there is no single method which gives all abundances with good accuracy. Rather, the abundances must be compiled from a large number of publications, and the values given must be used with scrutiny. Fortunately, several review papers exist doing just that: The classical paper of Anders and Grevesse [3] with two updates [21, 22], or the table in Landolt-Börnstein [31]. These tables have the advantage of being complete and readily available, and as such [3] has found very widespread use as a reference, in particular in the astronomical literature. This is of course ok if it is understood that the table lists reference, not necessarily solar abundances, as several values have been updated in the meantime.

The tables generally show very good agreement between solar and meteoritic abundances, which allows to infer gaps in the solar record from the meteoritic one (and vice versa). However, this immediately brings the danger of using circular arguments. For example, noble gases can be obtained by interpolating between s-process elements in the meteoritic record, or by inferring from the solar wind record. If the two methods give results which are in agreement we are tempted to attribute them a high degree of credibility, but they could still both be wrong.

Another point worth noting is the evolution of the oxygen abundance over the past decade: 8.93 on the scale where H≡12) in [3], 8.83 in [21], and 8.73 in [22], while the newest original paper even gives 8.69 ± 0.05 [2]. This decrease is of course not a solar effect, but reflects our improving knowledge and interpretation of the relevant lines in the solar spectrum (see also the paper of Holweger in this volume [24]).

In Table 1 we have summarised our conservative estimates of the relative accuracies that can be expected when deriving solar abundances from the different reservoirs discussed above.

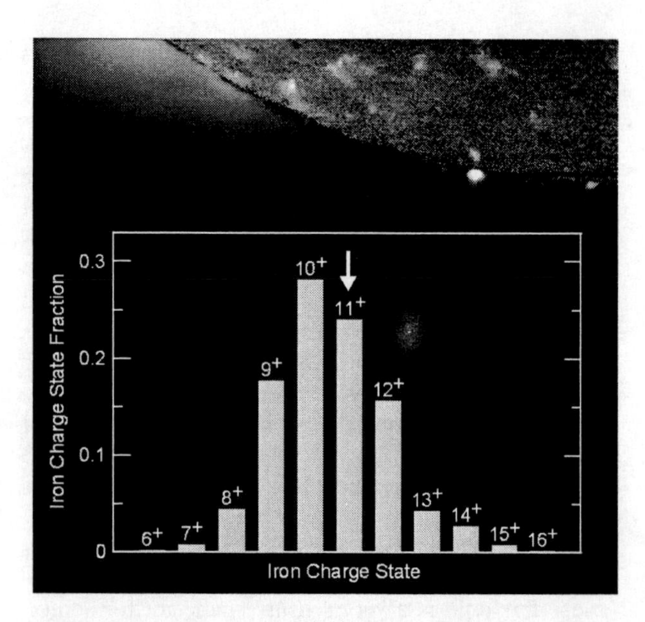

FIGURE 2. SOHO-SUMER picture in the Fe XII line, showing essentially no emission from the coronal hole, and Ulysses-SWICS Fe charge state spectrum in the fast solar wind, indicating a large fraction of the corresponding ion Fe^{11+} (arrow).

SOME UNANSWERED QUESTIONS

The Working Group has identified a number of unanswered questions (UQs), the answers to which have great potential to advancing our knowledge of the Sun and the solar system. These are:

UQ1: What are the solar isotope ratios, in particular of N, O, and Mg? The isotopes are the most convincing tool for taxonomy in the solar system. In the isotope system of oxygen, relative differences of the order of 10^{-4} are used to infer the parent of meteorites. Yet the solar ratios are known no better than to 15 %.

UQ2: What are the solar noble gas abundances? The noble gases, owing to their high volatility, are another important means for taxonomy in the solar system. Yet all we have are solar wind, not solar abundances. Even a relative knowledge of noble gas abundances among themselves would be helpful, but this only leads to the next question:

UQ3: Is the solar wind a faithful sample of the outer convective zone? Do we understand the FIP effect well enough to infer solar abundances? Although the FIP effect is not expected to cause substantial mass fractionation [42, 29], it may still be significant. Specifically, the cause of the added variability of the slow solar wind remains unknown (but see [16]).

UQ4: What causes the contradictory evidence between in-situ charge state and optical observations? Observations of solar wind charge states can be interpreted as a proxy for the coronal electron temperature at the site where the solar wind is formed. Results of Ulysses-SWICS in the fast solar wind indicate a coronal hole temperature profile with a maximum of $\approx$1.5 MK at an altitude of $\approx 1.5 R_\odot$ [18, 27]. On the other hand, SOHO-SUMER observations above the limb in coronal holes indicate electron temperatures barely reaching 1 MK near the limb and decreasing with altitude [51]. The discrepancy is made particularly apparent when looking at the charge states of iron: Whereas SUMER pictures in the 195-Å line of Fe XII are essentially black, SWICS observes a substantial fraction of this ion, $Fe^{11+}/Fe \simeq 25$ % (see Figure 2). The cause of this discrepancy is as yet unknown. It may equally well be rooted in the interpretation of optical observations, specifically in the assumption of ionisation equilibrium for the charge state fractions, or in the interpretation of charge state spectra observed in situ, which may be affected by differential streaming between ion species and/or by suprathermal tails of the electron distribution function when evolving over the first few solar radii [15], or both.

FUTURE OBSERVATIONS

In the near future our knowledge about solar abundances will undoubtedly be much improved by the Genesis mission, which is specifically designed to determine elemental and isotopic abundances with the highest feasible accuracy. Other missions will also carry in situ composition instrumentation, notably the Stereo, Solar Orbiter, and Solar Probe missions. In contrast to the Genesis mission, these missions are not specifically designed to measure the solar wind composition *per se*. Rather, the objective of these missions is to use the abundance measurements as a diagnostic tool to gain further insight into the ori-

TABLE 1. Table of the relative accuracy of solar abundances as obtained from different reservoirs.

Element	Interior	OCZ	Photosphere	Corona	Solar wind	SEP
H			0.3 %	$\lesssim$5 %	$\approx$3 %	
He	1 %	1 %	-	10 %	$\lesssim$10 %	5 %
Li			15 %			
CNO etc.			10 %	$\lesssim$50 %	$\lesssim$30 %	10 %
Nobles			-	$\lesssim$50 %	$\lesssim$30 %*	10 %
Trans-Fe			10 %	$\lesssim$50 %		$\gtrsim$100 %
Isotopes			15 %†	-	$\approx$10 %	$\approx$15 %

* better for trapped SW
† C,O,Mg

gin of the solar wind, in particular its acceleration and fractionation processes.

Genesis

Genesis is a 3-year NASA mission to capture an integrated sample of solar wind and return it to Earth for ground-based isotopic and elemental analysis in the style of the Apollo-era Solar Wind Composition 'sunshade' experiments. After its launch on 8 August, 2001, the spacecraft spends $\approx$2.5 years at the L1 point collecting solar wind before returning with a capsule-style reentry in September, 2004. A number of factors should lead to greatly expanded analytical capabilities relative to the Apollo foil collection experiments. These include much longer collection time (>2 years *vs.* 2 days), absence of lunar dust contamination, and higher purity collection materials, including Si, Ge, and SiC wafers, Au- and Al-on-sapphire, and CVD diamond. The Genesis payload also includes a solar wind concentrator aimed at obtaining 20$\times$ higher-concentration samples of elements lighter than Si, designed particularly for oxygen isotopes. Current technology suggests that abundances will be obtainable for many elements up to and including rare-earth elements. With these capabilities in mind, the science team laid out a large number of prioritised solar wind measurement objectives. The highest priorities are isotope ratios, including ^{16,17,18}O to $\pm$0.1 % 2-σ uncertainty, nitrogen isotopes to $\pm$1 %, carbon isotopes to 0.4 %, and noble gas elemental and isotopic ratios to varying uncertainties. Other priorities include checks on solar nebula solid-gas fractionation, heavy-light element comparisons, Li, Be, B measurements, determinations of ^{14}C and ^{10}Be, measurement of solar spallation-produced F, and general comparisons with terrestrial, chondritic, and s-process predicted abundances.

The required accuracies and precisions for the highest priority measurements were based on the need to distinguish between competing science claims. For example, for oxygen isotopes, although differences between Earth, Mars, and Vesta are significantly smaller, the Genesis measurement goal of $\pm$0.1 % will clearly distinguish between competing theories for the phenomenon of oxygen isotope heterogeneity. It should be noted that current solar and solar-wind isotopic measurements have accuracies only at the $\approx$10 % level. The oxygen isotope issue, and the rationales for the various Genesis measurement objectives, are discussed in greater detail in [50] and in the working group papers 4 and 5 elsewhere in this volume.

An obvious drawback is that the Genesis mission measures solar wind compositions, which are one step removed from the actual solar composition. To aid in understanding the corrections needed to obtain accurate solar abundances from solar wind data, the Genesis spacecraft is collecting separate samples of three different solar wind types: interstream, coronal hole, and coronal mass ejection material, selected by real-time on-board analysis of measurements by ion and electron spectrometers. By studying time-integrated samples of these three different solar wind regimes, it is expected that any isotopic fractionation between the photosphere and solar wind will for the first time be well constrained, and that general limits on elemental fractionation will lead to increased accuracy for solar elemental abundances as well.

Stereo

The principal scientific objective of the Stereo mission is to understand the origin and consequences of coronal mass ejections (CMEs). It will provide a totally new perspective on solar eruptions by moving away from our customary Earth-bound lookout point and providing stereoscopic views of solar eruptions, with one spacecraft leading Earth in its orbit and another lagging it. Each will carry a cluster of telescopes. When simultaneous telescopic images are combined with data from observatories on the ground or in low Earth orbit, the buildup of magnetic energy, and the lift off, and the trajectory of Earthward-bound CMEs can all be tracked in three dimensions. When a CME reaches Earth's orbit, magnetometers and plasma sensors on the Stereo spacecraft will sample the material and allow investigators to link the plasmas and magnetic fields unambiguously to their origins on the Sun. Launch is scheduled for December 2005.

One of the four instruments selected to fly on Stereo is the "Plasma And SupraThermal Ion Composition" (PLASTIC) instrument, which is the primary sensor for studying coronal-solar wind and solar wind-heliospheric processes. The PLASTIC measurements will be complemented at higher energies by the "In-situ Measurements of Particles and CME Transients" (IMPACT) – a suite of seven instruments that will sample the 3-D distribution of solar wind plasma electrons, the characteristics of the solar energetic particle (SEP) ions and electrons, and the local vector magnetic field.

Solar Orbiter

The key mission objective of the Solar Orbiter is to study the Sun from close-up (45 solar radii, or 0.21 AU) in an orbit tuned to solar rotation in order to examine the solar surface and the space above from a co-rotating vantage point at high spatial resolution. The Solar Orbiter will also provide images of the Sun's polar regions from heliographic latitudes as high as 38°. The mission was approved by ESA as a flexi-mission, to be implemented in the time frame 2008–2013.

The Solar Orbiter, while taking high-resolution images and making spectroscopic measurements of solar wind source regions that are magnetically linked to the spacecraft location and making simultaneous, in-situ measurements over the long co-rotation intervals, is ideally suited to address the critical issues of the chromospheric fractionation process. Also, the coronagraph on the Solar Orbiter may include a He channel, which would allow to determine for the first time the helium abundance (high FIP element) in those atmospheric layers where the acceleration of the slow and fast streams actually occurs, and where the charge states freeze in. The results expected will provide keys for the understanding of the processes at the origin of the solar wind and of the elemental composition in the heliosphere.

The strawman payload of the Solar Orbiter includes a Solar Wind Plasma Analyser (SWA), the three principal science goals of which are: (i) to provide observational constraints on kinetic plasma properties for a fundamental and detailed theoretical treatment of all aspects of coronal heating; (ii) to investigate charge- and mass-dependent fractionation processes of the solar wind acceleration process in the inner corona; (iii) to correlate comprehensive in-situ plasma analysis and compositional tracer diagnostics with space-based and ground-based optical observations of individual stream elements. Furthermore, the Solar Wind Plasma Analyser on the Solar Orbiter will investigate in detail ^{3}He and unusual charge states in CME-related flows, as well as the "recycling" of solar wind ions on dust grains in the distance range which has been located as the inner source [20]. Freshly produced pick-up ions from this inner source are specially suited as test particles for studying the dynamics of incorporation of these particles into the solar wind.

The SWA will measure separately the three-dimensional distribution functions of the major solar wind constituents: protons, alpha-particles and electrons. The basic moments of the distributions, such as density, velocity, temperature tensor, and heat flux vector will be obtained under all solar wind conditions and be sampled sufficiently rapidly to characterise fully the fluid and kinetic state of the wind. In this way we will be able to determine possible non-gyrotropic features of the distributions, ion beams, temperature anisotropies, and particle signatures of wave excitation and dissipation. In addition, measurements of representative high-FIP elements (the C, O, N group) and of low-FIP elements (such as Fe, Si or Mg) will be carried out in order to obtain their abundances, velocities, temperature anisotropies and charge states, to probe the wave-particle couplings (heavy-ion wave surfing), and to determine the freeze-in temperatures (as a proxy for the coronal electron temperature).

Solar Probe

Solar Probe is a NASA mission diving into the outer corona and thus will obtain an extreme close-up glimpse of the solar atmosphere. It will carry two complementary instrumentation packages, one for optical remote-sensing and another for in situ particle observations. Unfortunately, the selection process for these packages is currently interrupted and the prospects for the mission are unclear.

The optical measurements made at such a close distance of the Sun as 3 solar radii above the surface will allow for the determination of many elements at different ionisation stages and possibly at stages below the freeze-in temperatures. This unique feature could help to understand the discrepancies between in-situ freezing-in temperatures and temperatures derived from spectroscopic measurements, as discussed above. Open and closed regions will be explored as well. Table 1 of the Announcement of Opportunity provides a description of the quantities to be measured in-situ: (i) Spectra of energetic particles of H, ^{3}He, ^{4}He, C, O, Si, Fe with a sensitivity of 10 $\mathrm{cm^{-2}s^{-1}sr^{-1}keV^{-1}}$, a dynamic range of 10^7, and a time resolution of a few seconds in the spectral range of 0.02–2 MeV/nucleon. (ii) Time averaged distribution functions of H^+, $^3He^{++}$, $^4He^{++}$, C^{n+}, O^{n+}, Si^{n+}, and Fe^{n+} with a sensitivity of 10 km/s, a time resolution of 1 s (H, He, e^-) to 10 s (ions), a dynamic range of 2×10^7, in the spectral range of 0.05–10 keV/charge. It is envisaged to even have nadir view!

As far as remote sensing is concerned, there is no spectroscopic capability but a better understanding of the small-scale properties of the solar wind and, possibly, a determination of the acceleration process would help to understand the fractionation processes occurring in both the slow and fast solar wind.

CONCLUSIONS

We have seen that solar abundances can be measured using several different methods, any of which has its advantages and disadvantages. The accuracy of the results is reasonably good (depending on the specific question addressed), and future missions will improve on them significantly. Still, there remains the ultimate question: Is there any way to obtain *true* solar abundances? Most likely the answer is no: Neither with today's technology nor in the foreseeable future will it be possible to fly a mission to the Sun, obtain a sample of the OCZ, and return it to the laboratory for high-precision analysis of all its constituents. Our efforts should rather go into a different direction: As all observations of solar abundances, remote-sensing or in-situ, are affected by physical processes that may or may not alter those abundances, it is of utmost importance to identify all such processes and work towards a detailed, quantitative understanding of their effects. Such understanding can only be obtained and advanced through collaborations of observers, experimentalists, modelers, and theorists, which is exactly the goal of workshops such as this one ... and future ones of this kind.

ACKNOWLEDGMENTS

We gratefully acknowledge the effort of the scientific and local organisation committees, in particular of Robert F. Wimmer-Schweingruber, in making this workshop happen. We also thank Sylvie Vauclair and Robert F. Wimmer-Schweingruber for useful comments on the manuscript. JML was supported by NRL/ONR Solar Magnetism and the Earth's Environment 6.1 Research Option and by NASA Contracts W19473 and S137836.

REFERENCES

1. Aellig, M. R., Lazarus, A. J., and Steinberg, J. T., *Geophys. Res. Lett.* **28**, (2001) 2767–2770.
2. Allende Prieto, C., Lambert, D. L., and Asplund, M., *Astrophys. J. Lett.* **556**, (2001) L63–L66.
3. Anders, E., and Grevesse, N., *Geochim. Cosmochim. Acta* **53**, (1989) 197–214.
4. Bahcall, J. N., Pinsonneault, M. H., and Basu, S., *Astrophys. J.* **555**, (2001) 990–1012.
5. Banerjee, D., Teriaca, L., Doyle, J. G., and Lemaire, P., *Sol. Phys.* **194**, (2000) 43–58.
6. Basu, S., and Antia, H. M., *Mon. Not. R. Astr. Soc.* **276**, (1995) 1402–1408.
7. Benkert, J.-P., Baur, H., Signer, P., and Wieler, R., *J. Geophys. Res.* **98**, (1993) 13,147–13,162.
8. Bochsler, P., *Reviews of Geophysics* **38**, (2000) 247–266.
9. Bochsler, P., Geiss, J., and Maeder, A., *Sol. Phys.* **128**, (1990) 203–215.
10. Bochsler, P., Ipavich, F. M., Paquette, J. A., Weygand, J. M., and Wurz, P., *J. Geophys. Res.* **105**, (2000) 12,659–12,666.
11. Breneman, H. H., and Stone, E. C., *Astrophys. J.* **299**, (1985) L57–L61.
12. Brun, A. S., Turck-Chièze, S., and Zahn, J. P., *Astrophys. J.* **525**, (1999) 1032–1041.
13. Cook, W. R., Stone, E. C., and Vogt, R. E., *Astrophys. J.* **279**, (1984) 827–838.
14. DeForest, C. E., Plunkett, S. P., and Andrews, M. D., *Astrophys. J.* **546**, (2001) 569–575.
15. Esser, R., and Edgar, R. J., *Astrophys. J. Lett.* **532**, (2000) L71–L74.
16. Fisk, L. A., Zurbuchen, T. H., and Schwadron, N. A., *Astrophys. J.* **521**, (1999) 868–877.
17. Geiss, J., Bühler, F., Cerutti, H., Eberhardt, P., and Filleux, C., in *Apollo-16 Preliminary Science Report*, NASA-SP 315, 1972 pp. 14/1–14/10.
18. Geiss, J., Gloeckler, G., von Steiger, R., Balsiger, H., Fisk, L. A., Galvin, A. B., Ipavich, F. M., Livi, S., McKenzie, J. F., Ogilvie, K. W., and Wilken, B., *Science* **268**, (1995) 1033–1036.
19. Giordano, S., Antonucci, E., Noci, G., Romoli, M., and Kohl, J. L., *Astrophys. J. Lett.* **531**, (2000) L79–L82.
20. Gloeckler, G., and Geiss, J., *Space Sci. Rev.* **86**, (1998) 127–159.
21. Grevesse, N., and Sauval, A. J., *Space Sci. Rev.* **85**, (1998) 161–174.
22. Grevesse, N., and Sauval, A. J., *Adv. Space Res.* In press.
23. Heber, V. S., Baur, H., and Wieler, R., in *64th Meteoritical Society Conference*, Rome, Italy, 2001 .
24. Holweger, H., in *Solar and Galactic Composition*, edited by R. F. Wimmer-Schweingruber, vol. this volume, Woodbury, NY: AIP conference proceedings, 2001 .
25. Ipavich, F. M., Bochsler, P., Lasley, S., Paquette, J., and Wurz, P., *Eos Trans. AGU* **80** (17), (2000) S256.
26. Kern, O., Wimmer-Schweingruber, R. F., Bochsler, P., and Hamilton, D. C., in *Correlated Phenomena at the Sun, in the Heliosphere and in Geospace*, edited by A. Wilson, vol. 415 of *ESA SP*, Noordwijk, The Netherlands: ESA Publ. Div., 1997 pp. 345 – 348.
27. Ko, Y.-K., Fisk, L. A., Geiss, J., Gloeckler, G., and Guhathakurta, M., *Sol. Phys.* **171**, (1997) 345–361.

28. Leske, R. A., Mewaldt, R. A., Cohen, C. M. S., Cummings, A. C., Stone, E. C., Wiedenbeck, M. E., Christian, E. R., and von Rosenvinge, T. T., *Geophys. Res. Lett.* **26**, (1999) 2693+.
29. Marsch, E., von Steiger, R., and Bochsler, P., *Astron. Astrophys.* **301**, (1995) 261–276.
30. Murer, C., Baur, H., Signer, P., and Wieler, R., *Geochim. Cosmochim. Acta* **61**, (1997) 1303–1314.
31. Palme, H., and Beer, H., in *Landolt-Börnstein - Astronomy and Astrophysics*, vol. VI/3A of *New Series*, Heidelberg: Springer, 1993 pp. 199–222.
32. Paquette, J. A., Ipavich, F. M., Lasley, S. E., Wurz, P., and Bochsler, P., in *Solar and Galactic Composition*, edited by R. F. Wimmer-Schweingruber, vol. this volume, Woodbury, NY: AIP conference proceedings, 2001 .
33. Patsourakos, S., Habbal, S.-R., Vial, J.-C., and Hu, Y. Q., in *Solar and Galactic Composition*, edited by R. F. Wimmer-Schweingruber, vol. this volume, Woodbury, NY: AIP conference proceedings, 2001 .
34. Pedroni, A., and Begemann, F., *Meteoritics* **29**, (1994) 632–642.
35. Richard, O., Vauclair, S., Charbonnel, C., and Dziembowski, W. A., *Astron. Astrophys.* **312**, (1996) 1000–1011.
36. Rogers, F. J., and Iglesias, C. A., *Space Sci. Rev.* **85**, (1998) 61–70.
37. Signer, P., Baur, H., and Wieler, R., in *Alfred O. Nier Symposium on Inorganic Mass Spectrometry*, Durango, Colorado, 1993 pp. 181–202.
38. Thuan, T. X., and Izotov, Y. I., *Space Sci. Rev.* **84**, (1998) 83–94.
39. Turck-Chièze, S., *Space Sci. Rev.* **85**, (1998) 125–132.
40. Vauclair, S., *Space Sci. Rev.* **85**, (1998) 71–78.
41. von Steiger, R., *Space Sci. Rev.* **85**, (1998) 407–418.
42. von Steiger, R., and Geiss, J., *Astron. Astrophys.* **225**, (1989) 222–238.
43. von Steiger, R., Geiss, J., and Gloeckler, G., in *Cosmic Winds and the Heliosphere*, edited by J. R. Jokipii, C. P. Sonett, and M. S. Giampapa, Tucson: The University of Arizona Press, 1997 pp. 581–616.
44. von Steiger, R., Schwadron, N. A., Fisk, L. A., Geiss, J., Gloeckler, G., Hefti, S., Wilken, B., Wimmer-Schweingruber, R. F., and Zurbuchen, T. H., *J. Geophys. Res.* **105**, (2000) 27,217–27,236.
45. von Steiger, R., Wimmer-Schweingruber, R. F., Geiss, J., and Gloeckler, G., *Adv. Space Res.* **15**, (1995) (7)3–(7)12.
46. Weygand, J., Wurz, P., Bochsler, P., Paquette, J. A., and Ipavich, F. M., in *Solar and Galactic Composition*, edited by R. F. Wimmer-Schweingruber, vol. this volume, Woodbury, NY: AIP conference proceedings, 2001 .
47. Wieler, R., and Baur, H., *Meteoritics* **29**, (1994) 570–580.
48. Wieler, R., and Baur, H., *Astrophys. J.* **453**, (1995) 987–997.
49. Wieler, R., Kehm, K., Meshik, A. P., and Hohenberg, C. M., *Nature* **384**, (1996) 46–49.
50. Wiens, R. C., Huss, G. R., and Burnett, D. S., *Meteoritics and Planetary Science* **34**, (1999) 99–107.
51. Wilhelm, K., Marsch, E., Dwivedi, B. N., Hassler, D. M., Lemaire, P., Gabriel, A., and Huber, M. C. E., *Astrophys. J.* **500**, (1998) 1023–1038.
52. Wimmer-Schweingruber, R. F., Bochsler, P., and Wurz, P., in *Solar Wind Nine*, edited by S. R. Habbal, R. Esser, J. V. Hollweg, and P. A. Isenberg, AIP Conference Proceedings, Woodbury, NY: AIP Press, 1999 pp. 147–152.

Photospheric Abundances: Problems, Updates, Implications

H. Holweger

Institut für Theoretische Physik und Astrophysik, Universität Kiel, 24098 Kiel, Germany

Abstract.

Current problems encountered in the spectroscopic determination of photospheric abundances are outlined and exemplified in a reevaluation of C, N, O, Ne, Mg, Si, and Fe, taking effects of NLTE and granulation into account. Updated abundances of these elements are given in Table 2.

Specific topics addressed are (1) the correlation between photospheric matter and C I chondrites, and the condensation temperature below which it breaks down (Figure 1), (2) the question whether the metallicity of the Sun is typical for its age and position in the Galaxy.

INTRODUCTION

The Anders-Grevesse (1989) abundance table [1] has become a widely used reference for solar-system abundances. Follow-up compilations have appeared, the most recent one we are aware of being that by Grevesse and Sauval (1998) [2].

Most, if not all of the photospheric abundance determinations compiled rely on two basic approximations commonly made in abundance work on solar-type stars: local thermodynamic equilibrium (LTE) and onedimensional photospheric models.

Meanwhile deviations from LTE can be taken into account through detailed modeling of radiative and collisional processes, although important atomic data are still lacking (see below). Apart from NLTE abundance corrections, elaborate 2D and 3D hydrodynamical models of stellar convection coupled with radiative transfer are becoming available. It is the aim of the present contribution to provide a small data base for a few key elements that permits to assess the magnitude of the effects of NLTE and solar granulation in photospheric abundance determinations.

THE COMPLEXITY OF PHOTOSPHERIC LINE FORMATION

Judged from the simple Gaussian shape of most photospheric absorption lines, the solar photosphere might be regarded as an easy-to-model absorption tube. In reality it is a highly inhomogeneous and dynamic plasma permeated by an intense, anisotropic radiation field whose spectral energy distribution differs from that of a black body. It is not at all evident, although commonly assumed, that local thermodynamic equilibrium (LTE) should hold, i.e. that quantum-mechanical states of atoms, ions, and molecules are populated according to the convenient relations of Boltzmann and Saha, valid strictly only in thermodynamic equilibrium. Fortunately one can model this plasma with NLTE calculations, taking the essential microscopic processes into account. Such simulations allow to identify the processes that drive the plasma away from LTE, as well as those that act to thermalize it. It has become clear that departures from LTE arise from the above mentioned deviations from black-body radiation via bound-bound and bound-free radiative transitions, while the main thermalizing processes in solar-type stars are inelastic collisions with electrons and hydrogen atoms. Unfortunately accurate cross-sections are scarce. For the potentially dominating hydrogen collisions only a coarse estimate is available (see, e.g. [3]).

Absorption-line spectroscopy requires counting photons that were removed from the continuous spectrum by the species of interest. However, the picture of a 'reversing layer' that absorbs an underlying continuum is much too simple. Re-emission of both continuum and line photons occurs at all heights, although absorption generally dominates if temperature decreases outward. In order to determine elemental abundances one needs to know these processes in all atmospheric layers that contribute to the formation of the continuous and line spectrum. This implies knowledge of temperature and other variables as a function of height. In addition, horizontal and vertical fluctuations associated with convection and waves have, in principle, to be taken into account.

Another complication arises if the line considered is

CP598, *Solar and Galactic Composition*, edited by R. F. Wimmer-Schweingruber

strong enough to show effects of saturation, i.e. the linear relation between equivalent width and the product of f-value and abundance breaks down. The deviation from linearity depends on line-broadening mechanisms that have traditionally introduced substantial uncertainties in stellar spectroscopy: Doppler broadening by nonthermal velocity fields ('microturbulence'), mostly of convective origin, and broadening by collisions with ambient particles, primarily hydrogen atoms and electrons. A more detailed discussion of these problems may be found e.g. in [3].

UPDATED PHOTOSPHERIC ABUNDANCES OF SELECTED ELEMENTS

In this section we discuss and reevaluate recent data available for a few important elements including iron and oxygen.

Departures from LTE in the solar photosphere will be taken into account as outlined in [4], employing model atoms described in the respective sections. The HM photospheric model [5] is adopted as a default. This model has been used in many of the abundance determinations on which the above mentioned compilations are based. To illustrate the model-dependence of the results, calculations for iron and oxygen have also been carried out with the VAL model [6]. The latter is widely used as a reference model for chromospheric studies. However, its photospheric structure fails to reproduce the excitation equilibrium of temperature sensitive molecules like OH [7] [8] [9] and CO [10]. The NLTE abundance corrections are summarized in Table 2.

As a novel feature this analysis also considers the effects of photospheric temperature inhomogeneities associated with convection. Representing the mean vertical temperature structure by the empirical HM model, abundance corrections are applied to account for horizontal temperature inhomogeneities associated with granulation, following the approach described by Steffen [11], based on 2D numerical models of solar granulation and LTE line formation. For the purpose of this study abundance corrections were kindly made available by M. Steffen (private communication). They are summarized in Table 2.

We note that effects of NLTE and granulation are treated here as independent, second-order effects rather than attempting an ab-initio approach that includes everything. The smallness of the effects (Table 2) is consistent with this differential approach, which has the advantage of being computationally affordable and allowing to assess the importance of either effect in the Sun and in other solar-type stars.

In the following sections individual elements are discussed in a somewhat unconventional order, starting with iron and oxygen, two important cases where a more detailed report will be given.

Iron

In the Anders-Grevesse abundance table [1] the recommended photospheric iron abundance is $\log N_{\rm Fe} = 7.67 \pm 0.03$. In the recent compilation by Grevesse and Sauval [2] a value of $\log N_{\rm Fe} = 7.50 \pm 0.05$ is quoted. Note that the abundance has changed significantly, and the assessment of error limits is now more conservative.

Intriguingly, error limits quoted in the literature are defined in two different ways. The more optimistic version is the error of the mean abundance (derived from individual lines), while the standard deviation of individual lines is more conservative. Both differ by a factor $\sqrt{n}$, where n is the number of lines involved. The entry for iron (and for many other elements) in the Anders-Grevesse abundance table [1] conform to the former prescription, while the more conservative version is used throughout the present paper. In this version errors of individual lines are not assumed to be independent, with the reasoning that errors depending on wavelength, equivalent width, or excitation cannot be excluded.

The recommended iron abundance is based exclusively on lines of ionized iron. In the solar photosphere iron occurs mainly as Fe II, hence these lines are much less sensitive to the thermal structure of the photospheric model than Fe I lines are. In addition, a number of independent sources of f-values are available, both experimental and theoretical. Previous Fe II analyses [12] [13] have consistently inferred a lower iron abundance than that adopted in the Anders-Grevesse compilation, $\log N_{\rm Fe} = 7.48$. Recently, Schnabel et al. [14] (SKH99) have used improved lifetime measurements for Fe II levels in a new abundance analysis based on the line list of [12], inferring an even lower value, 7.42 ± 0.09. The quoted error limits conform to the more conservative prescription mentioned above. The mean abundance following from all 13 Fe II lines in Table 2 of SKH99 (entering the three infrared lines with half weight) is 7.419 with a standard deviation of 0.082.

For each of the 13 Fe II lines NLTE abundance corrections $\Delta \log N = \log N_{\rm NLTE} - \log N_{\rm LTE}$ have been calculated using a model atom described in [15]. NLTE effects in Fe II lines are extremely small, $|\Delta \log N_{\rm Fe}| \leq 0.001$, and mean LTE and NLTE abundances agree to within 0.001 dex. The photospheric iron abundance, including NLTE effects, but still without granulation corrections, becomes $\log N_{\rm Fe} = 7.419 \pm 0.082$.

If the HM model is replaced by the VAL model, the

mean abundance increases by 0.049 dex both in LTE and NLTE.

The effect of granulation on the iron abundance has been determined for a small number of Fe II lines representative of the SKH99 sample, leading to a mean correction of +0.029 dex. The recommended photospheric iron abundance is then obtained by combining the (1D) LTE result with corrections for effects of NLTE (1D) and granulation (2D): $\log N_{\rm Fe} = 7.448 \pm 0.082$.

Oxygen

The photospheric abundance listed in the Anders-Grevesse table [1] is $\log N_{\rm O} = 8.93 \pm 0.035$, while Grevesse and Sauval [2] recently have recommended $\log N_{\rm O} = 8.83 \pm 0.06$. Again, a significant change in the preferred abundance and a trend towards more conservative error limits is to be noted.

The oxygen abundance inferred here is derived from atomic lines, while the Anders-Grevesse value is based on an unpublished analysis of the CNO group using molecular lines. Like Fe II, O I is the dominant species, while molecules such as OH are trace constituents whose formation is quite temperature sensitive. Grevesse & Sauval [2] report CNO abundances which they have derived from molecular as well as atomic lines. Unfortunately no details are given and the authors regard the result (quoted above) as preliminary. Previous analyses based exclusively on atomic lines [16] [17] have led to closely coincident abundances of 8.86 and 8.87, respectively. However, this agreement is somewhat surprising in view of the different sets of lines and f-values that have been used. Moreover in [16] LTE is assumed and the HM model is used, while a detailed NLTE analysis is carried out in [17] in combination with a flux-constant ATLAS6 model. We feel that this situation needs clarification and have reevaluated the O I data as follows.

The oxygen lines analyzed are listed in Table 1. Our sample is based on the line list given in [18], supplemented by four additional infrared lines from [16]. In order to minimize blend problems, lines with uncertain profiles were omitted because they are likely to be contaminated by other atomic or molecular species whose contribution, if not properly accounted for, will inevitably lead to fictitiously high oxygen abundances. Specifically, all three lines marked ':' in Table 5 of [18] have been rejected ($\lambda\lambda$5577.3, 6156.8, and 6363.8 Å) together with the multiplet at 8446 Å which is strongly perturbed by Fe I. Our selection is in accordance with [19]. Of the infrared lines listed in [16] λ9262.8 Å is not accepted because its profile is badly blended. In our line formation calculations of multiplets 67 and 64 fine structure splitting is taken into account. In both cases the combined equivalent width of the multiplet is quoted in Table 1, and only one abundance value is assigned to each triplet. Equivalent widths refer to the center of the solar disk. All $\log gf$ values were taken from the NIST data base [20]; most of them are based on quantum-mechanical calculations [21]. LTE abundances derived from individual lines are presented in Table 1. Assigning equal weight to all lines, a mean LTE abundance of 8.780 results, with a standard deviation of 0.071.

Our LTE abundance is 0.08 dex lower than that inferred by Biémont et al. [16]. The line abundances quoted in Table 3 of [16] permit to trace the reason for this difference. Most of it is due to lines we have not used because of less reliable profiles; if the four lines in question ($\lambda\lambda$6156.8, 8446.3/8446.8 and 9262.8 Å) are excluded from the Biémont et al. sample and the remaining eight lines in their Table 3 given equal weight, the mean abundance decreases to 8.803. The residual 0.023 dex difference can be attributed to slightly different f values and equivalent widths. The four perturbed lines, if averaged separately, lead to a 0.120 dex higher abundance, illustrating the bias towards higher abundances that may arise if blends are included in abundance determinations.

The blend problem has also been addressed by Reetz [22] in the context of the two well-known [O I] lines ($\lambda\lambda$ 6300.3/6363.8 Å). These lines have been entered with high weight in many solar and stellar abundance determinations since they were considered as safe with respect to NLTE effects. While this is certainly true (see Table 1), their extreme faintness makes them very sensitive to blends. Indeed, Reetz argues that λ6363.8 (which we have not used) is more strongly perturbed by a CN line than realized previously, and that λ6300.3 may contain contributions by a Ni I line. This could explain why λ6300.3 yields the highest abundance of our sample. Detailed spectrum synthesis of the [O I] line regions using all available data bases may help to solve this problem.

NLTE calculations for oxygen have been carried out employing a model atom described in [23]. Deviations from LTE in O I are most important for the strongest lines, amounting to $\Delta\log N_{\rm O} = -0.065$ for λ7771.9 Å (Table 1). The mean NLTE correction is $\Delta\log N_{\rm O} = -0.028$. In all NLTE calculations inelastic collisions with H atoms have been taken into account, assuming a scaling factor $S_{\rm H} = 1.0$ in accordance with Takeda [17]. The photospheric oxygen abundance, including NLTE effects but without granulation corrections, becomes $\log N_{\rm O} = 8.752 \pm 0.078$.

The LTE and NLTE results quoted refer to the HM photospheric model. If the VAL model is adopted instead, the LTE abundance increases by 0.072 dex. NLTE effects are hardly different; the average NLTE abundance correction is now $\Delta\log N_{\rm O} = -0.027$ (VAL) instead of -0.028 (HM). Thus the NLTE abundance derived from

TABLE 1. Photospheric oxygen lines used for abundance analysis

Wavelength (Å)	Mult.	E.P.(eV)	log gf	W (mÅ)	log $N_{O,LTE}$	log $N_{O,NLTE}$	Δ_{gran}
6158.15	67	10.741	-1.841	...	...	...	...
6158.17	67	10.741	-0.996	...	...	...	...
6158.19	67	10.741	-0.409	5.0	8.780	8.775	−0.085
6300.30	92	0.000	-9.774	4.3	8.921	8.921	+0.015
7771.94	56	9.146	0.396	88.0	8.834	8.769	+0.015
7774.17	56	9.146	0.223	71.0	8.776	8.714	+0.007
7775.39	56	9.146	0.001	54.0	8.741	8.694	+0.000
9265.83	64	10.741	-0.719	...	...	...	...
9265.93	64	10.741	0.126	...	...	...	...
9266.01	64	10.741	0.712	35.6	8.718	8.695	−0.028
11302.4	63	10.741	0.076	14.0	8.697	8.684	−0.033
13164.9	76	10.989	-0.033	16.3	8.772	8.766	−0.017

the VAL model is 0.073 dex higher than the corresponding HM result.

Granulation corrections for the individual O I lines are given in the last column of Table 1. Note that corrections may have either sign, depending on excitation and line strength. On average, abundances derived from conventional 1D models have to be adjusted by -0.016 dex if the sample of Table 1 is adopted for analysis. The recommended photospheric oxygen abundance is the combination of the LTE result with corrections for effects of NLTE and granulation: $\log N_O = 8.736 \pm 0.078$.

Carbon

The photospheric abundance listed in the Anders-Grevesse table [1], $\log N_C = 8.56 \pm 0.04$, is based on the same analysis of CNO-group molecular lines used also for oxygen. Grevesse and Sauval [2] suggest $\log N_C = 8.52 \pm 0.06$ as a preliminary result of an unpublished revised evaluation of both molecular and atomic species.

As mentioned above, molecular lines are quite temperature sensitive. Our recommended value is based on published analyses of C I lines [24] [25]. The LTE carbon abundances derived from Table 2 of [24] and Table 2 of [25] are 8.583 ± 0.133 and 8.600 ± 0.098, respectively. The detailed NLTE calculations of the former paper [24] have shown that in C I NLTE effects depend strongly on equivalent width, and a mean NLTE correction of -0.05 dex was derived from this sample. For the sample of [25] we find a mean NLTE correction of -0.04 dex. Combining the respective LTE abundances and NLTE corrections and taking the average of both determinations leads to a NLTE abundance of $\log N_C = 8.571 \pm 0.108$.

Granulation corrections were determined for one representative line of either sample rather than for the dozens of lines. Its wavelength, excitation potential, and equivalent width were chosen as the mean value of each of the two samples. This resulted in two sets of line parameters, (9800 Å, 7.8 eV, 90 mÅ) for the sample of [24], and (12700 Å, 8.7 eV, 83 mÅ) for the [25] sample. The corresponding granulation corrections are +0.023 and +0.019 dex, respectively (M. Steffen, private communication).

The final photospheric carbon abundance is derived by combining the respective LTE abundances and corrections for NLTE and granulation, and taking the average of both samples. The recommended value is $\log N_C = 8.592 \pm 0.108$.

Nitrogen

The photospheric abundance listed in the Anders-Grevesse table [1] is $\log N_N = 8.05 \pm 0.04$ has resulted from the same (unpublished) analysis of CNO-group molecular used also for C and O. Grevesse and Sauval [2] suggest a significantly lower value, $\log N_N = 7.92 \pm 0.06$, derived from a preliminary analysis of atomic and molecular lines.

In analogy to carbon and oxygen our recommended value is based on a NLTE analysis of photospheric N I lines [15]. Table 2 of [15] leads to a mean NLTE correction of -0.032 dex and a NLTE abundance of $\log N_N = 8.049 \pm 0.097$. The f-values employed were taken from the Opacity Project data base. In the meantime a NIST compilation has appeared [20], which is used here. On average the new NIST $\log gf$ values are 0.048 dex larger than the OP values (for the line sample of [15]), implying a downward readjustment of the nitrogen abundance by the same amount for a given set of equivalent widths. Omitting the highly discrepant line at 10575.9 Å whose f-value is obviously in error, the new NLTE abundance becomes $\log N_N = 8.001 \pm 0.111$.

As in the case of carbon, granulation effects are derived for a 'mean' N I line with parameters representative of the sample used. The mean line is characterized by (9600 Å, 11.1 eV, 3.8 mÅ), implying that the lines are all weak, and of high excitation. Line of this type are quite sensitive to temperature inhomogeneities [11]. In the case of N I the granulation correction amounts to -0.070 dex. Taking this into account leads to the recommended photospheric abundance of $\log N_{\rm N} = 7.931 \pm 0.111$.

Magnesium

Identical entries for magnesium are found in the abundance tables of Anders-Grevesse [1] and Grevesse-Sauval [2], $\log N_{\rm Mg} = 7.58 \pm 0.05$. This value dates back to the 1984 review of Grevesse [26].

Our recommended value is based on an analysis of Mg II lines [27] because this one-electron system permits accurate quantum-mechanical calculations of f-values, while this is notoriously difficult for Mg I. The LTE abundance [27] is $\log N_{\rm Mg} = 7.54 \pm 0.06$. NLTE calculations for magnesium in the Sun [28] have focused on Mg I. The resulting departures from LTE are very small throughout the photosphere. Since magnesium is mostly present as Mg II, we expect even smaller departures for this dominant species. Hence zero NLTE corrections are assumed for the solar magnesium abundance, which is based exclusively on Mg II lines. This is supported by existing NLTE calculations for Mg II in A-type stars. Even though the radiation field of an A star with $T_{\rm eff}$ = 9500 K deviates from blackbody radiation much more than that of the Sun, NLTE corrections are typically only -0.04 dex [29].

The granulation correction for a representative Mg II line with mean parameters (8800 Å, 9.4 eV, 45 mÅ) was determined to be -0.002 dex. We note that this is not typical for high-excitation Mg II lines in general, but is due to cancellation of the dependence of granulation effects on excitation and on line strength. The recommended photospheric abundance is $\log N_{\rm Mg} = 7.538 \pm 0.060$.

Silicon

Like Mg, the Si abundance $\log N_{\rm Si} = 7.55 \pm 0.05$ listed in the Anders-Grevesse [1] and Grevesse-Sauval [2] tables was adopted from the 1984 review by Grevesse [26]. It is based on the LTE analysis of Si I and Si II by Becker et al. [30].

Recently Asplund [31] has carried out a re-analysis using a 3D hydrodynamical model. The result, $\log N_{\rm Si} = 7.51 \pm 0.04$, includes effects of granulation but rests on the assumption of LTE.

In an independent parallel study of essentially the same line sample Wedemeyer [32] has derived an LTE abundance of 7.560 ± 0.066 in close agreement with the abundance tables, using a conventional 1D model. His study also presents detailed NLTE calculations. For the combined Si I/II line sample an average NLTE correction of -0.010 dex is derived from standard 1D modeling, thus the NLTE result (without granulation effects) becomes 7.550 ± 0.056. The smaller standard deviation in NLTE as compared to LTE is mainly due to the good fit of Si I and Si II achieved by the NLTE calculations.

Wedemeyer [32] quotes a granulation correction of +0.021 dex, derived by Steffen [11] in the same way as the corrections for the other elements. The corresponding LTE abundance including granulation correction, is 7.581 ± 0.066. which may be compared with Asplund's [31] value of 7.51 ± 0.04. Although both values agree within their mutual error limits, the 0.07 dex difference is somewhat disturbing and deserves further investigation.

We adopt the mean of both determinations and assume the same NLTE correction of -0.010 dex for both analyses. The recommended photospheric abundance thus becomes $\log N_{\rm Si} = 7.536 \pm 0.049$.

Neon

The abundance listed in the Anders-Grevesse table [1] is $\log N_{\rm Ne} = [8.09 \pm 0.10]$; Grevesse and Sauval [2] recommend $\log N_{\rm Ne} = [8.08 \pm 0.06]$. Square brackets in the tables indicate that these values were not derived from photospheric absorption lines. Although Ne I does have lines in the visible part of the spectrum, these lines are of very high excitation and hence too weak to be observable in the solar photosphere.

The Anders-Grevesse value is based on solar particle and local galactic data, while Grevesse and Sauval have adopted the more recent result obtained by Widing (1997) [33] from EUV spectroscopy of emerging active regions.

We prefer the latter source over solar particle data because, as Widing [33] argues, emerging flux events most likely permit direct observation of unfractionated photospheric material. The abundance value of 8.08 was derived in [33] from the Ne/Mg ratio following from EUV data by combining it with the Anders-Grevesse photospheric Mg abundance of 7.58. We make two modifications: (1) we use also oxygen, in addition to magnesium, as the photospheric reference element, and (2) adopt the updated photospheric values derived above.

The abundances following from emerging active regions are quoted in Table 4 of [33] in terms of abundance ratios, Ne/Mg = 3.16 ± 0.07 (based on five observations) and O/Ne = 6.75 ± 0.65 (two observations). The Ne/Mg ratio then translates into a photospheric neon abundance of 8.038 while the O/Ne ratio yields 7.907. We take the mean, assigning weight 5 to the former and weight 2 to

TABLE 2. Recommended photospheric abundances ($\log N_{\mathrm{H}} = 12$)

Element	log N	Δ_{NLTE}	Δ_{gran}	δ_{AG89}	δ_{GS98}
6 C	8.592 ± 0.108	−0.045	+0.021	+0.03	+0.07
7 N	7.931 ± 0.111	−0.032	−0.070	−0.12	+0.01
8 O	8.736 ± 0.078	−0.028	−0.016	−0.19	−0.09
10 Ne	8.001 ± 0.069	...	...	−0.09	−0.08
12 Mg	7.538 ± 0.060	∼0	−0.002	−0.04	−0.04
14 Si	7.536 ± 0.049	−0.010	+0.021	−0.01	−0.01
26 Fe	7.448 ± 0.082	+0.000	+0.029	−0.22	−0.05

the latter. The recommended photospheric abundance becomes $\log N_{\mathrm{Ne}} = 8.001 \pm 0.069$.

Update Summary

The updated abundances are summarized in Table 2. Effects of NLTE and photospheric granulation are included in the abundances, and specified separately in the next two columns. Also given are the differences between the new abundances and those compiled by Anders and Grevesse [1] (δ_{AG89}) and Grevesse and Sauval [2] (δ_{GS98}) in the sense 'new minus old'.

SOLAR AND METEORITIC ABUNDANCES

The close correlation between solar and meteoritic abundances of non-volatile elements is well known. The unique mineralogy of CI chondrites qualifies their matter as an essentially unaltered condensate of the solar nebula. One may ask if there is a well-defined condensation temperature below which the compositional correlation breaks down.

This question is addressed in Figure 1. The quantity plotted is $\log(\mathrm{X/Si})_{\mathrm{CI}} - \log(\mathrm{X/Si})_{\odot}$, the depletion of various elements in CI chondrites with respect to the solar photosphere. Included are the elements up to the iron group and two volatile heavier elements, Cd and Pb. For meteorites we have adopted data from [1], for the Sun those listed in Table 2 supplemented by data from [2]. Not included are B, F, and Cl because their abundance must be regarded as highly uncertain (F and Cl are based on sunspot spectra). For the strongly depleted elements (hydrogen and noble gases) upper limits are given. Not shown is lithium, which is depleted in the Sun by 2.2 dex due to thermonuclear destruction.

Approximate condensation temperatures have been collected from the literature; actual values will depend on the conditions under which condensation occurs. Nevertheless, Figure 1 shows a strikingly close correlation down to the condensation temperature of Pb and Cd, the most volatile condensed elements with reliable photospheric data. For Cd, $T_{\mathrm{C}} \approx 420\,\mathrm{K}$ [34]. Even oxygen ($T_{\mathrm{C}} \approx 180\,\mathrm{K}$ [35]) obviously has condensed to a remarkable degree. It seems that the solar nebula was well-mixed at the time when bulk of the CI matter condensed from the gas phase, which is believed to have occurred at a distance of several AU from the central star.

THE SUN AND THE METALLICITY OF NEARBY STARS

Is the metallicity of the Sun typical for stars of solar mass and age? This question is addressed in a recent discussion by Gustafsson [36] of the comprehensive set of data on bright F and G main sequence stars presented by Edvardsson et al. [37]. Gustafsson emphasizes that their sample was chosen to be biased towards low metallicities: the stars were selected to be evenly distributed over the metal abundance range $-1.0 > [\mathrm{Me/H}] > +0.3$. In other words, compared to a volume limited sample the number of metal deficient stars was enhanced, hence the more metal-rich stars like the Sun appear less numerous. Furthermore, we note that the selection cutoff near [Me/H] = +0.3 of the Edvardsson et al. sample will cause an additional thinning out of the sample towards the metal-rich side. All this obviously has led some authors to believe that the age-metallicity relation depicted in Figure 14a of [37] implies that the Sun is metal-rich by about 0.2 dex. However, if the sample is corrected for the low-metallicity bias (which is not trivial) and restricted to nearby stars with ages between 4 and 5 Gyr, the solar 'anomaly' reduces to $[\mathrm{Fe/H}] = -0.09 \pm 0.22$ [36].

The Edvardsson et al. [37] analysis is strictly differential in the sense that the same instrument was used for recording stellar and solar (flux) spectra, and the same suite of model atmospheres was employed for analysis. Another strictly differential analysis was carried out by Fuhrmann [38]. Again no attempt was made to establish a volume limited sample, but unlike [37] no metallicity cutoff just above the solar level was imposed. Figure 11 of [38] suggests that [Fe/H] as well as [Mg/H] of the Sun is quite typical for nearby disk dwarfs.

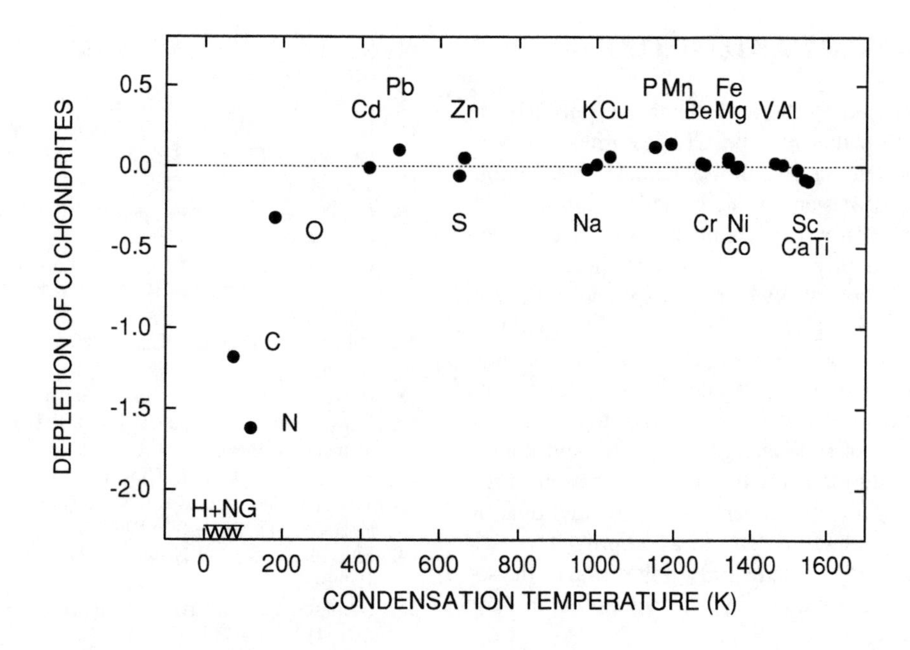

FIGURE 1. Comparison of abundances in CI chondrites with those in the solar photosphere

By contrast, another recent study of the age-metallicity relation [39] is not strictly differential but makes use of a metallicity calibration of *uvby* photometry [40] whose main source of abundance data is a compilation [41] which dates back to 1983 and includes a variety of earlier stellar and solar abundance determinations. While the clear trend of metallicity with age shown in Figure 13 of [39] is not affected by this uncertainty, the zero point of the [Fe/H] scale - which apparently indicates that the Sun is anomalously metal-rich - must be regarded as preliminary.

Probably the most accurate data set currently available is a by-product of the search for extrasolar planets. A sample of 77 single G dwarfs was used by Butler et al. [42] for comparison with planet-bearing stars. It represents a volume-limited set of field stars whose photometric metallicities were calibrated with strictly differential spectroscopic analyses. Figure 7 of [42] clearly shows that the Sun is about 0.1 dex more iron rich than nearby field stars (and that stars with detected planets are even more metal-rich). Since the age of the solar neighborhood is more than twice the age of the Sun [43], an enhanced solar metallicity fits qualitatively into the picture of galactic chemical evolution, although the metal enrichment during $\sim$ 5 Gyr predicted by most models is larger than 0.1 dex. Thus the Sun would appear to be metal-poor for its age, rather than metal-rich! Possibly the concept of a strict age-metallicity relation has to be revisited. A similar conclusion has been reached very recently [44] in view of the detection of super-metal-rich, $\sim$ 10 Gyr old stars.

THE SUN AND GALACTIC ABUNDANCE GRADIENTS

Does the Sun fit into the galactic abundance gradients determined from various nonstellar and stellar sources? While a general review of this topic is beyond the scope of this paper, we briefly mention some consequences of the updated solar abundances listed above (Table 2).

We may take advantage of the large body of observational data on galactic H II regions, planetary nebulae, and B stars recently compiled by Hou et al. [45]. In their Figure 6 - which may serve as a basis for our discussion - abundances of various elements are plotted versus galactocentric distance R_G, showing a rather large scatter and, in most cases, a more or less pronounced abundance gradient. The scatter possibly reflects real variations from object to object as well as uncertainties in the determinations. In any case it permits to assess whether the Sun (R_G = 8.5 kpc) fits into the overall picture emerging from recent data.

Close inspection of Figure 6 of [45], with the updated abundances of C, N, O, Ne, Mg, and Si taken into account, shows that the Sun is quite typical, even in the case of oxygen whose previous solar value seemed higher than typical. However, the solar carbon abundance appears to be ~0.4 dex higher compared to that derived from galactic B stars. It is highly improbable that the solar value is in error by this large amount. This important element deserves further study.

SUMMARY AND OUTLOOK

The elemental composition of the solar photosphere closely resembles that of type CI meteorites down to a condensation temperature of 400 K and agrees within 0.1 dex with that of stars in our galactic neighborhood. This qualifies solar-system abundances as an excellent reference for galactic and extragalactic studies.

Spectroscopic observations of photospheric matter can be carried out with high precision, yet the interpretation in terms of abundances is inevitably much more indirect than counting particles. While photon spectroscopy will remain indispensable for stars other than the Sun, the measurement of solar energetic particles and the solar wind - in combination with the determination of hydrogen and with a better understanding of fractionation processes - will hopefully lead to an improved knowledge of the isotopic and elemental composition of the solar photosphere, comparable in accuracy to that already achieved for meteorites.

ACKNOWLEDGMENTS

The author is grateful to Matthias Steffen for providing granulation abundance corrections in advance to publication. Support by ESA and by the Physikalisches Institut, University of Bern, is gratefully acknowledged.

REFERENCES

1. Anders, E., and Grevesse, N., *Geochim. Cosmochim. Acta*, **53**, 197–214 (1989).
2. Grevesse, N., and Sauval, A. J., *Space Science Reviews*, **85**, 161–174 (1998).
3. Holweger, H., *Physica Scripta*, **T65**, 151–157 (1996).
4. Steenbock, W., and Holweger, H., *A&A*, **130**, 319–323 (1984).
5. Holweger, H., and Müller, E. A., *Solar Physics*, **39**, 19–30 (1974).
6. Vernazza, J. E., Avrett, E. H., and Loeser, R., *ApJS*, **30**, 1–60 (1976).
7. Goldman, A., Murcray, D. G., Lambert, D. L., and Dominy, J. F., *MNRAS*, **203**, 767–776 (1983).
8. Grevesse, N., Sauval, A. J., and van Dishoek, E. F., *A&A*, **141**, 10–16 (1984).
9. Sauval, A. J., Grevesse, N., Brault, J. W., Stokes, G. M., and Zander, R., *ApJ*, **282**, 330–338 (1984).
10. Harris, M. J., Lambert, D. L., and Goldman, A., *MNRAS*, **224**, 237–255 (1987).
11. Steffen, M., "2D numerical simulation of stellar convection", in *Stellar Astrophysics*, edited by K. e. a. Cheng, Kluwer Academic Publishers, 2000, pp. 25–36.
12. Holweger, H., Heise, C., and Kock, M., *A&A*, **232**, 510–515 (1990).
13. Hannaford, P., Lowe, R. M., Grevesse, N., and Noels, A., *A&A*, **259**, 301–306 (1992).
14. Schnabel, R., Kock, M., and Holweger, H., *A&A*, **342**, 610–613 (1999).
15. Rentzsch-Holm, I., *A&A*, **312**, 966–972 (1996).
16. Biémont, E., Hibbert, A., Godefroid, M., Vaeck, N., and Fawcett, B. C., *ApJ*, **375**, 818–822 (1991).
17. Takeda, Y., *PASJ*, **46**, 53–72 (1994).
18. Müller, E. A., Baschek, B., and Holweger, H., *Solar Physics*, **3**, 125–145 (1968).
19. Lambert, D. L., *MNRAS*, **182**, 249–272 (1978).
20. Wiese, W. L., Fuhr, J. R., and Deters, T. M., *J. Phys. Chem. Ref. Data*, **7** (1996).
21. Hibbert, A., Biémont, E., Godefroid, M., and Vaeck, N., *J. Phys. B*, **24**, 3943–3953 (1991).
22. Reetz, J., *Sauerstoff in kühlen Sternen und die chemische Entwicklung der Galaxis*, Ph.D. thesis, Universität München (1998).
23. Paunzen, E., Kamp, I., Iliev, I. K., Heiter, U., Hempel, M., Weiss, W. W., Barzova, I. S., Kerber, F., and Mittermayer, P., *A&A*, **345**, 597–604 (1999).
24. Stürenburg, S., and Holweger, H., *A&A*, **237**, 125–136 (1990).
25. Biémont, E., Hibbert, A., Godefroid, M., and Vaeck, N., *ApJ*, **412**, 431–435 (1993).
26. Grevesse, N., *Physica Scripta*, **T8**, 49–58 (1984).
27. Holweger, H., "Abundances of the elements in the Sun", in *Les Elements et leurs Isotopes dans l'Univers*, Institut d'Astrophysique, Université de Liège, 1979, pp. 117–127.
28. Lemke, M., and Holweger, H., *A&A*, **173**, 375–382 (1987).
29. Gigas, D., *A&A*, **192**, 264–274 (1988).
30. Becker, U., Zimmermann, P., and Holweger, H., *Geochimica et Cosmochimica Acta*, **44**, 2145–2149 (1980).
31. Asplund, M., *A&A*, **359**, 755–758 (2000).
32. Wedemeyer, S., *A&A*, **373**, 988–1008 (2001).
33. Widing, K. G., *ApJ*, **480**, 400–405 (1997).
34. Meyer, J.-P., Drury, L. O., and Ellison, D. C., *ApJ*, **487**, 182–196 (1997).
35. Field, G. B., *ApJ*, **187**, 453–459 (1974).
36. Gustafsson, B., *Space Science Reviews*, **85**, 419–428 (1998).
37. Edvardsson, B., Andersen, J., Gustafsson, B., Lambert, D. L., Nissen, P. E., and Tomkin, J., *A&A*, **275**, 101–152 (1993).
38. Fuhrmann, K., *A&A*, **338**, 161–183 (1998).
39. Rocha-Pinto, H. J., Maciel, W. J., Scalo, J., and Flynn, C., *A&A*, **358**, 850–868 (2000).
40. Schuster, W. J., and Nissen, P. E., *A&A*, **221**, 65–77 (1989).
41. Cayrel de Strobel, G., and Bentolila, C., *A&A*, **119**, 1–13 (1983).
42. Butler, R. P., Vogt, S. S., Marcy, G. W., Fischer, D. A., Henry, G. W., and Apps, K., *ApJ*, **545**, 504–511 (2000).
43. Binney, J., Dehnen, W., and Bertelli, G., *MNRAS*, **318**, 658–664 (2000).
44. Feltzing, S., and Gonzalez, G., *A&A*, **367**, 253–265 (2001).
45. Hou, J. L., Prantzos, N., and Boissier, S., *A&A*, **362**, 921–936 (2000).

Elemental and Isotopic Abundances in Meteorites

P. Hoppe

Max-Planck-Institute for Chemistry, Cosmochemistry Department, P.O. Box 3060, D-55020 Mainz, Germany

Abstract. Abundance variations of refractory elements between different groups of undifferentiated meteorites (chondrites) are within a factor of 2. The elemental abundance pattern of CI chondrites matches that of the solar photosphere reasonably well except for Li, which is depleted in the Sun's convection zone, and the volatile elements H, C, N, O, and noble gases which are incompletely condensed in chondrites. Macroscopic-scale isotopic heterogeneities are largest for H (D/H varies by 8x) and N ($^{15}N/^{14}N$ varies by 2x). Bulk C- and O-isotopic compositions do not vary by more than a few percent and isotopic heterogeneities are even much smaller for refractory elements. The variations in bulk elemental and isotopic compositions are attributed to variations in the nebula environment from which the chondrites formed. Some meteoritic components (calcium-aluminum-rich inclusions and chondrules) also carry the decay products of now extinct short-lived radioactive nuclides which were either produced locally in the early solar system or in a stellar source shortly before seeding the solar nebula. Large isotopic anomalies with variations over more than four orders of magnitude are observed on a microscopic scale in chondrites. These anomalies are carried by presolar grains that formed mainly in the winds of red giant and asymptotic giant stars and in the ejecta of supernova explosions.

INTRODUCTION

Our solar system formed from the collapse of a molecular cloud about 4.57 Gy ago, possibly triggered by a nearby supernova (SN) explosion [see 1]. The release of gravitational energy led to the evaporation of a large fraction of dust grains present in the solar nebula and much of the nucleosynthetic memories carried by these grains was erased by chemical and isotopic equilibration. New minerals condensed out of the cooling nebula and an accretion disk formed around the young Sun. Formation of solids probably started about several 100,000 y after onset of molecular cloud collapse and lasted for at least several million years as inferred from the presence of the decay products of short-lived radionuclei, such as ^{41}Ca (half-life = 103,000 y) and ^{26}Al (half-life = 730,000 y) in primitive solar system matter [e. g. 2, 3]. According to modern models of planet formation [see 4] dust accumulates by low-velocity collisions and non-gravitational sticking mechanisms to km-sized bodies, the planetesimals. Mutual collisions of planetesimals finally led to the formation of the solar system planets and asteroids and fragments of those bodies eventually may reach the Earth as meteorites (Fig. 1). While most meteorites originate from asteroids some of them are from the Moon or from Mars.

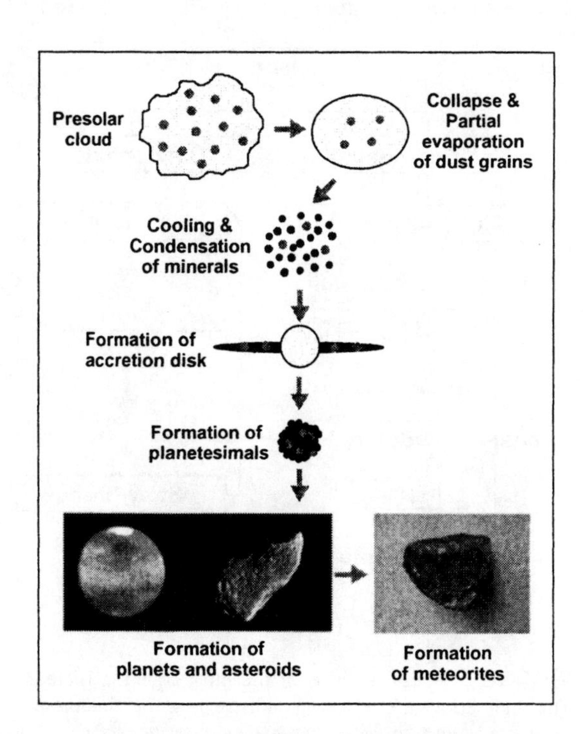

FIGURE 1. Origin and history of meteorites.

Although models of solar system formation do not support large chemical gradients in the inner solar system [5], elemental and isotopic heterogeneities may have existed among the formation locations of planetary bodies to some extent. The magnitude of such heterogeneities

CP598, *Solar and Galactic Composition*, edited by R. F. Wimmer-Schweingruber

in solar system matter depends on the mixing of compounds that formed at different locations or times in the solar nebula and on the possible admixture of molecular cloud material during the early evolution of the solar system. Isotopic heterogeneities among solar system materials are also expected from the presence of short-lived radioactive nuclides in the early solar system whose decay will leave characteristic imprints on the isotopic patterns of the daughter elements.

Meteorites represent a sample of solar system matter from different locations in the solar nebula that can be studied with high precision in the laboratory. Meteorites can be divided into differentiated and undifferentiated meteorites (Fig. 2). Differentiated meteorites are further subdivided into achondrites, irons, and stony-irons. These meteorites have experienced strong post-accretionary alteration and their elemental abundances are not representative of the bulk compositions of their parent bodies. Undifferentiated meteorites, the chondrites, on the other hand, have preserved the bulk elemental and isotopic compositions of their parent bodies. They thus provide information on the homogeneity of elemental and isotopic abundances over large distances (from Earth to Jupiter orbits) in the solar nebula at the time of chondrite formation.

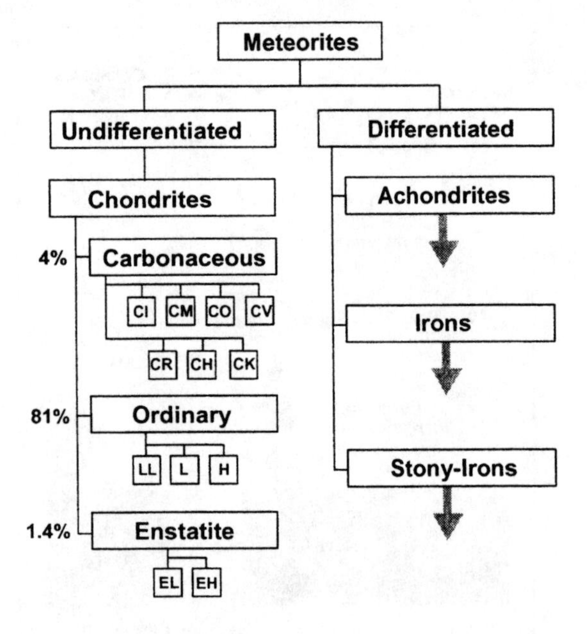

FIGURE 2. Classification of the most common meteorites. The numbers to the left of chondrites give the abundances of observed meteorite falls. The arrows indicate further subdivision of differentiated meteorites.

According to chemical composition, degree of oxydization, and degree of equilibration and metamorphic recrystallization the chondrites are divided into discrete groups. These include the carbonaceous, ordinary, and enstatite chondrites which are further subdivided into different types. The different carbonaceous chondrite types are named after a meteorite of each type (e.g., Ivuna for CI, Mighei for CM). Subdivision of ordinary and enstatite chondrites is according to Fe content (LL: low total Fe, low metallic Fe; L: low total Fe; H: high total Fe). Chondrites are mainly composed of chondrules and calcium-aluminum-rich inclusions (CAIs), mm- to cm-sized objects that experiencd high temperatures during formation in the solar nebula, and the finer grained matrix which may be considered as a glue. The relative abundances of those compounds vary from type to type. The matrix contains small amounts (ppb to ppm) of nm- to μm-sized refractory dust grains that are of presolar origin as indicated by large isotopic anomalies [6, 7]. The laboratory study of these rare objects allows to obtain a wealth of information on many astrophysical aspects.

In this paper I will discuss the elemental and isotopic homogeneity of meteorites both on a macroscopic (section 2 and 3) and microscopic scale (section 4) and to which extent meteorites may serve as a reference for bulk solar system matter (sections 2 and 3). An overview about the isotopic compositions of presolar grains will be presented and the origin of presolar grains will be briefly discussed (section 4).

ELEMENTAL ABUNDANCES

The most abundant elements in chondrites are O, Fe, Si, and Mg. Variations in elemental abundances among the different chondrites are generally small for refractory and moderately volatile elements (Fig. 3). These variations are typically within a factor of two. Highest refractory element abundances are observed in the carbonaceous chondrites and lowest in the enstatite chondrites (see compilation of abundance data in [8]). Larger variations are seen in the abundances of volatile elements such as H, N, C, and noble gases.

The small but noticable heterogeneity in refractory element abundances is attributed to variations in the nebula environment from which the chondrites formed [e. g. 9]. In contrast, the relatively large variations of highly volatile element abundances may be the result of different condensation behaviours at different locations in the nebula or of partial loss due to thermal metamorphism on the meteorite parent bodies [9].

The Si-normalized elemental abundance pattern of CI chondrites is remarkably similar to that observed in the solar photosphere [10]. The abundances of most elements agree within a factor of 1.5 and for many elements the agreement is even within a few percent (Fig. 4). Noticable exceptions are the volatile elements, such as H, C, N, O, and noble gases which are incompletely condensed in chondrites, and Li which is depleted in the solar pho-

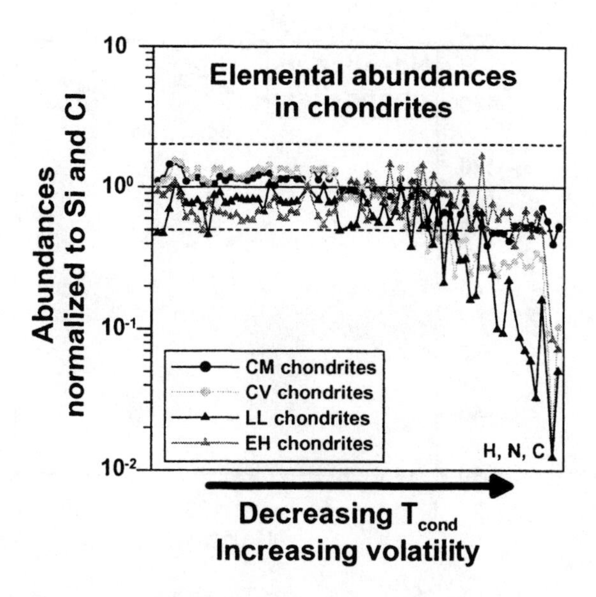

FIGURE 3. Si- and CI-normalized elemental abundances of CM, CV, LL, and EH chondrites. The elements are ordered according to decreasing condensation temperature (increasing volatility). For a pressure of 10^{-4} bar the condensation temperature on the left-hand side of this figure is 1800 K, on the right-hand side 80 K. Data from [8]. The dashed lines indicate a difference of a factor of 2 from CI abundances.

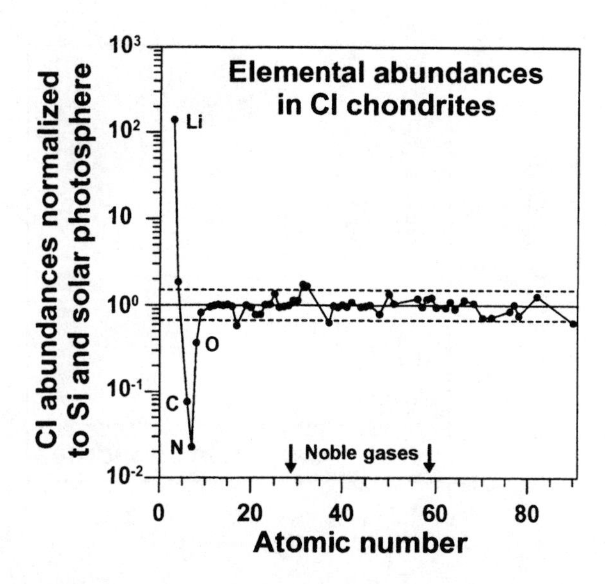

FIGURE 4. Elemental abundances in CI chondrites, normalized to Si and solar photospheric abundances [10]. The dashed lines represent a difference of a factor of 1.5 between CI and solar photospheric abundances.

tosphere due to nuclear reactions at the bottom of the Sun's convection zone. The good agreement between the two data sets justifies use of the elemental abundances of CI chondrites as chemical reference for bulk solar system matter except for H, C, N, O, and the noble gases. Their abundances in bulk solar system matter must be directly derived from the Sun or solar wind samples or, as for Kr and Xe, from s-process systematics and/or neighboring element abundances.

ISOTOPIC COMPOSITIONS

It is not the purpose of this paper to review the isotopic compositions of all elements in bulk meteoritic matter or major meteoritic compounds. In general, isotopic heterogeneities among the different chondrite groups are small for refractory elements. Heterogeneities are larger for volatile elements or if the decay products of short-lived radioisotopes or the products from cosmic ray spallation reactions add to the isotopic abundance pattern of specific elements. Here, I will briefly discuss the isotopic compositions of H, C, N, O, and Cr and the presence of now extinct ^{26}Al in meteorites.

Hydrogen

The D/H ratio in meteoritic water varies between 9×10^{-5} in clay minerals and 7×10^{-4} in chondrules. These variations are interpreted to be the result of progressive isotope exchange in the solar nebula between D-rich interstellar water and protosolar H_2 [11].

Carbon

Carbon concentrations are between 0.1 and several wt% in bulk chondrites. C-isotopic compositions vary by about 3% with $\delta^{13}C/^{12}C$ values (δ-values give the permil deviation of an isotopic ratio from a reference ratio) between 0 and -30 permil relative to terrestrial C (Fig. 5) [12, 13]. Each chondrite group has a restricted range in C-isotopic composition and C concentration. Highest C concentrations are seen in the CI chondrites. Ordinary chondrites have comparatively low C concentrations and exhibit the lowest $^{13}C/^{12}C$ ratios among the different chondrite groups.

Nitrogen

Nitrogen concentrations in bulk chondrites are much lower than C concentrations. On the other hand, N-isotopic variations are much larger than those in C. Nitrogen concentrations range from about 1 ppm to more

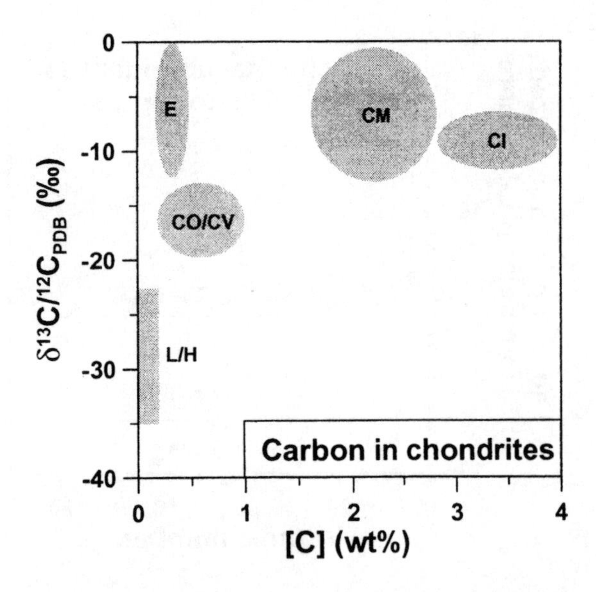

FIGURE 5. Bulk C-isotopic compositions given as permil deviation from the terrestrial PDB standard (the bulk Earth has $\delta^{13}C/^{12}C$ = -6.4 permil relative to PDB) and C concentrations of chondrites. Data taken from [12, 13].

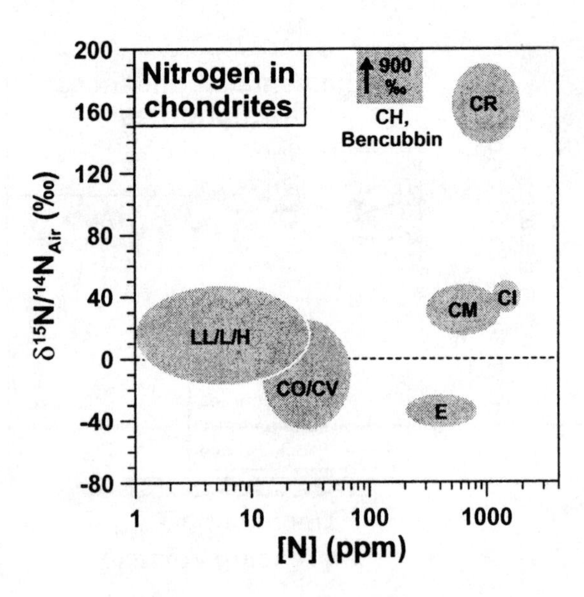

FIGURE 6. Bulk N-isotopic compositions given as permil deviation from the terrestrial air standard and N concentrations of chondrites. The data for CH chondrites and Bencubbin are off-scale with bulk $\delta^{15}N/^{14}N$ values of up to 900 permil. Data from [12, 13, 14, 15, 16].

than 1000 ppm. Most of the chondrites have $\delta^{15}N/^{14}N$ values between -40 and +40 permil relative to terrestrial N but some, such as the CH chondrites and the unique Bencubbin meteorite, show enrichments in ^{15}N by up to a factor of 2 (Fig. 6) [12, 13, 14, 15, 16]. These large variations are still not understood but might be caused by admixture of ^{15}N-rich interstellar (organic) material to protosolar N in different proportions [17, 18]. Although presolar grains of stellar origin exhibit huge variations in $^{15}N/^{14}N$ (see below), variable admixture of different presolar grain populations with different N-isotopic signatures is not likely to be the cause for the large heterogeneities seen in bulk chondrites as one would expect to see also large C-isotopic heterogeneities, contrary to the observation (see above). Because of the large spread in N-isotopic compositions of bulk meteorites, the protosolar N-isotopic composition can be hardly derived from the meteorite data. Distinctly lower $^{15}N/^{14}N$ ratios than those seen in meteorites and the Earth are derived for the Jupiter atmosphere ($\delta^{15}N/^{14}N$ = -374 +- 82 permil [19]). Further complication is introduced from measurements of solar wind N. In-situ spacecraft studies gave $\delta^{15}N/^{14}N$ = +360 +- 370 permil [20], measurements of solar wind implanted N in lunar samples $\delta^{15}N/^{14}N$ < -240 permil [17]. See also the contribution by Busemann et al. in these proceedings.

Oxygen

Similar to C, bulk chondrites have O-isotopic variations of about 20-30 permil and each chondrite group exhibits a characteristic O-isotopic signature (Fig. 7) [21]. As O has three stable isotopes, mass-dependent isotope fractionation processes can be distinguished from non-mass-dependent isotope fractionation processes. Mass-dependent fractionation in equilibrium and kinetically-controlled processes occurs along a line with slope 0.5 in a three-isotope representation (cf. TFL in Fig. 7). Imprints of mass-dependent fractionation is seen within individual chondrite groups. The difference between the chondrite groups points to formation from reservoirs with different O-isotopic signatures. The CAIs show considerable enrichments in ^{16}O of up to 5-7% relative to bulk chondrites and the O-isotopic compositions of different minerals plot on a mixing-line with slope 1 in a three-isotope-representation (Fig. 7). This feature is not yet understood and represents one of the major puzzles in the field of meteoritics. Possible explanations include non-mass-dependent chemical fractionation processes [22] and presence of ^{16}O-rich grains from supernovae (SN) (see below) in proto-CAIs. See also the contribution by Busemann et al. in these proceedings.

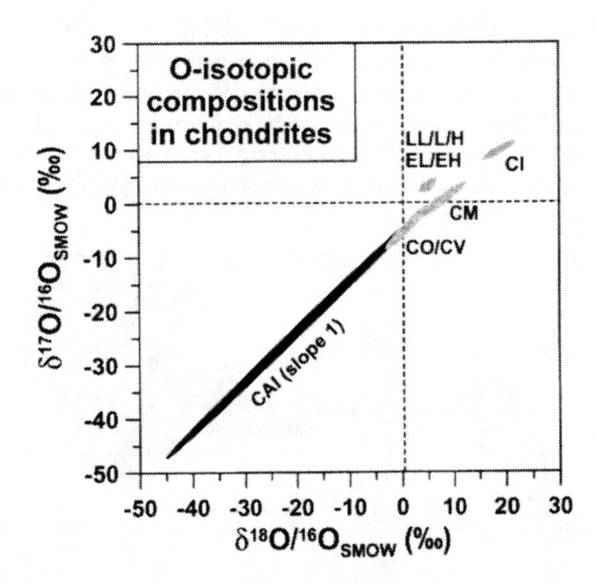

FIGURE 7. O-isotopic compositions of bulk chondrites and CAIs given as permil deviation from the terrestrial SMOW standard [21]. TFL = terrestrial fractionation line.

Aluminum-26

Another characteristic feature of CAIs are enrichments in ^{26}Mg from the decay of now extinct ^{26}Al. Aluminum-26 was either produced locally in the solar system by the interaction of an X-wind, emerging from the inner edge of the young Sun's accretion disk, with proto-CAIs [23, 24] or in a stellar source shortly before seeding the solar nebula [25]. In the latter scenario short-lived radioactive nuclides can be used as a chronometer to date events in the early solar system history. In this context CAIs are the first solids that formed in the solar system as they exhibit the highest level of ^{26}Al among solar system materials (except presolar grains) with inferred initial ^{26}Al/^{27}Al ratios of up to 5×10^{-5} [3].

Chromium

Heterogeneities in the solar nebula are also evident from the Cr-isotopic composition. Although variations in the ^{53}Cr/^{52}Cr ratio are less than 100 ppm in different solar system samples, this ratio apparently correlates with the heliocentric distance of the place of formation of planetary bodies. This is evidenced from Cr data for the Earth, martian (SNC) meteorites, and HED meteorites (achondrites) believed to originate from the asteroid Vesta (Fig. 8) [26, 27]. If this is generally true, then the ^{53}Cr/^{52}Cr ratio can be used to constrain the place of formation of solar system bodies. Enstatite chondrites would have formed close to the orbit of Mars and the ordinary chondrites close to that of Vesta [27]. The Cr data suggest that a radial heterogeneous distribution of radioactive ^{53}Mn (half-life = 3.7 My), which decays to ^{53}Cr, must have existed in the early inner solar system.

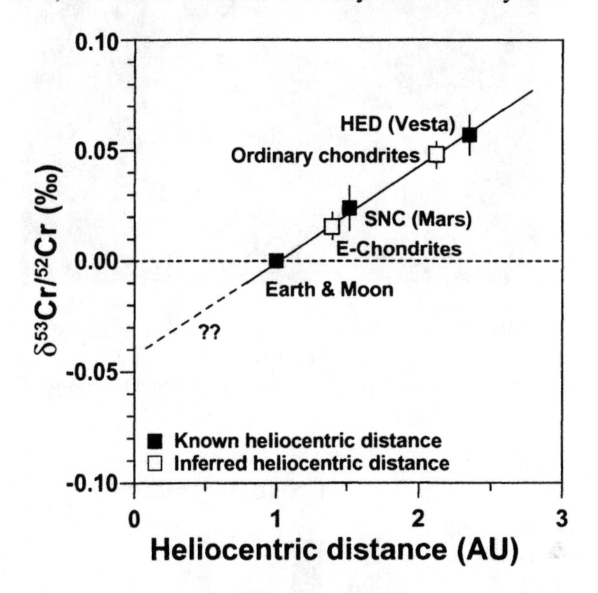

FIGURE 8. The ^{53}Cr/^{52}Cr ratio of the Earth and Moon, SNC meteorites, and HED meteorites given as permil deviation from terrestrial Cr as a function of heliocentric distance. The heliocentric distance of the place of formation of enstatite and ordinary chondrites is inferred from the Cr data, assuming that there is a linear relationship between ^{53}Cr/^{52}Cr and heliocentric distance [27].

PRESOLAR GRAINS

The topic of presolar grains has been reviewed in great detail in recent years and only a brief summary is given here. For more detailed informations and a complete list of references the reader is referred to the papers by [28, 29, 6, 7] and to the compilation of papers in "Astrophysical Implications of the Laboratory Study of Presolar Materials" [30]. Presolar grains identified to date in primitive meteorites include diamond, silicon carbide (SiC) (Fig. 9), graphite, silicon nitride (Si_3N_4), corundum (Al_2O_3), spinel ($MgAl_2O_4$), and hibonite ($CaAl_{12}O_{19}$). The presolar nature of those grains is indicated by large isotopic anomalies in the major and many trace elements contained in the grains (Fig. 10). The fact that the known presolar minerals are high temperature condensates implies that they formed in stellar outflows or in the ejecta of stellar explosions. These grains thus represent a sample of stardust that can be analyzed in the laboratory.

Figure 11 illustrates the path of presolar grains from their stellar sources to the laboratory. The isotopic compositions of presolar grains represent those in the stellar

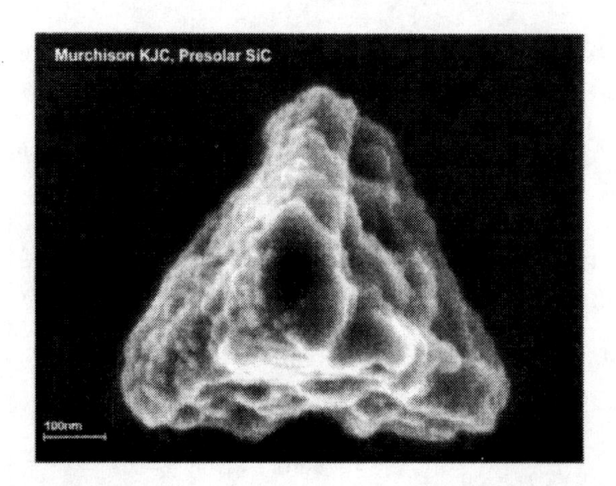

FIGURE 9. Presolar SiC grain from the Murchison meteorite. The scale bar is 100 nm. The size of $\approx 0.5\ \mu m$ represents a typical size of SiC grains from the Murchison meteorite.

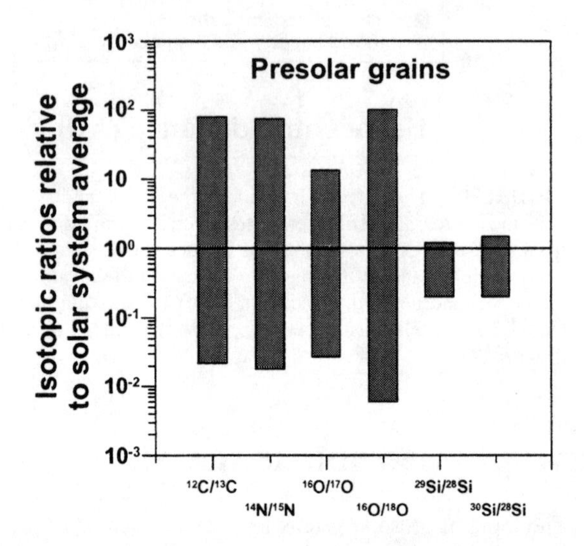

FIGURE 10. Range of isotopic ratios observed in presolar grains from primitive meteorites. All ratios are normalized to solar system reference ratios (terrestrial atmosphere for N, $^{14}N/^{15}N = 272$; terrestrial PDP for C, $^{12}C/^{13}C = 89$; terrestrial SMOW for O, $^{16}O/^{17}O = 2610$, $^{16}O/^{18}O = 499$; bulk meteorites and Earth for Si, $^{29}Si/^{28}Si = 0.05063$, $^{30}Si/^{28}Si = 0.03347$). For references see [6, 7].

atmosphere or in the ejecta of stellar explosions which in turn are determined by the compositions at the time the parent stars formed, and by the nucleosynthesis in and evolution of the parent stars. After passage through the interstellar medium such grains became part of the molecular cloud from which our solar system formed. They survived the events that led to the formation of the solar system inside small planetary bodies (asteroids) and comets. They are carried to the Earth by meteorites from which they can be separated by chemical and physical treatments which were invented by Edward Anders and co-workers at the University of Chicago in the 1980s. The laboratory study of presolar grains can provide important information on stellar nucleosynthesis and evolution, mixing in supernova (SN) ejecta, the galactic chemical evolution (GCE), grain formation in stellar winds or ejecta, and on the inventory of stars that contributed dust to the solar nebula.

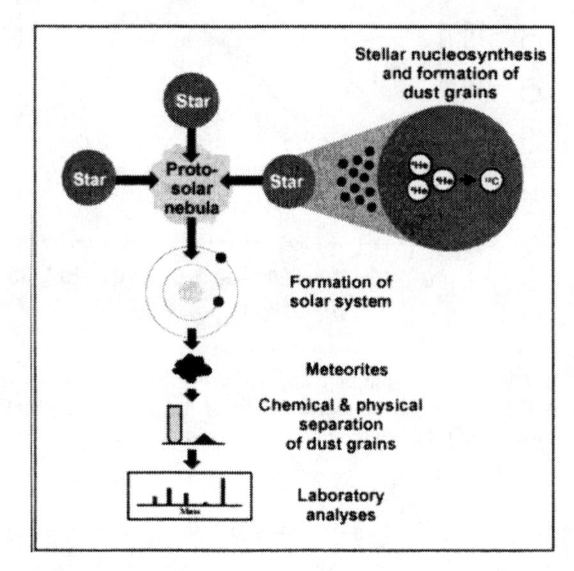

FIGURE 11. Path of presolar grains from their stellar sources to the laboratory.

Although most abundant (with concentrations of > 1000 ppm in the most primitive meteorites) the diamonds are least understood. The reason for this is their comparatively small size of only 2-3 nm that does not allow to measure the isotopic compositions of single grains. The other grain types are less abundant (at most a few ppm) but they have sizes of > 100 nm (Fig. 9) that allows to measure the isotopic compositions of many elements in single grains. A lot of isotopic information is available for SiC, graphite, and corundum. On the other hand, only very few presolar silicon nitride, spinel, and hibonite grains were identified so far and there are only few isotopic data available for those grains.

Silicon carbide, graphite, and silicon nitride

Based on the isotopic compositions of C, N, Si, and the abundance of radiogenic ^{26}Mg the SiC grains were divided into six distinct populations. Of particular importance are the so-called mainstream [31] and the X grains [32, 33, 34]. The mainstream grains make up the majority (> 90%) of the SiC grains. They are characterized by lower than solar system $^{12}C/^{13}C$ ratios (typically 40-80; bulk meteorites: 89-92) and higher than solar system $^{14}N/^{15}N$ ratios (up to 20,000; bulk meteorites and planets: 140-430) (Fig. 12) [e. g. 35, 36, 37]; heavier elements (e.g., Kr, Sr, Zr, Mo, Xe, Ba, Nd, Sm) show the signature of s-process nucleosynthesis [e. g. 38, 39, 40, 41, 42, 43]. From a comparison between the grain data and astronomical observations and stellar models, low-mass (1-3 $M_{\odot}$) asymptotic giant branch (AGB) stars are considered the most likely stellar sources of the mainstream grains.

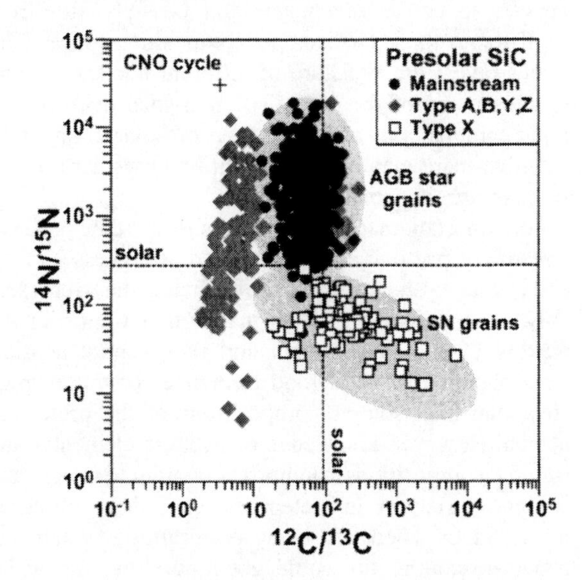

FIGURE 12. N- and C-isotopic compositions of different populations of presolar SiC grains. The dashed lines represent solar system reference ratios (N: terrestrial atmosphere; C: terrestrial PDB standard). The mainstream grains are believed to originate from AGB stars, the X grains from SN. For references see [6, 7].

Most mainstream grains have enrichments in the heavy Si isotopes of up to 20% relative to their solar system abundances (Fig. 13) [cf. 31]. In a Si-three-isotope-representation the data fall along a line with slope 1.3. In low-mass AGB stars Si is affected by the s-process in the He shell. He shell matter is mixed outward in the so-called third dredge-up, leading to enrichments of ^{29}Si and ^{30}Si in the star's envelope. However, the expected shifts in ^{29}Si/^{28}Si and ^{30}Si/^{28}Si are at most a few percent [e. g. 44] and evolution of the Si-isotopic composition in a Si-three-isotope representation is expected along a line with slope $\approx$ 0.5 [e. g. 45], at variance with the grain data. It is the preferred interpretation today that the slope 1.3 Si correlation line does not result from the dredge-up of He shell matter in AGB stars but reflects the GCE, both in time and space, of the Si isotopes and represents a range of Si starting compositions of a large number of AGB stars [46, 47, 44].

Most SiC X grains are characterized by isotopically light C (with ^{12}C/^{13}C of up to 7000), heavy N (with ^{14}N/^{15}N down to 13), and light Si (with enrichments in ^{28}Si of up to a factor of 5). Other isotopic features of X

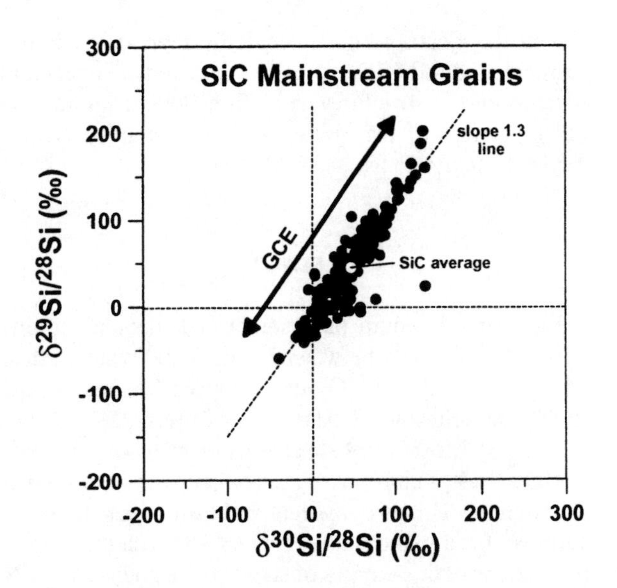

FIGURE 13. Si-isotopic ratios of SiC mainstream grains given as permil deviation from the solar system reference (terrestrial) ratios. Data are from [35]. The slope 1.3 line most likely reflects the GCE of the Si isotopes. The average Si-isotopic composition of presolar SiC (including all sub-types) is not solar, indicating that there may exist yet unidentified presolar mineral types with isotopically light Si on average.

grains are high inferred initial ^{26}Al/^{27}Al ratios and the presence of radiogenic ^{44}Ca from the decay of now extinct ^{44}Ti (half-life 60 y) in some grains [32, 33, 34]. On the basis of the enrichments in ^{28}Si and presence of ^{44}Ti at the time of grain formation, type II SN have been proposed as the most likely stellar sources of X grains. The same holds for the majority of the graphite grains and all silicon nitride grains. Most of these grains exhibit isotopic signatures that resemble those of X grains indicative for a close relationship between these types of presolar grains [e. g. 48, 49, 33, 50]. A small fraction of the SiC and graphite grains apparently is from novae [51] and for some of the graphite grains also a Wolf-Rayet star origin cannot be excluded [48].

The presolar SN grain data do not only allow to test models of nucleosynthesis in SN but also provide information on the mixing in SN ejecta. In the context of a type II SN origin, the isotopic compositions of presolar grains require mixing of matter from the innermost Ni- and Si-rich zones and the C-rich outer layers, indicative of deep mixing in the ejecta. This confirms similar conclusions derived from astronomical observations of SN light curves [e. g. 52, 53]. Although carbonaceous grains might form in the ejecta of type II SN explosions even while C/O < 1 [54], some kind of selective mixing, which limits contributions of matter from the intermediate, extremely O-rich zones to the condensation site in the ejecta, is indicated. This supports hydrodynamical models of SN explosions that predict fingers and mushroom-like structures rising from the interior into the outer portions of the ejecta as a result from Rayleigh-Taylor instabilities [e. g. 55, 56, 57].

Oxides

Presolar corundum (and spinel and hibonite) grains have $^{16}O/^{17}O$ ratios between 70 and 30,000 (meteorites: 2580-2740) and $^{16}O/^{18}O$ ratios between 200 and 50,000 (bulk meteorites and CAIs: 490-520) [e. g., 58, 59, 60, 61, 62, 63]. Most grains are characterized by enrichments in ^{17}O and depletions in ^{18}O as compared to solar system abundances. This is consistent with astronomical observations of red giant and AGB stars and with theoretical predictions of those types of stars. Oxide grains from SN are apparently rare among meteoritic stardust. Up to now only one corundum grain was found that shows the expected enrichment in ^{16}O for oxide grains from SN [61].

SUMMARY

Undifferentiated meteorites (chondrites) have preserved the bulk elemental and isotopic compositions of their parent bodies. Variations in elemental abundances between different chondrite groups are within a factor of 2 for the refractory elements and are attributed to variations in the nebula environment from which the chondrites formed. The CI chondrites, which represent the most primitive meteoritic matter, have elemental abundances that are compatible to those in the solar photosphere execpt for Li, which is destroyed in the Sun's convection zone, and the highly volatile elements H, C, N, O, and noble gases which are incompletely condensed in chondrites. This justifies use of CI chondrites as chemical reference for bulk solar system matter for most of the elements. Largest bulk isotopic heterogeneities are seen for H (variations in D/H of a factor of 8), followed by N ($^{15}N/^{14}N$ varies by 2x), and C and O (C- and O-isotopic ratios vary by a few percent). Much smaller isotopic variations are observed for more refractory elements. The $^{53}Cr/^{52}Cr$ varies by less than 100 ppm and this ratio appears to be correlated with the place of formation of planetary bodies. Short-lived radioactive nuclides such as ^{26}Al and ^{41}Ca existed live in the early solar system as evidenced from the presence of the radiogenic daughter nuclides in CAIs and chondrules. Provided that the radionuclides were injected into the solar nebula at the time of solar system formation, they can be used as a chronometer to date events in the early solar system.

Chondrites contain small quantities of presolar grains. These grains exhibit large isotopic anomalies in the major and many trace elements indicative of a circumstellar origin. Presolar minerals identified to date include diamond, silicon carbide, graphite, silicon nitride, corundum, spinel, and hibonite. Most of the grains apparently formed in the winds of red giant and AGB stars and in the ejecta of SN explosions. A small fraction of the grains appears to come from novae and possibly also from Wolf-Rayet stars. The isotopic compositions of presolar grains reveal the signature of different nuclear burning processes and of the GCE. The non-solar average isotopic compositions of Si and O in presolar grains indicate that there may be yet unidentified presolar mineral types in primitive meteorites.

An important merit of meteorites is that they preserve a record of presolar components and of processes in the early solar system. On the other hand, they represent only a small fraction of the matter that went into the making of the solar system and they cannot a priori be considered to be a good reference for the average elemental and isotopic compositions of the protosolar nebula. However, knowledge of average elemental and isotopic compositions is important in order to understand isotopic variations in meteorites, e.g., those observed in N and O. There are many opportunities where the meteorite community could get input from the wider SOHO/ACE community and these are outlined in the contribution of the results from the working group on "Applications in Cosmochemistry" (H. Busemann, these proceedings).

ACKNOWLEDGMENTS

I thank the organizers of the SOHO/ACE workshop on "Solar and Galactic Composition" for the invitation to present this paper. Critical and helpful reviews by P. Eberhardt and V. Heber are acknowledged.

REFERENCES

1. Boss, A., and Vanhala, H. A. T., "Triggering protostellar collapse, injection, and disk formation", in *From dust to terrestrial planets*, edited by W. Benz, R. Kallenbach, and G. W. Lugmair, Space Sciences Series of ISSI, Kluwer Academic Publishers, Dordrecht, 2000, pp. 13–22.
2. Srinivasan, G., Sahijpal, S., Ulyanov, A. A., and Goswami, J. N., *GCA*, **60**, 1823–1835 (1996).
3. MacPherson, G. J., Davis, A. M., and Zinner, E., *Meteoritics*, **30**, 365–386 (1995).
4. Benz, W., "Low velocity collisions and the growth of planetesimals", in *From dust to terrestrial planets*, edited by W. Benz, R. Kallenbach, and G. W. Lugmair, Space

Sciences Series of ISSI, Kluwer Academic Publishers, Dordrecht, 2000, pp. 279–294.
5. Cassen, P., *Icarus*, **112**, 405–429 (1994).
6. Zinner, E., *Ann. Rev. Earth and Planet. Sci.*, **26**, 147–188 (1998).
7. Hoppe, P., and Zinner, E., *JGR*, **275**, 10371–10385 (2000).
8. Lodders, K., and Fegley, Jr., B., *The Planetary Scientist's Companion*, Oxford University Press, New York, 1998.
9. Palme, H., "Are there chemical gradients in the inner solar system?", in *From dust to terrestrial planets*, edited by W. Benz, R. Kallenbach, and G. W. Lugmair, Space Sciences Series of ISSI, Kluwer Academic Publishers, Dordrecht, 2000, pp. 237–264.
10. Anders, E., and Grevesse, N., *GCA*, **53**, 197–214 (1989).
11. Robert, F., Gautier, D., and Dubrulle, B., "The solar system D/H ratio: Observations and theories", in *From dust to terrestrial planets*, edited by W. Benz, R. Kallenbach, and G. W. Lugmair, Space Sciences Series of ISSI, Kluwer Academic Publishers, Dordrecht, 2000, pp. 279–294.
12. Kung, C.-C., and Clayton, R. N., *EPSL*, **38**, 421–435 (1978).
13. Kerridge, J. F., *GCA*, **49**, 1707–1714 (1985).
14. Prombo, C. A., and Clayton, R. N., *Science*, **230**, 935–937 (1985).
15. Grady, M. M., Ash, R. D., Morse, A. D., and Pillinger, C. T., *Meteoritics*, **26**, 339–340 (1991).
16. Hashizume, K., and Sugiura, N., *GCA*, **59**, 4057–4070 (1995).
17. Hashizume, K., Chaussidon, M., Marty, B., and Robert, F., *Science*, **290**, 1142–1145 (2000).
18. Sugiura, N., Zashu, S., Weisberg, M., and Prinz, M., *MAPS*, **35**, 987–998 (2000).
19. Owen, T., Mahaffy, P. R., Niemann, H. B., Atreya, S., and Wong, M., *ApJ*, **553**, L77–L80 (2001).
20. Kallenbach, R., Geiss, J., Ipavich, F. M., Gloeckler, G., Bochsler, P., Gliem, F., Hefti, S., Hilchenbach, M., and Hovestadt, D., *ApJ*, **507**, L185–L188 (1998).
21. Clayton, R. N., *Ann. Rev. Earth Planet. Sci.*, **21**, 115–149 (1993).
22. Thiemens, M. H., and Heidenreich, J. E. I., *Science*, **219**, 1073–1075 (1983).
23. Shu, F., Shang, H., and Lee, T., *Science*, **271**, 1545–1552 (1996).
24. Shang, H., Shu, F. H., Lee, T., and Glassgold, A. E., "Protostellar winds and chondritic meteorites", in *From dust to terrestrial planets*, edited by W. Benz, R. Kallenbach, and G. W. Lugmair, Space Sciences Series of ISSI, Kluwer Academic Publishers, Dordrecht, 2000, pp. 153–176.
25. Cameron, A. G. W., and Truran, J. W., *Icarus*, **30**, 447–461 (1977).
26. Lugmair, G. W., and Shukolyukov, A., *GCA*, **62**, 2863–2886 (1998).
27. Shukolyukov, A., and Lugmair, G. W., "On the ^{53}Mn heterogeneity in the early solar nebula", in *From dust to terrestrial planets*, edited by W. Benz, R. Kallenbach, and G. W. Lugmair, Space Sciences Series of ISSI, Kluwer Academic Publishers, Dordrecht, 2000, pp. 225–236.
28. Anders, E., and Zinner, E., *Meteoritics*, **28**, 490–514 (1993).
29. Ott, U., *Nature*, **364**, 25–33 (1993).
30. Bernatowicz, T. J., and Zinner, E., editors, AIP Conference Proceedings 402, American Institute of Physics, Woodbury, New York, 1997.
31. Hoppe, P., and Ott, U., "Mainstream silicon carbide grains from meteorites", in *Astrophysical Implications of the Laboratory Study of Presolar Materials*, edited by T. J. Bernatowicz and E. Zinner, AIP Conference Proceedings 402, American Institute of Physics, Woodbury, New York, 1997, pp. 27–58.
32. Amari, S., Hoppe, P., Zinner, E., and Lewis, R. S., *ApJ*, **394**, L43–L46 (1992).
33. Nittler, L. R., Amari, S., Zinner, E., Woosley, S. E., and Lewis, R. S., *ApJ*, **462**, L31–L34 (1996).
34. Hoppe, P., Strebel, R., Eberhardt, P., Amari, S., and Lewis, R. S., *MAPS*, **35**, 1157–1176 (2000).
35. Hoppe, P., Amari, S., Zinner, E., Ireland, T., and Lewis, R. S., *ApJ*, **430**, 870–890 (1994).
36. Hoppe, P., Strebel, R., Eberhardt, P., Amari, S., and Lewis, R. S., *GCA*, **60**, 883–907 (1996).
37. Huss, G. R., Hutcheon, I. D., and Wasserburg, G. J., *GCA*, **61**, 5117–5148 (1997).
38. Ott, U., and Begemann, F., *ApJ*, **353**, L57–L60 (1990).
39. Prombo, C. A., Podosek, F. A., Amari, S., and Lewis, R. S., *ApJ*, **410**, 393–399 (1993).
40. Lewis, R. S., Amari, S., and Anders, E., *GCA*, **58**, 471–494 (1994).
41. Nicolussi, G. K., Davis, A. M., Pellin, M. J., Lewis, R. S., Clayton, R. N., and Amari, S., *Science*, **277**, 1281–1283 (1997).
42. Nicolussi, G. K., Pellin, M. J., Lewis, R. S., Davis, A. M., Amari, S., and Clayton, R. N., *GCA*, **62**, 1093–1104 (1998).
43. Nicolussi, G. K., Pellin, M. J., Lewis, R. S., Davis, A. M., Clayton, R. N., and Amari, S., *Phys. Rev. Lett.*, **81**, 3583–3586 (1998).
44. Lugaro, M., Zinner, E., Gallino, R., and Amari, S., *ApJ*, **527**, 369–394 (1999).
45. Gallino, R., Raiteri, C. M., Busso, M., and Matteucci, F., *ApJ*, **430**, 858–869 (1994).
46. Alexander, C. M. O. D., *GCA*, **57**, 2869–2888 (1993).
47. Timmes, F. X., and Clayton, D. D., *ApJ*, **472**, 723–741 (1996).
48. Hoppe, P., Amari, S., Zinner, E., and Lewis, R. S., *GCA*, **59**, 4029–4056 (1995).
49. Nittler, L. R., Hoppe, P., Alexander, C. M. O. D., Amari, S., Eberhardt, P., Gao, X., Lewis, R. S., Strebel, R., Walker, R. M., and Zinner, E., *ApJ*, **453**, L25–L28 (1995).
50. Travaglio, C., Gallino, R., Amari, S., Zinner, E., Woosley, S., and Lewis, R. S., *ApJ*, **510**, 325–354 (1999).
51. Amari, S., Gao, X., Nittler, L. R., Zinner, E., Jose, J., Hernanz, M., and Lewis, R. S., *ApJ*, **551**, 1065–1072 (2001).
52. Dotani, T., et al., *Nature*, **330**, 230–231 (1990).
53. Shigeyama, T., and Nomoto, K., *ApJ*, **360**, 242–256 (1990).
54. Clayton, D. D., Liu, W., and Dalgarno, A., *Science*, **283**, 1290–1292 (1999).
55. Arnett, D., Fryxell, B., and Müller, E., *ApJ*, **341**, L63–L66 (1989).
56. Ebisuzaki, T., Shigeyama, T., and Nomoto, K., *ApJ*, **344**, L65–L68 (1989).
57. Herant, M., and Woosley, S. E., *ApJ*, **425**, 814–828 (1994).

58. Huss, G. R., Fahey, A. J., Gallino, R., and Wasserburg, G. J., *ApJ*, **430**, L81–L84 (1994).
59. Nittler, L. R., Alexander, C. M. O. D., Gao, X., Walker, R. M., and Zinner, E., *Nature*, **370**, 443–446 (1994).
60. Nittler, L. R., Alexander, C. M. O. D., Gao, X., Walker, R. M., and Zinner, E., *ApJ*, **483**, 475–495 (1997).
61. Nittler, L. R., Alexander, C. M. O. D., Wang, J., and Gao, X., *Nature*, **393**, 222 (1998).
62. Choi, B.-G., Huss, G. R., Wasserburg, G. J., and Gallino, R., *Science*, **282**, 1284–1289 (1998).
63. Choi, B.-G., Wasserburg, G. J., and Huss, G. R., *ApJ*, **522**, L133–L136 (1999).

The Isotopic Composition of Solar Nitrogen and the Heterogeneity of the Solar System

Ko Hashizume*[†], Bernard Marty*, Marc Chaussidon* and François Robert*[¶]

*Centre de Recherches Pétrographiques et Géochimiques, CNRS, Rue Notre-Dame des Pauvres, BP 20, 54501 Vandoeuvre Cedex, France

[†]Department of Earth and Space Sciences, Osaka University, Toyonaka, Osaka 560-043, Japan

[¶]Muséum National d'Histoire Naturelle, CNRS, 61 rue Buffon, 75005 Paris, France

Abstract. The isotopic composition of solar nitrogen is a long standing issue that received recently new impetus. The analysis of nitrogen isotopes in lunar samples and in Jupiter show that solar nitrogen is depleted in ^{15}N by 30 % relative to terrestrial. The systematic enrichment of ^{15}N in terrestrial planets and bulk meteorites requires the contribution of ^{15}N-rich compounds to the total nitrogen in planetary materials. Most of these compounds are possibly of an interstellar origin that never equilibrated with the ^{15}N-depleted protosolar nebula N_2.

INTRODUCTION

The isotopic composition of nitrogen ($^{15}N/^{14}N$, expressed as ($\delta^{15}N = [(^{15}N/^{14}N)_{sample}/(^{15}N/^{14}N)_{air} - 1] \times 1000$, in parts per mil, or ‰) is extremely variable among solar system objects such as meteorites, planetary atmospheres, interplanetary dust particles (IDPs) (e.g., [1, 2] and refs. therein), see Fig. 1. Because nitrogen has only two isotopes, it is unclear if such variations are due to isotopic fractionation in the proto-solar nebula, or in the interstellar medium, or if they result from mixing of pre-solar components processed in different stars, or from a combination of such processes.

The basic problem with this element is that the isotopic composition of solar nitrogen, the main reservoir of N in the solar system, has not been measured directly with precision, preventing an assessment of the solar system reference value for $^{15}N/^{14}N$. We have recently proposed that solar N is drastically depleted in ^{15}N, based on newly developed analyses of lunar soils and we discuss briefly some of the implications of this isotope heterogeneity for the evolution of the solar system.

SOLAR NITROGEN

A direct determination of the solar wind nitrogen composition will be possible when samples having been exposed in space during the

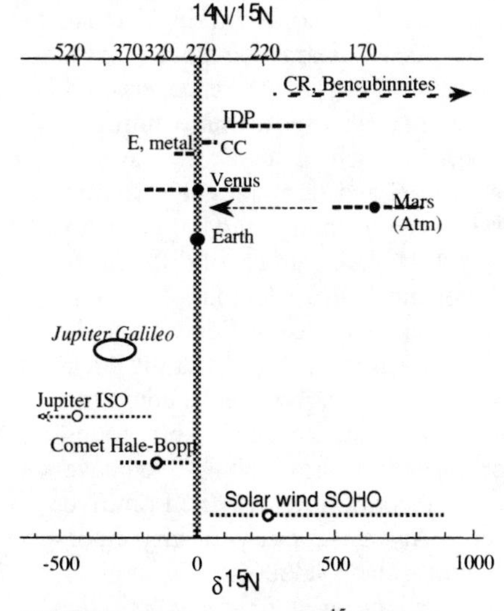

FIGURE 1. Variation of $\delta^{15}N$ in the solar system Different types of chondrites (E, CR) are shown together with measurements of the atmosphere of Venus and of Mars.

Genesis Discovery Mission will be returned to laboratories. Studies of the lunar regolith, which has been irradiated by the solar corpuscular emission for long periods of time, have allowed to identify and measure precisely the isotopic composition of solar wind rare gases (e.g., [3]), which agrees well with direct measurement of solar wind rare gases in aluminium foils exposed

CP598, *Solar and Galactic Composition*, edited by R. F. Wimmer-Schweingruber

on the Moon. The abundance of nitrogen (and carbon) of different soils correlates fairly well with that of rare gases, and this has been taken as a firm evidence for a solar origin of N.

However, the N/Ar ratio of lunar soils is on an average one order of magnitude higher than Solar, and the $\delta^{15}N$ lunar soil values range over 300 ‰ (e.g., [1] and refs. therein). These observations were interpreted following two different types of models, not necessarily exclusive : (i) the isotopic composition of nitrogen in solar emission varied with time [4], contrary to that of rare gases (e.g., [5]), or (ii) several exotic components contributed nitrogen to lunar soils [6]. The first possibility has been discarded by Geiss and Bochsler [6] on the ground that no known spallation or thermonuclear reaction could produce enough ^{15}N and these authors concluded that the N isotope ratio at the solar surface has been constant during the last 4 billion years. They instead proposed that variations of the $^{15}N/^{14}N$ ratio is due to mixing between a ^{15}N-rich solar component and a very light N component admixed in various amount to planetary matter. The analysis of N abundance together with that of Ar isotopes in single lunar soil grains by laser extraction-static mass spectrometry developed in Nancy, France, showed that lunar soils contained two or more nitrogen components [7]. In such a case the overall correlation between N and rare gases in different soils does not imply a common origin but is due to the implantation at the grain surface of different N and Ar components. In addition to solar corpuscular radiation, possible candidates include meteoritic N since it has been shown that the lunar regolith contains 1-2 % of carbonaceous chondrite debris [8] and that some of lunar soil grains present vapor-deposited rims though to have originated from meteoritic impacts [9]. Primitive meteorites have N/Ar ratios 5 orders of magnitude higher than the solar value so a limited contribution of such material only be seen for nitrogen, leaving the solar rare gas component essentially unaltered. A cometary contribution would also be possible but such case would imply either contribution of cometary silicates, or preferential trapping of nitrogen relative to rare gases if these volatiles were supplied by cometary ice.

The occurrence of several N components was further confirmed when it became possible to determine also the N isotopic composition of single lunar soil grains [10, 11] for soil sampled at different sites. These analyses demonstrated that

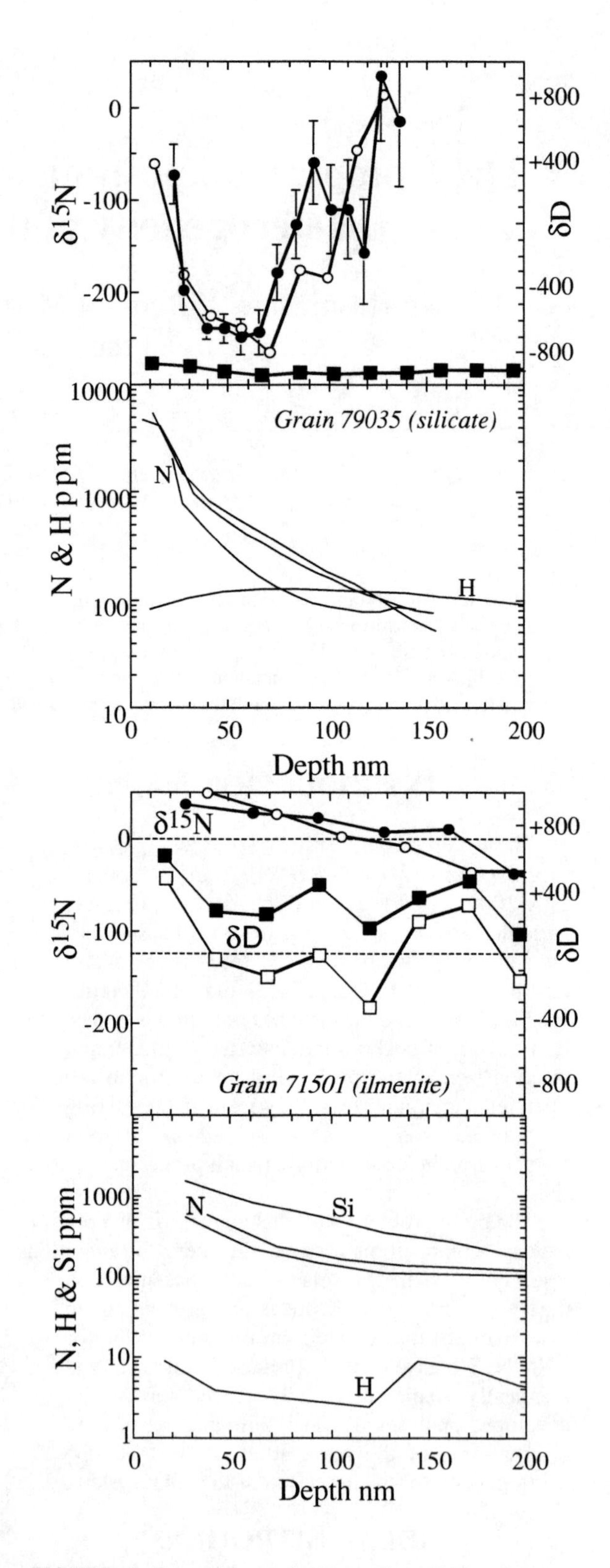

FIGURE 2. Examples of depth profiles of N and H concentrations and isotope compositions (dots and squares for H and N, respectively) in grains from soil 79035 and soil 71501 (adapted from ref. [2]).

solar wind nitrogen is not enriched in ^{15}N contrary to what was proposed by [6] since ^{15}N-rich grains are depleted in solar wind argon and $\delta^{15}N$ values become negative with increasing contribution of solar wind argon. A lower limit of – 200 ‰ for solar $\delta^{15}N$ could be set based on $\delta^{15}N$-Ar/N relationships.

A new method of ion probe rastering analysis developed at CRPG, Nancy [12] allowed to measure the depth profile isotopic variation of H and N on soil grain surfaces with a depth resolution of 10 nm for two soils of different antiquities (the antiquity refers to the epoch at which a given soil was exposed at the Moon's surface, based on indirect, semi-quantitative soil characteristics).

Results [2] from soil 79035, a regolith presumably exposed for > 500 Ma, show that a light N component depleted in ^{15}N by at least 240 ‰ is implanted in the depth range of few tens of nanometers that corresponds to implantation of SW ions. The light N component is associated with solar wind, D-free hydrogen (Fig. 2), confirming the suspicion from single grain analysis that solar wind N is strongly depleted in ^{15}N relative to planetary N.

Other measurements aimed to determine the solar N composition are sparse. Kallenbach et al. [13] reported a $\delta^{15}N$ value of 360^{+520}_{-290} ‰ for the present -day solar wind using the SOHO TOF mass spectrometer facility. Fouchet el al. (Icarus 143, 223, 2000) reported a range of -480^{+240}_{-280} ‰ for ammonia in the Jovian atmosphere using the ISO short wavelength spectrometer. Since the N/C/Ar ratios of the Jupiter atmosphere are close to solar [14], the Jovian N isotopic composition should reflect closely that of the proto-solar nebula and the discrepancy between the two measurements may be analytical. Recently, the Galileo team has published the results of the mass spectrometric measurements of N isotopes during the Jovian atmosphere entry and proposed a $\delta^{15}N$ value of -375+/-80 ‰ for Jupiter [14], in good agreement with the Fouchet et al's determination and with the upper limit of –240 ‰ for solar wind nitrogen on the Moon proposed by Hashizume et al. [2].

Since Jupiter and comets have sampled proto-solar N 4.5 Ga ago, and soil 79035 has been presumably exposed to the solar wind several hundreds of Ma ago, this would still leave space in principle for secular evolution of solar wind N towards a more positive $\delta^{15}N$ value at present as proposed by [4]. We think that this is not the case, for the reasons exposed by [6] and because we have recently analyzed a Luna-24 soil showing evidence for recent exposition (notably a low $^{40}Ar/^{36}Ar$ ratio of 0.5) which shows also depletion of ^{15}N with increasing solar wind ^{36}Ar contribution [11].

EVIDENCE FOR N ISOTOPE HETEROGENEITY IN THE SOLAR SYSTEM

All solar system objects analyzed so far present enrichment in ^{15}N relative to solar and Jupiter (Fig. 1). Such enrichment could be in some cases the result of atmospheric processes (e.g., Mars and Titan) but a contribution of ^{15}N-rich component(s) is needed for primitive objects as well as planetary interiors. Mixing between solar N and other(s) non-solar N source(s) is evident at the Moon's surface. Analysis of grains from soil 71501 which was presumably exposed recently [5] display a ^{15}N-rich component having deuterium-rich (non-solar) hydrogen. Positive $\delta^{15}N$ values were also found at the very surface of 71501 grains as well as grains and data taken together define a range of values for non-solar N between -50 and +130 ‰ [2], essentially similar to that defining the upper limit of bulk soil spallation-corrected data (e.g., [1]). The depth profile analysis of major elements of 71501 ilmenite grains showed that the ^{15}N-rich component is associated with silicon coatings [2]. The association of Si-rich coatings with ^{15}N-rich nitrogen and D-rich hydrogen strongly suggests that such coatings originated from exotic material having impacted the Moon's surface. We can evaluate comets and meteorites as potential sources. The single existing cometary N isotope measurement was done on HCN in comet Hale-Bopp [15] and gave a $\delta^{15}N$ value of -157^{+136}_{-109} ‰. If representative of the cometary reservoir, this measurement would rule out comets as the unique source of non-solar N on the Moon. Micrometeorites and meteorites are better candidates. The range of $\delta^{15}N$ values observed in most meteorites bracket the terrestrial ratio (Fig. 1) and micrometeorites share similarities with the CM carbonaceous chondrite clan which contains on an average 10^3 ppm of nitrogen with $\delta^{15}N$ around +40 ‰. Since heavier N is observed in the Moon, other sources of N are required and IDP-like matter could fit since IDPs have $\delta^{15}N$ values up to 480 ‰ [16]. In such case the secular N isotopic variations for soils having different antiquities reflect variations in the solar versus meteoritic/cometary contributions through time and would provide an unique way to investigate the variability of meteoritic bombardment of the Earth through geological time. Preliminary mass balance calculations are very encouraging with this respect as they show a good consistancy between the required metoritic flux on the Moon to account for N isotope variability in lunar regolith and the micrometeoritic flux estimated for the Earth (Hashizume et al., in prep.).

The systematic enrichment of ^{15}N in terrestrial planets and bulk meteorites requires the contribution of ^{15}N-rich compounds to the total nitrogen in planetary materials. Most of these compounds are possibly of an interstellar origin that never equilibrated with the ^{15}N-depleted protosolar nebula N_2. Isotopic fractionation of nitrogen will enrich HCN in ^{15}N by up to 30% relative to N_2 following ion-molecule reactions in interstellar clouds, qualitatively consistent with the difference in $^{15}N/^{14}N$ between solar and planetary N.

The $\delta^{15}N$ value at the Earth's surface (atmosphere plus sediments plus crust) is + 1.5 ‰ and the upper mantle value is around – 5 ‰. Thus terrestrial nitrogen does not show evidence of solar gas origin, contrary to mantle neon. Taken at face value, terrestrial nitrogen could have been supplied by both cometary- and meteorite-like matter, however the terrestrial D/H ratio rules out significant contribution from comets [17]. In addition, the volatile element pattern of the terrestrial mantle is chondritic [18], which leaves little doubt that nitrogen has been supplied to the Earth under the form of meteoritic compounds rather than N_2.

ACKNOWLEDGMENTS

Discussions with Rainer Wieler, Paul Mahaffy, John Kerridge, Kurt Marti, Richard Becker and Toby Owen were greatly appreciated, although this work does not necessarily reflects the view of some of these scientists. This work was supported by grants from the Programme National de Planétologie, Institut National des Sciences de l'Univers, the Japanese Ministry of Education, Science, Sports and Culture, the Centre National de la Recherche Scientifique and the Région Lorraine.

REFERENCES

1. Kerridge, J.F., *Rev. Geophys.* **31**, 423-437, (1993).
2. Hashizume, K., Chaussidon, M., Marty, B., and Robert, F., *Science* **290**, 1142-1145 (2000).
3. Benkert, J.P., Baur, H., Signer, P., and Wieler, R., *J. Geophys. Res.* **98**, 13,147-13,162 (1993).
4. Kerridge, J.F., *Science* **188**, 162-164 (1975)
5. Wieler, R., and Baur, H., *Meteoritics* **29**, 570-580 (1994).
6. Geiss, J., and Bochsler, P., *Geochim. Cosmochim. Acta* **46**, 529-548 (1982).
7. Wieler, R., Humbert, F., and Marty, B., *Earth Planet. Sci. Lett.* **167**, 47-60 (1999).
8. Keays, R.R., Ganapathy, R., Laul, J.C., Anders, E., Herzog, G.F., and Jeffery, P.M., *Science* **167**, 490-493 (1970).
9. Keller, L.P., and McKay, D.S., *Geochim. Cosmochim. Acta* **61**, 2331-2342 (1997).
10. Hashizume, K., Marty, B., and Wieler, R., in *LPS XXX* CD-ROM, pp. 1567, LPI, Houston (1999).
11. Assonov, S.S., Marty, B., Shukolyukov, Y.A., and Semenova, A.S., in *LPS XXXII*, CD-ROM, pp. 1798, LPI, Houston (2001).
12. Chaussidon, M., and Robert, F., *Nature* **402**, 270-273 (1999).
13. Kallenbach, R., et al., *Astrophys. J.* **507**, L185-L188 (1998).
14. Owen, T., Mahaffy, P.R., Niemann, H.B., Atreya, S., and Wong, M., *Astrophys. J.* **553**, L77-L79 (2001).
15. Jewitt, D.C., Matthews, H.E., Owen, T., and Meier, R., *Science* **278**, 90-93 (1997).
16. Messenger, S., *Nature* **404**, 968-971 (2000).
17. Dauphas, N., Robert, F., and Marty, B., *Icarus* **148**, 508-512 (2000).
18. Marty, B., and Zimmermann, J.L., *Geochim. Cosmochim. Acta* **63**, 3619-3633 (1999).

The SUMER Spectral Atlas of Solar-Disk Features

W. Curdt[1], P. Brekke[2], U. Feldman[3], K. Wilhelm[1], B.N. Dwivedi[1,4], U. Schühle[1], and P. Lemaire[5]

[1]Max-Planck-Institut für Aeronomie, 37191 Katlenburg-Lindau, Germany
[2]ESA Space Science Department / NASA Goddard Space Flight Center, Greenbelt, Md. 20771, USA
[3]E. O. Hulburt Center for Space Research, Naval Research Laboratory, Washington D.C., 20375, USA
[4]Department of Applied Physics, Banaras Hindu University, Varanasi 221005, India
[5]Institut d'Astrophysique Spatiale, Unité Mixte CNRS - Université de Paris XI, 91405 Orsay, France

EXTENDED ABSTRACT

A far-ultraviolet and extreme-ultraviolet (FUV, EUV) spectral atlas of the Sun between 670 Å and 1609 Å in first order of diffraction has been derived from observations obtained with the SUMER (Solar Ultraviolet Measurements of Emitted Radiation) spectrograph on the spacecraft SOHO (Solar and Heliospheric Observatory) [1]. The atlas contains spectra of the average quiet Sun, a coronal hole and a sunspot on disk. Different physical parameters prevalent in the bright network (BN) and in the cell interior (CI) - contributing in a distinct manner to the average quiet-Sun emission - have their imprint on the BN/CI ratio, which is also shown for the entire spectral range. With a few exceptions, all major lines are given with their identifications and wavelengths. Lines that appear in second order are superimposed on the first order spectra, but below 500 Å the responsivity of the normal-incidence optical system is very low. The spectra include emissions from atoms and ions in the temperature range $6\ 10^3$ K to $2\ 10^6$ K, i.e., continua and mission lines emitted from the lower chromosphere to the corona. This spectral atlas, with its broad wavelength coverage, provides a rich source of new diagnostic tools for studying the physical parameters in the chromosphere, the transition region and the corona. In particular, the wavelength range below 1100 Å as observed by SUMER represents a significant improvement over the spectra produced in the past. In view of the manifold appearance and temporal variation of the solar atmosphere it is obvious that our atlas can only be a - hopefully typical - snapshot. The spectral radiances are determined with a relative uncertainty of 0.15 to 0.30 (1σ), and the wavelength scale is accurate to typically 10 mÅ, which is the level achievable with semi-automatic processing.

The SUMER solar-disk spectral atlas will be published in the near future by Curdt et al. [2]. It includes profiles of the average quiet Sun, an equatorial coronal hole, and a sunspot. As an example we show in Fig. 1 the spectral range from 1300 Å to 1342 Å with the prominent O I and C II lines. Resolved emission lines are indicated by a mark, the measured wavelength in angstrom, and the identification, if available. Marks point to line lists available in the literature, where additional information about a specific line can be found [3-7]. New lines or identifications are indicated. Lines observed in first order and in second-order of diffraction are distinguished. Only the three least-significant digits of the wavelength values are given. If available, unidentified lines are characterized by the temperature classification defined in [3] (a: $T_e < 3\ 10^5$; b: $T_e \approx 3\ 10^5$; c: $T_e \approx 4\ 10^5$; d: $6\ 10^5 < T_e < 9\ 10^5$; e: $T_e \approx 1.4\ 10^6$; f: $T_e \approx 1.8\ 10^6$). The vertical axes are scaled to spectral radiance in units of mW sr^{-1} m^{-2} Å^{-1}; on the left the radiometric calibration for first order lines is given, on the right for second order lines. Note, that second order lines are always superimposed on a first order background. We have taken care of the type of photocathode (bare or KBr) when applying the radiometric calibration to different sections of the spectrum. Also displayed in green is the BN/CI ratio in an attempt to characterize the quiet-Sun chromospheric network structure. A pre-print of the SUMER spectral atlas and a line list is available at http://www.linmpi.mpg.de/~curdt. All SUMER telemetry data and routine-type data handling procedures are available from

CP598, *Solar and Galactic Composition*, edited by R. F. Wimmer-Schweingruber

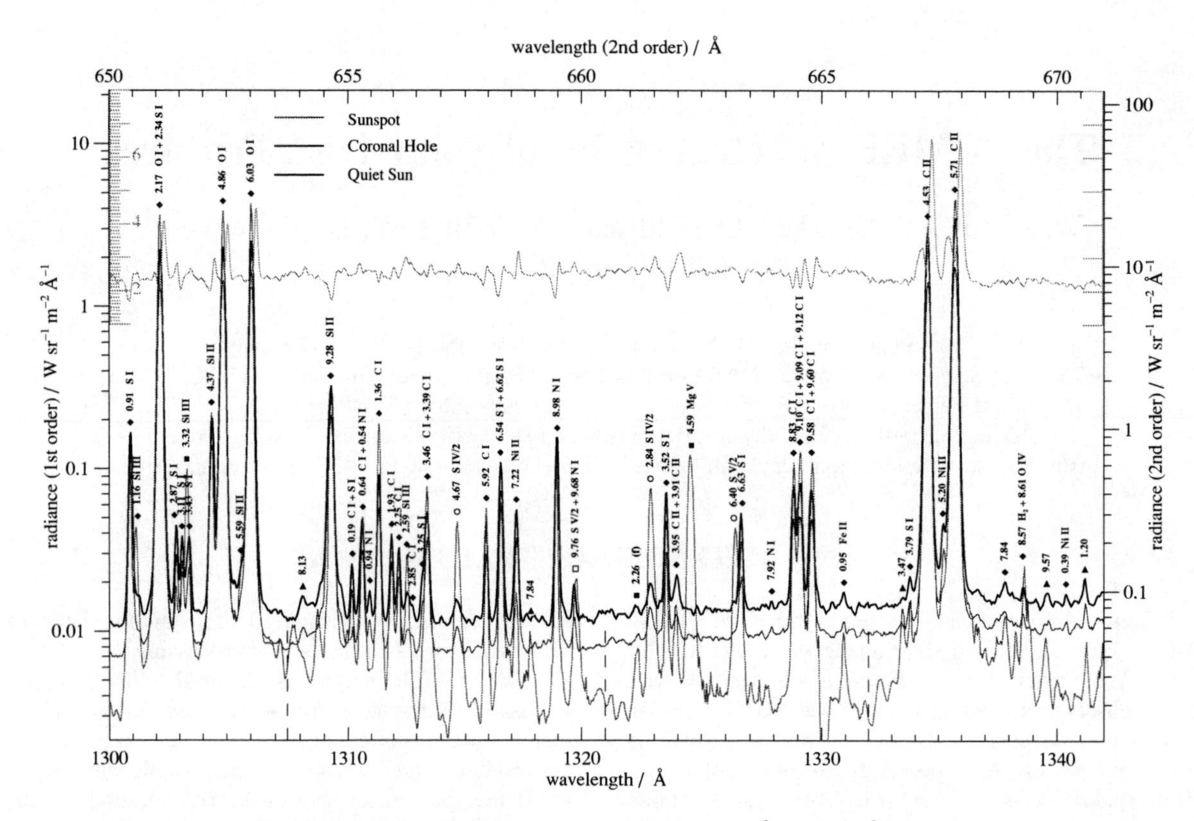

Figure 1. The SUMER spectral atlas in the wavelength range from 1300 Å to 1342 Å as an example of the full range extending from 670 Å to 1609 Å in first order of diffraction.

http://www.linmpi.mpg.de/english/projekte/sumer/FILE/SumerEntryPage.html for scientific evaluation.

REFERENCES

1. Wilhelm, K., Curdt, W., Marsch, E., et al., *Sol. Phys.* **162**, 189-231 (1995).
2. Curdt, W., Brekke, P., Feldman, U., et al., *Astron. Astrophys.*, in press (2001).
3. Feldman, U., Behring, W. E., Curdt, W., et al., *Astrophys. J. Suppl. Ser.* **13,** 195- 219 (1997).
4. Curdt, W, Feldman, U., Laming, J. M., et al., *Astron. Astrophys. Suppl. Ser.* **126,** 281-296 (1997).
5. Kelly, *J. Phys. Chem. Ref. Data* **16**, 1 (1987).
6. Cohen, L., Feldman, U., Doschek, G. A., *Astrophys. J. Suppl. Ser.* **37,** 393-405 (1978).
7. Sandlin, G. D., Bartoe, J.-D. F., Brueckner, G. E., et al., *Astrophys. J. Suppl. Ser.* **61,** 801-898 (1986).

Coronal and Solar Wind Composition

Coronal and Solar Wind Elemental Abundances

J. C. Raymond*, J. E. Mazur†, F. Allegrini**, E. Antonucci‡, G. Del Zanna§, S. Giordano‡, G. Ho¶, Y.-K. Ko*, E. Landi||, A. Lazarus††, S. Parenti‡‡, G. Poletto§§, A. Reinard¶¶, J. Rodriguez-Pacheco***, L. Teriaca§§, P. Wurz** and L. Zangrilli†††

*Center for Astrophysics, 60 Garden St., Cambridge, MA 02138, USA
†The Aerospace Corporation, El Segundo, CA 90245, USA
**Physikalisches Institut, University of Bern, sidlerstrasse 5, CH-3012 Bern, Switzerland
‡Osservatorio Astronomico di Torino, 10025 Pino, Torinese, Italy
§DAMTP, University of Cambridge UK, Cambridge, UK
¶Johns Hopkins University, Applied Physics Laboratory, Johns Hopkins Road, Laurel, MD 20723 USA
||Naval Research Laboratory, Washington, DC USA
††MIT, Room 37-687, 77 Massachusetts Avenue, Cambridge, MA 02139 USA
‡‡Centre for Astrophysics, University of Central Lancashire, Preston, UK
§§Arcetri Astrophysics Observatory, Largo Enrico Fermi 5, Firenze 50125, Italy
¶¶Department of Atmospheric, Oceanic, and Space Science, University of Michigan, Ann Arbor, MI 48109 USA
***Departamento Fisica - Universidad Alcala, Alcala de Henares, Madrid 28871 Spain
†††Dipartimento di Astronomia e Scienza dello Spazio, Universita di Firenze, Largo Enrico Fermi 5, Firenze 50125 Italy

Abstract. Coronal elemental abundances, as compared with abundances in the solar wind and solar energetic particles, provide the means for connecting solar wind gas with its coronal source. Comparison of coronal abundances with photospheric values shows fractionation with the ionization potential of the atom, providing important, though not yet fully understood, information about the exchange of material between corona and chromosphere. Fractionation due to gravitational settling provides clues about flows within the corona.

In this paper, we discuss the uncertainties of abundance determinations with spectroscopic techniques and *in situ* measurements, we survey the ranges of abundance variations in both the corona and solar wind, and we discuss the progress in correlating solar wind features with their coronal sources.

1. INTRODUCTION

Elemental abundances in the solar corona are the basis of comparison for investigations of the coronae of other stars and for abundances measured in the solar wind. They differ from solar photospheric abundances by as much as an order of magnitude, and they vary from place to place and time to time. Fractionation according to the ionization potential of the neutral atom (First Ionization Potential, or FIP) is ubiquitous, and evidence for gravitational settling in quiescent streamers has been reported. Measurements of abundances in the solar wind show similar variations, strongly correlated with the speed of the wind. Abundances of solar energetic particles show variations in isotopic ratios indicating strong mass fractionation.

Abundance variations are useful as means of connecting features in the corona with solar wind structures that arise from them. They have enormous potential as tools for understanding the physical processes in the chromosphere that give rise to the wind through analysis of the FIP effect. Preferential heating of various ion species [1] and the effects of ion drag in the accelerating wind [2] will eventually be useful for determining the physical conditions and processes in the region out to a few solar radii where the wind forms.

To realize this potential, we must understand and exploit the variations of coronal and solar wind abundances. We must understand and reduce the uncertainties in both the analysis of spectroscopic observations of the corona and the measurement of abundances in the solar wind. We must be able to reliably connect structures in the solar wind with their coronal sources. The following sections discuss the uncertainties, the variations and the connection between solar and solar wind features, respectively.

CP598, *Solar and Galactic Composition*, edited by R. F. Wimmer-Schweingruber

2. UNCERTAINTIES IN SPECTROSCOPIC ESTIMATES OF CORONAL ABUNDANCES

Before SOHO, remote sensing instruments revealed that different solar features had very different elemental compositions. Most of these results were obtained using the Skylab S0-82A observations (see e.g., [3]) that were limited by the fact that only a few lines could be used, given the characteristics of the overlapping spectroheliograms. Now, the spectroscopic instruments on board SOHO provide new opportunities to study in much greater detail the chemical composition of the solar transition region and corona. These instruments provide better radiometric calibration, much higher spatial and spectral resolution, and coronagraphic capabilities. However, it is important to be aware of the other factors that limit the accuracy of coronal abundance determinations.

Del Zanna et al. [4] reviewed the main factors that affect the determination of the element abundances from spectroscopic instruments that observe the low corona, such as CDS and SUMER on board SOHO. The element abundances that are derived depend on:

a) The spectroscopic method used. The method of estimating the emission measure distribution as a function of temperature can affect the results, particularly if only a few spectral lines are available.

b) The atomic data and the ionization equilibrium calculations that are adopted. We discuss these below.

c) Temperature and density effects. Many analyses use, for instance, an ionization model computed for the low density limit, while at transition region densities some dielectronic recombination ratios are reduced by about a factor of 2. In the case of spectroscopic instruments that observe the outer corona (such as UVCS) temperature and density effects are reduced because the coronal plasma is nearly isothermal and the density is low. On the other hand, other effects such as the photo-excitation become important, and have to be taken into account.

d) Instrument calibration. For many purposes only the relative sensitivity as a function of wavelength needs to be known. For modern UV and EUV instruments, this should be calibrated to 15% or better before launch.

Several approximations are commonly made, often tacitly. Proton collisional excitation and de-excitation processes are usually neglected. They are generally most important for fine structure transitions. Stimulated emission and absorption of ambient radiation are often neglected, though they can be important for hydrogen thanks to its metastable 2s level [5]. The optical depth to resonance scattering is generally ignored because of the geometrical complexity introduced by scattering, but it may be important in some cases [6, 7].

Various authors (see, e.g., [8, 9, 10]) have pointed out that all of these factors may have led to inaccurate determinations of the element abundances in the past. Del Zanna et al. [4] show examples where in particular cases these factors do indeed lead to variations of a factor of two or more in the derived element abundances. Since each of these variations is of the same order as the FIP effect that we want to measure, these factors should be given full consideration.

2.1 Atomic Rates

Any plasma diagnostic technique involving EUV line intensities requires the knowledge of a large amount of atomic data and transition probabilities in order to be carried out; these are necessary to calculate the theoretical line intensities for a given ion to be compared with the observations. Any uncertainty or inaccuracy in these atomic data can have significant impact on the diagnostic results.

The number of photons emitted in an optically thin spectral line $i \rightarrow j$ is given by

$$I_{ij} = \frac{1}{4\pi}\int_h N_j(X^{+m})A_{ji}dh \qquad ph\ cm^{-2}\ s^{-1}\ sr^{-1} \qquad (1)$$

The *Contribution Function* $G(T,N_e)$ of the line is defined as

$$G(T,\lambda_{i,j}) = \frac{N_j(X^{+m})}{N(X^{+m})}\frac{N(X^{+m})}{N(X)}\frac{N(X)}{N(H)}\frac{N(H)}{N_e}\frac{A_{ji}}{N_e} \qquad (2)$$

where $\frac{N_j(X^{+m})}{N(X^{+m})}$ is the relative upper level population; $\frac{N(X^{+m})}{N(X)}$ is the relative abundance of the ion X^{+m} (*ion fraction*); $\frac{N(X)}{N(H)}$ is the abundance of the element X relative to hydrogen; $\frac{N(H)}{N_e}$ is the hydrogen abundance relative to the electron density; and, A_{ji} is the Einstein coefficient for spontaneous emission.

The *Differential Emission Measure* (DEM) is defined so that

$$\varphi(T) = N_e^2\frac{dh}{dT} \qquad (3)$$

the number of photons emitted in a spectral line may be expressed as

$$I_{ij} = \frac{1}{4\pi}\int_T G(T,\lambda_{i,j})\varphi(T)dT \qquad (4)$$

>From Equation 2 it is possible to identify the main sources of uncertainties in the evaluation of the $G(T,N_e)$ functions.

2.2 Relative upper level population

The relative population $\frac{N_j(X^{+m})}{N(X^{+m})}$ must be calculated by solving the statistical equilibrium equations for a number of low lying levels and including all the important collisional and radiative excitation and de-excitation mechanisms.

Stimulated and spontaneous radiative transition probabilities are mostly obtained from *ab initio* theoretical calculations, using programs such as SUPERSTRUCTURE [11], or CIV3 [12].

The electron collisional excitation rate coefficient ($cm^3\ s^{-1}$) for a Maxwellian electron velocity distribution with a temperature T_e (K), is given by:

$$C^e_{i,j} = \frac{8.63 \times 10^{-6}}{T_e^{1/2}} \frac{\Upsilon_{i,j}(T_e)}{\omega_i} e^{(-E_{i,j}/kT_e)} \quad (5)$$

where ω_i is the statistical weight of level i; $E_{i,j}$ is the energy difference between levels i and j; k is the Boltzmann constant and $\Upsilon_{i,j}$ is the thermally-averaged collision strength:

$$\Upsilon_{i,j}(T_e) = \int_0^\infty \Omega_{i,j}\, e^{(-E_j/kT_e)} d(E_j/kT_e) \quad (6)$$

where Ω is the collision strength and E_j is the energy of the scattered electron relative to the final energy state of the ion. Non-Maxwellian electron distributions can also be considered. However, the effect of a non-Maxwellian tail on collisional excitation rates is generally small, because the excitation rate is usually dominated by electron energies below 2 or 3 kT.

Collision strengths are obtained from theoretical calculations. The solution of the electron-ion scattering problem is complex and requires extensive computing resources. The accuracy of a particular calculation depends on two main factors. The first is the representation which is used for the target wavefunctions, the second is the type of scattering approximation chosen. The target must take account of configuration interaction and allow for intermediate coupling for the higher stages of ionization. The main approximations used for electron-ion scattering are *Distorted Wave* (DW) [13], *Coulomb Bethe* (CBe) [14] and the more elaborate *Close-Coupling* (CC) [15]. Laboratory measurements of the cross sections are available for some transitions, and they provide an important benchmark for the theoretical calculations.

Comprehensive databases that include the best atomic data are becoming available. Besides making new data quickly available to researchers, they promote consistency among various analyses. In solar physics, CHIANTI [16] is most popular. Recently, CHIANTI has been extended to cover the X-ray wavelengths [18]. Other databases such as APEC/APED [17] emphasize X-ray wavelengths.

The accuracy of theoretical calculations for the radiative and collisional transition probabilities is of crucial importance for the determination of relative level populations. However, there is no direct way of assessing the accuracy of these theoretical calculations, and in the literature the authors limit themselves to quoting an accuracy of usually 10% for radiative transition probabilities, and of 30% for collisional transition probabilities.

However, the accuracy quoted for the collisional transition probabilities is sometimes very optimistic. In fact, the increasing computing power of modern computers allows the inclusion of more and more complex atomic models and to take into account interactions between a larger number of terms. The presence of these previously neglected configurations, terms and interactions can lead to very different results in the collision rates. Examples are given for Fe XII [19], for Fe XIV [20] and for Fe IX [21]. In these cases, the effects of more accurate calculations go beyond the 30% accuracy quoted by previous authors, and have a strong impact in the diagnostic results obtained from line intensity ratios. However, Fe IX, XII and XIV are probably the worst cases. Cascades from higher-lying levels are also important in many cases, such as Fe XVII [22].

One way to evaluate the consistency of the atomic data is to model a set of observational data containing many spectral lines and examine the scatter in ratios of observed to predicted fluxes. The scatter will include measurement and calibration errors, along with any inappropriate assumptions of the model, but it gives a general indication of the accuracy of the atomic rates. Spectra from SERTS observations of a solar active region [23] between 171 and 445 Å have been especially useful in this regard [24, 25, 26], and they demonstrate the improvements in recent years incorporated into the CHIANTI database. More recently, [27] and [10] have carried out a similar comparison using an off-limb observation of a streamer obtained with SUMER [29] between 800 and 1600 Å. This comparison involves mainly intercombination and forbidden lines within the ground configuration for a number of coronal ions. Landi & Feldman [27] find in general good agreement between the CHIANTI database and observed line intensities.

2.3 Ion fractions

It is usually assumed that the plasma is in ionization equilibrium; this assumption, which can be misleading in highly dynamic plasma, allows the use of the ion fraction datasets found in the literature. It is not difficult to compute time-dependent ionization states for any specified model of temperature and density (e.g., [30]), but there is rarely a unique model available for a given data set. The equilibrium ion fractions available in the literature have been calculated using the state-of-the-art ionization and recombination rates available at the time of publication. However, progress in the theoretical models for ionization and recombination processes has led to significant changes in ionization and recombination rates. The tables of Mazzotta et al. [31] reflect the best information available to date.

An assessment of the quality of ion abundance computations is not an easy task, due to the complicated and largely unknown temperature structure of the solar atmosphere as seen along the line of sight. In the recent past, Masai [32] investigated the impact of uncertainties in the ionization and recombination rates on X-ray spectral analyses, finding that differences in the rates led to significant differences in iron abundance and plasma temperature measurements. Phillips & Feldman [33] have used *Yohkoh* flare observations to check the ion fractions of He-like ions, concluding that the observed spectra were consistent with the adopted ion fractions at the 50% level of precision, and this led to changes to the plasma diagnostic results. Phillips & Feldman suggest that Arnaud & Rothenflug [34] ion abundances for [S XV] and [Ca XIX] need to be improved.

Young & Mason [28] used SOHO/CDS spectral data to study element abundance variations. They found enhancements of up to a factor of 5 for Ca X, but suggested that this was due to inaccurate ionization fraction calculations. Del Zanna [26] has also used SOHO/CDS observations to point out disagreements between lines of the Li and Na isoelectronic sequences, compared to the lines of all the other sequences. Del Zanna [26] has also shown that some of these discrepancies, including the Ca X one, are indeed related to inaccurate ionization equilibrium calculations, since they are resolved with the use of more recent calculations [31].

More recently, [35] investigated the effect of changing ion fractions datasets on temperature diagnostics. They used the spectra from an isothermal region outside the solar limb, and adopted in turn the ionization equilibria from Shull & Van Steenberg [36], Arnaud & Rothenflug [34] incorporating the latest revisions to the iron ions by Arnaud & Raymond [37] and Mazzotta et al. [31]. They found negligible differences in the results, showing that in the case of an isothermal spectrum the choice of the ion fractions has little effect on the temperature diagnostic results. Unfortunately, they did not investigate the effect on abundance diagnostics. Gianetti et al. [38] carried out a similar study using active region observations on the disk. They found that the choice of the ion fractions has a large effect on the diagnostic results, both in terms of DEM and element abundance measurements.

It is important to recognize that improved abundance determinations in recent years result in almost equal measure from the efforts of laboratory and theoretical atomic physicists, the calibration efforts of the SOHO instrument teams, and the careful planning and data analysis of SOHO observers.

3. UNCERTAINTIES OF SOLAR WIND AND SOLAR ENERGETIC PARTICLE ABUNDANCE MEASUREMENTS

Abundances can be measured both in the normal solar wind and in Solar Energetic Particle (SEP) events. As for spectroscopic measurements, instrument calibration is a crucial part of the analysis. Modern instruments such as SWICS/ULYSSES [39], CELIAS/SOHO [40] and SWIMS and SWICS/ACE [41] can measure solar wind abundances with high sensitivity and resolution of mass and charge. Instruments such as ACE/ULEIS and ACE/SIS measure higher energy particles (100 keV/nuc to 10 MeV/nuc), again with better sensitivity and resolution than earlier generations of instruments. With instruments such as these it is possible to measure the abundances of rare species and to measure isotopic abundance ratios. The sensitivity to particles of different mass is calibrated before launch, and sometimes extensive calibration campaigns are performed even after launch using the flight-spare instruments. In recent publications on solar wind abundances the quoted errors are around 10% (or larger for elements of very low abundance) and as low as 5% for isotopic ratios.

One important limitation, as for spectroscopic instruments, is dynamic range. This is especially an issue when comparing hydrogen and helium to other species, as protons and alpha particles are usually measured with a different sensor than that used for the heavy elements. Thus most papers quote ratios of different elements to oxygen or silicon, but the ratios relative to hydrogen are not often determined directly. Ratios relative to hydrogen are given by Wimmer-Schweingruber [42] for the fast and slow solar wind. They imply that the FIP effect is an enhancement of low-FIP elements rather than a depletion of high-FIP elements relative to hydrogen. For gradual SEP events corresponding to particles accelerated by Coronal Mass Ejection (CME) shocks in the corona, Reames [43] gives a H/O ratio of 1570 $\pm$220 (statistical uncertainty), within about 16% of the photospheric value. For impulsive SEP events (corresponding to particles accelerated in flares) the ratio is somewhat lower, but less well determined.

SEPs can be measured with very good mass resolution, as low as $\sigma_m > 0.3$ amu [44]. Isotopic abundance measurements can be reliably performed. The main limitation is that acceleration processes and transport effects lead to fractionation in mass or mass/charge (Reames [43]; see section 6).

4. ABUNDANCE VARIATIONS IN THE CORONA

Most astronomers and solar wind physicists tend to assume that there is one set of photospheric abundances and one set of coronal abundances, differing by a FIP fractionation of a factor of 3-4. The tables of [45] are the current standard. A single set of photospheric abundances is probably a good approximation, though the 'standard' abundance set continues to evolve. This is especially troublesome for elements such as neon which lack strong lines at optical wavelengths. The most important change in recent years is the decrease in O abundance by 0.2 dex [45], as many relative abundance measurements are scaled to oxygen. (See also Holweger, these proceedings, and references therein.)

Substantial FIP abundance variations within the corona were established some time ago, but the fragmentary nature of the results (often just the ratio of one element to one other), disputes about whether erroneous atomic data might account for some of the apparent variations, and the large range of abundances without a clear pattern have clouded the issue. Abundance variations in solar flares (e.g., [46]) are believed to result from evaporation of chromospheric material into loops filled with pre-existing coronal plasma. The large number of spectral lines available to the spectrographs aboard SOHO have begun to clarify the situation. Some results for the FIP bias are listed in Table 1. While the expected factor of about 3-4 is often observed in the quiet Sun, the FIP bias is only half as large in coronal holes, and it can be up to twice as large in active regions. Coronal hole plumes do not show extreme FIP effects [47]. Earlier reports of extreme FIP enhancement can be explained in terms of their emission in a narrow temperature band [4].

Off-limb observations in the UV open the possibility of measuring abundances relative to hydrogen. UVCS observations of a strong depletion of O in the core of the equatorial streamer at solar minimum were interpreted in terms of gravitational settling [48], and SUMER observations of a strong decline of iron lines with height compared with the decline of silicon lines support this idea. Parenti et al. [49] derived the oxygen abundances in a streamer core and edges from UVCS data taken in June 2000. The absolute abundance of oxygen turned out to be $\sim$ 8.6 on the edges and 8.4 in the core. The electron temperature derived from line ratio techniques appears to decrease by 15% from the core to the edges. Zangrilli, Poletto and Biesecker [50] studied the oxygen abundance in streamers in July 1996 between 2.5 and 4 $R_{\odot}$. Two streamers at the minimum of solar activity were analyzed (6 and 11 July 1996) and the oxygen abundance was found to vary from about 8.0 in the core to 8.4 in the legs of the streamer (about 10° above the equator). Zangrilli et al. also found a hint of abundance decrease with height (about 50% between 2.5 and 5 $R_{\odot}$) in the core. At these heights there may be some departure from ionization equilibrium. Marrocchi et al [51] found a 40% decline in oxygen abundance in a streamer between 1.5 and 2.2 $R_{\odot}$.

It is more difficult to obtain abundances in coronal holes, because the lines are fainter and because the assumption of ionization equilibrium is problematic. An attempt to determine the oxygen abundance in interplume lanes has been performed by Teriaca et al [52]. From coronal hole O VI doublet data taken by SUMER in 1996, the authors infer an oxygen abundance $\geq$ 8.50 depending on the adopted density profile. Antonucci et al. [53] used UVCS observations of solar minimum coronal holes to infer an oxygen abundance consistent with photospheric and ULYSSES values.

Among the most difficult, but most important, measurements is the helium abundance. Laming and Feldman [54] find He/H ratio of 0.05 in both a streamer and coronal hole near solar minimum.

5. ABUNDANCE VARIATIONS IN THE SOLAR WIND

The variations in He/H and FIP bias between fast and slow solar wind are well established [67, 68], and solar physicists tend to view solar wind abundances as bimodal. The FIP bias is modest (around 1.5) in the fast wind and 3-4 in the slow wind. While this does seem to be a good approximation for the average abundances, there are short time-scale variations.

Heavy ions in the solar wind (e.g., O, Si, Fe) and helium generally show a much larger variability in their densities than the protons. Typically the variability is larger by a factor of 10. This variability has been observed for helium with respect to protons already a while ago as has been reviewed by Neugebauer [69]. She found that the ratio of protons to alpha particles in the solar wind (n_p/n_a) varies in the range from 8.1×10^{-4} to 4.17×10^{-1}, a variation by a factor of 500. For heavy elements this variability is illustrated in Figure 1 where two commonly used abundance ratios Fe/O and Si/O are plotted for a period of 80 days during solar minimum in 1996. This time period also contains short coronal hole solar wind sections of about one day since the "elephant trunk" coronal hole passed three times during that period. One can easily see the large fluctuations in the abun-

TABLE 1.

Structure	Paper	Height	Instrument	FIP Bias
Coronal Hole	Doschek et al. 98 [55]	1.0-1.2 $R_\odot$	SUMER	< 2
Coronal Hole	Feldman et al. 98 [56]	1.03-1.5 $R_\odot$	SUMER	< 1.3
Coronal Hole Plume	Young et al. 99 [57]	1.0-1.1 $R_\odot$	CDS	1.5
Quiet Sun	Warren 99 [58]	1.05-1.35 $R_\odot$	SUMER	2.3 ± 0.7
Quiet Sun	Laming et al. 99 [59]	1.1 $R_\odot$	SUMER	3 - 4
Quiet Sun	Young & Mason 98 [28]	Disk	CDS	2
EQ Streamer core	Raymond et al. 97 [48]	1.5 $R_\odot$	UVCS	3 - 4
EQ Streamer legs	Raymond et al. 97 [48]	1.5 $R_\odot$	UVCS	3 - 4
EQ Streamer base	Feldman et al. 99 [56]	1.03-1.5 $R_\odot$	SUMER	4
EQ Streamer	Parenti et al. 00 [61]	1.6 $R_\odot$	UVCS	3
EQ Streamer base	Parenti et al. 00 [61]	1.02-1.19 $R_\odot$	CDS	1.1
Active Region	Young & Mason 98 [28]	1.5 $R_\odot$	CDS	1 - 9
Active Region	Rank et al. 99 [62]	Disk	CDS	5-9
Active Region	Dwivedi et al. 99 [63]	1.04-1.11 $R_\odot$	SUMER	8
Active Region	Ko et al. 01 [64]	1.3-1.7 $R_\odot$	UVCS	4
CME	Ciaravella et al. 97 [65]	1.5 $R_\odot$	UVCS	1
Prominence	Spicer et al. 98 [66]	1.5 $R_\odot$	SUMER	2

dance ratios exceeding the values for the FIP fractionation by an order of magnitude. Of course, some of the spikes in Figure 1 are of instrumental or statistical nature. This high variability averages out when investigating longer time periods. Analysis of the data presented in Figure 1 showed that the established heavy element abundances for Fe/O and Si/O are obtained including the FIP fractionation pattern when sufficiently long averages are performed [70]. These short-time variations in the abundance of heavy ions in the solar wind are thought to be caused in the corona and are believed to be of temporal and spatial nature [71]. Similar short-time fluctuations are also observed for the charge states of heavy elements [72, 73], although the amplitude of the variations is smaller.

6. ABUNDANCE VARIATIONS IN SOLAR ENERGETIC PARTICLE EVENTS

Transport effects can be a significant source of abundance variations in individual SEP events associated with interplanetary shocks. The basic parameter that describes a particle's motion in the interplanetary medium is its rigidity, or momentum per unit charge. Rigidity is proportional to the mass to charge (A/Q) ratio, and it is the case that A/Q effects seem to organize the abundance variations at the onsets of some SEP events, e.g., Tylka et al. [74] Recently Ng et al. [75] developed a model of particle escape from the Alfvèn wave turbulence that energetic protons create at such a shock. Also, Zank et al. [76] have begun to model the rigidity-dependent acceleration and escape processes in a 3-dimensional model of a traveling shock. Models such as these are providing valuable insight into the SEP accelerator and may develop into more powerful tools for describing the event-to-event variations recently seen, e.g., Boberg and Tylka [77].

In addition to transport effects, the seed population available for acceleration at the CME/shock front includes suprathermal particles from various sources such as the solar wind, previous CME events, and ^{3}He-rich impulsive flares, e.g., Mason et al. [78]. The actual mixture of suprathermal particles available for injection at the shock is unknown for any given event, but the abundance of ^{3}He may serve as a tracer for the flare contribution. The changing magnetic connection to a propagating shock implies that the observer samples the acceleration within a wide range of heliolongitudes. This time-dependent connection combines with a non-uniform mixture of source populations to create a complex picture of the source population for the SEPs.

Nevertheless, previous studies have gone ahead and averaged SEP abundances in samples of many (>10) events [79, 80, 81, 43]. In this sense, the SEPs might be a measure of the time-averaged and longitude-averaged abundance of the interplanetary suprathermal particles, part of which includes the fast and slow coronal flows that are sampled directly and remotely via spectroscopic methods. The SEP event-to-event variations then arise from the transport and seed population effects discussed above. Reames [43] lists such an average of 50 events and suggests that the residual Q/A effects have been averaged out to "at least an accuracy of about 10%". Measurements of the SEP abundances will continue in the current solar maximum with more sensitive instruments on platforms such as Wind and ACE. As the new mea-

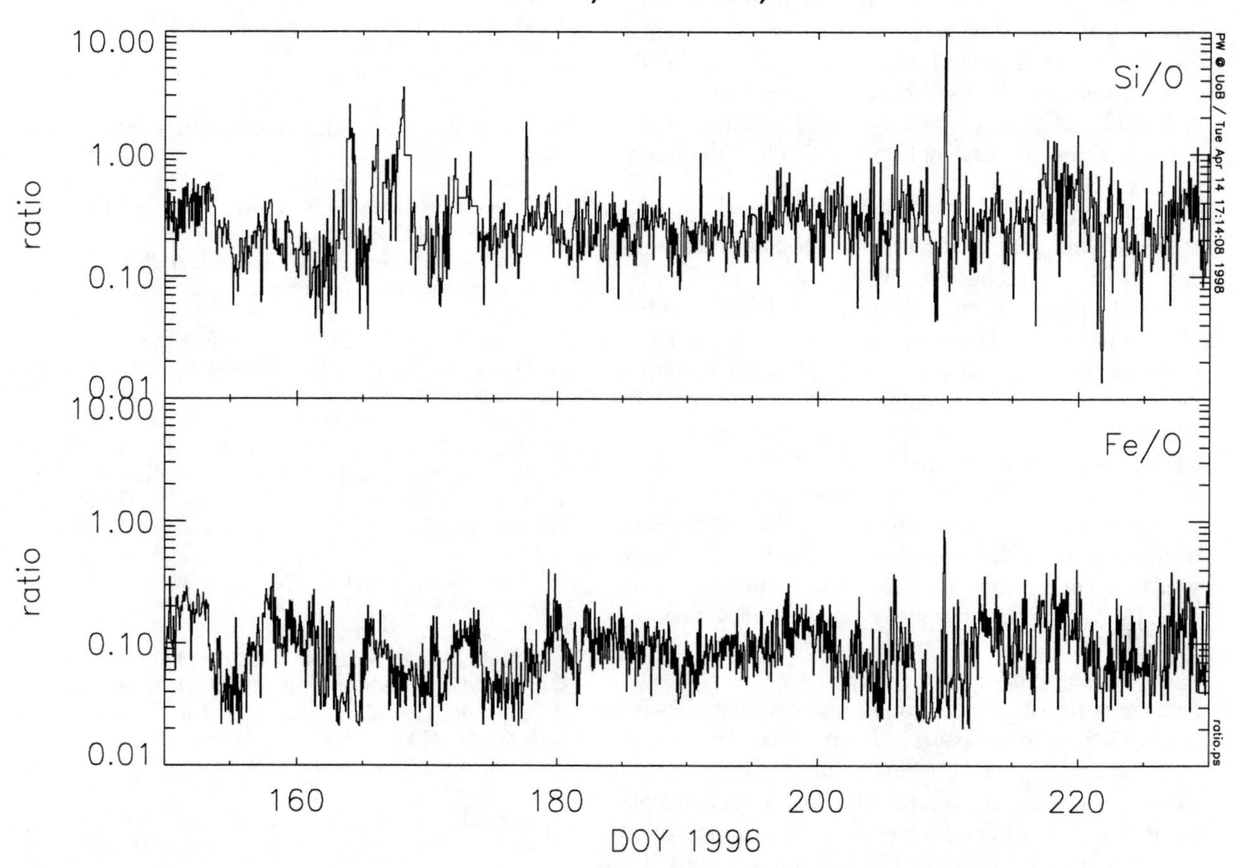

FIGURE 1. Abundance ratios of silicon and iron to oxygen. Thirty minute running averages of 5 minute data from CELIAS/MTOF aboard SOHO.

surements accumulate, we expect that new SEP abundance compilations along the lines of Reames [43] will address the issue of event-averages over a broader energy range and with high sensitivity.

Another broad class of SEP events have characteristics that suggest an origin not in interplanetary space but at a flare site. We list several of these observables here: enrichments of ^{3}He by a factor of 10-1000 compared to the solar wind; enrichments of Ne-Si and Fe by factors of 3-5 and 10, respectively; association with streaming 10-100 keV electrons; scatter-free propagation; high ionization states appropriate to a 5-10 MK source. An observer at 1 AU sees the SEPs from these impulsive solar flares arrive with a velocity dispersion when there is a good magnetic connection between 1 AU and an active region at western solar longitudes, e.g., [82, 83]. Different wave-particle resonance modes are thought to cause the large abundance enhancements of the ^{3}He and heavy ions. New measurements at 1 AU, e. g. Ho et al. [84] are revisiting the possible link between the energetic electrons and the ^{3}He enrichments.

Current instrumentation can readily resolve individual particle injections of ^{3}He-rich events at 1 AU [85], but it is difficult to tie the in-situ particles to observations of the same flare via photons. The X-ray events are often low intensity and more numerous, so correlations are difficult. The best hope for correlated optical and particle measurements lies in the broad line emissions from SEPs that precipitate into the solar atmosphere during a solar flare. Such correlations may become possible with current particle instrumentation and X-ray imaging spectroscopy from the HESSI mission [86].

7. CONNECTION BETWEEN CORONAL AND SOLAR WIND STRUCTURES

The general connection between fast solar wind and coronal holes is well established. Comparison of elemental abundances in streamer legs with slow solar wind abundances seems to show that the slow wind does arise from the edges (not the cores) of streamers.

There remains a troubling question as to what exactly the streamer legs are. Observationally, they are high density, low outflow speed structures very similar to the streamer cores except in abundance relative to hydrogen. While their appearance suggests that they correspond to the open field lines closest to the closed field structures of the streamer cores, this has not been verified by sufficiently detailed field models.

The next challenge is to identify specific features in the solar wind with their sources at the Sun. Traditionally, the solar wind has been traced back to the Sun by means of simple ballistic outflow models. More recently, global magnetic field models and MHD outflow solutions have been applied. Ko et al. [87] present a promising attempt to relate solar wind properties measured with ACE/SWICS to structures observed by UVCS at several heights based on a 3-D MHD code. The energetic particles measured within magnetic clouds originate in impulsive flares (i.e., ^{3}He-rich and Fe/O$\sim$1). If cases can be found where a shock forms in the corona and rapidly weakens, it may eventually be possible to infer the region (e.g., legs of coronal streamers, etc.) where they were injected.

An exciting prospect is the measurement of energetic particles within Magnetic Clouds. Energetic particles inside MCs that are not associated with IP shocks (or with very weak ones), should show composition rather the same as that in the solar corona. Thus they would indicate the region (legs of coronal streamers, etc.) from which they were injected inside the CME. Moreover, their ^{3}He and Fe abundances could provide additional information about the problematic flare-CME connection.

REFERENCES

1. Kohl, J.L. et al. 1997, Sol. Phys. 175, 613
2. Ofman, L. 2000, submitted to ApJ
3. Sheeley, N.R. Jr., 1996, ApJ, 469, 423.
4. Del Zanna, G., Bromage, B.J.I., and Mason, H.E., 2001, this issue.
5. Gabriel, A.H. 1971, Sol. Phys., 21, 392
6. Schrijver, C.J., and McMullin, R.A. 2000, ApJ, 531, 1121
7. Wood, K., and Raymond, J. 2000, ApJ, 540, 563
8. Mason, H.E., 1992, ESA SP-348, 297
9. Phillips, K. J. H., 1997, Adv. Space Res. 20, 79.
10. Young, P.R., Landi, E., Thomas, R.J., 1998, A&A 329, 291
11. Eissner, W., Jones, M., Nussbaumer, H., 1974, Comp.Phys.Comm., 8, 270
12. Hibbert, A., 1975, Comp.Phys.Comm., 9, 141
13. Eissner, W., Seaton, M.J., 1972, J.Phys.B, 5, 2187
14. Burgess, A., Sheorey, V.B., 1974, J.Phys.B, 7, 2403
15. Burke, P.G., Hibbert, A., Robb, W.D., 1971, J.Phys.B, 4, 153
16. Dere, K.P., Landi, E., Mason, H.E., Monsignori Fossi, B.C., Young, P.R., 1997, A&ASS, 125, 149
17. Smith, R.K., Brickhouse, N.S., Lieddahl, D.A., and Raymond, J.C. 2001, R, ApJL, in press
18. Dere, K.P., Landi, E., Young, P.R., and Del Zanna, G., 2001, ApJSS, 134, 331.
19. Binello, A.M., Landi, E., Mason, H.E., Storey, P.J., Brosius, J.W., 2001, A&A, 370, 1071
20. Storey, P.J., Mason, H.E., Young, P.R., 2000, A&AS, 141, 285
21. Storey, P.J. and Zeippen, C.J., 2001, MNRAS, 324, L7
22. Smith, B.W., Mann, J.B., Cowan, R.D., and Raymond, J.C. 1985, ApJ, 298, 898
23. Thomas, R.J., Neupert, W.M., 1994, ApJSS, 91, 461
24. Brickhouse, N.S., Raymond, J.C., and Smith, B.W. 1995, ApJS, 97, 551
25. Landi, E., 1998, PhD Thesis - University of Florence
26. Del Zanna, G., 1999, PhD Thesis - University of Central Lancashire.
27. Landi, E., Feldman, U., 2001, ApJ, in preparation
28. Young,P.R. and Mason H.E., 1998, SSRv. 85, 315.
29. Wilhelm, K., et al., 1995, Solar Phys. 162, 189
30. Akmal, A. Raymond, J.C., Vourlidas, A., Thompson, B., Ciaravella, A., Ko, Y.-K., Uzzo, M., and Wu, R. 2001, ApJ, 553, 922
31. Mazzotta, P., Mazzitelli G., Colafrancesco S., Vittorio N., 1998, A&AS 133, 403
32. Masai, K., 1997, A&A, 324, 410
33. Phillips, K.J.H., Feldman, U., 1997, ApJ, 477, 502
34. Arnaud, M., Rothenflug, R., 1985, A&AS 60, 425
35. Allen, R., Landi, E., Landini, M., Bromage, G.E., 2000, A&A, 358, 332
36. Shull, J.M., Van Steenberg, M., 1982, ApJS 48, 95 and ApJS 49, 351
37. Arnaud, M., Raymond, J.C., 1992, ApJ 398, 394
38. Gianetti, D., Landi, E., Landini, M., 2000, A&A, 360, 1148
39. Gloeckler, G., Geiss, J., Balsiger, H., Bedini, P., Cain, J.C., Fisher, J., Fisk, L.A., Galvin, A.B., Gliem, F., and Hamilton, D.C., 1992 A & A Suppl., 92, 267
40. Hovestadt, D., et al. 1995, Solar Phys., 162, 441
41. Gloeckler, G., Bedini, P., Bochsler, P., Fisk, L.A., Geiss, J., Ipavich, F.M., Cani, J., Fischer, J., Kallenbach, R., Miller, J., Tums, O., and Winner, R. 1998, SSRv, 86, 495
42. Wimmer-Schweingruber, R. 1994, PhD Thesis, University of Bern
43. Reames, D.V. 1999, SSRv, 90, 413
44. Stone, E.C., et al., 1998, SSRv, 86, 1
45. Grevesse, N., and Sauval, A.J. 1998, SSRv, 85, 161
46. Fludra, A., and Schmelz, J.T. 1999, A& A 348, 286
47. Del Zanna, G., and Bromage, B.J.I. 1999, SSRv, 87, 169
48. Raymond, J.C., et al. 1997, Sol. Phys., 175, 645
49. Parenti, S., Bromage, B.J.I., Poletto, G., Suess, S., Raymond,J.C., Noci, G., and Bromage, G.E. 2001, this volume
50. Zangrilli, L., Poletto, G., and Biesecker, D. 2000, this volume
51. Marrocchi, D., Antonucci, E., and Giordano, S. 2001, Annales Geophysicae, 19, 135
52. Teriaca, L., Poletto, G., Falchi, A., and Doyle, J.G. 2001, this volume
53. Antonucci, E., Giordano, S., and Marocchi, D. 2001, this volume
54. Laming, J.M., and Feldman, U. 2001, this volume
55. Doschek, G.A., Laming, J.M., Fledman, U., Wilhelm, K.,

Lemaire, P., Schüle, U., and Hassler, D.M. 1998, ApJ, 504, 573
56. Feldman, U., Schüle, U., Widing, K.G., and Laming, J.M. 1998, ApJ, 505, 999
57. Young, P.R., Klimchuk, J.A., and Mason, H.E. 1999, A& A, 350, 286
58. Warren, H.P. 1999, Sol. Phys., 190, 363
59. Laming, J.M., Feldman, U., Drake, J.J., and Lemaire, P. 1999, ApJ, 518, 926
60. Feldman, U., Doschek, G.A., Schüle, U., and Wilhelm, K. 1999, ApJ, 518, 500
61. Parenti, S., Bromage, B.J.I., Poletto, G., Noci, G., Raymond, J.C., and Bromage, G.E. 2000, A& A, 363, 800
62. Rank, G., Czaykowska, A., Bagalá, L.G., and Haerendel, G. 1999, ESA-SP-448, p. 349
63. Dwivedi, B.H., Curdt, W., and Wilhelm, K. 1999, ApJ, 517, 516
64. Ko, Y.-K., Raymond, J.C., Li, J., Ciaravella, A., Michels, J., Fineschi, S., and Wu, R. 2001, in preparation
65. Ciaravella, A., et al. 1997, ApJL, 491, 59
66. Spicer, D. Feldman, U., Widing, K.G., and Rilee, M. 1998, ApJ, 494, 450
67. Feldman, W.C., Asbridge, J.R., Bame, S.J., and Gosling, J.T. 1978, JGR, 83, 2177
68. Geiss, J., Gloeckler, G., von Steiger, R., Balsiger, H., Fisk, L.A., Galvin, A.B., Ipavich, F.M., Livi, S., McKenzie, J.F., Ogilvie, K.W., and Wilken, B. 1995, Science, 268, 1033
69. Neugebauer, M., 1981, in *Fundamentals of Cosmic Physics*, Gorden and Breach Science Publishers, Inc., 1981, 7, 131-199.
70. Wurz, P., Aellig, M.R., Bochsler, P., Hefti, S., Ipavich, F.M., Galvin, A.B., Gruenwaldt, H., Hilchenbach, M., Gliem, F., Hovestadt, D. 1999, Phys. Chem. Earth (C), 24(4), 421
71. Geiss, J., Gloeckler, G., and von Steiger, R. 1995, SSRv, 72, 49
72. Aellig, M.R., Hefti, S., Grünwaldt, H., Bochsler, P., Wurz, P., Ipavich, F.M., Hovestadt,D. 1999, JGR, 104, 24769
73. Hefti, S., Grünwaldt, H., Ipavich, F.M., Bochsler, P., Hovestadt, D., Aellig, M.R., Hilchenbach, M., Kallenbach, R., Galvin, A.B., Geiss, J., Gliem, F., Gloeckler, G., Klecker, B., Marsch, E., Moebius, E., Neugebauer, M., Wurz,P. 1998, JGR, 103, 29697
74. Tylka, A. J., Boberg, P.R., McGuire, R.E., Ng, C.K., and Reames, D.V. 2000, Acceleration and Transport of Energetic Particles Observed in the Heliosphere, ed. R. A. Mewaldt et al., AIP Conf. Proc., 528, 147
75. Ng, C. K. et al. 1999, Geophys. Res. Lett. 26, 2145
76. Zank, G. P., Rice, W.K.M., and Wu, C.C. 2000, J. Geophys. Res. 105, 25079
77. Boberg, P. R., and Tylka, A.J. 2000, Acceleration and Transport of Energetic Particles Observed in the Heliosphere, ed. R. A. Mewaldt et al., AIP Conf. Proc., 528, 115
78. Mason, G. M., Mazur, J.E., and Dwyer, J.R. 1999, ApJL, 525, L133
79. Cook, W. R., Stone, E.C., and Vogt, R.E. 1984, ApJ, 279, 827
80. Breneman, H. H. and E. C. Stone 1985, ApJL, 299, L57
81. Mazur, J. E. et al. 1993, ApJ, 404, 810
82. Kahler, S. W. et al. 1987, Sol. Phys., 107, 385
83. Mazur, J. E. et al. 2000, ApJL, 532, L79
84. Ho, G.C., Roelof, E.C., Hawkins, S.E. III, Gold, R.E., Mason, G.M., Dwyer, J.R., and Mazur, J.E. 2001, Astrophys. J.,552, 863
85. Mason, G. M., Dwyer, J.R., and Mazur, J.E. 2000, ApJL, 545, L157
86. Lin, R. P. 2000, in High Energy Solar Physics: Anticipating HESSI, ASP Conf. Proc. 206, 1
87. Ko, Y.-K., Zurbuchen, T., Strachan, L., Riley, P., and Raymond, J.C., this volume

Elemental abundances of the low corona as derived from SOHO/CDS observations

G. Del Zanna*, B.J.I. Bromage† and H.E. Mason*

*DAMTP, University of Cambridge, Cambridge, UK
†Centre for Astrophysics, University of Central Lancashire, UK

Abstract. Some of the main factors that affect the determination of the element abundances from EUV spectra are reviewed. The ionization balance, the selection of lines, and the spectroscopic method used can each account for a variation of a factor of two or more in the derived element abundances, in particular cases. Diagnostic techniques are applied to Skylab/HCO and SOHO/CDS observations of solar coronal holes and plumes, in order to derive their relative element abundances. It is confirmed that coronal holes have photospheric abundances, while plumes only show a small FIP effect, contrary to what has long been thought. It is shown that the plume characteristics can mainly be explained in terms of their temperature structure rather than a large FIP effect.

INTRODUCTION

A correlation between the coronal abundances of some solar regions and the first ionization potential (FIP) of the various elements (e.g., see the review of Raymond et al. [1]) has been found by many authors. A large variety of coronal abundances have been reported, with differences from the photospheric values that usually range from 2 to 4, except for extreme cases.

In the past, using Skylab S-082A observations, only a limited number of lines could be used unambiguously, due to the characteristics of the overlapping spectroheliograms. Most results were based on observations of Mg VI and Ne VI lines, selected as representatives of a low- and a high-FIP element, respectively. Although the observed intensity ratios of lines from these ions do show large variations for different coronal structures, it is not straightforward to deduce variations in relative element abundances. In fact, various other effects can change the observed ratios, as discussed below.

Recently, the spectroscopic instruments on board SOHO, have provided a new opportunity to study in detail the chemical composition of the solar transition region and corona. Here, we are primarily concerned with spectroscopic measurements of relative abundances, and with the various factors that can affect the determination of the element abundances. These factors can be broadly grouped into the following classes:

1. The diagnostic method used.
2. The selection of spectral lines used.
3. The assumption of ionization equilibrium and the ionization balance used.
4. The atomic data used for each ion.
5. Temperature and density effects.
6. Instrument calibration.

Previously, other authors [2, 3, 4] have pointed out that some of these factors may have led to inaccurate determinations of the element abundances. The problems summarised here are general, that is are not instrument dependent, since they have been found in Skylab, SOHO, and other data. More details can be found in [5]. Here, only a few examples are given, to point out that some of these factors can account for large variations (of a factor of 2-3) in the derived element abundances.

In this paper we present results from the Coronal Diagnostic Spectrometer (CDS) on SOHO [6] which consists of two spectrometers (a Normal Incidence, NIS; and a Grazing Incidence, GIS) and six channels, covering almost entirely the 151-785 Å wavelength region. The CDS observations have many advantages in reducing some of the uncertainties listed above. One of the key issues is the CDS ability to observe many emission lines from a large number of highly ionized ions of the most abundant elements. These cover a large range of temperatures and isoelectronic sequences. The CDS radiometric calibration was uncertain during the first period of the mission. This produced large uncertainties (factor of 2-3) in some earlier element abundance measurements (e.g., Mg/Ne, see [7]). However, the CDS instrument is now well calibrated [8] within 20-30%, which is of the same order as the accuracy of the atomic data.

CP598, *Solar and Galactic Composition*, edited by R. F. Wimmer-Schweingruber

THE DIAGNOSTIC METHODS

The intensity $I(\lambda_{ij})$, of a spectral line emitted by an optically thin plasma can be written as in [1]:

$$I(\lambda_{ij}) = A_b(X) \int C(T,\lambda_{ij},N_e)\, N_e\, N_H\, dh \qquad (1)$$

by asuming that the elemental abundance $Ab(X)$ is constant over the line of sight h. $C(T,\lambda_{ij},N_e)$, usually called the *contribution function*, is mostly a function of the electron temperature, and contains all the atomic parameters, and in particular the ionization fraction. N_H, N_e are the hydrogen and electron number densities. If we define the differential emission measure $DEM(T) = N_e N_H (dh/dT)$ we have:

$$I(\lambda_{ij}) = A_b(X) \int C(T,\lambda_{ij},N_e)\, DEM(T)\, dT \qquad (2)$$

from which in principle the relative element abundance $A_b(X_1)/A_b(X_2)$ of two elements X_1 and X_2 can be deduced from the observed intensity ratio I_1/I_2.
The best available atomic data, stored in the CHIANTI database (v.2, [9]) have been used here.

Most authors use various approximations to express the above equation in an even simpler way. These approximations were introduced about 30 years ago, when the uncertanties in the instrument's calibrations and in the atomic data were much larger than they are now. These approximations had the advantage of being computationally simple. In what follows, we briefly review these approximations, and outline the more correct approach that has been adopted here.

Following [10], many authors have approximated the expression for the line intensity by removing an averaged value of C(T) from the integral:

$$I = A_b(X)\ < C(T) > \int N_e N_H dh \qquad (3)$$

An emission measure $EM_L = I_{\rm ob}/(A_b(X) < C(T) >)$ can therefore be immediatly calculated for each observed line of intensity $I_{\rm ob}$. We define here the EM_L the line emission measure. The values $A_b(X)\ EM_L$ are plotted at the temperature $T_{\rm max}$ (defined as the temperature where $C(T)$ has a maximum), and the relative abundances derived in order to have all the points of the various ions lie along a common smooth curve. The differences between the various methods are related to the way the average $< C(T) >$ is calculated. Most authors follow the approximation given in [11].

A different approximation was proposed by [12]. The idea is to define for each observed line of intensity $I_{\rm ob}$ a single DEM value, that for clarity we define here as the *line* Differential Emission Measure DEM_L:

$$DEM_L \equiv \left\langle N_e N_H \frac{dh}{dT} \right\rangle \equiv \frac{I_{\rm ob}}{A_b(X) \int C(T) dT} \qquad (4)$$

A plot of the $A_b(X)\ DEM_L = I_{\rm ob}/\int C(T)dT$ values displayed at the temperatures $T_{\rm max}$ is used to derive the relative element abundances, by adjusting them in order to have a continuous sequence of values.

The more accurate approach adopted here is to use the largest possible number of lines, calculate the contribution functions at the measured densities, and then determine the relative element abundances and the DEM at the same time. In this way, most of the uncertainties will be reduced, and systematic effects highlighted.

Since the lines observed by CDS are produced by many ions covering a large temperature range, it is possible to deduce a DEM curve for each element, and to determine relative element abundances, by normalizing the DEM curves of the different elements. It should be mentioned that the determination of the DEM distribution is an ill-conditioned problem [see, e.g., 13, and references therein] where solutions are not unique, thus producing some added uncertainty. However, the conditioning can be improved with an appropiate line selection. All the low temperature lines (T $\leq 10^{5.5}$ K) observed by CDS/NIS are from high-FIP elements (N, O, Ne) while all the high temperature lines (T $\geq 10^{5.9}$ K) are from low-FIP elements (Mg, Ca, Fe, Si, Al). It is therefore possible to deduce the relative abundances within these two groups of lines. The scaling between the high and low-FIP elements can be done (see [14] for applications to quiet Sun and coronal hole observations) using lines that overlap in temperature. CDS/GIS observations are important, because GIS observes many other lines and ions that extend the overlapping region between the high- and low-FIP ions. This is particularly important, since the use of different ions from different isoelectronic sequences can help to indicate where the atomic physics is in disagreement with the observations (see, e.g., [4]).

An example

Here, the Skylab coronal hole cell-centre data of [15] are used as an example. The photospheric [16] abundances were used, together with the [17] ionization balance. The DEM distribution was derived using an inversion technique. The DEM method indicated the need to modify the O, Na, and Ca abundances by -0.2, +0.4, +0.2 (dex), respectively. Figure 1 shows the DEM_L values (left), and the DEM distribution (right). The DEM_L values are plotted at the temperature $T_{\rm max}$, while the points in the DEM curve are plotted at the effective temperature log $T_{eff} = \int C(T)\ DEM(T)$ log $T\ dT/(\int C(T)\ DEM(T)\ dT)$, that is an average temperature more indicative of where the bulk of the emission is. In both cases the points have been calculated with photospheric abundances [16], to show the sensitivity of these

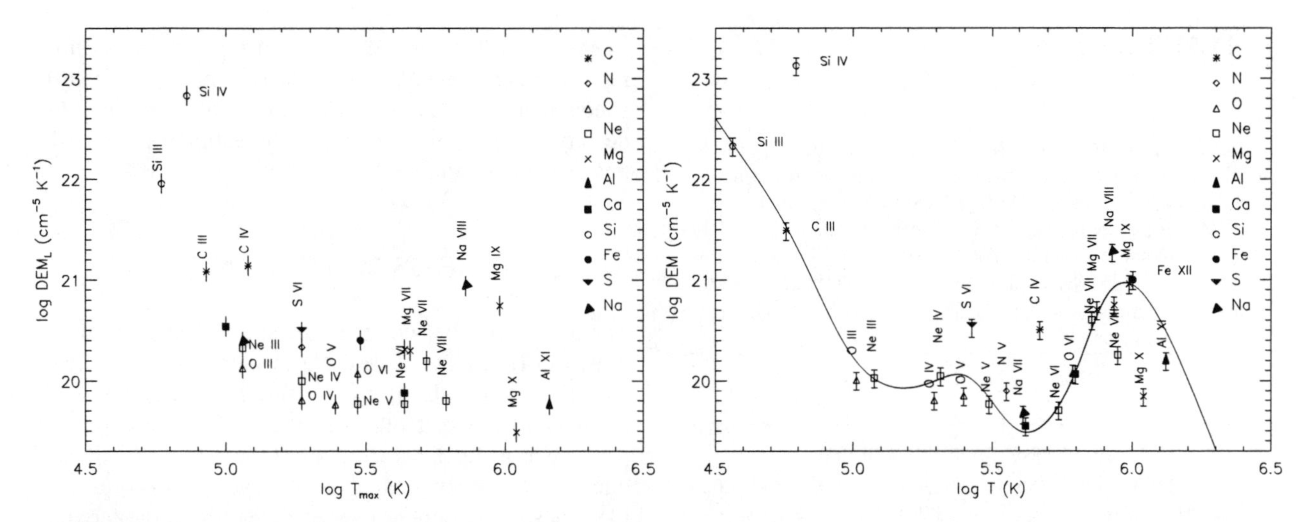

FIGURE 1. Left: line differential emission measures DEM_L of the selected lines plotted at the temperature $T_{\max}$. Right: the DEM distribution, with each experimental data point plotted at the effective temperature T_{eff} and at a value equal to the product $DEM(T_{\text{eff}})\times(I_{ob}/I_{th})$. The error bars represent an indicative 20% error on the observed intensities [15].

methods in measuring relative abundances.

For some lines, the DEM_L and DEM methods are in agreement. For example, they both clearly indicate the need to decrease the adopted O/Ne abundance (see e.g., the O IV and Ne IV points), and the fact that the Ne VIII and Mg X points are in total disagreement with the others. It is interesting to note that the photospheric abundance of O cited by [16] is 8.93 (log value), and if we assume fixed the Ne abundance, both methods indicate an oxygen abundance of 8.73, exactly the same value that has only recently been revised by [18].

However, in other cases the two methods produce very different results. For example, the DEM_L method does not indicate any need to modify the Na abundance, since the Na VIII point lies along a common smooth curve (neglecting Ne VIII, see below). On the other hand, the intensity of the Na VIII line, calculated with the DEM, is lower than the observed ones, by a factor of more than 2. How can the DEM and DEM_L methods differ by factors of more than 2 ? Only when the two lines have similar $C(T)$ and are emitted over a similar range of temperatures, can one assume the DEM to be constant and write:

$$\frac{\int C_1(T,N_e)\,DEM(T)\,dT}{\int C_2(T,N_e)\,DEM(T)\,dT} = \frac{\int C_1(T,N_e)\,dT}{\int C_2(T,N_e)\,dT} \qquad (5)$$

If the above equality holds, then it is possible to deduce the relative abundances directly from the observed intensities and the contribution functions, because:

$$\frac{A_b(X_1)}{A_b(X_2)} = \frac{I_1 \cdot \int C_2(T,N_e)\,dT}{I_2 \cdot \int C_2(T,N_e)\,dT} = \frac{DEM_L(X_2)}{DEM_L(X_1)} \qquad (6)$$

i.e., the DEM and DEM_L methods are equivalent. In any other case (defined here as the DEM effect), the two methods obviously produce different results. The DEM effect is particularly important when the emission $C(T)\ DEM(T)$ of a line peaks at temperatures where there is a non-negligible DEM gradient and Equation 5 does not hold. In the example produced here, the DEM distribution is such that the Na abundance estimate is mostly affected. However, other solar region observed have very different DEM distributions, and is impossible to know *a priori* if the approximation proposed by [12] is valid or not. These authors used their method to derive the Mg/Ne abundance of an erupting prominence, using Mg VI and Ne VI lines. A DEM analysis of this observation, performed by [5], has shown that the DEM peaks at $T = 10^{5.7}$ K, i.e. at the same temperature where the $C(T)$ of the Mg VI and Ne VI lines peaks. There is no DEM effect in this case, and the approximation used by [12] is therefore perfectly valid.

Another difference in the DEM and DEM_L methods is in the use of a different temperature at which the points are plotted. Note that there are substantial differences between $T_{\max}$ and T_{eff} for some lines. This can occur for example when the bulk of the emission comes from plasma at temperatures far from $T_{\max}$ (e.g. when there is a strong DEM gradient) or when the observed lines are blends of spectral lines that have $C(T)$ that peak at different temperatures.

TEMPERATURE (*DEM*) AND DENSITY EFFECTS

TABLE 1. Table of two Mg/Ne theoretical intensity ratios, calculated assuming A_b (Ne/Mg) = 0.5, for two densities N_e and: a) with a constant *DEM*; b) with a coronal hole plume *DEM* [19]; c) with a quiet Sun (network) *DEM* [14]. The values calculated in [20] (W F) for $N_e = 10^{10}$ and those presented by [21] (S) are also displayed for comparison in the last two columns. Note that the Mg VI 403.3 Å line is blended with a Ne VI line.

	$N_e = 10^8$	$N_e = 10^{10}$	$N_e = 10^{10}$ WF	S
Mg VI 400.666 Å /Ne VI 401.926 Å				
a) No *DEM*	1.50	0.90	0.97	1.00
b) Plume *DEM*	3.03	1.76	-	-
c) QS *DEM*	2.28	1.34	-	-
Ne VI 401.926 Å /Mg VI (+ Ne VI) 403.3 Å				
a) No *DEM*	0.40	0.65	-	0.61
b) Plume *DEM*	0.21	0.36	-	-
c) QS *DEM*	0.28	0.46	-	-

It is well known [3] that the Mg VI contribution functions are slightly skewed towards higher temperatures, when compared to the Ne VI ones. It is interesting to see the importance of the *DEM* effect here. Table 1 present two Mg VI / Ne VI theoretical intensity ratios, calculated assuming a constant *DEM* and using two *DEM* distributions, of a coronal hole plume and a quiet Sun. The Mg VI and Ne VI lines in Table 1 have been widely used by many authors [see, e.g., 21], because they are close in wavelength and because they have similar $C(T)$. The values in Table 1 show that if the *DEM* effect is neglected, the Ne/Mg relative abundance can be substantially underestimated, thus overestimating the FIP effect up to a factor of 3 (in the case of the plume). The *DEM* effect is much more pronounced when other line ratios such as Ca IX / Ne VII and Mg VII / Ne VII are considered, because their $C(T)$ peak at temperatures ($\log T = 5.9$) where the *DEM* gradient is usually large. The small differences in their $C(T)$ are amplified when forming the integrals. Most of previous works on element abundances have neglected the shape of the *DEM* distribution when calculating the relative abundances, and it is therefore possible that some previous estimates were wrong by factors of 3 or more.

The Mg VI lines considered here are slightly density-dependent. Density variations can therefore change the observed Mg VI / Ne VI intensity ratios aswell. Table 1 also shows the effect that different densities have on the Mg/Ne intensity ratios. Transition region densities are difficult to measure, and usually different line ratios produce different values [see, e.g., 20]. The densities adopted for the calculation should be considered as extreme values, in the sense that measured transition region densities are $N_e = 1 \pm 0.5\text{x}10^9$ cm^{-3} [5]. Table 1 shows that the *DEM* effect is more important than the density effect. However, inaccurate estimates of densities can lead to non-negligible effects, up to 50%.

PROBLEMS WITH MANY IONS

The anomalous behaviour of the spectral lines of many ions, mostly of the Li and Na isoelectronic sequences was discussed in detail by [5] using Skylab data as well as SOHO/CDS and other data. If a *DEM* analysis is performed using lines from any other isoelectronic sequences, the theoretical intensities of the Li- and Na-like lines are under- or over-estimated by large factors, ranging from 2 up to 10. These discrepancies cannot be ascribed to element abundance anomalies, and actually give a strong warning against the use of these lines for *DEM* or element abundance analyses.

Anomalous behaviour of the Li-like and Na-like ions, was first reported by [22], using OSO-IV quiet Sun spectra. Such problems were not reported by [23] who used Skylab data. This can be explained by the fact that [23] mainly used Li-like lines (O VI, Ne VIII, Mg X, Si XII, S XIV) to constrain the *DEM* at high temperatures.

In the example presented here, the lines of the Li-like N V and C IV are underestimated by factors of 3 and 10, while those of Ne VIII and Mg X are overestimated by factors of 5 and 10, respectively. The S VI 933.3 Å (Na-like) is also underestimated by a factor of 3.

A *DEM* analysis of a rocket solar spectrum was presented by [24]. They found ‘very significant and systematic differences’ between the line intensities (by factors of 2 to 5) of the Li and Na isoelectronic sequences. A possible cause for this effect is a departure from ionization equilibrium, which can be explained with the long timescales of the dielectronic recombination from the He-like ions. Another possible cause could be due to inaccurate ionization equilibrium calculations.

A comparison between different ionization equilibrium calculations was reported by [5] for CDS observations of a simple quasi-isothermal region. Large differences were found, showing that the ionization balance plays a major role in the derivation of any element abundances, confirming the suggestions by [4]. In particular, [5] showed that if the more recent calculations of [25] are used instead of [17], the theoretical intensities of Ca IX and Ca X lines increase by factors of more than 3. If the ionization balance of [25] is used for the example shown here, significant differences for some of the ions are also found.

EXAMPLE OF CORONAL HOLE ABUNDANCES

Figure 2 shows ratios of selected lines of a CDS/GIS E-W scan across the Elephant's Trunk [14] coronal hole, when it was near disc centre, on 1996 August 27. The Mg VI / Ne VI and Ca IX / Ne VII ratios present variations that follow the cell-centre network pattern. The Mg VI/Ne VI values indicate, if no density and/or *DEM* effect are accounted for, an almost photospheric Ne/Mg abundance in the network regions (at Solar X=35, 70, 110 arcsec, where the ratio $\simeq$ 0.5), with smaller values in the cell-centre regions (where the ratio $\simeq$ 0.9).

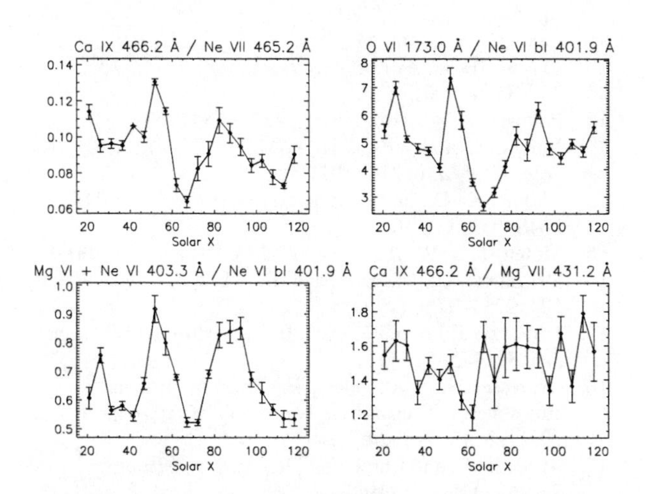

FIGURE 2. Intensity ratios (energy units) of selected lines of a CDS/GIS E-W scan across a coronal hole. The higher Mg VI / Ne VI values are located at the cell centres, at Solar X=50,90 arcsec

Can the higher Mg VI/Ne VI intensities in the cell-centres be explained instead by a density effect? Not really, since O IV measurements [5, 14] have indicated that the cell-centres have a higher electron density by about a factor of two. If this is true also for the heights where Mg VI is formed, then the Ne/Mg abundance would be slightly lower (and the FIP effect larger), since the Mg VI emissivity of the 403.3 Å line decreases with density. On the other hand, the higher Mg VI / Ne VI intensities can partly be explained by a temperature effect. Indeed, as shown in [14], the *DEM* distributions of the network and cell-centre regions are different, with the cell-centres having a steeper increase towards coronal temperatures.

However, an inspection of other combinations (Ca IX / Mg VII, O VI / Ne VI in Figure 2) suggests that most of these variations are probably due to a decreased Ne abundance in the cell centres which appears to occur relative to both low-FIP elements (Mg, Fe, Ca) and high-FIP ones (O). Variations of the Ne abundance, also relative to other high-FIP elements (such as O) have already been reported in a number of cases [19, 26].

CORONAL HOLE PLUMES

The most striking example in terms of a large Mg VI/Ne VI intensity ratio is given by coronal hole plumes. A large FIP bias (factor of 10) was derived by [27] from a Skylab off-limb EUV observation of a bright plume, using the DEM_L method and has long been thought that plumes have a large FIP effect. A *DEM* analysis was performed on the data tabulated in [27]. It showed that the plume had an isothermal distribution, similar to that one derived by [19] for an equatorial plume, and used as example in Table 1. The peak of the *DEM* was at $\log T = 5.9$, with a strong gradient where the $C(T)$ of the Ne VI and Mg VI lines differ most. The *DEM* effect here is so large that a photospheric Ne/Mg abundance can explain the Ne VI and Mg VI lines.

The fact that the large Mg VI/Ne VI intensity ratios observed in plumes are not indicative of a large FIP effect was also shown by [19], using SOHO/CDS observations. Here, we present further on-disc SOHO/CDS observations of a coronal hole plume to confirm this result. This plume was observed by CDS during the second week of October 1997 in the north polar hole. More details can be found in [5].

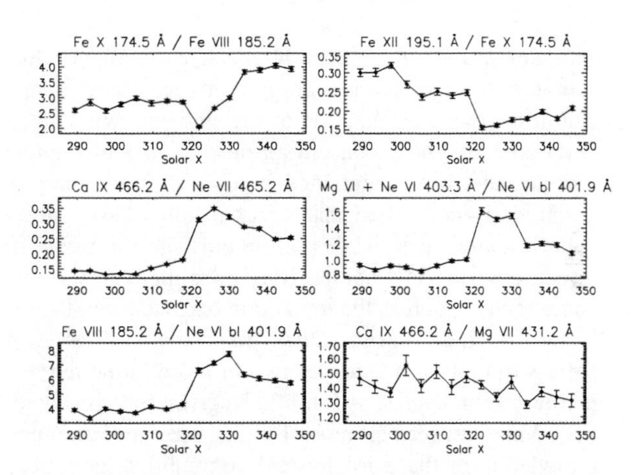

FIGURE 3. Intensity ratios (energy units) of few GIS lines during an E-W scan across a coronal hole plume (Solar X=320-330 arcsec).

A GIS scan was performed across the plume. Figure 3 shows how the intensity ratios of few GIS lines varies across the plume. The upper transition region lines have an increased intensity by a factor of about 4 in the plume area, while the high-temperature lines show a decreased intensity, indicating lower emission measures. There are indications of a density increase inside the plume area, as well as a decreased temperature. The Mg VI / Ne VI and Ca IX / Ne VII ratios show undoubtedly a large increase in the plume, and therefore a possibly large FIP effect. Nevertheless, an inspection of Fe VIII/Ne VI and

Ca IX / Mg VII ratios indicates that most of the observed variations are to be attributed to abundance variations of Ne only, as was observed for the equatorial plume [19]. Other ratios examined (e.g., Mg/Si) indicate that the relative abundances between the low-FIP elements remain almost unchanged.

If no density and *DEM* effects are considered, then the Mg/Ne abundance can be derived directly from the value of $\simeq$ 1.6 of the Mg VI 403.3 Å / Ne VI 401.9 Å ratio (Figure 3). From Table 1 one derives (the inverse Ne VI/Mg VI value being 0.62) a Ne/Mg abundance of 0.5, and a large FIP effect of 6.8. The transition region density of that plume (as derived from Mg VII) was not much higher than the adjacent coronal hole network region, and therefore a density effect (that would increase the FIP effect) can be excluded.

A *DEM* analysis was performed on the plume area (Figure 3, $SolarX = 326''$), in order to determine its elemental abundance. The *DEM* peaks at $T = 7\text{x}10^5$ K with a quasi-isothermal distribution at these heights. The resulting FIP effect is less than 2, similar to the values found by [28, 29].

CONCLUSIONS

The derivation of element abundances from spectroscopic measurements is a complex issue. Many factors can affect the determination of the element abundances. Only some of those concerning observations of the low corona have been mentioned here, with few examples given. It is confirmed that coronal holes have photospheric abundances, while plumes only show a small FIP effect, contrary to what has long been thought. Clearly, some factors such as the ionization balance used, the selection of lines, and the *DEM* effect can each account for a variation of a factor of two or more in the derived element abundances, and should be given full consideration. Many problems, some of which are not of common knowledge in the astrophysical community, have been highlighted. If the problem with the Li-like ions is related to departures from ionization equilibrium, then it is likely that a large amount of work based on these ions in solar and stellar coronal physics will have to be revisited.

ACKNOWLEDGMENTS

Financial support from PPARC is acknowledged. We thank the CDS team for their support in the instrument operations. SOHO is a project of international cooperation between the European Space Agency and NASA.

REFERENCES

1. Raymond, J. C., et al., *this issue* (2001).
2. Mason, H. E., "Abundance determination in the quiet corona", in *Proceedings of the First SOHO Workshop*, 1992, pp. 297–304.
3. Phillips, K. J. H., *Advances in Space Research*, **20**, 79 (1997).
4. Young, P. R., and Mason, H. E., *Space Science Reviews*, **85**, 315 (1998).
5. Del Zanna, G., Ph.D. thesis, Univ. of Central Lancashire, UK (1999).
6. Harrison, R. A. et al., *Sol. Phys.*, **162**, 233 (1995).
7. Young, P. R., and Mason, H. E., *Sol. Phys.*, **175**, 523–539 (1997).
8. Del Zanna, G., Bromage, B. J. I., Landi, E., and Landini, M., *A&A*, submitted (2001).
9. Landi, E., Landini, M., Dere, K. P., Young, P. R., and Mason, H. E., *Astron. Astrophys. Suppl. Ser.*, **135**, 339–346 (1999).
10. Pottasch, S. R., *Astrophys. J.*, **137**, 945 (1963).
11. Jordan, C., and Wilson, R., *ASSL Vol. 27: Physics of the Solar Corona*, p. 219 (1971).
12. Widing, K. G., and Feldman, U., *Astrophys. J.*, **344**, 1046–1050 (1989).
13. McIntosh, S. W., *Astrophys. J.*, **533**, 1043–1052 (2000).
14. Del Zanna, G., and Bromage, B. J. I., *J. Geophys. Res.*, **104**, 9753–9766 (1999).
15. Vernazza, J. E., and Reeves, E. M., *Astrophys. J. Suppl. Ser.*, **37**, 485–513 (1978).
16. Grevesse, N., and Anders, E., Solar interior and atmosphere. Tucson, AZ, University of Arizona Press, 1991, pp. 1227–1234.
17. Arnaud, M., and Rothenflug, R., *Astron. Astrophys. Suppl. Ser.*, **60**, 425–457 (1985).
18. Grevesse, N., *Adv. Space Res.*, in press (2001).
19. Del Zanna, G., and Bromage, B. J. I., *Space Science Reviews*, **87**, 169–172 (1999).
20. Widing, K. G., and Feldman, U., *Astrophys. J.*, **416**, 392 (1993).
21. Sheeley, N. R., *Astrophys. J.*, **469**, 423 (1996).
22. Dupree, A. K., *Astrophys. J.*, **178**, 527–542 (1972).
23. Raymond, J. C., and Doyle, J. G., *Astrophys. J.*, **247**, 686–691 (1981).
24. Judge, P. G., Woods, T. N., Brekke, P., and Rottman, G. J., *Astrophys. J. Letters*, **455**, L85 (1995).
25. Mazzotta, P., Mazzitelli, G., Colafrancesco, S., and Vittorio, N., *Astron. Astrophys. Suppl. Ser.*, **133**, 403–409 (1998).
26. Schmelz, J. T., Saba, J. L. R., Ghosh, D., and Strong, K. T., *Astrophys. J.*, **473**, 519 (1996).
27. Widing, K. G., and Feldman, U., *Astrophys. J.*, **392**, 715–721 (1992).
28. Young, P. R., Klimchuk, J. A., and Mason, H. E., *Astron. Astrophys.*, **350**, 286–301 (1999).
29. Wilhelm, K., and Bodmer, R., *Space Science Reviews*, **85**, 371–378 (1998).

Oxygen Abundance in Polar Coronal Holes

L. Teriaca*, G. Poletto*, A. Falchi* and J. G. Doyle†

*Osservatorio Astrofisico di Arcetri, Largo E. Fermi 5, 50125 Firenze, Italy
†Armagh Observatory, College Hill, BT61 9DG Armagh, N. Ireland

Abstract.
Fast solar wind is known to emanate from polar coronal holes. However, only recently attention has been given to the problem of where, *within* coronal holes, fast wind originates. Information on whether the fast solar wind originates from plumes or interplume regions may be obtained by comparing the elemental abundances in these regions with those characterizing the fast wind. Here we present a first attempt to determine the oxygen abundance in the interplume regions by using spectra taken at times of minimum in the solar cycle (when it is easier to identify these structures) by the SUMER spectrograph aboard SoHO. To this end, we analyze spectra taken in 1996 in polar regions, at altitudes ranging between 1.05 and 1.3 $R_{\odot}$, finding a value ≥ 8.5 for the oxygen abundance in the interplume regions. From the analysis of the O VI 1032 to 1037 line intensity ratio we also find no evidence of outflow velocities below 1.2 solar radii in interplume regions, while there are indications that outflow motions start to be significant above 1.5 solar radii. The method used and the assumptions made are discussed in light of the derived values. Our values are compared with previous determinations in the corona and solar wind.

INTRODUCTION

Polar coronal holes have long been recognized to be the sources of the high speed solar wind, but only recently it has been identified where, within coronal holes, the solar wind originates. UVCS and SUMER measurements of line widths and of the ratio of the O VI doublet lines at 1032 and 1037 Å in the low corona (at heliocentric distances lower than 2.5 $R_{\odot}$) seem to indicate that the low density background plasma, rather than the high density plume plasma, is the site where high speed wind originates. Recent analyses of SUMER and UVCS data [see e. g. 1, 2, 3] have shown that the width of UV lines is larger in interplume than in plume regions, hinting to interplumes as the site where energy is preferentially deposited and, possibly, fast wind emanates. Moreover, the analysis of fast polar wind performed by [4] seems to favor the low density regions as sources of the fast wind streams, while a work by [5] shows that there is no evidence of outflow motions in bright points and plumes observed within a coronal hole in the Ne VIII 770 Å line.

A further factor that may differentiate plume and interplume regions is their elemental abundance: a FIP bias (overabundance of particles with low First Ionization Potential over photospheric abundances) in plumes of ~ 15 has been found by [6]. Because only small FIP bias of up to 2 is found in the fast wind [7], we may expect that interplumes would not show large FIP bias either.

Ulysses observations have identified a difference in the elemental abundances of the fast and slow wind: low FIP elements being more enriched, with respect to their photospheric abundance, in the slow speed wind than in the high speed wind. Hence, if interplume regions are really the sources of the fast wind, we may possibly find a difference in the elemental composition of plume and interplume plasma at coronal levels.

The purpose of the present work is to shed some light on the interplume composition by determining which oxygen abundance is consistent with the O VI 1032, 1037 Å line intensities measured in the range between 1.05 and 1.3 $R_{\odot}$. It is also necessary to point out that the FIP bias usually refers to the comparison between high FIP and low FIP elements. This paper deals with absolute abundance instead, which actually is a more appropriate parameter in distinguishing the coronal origin of the solar wind.

OBSERVATIONS

The observations discussed here were acquired by SUMER on 3 June 1996 and comprise two rasters of the North Polar Coronal Hole (NPCH). The first part is a raster scan consisting of 75 spectra obtained using the $4'' \times 300''$ slit with an exposure time of 60 s and covering an area of $\sim 278'' \times 300''$ centered at helio-

CP598, *Solar and Galactic Composition*, edited by R. F. Wimmer-Schweingruber

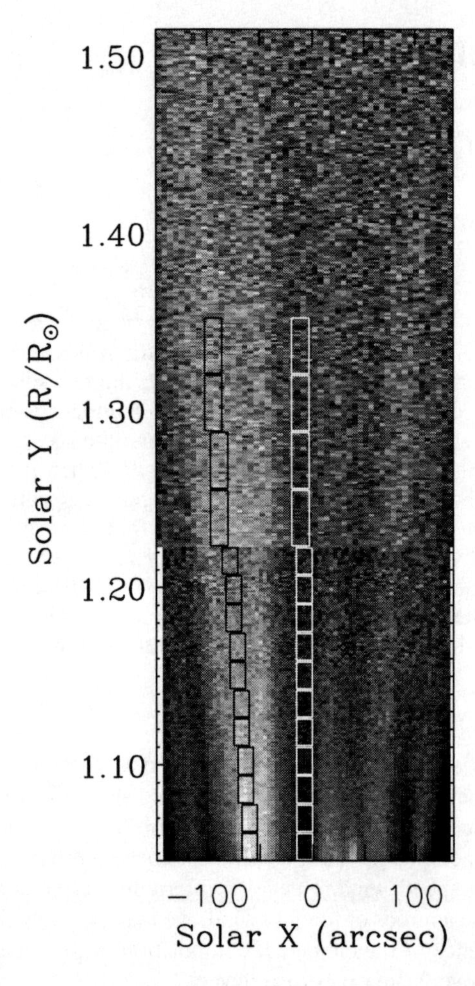

FIGURE 1. Intensity map of the north polar coronal hole as seen by SUMER in O VI on June 3rd 1996. Strong plumes and inter plume lanes can be identified in this diagram out to 1.5 $R_\odot$ above the limb. The contrast was enhanced by normalizing to the average intensity profile in the y-direction. Areas enclosed by white (black) solid lines are the ones over which an integration was performed in order to obtain the the averaged line intensities in plumes (interplumes) regions discussed in the text. The averaged areas measure $\sim 19'' \times 15''$ and $\sim 23'' \times 30''$ in the lower and upper dataset, respectively

centric coordinates $X = -4''$, $Y = 1150''$. An exposure time of 150 s and the $4'' \times 300''$ slit were used for the second part, consisting of 52 spectra covering a region of $\sim 287'' \times 300''$ centered at $X = -6''$, $Y = 1300''$ (see Fig. 1). These rasters were aimed to obtain high signal-to-noise line profiles in the plume and interplume regions above the limb out to 1.5 R$_\odot$. Data were reduced using various IDL routines from within the SUMER software tree.

Polar plumes are linear structures that are apparent over the solar poles in visible light, in extreme ultraviolet and in soft X-rays [see 8, and references therein]. Despite the fact that bright plumes are striking features within coronal holes that can extend to more than 30 solar radii from the Sun [8, 9], they only account for a tiny fraction of the solid angle subtended by the polar coronal holes. Plumes are well visible in O VI lines and our rasters show at least three of these structures. After selecting a plume and an interplume region, data within these regions were binned over 5 × 15 pixels (in the x and y direction, respectively) for the first (lower) scan, and over 4 × 30 pixels for the second scan. In such a way, line profiles were determined at 15 different locations above the solar limb for both plume and interplume (see Fig. 1). Further details on this dataset can be found in [2].

DATA ANALYSIS

Spectral profiles were carefully fitted in order to obtain C II 1037, O VI 1032 and O VI 1037 Å line intensities.

Despite the very high quality of the telescope mirror, the level of stray light is not negligible when observations of lines that are bright on disk are carried out above the limb [10, 11].

A chromospheric line, such as C II 1037, can be assumed not to have any coronal contribution and, hence, to be entirely due to stray light. The O VI 1032 stray component at a certain altitude above the limb will be, hence, given by the product of the C II 1037 intensity at that altitude times the ratio of the disk-averaged O VI 1032 and C II 1037 line intensities. The above ratio was evaluated using observations at 1.8 R$_\odot$, where the SUMER spectrum is entirely due to stray light [10, 12, 2].

Errors on line intensities were calculated through Poissonian statistics and, finally, their propagation in the O VI 1032/1037 line ratio was established.

RESULTS AND DISCUSSION

In the case of allowed transitions from the ground level of ions without metastable levels (such as Li-like ions), only the ground level g and the excited upper level u responsible for the line emission are important for calculating the line intensity and the transition is, hence, well described by a two level atomic model. This is the case of the O VI 1032 and 1037 Å lines. In the solar corona, these lines are formed by electron impact excitation (collisional component) and by resonant absorption of the O VI radiation from the transition region (radiative com-

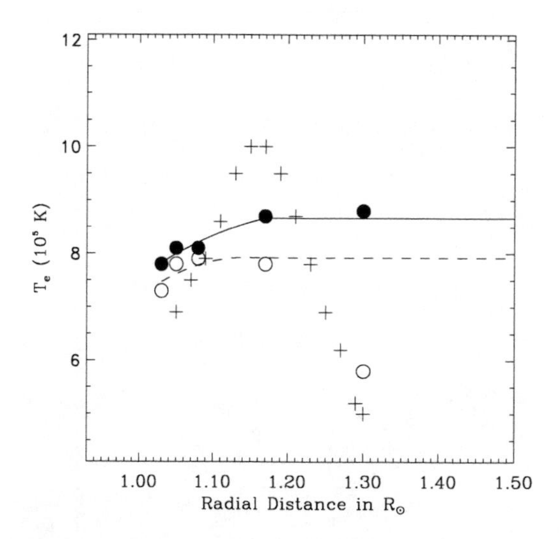

FIGURE 2. Electron temperature as a function of altitude in polar coronal holes. Measurements obtained in interplume (*filled circles*) and plume (*open circles*) regions, respectively [3]. Values obtained without distinguishing between plume and interplume: *crosses* [11]. Dashed and solid lines represent the adopted temperature profiles in plume and interplume, respectively.

ponent),

$$I_{obs} = I_{Coll.} + I_{Rad.} = 0.85 \frac{\Delta E_{gu}}{4\pi} A_{O/H} \times \times [\langle q(T_e)R(T_e)N_e^2\rangle + B_{gu} I_\odot \langle D(T_O,W)R(T_e)N_e\rangle] \quad (1)$$

where ΔE_{gu} is the energy of the transition from g to u, $A_{O/H}$ is the oxygen abundance relative to hydrogen, $q(T_e)$ the collisional excitation rate coefficient, B_{gu} is the Einstein absorption coefficient, $I_\odot$ the O VI disk-averaged line intensity, $D(T_O,W)$ accounts for Doppler dimming and geometrical dilution factors and is function of the oxygen temperature T_O (consisting of the components parallel $T_o^{\parallel}$ and perpendicular $T_o^{\perp}$ to the direction of the magnetic field lines) and the wind velocity W. N_e is the electron number density, $R(T_e)$ is the oxygen ionic fraction calculated in ionization equilibrium and 0.85 is the value of the hydrogen to electron number density ratio for a fully ionized plasma with composition given by [13]. The quantities in brackets $\langle ... \rangle$ are integrated along the line of sight. We are interested in obtaining information on the abundance through the comparison of observed O VI line intensities with values computed using Eq. 1. Both radiative and collisional components are strongly dependent on the electron density (N_e), electron temperature (T_e), and oxygen elemental abundance (A_{el}), while the radiative component depends also on the adopted values of $I_\odot$, $T_o^{\parallel}$, $T_o^{\perp}$ and W. We are also dealing with the assumption of ionization equilibrium (implied in the calculation of $R(T_e)$) and with the uncertainties in the knowledge of the atomic data (see Raymond *et al.*, present proceedings).

$I_\odot$ was evaluated assuming a $1/cos(\theta)$ center-to-limb line intensity variation ([19]) and adopting an intensity at disk centre of 280 mW m^{-2} St^{-1} for the O VI 1032 Å line. The above value was obtained from a disk centre spectrum obtained on June 4th 1996.

In the case of the observations here discussed, we can assume the magnetic field lines to be perpendicular to the line of sight. This allow us to identify $T_o^{\perp}$ with the effective temperature T_{eff} defined through the equation:

$$v_{1/e}^2 = \frac{2kT_{eff}}{m} \quad (2)$$

where m is the ion mass and $v_{1/e}$ is the Doppler width (in km s^{-1}) or most probable speed of the unresolved (thermal + non-thermal) motions distribution, given by

$$v_{1/e} = \frac{\Delta\lambda_D}{\lambda} c, \quad (3)$$

where λ is the wavelength, c the velocity of light and $\Delta\lambda_D$ the measured half width of the line at $1/e$ of the peak intensity. I_{Rad} is not a strong function of $T_o^{\perp}$ and, hence, a $v_{1/e}$ value of 65 km s^{-1} was assumed, for the present work, as representative of both plume and interplume. The above value was obtained averaging the results published by [2]. Furthermore, $T_o^{\parallel}$ has been assumed equal to T_e.

The determination of the electron temperature profile in the solar corona is particularly controversial. The discrepancy between the electron temperatures obtained in the inner corona through spectroscopic diagnostics [see e. g. 3] and the values deduced from ion charge composition measurements obtained *in-situ* in the high-speed solar wind by Ulysses [see e. g. 20], is well known. Spectroscopically derived T_e values never exceed 10^6 K in the range 1.05–1.3 $R_\odot$ (see Figure 2). Values by [3] are particularly accurate since they use the ratio of lines of the same ion (Mg IX 706/750), thus avoiding problems with elemental abundances, ionic fractions, and differential flow speeds between different ions [21]. The closeness in wavelength minimize also the errors arising from the instrument calibration. Moreover, these are the only measurement of T_e in plume and interplume. It has been shown by [21] that in the region 1.05–1.5 $R_\odot$ this ratio is scarcely affected by the presence of a small percentage of electrons following non-Maxwellian distribution functions. From the above discussion we adopted the T_e profiles shown in Figure 2.

The O VI 1032/1037 line ratio is insensitive to T_e [22] and does not depend on the elemental abundance. It is, however, strongly dependent on the electron density N_e

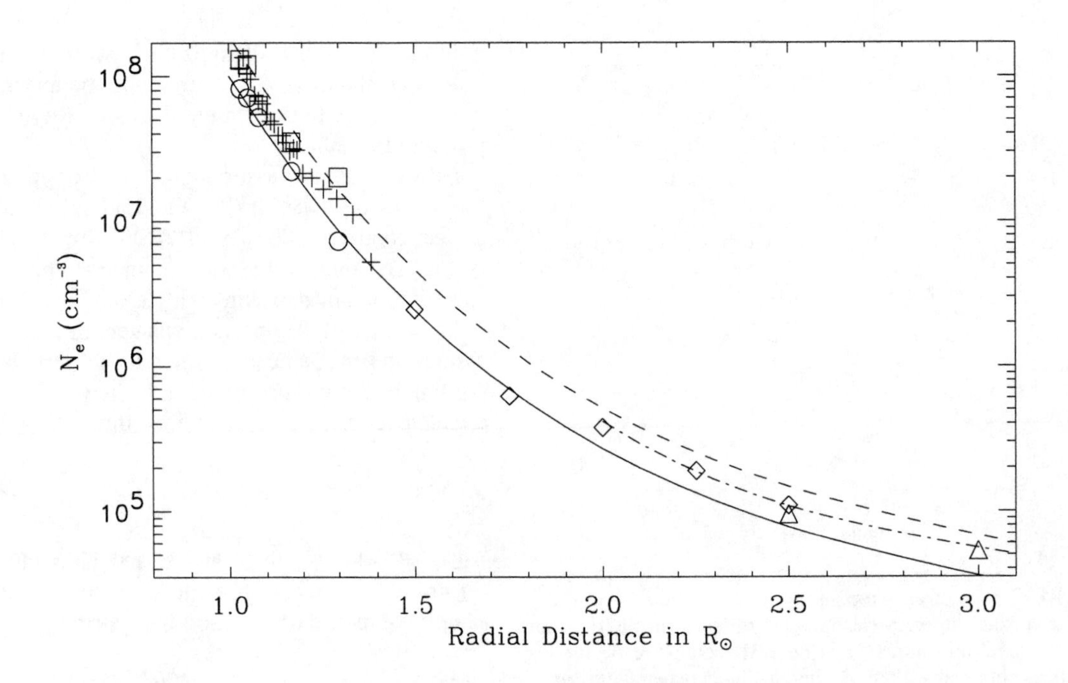

FIGURE 3. Electron density as a function of altitude in polar coronal holes. Measurements obtained in interplume: *crosses* [14], *open circles* [3]. Measurements obtained in plumes: *open squares* [3]. Measurements obtained without distinguishing between plume and interplume: *open diamonds* [15], *open triangles* [16]. Dashed-dotted line shows the expression, valid from 2 to 4 solar radii, given by [17]. Dashed and solid lines represent the adopted density profiles in plumes and interplumes, respectively.

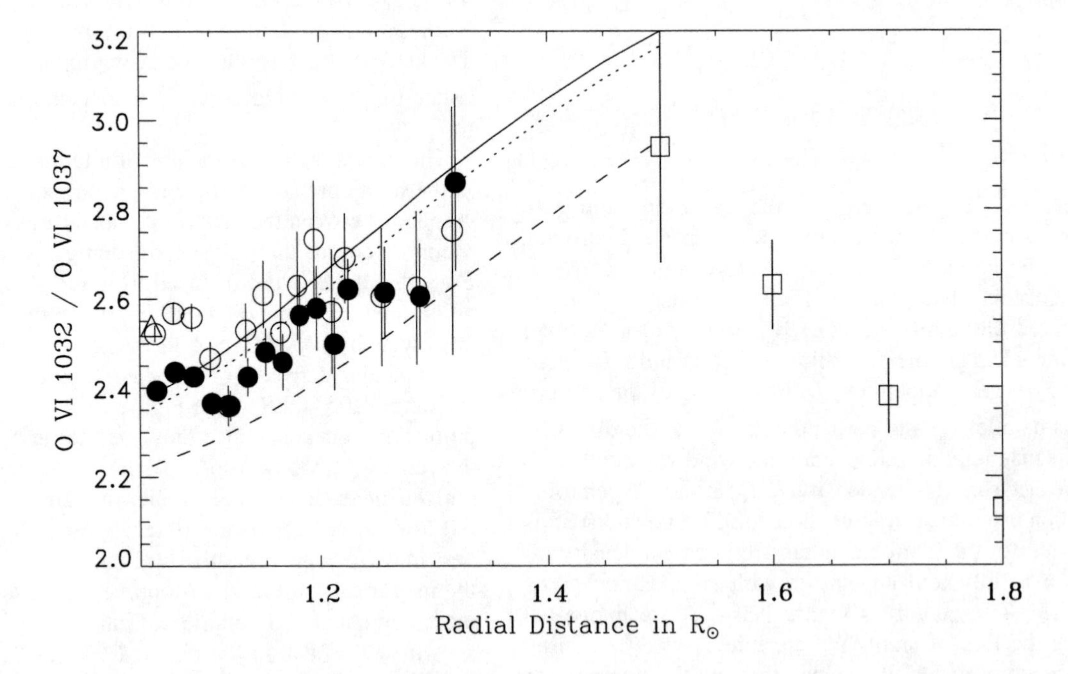

FIGURE 4. O VI 1032 to O VI 1037 intensity line ratios as a function of altitude in polar coronal holes. Values obtained in interplume regions: *open circles* [present paper], *open triangle* [18], *open squares* (UVCS, 21 May 1996) [4]. Values obtained in plumes: *filled circles* [present paper]. Solid and dashed lines represent models for interplume and plume conditions, respectively. Dotted line represents a mixed model comprising plume and interplume lanes.

as well as on the wind speed W. In order to reproduce the observed line ratio, the N_e profile as a function of height needs to be known, together with the exciting transition region radiation. Densities in coronal holes have been evaluated by several authors [e. g. 17, 23]. However, in the present analysis only data obtained in 1996 (*i. e.* closer in time to our data) were used.

Figure 3 shows N_e as a function of altitude in coronal holes from previous measurements, together with the adopted density profiles in plumes and interplumes. Using those electron density and temperature profiles the O VI 1032 and 1037 line intensities were computed using Eq. 1 and the O VI 1032 to 1037 line ratio was then evaluated. Figure 4 shows that our choice of N_e is able to reproduce the observed line ratios in interplumes with negligible outflow velocity up to 1.3 $R_\odot$. Above this altitude, the comparison with UVCS data obtained on 21 May 1996 in interplume lanes by [4] shows that the region where the fast solar wind is accelerated lies above 1.3 solar radii, as already suggested by [2]. An upper limit to the outflow speed, possibly present at 1.3 $R_\odot$, can be derived by computing the outflow speed which would bring the O VI 1032/1037 theoretical line intensity ratio down to the lowest value compatible with the error bar of the observed ratio. It turns out an outflow velocity of 75 km s^{-1}. We notice, however, that such an high outflow speed is not consistent with the O VI line ratio observed by UVCS at 1.5 $R_\odot$.

Note that $W \sim 0$, together with the small temperature gradient, implies the assumption of ionization equilibrium being valid at low altitude above the limb. We are, hence, confident in using Eq. 1 for calculating line intensities in the low corona.

Using the adopted interplume T_e and N_e profiles, we are able to reproduce the observed intensities in interplumes (Figure 5) with an oxygen abundance of 8.5. However, this value is lower than that measured in the fast solar wind, where an abundance of $\sim 8.65 \pm 0.18$ dex is observed in the fast wind around 700 km s^{-1} ([see 24]). It is interesting to note that the latter value agrees very well with the recently measured value of 8.69 ± 0.05 dex of the photospheric oxygen abundance ([25]). We point out that a higher value of T_e would translate in a higher value of abundance, thus allowing us to consider the value of 8.5 as a lower limit to the value of the oxygen abundance in the interplume lanes of this polar coronal hole. Similar considerations hold also in the case of the electron density. Higher values of N_e would result in a theoretical ratio *lower* than the observed one, while a lower N_e would result in a small amount of outflow velocity and a higher value of the oxygen abundance.

A realistic attempt to model our plumes observations requires a combination of plume and interplume emissivities along the line of sight, thus introducing additional free parameters such as the plume filling factor and oxygen abundance. A mixed plume-interplume model has been, hence, created assuming a single plume embedded in interplume plasma. The line of sight fraction occupied by the plume was estimated from Fig. 1 while the same oxygen abundance as in interplume was adopted. Despite the fact that we do not feel to attach a physical significance to adopted plume abundance, it is interesting to note that the introduction of the surrounding interplume

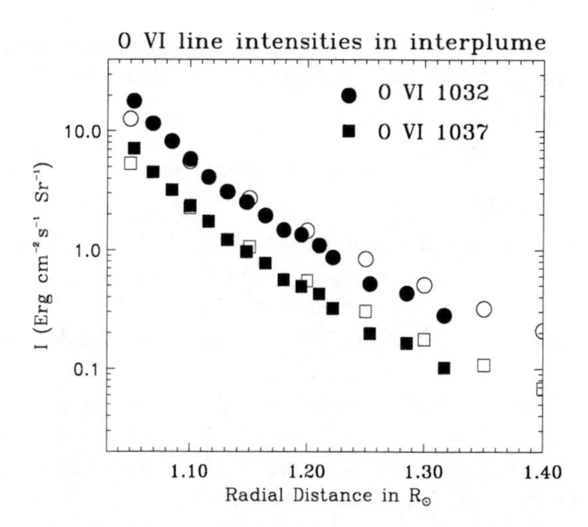

FIGURE 5. O VI line intensities as a function of altitude in an interplume region of the North polar coronal hole. Filled circles and squares represent the O VI 1032 and O VI 1037 observed line intensities. Open circles and squares represent the O VI 1032 and O VI 1037 modeled line intensities for the adopted interplume conditions.

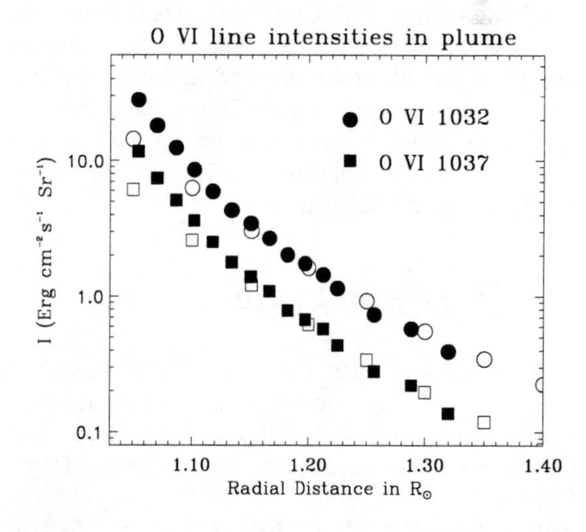

FIGURE 6. O VI line intensities as a function of altitude in a plume region of the North polar coronal hole. Filled circles and squares represent the O VI 1032 and O VI 1037 observed line intensities. Open circles and squares represent the O VI 1032 and O VI 1037 modeled line intensities obtained by combining plume and interplume models.

plasma is necessary for reproducing the observed line ratios and intensities (see Figures 4 and 6).

CONCLUSIONS

Our measurements of the O VI 1032/1037 line intensity ratio shows that negligible outflow velocities are consistent with the vaues measured below 1.3 $R_\odot$, while an oxygen abundance of 8.5 is required to reproduce the observed O VI line intensities in interplume. This value is lower than the oxygen abundance measured in the fast solar wind ($\sim 8.65 \pm 0.18$). In order to reproduce the abundances measured *in situ*, lower N_e values would be required. Simulations show that densities should be lower by more than a factor of two than those adopted. However, such lower N_e profile may imply the solar wind to be accelerated just above ~ 1.2 solar radii. An indication of this could be found in the apparent decrease of the observed O VI ratio in interplume above 1.2 $R_\odot$ but is apparently inconsistent with the UVCS observations, which, however, refer to structures observrd about 10 days before the SUMER measurements discussed here. Because densities play a fundamental role in determining where the solar wind starts being accelerated and in determining the correct value of the oxygen abundance, we like to point out that it is of the utmost importance to have *simultaneous* determinations of electron density, temperature and line intensity.

ACKNOWLEDGMENTS

L. T. and G. P. are partially supported by MURST. The authors would like to thank the anonymous referee for useful comments and suggestions. Research at Armagh Observatory is grant-aided by the N. Ireland Dept. of Culture, Arts and Leisure. The SUMER project is financially supported by DLR, CNES, NASA, and PRODEX. SUMER is part of SoHO, the Solar and Heliospheric Observatory of ESA and NASA.

REFERENCES

1. Giordano, S., Antonucci, E., Noci, G., and *others*, *Astrophysical Journal*, **531**, L79–L82 (2000).
2. Banerjee, D., Teriaca, L., Doyle, J. G., and Lemaire, P., *Solar Physics*, **194**, 43–58 (2000).
3. Wilhelm, K., Marsch, E., Dwivedi, B. N., and *others*, *Astrophysical Journal*, **500**, 1023–1038 (1998).
4. Antonucci, E., Dodero, M. A., and Giordano, S., *Solar Physics*, **197**, 115–134 (2000).
5. Wilhelm, K., Dammasch, I. E., Marsch, E., and Hassler, D. M., *Astronomy and Astrophysics*, **353**, 749–756 (2000).
6. Widing, K. G., and Feldman, U., *Astrophysical Journal*, **392**, 715–721 (1992).
7. Gloeckler, G., and Geiss, J., "The abundances of elements and isotopes in the solar wind", in *Proceedings of the AIP Conference: Cosmic abundances of matter*, AIP Conference Proceedings 183, American Institute of Physics, New York, 1989, pp. 49–71.
8. DeForest, C. E., Hoeksema, J. T., Gurman, J. B., and *others*, *Solar Physics*, **175**, 393–410 (1997).
9. DeForest, C. E., Plunkett, S. P., and Andrews, M. D., *Astrophysical Journal*, **546**, 569–575 (2001).
10. Lemaire, P., Wilhelm, K., Curdt, W., and *others*, *Solar Physics*, **170**, 105–122 (1997).
11. David, C., Gabriel, A. H., Bely-Dubau, F., and *others*, *Astronomy and Astrophysics*, **336**, L90–L94 (1998).
12. Hassler, D. M., Wilhelm, K., Lemaire, P., and Schüle, U., *Solar Physics*, **175**, 375–391 (1997).
13. Feldman, U., Mandelbaum, P., Seely, J. F., and *others*, *Astrophysical Journal Supplement*, **81**, 387–408 (1992).
14. Doyle, J. G., Teriaca, L., and Banerjee, D., *Astronomy and Astrophysics*, **349**, 956–960 (1999).
15. Zangrilli, L., Private communication (2001), unpublished.
16. Lamy, P., Quemerais, E., Liebaria, A., and *others*, "Electronic Densities in Coronal Holes from LASCO-C2 Images", in *Fifth SOHO Workshop: The Corona and Solar Wind Near Minimum Activity*, edited by A. Wilson, ESA SP 404, ESA Publications Division, ESTEC, Noordwijk, The Netherlands, 1997, pp. 491–494.
17. Kohl, J. L., Noci, G., Antonucci, E., and *others*, *Astrophysical Journal*, **501**, L127–L131 (1998).
18. Patsourakos, S., and Vial, J. C., *Astronomy and Astrophysics*, **359**, L1–L4 (2000).
19. Wilhelm, K., Lemaire, P., Dammasch, I. E., and *others*, *Astronomy and Astrophysics*, **334**, 685–702 (1998).
20. Ko, Y. K., Fisk, L. A., Geiss, J., and *others*, *Solar Physics*, **171**, 345–361 (1997).
21. Esser, R., and Edgar, R. J., *Astrophysical Journal*, **532**, L71–L74 (2000).
22. Li, X., Habbal, S. R., Kohl, J. L., and Noci, G., *Astrophysical Journal*, **501**, L133–L137 (1998).
23. Fisher, R., and Guhathakurta, M., *Astrophysical Journal*, **447**, L139–L142 (1995).
24. von Steiger, R., Fisk, L. A., Gloeckler, G., and *others*, "Composition Variations in Fast Solar Wind Streams", in *Solar Wind Nine, Proceedings of the Ninth International Solar Wind Conference*, edited by S. R. Habbal, R. Esser, J. V. Hollweg, and P. A. Isenberg, AIP Conference Proceedings 471, American Institute of Physics, New York, 1999, pp. 143–149.
25. Prieto, C. A., Lambert, D. L., and Asplund, M., *Astrophysical Journal Letters*, **556**, L63–L66 (2001).

Oxygen abundance in streamers above 2 solar radii

Zangrilli, L.*, Poletto, G.†, Biesecker, D.** and Raymond, J. C.‡

*Dipartimento di Astronomia e Scienza dello Spazio, Università di Firenze, Italy
†Osservatorio di Arcetri, Largo Fermi, 5, Firenze, Italy
**Emergent-IT, NASA/GSFC, Greenbelt, MD, USA
‡Harvard–Smithsonian Center for Astrophysics, Cambridge, MA 02138, USA

Abstract. The oxygen abundance in streamers has been evaluated by several authors [see *e.g.* 1, 2, 3] who found, in the core of streamers, an oxygen abundance lower by a factor 3-4 than in the lateral branches (legs). All estimates were made at heliocentric distances $h \leq 2.2$ $R_\odot$. In this paper we analyze UVCS observations of two streamers, observed during solar minimum at altitudes $h \geq 2.4$ $R_\odot$ to derive the oxygen abundance, relative to hydrogen, and its latitude dependence within streamers, in the range $2.4 \leq h \leq 4$ $R_\odot$. To this end, electron densities have been derived from LASCO data, taken at the time of the UVCS observations, and the radial temperature profile has been taken from literature. These parameters allow us, after the collisional contribution to the O VI 1032, 1037 Å line intensities has been identified, to determine the oxygen abundance that reproduces the observed collisional components. Our results are compared with previous abundance determinations and the relationship between coronal and *in situ* abundances is also discussed.

INTRODUCTION

Observations made with the SOHO/UVCS spectrometer have shown that the streamer morphology in minor ion emission lines can be markedly different from the streamer morphology in H Lyα. Whereas the brightness of the Lyα peaks in what would be the streamer core in the case of a global dipolar magnetic configuration, the minor ion emission peaks in lateral structures. These structures, in the dipolar configuration, represent the legs of the streamer, or, in a multipolar magnetic field not identifiable in Lyα radiation, correspond to lobes of the magnetic configuration. This difference is not always present: whether there are two distinct types of streamers, or whether projection effects sometimes mask the true streamer morphology, is still an open question.

The O VI depletion in the streamer core, and O VI relative enhancement in the streamer legs, is most easily interpreted in terms of a variation of the oxygen abundance across the streamer [see *e.g.* 4]. Raymond *et al.* [1, 5], from UVCS data analysis, derived an oxygen abundance higher in the legs, than in the core of streamers. Similar results have been found by [2] and [3]. All these studies are based on data taken at heliocentric altitudes ≤ 2.2 $R_\odot$.

The issue has a strong impact on the problem of the origin of the slow wind. Slow wind emerges from low latitude solar regions, where streamers are mostly rooted. A comparison of the elemental composition of the streamers with the elemental composition of the slow wind may help identifying the site where the slow wind originates.

In this contribution we extend the analysis of the O VI abundance in streamers to altitudes ($2.4 \leq h \leq 4$ $R_\odot$) higher than those addressed by previous works. This will allow us to see whether the oxygen abundance varies across the streamer at such high levels and whether the decrease of the oxygen abundance with height, found by other authors [see *e.g.* 2], is still detectable. We also considered two streamers – possibly pertaining to the two classes previously described – with the purpose of checking whether there is any difference in their composition.

DATA AND ANALYSIS TECHNIQUE

The observations

For the present study we selected two streamers observed by UVCS in 1996, on July 6 and 11, in the equatorial region along the East direction. These two streamers are representative of the two classes described above. However, we note that their morphology, as discussed in the following, leaves room for a different interpretation. UVCS data were acquired in the range 2 to 4.5 $R_\odot$ on July 6, 1996, and 1.6 to 4 $R_\odot$ on July 11, 1996. The slit width was 300 μm, a spatial binning of 2 pixels and a six

CP598, *Solar and Galactic Composition*, edited by R. F. Wimmer-Schweingruber

pixel spectral binning were adopted. Spectra have been taken at increasing heliocentric distances, every 0.25, or 0.5 $R_\odot$, with the slit 5° above the equator, on July 6, and normal to the axis of the streamer, on July 11.

Images of the two streamers in Ly_α and O VI have been built by integrating over the line width and interpolating between data taken at contiguous altitudes, and are shown in In Fig. 1. Spectra at each altitude were integrated over 14 spatial bins, to improve the count rate statistics. Data acquired by LASCO/C2 over the same streamers have also been used in this work. Observations of polarized brightness (pB) by C2 start at ≈ 2 $R_\odot$, hence, as we need to use data taken at the same position by UVCS and LASCO, we did not use UVCS observations taken at lower heights.

How to evaluate oxygen abundances

UV lines in the extended corona form by collisional excitation and resonant scattering of solar disc radiation. The ratio between the collisional components of the O VI doublet lines is 2, while provided the plasma where the lines originate has a low enough flow speed and kinetic temperature, the ratio between the radiative components is 4. Thus a simple relationship holds between the total intensities of the 1032 and 1037 lines and the collisional component of the 1037Å line [see *e. g.* 6]:

$$I_{c,1037} = 2I_{tot,1037} - \frac{1}{2}I_{tot,1032} \qquad (1)$$

Here $I_{tot,1032}$ ($I_{tot,1037}$) is the total intensity of the O VI 1032 Å (1037) line, made up of a collisional $I_{c,1032}$ ($I_{c,1037}$) and a radiative component $I_{r,1032}$ ($I_{r,1037}$). The collisional component of the O VI 1037 Å line can be derived from (1) and, as a consequence, also the components of the O VI 1032 Å line can be easily obtained.

The collisional component of the O VI lines is given by:

$$I_c = \frac{0.8}{4\pi} h\nu_{k1} A_{el} \int_{LOS} R_{ion}(T_e) n_e^2 C_{1k}(T_e) dx \qquad (2)$$

where A_{el} and $R_{ion}(T_e)$ are the oxygen and ion abundance; T_e, n_e are the electron temperature and density; $C_{1k}(T_e)$ is the collisional excitation coefficient, the indices 1 and k indicate the ground and the upper level of the transition and the integration is extended along the line of sight (LOS).

Relationship (2) allow us to derive the element abundance A_{el}, provided we know I_c, n_e, T_e and the geometry of the structures where the O VI emission originates. The following subsection illustrates the physical parameters we used in the abundance determination.

Physical parameters in streamers at $h \geq 2.4$ $R_\odot$

As we mentioned, in order to apply the method described in the previous subsection, we need to know the density and electron temperature throughout the region where we evaluate the oxygen abundance. Densities have been derived from LASCO/C2 pB data using the standard Van de Hulst inversion technique [7], which is based on the assumption of spherical symmetry. Hence we are modeling a streamer belt, not an isolated feature: this motivated our choice of features observed in 1996, at minimum activity of the solar cycle, when the spherical symmetry assumption is more likely to be tenable.

From LASCO data, densities along a grid of radial directions, 4 degrees apart, have been evaluated throughout the streamers. These are given in the bottom panel of Figs. 2 and 3. Densities at ≈ 2.4 $R_\odot$ seem to be practically constant, throughout the core of the streamer and to decrease towards the streamer legs. All densities decrease with height, with a steeper profile as one moves from the streamer core and enters in the adjacent coronal hole at progressively lower altitudes. Densities in the two streamers at 2.4 $R_\odot$ agree fairly well with those given by [8] for a streamer observed in August 1996 and, at ≈ 4 $R_\odot$, lie on the upper edge of the strip of values given by those authors at that height.

There are a few determinations of electron temperature T_e in streamers, at ≈ 1.5 $R_\odot$. In the altitude range we are dealing with, there is only the determination of Fineschi *et al.* [9] who give $T_e = 1.1 \times 10^6 \pm 0.25 \times 10^6 K$, at $r = 2.7$ $R_\odot$. This value is in agreement with the streamer scale height temperature derived by Gibson *et al.* [8] at that altitude. Hence, we assumed T_e to decrease with altitude with the profile given by Gibson *et al.*

The variation of the electron temperature across a streamer is essentially unknown. The results of Parenti *et al.* (these Proceedings) [10] seem to favor a decrease of the electron temperature towards the streamer legs. However, the Parenti *et al.* work, which refers to much lower altitudes (1.6 $R_\odot$), does not give a profile of temperature across a streamer: it only shows that the electron temperature, evaluated in different days at only one position, is slightly lower at positions which might correspond to the edge of streamers rather than at positions within the streamer core. Possibly, the results of Parenti *et al.*, with a very limited statistics, refer to streamers which have a different T_e, constant across the structure. Hence, we assumed T_e to be only a function of the radial distance: $T_e = T_e(r)$.

A further relevant parameter is the plasma flow speed. This parameter affects the identification of the collisional component of the O VI lines, and makes the determination of I_c from eq. (1) inaccurate.

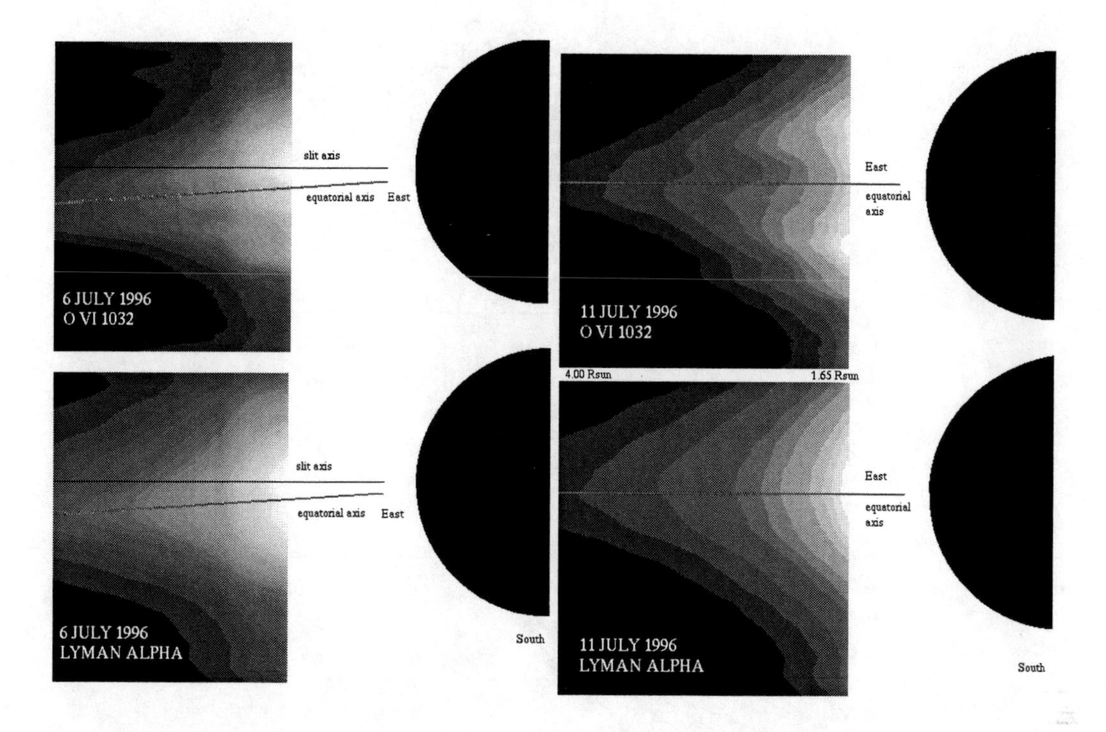

FIGURE 1. Images of the two streamers in O VI and Lyα for July 6 and 11, 1996. The spectral images for July 6 have been taken with the slit axis perpendicular to the radius at 5° above the equator. The dotted lines indicate the position of the equator.

Another parameter affecting the identification of I_c is the line width. This because the percentage contribution of the radiative component to the total line intensities (hence also the collisional contribution) depends also on the width of the absorption profile. The present observations have been acquired with too large a slit width, and too coarse a spectral resolution, to allow us to measure line widths from the data: hence in the following we took the O VI lines width from literature [11]. Then we calculated, from the above densities, temperature, line width, and the value of oxygen abundance derived from eqs. (1) and (2), the plasma speed which reproduces the observed individual O VI line intensities, taking into account the Doppler dimming effect. Once the plasma speed is known, we may also calculate the error made deriving I_c from eq. (1) and recalculate the oxygen abundance from the revised value of I_c. This procedure is iterated until a convergence is reached. The correction factor to the collisional component derived from eq. (1) is from about 0.7 to 1.0, according to the local kinetic temperature, and the derived outflow speed, which turns to be in the interval $50-100$ $km\ s^{-1}$, in agreement with the values given in the literature [11].

RESULTS

Oxygen abundances have been derived along four radial directions through the streamers bodies. In Figs. 2 and 3 we show the results of our analysis for, respectively, the July 6 and 11, 1996, streamers. In both figures the upper panel gives isophotes of the intensity of the O VI 1032 Å line, integrated over the line width, as a function of latitude and heliocentric distance. The morphology of the streamers, at least above ≈ 2 $R_\odot$, shows that the July 11 feature has a double peaked appearance which disappears at ≈ 3 $R_\odot$: above this altitude the O VI intensity peaks at the latitude where, at lower levels, the intensity has a local minimum in between the two peaks. The July 6 isophotes, on the contrary, do not provide any clear evidence of a double peaked structure, although we cannot discard the hypothesis that the small secondary intensity enhancement, which appears at southern latitudes, is the remnant of a peak which is partially masked by projection effects.

In the top panel, different symbols, drawn at different heliocentric distances along radial cuts through the northern latitudes $2, 6, 10°$ and the southern latitude 2°, indicate the positions where the oxygen abundance has been evaluated. In the July 11 streamer, the cut at 2° passes through the O VI intensity drop, while the cuts through $-2°$ and $6°$ pass through the O VI intensity peaks sideways of the intensity minimum. Values of the

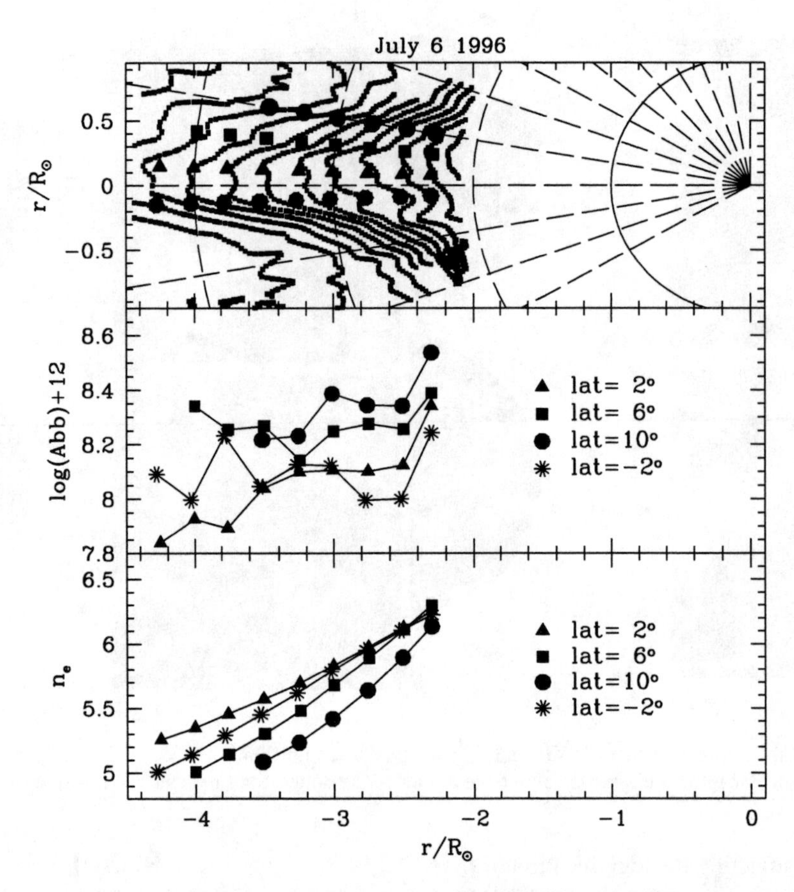

FIGURE 2. Results for the July 6, 1996 streamer. Top panel: isophotes of the O VI 1032 total line intensity *vs.* latitude and heliocentric distance. The lowest isophote corresponds to $logI_{O\ VI,1032} = 9.0$, the highest isophote corresponds to $logI_{O\ VI,1032} = 6.8$. Symbols drawn along different radial cuts indicate the positions where the oxygen abundance has been evaluated. Middle panel: oxygen abundance at the positions given in the top panel. Bottom panel: densities at the positions given in the top panel.

oxygen abundance at these positions are given in the middle panel of Figs. 2, 3 and the bottom panel shows the density values adopted in the oxygen abundance determination. Although, as we already mentioned, densities are about constant, within the streamer core (at the lowest heliocentric altitude), the bottom panel hints at a decrease of densities with latitude while the opposite behavior is shown by the oxygen abundance, which increases as one moves away from the streamer axis. The anticorrelation between densities and oxygen abundance has already been suggested by [1] and by [3], who derived abundances with a completely different technique than we adopted. The error in oxygen abundances, taking into account only uncertainties originating from the intensity count rate statistics, is 0.05 dex, at heliocentric distances ≤ 3 R$_\odot$ and 0.1 dex at larger heights. Assuming an uncertainty of about 30 % in the kinetic temperatures, we estimate a variation in the abundance determination of about 0.005 dex at 2.75 R$_\odot$, and 0.06 dex at 3.75 R$_\odot$.

The average value of the oxygen abundance in the streamers' cores at $r/R_\odot = 2.5$, for latitudes within $\pm 2°$ from the equator, is 8.11, in units of $\log(Abb) + 12$, with $Abb = N(O)/N(H)$. Marocchi *et al.* [3] give 8.04, with an uncertainty ≈ 0.08 dex, at 2.2 R$_\odot$. We point out that Marocchi *et al.* used a temperature of $1.58 \times 10^6 K$, while we have, at 2.5 R$_\odot$, $T = 1.18 \times 10^6 K$. The two values appear to agree, within the uncertainties, and indicate that the abundance is constant with altitude, with no evidence for gravitational settling. The same conclusion seems to be consistent with the results we obtained at higher altitudes, at least within the streamer cores. Qualitatively, if T_e decreases more quickly than we have assumed by a factor of two, the oxygen abundance at the largest heights would be reduced of ≈ 0.48 dex. This is because the O VI emissivity is higher at lower T_e.

The average value of the oxygen abundance in the streamers' lateral branches – latitudes $\geq 6°$ from the equator, is on the order of 8.30 – 8.35, with an ≈ 0.23 dex increase, with respect to the streamer core. A higher variation has been found by Raymond *et al.* [1], who give an increase of 0.60 dex, at an altitude of ≈ 1.5 R$_\odot$, and Marocchi *et al.* [3], who find an increase slightly larger

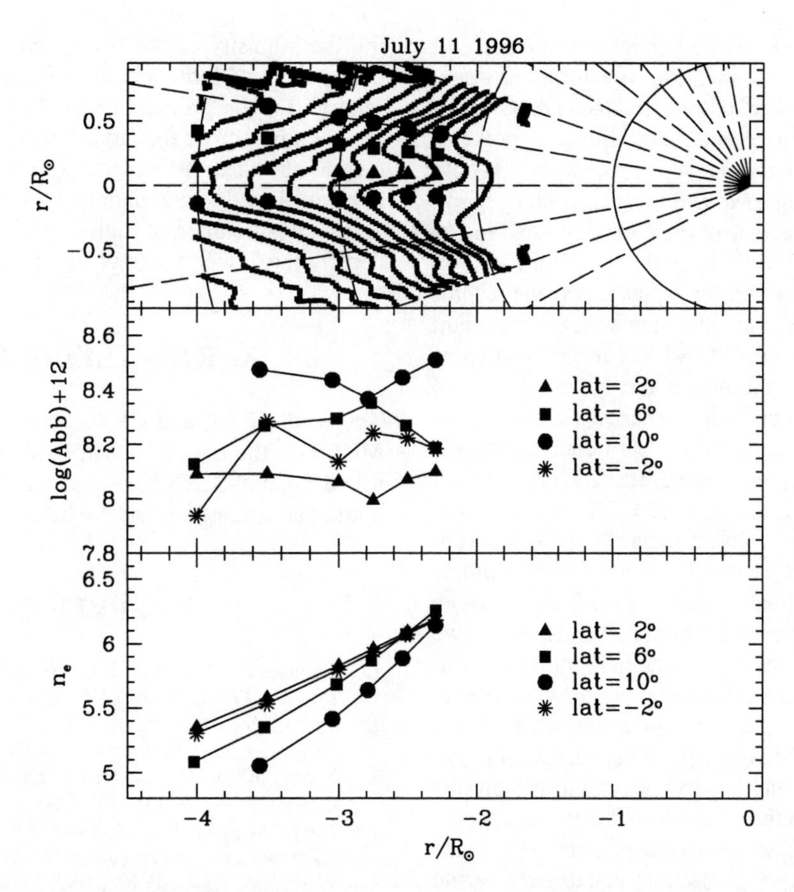

FIGURE 3. Results for the July 11, 1996 streamer. Top panel: isophotes of the OVI 1032 total line intensity *vs.* latitude and heliocentric distance. The lowest isophote corresponds to $logI_{\mathrm{O\ VI},1032} = 9.0$, the highest isophote corresponds to $logI_{\mathrm{O\ VI},1032} = 6.8$. Symbols drawn along different radial cuts indicate the positions where the oxygen abundance has been evaluated. Middle panel: oxygen abundance at the positions given in the top panel. Bottom panel: densities at the positions given in the top panel.

than 0.30 dex, from the streamer core to the legs, at an altitude of 2.2 $R_\odot$. Possibly, higher ratios are found at lower heights because projection effects are less important, and core and legs are better separated.

Abundances in the streamer of July 6, 1996, are slightly lower than abundances in the July 11, 1996 streamer, although values are within the estimated uncertainties. Hence, there is apparently no difference, in this respect, between the two class of streamers although we stress once more that the two streamers do not unambiguously belong to different classes.

DISCUSSION AND CONCLUSIONS

A comparison of our study with others supports previous results indicating that streamers show an enhanced oxygen abundance in their lateral branches. On this basis, because the oxygen abundance in slow wind is higher than the that in the streamer core, Raymond *et al.* [1] suggested streamer legs as the site where slow wind originates. On the theoretical side, Ofman [12] recently showed, through a 2.5 D numerical MHD model, that the enhanced oxygen O VI emission in the legs of streamers may be caused by Coulomb friction with outflowing protons and thus may trace the source regions of slow wind.

From our results, the abundance values seem to decrease slightly along different radial cuts, in particular in the streamer of July 6. The difference of abundance between core and legs of streamers persists with the heliocentric distance in both cases. The core depletion is usually ascribed [1, 5] to gravitational settling: the core of the streamer, being made of closed loops, may constitute a medium where oxygen ions, trapped for a long time, settle down. In the streamer legs, plasma outflows may make this mechanism less efficient. In this case, we may surmise that, as we move to higher heights, we enter regions where loops tend to open up and, as a consequence, the difference between legs and core gradually disappears. We notice, however, that this scenario implies that the oxygen abundance at high levels is approximately the same as in the legs of the streamer, but there is no indication for such a behavior in our data.

We would like to call attention to a different issue. The results we, and other authors, obtained are based on the assumption of ionization equilibrium: an hypothesis which is fairly often adopted (even in recent coronal hole models) without being adequately tested. A comparison between the expansion time of the coronal plasma in our streamer and the O VI ion recombination time, shows that the expansion time becomes smaller than the recombination time below 2 solar radii. However, the important rate is not recombination of O VI, but recombination of O VII, whose rate coefficient is about 10^{-12} cm^3s^{-1}; so the time scale is 10^6 s at 2.5 $R_\odot$, where the electron density is 10^6 cm^{-3}. Since the initial O VI concentration is only about a per cent, only a small fraction of the O VII needs to recombine to change the O VI concentration. Nevertheless, the gas is almost certainly out of equilibrium at $3-4$ $R_\odot$, and the result is that we underestimate the abundance. This tends to cancel the effect of a more rapid drop in T_e that we talked about earlier. However, we point out that, independently of whether ionization equilibrium is tenable or not, the behavior across the streamer is maintained, because it depends on the assumption of a constant T_e across the streamer, rather than on the absolute value of T_e. Only large variations in the plasma outflow speed within the streamer will affect the profile of abundances normal to the streamer's axis.

As a first attempt to guess how abundances would behave in a non-equilibrium medium, we re-calculated oxygen abundances assuming the electron temperature ($T_e = 1.18 \times 10^6$ K) at $r = 2.3$ $R_\odot$, to be the freezing-in temperature of O VI ions. Results lead to a slight increase of the abundance with height.

Whether oxygen is frozen-in in streamers, and at which level oxygen starts being frozen-in, has a potentially strong impact on the relationship between oxygen abundance in the slow wind and oxygen abundance in streamer legs. Marocchi *et al.* [3] pointed out that the slow wind is richer in oxygen than streamers, the difference being not significant possibly only if streamers abundances at 1.5 $R_\odot$ are considered. Our calculations show, although in a very approximate way, that non-equilibrium may act in the correct direction, that is, keeping up abundances at a value possibly consistent with that measured in slow wind. Also, we notice that streamer abundances should match those of the slow wind only if the oxygen and hydrogen outflow speeds are the same.

The analysis presented in this paper is preliminary. The problems we plan to address for a better data analysis include a study of the ionization state in the streamer and an extension of the analysis to the whole structure of the streamer. This implies the evaluation of the plasma flow speed in streamers through a technique which is at present under development. A further check of the behavior of T_e in the streamer may be given by simulating the intensity of the H Ly_α and Ly_β lines and comparing them with the measured values. This is especially relevant for the problem of the variation of the electron temperature across the streamer. Although at low heliocentric altitudes the oxygen core depletion is too large to be attributed to a variation of the electron temperature across the streamer, at higher levels where possibly the depletion is lower, the issue deserves further study.

ACKNOWLEDGMENTS

The work of LZ and GP has been partially funded by MURST – the Italian Ministry for University and Scientific Research. SOHO is a mission of international cooperation between ESA and NASA.

REFERENCES

1. Raymond, J. C., Kohl, J. L., Noci, G., and *others*, *Solar Physics*, **175**, 645–665 (1997a).
2. Parenti, S., Bromage, B. J. I., Poletto, G., and *others*, *Astronomy and Astrophysics*, **363**, 800–814 (2000).
3. Marocchi, D., Antonucci, E., and Giordano, S., *Annales Geophysicae*, **19**, 135–145 (2001).
4. Noci, G., Kohl, J. L., Antonucci, E., and *others*, "The quiescent corona and slow solar wind", in *Fifth SOHO Workshop*, ESA-SP 404, ESA Publication Division, Noordwijk, The Netherlands, 1997, pp. 75–84.
5. Raymond, J., Suleiman, R., van Ballegoijen, A., and Kohl, J. L., "Absolute Adundances in Streamers from UVCS", in *Correlated Phenomena at the Sun, in the Heliosphere and in Geospace*, edited by A. Wilson, ESA-SP 415, ESA Publication Division, Noordwijk, The Netherlands, 1997b, pp. 383–386.
6. Zangrilli, L., Poletto, G., Nicolosi, P., and Noci, G., "Latitudinal Dependence of the Outflow Speed of the Solar Wind from UVCS Observations", in *Plasma Dynamics and Diagnostics in the Solar Transition Region and Corona*, edited by J.-C. V. . B. Kaldeich-Schürmann, ESA-SP 446, ESA Publication Division, Noordwijk, The Netherlands, 1999, pp. 721–726.
7. van de Hulst, H. C., *Bulletin of the Astronomical Institute of the Netherlands*, **11**, 135–150 (1950).
8. Gibson, S., Fludra, A., Bagenal, F., and *others*, *Journal of Geophysical Research*, **104**, 9691–9700 (1999).
9. Fineschi, S., Gardner, L. D., Kohl, J. L., and *others*, "Grating stray light analysis and control in the UVCS/SOHO", in *Proc. SPIE Int. Soc. Opt. Eng., X-Ray and Ultraviolet Spectroscopy and Polarimetry II*, edited by S. Fineschi, 3443, 1998, pp. 67–74.
10. Parenti, S., Poletto, G., Bromage, B. J. J., and *others*, "Preliminary results from coordinated CDS-UVCS-Ulysses observations", ESA SP, American Institute of Physics, New York, 2001.
11. Kohl, J. L., Noci, G., Antonucci, E., and *others*, *Solar Physics*, **175**, 613–644 (1997).
12. Ofman, L., *Geophysical Research Letters*, **27**, 2885–2888 (2000).

Oxygen Abundance in the Extended Corona at Solar Minimum

Ester Antonucci* and Silvio Giordano*

*Osservatorio Astronomico di Torino, 10025 Pino Torinese, Italy

Abstract.

We present a study on the abundance of oxygen relative to hydrogen in the solar minimum corona and for the first time we measure this quantity in polar coronal holes. The results are derived from the observations of the extended corona obtained with the Ultraviolet Coronagraph Spectrometer (UVCS) on SOHO. The diagnostic method used to obtain the oxygen abundance is based on the resonant components of the O VI 1032 Å and HI 1216 Å emission lines. This method fully accounts for the effects of the outflow velocity of the solar wind, which can be determined through Doppler dimming, and of the width of the absorbing profiles of the coronal ions or neutral atoms involved in resonant scattering. The oxygen abundance is higher in the polar coronal hole regions, where the fast wind is accelerated, than in the streamer belt. In the polar regions the observed oxygen abundance is consistent with the photospheric value and with the composition results obtained with Ulysses for the fast wind. The oxygen abundance values derived with UVCS suggest that the plasma remains substantially contained in quiescent streamers, that therefore do not contribute significantly to the solar wind.

INTRODUCTION

The capability of determining elemental abundances in the extended corona, the region marking the transition between the inner solar atmosphere and the heliosphere, has been achieved only recently with the Ultraviolet Coronagraph Spectrometer (UVCS) onboard SOHO. This instrument has opened the possibility of determining the ultraviolet emissions of neutral hydrogen and minor ions beyond 1.5 $R_\odot$, and, on the basis of these quantities, the abundance of minor elements relative to hydrogen.

Most of the attention has been focussed up to now on the abundance of oxygen, since the OVI 1032 and 1037 are, with the HI Ly α at 1216, the strongest lines in the extended corona [e. g. 17]. Moreover, during solar minimum the oxygen abundance has been measured first in streamers where emission lines are brighter.

A new phenomenon was immediately evident when imaging the extended corona with UVCS. Solar minimum quiescent streamers are characterized by a depletion of oxygen which is particularly strong in their core (of roughly an order of magnitude with respect to the photosphere) and less severe in the lateral branches [16, 17]. This result, coupled with the fact that Feldman et al. [7] and Feldman [8] found an oxygen abundance value close to the limb, at 1.03 $R_\odot$, consistent with the photospheric value, led Raymond et al. [17] to the suggestion that gravitational settling of heavier ions is occurring in the closed field line region present in the center of streamers. The magnetic configuration, in this case, is roughly that of Figure 1. The OVI core dimming, however, can also be related to the origin of a slow wind flowing outward along open magnetic field lines that separate the OVI bright structures existing within a streamer, that in this case are interpreted as evidence for substreamers (Figure 2). Therefore in this case the depletion of oxygen is related to the outflows in the magnetically complex structure of the equatorial belt at solar minimum [16].

A statistical study of the quiescent streamer belt performed by Marocchi, Antonucci and Giordano [15] has, on the one hand, confirmed the deficiency of oxygen in the core of the streamer belt, with an abundance value relative to hydrogen close to 1.0×10^{-4} (8.0); on the other hand, it has shown that the oxygen depletion is dependent on heliodistance, and the dependence is more pronounced in the lateral branches where the abundance decreases from 3.9×10^{-4} (8.59) at 1.5 $R_\odot$ to 2.2×10^{-4} (8.34) at 2.2 $R_\odot$.

In the solar wind the oxygen abundance remains close to photospheric values. Although the SWICS/Ulysses instrument has detected a tendency to a systematic variation of the oxygen abundance between 5.3×10^{-4} (8.7) and 6.3×10^{-4} (8.8), related to the variation of wind speed between 400 km s^{-1} and 800 km s^{-1}, during the period of declining activity between mid–1992 and

CP598, *Solar and Galactic Composition*, edited by R. F. Wimmer-Schweingruber

mid–1993, this difference might not be fully significant [19]. The large discrepancy in the oxygen abundance of streamers and wind suggests that it is unlikely that the streamer plasma does significantly contribute to the wind, and in particular to the slow wind confined in the low–latitude/equatorial heliospheric regions during solar minimum. Hence, a possible hypothesis is that the slow solar wind would not originate primarily from quiescent streamers but from the layers surrounding the streamers, where the plasma is flowing outward channeled in flux tubes with large expansion factors [15]; that is, in a region of transition between streamers and the core of coronal holes. The above hypothesis is reasonable since quiescent streamers are mainly characterized by closed field lines. If instead the magnetic topology is such that a streamer is structured in substreamers, as suggested by the OVI images, there is the possibility of leakage of coronal material flowing along the open magnetic field lines separating adjacent substreamers (Figure 2).

Active streamers might certainly behave differently by significantly contributing to the slow wind; but the alternating slow and fast wind streams observed with SWICS/Ulysses persists substantially unmodified independently of the level of activity of a streamer. Therefore the degree of change in the magnetic topology of the equatorial streamer belt, in principle, should not be determinant in the formation of the slow wind.

If the interpretation of the data analyzed up to now is correct we then have to measure the oxygen abundance in the coronal regions outside the streamer belt, predominantly characterized by open field lines, in order to identify the abundance signature of the solar wind. Furthermore we might attempt to identify possible oxygen variations in the slow and fast wind streams. This task might be not so simple when studying oxygen since this is a high FIP element and therefore its concentration is not so variable from slow to fast wind streams as expected for the low FIP element concentrations [e. g. 21]. Here we address the first problem, that of determining the oxygen abundance close to the Sun in a polar coronal hole, with a new diagnostic method that allows to derive such quantity in a region of significant outflows related to the acceleration of the fast wind.

OXYGEN ABUNDANCE DIAGNOSTICS IN AN EXPANDING CORONA

In order to determine the oxygen abundance of the extended corona outside the streamer belt, we need to adopt a diagnostic method that accounts for the presence of outward plasma flows related to the expansion of the solar corona along open field lines. The radial flows induce a dimming in the emission of the oxygen and hydrogen

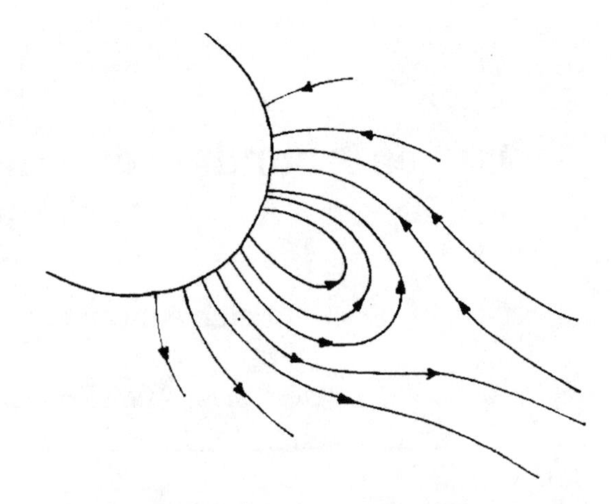

FIGURE 1. Dipolar magnetic configuration of a streamer.

lines used in the analysis. This is because the main mechanism of line formation in the tenuous extended corona is resonant scattering of photons originating in the lower atmosphere, since the importance of the radiative relative to the collisional contribution to the intensity of a coronal line is increasing with heliodistance (the radiative component decreases with density, while the collisional one decreases as density squared), and the resonant emission, in presence of a radial outward flow, undergoes Doppler dimming. This effect is due to the red shift of the incident radiation in the frame of reference of the expanding plasma. The oxygen coronal abundance relative to hydrogen is derived from the ratio of the bright oxygen line, the OVI 1032 line and the dominant coronal hydrogen line, H I Lyα 1216. Since in the corona the latter is formed almost exclusively by resonant scattering, the abundance is derived by computing the ratio of the radiative component of O VI 1032, and the H I Lyα 1216 line, according to the formula derived by Antonucci and Giordano [2]:

$$\left(\frac{N_O}{N_H}\right)_r \sim \frac{I_{r,OVI}}{I_{r,HI}}\frac{b_{HI}}{b_{OVI}}\frac{B_{12,HI}}{B_{12,OVI}}\frac{\lambda_{0,HI}}{\lambda_{0,OVI}}\frac{F_{D,HI}(w)}{F_{D,OVI}(w)}\frac{n_{HI}/n_H}{n_{OVI}/n_O}, \tag{1}$$

where $I_{r,OVI}$ and $I_{r,HI}$ are the intensities of the radiative components of O VI and the HI line, b_{HI} and b_{OVI} are the branching ratios of the considered transitions, $B_{12,HI}$ and $B_{12,OVI}$ are the Einstein coefficients for stimulated emission, $\lambda_{0,HI}$ and $\lambda_{0,OVI}$ are the reference wavelengths of the transitions, and $\frac{n_{OVI}}{n_O}$, $\frac{n_{HI}}{n_H}$ are the concentrations of the OVI ions and HI atoms, respectively.

The quantities $F_D(w)$ are the functions which take into account the line intensity dimming due to the outflow velocity of coronal plasma, w. They also account for the width of the coronal absorbing profiles along the direction of the incident radiation. Coronal absorbing profiles are usually much wider than thermal profiles in regions of open magnetic field lines, and their width may

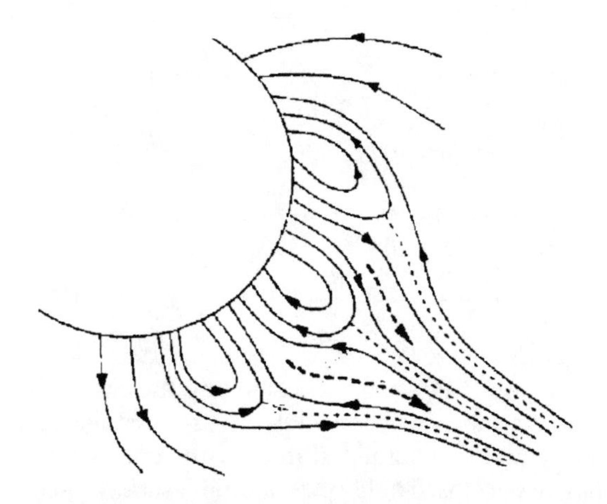

FIGURE 2. Magnetic topology of a complex streamer consisting of substreamers separated by open field lines. The dashed lines with arrows indicate slow plasma flows along the open magnetic field regions [16].

vary depending on the direction, since in polar coronal holes the velocity distribution of the oxygen ions are bi-Maxwellian above 1.8 $R_\odot$ [14].

The separation of the radiative and collisional component of the OVI 1032 is performed by taking into account the OVI 1037 line [3].

RESULTS ON THE OXYGEN ABUNDANCE IN A POLAR CORONAL HOLE DURING SOLAR MINIMUM

The oxygen abundance is determined in a polar coronal hole observed at North on May 21, 1996. The intensities of the OVI 1032 and 1037 and HI 1216 lines are averaged over an instantaneous field of view (30 arcmin x 84 arcsec) perpendicular to, and centered on, the radial direction at 1.6 $R_\odot$. Therefore the average height of the measurement is 1.64 $R_\odot$. The local outflow velocity, $w = 80$ km s^{-1}, is derived from the Doppler dimming of the OVI doublet by means of the the OVI 1037, 1032 line ratio diagnostics. The code used in the analysis is discussed in the paper by Dodero et al. [6] and Antonucci, Dodero and Giordano [1]. The ion velocity distribution is considered to be Maxwellian since at the coronal level of 1.6 R there is no evidence for anisotropy in the ion velocity distribution. In order to derive Doppler dimming we have assumed a local electron temperature, 8×10^5 K, compatible with the low temperature values determined in the same coronal hole by David et al. [5] and the electron density as derived by Guhathakurta et al. [12]. The same electron temperature is assumed to compute the concentrations of the HI atoms and OVI ions. In these conditions we find an oxygen abundance relative to hydrogen of $6.0 \pm 1.1\ 10^{-4}$. The error is due to the low statistics found in the faint coronal hole regions. Possible systematic errors ($\leq 30\%$) can be due to uncertainties in the atomic parameters.

The electron density and temperature values used correspond to the model of tenuous, cool corona considered by Antonucci, Dodero and Giordano [1], to derive the outflow velocity in coronal holes. As discussed in that paper, the still existing discrepancies in the observational results on the physical conditions in coronal holes allow also a model of relatively denser and hotter plasma, if we use the temperature derived by Ko et al. [13], on the basis of the in situ composition measurements of the solar wind performed with SWICS/Ulysses, and the density derived from the visible light data obtained with UVCS/SOHO by Kohl et al. [14]. In this case, at 1.6 $R_\odot$ the outflow velocity is negligible and the corresponding oxygen abundance is a factor 2.4 larger than that derived for a tenuous and cool corona (first model). Since this value is clearly unsound, the cool, tenuous coronal model is preferred, in agreement with other indications in favor of this model (for instance, the electron density value as derived for the extended corona with the spectroscopic method discussed by Antonucci and Giordano [3] is consistent with the Guhathakurta et al. [12] density curve derived from visible light data). For what concerns the electron temperature, although the results obtained by David et al. [5] do trace this quantity only out to roughly 1.3 $R_\odot$, its values remain below 1×10^6 K and there is a tendency to decrease outward when approaching 1.3 $R_\odot$. Therefore, considering this tendency and that the interplanetary value is of the order of 10^5 K, we do not expect significant temperature variation from the inner corona out to 1.6 $R_\odot$, where we assume as a value 8×10^5 K. The slow variation of temperature and relatively low expansion velocity ensure that, even in presence of an expanding corona, we can determine the concentrations of neutral hydrogen and oxygen ions which do not vary significantly out to 1.6 $R_\odot$.

The oxygen abundance derived for a polar coronal hole in this analysis is consistent with the fast wind values obtained with SWICS/Ulysses, 6.3×10^{-4}, and the photospheric value as shown in Table I, where we report both the photospheric value, 6.7×10^{-4} obtained by Grevesse et al [10] and the value recently re-evaluated by Grevesse [11], which is somewhat lower, 5.9×10^{-4}. Table I summarizes the oxygen concentration values resulting from the UVCS observations of the extended corona, compared with the heliospheric data measured with SWICS and the photospheric data.

TABLE 1. Oxygen Abundance.

Extended Corona	streamer core	1×10^{-4}
	streamer lateral structures	$3.5–2.2 \times 10^{-4}$
	coronal holes	6.0×10^{-4}
Heliosphere	slow wind	5.3×10^{-4}
	current sheet	$\leq 4.0 \times 10^{-4}$
	fast wind	6.3×10^{-4}
Photosphere		$6.7 - 5.9 \times 10^{-4}$

DISCUSSION

Our analysis indicates that, on the basis of the oxygen abundance, the plasma of coronal holes is the same as that observed in the fast wind streams. In addition, this plasma, accelerated in the extended corona to form the fast wind, retains the photospheric abundance of a high-FIP element such as oxygen. Therefore the present result, that provides the first determination of oxygen abundance in coronal holes, shows that this quantity is preserved from the convection zone to the heliosphere along the open field lines that are channeling the fast wind.

Regarding the low–latitude/equatorial regions, in the introduction it was pointed out that the oxygen abundance of quiescent streamers derived in previous studies is significantly lower than the slow wind abundance. In fact in the bright lateral structures of streamers oxygen is depleted from 3.5×10^{-4} at 1.5 $R_{\odot}$ to 2.2×10^{-4} at 2.2 $R_{\odot}$ and in the core of a streamer from 1.3×10^{-4} to 0.8×10^{-4}, in the same range of heliodistance [15].

In order to reconcile the abundance values in the outer corona and heliosphere to the observational evidence pointing to a general association of slow wind and streamer equatorial belt, Raymond et al. [18] suggested the following interpretation. Whereas the core plasma is magnetically confined for long intervals of time, and thus the confinement is causing the observed severe oxygen depletion due to gravitational settling, the slow wind might originate in the lateral branches where the plasma might be confined for shorter time and therefore partially escape outward (Figure 1). However even considering possible systematic errors in the spectroscopic analysis of the oxygen abundance ($< 30\%$), the oxygen depletion in the streamer lateral structures is significant; whereas the wind abundance, even in the case of the slow wind which is showing a tendency for a decrease relative to fast wind, does not differ significantly from the photospheric abundance on the basis of the statistical errors of the available data. The only channels for plasma leakage in the equatorial belt, therefore, remain the open field lines between substreamers present in the magnetic configuration proposed by Noci et al. [16], as shown in (Figure 2). This plasma might contribute, but only to the heliospheric region close to the interplanetary sector boundary, that indeed is noticeably poor in oxygen (Table I). In the case of this magnetic topology, the clear heliodistance dependence observed in the streamer lateral structures could be interpreted in terms of gravitational settling in the closed field lines outlining substreamers.

In this scenario the most probable source of the slow wind is likely to be found in a region of transition between streamers and the core of coronal holes. However, whether the oxygen abundance signature can contribute to the identification of the source of the slow wind remains an open question. In fact, slow and fast wind compositions certainly differ for what concerns the 'first ionization potential' (FIP) effect, which is enhanced by a factor of 1.5–2 in the slow wind relative to the fast wind, with a quite sharp transition of the low FIP element abundance at the fast/slow wind boundary [21]. Although oxygen is a high FIP element, and therefore it is probably not the best candidate to identify the transition between slow and fast wind in the corona, future studies might, however, provide information on possible variations in composition in the open field line regions outside streamers, that might also exist in the heliosphere but be masked by the statistical uncertainty.

ACKNOWLEDGMENTS

This work was supported by ASI (Italian Space Agency) and MURST (Ministero dell'Università e della Ricerca Scientifica e Tecnologica) contracts.

REFERENCES

1. Antonucci, E., Dodero, M.A. and Giordano, S. *Sol. Phys.*, **197**, 115–134 (2000).
2. Antonucci, E. and Giordano, S., *in preparation*, (2001).
3. Antonucci, E. and Giordano, S., *A&A*, *submitted*, (2001).
4. Bame, S.J., Asbridge, J.R., Feldman, W.C., Montgomery, M.D. and Kearny, P.D., *Sol. Phys.*, **43**, 463–473 (1975).
5. David, C., Gabriel, A.H., Bely-Dubau F., Fludra, A., Lemaire, P. and Wilhelm, K., *A&A*, **336**, L90–94 (1998).
6. Dodero, M.A., Antonucci, E., Giordano, S and Martin., *Sol. Phys.*, **183**, 77–90 (1998).
7. Feldman, U., Schüle, U., Widing, K.G. and Laming, J.M., *ApJ*, **505**, 999–1006 (1998).
8. Feldman, U., *Space Science Reviews*, **85**, 227–240 (1998).

9. Geiss, J., Gloeckler, G., von Steiger, R., Balsiger, H., Fisk, L.A., Galvin, A.B., Ipavich, F.M., Livi, S., McKenzie, J.F., Ogilvie, K.W. and Wilken, B., *Science*, *268*, 1033–1036, 1995.
10. Grevesse, N. and Sauval, A.J., *Space Science Reviews*, **85**, 161–174 (1998).
11. Grevesse, N.,*COSPAR 2000, private comunication* (2000).
12. Guhathakurta, M., Fludra, A., Gibson, S. E., Biesecker, D. and Fisher, R., *JGR*, **104**, 9801–9808 (1999).
13. Ko, Y.–K., Fisk, L.A., Geiss, J., Gloeckler, G. and Guhathakurta, M., *Sol. Phys.*, **171**, 345–361 (1997).
14. Kohl, J.L., Noci, G., Antonucci, E., Tondello, G., Huber, M.C.E., Cranmer, S.R., Strachan, L., Panasyuk, A.V., Gardner, L.D., Romoli, M., Fineschi, S., Dobrzycka, D., Raymond, J.C., Nicolosi, P., Siegmund, O.H.W., Spadaro, D., Benna, C., Ciaravella, A., Giordano, S., Habbal, S.R., Karovska, M., Li, X., Martin, R. , Michels, J.G., Modigliani, A., Naletto, G., O'Neal, R. H., Pernechele, C., Poletto, G., Smith, P.L. and Suleiman, R. M., *ApJ*, **501**, L127–131 (1998).
15. Marocchi, D., Antonucci, E. and Giordano, S., *Annales Geophysicae*, **19**, 135–145 (2001)
16. Noci, G., Kohl, J.L., Antonucci, E., Tondello, G., Huber, M.C.E., Fineschi, S., Gardner, L.D., Korendyke, C.M., Nicolosi, P., Romoli, M., Spadaro, D., Maccari, L., Raymond, J.C., Siegmund, O.H.W., Benna, C., Ciaravella, A., Giordano, S., Michels, J., Modigliani, A., Naletto, G., Panasyuk, A., Pernechele, C., Poletto, G., Smith, P.L. and Strachan, L., *ESA SP–*, **404**, 75–84 (1997).
17. Raymond, J.C., Kohl, J.L., Noci, G., Antonucci, E., Tondello, G., Huber, M.C.E., Gardner, L.D., Nicolosi, P., Fineschi, S., Romoli, M., Spadaro, D., Siegmund, O.H.W., Benna, C., Ciaravella, A., Cranmer, S., Giordano, S., Karovska, M., Martin, R. , Michels, J., Modigliani, A., Naletto, G., Panasyuk, A., Pernechele, C., Poletto, G., Smith, P.L., Suleiman, R. M. and Strachan, L., *Sol. Phys.*, **175**, 645–665 (1997a).
18. Raymond, J.C., Suleiman, R. M., van Ballegooijen, A. and Kohl, J.L., *ESA SP–*, **415**, 383–386 (1997b).
19. von Steiger, R., Wimmer Schweingruber, R. F., Geiss, J. and Gloeckler, G., *Adv. Space Res.*, **15**, 3–12 (1995).
20. von Steiger, R., *Space Science Reviews*, **85**, 407–418 (1998).
21. von Steiger, R., Schwadron, N.A., Geiss, J., Gloeckler, G., Fisk, L.A., Hefti, S., Wilken, B., Wimmer Schweingruber, R. F. and Zurbuchen, T.H., *JGR*, **105**, 217–238 (2000).
22. Wimmer Schweingruber, R. F., *Ph.D. Thesis, University of Bern*, (1994).

Preliminary results from coordinated SOHO-Ulysses observations

S. Parenti*, G. Poletto†, B.J.I. Bromage*, S.T. Suess**, J.C. Raymond‡, G. Noci§ and G.E. Bromage*

*Centre for Astrophysics, University of Central Lancashire, Preston, UK
†Osservatorio di Arcetri, Firenze, Italy
**Marshall Space Flight Center, Huntsville, AL, USA
‡Harvard-Smithsonian Center for Astrophysics, Cambridge, MA, USA
§Università di Firenze, Firenze, Italy

Abstract. SOHO-Ulysses quadratures occur at times when the SOHO–Sun–Ulysses angle is 90° and offer a unique possibility to compare properties of plasma parcels observed in the low corona with properties of the same parcels measured, in due time, *in situ*. The June 2000 quadrature occurred at a time Ulysses was at 3.35 AU and at a latitude of 58.2 degrees in the south–east quadrant. Here we focus on the UVCS observations made on June 11, 12, 13, 16. UVCS data were acquired at heliocentric altitudes ranging from 1.6 to 2.2 solar radii, using different grating positions, in order to get a wide wavelength range. The radial direction to Ulysses, throughout the 4 days of observation, traversed a region where high latitude streamers were present. Analysis of the spectra taken by UVCS along this direction shows a variation of the element abundances in the streamers over our observing interval: however, because the radial to Ulysses crosses through different parts of streamers in different days, the variation could be ascribed either to a temporal or to a spatial effect. The oxygen abundance, however, seems to increase at the edge of streamers, as indicated by previous analyses. This suggests the variation may be a function of position within the streamer, rather than a temporal effect. Physical conditions in streamers, as derived from UVCS observations, are also discussed.

INTRODUCTION

SOHO-Ulysses quadratures occur when the SOHO–Sun–Ulysses angle is 90°. At such times, we may observe the plasma parcels that leave the corona in the direction of Ulysses first, remotely, with SOHO, and later on, *in situ*, with Ulysses. This geometric configuration occurs twice per year and a number of coordinated SOHO Ulysses campaigns have been run at those times.

The June 2000 quadrature occurred when Ulysses was at 3.35 AU and at a latitude of 58.2 degrees in the south–east quadrant. A JOP – JOP 112 – was running at that time, with the participation of CDS, SUMER, UVCS, LASCO on SOHO and SWOOPS and SWICS experiments on Ulysses. The JOP aimed at deriving element abundances along the radial to Ulysses, at different heliocentric distances, with the purpose of establishing whether and how the element abundance varies with altitude and compare coronal and in situ abundances. CDS data are composed by normal incidence telescope (NIS) rasters of $120'' \times 120''$ centred at altitudes that reach up to 1.18 solar radii, and grazing incidence telescope (GIS) $4'' \times 4''$ rasters within the same field of view, out to 1.2 solar radii. SUMER data were acquired in the range between 1.04 and 1.6 solar radii. UVCS will be discussed extensively in the following. LASCO data were providing the overall coronal configuration at the time of the quadrature.

It is hard to overestimate the role that studies of element abundances have played in our understanding of the coronal-solar wind relationship: here it suffices to remind the reader of those analyses [e. g. 1], which, from the depletion of the He/H ratio, at magnetic sector boundaries, lead the authors to suggest streamers as plausible source regions of the wind measured at those positions. Where, within streamers, does slow wind emerge from, is still a matter of debate: Raymond et al. [2], from the oxygen abundance measured in the streamer legs, propose streamers lateral branches as the site where slow wind originates, Noci et al. [3], on the other hand, point to a more complicated configuration where the open regions between substreamers are possible sources of the slow wind. Because JOP 112 occurred at a phase of maximum activity in the solar cycle, we may expect to ob-

CP598, *Solar and Galactic Composition*, edited by R. F. Wimmer-Schweingruber

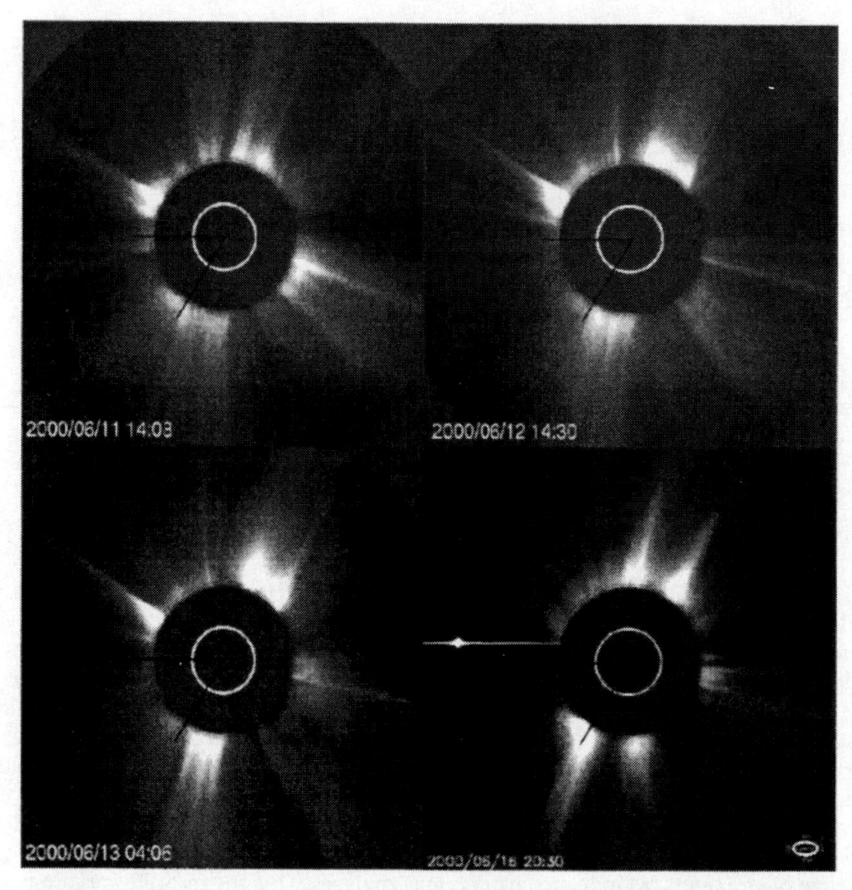

FIGURE 1. LASCO C2 images of the white light corona taken on 11, 12, 13, 16 June 2000. The inner and outer circles indicate, respectively, the solar disk and the lower edge of the C2 coronagraph. The radial to Ulysses, at -58.2° in the south-west quadrant, has been superposed on the images. The UVCS slit was set normal to the radial to Ulysses; the CDS FOV is shown in figure 2; the SUMER slit, which can be moved only along the North-South direction, was set at altitudes allowing SUMER to observe areas along the Sun-Ulysses direction.

serve streamers, in the direction pointing to Ulysses and relate abundances at coronal levels to *in situ* abundances and wind speed.

In this contribution we present preliminary results from JOP 112, focussing on an analysis of UVCS data. After a description of the observations we made and of the morphology of the regions traversed by the radial to Ulysses, we describe the techniques used to derive element abundances, and discuss the results we obtained. We conclude by outlining future work.

THE OBSERVATIONS

The June 2000 quadrature campaign extended for two weeks, around the quadrature date, which occurred on June 13. JOP 112 has been run during the second week of the campaign. Here we focus mainly on observations taken on 11, 12, 13 and 16 June. Fig. 1 shows the morphology of the corona on those days from LASCO C2 observations. As we might expect at this phase of the solar cycle, streamers extend to high latitudes, and the radial to Ulysses either runs at the edge of streamers or crosses the streamer core.

UVCS acquired data at 1.6 and 1.9 $R_\odot$. The slit (100 μm wide) was set normal to the radius of the Sun, with its center along the direction to Ulysses. Spectra have been obtained using different grating positions (usually 5), with a spatial resolution of $\approx 70''$ and a 2 pixel spectral binning (1 pixel = 9.25 mÅ). Observing times at each grating positions were on the order of 90 to 120 minutes. Spectra span the 951 Å - 1117 Å interval, in the O VI primary channel and the 1180 Å to 1250 Å interval in the redundant channel and include the H I Ly_α, Ly_β, Ly_γ lines, the O VI 1032 and 1037 Å lines, several lines from minor ions in different ionization stages, e.g. Fe X through Fe XIII, S X and , Ar XII and Ar XIV, in addition to Si XII, Mg X, Ca X and N V lines.

To show how coordinated observations have been made during JOP 112, we give in Fig. 2 the position

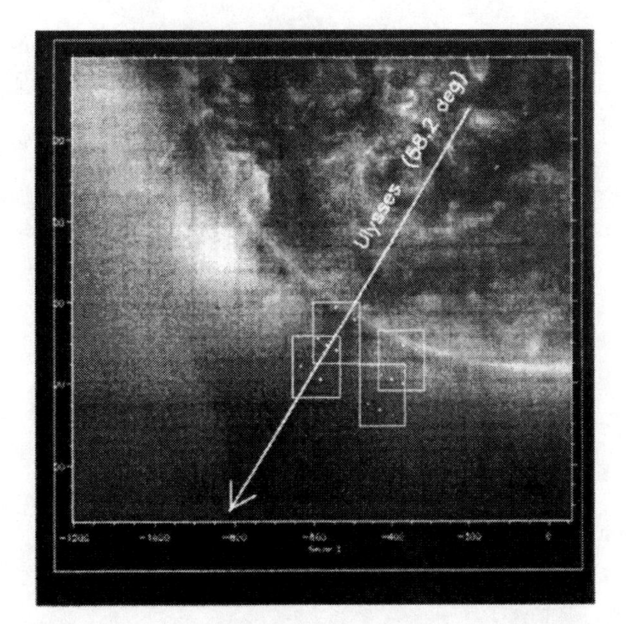

FIGURE 2. An excerpt from an EIT image in 195 Å taken on 13 June, showing the direction to Ulysses and the positions where CDS made observations with the NIS and GIS spectrometers. The CDS SUMER slits are along the north-south direction. NIS the full spectra are taken in 120″ × 150″ rasters; GIS spectra are acquired at several positions within the NIS rasters in a 4″ by 4″ pixel.

of the CDS field of view: as we mentioned, CDS – and SUMER – made spectra along the radial to Ulysses (or at nearby positions) at altitudes $1.02 \leq r/R_\odot \leq 1.6$. These observations will allow us to find the profile of abundances vs. height, as well as the density and electron temperature profile vs. height [see e. g. 4], over several days at the different positions traversed by the radial to Ulysses. In this preliminary analysis, however, we report only on results from the analysis of 4 days of UVCS observations. In the next section diagnostic techniques to derive density, temperature and oxygen abundance will be illustrated.

DIAGNOSTIC TECHNIQUES

We give now a short description of the diagnostic techniques used to derive density, electron temperature and oxygen abundance. We refer the reader to Raymond [5] for a more detailed discussion of these methods.

a) Electron densities
In the hypothesis that the plasma speed is negligible at the positions where UVCS took data (heliocentric altitudes: 1.6 and 1.9 $R_\odot$, in streamer areas) we can make a crude evaluation of the electron density N_e in the streamer from O VI lines, assuming that the major contribution to the line intensity is given by plasma in the plane of the sky. In this case we can write [6]

$$\frac{I_{12}}{I_{13}} = \frac{C_{12}N_1 + 4\pi\frac{j_{12}}{h\nu_{12}}}{C_{13}N_1 + 4\pi\frac{j_{13}}{h\nu_{13}}} \qquad (1)$$

where 1, 2, 3, indicate, respectively, the ground level and the lower and upper level of the transitions from which the O VI doublet lines originate and j_{12}, j_{13} are the line emissivities. Eq. (1) can be rewritten as

$$\frac{I_{12}}{I_{13}} = \frac{g_2}{g_3}\frac{1+\frac{g_2}{g_3}\theta}{1+\theta} \qquad (2)$$

with g statistical weights of the levels and the ratio $\theta = \frac{I_{13,rad}}{I_{13,coll}}$ between the radiative and collisional component of the O VI 1032 Å line given by

$$\theta = 5.75\ 10^2 \frac{\lambda_{13}^2 exp\frac{E_{13}}{kT_e}\sqrt{T}_e \int_{13} I_{ex}(\lambda)d\lambda}{\bar{g}N_e(\Delta\lambda_{cor}^2 + \Delta\lambda_{ex}^2)^{\frac{1}{2}}}\left(\frac{R_{sun}}{r}\right)^2 h(r) \qquad (3)$$

where I_{ex} is the exciting chromospheric radiation, other symbols have their usual meaning and $h(r)$ is a geometrical factor.

b) Electron temperatures
As we mentioned, our spectra contain lines from Fe in different ionization stages, as well as H I Lyman lines. Then, if we identify the collisional component of the Lyβ line [1], calculate the ratio R_{obs} between the observed Fe ion intensities to the observed Lyβ component and the ratio R_{th} between the predicted Fe ion intensities and the predicted Lyβ component at different temperatures, we may build a plot of log R = log $(\frac{R_{th}}{R_{obs}})$ vs. log T, for all the Fe ions of the spectra [7]. The common intersection of the curves built from different ions gives a good indication of the plasma electron temperature. Clearly, if the plasma is not isothermal, curves intersect at different temperatures.

Temperatures can be derived also via the DEM (Differential Emission Measure) technique: a plot of the DEM of different from different lines vs. T indicate the temperature at which the peak(s) in emission occur(s). We refer the reader to Raymond et al., this volume, for a definition of the DEM.

c) Element abundances
Element abundances can be derived, at the same time as T_e is derived, from the techniques briefly described

[1] this can be evaluated from the measured intensities of the H I lines, taking into account the predicted ratios between their collisional and radiative components, see [2].

above. For instance, iron abundance is derived from the $\log R$ vs. $\log T$ curves: at the temperature where the curve intersect, $\log R = 0$, if the abundance used in the theoretical calculation were corresponding to the abundance in the regions where observations were taken. The shift in the $\log R$ vs. $\log T$ profiles, required to make $\log R = 0$, gives the correct value of the iron abundance.

The above method cannot be applied to oxygen ions, though, as we do not have oxygen ions from different ionization stages. Hence, we used an alternative technique, which requires first the identification of the radiative and collisional components of the O VI doublet lines.
From the radiative components we have

$$\left(\frac{N_{\rm O}}{N_{\rm H}}\right)_{rad} = \frac{I_{rad}(1032)}{I_{rad}({\rm Ly}\beta)}\frac{C_{HI}}{C_{OVI}}\frac{B_{{\rm Ly}\beta}}{B_{OVI}}\frac{f_{{\rm Ly}\beta}}{f_{1032}}\frac{I_{disk}({\rm Ly}\beta)}{I_{disk}(1032)}\frac{\delta\nu_{OVI}}{\delta\nu_{HI}} \quad (4)$$

while from the collisional components we have

$$\left(\frac{N_{\rm O}}{N_{\rm H}}\right)_{col} = \frac{I_{col}(1032)}{I_{col}({\rm Ly}\beta)}\frac{C_{HI}}{C_{OVI}}\frac{B_{{\rm Ly}\beta}}{B_{OVI}}\frac{q_{{\rm Ly}\beta}}{q_{1032}} \quad (5)$$

where $(N_{\rm O}/N_{\rm H})_{col}$ and $(N_{\rm O}/N_{\rm H})_{rad}$ indicate the oxygen abundance value derived from the collisional and radiative components of the line intensity; I_{col} and I_{rad} are the collisional and radiative components of the intensity in photon cm^{-2} s^{-1} sr^{-1}; $C_{\rm OVI}/C_{\rm HI}$ is the ratio of ion concentration (which, in the $\log T$ interval 6.1-6.3 changes by $\leq 16\%$); f is the oscillator strength; B is the branching ratio; I_{disk} is the disk intensity in O VI/Lyβ lines; $\delta\nu$ is the line width and q the excitation rate.

RESULTS

Figure 3 gives an example of the application of the technique described in b) of the previous Section to observations acquired at $1.6R_\odot$ on June 16, 2000. The plot shows that plasma is isothermal, at the position to which data refer, as the $\log\frac{R_{th}}{R_{obs}}$ vs. $\log T$ curves for different ions intersect at practically the same temperature. Figure 4 gives an example of the DEM technique, mentioned in b) of the previous Section, applied to data taken on the same day as in Figure 3. Both methods yields approximately the same temperature. Results for the four days we are analyzing appear in Table 1. Temperature and densities are given along the radial to Ulysses at 1.6 and 1.9 $R_\odot$. On June 11 we made observations only at the lower altitude, hence physical parameters are given only at the lower height. The error in temperature values is on the order of 50 10^3 K, in densities is on the order of 20%. Temperatures at 1.6 $R_\odot$ are $\approx 1.2\ 10^6\ K$, for all days but June 16. It also looks like temperature decreases with altitude, with the exception of June 12. This day may have an anomalous behaviour also as far as densities are concerned: while its density at 1.6 $R_\odot$ is lower, its density gradient is less steep than in any other day. We notice, however, that an error of 20% will raise the density at 1.6 $R_\odot$ to values found on other days.

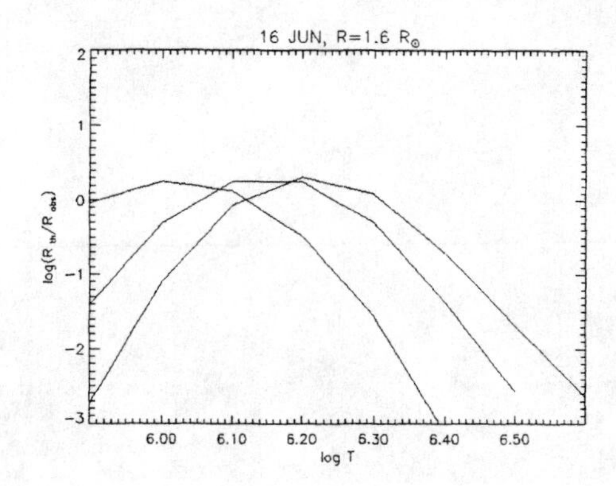

FIGURE 3. Plot of $\log\frac{R_{th}}{R_{obs}}$ vs. $\log T$ for June 16, 2000, at the heliocentric altitude of $1.6R_\odot$. Data refer to the UVCS pixel lying along the radial to Ulysses. Lines from three ionization stages of Ulysses – Fe X, Fe XII, Fe XIII– have been used to build the plot, theoretical line emissivities have been taken from [8].The streamer appears to be approximately isothermal.

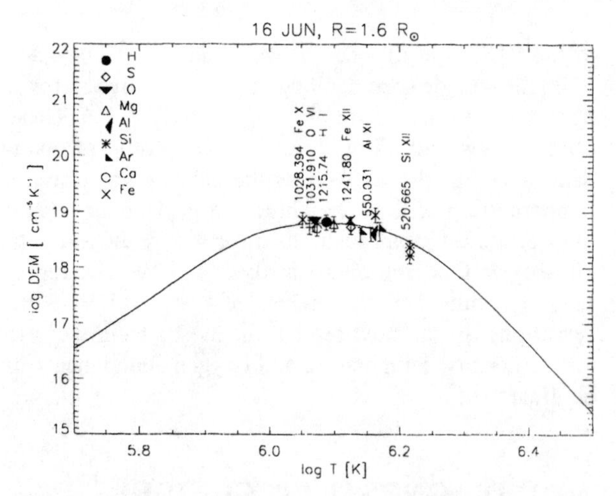

FIGURE 4. Plot of $\log DEM$ vs. $\log T$ for June 16, 2000, at the heliocentric altitude of $1.6R_\odot$. Data refer to the UVCS pixel lying along the radial to Ulysses. Lines from different ions have been used (see symbols); the temperature at which the DEM peaks is approximately the same as given by the plot of $\log\frac{R_{th}}{R_{obs}}$ vs. $\log T$ (Figure 3).

In order to interpret our results we need to understand which feature we are observing. To this end we cannot rely on LASCO images, as they refer to higher altitudes than our data. Hence we made plots of the total intensity of the O VI 1032 Å line vs. distance along the UVCS slit, at 1.6 and 1.9 $R_\odot$. Because the UVCS slit

TABLE 1. Electron temperature, T_e, and density, N_e, along the radial to Ulysses

		T_e/ **MK**	N_e/ $\mathbf{10^6}$ **cm**$^{-3}$
11 *June*	$1.6R_\odot$	1.17	5.00
12 *June*	$1.6R_\odot$	1.20	4.62
	$1.9R_\odot$	1.48	1.73
13 *June*	$1.6R_\odot$	1.23	5.60
	$1.9R_\odot$	1.03	1.52
16 *June*	$1.6R_\odot$	1.38	5.58
	$1.9R_\odot$	1.26	2.50

center lies along the radial to Ulysses, those plots show that the direction to Ulysses traverses a region with an approximately flat intensity distribution, i.e. through the streamer body, on June 16, while it is adjacent to a region of very steep intensity gradient, i.e. intersects a streamer edge, on June 11 and 13. On June 12, however, there is an ambiguous situation, because we have contrasting information at the two altitudes: it looks like we observe the streamer edge, at the low heliocentric distance, and the streamer body at high heliocentric distance. Possibly, projection and/or temporal effects do not allow us to identify clearly where the radial to Ulysses is lying. Hence, from now on, we consider data from June 11 and 13 as representative of conditions of the streamer's legs, data from June 16 as representative of conditions of the streamer's "body" and leave the June 12 case apart. In a later study, we plan to check whether coronagraph observations at lower heights (e. g., from Mauna Loa) provide any useful indications to understand the morphology of the June 12 case.

In this scenario, we may conclude that the electron temperature is lower at the streamer's edges, than in streamer's cores. This conclusion should be further tested calculating the electron temperature distribution across a streamer, because we cannot rule out, so far, the possibility that the lower T_e of June 11 and 13, is lower *throughout the streamer* than the T_e derived for the streamer's body on June 16. However, should the present indication be confirmed by a more detailed analysis, the abundance obtained in streamer's legs by different authors should be rediscussed in terms of a core-leg variation of the streamer electron temperature.

Densities, on the other hand, appear to have, at 1.6 $R_\odot$, the same value independent of whether they refer to the streamer edge or to the streamer body. In the streamer body the density gradient is possibly less steep than in the legs, but once more we need to confirm these indications by a more extended analysis. We notice, however, that densities inferred by LASCO at higher altitudes seem to give analogous indications (see Zangrilli et al., this volume [9]).

In order to derive the oxygen abundance with the technique described in c) of the previous Section we need to know the disk intensities in the H Ly$_\beta$ and O VI 1032 Å lines and to identify the collisional and radiative components of the O VI and hydrogen lines. As disk intensities vary with time, a precise knowledge of these parameters is essential to a good determination of the oxygen abundance. Also, to identify correctly the collisional and radiative components of the Ly$_\beta$ line, we need to know the ratio between the radiative and collisional components of the Ly$_\alpha$ and Ly$_\beta$ lines, which depend as well from the chromospheric illumination.

There is no room here for a detailed description of how we evaluated these factors. Briefly, we can say that, as we do not have disk intensities measured at the time our data have been acquired, we estimated disk intensities starting from measured values (SOLSTICE values for Ly$_\alpha$ and UVCS Ly$_\alpha$ and O VI 1032 disk measurements in 1996 and 1997) and extrapolated these to the time of our observations on the basis of the temporal increase predicted by [10] for the Lyman continuum and the NeVIII and NV lines, assumed to be representative of the Ly$_\alpha$, and of the Ly$_\beta$ and O VI lines, respectively. Extrapolated intensities agree with indications given by [11] for the variation between maximum and minimum of the solar cycle for Ly$_\alpha$ and other lines.

As for the separation of the collisional and radiative components of the lines, Raymond et al. [2] give values for these ratios. But, as we mentioned, these as well depend on disk intensities: as a consequence, we estimated new ratios, from the disk intensities we predicted. However, in evaluating abundances from Eq. 4 and Eq. 5, we let the ratios vary in between Raymond et al. values and the newer estimates, until $(\frac{N_O}{N_H})_{rad} = (\frac{N_O}{N_H})_{coll}$. In this way, we sort of took care of uncertainties in the determination of disk intensities, as well as of the unknown short temporal variation they may go through over the time interval when data were acquired. Alternatively, as more usually done, we might have considered average values from the collisional and radiative abundance determination. Practically, the two technique are equivalent, as values from the two determinations are pretty close: the *maximum* discrepancy we had, using the old vs. the new factors, is on the order of 35%.

Oxygen abundance values are given in Table 2. As with other parameters, the behavior of June 12 is anomalous, as abundances increase with altitude. In the other days, we find an higher abundance value, when observing the streamer edges (June 11 and 13), than when observing the streamer body (June 16). This result confirms previous determination by [2, 12], at altitudes comparable to those of this work, and by Zangrilli et al. [9] at higher heliocentric distances.

TABLE 2. Oxygen abundances along the radial to Ulysses

		Oxygen Abundance (Phot. = 8.93)
11 *June*	$1.6R_\odot$	8.61
12 *June*	$1.6R_\odot$	8.38
	$1.9R_\odot$	8.53
13 *June*	$1.6R_\odot$	8.67
	$1.9R_\odot$	8.55
16 *June*	$1.6R_\odot$	8.43
	$1.9R_\odot$	8.49

SUMMARY AND CONCLUSIONS

In this paper we have presented results from a preliminary analysis of UVCS data taken in June 2000, at the time of an Ulysses-SOHO-Sun quadrature. Our analysis provides further evidence in favor of an enhanced oxygen abundance in the legs of streamers, with respect to the streamer's core. This enhancement favors the hypothesis that streamer legs are sources of the slow solar wind: because abundances in the core are too low to be compatible with abundances in the wind, as measured by *in situ* experiments. It may be worth pointing out that the $\frac{H}{O}$ ratio in the slow wind, takes the value $\frac{H}{O} = 1890 \pm 600$ while in the fast wind $\frac{H}{O} = 1590 \pm 500$, [13]. The oxygen abundances we derived from June 11 and 13 are consistent with the slow wind estimates, while both the June 12 and 16 values are inconsistent with reported values.

One of the aims of quadrature campaigns is the comparison between coronal and *in situ* data. Hopefully, we might be able to check values of $\frac{H}{O}$ from data measured by Ulysses versus our coronal values. However, the two values quoted above for slow and fast wind are only marginally different and it may be difficult to draw definite conclusions from this kind of comparison.

Obviously, we need to extend the analysis of quadrature observations to the other days when we acquired data. Also, we are planning to extend the analysis of element abundances to include other elements as well. Estimates of the Fe abundance lead to values higher than found by [2]. For instance, Figure 3 shows that the iron abundance is photospheric on May 16, at 1.6 $R_\odot$. Fe is a low FIP element: it will be interesting to check whether this overabundance is common to other low FIP elements. Also, analysis of CDS and SUMER data will allow us to compare the values we find from UVCS with values derived from other data sets, at different altitudes.

Another interesting result we got is related to the electron temperatures in streamers. There have been a few estimates of T_e in streamers, but no attempt has been made, yet, to derive the profile of T_e across a streamer. We plan to extend our analysis to different location than so far examined, to check whether the indications from this work, which favor an increase of T_e towards the streamer edge is confirmed. This would allow a better evaluation of the abundance profile across streamers.

ACKNOWLEDGMENTS

The work of GP has been partially funded by MURST – the Italian Ministry for University and Scientific Research. SP acknowledges support from ASI, the Italian Space Agency. SOHO is a mission of international cooperation between ESA and NASA.

REFERENCES

1. Borrini, G., Wilcox, J. M., Gosling, J. T., and *coauthors*, *Journal Geophysical Research*, **86**, 4565–4573 (1981).
2. Raymond, J. C., Kohl, J. L., Noci, G., and *coauthors*, *Solar Physics*, **175**, 645–665 (1997a).
3. Noci, G., Kohl, J. L., Antonucci, E., and *coauthors*, "The quiescent corona and slow solar wind", in *Fifth SOHO Workshop*, ESA-SP 404, ESA Publication Division, Noordwijk, The Netherlands, 1997, pp. 75–84.
4. Parenti, S., Bromage, B. J. I., Poletto, G., and *coauthors*, *Astronomy and Astrophysics*, **363**, 800–814 (2000).
5. Raymond, J. C., "Radiation from hot, thin plasmas", in *Hot Thin Plasmas in Astrophysics*, edited by R. Pallavicini, ASI 249, Kluwer, Dordrecht, The Netherlands, 1998, pp. 3–20.
6. Noci, G., Kohl, J. L., and Withbroe, G. L., *Astrophysical Journal*, **315**, 706–715 (1987).
7. Raymond, J. C., Suleiman, J. L., Kohl, J. L., and Noci, G., *Space Science Review*, **85**, 283–288 (1998).
8. Mazzotta, P., Mazzinelli, G., Colafrancesco, S., and *coauthors*, *Astronomy and Astrophysics*, **133**, 403–409 (1998).
9. Zangrilli, L., Poletto, G., Biesecker, D., and Raymond, J. C., "Oxygen abundance in streamers above 2 solar radii", AIP Conference Proceedings, American Institute of Physics, New York, 2001.
10. Schuhle, U., Wilhelm, K., Hollandt, J., and *coauthors*, *Astronomy and Astrophysics*, **354**, L71–L74 (2000).
11. Tobiska, W. K., and Eparvier, F. G., *Solar Physics*, **177**, 147–159 (1998).
12. Raymond, J., Suleiman, R., van Ballegoijen, A., and Kohl, J. L., "Absolute Adundances in Streamers from UVCS", in *Correlated Phenomena at the Sun, in the Heliosphere and in Geospace*, edited by A. Wilson, ESA-SP 415, ESA Publication Division, Noordwijk, The Netherlands, 1997b, pp. 383–386.
13. Wimmer-Schweingruber, R. F., , Ph.D. thesis, University of Bern (1994).

The Solar Wind Helium Abundance: Variation with Wind Speed and the Solar Cycle

Matthias R. Aellig*, Alan J. Lazarus* and John T. Steinberg†

*MIT Center for Space Research, Cambridge, MA 02139
†Los Alamos National Lab., Los Alamos, NM 87545

Abstract. We investigate the helium abundance in the solar wind and variations thereof on a time scale of years. Data from the WIND/SWE experiment gathered between the end of 1994 and early 2000 are analyzed. In agreement with similar work for previous solar cycles, we find a clear dependency of the He/H ratio in the solar wind on the solar cycle. In the slow solar wind, the average He/H rises from a minimum of less than two percent around solar minimum to about 4.5% in early 2000. The solar cycle dependency is stronger the lower the speed that is used to sort the data. We observe the strongest dependency of the He/H ratio on the solar wind speed around solar minimum and it weakens as the solar activity increases. We speculate that the expansion factor of the magnetic field close to the Sun changes over the solar cycle and thereby changes the efficiency of the Coulomb drag. Inefficient Coulomb drag leads to a low helium abundance in the solar wind.

INTRODUCTION

This contribution is a somewhat expanded version of a poster given at the SOHO-ACE Workshop 2001 in Bern. A formal paper on our results will be found in *Aellig et al.* [1]. Some of the figures and text from that paper have been used by permission.

The solar wind helium abundance was first reported to vary over the solar cycle by *Ogilvie and Hirshberg* [2]. Combining data sets from several spacecraft they found a solar cycle variation of n_α/n_p of 0.01 ± 0.01 with the average helium abundance being higher during periods of higher activity. Adding data from Imp 6, 7, and 8, *Feldman et al.* [3] confirmed the solar cycle dependency, but they found that the maximum difference in helium abundance between the low- and high-speed wind occurred near solar minimum in contradiction to the results reported by *Ogilvie and Hirshberg* [2] who reported the maximum speed variation near the peak of solar activity. Analyzing ISEE-3 data from the ICI instrument, *Ogilvie et al.* [4] reported the solar cycle dependency of the solar helium abundance for solar cycle 21. They suggested that the smallest dependence on speed occurs during the period of decreasing activity but before solar minimum. In this paper we report the abundance variation with wind speed from one year before solar minimum to nearly solar maximum. An overview of previously reported He/H in the solar wind plus the work described in this paper is shown in Figure 1.

DATA

For this study we used data from the Faraday Cup portion of the SWE instrument on the Wind spacecraft, *Ogilvie et al.* [5]. The Faraday Cup data allow the determination of the solar wind proton velocity, thermal speed and density. The same parameters can be derived for solar wind alpha particles under most circumstances.

Isotropic Maxwellian velocity distributions are assumed for both the protons and the alpha particles, and every energy/charge sweep (lasting approximately 90 seconds) is fitted individually. Depending upon the quality of the data and how well the two ion populations are separated, different fitting approaches are taken or some parameters may be subject to a constraint. In all cases that are used, the distinction between protons and alphas is clear.

The data analyzed cover the time period from the launch of the Wind mission in late 1994 to April 2000, i.e., about half a solar cycle. The average helium abundance over a certain period was calculated as the ratio of the integrated number densities of helium and protons. This corresponds to a weighted average of the He/H ratios where the weights are the proton densities. As a measure of the variability of the individual values we use the central interval in which 68% of all measurements fall. We stress that CMEs, current sheet crossings, and shocks have not been excluded from the analysis since it is our intention to characterize the solar wind He/H at large.

As a proxy for the solar activity we used the monthly

CP598, *Solar and Galactic Composition*, edited by R. F. Wimmer-Schweingruber

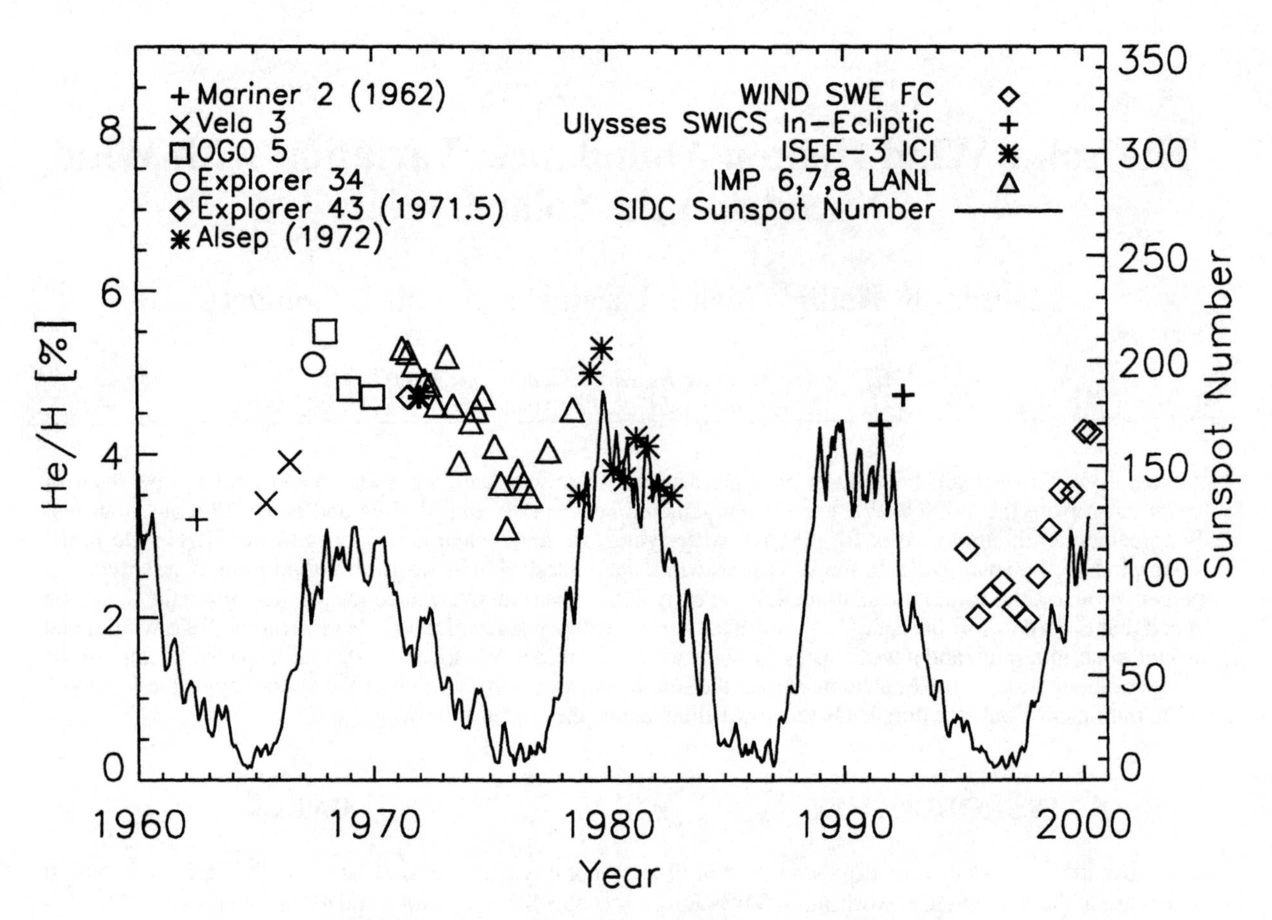

FIGURE 1. Solar wind He/H over several solar cycles (from [1]).

sunspot number provided by the Sunspot Index Data Center (SIDC).

The He/H data that fall into a certain speed range were averaged over 250 days. The speed bins cover the speed below 350 km/s and then bins 50 km/s wide for higher speeds up to 600 km/s. Typically those averages are based on several tens of thousands of measurements. Since the observation period contains the solar minimum around 1996, faster streams are not that prominently represented in our study particularly around 1996. Nevertheless averages between 550 km/s and 600 km/s contain at least 1600 data points (see Table 1).

RESULTS

The result of our analysis is shown in Figure 2. For the slowest wind considered, i.e., for wind speeds below 350 km/s the average He/H at the beginning of the observation period is 1.8 % and then drops to as low as 1.4 % close to solar minimum. Then, the average He/H in that speed range starts to increase and reaches 4 % by the beginning of the year 2000, tripling within three years. For larger speeds, the average value at the start of the observation period increases as well as does the value observed around solar minimum. Furthermore, the change over the observation period becomes smaller. While the uncertainties of the mean values of He/H (indicated by error bars) in the speed range below 350 km/s shows a cycle dependency with the smallest uncertainty around solar minimum, no obvious pattern is observed for the speeds from 550 km/s to 600 km/s. We also analyzed the speed range between 600 and 850 km/s in which a slight linear increase of the average He/H is observed from 1995 to 2000 although there are only relatively few measurements in that speed range, especially around solar minimum.

Because the number of observations is large, the mean values of the He/H ratio are well-determined. Nevertheless, the range of values contributing to each mean value is also large: the long error bars in Figure 2 show the ranges of the ratios corresponding to the 16ãnd 84q̃uantiles of the measured He/H The range spanned by those values contains the central range into which 68% of the He/H determinations fall.

In summary, throughout the observation period increasing solar wind speed implies increasing helium abundance in the solar wind. The strength of that speed dependence also varies over the solar cycle. Using a linear fit to the observed He/H ratio as a function of speed, we get

$$\mathrm{He/H} = a + b\, v_{sw} \qquad [\mathrm{He/H}] = \% \qquad (1)$$

where the parameter b indicates the strength of the speed dependence. The resulting values for a and b derived with a robust fit are shown in Table 2. Based on the variation of b over the solar cycle, the speed dependence is four times stronger around solar minimum than during 1999 and early 2000.

Another way to display the increasing He/H ratio for lower speed wind as solar activity increases is shown in Figure 3 which displays a histogram of the measured ratios in 1996 and 2000.

SUMMARY

We have found a strong variation of the solar wind helium abundance between the end of 1994 and early 2000. The lowest values are observed around solar minimum in 1996. Those finding are in agreement with previous reports about the solar cycle dependency during earlier activity cycles [2,3,4,6], We find the dependence of the helium abundance on the solar wind speed to be strongest during solar minimum which agrees with results reported by *Feldman et al.* [3] but that solar cycle speed dependence disagrees with those reported by *Ogilvie and Hirshberg* [2] and *Ogilvie et al.* [4]. Our value for He/H that was averaged over all speeds around solar minimum is considerably lower than the speed-averaged helium abundance around previous solar minima. It is not clear, however, to what degree this is a real effect and to what degree it is a systematic difference between different sensors.

DISCUSSION

In the following we discuss a possible explanation for the solar cycle dependency of the helium abundance that also would explain the observed speed dependency.

Many theoretical studies indicate that the proton flux is the most important parameter regulating the behaviour of the alpha particles in the solar wind *Bürgi* [7] and references therein. Coulomb collisions between the protons and the alpha particles and the minor ions play an important role in accelerating the latter two constituents *Geiss et al.* [8]. The proton flux transfers momentum to helium ions and accelerates them out of the corona. The larger the proton flux, the more efficiently helium is accelerated and the more helium is observed in the solar wind. It turns out that, under coronal conditions, among all heavy ions ^{4}He couples the least efficiently to the protons.

Although the proton flux at 1 AU varies only by little over the solar cycle *Wang* [9], the proton flux in the acceleration region of the solar wind does not necessarily have to be constant as we shall now discuss. In a stationary case, the continuity equation along a flux tube with

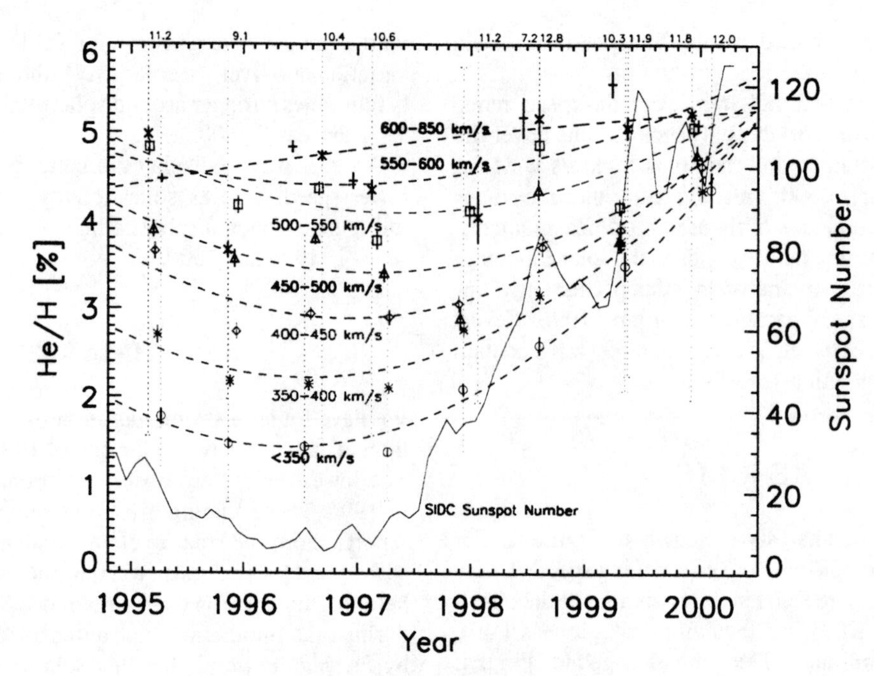

FIGURE 2. The solar wind helium abundance derived from the Wind/SWE Faraday Cup data. The solid error bars shown for each point represent the uncertainty of the plotted mean value. The dotted error bars indicate the range in which 68/the observations fall; see text. The numbers across the top of the figure are the values of range error bars when they fall outside the figure. The figure is from [1]

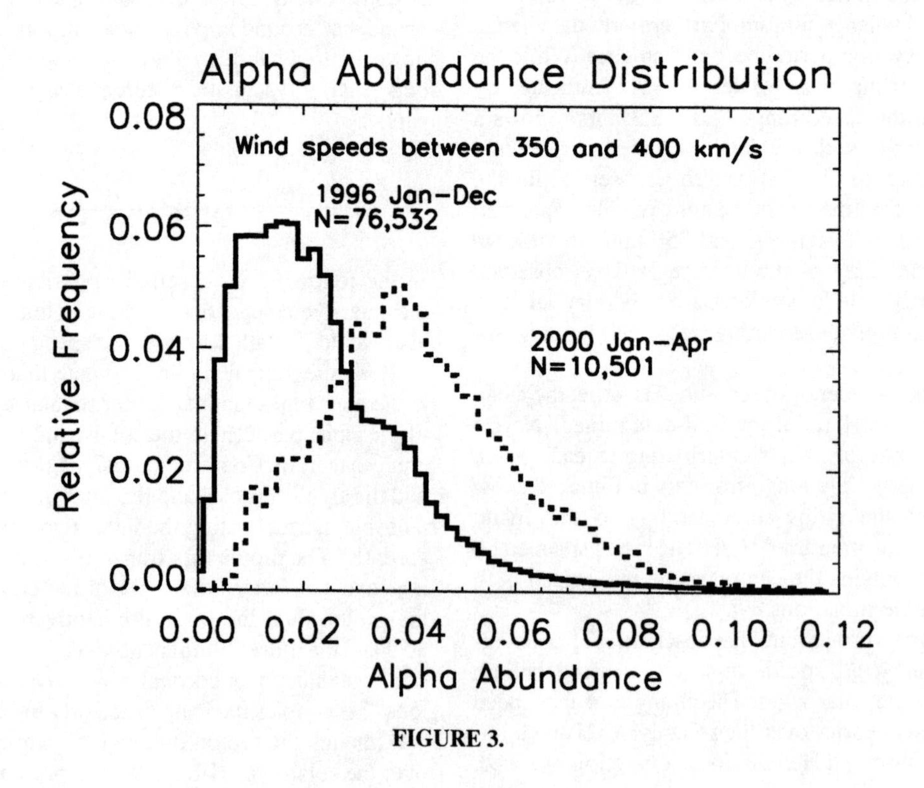

FIGURE 3.

TABLE 1. The number of spectra contributing to each speed range for the year centered about the average year indicated. Note that speeds are given in km/s.

Speed range \ Year	1995.3	1996.5	1997.9	1998.6	1999.4	2000.0
< 350	56238	35432	88786	34608	40376	21012
350 – 400	44296	74580	67348	55370	46556	35934
400 – 450	36686	52124	41666	45650	41168	30544
450 – 500	21150	31590	14940	25290	21420	20790
500 – 550	15064	15624	6440	13440	13216	17248
550 – 600	14600	8800	1600	9160	8200	16800
600 – 850	24849	3597	132	7722	7392	15147

TABLE 2. Parameters for the speed dependence of He/H given in eq. 1. The mean absolute deviation between the measured He/H and the linear fit is denoted with δ.

Year, Month	a [%]	b [% s/km]	δ [%]
1995, Mar	-1.4	$1.07\ 10^{-2}$	0.19
1995, Nov	-2.4	$1.20\ 10^{-2}$	0.21
1996, Aug	-2.8	$1.30\ 10^{-2}$	0.09
1997, Mar	-1.9	$1.05\ 10^{-2}$	0.09
1997, Dec	0.63	$0.54\ 10^{-2}$	0.19
1998, Aug	-0.62	$0.96\ 10^{-2}$	0.08
1999, May	2.36	$0.30\ 10^{-2}$	0.18
2000, Jan	3.03	$0.32\ 10^{-2}$	0.07

the areal expansion $A(r) = r^2 f(r)$ reads

$$n(r)v(r)A(r) = \text{const.} \tag{2}$$

From the measured proton flux at $R = 1$ AU, the proton flux $\Phi(r)$ at any given radial distance r can be inferred using

$$\Phi(r) = n(r)v(r) = \Phi(R)\,\frac{A(R)}{A(r)} \tag{3}$$

The ratio of the expansion factors of the magnetic field is likely to change over the solar cycle since the large scale magnetic structure of the corona changes from a well-organized morphology with polar coronal holes and an equatorial streamer belt around solar minimum to a much more complicated structure with small coronal holes and small streamers at all solar latitudes. The expansion factors of the magnetic field can be inferred using solar magnetograms and the assumption of a potential field model, either with or without a heliospheric current sheet *Wang and Sheely* [10]. *Wang* [9] gives two expansion profiles $f(r)$ of the magnetic field for the slow solar wind during solar minimum and solar maximum. The expansion profile for solar minimum shows a very strong over-expansion around 2.5 $R_\odot$ before it decreases at larger radii and levels off at about 5. Contrarily, the solar maximum expansion profile monotonically increases to about 20 at large heliocentric distances. Using eq. 3 and Wang's [9] profiles to calculate the proton flux at 2.5 $R_\odot$, we find that during solar minimum it is lower by a factor of nine assuming a cycle-independent flux at 1 AU. A quantitative study showed that indeed the proton flux at 2.5 $R_\odot$ varies by about a factor of ten over the solar cycle between 1977 and 1994 [Y.-M. Wang, private communication, 2000] when near-Earth wind speeds of less than 400 km/s were analyzed. A parameter study by *Bürgi* [11] has shown that with decreasing proton flux close to the Sun (induced by a locally increasing over-expansion) the helium abundance in the solar wind decreases because of the decreased momentum transfer via Coulomb collisions. The change of the expansion profile, i.e., the magnetic structure, over the course of the solar cycle leads to a decreased proton flux at around 2.5 $R_\odot$ during solar minimum. We suggest that this reduction of the proton flux close to the Sun decreases the efficiency of the Coulomb drag and therefore reduces the He/H ratio in the solar wind around solar minimum.

Our measurements during solar minimum were taken close to the heliospheric current sheet. It is known that the helium abundance drops considerably upon current sheet crossings *Borrini et al.* [12]. Indeed, their observation fits the idea of strong reduction of the proton flux close to the Sun and the subsequent depletion of helium ions because of inefficient Coulomb drag. Solar wind streams close to the current sheet have very large expansion factors close to the Sun *Wang and Sheely* [13] which lead to a low helium abundance in the solar wind *Bürgi* [11]. So far, we have discussed a potential explanation for the solar cycle variation in a low speed bin. We can apply the same concept of flux tube expansion close to the Sun to the question why during solar minimum we observe a strong speed dependence of the helium abundance. *Wang and Sheeley* [13] reported an inverse correlation between the rate of magnetic flux-tube expansion and the solar wind speed at 1 AU. With increasing solar wind speed at 1 AU they associate smaller expansion factors close to the Sun and therefore a larger proton flux in corona. As discussed above, that leads to a higher He/H in the solar wind which is observed in Figure 2 around solar minimum.

We tried to verify our hypothesis also for the fast wind. In the time interval between 1977 and 1994 there was also a solar cycle dependence of the proton flux at 2.5 $R_\odot$ for near-Earth wind speeds larger than 550 km/s [Y.-M. Wang, private communication, 2000]. Following the arguments given above for the slow wind, we should therefore also expect a solar cycle dependence of the solar wind He/H for the fast wind, which we do *not* observe. Still, it can be argued, that the fast solar wind is wave-driven and that the helium ions are not accelerated by Coulomb drag but by wave-particle interaction. In that case, the fast wind cannot serve as a test for the hypothesis.

If indeed variations of the proton flux close to the Sun via Coulomb drag cause the observed helium abundance variation over the solar cycle, the same effect should be present in other elemental or even isotopic ratios. In those ratios, we expect a considerably less pronounced effect since ^{4}He has the strongest dependency on the proton flux *Bodmer and Bochsler* [14].

ACKNOWLEDGMENTS

We are grateful to the many individuals who contributed to the success of the Solar Wind Experiment (SWE) on the Wind spacecraft. Furthermore, we thank Yi-Ming Wang for valuable suggestions and discussions. This work was supported in part by NASA Grant NAG5-7359. The figures, some text, and tables from [1] are copyrighted 2001 by the AGU and are used with permission.

REFERENCES

1. Aellig, M. R, A. J. Lazarus, and J. T. Steinberg, The solar wind helium abundance: variation with wind speed and the solar cycle, *Geophys. Res. Lett.*, **28,** 2767–2770 (2001).
2. Ogilvie, K. W. and J. Hirshberg, The solar cycle variation of the solar wind helium abundance, *J. Geophys. Res.*, **79,** 4595–4602 (1974).
3. Feldman, W. C., J. R. Asbridge, S. J. Bame, and J. T. Gosling, Long-term variations of selected solar wind properties: Imp 6, 7, and 8 results, *J. Geophys. Res.* **83,** 2177–2189 (1978).
4. Ogilvie, K. W., M. A. Coplan, P. Bochsler, and J. Geiss, Solar wind observations with the Ion Composition Instrument aboard the ISEE-3/ICE spacecraft, *Sol. Phys.* **124,** 167–183 (1989).
5. Ogilvie, K. W. et al., SWE, a comprehensive plasma instrument for the Wind spacecraft, *Space Science Reviews* **71,** 55–77, 1995.
6. Neugebauer, M., Observations of solar wind helium, *Fundamentals of Cosmic Physics* **7,** 131–199 (1981).
7. Bürgi, A., Proton and alpha particle fluxes in the solar wind: results of a three-fluid model, *J. Geophys. Res.* **97,** 3137–3150 (1992).
8. Geiss, J., P. Hirt, and H. Leutwyler, On acceleration and motion of ions in corona and solar wind, *Sol. Phys.* **12,** 458–483 (1970).
9. Wang, Y.-M., Two types of slow solar wind, *Ap. J. Lett.* **437,** L67–L70 (1994).
10. Wang, Y.-M. and N. R. Sheeley Jr., The solar origin of long-term variations of the interplanetary magnetic field strength, *J. Geophys. Res.* **93,** 11227–11236 (1988).
11. Bürgi, A., Dynamics of alpha particles in coronal streamer type geometries, in *Solar Wind Seven*, edited by E. Marsch and R. Schwenn, pp. 333–336, Pergamon Press, Tarrytown, N. Y., (1992).
12. Borrini, G., J. T. Gosling, S. J. Bame, W. C. Feldman, and J. M. Wilcox, Solar wind helium and hydrogen structure near the heliospheric current sheet: a signal of coronal streamers at 1 AU, *J. Geophys. Res.* **86,** 4565–4573 (1981).
13. Wang, Y.-M. and N. R. Sheeley Jr., Solar wind speed and coronal flux-tube expansion, *Ap. J.* **355,** 726–732 (1990).
14. Bodmer, R. and P. Bochsler, Influence of Coulomb collisions on isotopic and elemental fractionation in the solar wind acceleration process, *J. Geophys. Res.* **105,** 47–60, 2000.

The Relative Abundance of Chromium and Iron in the Solar Wind

J.A. Paquette*, F.M. Ipavich*, S.E. Lasley*, P. Bochsler† and P. Wurz†

*Department of Physics, University of Maryland, College Park, MD, USA 20742
†Physikalisches Institut der Universität Bern, Switzerland, CH3012

Abstract. Chromium and iron are two heavy elements in the solar wind with similar masses. The MTOF (Mass Time Of Flight) sensor of the CELIAS investigation on the SOHO spacecraft easily allows these two elements to be resolved from one another. Taking the ratio of the densities of these two elements - as opposed to considering their absolute abundances - minimizes the effects of uncertainties in instrument efficiency. Measurements of the abundance ratio are presented here. The First Ionization Potential (FIP) of chromium is 6.76 eV, while the FIP of iron is 7.87 eV. Since Cr and Fe have similar FIPs the ratio of their abundances should not be biased by the FIP effect which is well known in different solar wind flows. Therefore the Cr/Fe ratio from the MTOF data should give a good measure of the photospheric abundance ratio. We also compare the ratio measured in this work to the meteoritic value.

INTRODUCTION

Chromium and iron are two heavy elements in the solar wind with similar properties. In this paper we present a preliminary study of the chromium to iron abundance ratio in the solar wind using data from the MTOF (Mass Time of Flight) sensor on SOHO. This is the first measurement of chromium in the solar wind. The abundance ratio for three short time periods (corresponding to three different solar wind flow types) and three long ($\approx$ 1 year duration) time periods are derived from the data. The abundance ratio of these two elements can be measured with sufficient precision using MTOF to allow comparison to photospheric and meteoritic values.

The three short periods were chosen for their relatively stable bulk speeds, which makes measurment simpler, as the instrument efficiency is a function of speed. No variation with solar wind flow type in the chromium to iron abundance ratio was expected *a priori*, but the choice of three different flow types allowed the possibility to be investigated. The three longer time periods were not explicitly differentiated by flow type. They consisted of all data during approximately year-long periods in which the bulk speed was in a narrow range. The study of these longer periods was intended to test the supposition that the abundance ratio measured for the shorter periods was not unusual or atypical.

INSTRUMENTATION

The MTOF sensor, which is a part of the CELIAS (Charge Element and Isotope Analysis System) Investigation on SOHO, is shown in Figure 1. It consists of an energy per charge filter (the WAVE) and a high mass-resolution ($M/\Delta M > 100$) spectrometer (the VMass). A solar wind ion with the requisite E/q to negotiate the WAVE encounters the carbon foil. After passing through the foil, a large fraction (the exact value is a function of both energy and species, but typically from 70% to more than 99%) of ions will either have been neutralized, or left with charge state +1. Neutrals will travel in a straight line to the neutral microchannel plate (MCP) detector. Positive ions will be bent towards the ion MCP by the electric field inside the VMass. In the harmonic electric potential inside the VMass, the time of flight (TOF) of the ions is proportional to the square root of M/q^* where q^* is the charge state after the foil. So in the case in which $q^* = 1$, the TOF is proportional to $\sqrt{M}$. MTOF and CELIAS are discussed in greater detail in [1].

In addition, the PM (Proton Monitor) - which is a subsensor of MTOF - provides solar wind parameters on a near real time basis. It was used to aid in the analysis of the MTOF data. The proton monitor consists of a three-box electrostatic deflection system followed by a position sensing Microchannel Plate. The proton monitor is described in detail in [2].

Instrumental fractionation in MTOF can result from several sources. Ions whose E/q values are close the up-

CP598, *Solar and Galactic Composition,* edited by R. F. Wimmer-Schweingruber

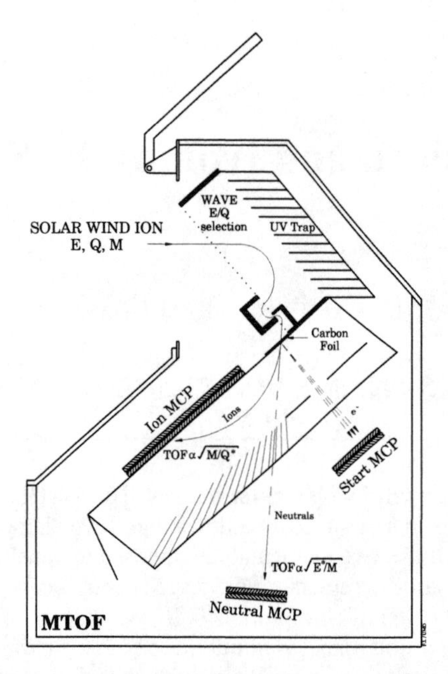

FIGURE 1. Schematic of the MTOF (Mass Time of Flight) Instrument. Solar wind ions enter through an energy/charge filter (the WAVE) and pass through a carbon foil, leaving them with a charge state q^*. They then enter a harmonic potential region in which those with $q^* > 0$ are electrostatically deflected down to a microchannel plate detector. For an ion of mass M, the time of flight in this region is $\propto \sqrt{M/q^*}$.

per or lower limits of the WAVE's passband are less efficiently transmitted than other ions, but this effect is well-understood and can be corrected for. Another fractionation effect is introduced by the carbon foil. The portion of ions which exit the foil with charge state +1 is a function of both energy and of species. The charge state distribution after exiting a carbon foil has been measured for many species [3], [4], [5]. While the distribution of charge states after a carbon foil has been measured for iron, it has not been done for chromium, and this is therefore a source of uncertainty. Finally, the efficiency of ion transmission through the VMass region is a function of energy, since sufficiently energetic ions cannot be turned back by the electric field and will strike the hyperbola. Ions of different species in the solar wind (all having roughly the same speed) will of course have different energies and hence will be fractionated in the VMass. Calibration of MTOF has provided a good understanding of the details of VMass transmission, although some uncertainty remains.

CHROMIUM AND IRON

Chromium and iron are close to each other in mass (the most common isotopes being 52 and 56, respectively). It may be surmised that these two metals have similar charge state distributions in the solar wind. Hence Cr ions and Fe ions are likely to have similar E/q distributions at any given time, and so fractionation by the WAVE will not be important. The small difference in the masses of the two elements likewise implies similar values for VMass transmission. While the charge state distributions after encountering a carbon foil were measured for iron, they have not been measured for chromium. This is a source of uncertainty, but it is unlikely that their charge states after the carbon foil will be very different from each other. Therefore, it is likely that instrument fractionation will be a small effect. It is also likely that any uncertainty in the ratio of the instrument efficiencies for the two elements will be less than the uncertainty in either element's absolute efficiency.

Another consideration in selecting chromium and iron for study is that the two elements have have similar First Ionization Potentials (FIPs). Chromium's FIP is 6.76 eV, while the FIP of iron is 7.87 eV. Since both elements have low FIP, the well-known FIP effect (see, for example, [6]) will not introduce any fractionation, and the abundance ratio measured in the solar wind should be directly comparable to the photospheric abundance ratio. Because both metals have low volatility, the ratio in the solar wind should also be directly comparable to the meteoritic ratio.

ANALYSIS TECHNIQUE

The TOF spectra from MTOF are fit using a maximum likelihood algorithm. The model function has 24 parameters, but fourteen of these are the heights of the peaks of the various species with masses between 48 and 64. Of the remaining 10, two describe a linear background, two are needed to convert from time of flight channel to mass, two are associated with small subsidiary peaks caused by electronic "ringing" in the instrument, and the the remaining 4 describe the shape and width of the peaks. The peak shape (which is identical for all 14 peaks) is a modified asymmetric Lorenztian. Although this work is only concerned with masses in the range 52 to 57, a wide range of time of flight channels was chosen to allow an adequate fit of the background and other parameters not related to the heights of the mass peaks.

Figure 2 shows the MTOF data for the relevant range, together with the model function for the same interval. The various peaks are labeled with the corresponding masses. The chromium peak is at mass 52, and the iron creates peaks at 54, and 56, with a shoulder at 57. The peak at 48 is titanium, and the peaks at 58, 60, and 62 are primarily nickel.

Two of the parameters provided by the fitting program

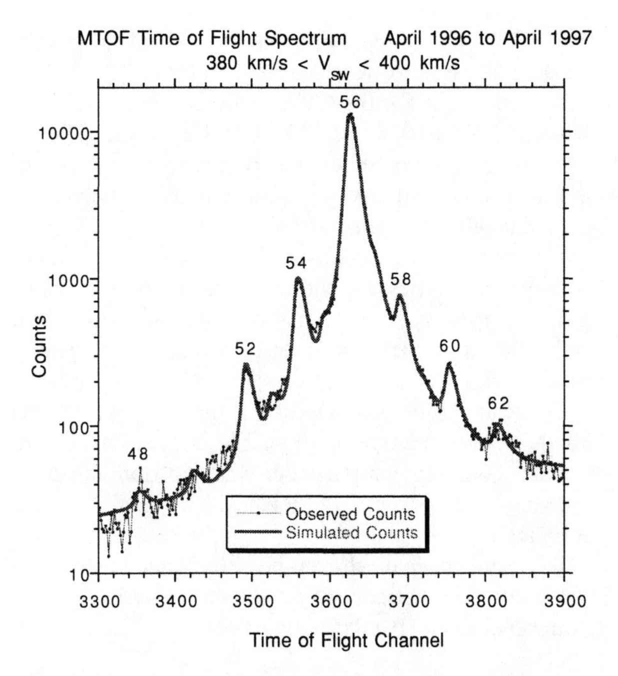

FIGURE 2. A portion of a TOF spectrum from MTOF. The points connected by a thin line are the observed counts as a function of TOF channel, and the thick line is the model fit to the data. The various peaks are labeled with the masses to which they correspond. The peak at mass 48 is titanium, the 52 peak is chromium, the 54 and 56 peaks are iron, and the 58, 60, and 62 peaks are primarily nickel.

correspond to the peak heights of the Cr^{52} and the Fe^{56} peaks. The ratio of the peak values for the 52 peak to the 56 peak provided by the fitting program is corrected to account for the small (typically a few percent) fractionation effect in the instrument, and for the isotopic fractions of each species to give the elemental abundance ratio.

THE CR TO FE RATIO IN DIFFERENT FLOW TYPES

Three time periods were selected, each corresponding to a different solar wind flow type. For ease of analysis, time periods without large variations in bulk speed were chosen.

An interstream time period is shown in Figure 3. The shaded region on the plot is the time period selected for analysis. This period has a nearly constant velocity.

A coronal hole time period is shown in Figure 4. The speed of around 450-500 km/s is low for a coronal hole associated wind, but coronal hole flows with speeds in this range have previously been observed (see, e.g. [7]). The time period has the morphology of a coronal hole, with an initial density increase, followed by a subsequent decrease and a simultaneous increase in both bulk speed

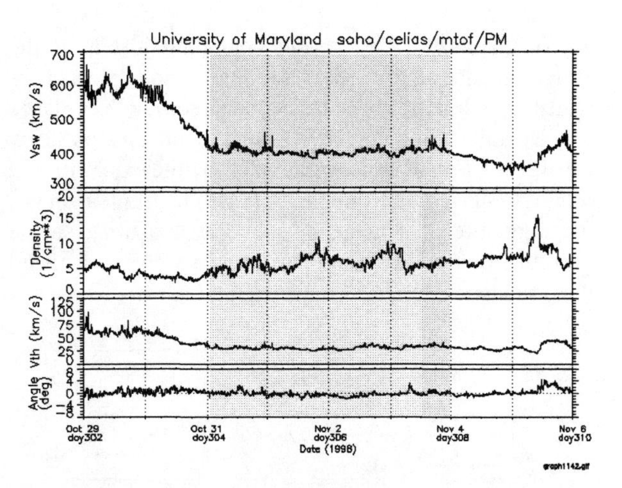

FIGURE 3. An interstream time period. The solar wind parameters plotted here are from the CELIAS/MTOF/PM (Proton Monitor) subsensor on SOHO. The top panel is the bulk speed of the solar wind, the second panel is the density, the third is the thermal speed ($\sqrt{2kT/m}$), and the bottom panel is the out-of-ecliptic flow angle. The shaded region is the fraction of the time period shown for which the MTOF data was used to derive the chromium to iron abundance ratio.

and thermal speed. An examination of the soft X-ray data from the Yohkoh spacecraft for the few days preceding this time shows a coronal hole extending to relatively low latitude that could have been the source of this flow [8]. In addition, the coronal hole maps produced by the NSO (National Solar Observatory) at Kitt Peak also show an equatorward extension of the south polar coronal hole during that time. Again, the shaded region was the time period that was analyzed.

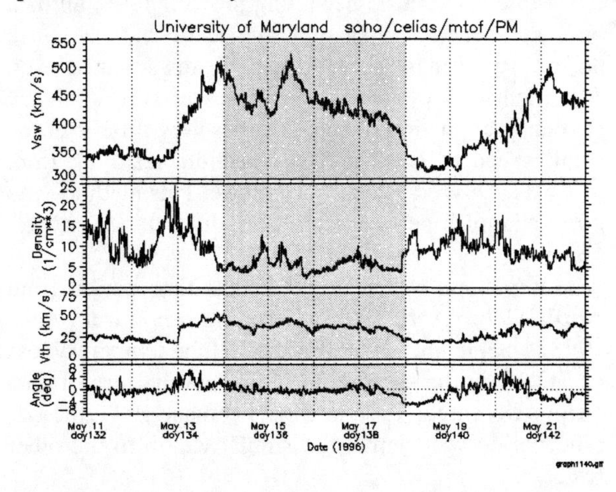

FIGURE 4. PM solar wind parameters for a time period with coronal hole associated flow.

A probable CME associated time period is shown in Figure 5. LASCO saw a halo CME on May 12 1997 at about 04:35 UT [9], and the shock driven by the CME arrived at SOHO slightly before 01:00 UT on May

15. The nearly 3-day period beginning just after the shock (shaded in gray) was the time period selected for analysis. While this time period bears some resemblance to fast solar wind, the NSO coronal hole maps show no likely coronal hole within 50° of the equator, and magnetometer data from the ACE MFI instrument shows the signature of a magnetic cloud at this time, so the identification of this time period as a CME associated one is reasonable.

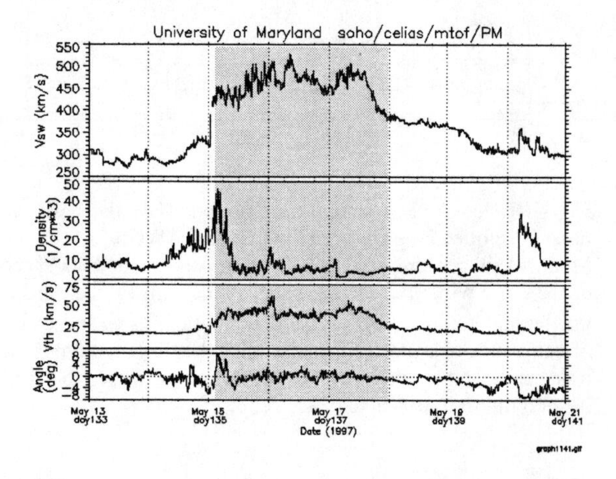

FIGURE 5. PM solar wind parameters for a time period with CME associated solar wind. LASCO saw a halo CME about 66 hours prior to the arrival of the shock near the beginning of May 12.

ONE YEAR PERIODS

For a one year period beginning in April 1996, all data with the bulk speed (as determined by the PM) in a narrow speed range (380-400 km/s) was summed over. This method gives very good statistics, and the efficiency (which is a function of speed) varies very little over the small speed range. The TOF spectrum resulting from this was analyzed as described above. This method does not distinguish between data from different flow types, however.

This process was repeated for the time period from April 1997 to April 1998, and for the period from April 1998 to September 1999. Because of the temporary loss of SOHO in the summer of 1998, and the loss of data during December 1998 to early February 1999, the last time period was actually of similar length to the other two.

RESULTS

For the the interstream time period, a value of the chromium to iron abundance ratio of $1.65 \times 10^{-2} \pm 0.30 \times 10^{-2}$ was found. For the coronal hole time period, the abundance ratio was measured to be $1.74 \times 10^{-2} \pm 0.31 \times 10^{-2}$, and for the CME-associated time period, a value of $1.59 \times 10^{-2} \pm 0.29 \times 10^{-2}$ was found. The errors estimates are based on uncertainty in instrument response, primarily due to uncertainty in the +1 yields for chromium after the carbon foil.

The results for the comparison of the three events of different flow types are shown in Figure 6. As can be seen, the values are all consistent, within errors, with each other and with both meteoritic and photospheric values.

The result of the average of the three 1-year periods was $1.73 \times 10^{-2} \pm 0.30 \times 10^{-2}$. The error estimate here includes counting statistics, but was still dominated by uncertainty in instrument efficiency. This is also shown in Figure 6.

The results from the three time periods and the average of the one year periods are all shown in Table 1, along with values from [10] for comparison.

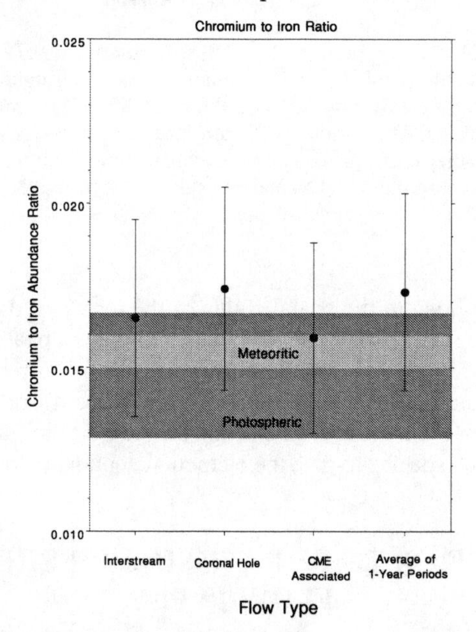

FIGURE 6. Comparison of the chromium to iron abundance ratio in different solar wind flow types and for the average of three one-year periods to the photospheric and meteoritic values. In this figure, the lighter shaded region represents the 1-σ range of the meteoritic abundance ratio reported by [10]. The darker shaded region is the analogous photospheric abundance ratio also from [10]. Note that the wider error range of the photospheric value completely encompasses the meteoritic value with its error range. The points shown on the plot are for the average of three periods of aproximately year-long periods and for three much briefer events; one an interstream flow, another a coronal hole flow, and the last a CME-associated flow.

TABLE 1. Chromium to Iron Abundance Ratio

Cr/Fe Abundance Ratio	
Photospheric (Grevesse and Sauval, 1998)	$1.48 \times 10^{-2} \pm 0.19 \times 10^{-2}$
Meteoritic (Grevesse and Sauval, 1998)	$1.55 \times 10^{-2} \pm 0.051 \times 10^{-2}$
This Work (Interstream)	$1.65 \times 10^{-2} \pm 0.30 \times 10^{-2}$
This Work (Coronal Hole)	$1.74 \times 10^{-2} \pm 0.31 \times 10^{-2}$
This Work (CME associated)	$1.59 \times 10^{-2} \pm 0.29 \times 10^{-2}$
This Work (1 year beginning Apr 1996)	1.77×10^{-2}
This Work (1 year beginning Apr 1997)	1.75×10^{-2}
This Work (1 year beginning Apr 1998)	1.67×10^{-2}
This Work (Average of three 1-year periods)	$1.73 \times 10^{-2} \pm 0.30 \times 10^{-2}$

CONCLUSIONS

As can be seen from Table 1, all of the abundance ratios are consistent (within errors) with the photospheric and meteoritic values. It is also true that all of the values are slightly higher than both the meteoritic and the photospheric values, but without increased accuracy in the measurement of the ratio, little can be concluded from this. It is possible that a small systematic error was introduced in the assumed chromium +1 yields.

Since both chromium and iron are low FIP elements, the agreement of the values from various flow types with each other is to be expected. The fact that they also agree with photospheric values is consistent with the "plateau" of constant fractionation that is often shown (e.g in [6]) for low FIP elements.

Further study of this question might include consideration of more events of each flow type. Since most of the uncertainty in the measurement of the ratio is due to uncertainty in the chromium efficiency, precision could be improved either by measurements of the post-carbon-foil charge state distributions for chromium, or by calibration of the the MTOF spare instrument to better determine the overall chromium efficiency.

ACKNOWLEDGMENTS

This research was supported by NASA grants NAG5-7678 and NAG5-9282. We would also like to thank the many individuals at the University of Maryland and at other CELIAS institutions who contributed to the success of MTOF.

NSO/Kitt Peak data used here are produced cooperatively by NSF/NOAO, NASA/GSFC, and NOAA/SEL.

REFERENCES

1. Hovestadt, D., Hilchenbach, M., Bürgi, A., Klecker, B., Laeverenz, P., Scholer, M., Grünwaldt, H., Axford, W. I., Livi, S., Marsch, E., Wilken, B., Winterhoff, H. P., Ipavich, F. M., Bedini, P., Coplan, M. A., Galvin, A. B., Gloeckler, G., Bochsler, P., Balsiger, H., Fischer, J., Geiss, J., Kallenbach, R., Wurz, P., Reiche, K.-U., Gliem, F., Judge, D. L., Ogawa, H. S., Hsieh, K. C., Möbius, E., Lee, M. A., Managadze, G. G., Verigin, M. I., and Neugebauer, M., *The SOHO Mission*, Kluwer Academic Publishers, Dordrecht, 1995, pp. 441–481.
2. Ipavich, F. M., Galvin, A. B., Lasley, S. E., Paquette, J. A., Hefti, S., Reiche, K.-U., Coplan, M. A., Gloeckler, G., Bochsler, P., Hovestadt, D., Grünwaldt, H., Hilchenbach, M., Gliem, F., Axford, W. I., Balsiger, A., Bürgi, A., Geiss, J., Hsieh, K. C., Kallenbach, R., Klecker, B., Lee, M. A., Managadze, G. G., Marsch, E., Möbius, E., Neugebauer, M., Scholer, M., Verigin, M. I., Wilken, B., and Wurz, P., *Journal of Geophysical Reasearch*, **103**, 17205–17213 (1998).
3. Oetliker, M., *Charge state distribution, scattering, and residual energy of ions passing through thin carbon foils; Basic data for the MTOF mass spectrometer on the space mission SOHO*, Master's thesis, University of Bern, Switzerland CH3012 (1989).
4. Gonin, M., *Interaction of low energy ions with thin carbon foils; charge exchange, energy loss, and angular scattering*, Master's thesis, University of Bern, Switzerland CH3012 (1991).
5. Gonin, M., *Ein semiempirisches Modell des Ladungsaustauches non niederenergetischen Ionen beim Durchang durch dünne Folien, zur Eichung von isochronen Flugzeit-Massenspektrometern*, Ph.D. thesis, University of Bern, Switzerland CH3012 (1995).
6. Geiss, J., *Space Science reviews*, **85**, 241–252 (1998).
7. Bohlin, J. D., *Coronal Holes and High Speed Streams*, Colorado Associated University Press, Boulder, Colorado, 1977, pp. 27–69.
8. Solar geophysical data prompt reports, Tech. Rep. 623-Part I, National Geophysical Data Center, Boulder CO 80303 (1996).
9. Plunkett, S. P., Thompson, B. J., Howard, R. A., Michels, D. J., Cyr, O. C. S., Tappin, S. J., Schwenn, R., and Lamy, P. L., *Geophysical Research Letters*, **25**, 2477–2480 (1998).
10. Grevesse, N., and Sauval, A. J., *Space Science Reviews*, **85**, 161–174 (1998).

Determination of the Ar/Ca solar wind elemental abundance ratio using SOHO/CELIAS/MTOF

J. M. Weygand*, F. M. Ipavich†, P. Wurz**, J.A. Paquette† and P. Bochsler**

*IGPP/UCLA 3845 Slichter Hall, Los Angeles, CA, 90095-1567, USA
†University of Maryland, College Park, MD 20742, USA
**Physikalisches Institute, University of Bern, Sidlerstrasse 5, CH-3012, Bern, Switzerland

Abstract.
This study examines the first direct measurements of the solar wind (SW) Ar/Ca elemental abundance ratio with the Mass Time-Of-Flight (MTOF) sensor of the Charge, Element, and Isotope Analysis System (CELIAS) instrument on the Solar and Heliospheric Observatory (SOHO) spacecraft. Two mass spectra are compiled for interstream (IS) associated SW and coronal hole (CH) associated SW. A detailed analysis of over 3.6 days of non-consecutive IS associated SW speed (395 $\pm$25 km/s) places the Ar/Ca elemental ratio at 0.38$\pm$0.05. A similar analysis for CH associated SW speed (525 $\pm$ 25 km/s), with about 5.6 non-consecutive days of CH associated SW speed data, derived an Ar/Ca ratio of 0.59$\pm$0.07. The results of this study are consistent with most previously published Ar/Ca values from gradual solar energetic particle events and spectroscopy studies.

INTRODUCTION

It has been observed in the solar wind (SW) that elements with a low first ionization potential (FIP <10 eV) are enriched relative to elements with a high FIP (i.e., FIP >10 eV) as compared to elements in the photosphere using oxygen as a reference ($[X/O]_{SW}/[X/O]_{photosphere}$. The enrichment in the interstream (IS) associated SW is about two to five for the low FIP elements and around one for the high FIP elements. It is generally believed that the relative enrichments are due to a fractionation process in the upper chromosphere and lower transition zone and are the result of neutral-ion separation with ionization most likely caused by EUV radiation from the solar corona.

The high FIP element argon (FIP = 15.76 eV) is highly volatile and of low abundance in the SW. Because of argon's volatility, inferences about the solar and solar system argon abundance from planetary samples are problematic. Calcium is a low FIP (FIP = 6.11eV) species and a refractory element, which is moderately abundant in the SW compared to argon. The present measurements are selected with the goal of obtaining accurate information on the present day solar atmosphere elemental composition. Furthermore, the Sun represents 99.9% of the solar system's matter, and since the Sun's central temperature has never been high enough to alter the elemental composition of the heavy elements by nuclear reactions, an accurate elemental composition measurement should reflect the pre-solar nebula elemental composition. Until recently, direct measurements of the IS and coronal hole (CH) associated SW argon and calcium composition have not been possible. Previously reported Ar/Ca elemental abundance ratio values were determined through 1) gradual SEP event measurements, 2) the corona and photosphere emission spectra in a variety of wavelength regimes, 3) the Apollo missions foil experiments, and 4) through meteoritic samples [1, 2, 3, 4]. While the above methods are accurate for the measurements made, those methods do not directly sample the IS and CH associated SW nor do these measurements tie the argon abundance to a refractory element. Direct measurements may provide limits to the variability of elemental abundance ratios within different SW regimes, and hence, obtain a clue on the importance of fractionation effects occurring in the SW. As of the late 1990's, several different composition instruments, such as WIND/SMS, SOHO/CELIAS, ACE/SWIMS, and ACE/SWICS [5, 6, 7], made it possible to determine the solar elemental composition quite reliably from direct SW observations [8, 9, 10]. Recent argon isotopic ratio measurements have demonstrated that both ^{36}Ar and ^{38}Ar can be successfully detected in the SW [11]. Calcium solar elemental composition measurements, on the other hand, have been extensively reported, but only recently have measurements come directly from the SW [8, 9].

CP598, *Solar and Galactic Composition,* edited by R. F. Wimmer-Schweingruber

Recent statistical measurements of the Fe/O elemental abundance ratio in the SW over all SW speeds have shown three compositional regions [10, 9]. The first of these regions is a slow speed SW for speeds less than 400 km/s, where the elemental abundance ratio appears to plateau. The second region is the highest speed bin for speeds above 500 km/s, where the elemental abundance ratio again looks constant. These two regimes are IS associated SW and CH associated SW, respectively. For the speeds in between (i.e., 400 km/s to 500 km/s), the elemental abundance ratio seems to monotonically decrease from slow speeds to high speeds. Furthermore, it was demonstrated that there is approximately a factor of two difference in the Fe/O ratio between the IS associated SW (i.e., < 400 km/s) and the CH associated SW (i.e., > 500 km/s) [10, 9].

This study determines the Ar/Ca ratio with the Mass Time-Of-Flight (MTOF) sensor of the Charge, Element, and Isotope Analysis System (CELIAS) instrument on the Solar and Heliospheric Observatory (SOHO) spacecraft for both the IS associated SW and the CH associated SW. For details on the MTOF sensor and CELIAS instrument see the work of Hovestadt et al. [6].

PROCEDURE

The most important criterion for the data selection is a high signal-to-noise ratio (SNR) for the argon. Unlike previous CELIAS/MTOF investigations, where the SNR is anywhere from about four to a few orders of magnitude, the SNR for ^{36}Ar is around 1.1 and about 1.8 for ^{40}Ca. To further complicate matters, additional constraints had to be imposed on the data set. The first of these constraints concerns data binning into either IS or CH associated SW. The IS associated SW sample contains SW speeds between 370 km/s and 420 km/s, and the CH associated SW bin lies between 500 km/s and 550 km/s. The two SW ranges precise location is derived from instrumentation requirements and from previous composition studies [10, 9, 12]. The two SW bins are selected such that they mostly lie in the two compositionally separate SW regimes previously observed [10, 9], which, in this study, we refer to as IS associated SW (SW < 400 km/s) and CH associated SW (SW > 500 km/s). While it is true that the IS associated SW speed bin is not entirely below 400 km/s, some leeway is allowed to improve the argon peak's SNR. Finally, the SW bins are also selected to be similar to those of the Weygand et al. [11] study so we may unambiguously determine the total argon counts in the ^{38}Ar peaks. In addition to restricting the speed, the SW's duration is required to be a minimum of 12 hours for the IS associated bin and a minimum of 6 hours for the CH associated SW bin. This minimum duration criterion limits the data set to large-scale and uniform flows as well as excludes variable speed flows, which would possibly mix results from the two SW regimes. Also not included in the mass spectra are data associated with CME's, shocks, and corotating interaction regions, since the passage of a CME can have markedly different elemental composition [13, 14, 15].

For the second constraint, the only instrument settings considered are those which allowed argon and calcium to pass through the MTOF sensor in a very similar manner. The two bins need to be in a region of high transmission for the time of flight (TOF) portion of the MTOF sensor as well as in the region of maximum effective carbon foil area as determined by the wide-angle, variable energy/charge (WAVE) entrance system. While these constraints restrict the total amount of data, they guarantee that Ar and Ca are treated nearly identically by the instrument.

SW argon and calcium enter the MTOF sensor through the WAVE system. The WAVE section consists of three energy over charge selection chambers. The entrance system allows a large energy range of ions to enter the sensor through a wide angular acceptance cone. Particles, which pass through this region, cross through a carbon foil, which reduces their charge from some initial SW charge distribution (i.e., +8, +9, and +10 for argon and +9, +10, +11, and +12 for calcium) to neutral, +1, or +2, where the most probable ion charge is +1 [16]. Argon and calcium essentially pass through the WAVE system in the same way with one assumption. If the argon and calcium ion's speed is about the same, then their nearly identical passage is determined by their M/Q SW value, which is four for both ions when we use their most probable charge state in a SW (i.e., +9 for argon and +10 for calcium) of a freeze-in temperature of 1.4 MK. The results vary at maximum by about 5% when other less probable charge states are examined. This 5% will be used as our uncertainty in the WAVE portion of the MTOF sensor. After the carbon foil, the ions follow a hyperbolic path through the TOF portion of MTOF to the ion MCP detector where they are recorded as counts.

Figure 1 is a plot of the argon and calcium total transmission through the TOF section of MTOF, where the transmission is calculated with a forward model [12]. The transmission curve has been compared with ground calibration data and there is a good agreement between the two. The dashed curve is the transmission for calcium over a range of SW values and the solid curve is the transmission for argon. The first pair of vertical lines indicates the IS associated SW speed range and the second set of vertical lines indicates the transmission for the CH associated SW speed range. Clearly, the total transmission for calcium is much higher than the total transmission for argon in the two SW speed ranges. This dif-

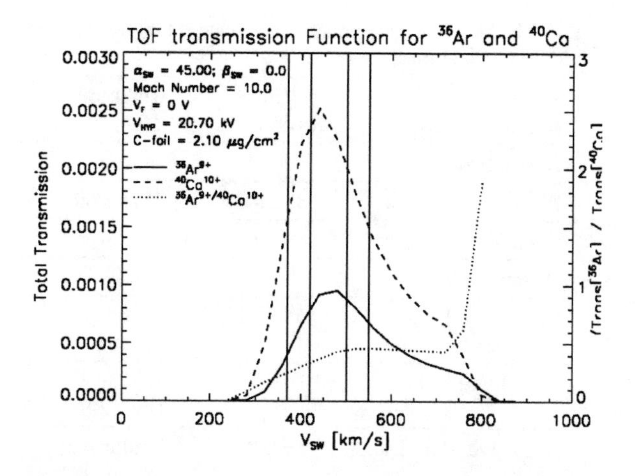

FIGURE 1. The isochronous TOF mass spectrometer total transmission is plotted for argon and calcium investigated in this study. The approximate MTOF sensor settings and the general SW values are located in the upper left-hand corner. The total transmission of argon is indicated with the solid line and calcium is represented with the dashed line. The first pair of vertical lines indicates the IS associated SW bin while the second set of vertical lines indicates the CH associated SW bin. The dotted curve gives the transmission ratio of ^{36}Ar to ^{40}Ca.

ference is the result of the larger yield of positive ions after the carbon foil from calcium than argon. The dotted curve indicates the transmission ratio of ^{36}Ar to ^{40}Ca. This curve graphically shows how calcium is more efficiently detected than argon. This difference in transmissions is accounted for by taking the mean transmission ratio of calcium to argon in both speed ranges, which is 3.45 for the IS associated SW bin and 2.22 for the CH associated SW range.

Table 1 lists the instrument settings required for the selection of data for the two different SW speed ranges. Again, these settings are selected to maximize the SNR for the argon peaks in the Weygand et al. [11] study, to examine the two compositionally different SW regimes, and to maximize the particle's transmission through the instrument.

Data from 1996 to the first half of 1999 are examined and the selected mass spectra are summed together to make one total mass spectrum for each SW speed range. For the IS associated SW the summed spectra amounts to 3.6 days worth of data. For the CH associated SW the summed spectra is for 5.6 days worth of data. The total number of days appears to be low because the entrance system settings are not always ideal and a minimum of 12 hours worth of IS associated solar wind (or a minimum of 6 hours of CH associated SW) are required to guarantee a pure sample. Figure 2 shows the compiled mass spectrum for the IS associated speed SW and Figure 3 displays the mass spectrum for the CH associated SW, respectively. On the figure's top half within the spectra seven mass

TABLE 1. MTOF sensor instrumental settings for different SW bins.

Portion of MTOF Sensor	Interstream Associated SW 370 to 420 km/s	CH Associated SW 500 to 550 km/s
Acceleration Voltage (V_f)	0.0	0.0
Hyperbola Voltage	20.4 kV	20.4 kV
Entrance System Voltage	4.9 to 7.9 kV	8.0 to 12.4 kV
Min SW Speed Duration	12 hr	6 hr
N/S SW Angle	±8°	±8°
Mach Number	≥10.0	≥0.0

peaks are visible, which are ^{34}S, ^{35}Cl, ^{36}Ar, ^{37}Cl, ^{38}Ar, ^{39}K, and ^{40}Ca. Along the x-axis is the TOF channel number, along the y-axis is the total number of counts per channel, the thin black line indicates the raw data, and the thick curve is a 16 parameter Maximum Likelihood fit (MLF) to the spectrum. The MLF is explained in detail in the study of Bochsler et al. [17]. The difference in this study is we have only seven mass peaks and we fit a quadratic background. The bottom part of each figure is a plot of the relative difference of the raw spectrum from the fit spectrum ((Raw − Fit)/Raw). It should be clear from the lower plot how good the fit is and relative difference variations above and below zero indicates that we are not consistently fitting above or below the raw data. The residuals are basically random, thus no systematic error in the fit and the fit function selection has been made. As a MLF consistency check, the Ar/Ca fit ratio was compared with the Ar/Ca raw count ratio and the ratios are consistent within the uncertainty.

RESULTS AND DISCUSSION

Displayed in Table 2 are the argon to calcium ratios derived from the IS and CH associated SW spectra along with the results of other studies. An examination of the previously published Ar/Ca ratios shows the scatter of the Ar/Ca values as well as the large uncertainty of these determinations. The ratios for this study represent the ratio of the area under the argon peaks and the calcium peaks. While only the area under the ^{36}Ar and ^{40}Ca peaks is determined, the various isotopes of calcium and argon are accounted for by using the isotopic abundance ratios. In addition to the ^{36}Ar peak there is an ^{38}Ar peak, which is difficult to see in Figure 2. ^{40}Ar also exists within the SW, but its overall contribution is negligible. An argon isotopic ratio (^{36}Ar/^{38}Ar) for the IS associated SW of 5.6 is used to determine the area under the ^{38}Ar peak and an argon isotopic ratio of 5.5 is used for the CH associ-

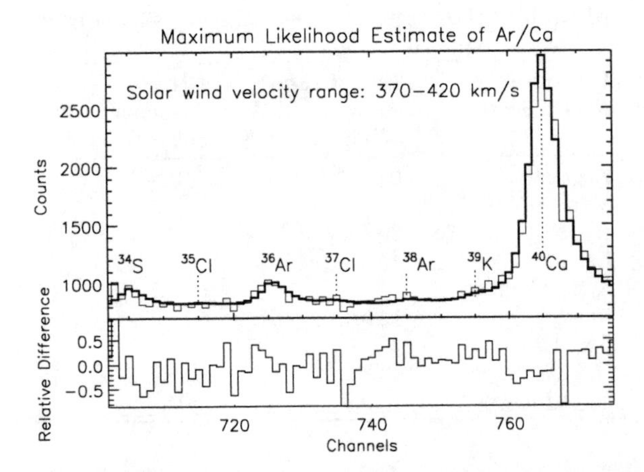

FIGURE 2. The plot's top half is the sum of the five minute mass spectra between 1996 and 1999 for the IS associated SW range of 370 km/s to 420 km/s. The thin black line represents the raw data and the thick black line is the MLF of the argon peaks. Along the Y-axis is the counts, along the X-axis are the TOF channel, and each isotope is labelled above each peak. The dotted line indicates the peaks position. The plot's bottom half is the relative difference between the fit and raw counts.

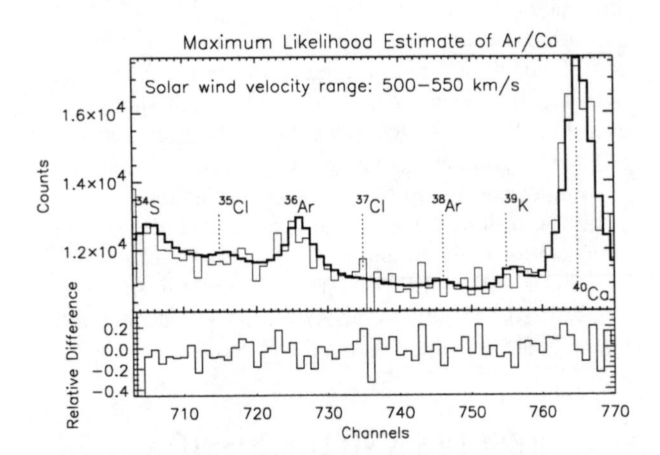

FIGURE 3. Same as Figure 2, but for SW speeds from 500 km/s to 550 km/s.

TABLE 2. Ar/Ca elemental abundance ratios for this work and previous studies.

Measured Regime	Ar/Ca Ratio	Reference
In situ SW		
IS SW	0.38±0.05	This Work
CH SW	0.59±0.07	This Work
Optical (Flare)		
EUV Flare Data	0.55±0.21	Widing & Spicer [19]
EUV Flare Data	1.10±0.25	Widing & Hiei [20]
Impulsive Flare Data	1.5±0.8	Feldman & Widing [21]
Flare Data	$0.79\ ^{+0.54}_{-0.45}$	Veck & Parkinson [22]
Optical (Other)		
Photosphere, Coronal SEP data Spectroscopy	1.10±0.14	Grevesse & Sauval [23]
CI Condrites (Ca) Photosphere Spectroscopy (Ar)	1.65±0.15 1.58±0.41	Anders & Grevesse [3]
Coronal Spectra	0.85±0.20	Young et al. [24]
Energetic particles		
SEP Derived Corona	$0.30\ ^{+0.74}_{-0.63}$	Breneman & Stone [2]
SEP Derived Corona	0.31±0.02	Reames [4]
Mass Unbiased Baseline SEP	0.71±0.70	Meyer, [25]

ated SW bin [11]. Like argon, calcium has more than one peak, ^{42}Ca and ^{44}Ca, which are not shown in either Figure 2 or 3. Calcium isotopic ratios are used to calculate the area under those calcium peaks [18]. An additional correction is made to account for the difference in calcium and argon transmission through the TOF portion of the instrument as discussed in the procedure section (see also Figure 1).

At this time, it is appropriate to explain the method of determination for the Ar/Ca ratio absolute uncertainty. The IS associated SW background for both elements is estimated to have a relative uncertainty of 3.6% ($\sqrt{n}/n$, where n is the background counts). For the CH associated SW the background is 0.9% for argon and 1.0% for the calcium. The instrumental relative uncertainty for the entrance system of MTOF, according to carbon foil area variation for different elemental charges, is approximately 5.0% for the IS associated SW and about 3.8% for the CH associated SW. The instrumental relative uncertainty for the TOF portion of MTOF is 7% in the elemental discrimination due to the release of start electrons, 5% for the uncertainty in the elemental discrimination due to charge state conversion in the carbon foil, and 7% in the elemental discrimination due to the stop detection efficiency. The area under the IS associated SW peaks has a relative uncertainty of approximately 3.2% for the ^{36}Ar peak and 2.6% for the ^{40}Ca peak. For the area under the coronal hole associated SW spectrum peak we find about 1.2% and 0.6% relative uncertainty for the ^{36}Ar and ^{40}Ca peak, respectively. A final uncertainty of 12% and 10% is included for the ^{36}Ar/^{38}Ar isotopic ratio in the IS and

CH associated speed SW regimes, respectively. The sum of the squares of the uncertainties results in a total relative uncertainty of about 13.3% for the IS associated SW value and about 11.8% for the CH associated SW value. Unfortunately, these uncertainties are not expected to improve much with the addition of more MTOF data. Only a moderate improvement could be made by further restricting the sensor settings and with the accumulation of a much larger number of spectra.

The IS associated SW Ar/Ca ratio of this work 0.38 ± 0.05 is consistent with the results of the Breneman and Stone [2], Meyer [25], and Reames [4] studies. These results, which are derived from SEP measurements, are believed to reflect the mean composition for the IS associated SW and the coronal composition. The IS associated SW value of this work, as well as the other SEP results, are not consistent with the work of Young et al. [24].

The results of this work for the CH associated SW Ar/Ca ratio 0.59 ± 0.07 is consistent with the Widing and Spicer [19] work's results as well as the Veck and Parkinson [22] studies. These previously published studies are believed to reflect the mean composition for the CH associated SW. We should also point out that the CH associated SW value of this study falls within the uncertainty of the Meyer [25] and the Young et al. [24] works, which are believed to reflect IS associated SW results. The purpose of this comment is to emphasize the highly variable values and large uncertainties associated with previous studies. The IS associated SW value of this work is not consistent with the work of Grevesse and Sauval [23]. The reason our value may be different from the Grevesse and Sauval [23] study is that they use sunspot data, solar corona data, and SW particle data to derive an argon photospheric abundance and that work does not have a direct photospheric value.

It is interesting to note that the Ar/Ca published values averaged together for the composition representative of the photosphere (1.16) is greater than the Ar/Ca average ratio (0.54) representative of the published interstream associated SW by a factor of about 2.1. Similarly, the IS associated SW ratio of this study is less than the CH associated SW Ar/Ca ratio by a factor of also 1.6. This difference is most likely due to FIP type fractionation which occurs in the solar atmosphere. The details of this process are still unclear at this time. This ratio of CH associated SW to the IS associated SW is similar in nature to that factor found in the Aellig et al. [10] and Wurz et al. [9] studies, where the O/Fe elemental ratio in the IS associated SW is about a factor of 1.82 less than in the CH associated SW. For this work, we have inverted the Aellig et al. [10] study Fe/O elemental abundance ratio to simplify the comparison of a high FIP element to a low FIP species. For the Bochsler et al. [17] study, a FIP factor of 2.1 is reported, where the difference between the Al/Mg elemental abundance ratio in the IS and CH associated SW is examined. While it is known that the $[X/O]_{ISSW}/[X/O]_{CHSW}$ varies for each element, the purpose of the above discussion is to indicate that we find FIP fractionation of 1.6, which is a typical FIP fractionation pattern between the IS and CH associated SW.

The difference between the Ar/Ca IS and CH associated SW has one of three implications. The first option is the argon abundance is constant in all SW regimes and the calcium is depleted in the CH associated SW, which is suggested in the research of Meyer [26], Veck and Parkinson [22], and Raymond et al. [27, 28]. This theory is supported by the results of Wurz [12], which demonstrates a decrease in the calcium abundance on the order of 60% in the SW. This abundance decrease is sufficient to explain the factor of 1.6 difference in the MLF counts. The second option is that the argon abundance is depleted in the IS associated SW and the calcium abundance remains unchanged. This has been suggested in the work of Marsch et al. [29] and Peter [30]. The third option is both abundances change in the SW, with the abundance of the low-FIP element stronger than the high-FIP element. This has been observed for O, Si, and Fe abundances [12].

CONCLUSIONS

The Ar/Ca elemental ratios for the IS and CH associated SW of this study are consistent with many previously published values determined from gradual SEP events, optical spectroscopy, the Apollo foil experiment results, and from meteorites. The IS associated SW Ar/Ca elemental ratio of 0.38 ± 0.05 is a factor of 1.6 larger than the CH associated SW Ar/Ca elemental ratio of 0.59 ± 0.07. This difference can be easily understood in terms of the FIP fractionation process. The difference between the two SW bin's Ar/Ca ratios suggest that either the argon is enhanced in the CH associated SW or the calcium is depleted in the CH associated SW with respect to the IS associated SW.

The uncertainty for these values is generally smaller than the Ar/Ca ratio found in the gradual SEP event studies and significantly smaller than the uncertainty related to spectroscopic Ar/Ca ratios. Slight improvements to the uncertainty can be made in the future by collecting more spectra as CELIAS/MTOF gathers more data and by tightening the data selection. Unfortunately, the measurement uncertainty for the ratios within this work are large and are not expected to improve by more than a few percent, because of the significant instrumental uncertainty. The most significant improvement of these observations over the previous ones is these measurements represent direct samples from two different SW regimes, and they tie a volatile element to a refractory element, as opposed to previous investigations that determined the ratio indirectly from the solar wind or the solar atmosphere. Furthermore, we place limits on the Ar/Ca elemental abundance ratio variability directly in the contemporary SW under different SW conditions.

ACKNOWLEDGMENTS

This work is supported by the Swiss National Science Foundation. CELIAS is a joint effort of five hardware institutions under the direction of the Max-Plank Institut für Extraterrestrische Physic (pre-launch) and the University of Bern (post-launch).

REFERENCES

1. Cerutti, H., *Die Bestimmung des Argons im Sonnenwind aus Messungen an den Apollo - SWC - Folien*, Ph.D. thesis, University of Bern, Bern, Switzerland (1974).
2. Breneman, H., and Stone, E., *Astrophys. J.*, **299**, L57–L61 (1985).
3. Anders, E., and Gevesse, N., *Geochim. Cosmochim. Acta*, **53**, 197–214 (1989).
4. Reames, D., *Space Sci. Rev.*, **85**, 327–340 (1998).
5. Gloeckler, G., Balsiger, H., Bürgi, A., Bochsler, P., Fisk, L. A., Galvin, A. B., Geiss, J., Gliem, F., Hamilton, D. C., Holzer, T. E., Hovestadt, D., Ipavich, F. M., Kirsch, E., Lundgren, R. A., Ogilvie, K. W., Sheldon, R. B., and Wilken, B., *Space Sci. Rev.*, **71**, 79–124 (1995).
6. Hovestadt, D., Hilchenbach, H., Bürgi, A., Laeverenz, B. K. P., Scholer, M., Grünwaldt, H., W. I. Axford, S. L., Marsch, E., Wilken, B., Winterhoff, P., Ipavich, F. M., Bedini, P., Coplan, M. A., Galvin, A. B., Gloeckler, G., Bochsler, P., Balsiger, H., Fischer, J., Geiss, J., Kallenbach, R., Wurz, P., Reiche, K. U., Gliem, F., Judge, D. L., Hsieh, K. C., Möbius, E., Lee, M. A., Managadze, G. G., Verigin, M. I., and Neugebauer, M., *Solar Physics*, **162**, 441–481 (1995).
7. Gloeckler, G., Cain, J., Ipavich, F. M., Tums, E. O., Bedini, P., Fisk, L. A., Zurbuchen, T., Bochsler, P., Fischer, J., Wimmer-Schweingruber, R. F., Geiss, J., and Kallenbach, R., *Space Sci. Rev.*, **86**, 497–539 (1998).
8. Kern, O., Wimmer-Schweingruber, R. F., Bochsler, P., Gloeckler, G., and Hamilton, D. C., *ESA*, **SP-415**, 345–348 (1997).
9. Wurz, P., Aellig, M. R., Ipavich, F. M., Hefti, S., Bochsler, P., Galvin, A. B., Grünwaldt, H., Hilchenbach, M., Gliem, F., and Hovestadt, D., *Phys. Chem. Earth*, **24**, 421–426 (1999).
10. Aellig, M. R., Hefti, S., Grünwaldt, H., Bochsler, P., Wurz, P., Ipavich, F. M., and Hovestadt, D., *J. Geophys. Res.*, **104**, 24769–24780 (1999).
11. Weygand, J. M., Ipavich, F. M., Wurz, P., Paquette, J. A., and Bochsler, P., *Geochim. Cosmochim.* (2001), (in press).
12. Wurz, P., *Heavy Ions in the solar wind: Results from SOHO/CELIAS/MTOF*, Habilitation thesis, University of Bern, Bern, Switzerland (1999).
13. Neukomm, R. O., *Composition of coronal mass ejections derived with SWICS/Ulysses*, Ph.D. thesis, University of Bern, Bern, Switzerland (1998).
14. Wurz, P., Ipavich, F. M., Galvin, A. B., Bochsler, P., Aellig, M. R., Kallenbach, R., Hovestadt, D., Grünwaldt, H., Hilchenbach, M., Axford, W. I., Balsiger, H., Bürgi, A., Coplan, M. A., J. Geiss, F. G., Gloeckler, G., Hefti, S., Hsieh, K. C., Klecker, B., Lee, M. A., Livi, S., Managadze, G. G., Marsch, E., Möbius, E., Neugebauer, M., Reiche, K. U., Scholer, M., Verigin, M. I., and Wilken, B., *Geophys. Res. Lett.*, **25**, 2557–2560 (1998).
15. Wurz, P., Wimmer-Schweingruber, R., Issautier, K., Bochsler, P., Galvin, A., and Ipavich, F., (these preceedings).
16. Bürgi, A., Gonin, M., Oetliker, M., Bochsler, P., Geiss, J., Lamy, T., Brenac, A., Andrä, H. J., Roncin, P., Laurent, H., and Coplan, M. A., *J. Appl. Phys.*, **73**, 4130–4139 (1993).
17. Bochsler, P., Ipavich, F. M., Paquette, J. A., Weygand, J. M., and Wurz, P., *J. Geophys. Res.*, **105**, 12659–12666 (2000).
18. Kallenbach, R., Ipavich, F. M., Bochsler, P., Hefti, S., Wurz, P., Aellig, M. R., Galvin, A. B., Geiss, J., Gliem, F., Gloeckler, G., Grünwaldt, H., Hilchenbach, H., Hovestadt, D., and Klecker, B., *Astrophys. J.*, **498**, L75–L78 (1998).
19. Widing, K., and Spicer, D., *Astrophys. J.*, **242**, 1243–1256 (1980).
20. Widing, K., and Hiei, E., *Astrophys. J.*, **281**, 426–434 (1981).
21. Feldman, U., and Widing, K., *Astrophys. J.*, **363**, 292–298 (1990).
22. Veck, N. J., and Parkison, J. H., *MNRAS*, **197**, 41–55 (1981).
23. Grevesse, N., and Sauval, A. J., *Space Sci. Rev.*, **85**, 161–174 (1998).
24. Young, P. R., Mason, H. E., Feenan, F. P., and Widing, K. G., *Astron. Astrophys.*, **323**, 243–249 (1997).
25. Meyer, J. P., *Ap. J. Suppl. Ser.*, **57**, 151–171 (1985a).
26. Meyer, J. P., *Ap. J. Suppl. Ser.*, **57**, 173–204 (1985b).
27. Raymond, J. C., Kohl, J. L., Noci, G., Antonucci, E., Tondello, G., Huber, M. C. E., Gardner, L. D., Nicolosi, P., Fineschi, S., Romoli, M., Spadaro, D., Siegmund, O. H. W., Benna, C., Ciaravella, A., Cranmer, S., Giordano, S., Karovska, M., Martin, R., Michels, J., Modigliani, A., Naletto, G., Panasyuk, A., Pernechele, C., Poletto, G., Smith, P. L., Suleiman, R. M., Strachan, L., and van Ballegooijen, A. A., *Solar Physics*, **175**, 645–665 (1997).
28. Raymond, J. C., Suleiman, R., Kohl, J. L., and Noci, G., *Space Sci. Rev.*, **85**, 283–289 (1998).
29. Marsch, E., von Steiger, R., and Bochsler, P., *Astron. Astrophys.*, **301**, 261–276 (1995).
30. Peter, H., *Astron. Astrophys.*, **312**, L37–L40 (1996).

Solar Coronal Abundances of Rare Elements Based on Solar Energetic Particles

C. M. S. Cohen*, R. A. Mewaldt*, E. R. Christian†, A. C. Cummings*, R. A. Leske*, P. L. Slocum**, E. C. Stone*, T. T. von Rosenvinge† and M. E. Wiedenbeck**

*California Institute of Technology, MC 220-47, Pasadena, CA 91125
†NASA/Goddard Space Flight Center, Code 661, Greenbelt, MD 20771
**Jet Propulsion Laboratory, 4800 Oak Grove Dr., Pasadena, 91109

Abstract. Although solar energetic particle (SEP) abundances vary from event to event, it has been shown that by accounting for these variations it is possible to use SEP data to obtain reliable estimates of elemental abundances for the solar corona. We analyze $\sim$ 20 to 65 MeV/nucleon measurements from the Solar Isotope Spectrometer on ACE in large SEP events observed from November 1997 to January 2001 to obtain new values of the average SEP composition of rare species, P, Cl, K, Ti, Mn, Cr, Co, Cu, and Zn, which have had limited statistical accuracy in SEPs in the past. The measured SEP abundances are compared with other sources of solar-system composition data.

INTRODUCTION

It is clear that the composition of matter in many regions of the corona differs significantly from that of the photosphere. The first ionization potential (FIP) of the elements is commonly used to organize the differences, resulting in a step-like function where elements with FIPs less than $\sim$10 eV are enhanced in the corona over elements with higher FIPs when compared to photospheric abundances (see, e.g., [1]). Although the abundances of nearly all the elements from He to Ga are relatively well known for the photosphere this is not the case for the solar corona. Because the coronal conditions are typically far from local thermal equilibrium, calculations of ionization equilibrium are difficult and significantly affect the accuracy of spectroscopically measured abundances [1]. Additionally, the charge state distribution of most elements is broad in the hot corona which results in the abundance of a particular element being distributed over many charge states, often yielding weaker spectral lines. This makes spectroscopic measurements of the abundance of elements such as P, Cl, Co, and Mn particularly difficult. Thus, another way of determining these rare abundances is needed.

Although the FIP-fractionation is understood to be a ion-neutral separation process that occurs in the chromosphere/transition region [2], the specific mechanism involved is not known. In fact, arguments have been put forth that the elements' first ionization time (FIT) is a better organizing parameter [3]. Further, it has been observed that the FIP step height is not the same everywhere in the corona. A large step height has been inferred from the Ne/Mg ratio measured in polar plumes and solar active regions [4] while the same ratio in coronal holes and the Si/C ratio measured over sunspots indicate a small step height in these regions [5, 6]. Thus deriving the elemental composition of the corona from that of the photosphere is not a simple matter.

Abundances measured in the solar wind are known to be similar to those obtained spectroscopically in the corona and so are a natural proxy for coronal composition [1]. Unfortunately, there are no measurements of the rare elements such as those mentioned above, although the first solar wind measurement of Cr/Fe was recently made using the MASS instrument on SoHO [7]. It is possible that the relative abundances of Mn, Ni and Zn will follow from the same data set in the near future (F. Ipavich, private communication), but currently such abundances are unavailable for the solar wind.

Solar energetic particle (SEP) events are believed to arise from two distinct origins [8]. Impulsive events contain flare accelerated material, typically last for hours and tend to be smaller in intensity than gradual events. Gradual events contain material that is accelerated out of the corona and solar wind by shocks driven by coronal mass ejections and last for days. Given their origin, it is natural to expect the composition of gradual SEP events to be similar to that of the corona and perhaps a useful measure

CP598, *Solar and Galactic Composition*, edited by R. F. Wimmer-Schweingruber

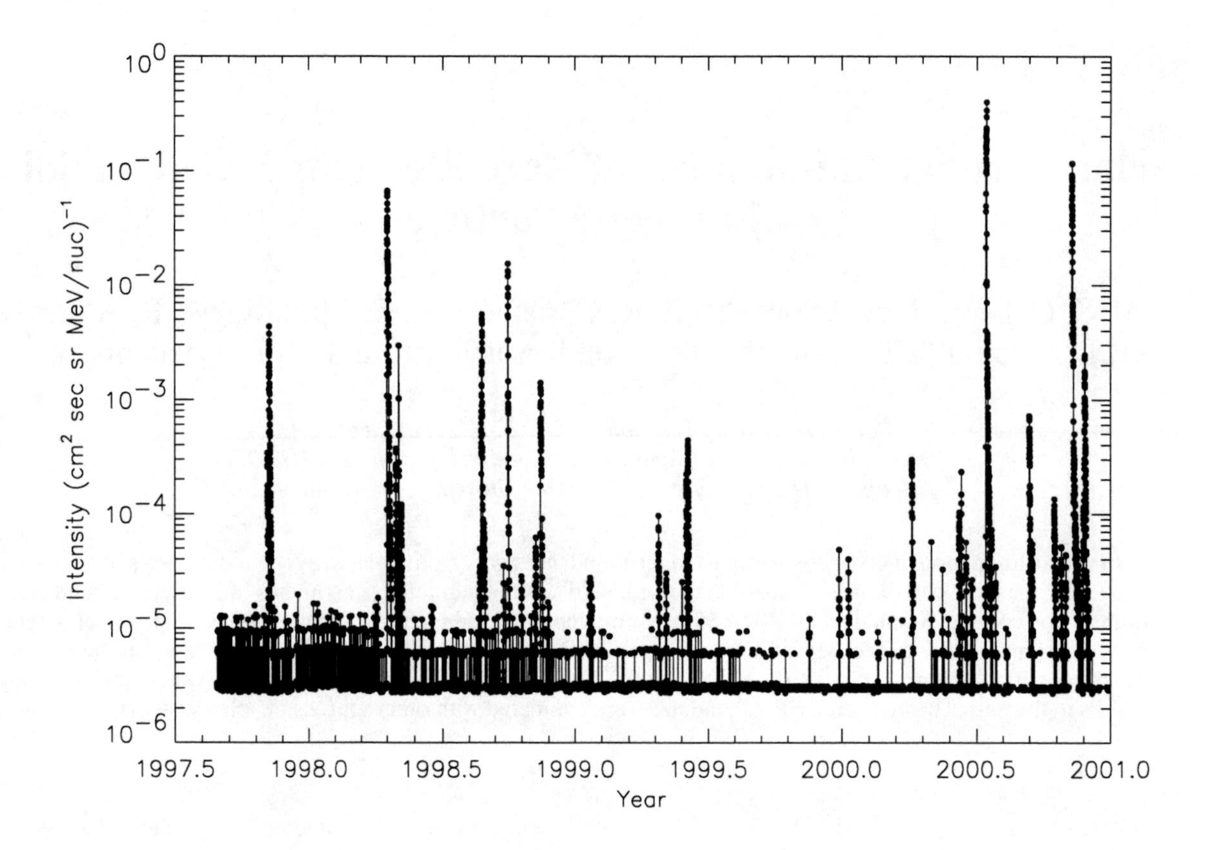

FIGURE 1. Hourly-averaged intensities of $\sim$ 14 MeV/nucleon oxygen from launch in August 1997 to the beginning of 2001.

of the coronal composition.

However, the elemental composition of gradual SEP events is known to vary from event to event due to transportation/acceleration effects which often can be characterized as a charge-to-mass (Q/M) dependent fractionation [9]. The effect of the Q/M fractionation on the elemental composition can be accounted for to obtain an average coronal composition. Garrard and Stone (hereafter G&S) [10] empirically corrected for the Q/M fractionation by comparing the measured abundances of low-FIP elements to photospheric abundances as a function of Q/M and fitting a power law to the data. Reames [8] states that the event-to-event variation in the Q/M fractionation tend to cancel out when many SEP events are averaged together resulting in an SEP composition that is representative of the average corona.

Additionally, while the FIP fractionation apparent in the spectroscopic coronal abundances is present in SEP events, the degree of fractionation varies from event to event [11, 12]. Since the FIP fractionation is an ion-neutral separation process and the SEPs are all ions, the variations observed in SEP events are reflections of the corona itself rather than a result of the acceleration/transport of the energetic particles. However, it is unclear how to directly relate the observed variation in FIP fractionation in SEP events to that observed spectroscopically on the Sun. Thus, we are left with the hope that averaging over many SEP events will result in a FIP step height that is 'typical' of the corona.

Although there are significant difficulties in using SEP data to obtain the composition of the solar corona, many rare elements that are not available through spectroscopy or the solar wind instruments can be measured with reasonable accuracy. With the increased sensitivity of today's energetic particle detectors, it is possible to measure the rare elements with an accuracy that is much improved over previous studies. Additionally, solar cycles are known to differ in character and analyzing the composition of events from solar cycle 23 is useful in furthering our understanding of both the solar corona and solar activity and their variations.

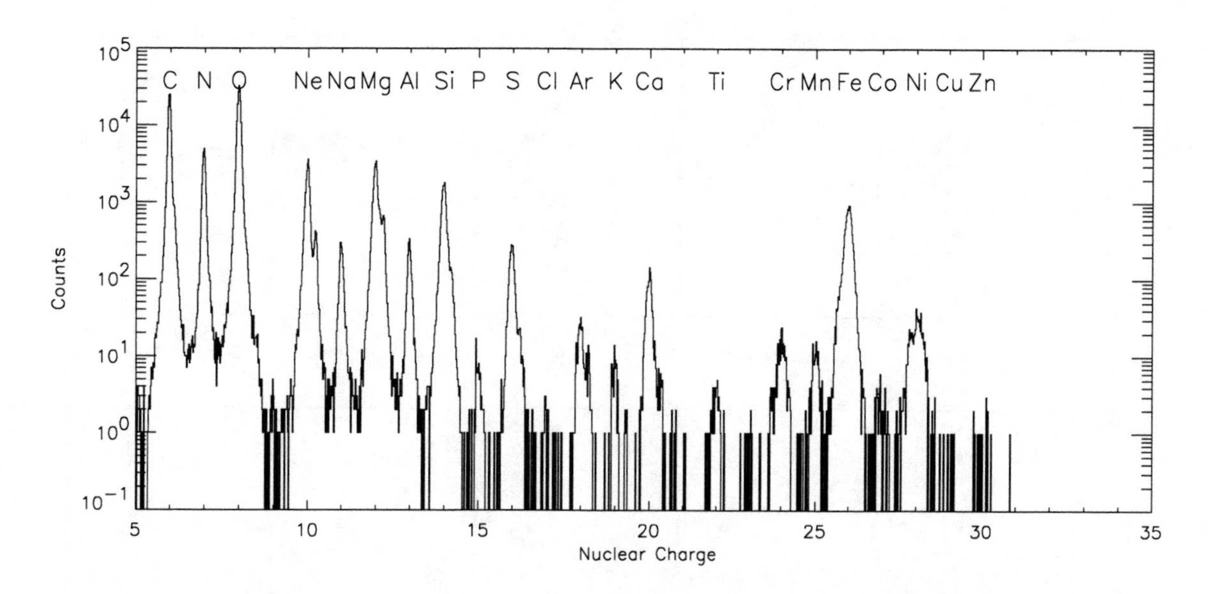

FIGURE 2. Histogram of events as a function of nuclear charge for particles stopping in the 3rd stack detector of SIS (corresponding to ~19 MeV/nucleon for oxygen).

INSTRUMENTATION AND DATA ANALYSIS

The data presented here are from the Solar Isotope Spectrometer (SIS; [13]) on the ACE spacecraft. The SIS sensor contains two identical telescopes, each of which has two position-sensing matrix detectors followed by a stack of large-area silicon detectors. The matrix detectors allow the particle trajectories to be determined, which combined with the deposited energies measured by the stack detectors allow accurate determinations to be made of nuclear charge, Z, mass, M, and total kinetic energy, E. The large geometry factor (~38 cm^2-sr) of SIS is key in obtaining statistically significant measurements of the rare elements that are the focus of this study.

We combine the data from 27 large SEP events which occurred between November 1997 and January 2001 (Figure 1). Combining the data makes it possible to identify statistically significant peaks in a nuclear-charge histogram which correspond to rare elements, something not possible on an event-by-event basis. Peaks for P, Cl, K, Ti, Cr, Mn, Co, Cu, and Zn can be clearly identified in Figure 2 with little apparent contamination from neighboring, more abundant elements.

The 27 events examined varied not only in intensity but in composition, spectral shape, and in FIP and Q/M fractionation. By taking ratios of elements with similar FIP and Q/M values, the fractionation effects can be minimized. Any residual fractionation is further reduced by combining the data from many events, resulting in an 'average' abundance that is reasonably representative of a global average of the solar corona. Such an averaging technique was used by Reames [14] to infer coronal abundances from 5-12 MeV/nucleon energetic particles.

From the combined data set the differential-intensity spectrum of each element is determined. At several tens of MeV/nucleon the contribution of galactic cosmic rays (GCRs) can be substantial and must be subtracted from the spectra before calculating abundances. The GCR spectra are determined by analyzing a data set that consists of 252 days of low solar activity (as defined by the 1.0-4.75 MeV/nucleon daily-averaged proton flux measured by the Electron, Proton, Alpha Monitor on ACE being less than 0.16 $(cm^2\text{-sec-sr-MeV})^{-1}$). The low-solar-activity days were spread throughout the November 1997 to January 2001 time period to account for variations in the GCR intensities due to changing levels of solar modulation during this time period. Contributions from anomalous cosmic rays (ACRs) are present and are also accounted for through the low-activity data.

Both the SEP spectra and the low-activity spectra for P, Cl, K, Ti, Cr, Mn, Co, Ni, Cu, and Zn are shown in Figure 3. For Co, Cu and Zn there are insufficient statistics in the low-activity data set to obtain intensities for all energy bins. For these elements the low-activity spectra are estimated by scaling the Fe low-activity spectrum by the Co/Fe, Cu/Fe, and Zn/Fe GCR abundances as measured by the Cosmic Ray Isotope Spectrometer on ACE (M. Wiedenbeck, private communication). These spectra are shown in Figure 3 as solid lines and are in agreement with the available intensities measured with SIS. After subtracting the low-activity spectra, the re-

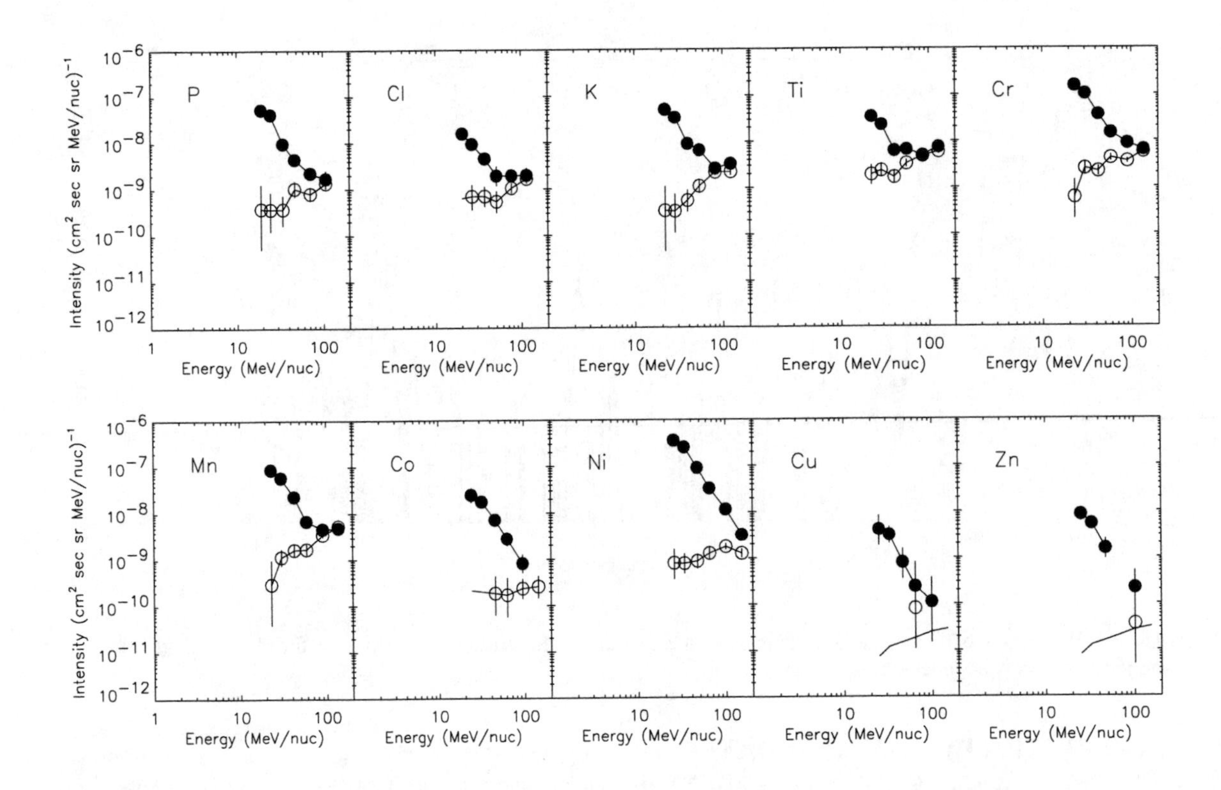

FIGURE 3. Spectra for different elements. Filled circles are SEP data, open circles are low-activity data. The solid lines are approximations of the low-activity spectra, obtained by scaling the Fe low-activity spectrum by the GCR abundances relative to Fe.

sulting corrected spectra are integrated from 20 to 65 MeV/nucleon to obtain relative abundances. The high end of this energy range was chosen such that the GCR corrections were not substantial (less than 50% contribution) while the low end was chosen based on where the nuclear charge resolution of the instrument was adequate for resolving the rare elements of interest.

RESULTS

The abundances of P, Cl, K, Ti, Cr, Mn, Co, Ni, Cu, and Zn (relative to Si, Ca or Fe) are presented in Figure 4 (and in Table 1). For comparison the SEP measurements of Reames (hereafter DVR) [14] and Breneman and Stone (hereafter B&S) [9] are given as well. The DVR values were obtained from a combined data set of 49 gradual SEP events and measured over 5-12 MeV/nucleon, while the B&S values were from 3.5-50 MeV/nucleon SEPs. The B&S values were then used by G&S to derive coronal abundances by explicitly correcting for the apparent Q/M fractionation by fitting a power law in Q/M to the low-FIP elements. These results are also given in Figure 4 and Table 1 along with the abundance ratios as measured in the photosphere and meteorites [15]. The uncertainty in the fractionation correction is included in the uncertainties plotted for the G&S values, whereas the uncertainties for the other SEP values are statistical only. By examining ratios of neighboring elements, the Q/M fractionation is minimized, as can be seen from the small differences between the B&S observed SEP and G&S SEP-derived coronal values.

Generally all the SEP abundances agree well with each other except for Cr/Fe and Ni/Fe. In the case of Mn and Cu there are no reported values by DVR and values with large uncertainties from B&S. There are no other SEP values for Co. While the Ni/Fe is higher in this work than that obtained by DVR and B&S, it is close to the photospheric and meteoritic values. Cr/Fe and Co/Fe, however, are 30% and 50% higher than the meteoritic abundances. It does not appear from Figure 2 (and from examination of a similar histogram of particles stopping in the second stack detector; not shown) that contamination from the Fe peak is the cause of this discrepancy as the peaks are well separated and there is little apparent background.

By choosing the ratios appropriately, the FIP fractionation is not a significant contribution to any difference between the photospheric (or meteoritic) and SEP abundances. Note that since P has a FIP value of 10.5 eV it, like S, is a transition element which does not appear to

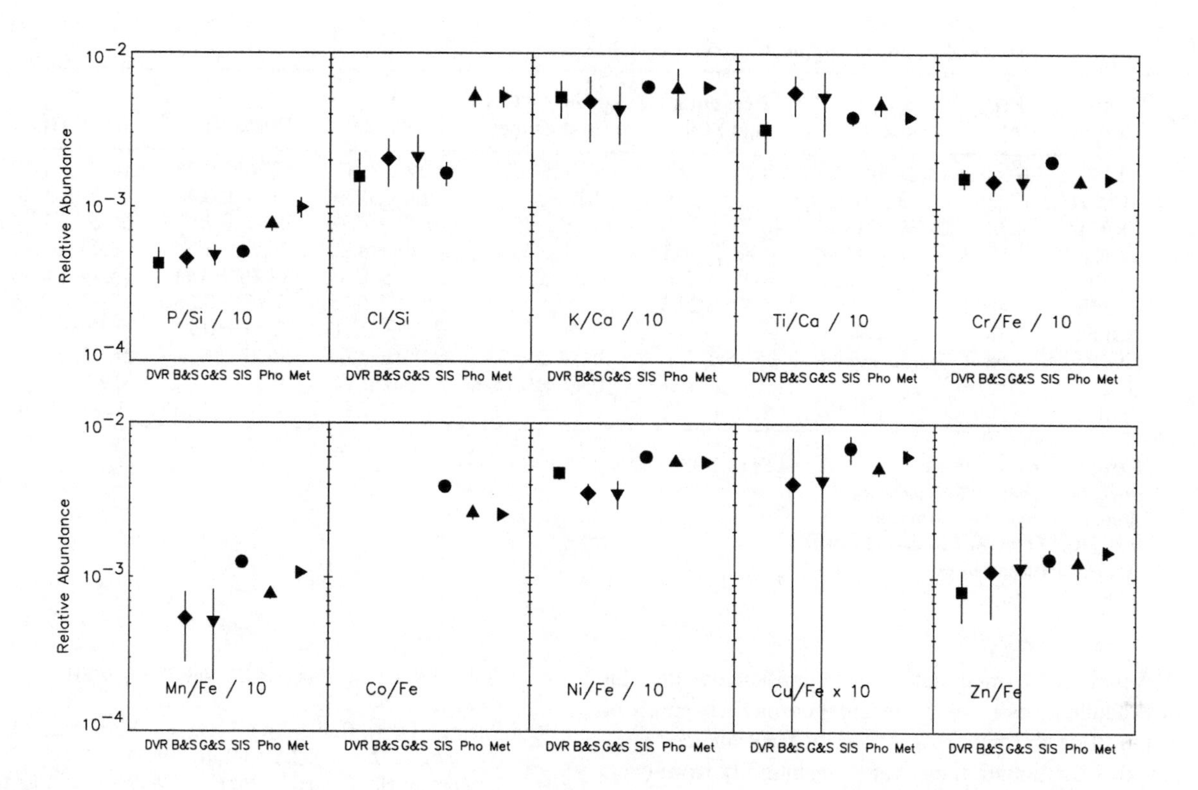

FIGURE 4. Resulting elemental ratio abundances for 20-65 MeV/nucleon as compared to SEP measurements by Reames (DVR) [14] and Breneman and Stone (B&S) [9]; SEP-derived coronal abundances by Garrard and Stone (G&S) [10]; and photospheric (Pho) and meteoritic (Met) abundances by Grevesse and Sauval [15].

belong to either the high-FIP or low-FIP element groups, and so the P/Si ratio is only partially depleted relative to photospheric values. The Cl/Si ratio is another exception in that Cl is a high-FIP element while Si is a low-FIP one. A more appropriate ratio would be Cl/Ar (two high-FIP elements), but both photospheric and meteoritic abundances for Ar are unavailable (note, the Ar entry in most tables for photospheric abundances is based on the Ar measurements from solar wind or SEPs [15]).

A modification of the analysis is required to extend the present study to a comprehensive one of all the elements, C-Zn, measured by SIS. By combining the data from SEP events of significantly different intensities, the resulting data set is skewed towards the larger intensity events. Such an averaging technique does not seem appropriate for element ratios that vary significantly between events due to Q/M or FIP fractionation effects.

Certainly such weighting issues are less of a concern when there is little event-to-event variability in the SEP composition. Evaluation of the variation of abundances as a function of energy has shown that the event-to-event abundance variation at 1 MeV/nucleon is less than it is above 20 MeV/nucleon [16]. Thus the averaging technique of creating and analyzing a combined data set from many SEP events may be more appropriate at energies near 1 MeV/nucleon.

In order to derive a complete set of meaningful coronal abundances from several tens of MeV/nucleon SEP data, it may be necessary to make event-specific Q/M fractionation corrections to the data prior to calculating average abundances. While this poses a statistical problem for the rare species a compromise may be possible. By characterizing each event by its Q/M fractionation (possibly using the Fe/Si ratio as a proxy [12]), events of similar fractionation can then be combined to increase statistical accuracy and then corrected for the Q/M fractionation before taking an average of the results of different event groups. Such an averaging technique will be investigated in future work.

CONCLUSIONS

By combining the SIS data from 27 large SEP events, we obtain statistically accurate measurements of P, Cl, K, Ti, Cr, Mn, Co, Cu and Zn. These measurements are significant improvements over previous SEP determinations. Since these elements are extremely difficult to measure spectroscopically in the corona, SEP measurements are currently the best determined sample of coronal material.

TABLE 1. Elemental Ratios Multiplied by 1000

Element Ratio	FIP (eV)*	Reames†	Breneman and Stone**	Garrard and Stone‡	This Work§	Photosphere¶	Meteoritic¶
P/Si	10.49	4.28 ± 1.12	4.61 ± 0.65	4.89 ± 0.75	5.12 ± 0.47	7.76 ± 0.75	10.00 ± 1.48
Cl/Si	13.00	1.58 ± 0.66	2.05 ± 0.70	2.12 ± 0.81	1.66 ± 0.29	5.24 ± 0.78	5.25 ± 0.78
K/Ca	4.34	51.89 ± 14.15	48.53 ± 22.06	43.25 ± 17.56	60.87 ± 4.78	58.87 ± 20.56	60.26 ± 2.84
Ti/Ca	6.82	32.08 ± 9.43	55.88 ± 16.18	52.48 ± 23.20	38.79 ± 4.22	46.74 ± 6.97	38.90 ± 1.83
Cr/Fe	6.77	15.67 ± 2.24	14.91 ± 2.71	14.86 ± 3.30	20.06 ± 0.85	14.76 ± 1.08	15.49 ± 0.36
Mn/Fe	7.44		5.42 ± 2.61	5.68 ± 3.34	12.55 ± 0.68	7.76 ± 0.56	10.72 ± 0.25
Co/Fe	7.86				3.88 ± 0.31	2.63 ± 0.26	2.57 ± 0.06
Ni/Fe	7.64	47.76 ± 4.48	35.25 ± 5.21	35.24 ± 7.32	60.89 ± 1.42	56.23 ± 1.31	56.23 ± 1.31
Cu/Fe	7.73		0.41 ± 0.51	0.47 ± 0.75	0.69 ± 0.14	0.51 ± 0.05	0.62 ± 0.06
Zn/Fe	9.39	0.82 ± 0.30	1.11 ± 0.55	1.18 ± 1.31	1.33 ± 0.21	1.26 ± 0.26	1.48 ± 0.14

* FIP of numerator; FIP values for Si, Ca, and Fe are 8.12, 6.09, and 7.83 eV respectively
† from [14] average SEP abundances
** from [9] average SEP abundances
‡ from [10] SEP-derived coronal abundances
§ average SEP abundances
¶ from [15]

However, Q/M fractionation can significantly alter the SEP abundances from the original coronal composition and must be accounted for. At tens of MeV/nucleon energies this fractionation can vary substantially from event to event and may need to be corrected on an event-by-event basis. Balancing this with the need for statistical accuracy in obtaining the abundances of rare elements will be an aspect of future work. These studies will be aided by direct examination of the event-to-event variability for some elements such as Cr. In this work, we have dealt with these fractionation issues by examining abundance ratios of neighboring elements with similar FIP and Q/M values.

ACKNOWLEDGMENTS

This work was supported by NASA at Caltech (under grant NAG5-6912), Jet Propulsion Laboratory, and Goddard Space Flight Center.

REFERENCES

1. Meyer, J. P., *Astrophys. J. Supp.*, **57**, 173–204 (1985).
2. Henoux, J. C., *Space Sci. Rev.*, **85**, 215–226 (1998).
3. Geiss, J., *Space Sci. Rev.*, **85**, 241–252 (1998).
4. Widing, K., and Feldman, U., *Astrophys. J.*, **416**, 392–397 (1993).
5. Feldman, U., *Space Sci. Rev.*, **85**, 227–240 (1998).
6. Dere, K. P., Bartoe, J.-D. F., and Brueckner, G., *Astrophys. J.*, **259**, 366–371 (1982).
7. Paquette, J. A., Ipavich, F., Lasley, S., Bochsler, P., and P.Wurz, "The Relative Abundance of Chromium and Iron in the Solar Wind", in *Proc. Joint SoHO–ACE Workshop 2001*, AIP Conf. Proc., AIP, New York, 2001, this volume.
8. Reames, D. V., Richardson, I. G., and Barbier, L. M., *Astrophys. J. Lett.*, **382**, L43–L46 (1991).
9. Breneman, H. H., and Stone, E. C., *Astrophys. J. Lett.*, **299**, L57–L61 (1985).
10. Garrard, T. L., and Stone, E. C., *Proc. 23rd Internat. Cosmic Ray Conf. (Calgary)*, **3**, 384–387 (1993).
11. Garrard, T. L., and Stone, E. C., *Adv. Space Res.*, **14**, (10)589–(10)598 (1994).
12. Mewaldt, R. A., et al., "Variable Fractionation of Solar Energetic Particles According to First Ionization Potential", in *Acceleration and Transport of Energetic Particles Observed in the Heliosphere: ACE 2000 Symposium*, edited by R. A. Mewaldt et al., AIP Conf. Proc. 528, AIP, New York, 2000, pp. 123–126.
13. Stone, E. C., et al., *Space Sci. Rev.*, **86**, 357–408 (1998).
14. Reames, D. V., *Adv. Space Res.*, **15**, (7)41–(7)51 (1995).
15. Grevesse, N., and Sauval, A. J., *Space Sci. Rev.*, **85**, 161–174 (1998).
16. Mazur, J. E., Mason, G. M., Klecker, B., and McGuire, R. E., *Astrophys. J.*, **404**, 810–817 (1993).

Isotopic Composition Measured In-Situ in Different Solar Wind Regimes by CELIAS/MTOF on board SOHO

R. Kallenbach

International Space Science Institute, Hallerstrasse 6, CH-3012 Bern, Switzerland

Abstract. The Sun is the largest reservoir representing the matter of the early solar nebula. Its isotopic composition for the elements N, Ne, Mg, Si, Ar, and Fe has been determined by measuring solar wind abundances with the CELIAS experiment on board SOHO and other spacecraft instruments. These measurements and theoretical considerations indicate that the solar wind, in particular the coronal-hole type solar wind, is much less isotopically fractionated than solar energetic particles. The data give evidence that the isotopic abundance ratios typically vary by only $\sim$1 - 2% per amu in different solar wind regimes such as coronal-hole type and streamer-belt associated solar wind, ejecta of coronal mass ejections (CMEs), or so-called 'blobs'.

INTRODUCTION

The elemental and isotopic composition of solar system samples is usually compared to the composition of terrestrial material as a reference. However, due to the distribution of mass, any non-nuclear fractionation process during the formation of the inner solar system from the early solar nebula could have had impact on the terrestrial composition, but practically not on the solar composition. The relevant fractionation processes are related to the typical size-scale of the dominating bodies present in the protosolar disk at different times [1]:

1. The first μm- to mm-sized solids, such as dust grains, chondrules, and chondrites, found today in meteoritical inclusions, formed within about ten million years in the protosolar disk. Due to heterogeneities in pressure and temperature or because of the interaction of the accretion flow with the magnetosphere of the young Sun, isotopic heterogeneities have been created by reprocessing and redistribution of solids. These heterogeneities are much less pronounced for refractory elements than for volatiles. Isotopic fractionation of the total matter in the protosolar disk with respect to the molecular cloud is assumed to be minor. The protosolar disk collapsed from the molecular cloud within only a few hundred thousand years. Most of the disk accretion occured in even shorter episodes of a few ten thousand years with rates up to 10^{-5} $M_{\odot}$ y^{-1}.
2. The planetesimals, up to several hundred km in size, accreted from smaller bodies and differentiated into core, mantle, and crust within one to two million years. The differentiation has not influenced very much the mean composition of the planetesimals, which is basically given by the assemblage of chondritic types.
3. The final accretion of planets, hundreds to thousands of kilometers in size, occurred by collisions between planetesimals and 'planetary embryos' on timescales of ten to hundred million years. Impacts play a crucial role in the formation of the terrestrial planets' atmospheres as they trigger losses of volatiles. Additionally, hydrodynamic escape continued during later phases of the inner solar system's history.

This scenario creates Earth with an isotopic composition of refractory elements similar to that of the Sun, but with compositional deviations for the volatile isotopes.

The goal of this summary is to constrain isotopic fractionation processes in the inner corona to infer the solar isotopic composition from solar wind isotopic abundance ratios. The variations of these ratios have been observed in different types of solar wind streams such as streamer-belt ('slow'), coronal-hole associated ('fast'), and CME-related solar wind, as well as in 'blobs', which are radially elongated structures of high density, forming near the heliospheric current sheet at the tip of helmet streamers beyond heliocentric distances of $\sim$2.5 $R_{\odot}$ and passively tracing the outflow of the 'slow' solar wind [2]. The isotopic abundance variations in the solar wind have been compared to theoretical considerations on variable Coulomb friction and wave heating in the solar wind, and on gravitational settling in closed coronal magnetic structures [3].

CP598, *Solar and Galactic Composition*, edited by R. F. Wimmer-Schweingruber

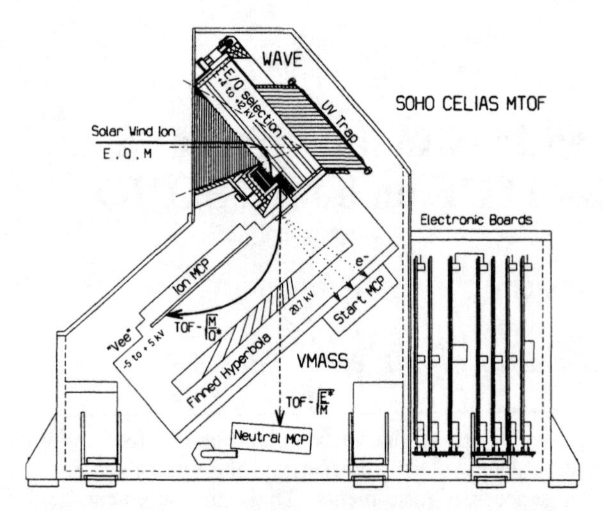

FIGURE 1. Schematic view of the MTOF sensor. Highly charged solar wind ions enter the instrument through the so-called WAVE (Wide Angle Variable Energy) entrance system that has an energy-per-charge (E/q) acceptance bandwidth of about half a decade, and a conic field of view of $\pm 25°$ width. When passing through the thin ($\sim 3\ \mu g\ cm^{-2}$) carbon foil of the V-shaped 'VMASS' spectrometer, the ions release secondary electrons to trigger a start signal on a micro channel plate detector and exchange charge. The fractions of charge states have been calibrated for 16 solar wind elements as a function of their speed when leaving the foil [4]. Typically, the ions leave the foil as neutrals and singly charged, but with increasing speed also multiply charged. After a time of flight proportional to $\sqrt{M/Q^*}$ (M: mass, Q^*: charge after passage through the carbon foil), the ions hit a position sensing stop detector.

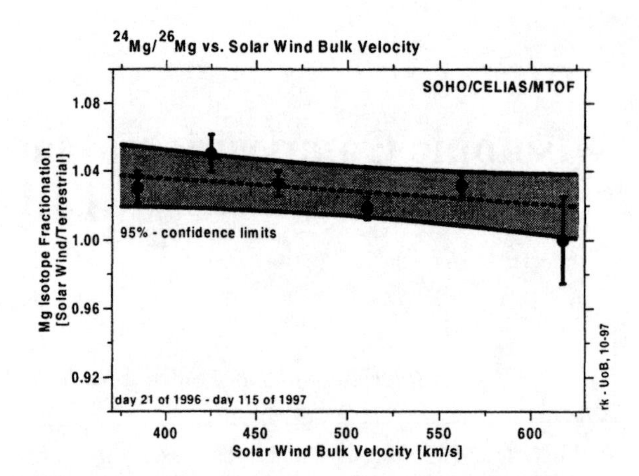

FIGURE 2. Solar wind $^{24}Mg/^{26}Mg$ ratio divided by the terrestrial $^{24}Mg/^{26}Mg$ value versus the solar wind bulk speed. A trend can be seen in the sense that the heavier isotope is depleted by a few percents in the slow solar wind. Data from [7].

INSTRUMENTATION

The data reported in this work mainly originate from the Charge, Element, and Isotope Analysis System (CELIAS) [5] on board SOHO, and in particular from the MTOF sensor (Figure 1). MTOF is an isochronous time-of-flight mass spectrometer with a resolution $M/\Delta M$ of about 100. It provides the possibility of resolving the different isotopes in the mass range 3 to 60 amu of the elements in the solar wind at bulk speeds of 300 to 1000 km/s. Mass spectra, instrument settings, and solar wind parameters from the proton monitor, a subsystem of MTOF, are available every five minutes. Therefore, the flight data can reliably be classified according to solar wind parameters, detection efficiencies of the time-of-flight (VMASS) spectrometer, and the ion optical instrument discrimination of the entrance system. Besides statistical uncertainties, the isotopic abundance measurements have a precision of $\sim 1.5\%$ per amu in the mass range above 20 amu, as evaluated in detail for the Ne isotopes [6]. Additional uncertainties arise for some isotopic species due to interferences with isotopes that have the same time of flight, i.e. the same M/Q^*, in the VMASS spectrometer. In particular, it was difficult to evaluate the $^{15}N/^{14}N$ ratio [8] because of the interference of $^{30}Si^{2+}$ with $^{15}N^+$ in the VMASS and because of the very low abundance of ^{15}N in the solar wind. Precise measurements have been made for Mg, as the minor isotopes ^{25}Mg and ^{26}Mg are fairly abundant in the solar wind and do not suffer interferences in the VMASS. Therefore, the Mg isotopic abundances were mainly used to constrain fractionation processes in the inner corona.

OBSERVATIONS

Figure 2 shows the solar wind $^{24}Mg/^{26}Mg$ ratio divided by the terrestrial $^{24}Mg/^{26}Mg$ value versus the solar wind bulk speed [7]. Time-of-flight spectra of the time period from day 21 in 1996 to day 115 in 1997 have been filtered to restrict the instrumental isotopic fractionation to at most $\pm 15\%$. As mentioned above, this procedure leads to $^{24}Mg/^{26}Mg$ ratios with an absolute experimental uncertainty of about 3%. Evaluating the variation of $^{24}Mg/^{26}Mg$ with solar wind speed, the experimental uncertainty is mainly statistical. Most of the instrumental fractionation arises from the energy-per-charge (E/q) analyzer WAVE. Its voltage U is stepped to cover the E/q range of all solar wind ions for any bulk speed. For long-time accumulations with variable solar wind speeds the measured data are spread in a similar way into the different classes of instrument fractionation as the ion optics of the WAVE only depend on $E/(qU)$. Thus, systematic instrumental uncertainties cancel out to some extent. The data suggest a variation of the $^{24}Mg/^{26}Mg$ ratio as a function of solar wind bulk speed with a slope of $(-2.7\pm 2.0)\times 10^{-5}$s/amu/km. Similar trends have been observed for the solar wind Ne and Si isotopes [6, 9].

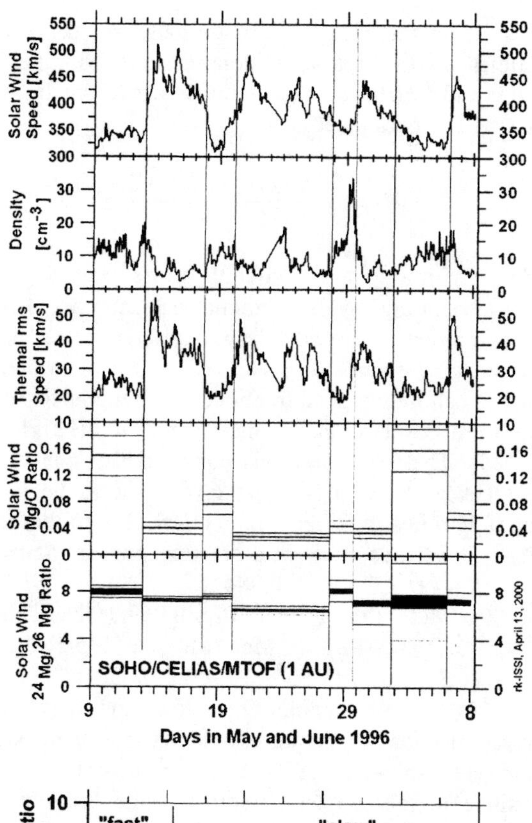

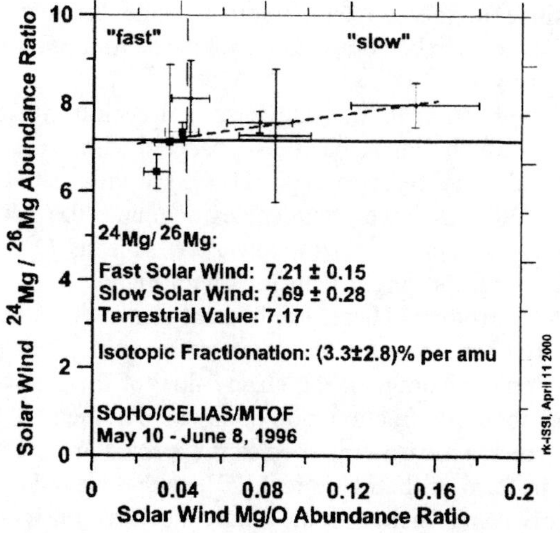

FIGURE 3. Time series of solar wind speed, density, and temperature. The Mg/O ratio and the ^{24}Mg/^{26}Mg ratio are averaged over time periods during which the solar wind was either of 'slow' or of 'fast' type. The width of the center lines for the Mg/O and the ^{24}Mg/^{26}Mg data indicate the statistical uncertainty, and the additional lines show the uncertainty in the instrument efficiencies.

This result has been cross-checked by analyzing the correlation of the variation in the ^{24}Mg/^{26}Mg isotopic ratio with other indicators to distinguish streamer-belt ('slow') and coronal-hole type ('fast') solar wind. A better indicator than the solar wind speed is the First Ionization Potential (FIP) effect [10]. Low-FIP elements are enriched by up to a factor four in the streamer-belt solar wind but to much less extent in the coronal-hole type solar wind. Furthermore, streamer-belt solar wind has typically higher density, lower temperature, and a lower ^{4}He/^{1}H flux ratio than coronal-hole type solar wind. The ^{4}He/^{1}H ratio may vary from typically $\sim$0.05 in the 'fast' solar wind to $\sim$0.02 in the 'slow' solar wind, because in the strongly superradially expanding source regions of the 'slow' wind with a rapidly decreasing flux density of the protons, the Coulomb friction with the protons is less efficient in dragging away the $^4He^{2+}$ ions from the solar surface than in the 'fast' wind [11]. Approximately the same variation in the solar wind ^{24}Mg/^{26}Mg ratio has been observed in correlation to the ^{4}He/^{1}H ratio as in correlation to the bulk speed [12].

Recently, time series on the solar wind abundance ratios at 1 AU of low- and high-FIP elements have become available, such as Fe/O from the Charge Time-of-Flight (CTOF) sensor of the CELIAS experiment [13] and Mg/O from MTOF (this work). Figure 3 shows preliminary results from a particular time period in May and June 1996 on the correlation of the ^{24}Mg/^{26}Mg ratio with several solar wind parameters. The solar wind has been divided into two classes, one with rather high Mg/O ratio, higher density, but lower speed and temperature, and one with the opposite characteristics. At least one of the parameters shows a step-like transition when switching from one class to the other. For each of the time intervals the average isotopic abundance ratio ^{24}Mg/^{26}Mg is shown. Unlike for Figure 2, the instrumental uncertainties do not cancel out because no data with large instrument fractionation (more than $\pm$15%) have been discarded, and because solar wind and instrument parameters are not evenly distributed in $E/(qU)$ for the short time period evaluated.

Nonetheless, the data of Figure 3 confirm the result of Figure 2 that there is a fractionation of about 1.5% per amu, depleting the heavier isotopes in the streamer-belt solar wind relative to their abundance in the coronal-hole associated solar wind. Similar results have been obtained when correlating ^{24}Mg/^{26}Mg with the Fe/O elemental abundance ratio in the solar wind [14].

The May 29 data show a particular structure with very high density resembling the so-called 'blobs' [2], which have initial sizes of $\sim$1 $R_\odot$ in the radial direction and 0.1 $R_\odot$ in the transverse direction. For this structure, the MTOF data show low Mg/O, low speed, and low temperature. The very low ^{26}Mg/^{24}Mg ratio may indicate that the 'blobs' are released from gravitationally stratified helmet streamers. It is necessary to search the full MTOF data set to possibly identify more of these structures and to exclude them from the analysis.

The CME ejecta belong to a similar class of special plasma. Isotopic fractionation of up to a factor 2 - 3 has

TABLE 1. Average solar wind isotopic abundance ratios from in-situ spacecraft measurements on board SOHO, WIND, and ULYSSES, compared to solar system values [15]. If not noted otherwise, values are measured with SOHO/CELIAS/MTOF [3]. For the ^{3}He/^{4}He ratio the values [16] for the 'fast' (f) and the 'slow' (s) solar wind are given separately, as the fractionation appears to be much larger than the experimental uncertainty. See also the review, Ref. [17].

Ratio	Solar wind	Solar System
^{3}He/^{4}He$^{(s)}$	0.00041±0.000025	0.000488
^{3}He/^{4}He$^{(f)}$	0.00033±0.000027	0.000488
^{15}N/^{14}N	0.005±0.0014	0.00368
^{18}O/^{16}O	0.0022±0.0006 [18]	0.0020
^{21}Ne/^{20}Ne	0.0023±0.0006	0.0024±0.0003
^{22}Ne/^{20}Ne	0.0728±0.0013	0.073±0.00007
^{25}Mg/^{24}Mg	0.1260±0.0014	0.12658
^{26}Mg/^{24}Mg	0.1380±0.0031	0.13947
^{29}Si/^{28}Si	0.03344±0.00024	0.033612
^{30}Si/^{28}Si	0.05012±0.00072	0.050634
^{36}Ar/^{38}Ar	0.183±0.008 [19]	0.1880
^{42}Ca/^{40}Ca	0.00657±0.00017	0.006621
^{44}Ca/^{40}Ca	0.0209±0.0011	0.021208
^{54}Fe/^{56}Fe	$0.085^{+0.005}_{-0.022}$ [20]	0.063236

been observed in solar energetic particles associated with the November 1997 events [21]. Data from ACE/SWIMS have constrained the isotopic fractionation in the bulk solar wind during the same events to less than 5% per amu [22]. This implies that the energetic particles are fractionatied by injection or acceleration mechanisms.

However, as the isotopic composition in CME ejecta has not yet been studied rigorously, any identified CME-related solar wind has been excluded from the MTOF data analysis. It remains uncertain, though, how many 'blobs' have contributed to the analysis. Despite this uncertainty, the solar wind isotopic composition of elements heavier than ~20 amu can be identified with the solar isotopic composition within ~1% precision. Table I summarizes the solar wind isotopic composition measured by spacecraft instruments. Note, that concordant values for the He and Ne isotopic composition of the solar wind have been measured before with the Apollo Solar Wind foil Collection (SWC) experiment [23].

THEORETICAL CONSIDERATIONS

As potential processes that fractionate isotopes in the solar wind, variable Coulomb friction and wave heating in the inner corona, as well as gravitational settling in closed magnetic coronal loops are discussed.

Variable Coulomb friction in the inner corona has been studied in detail [24, 25]. The fractionation model is based on the idea that the protons, and the alpha-particles, are the fastest and dominant species in the hot corona and drag the heavier elements away from the solar surface. The parameter

$$H_i = \frac{2A_i - Z_i - 1}{Z_i^2}\sqrt{\frac{A_i+1}{A_i}}$$

orders the fractionation strength of elements and isotopes depending on their atomic mass number A_i and charge number Z_i in the solar wind [25]. The strongest fractionation occurs for $^4He^{2+}$ and $^3He^{2+}$, which is taken to calibrate typical fractionation strengths for other ions from observations on the solar wind ^{4}He/^{3}He ratio [25]. These observations constrain the typical ratio $(^4He/^3He)_s$ ratio in the 'slow' wind to be larger than 0.75 times the typical ratio $(^4He/^3He)_f$ in the 'fast' wind. This leads to typical fractionation strengths of $(^{13}C/^{12}C)_s / (^{13}C/^{12}C)_f = 0.984$, $(^{15}N/^{14}N)_s / (^{15}N/^{14}N)_f = 0.985$, $(^{18}O/^{16}O)_s / (^{18}O/^{16}O)_f = 0.969$, $(^{22}Ne/^{20}Ne)_s / (^{22}Ne/^{20}Ne)_f = 0.981$, $(^{26}Mg/^{24}Mg)_s / (^{26}Mg/^{24}Mg)_f = 0.989$, $(^{30}Si/^{28}Si)_s / (^{30}Si/^{28}Si)_f = 0.992$, and $(^{34}S/^{32}S)_s / (^{34}S/^{32}S)_f = 0.989$, comparing 'slow' and 'fast' wind with about equal temperatures of heavy ions in the 'slow' wind (see Tab. 4 of Ref. [25]; their results are not very sensitive to the ion temperatures in the 'fast' wind). This appears to be compatible with the measurements reported above.

Wave heating in a strong form, ion cyclotron damping of Alfvén waves, has been observed with the Ultraviolet Coronal Spectrograph (UVCS) in the 'fast' solar wind [26]. As lower frequencies dominate the Alfvén wave spectrum, ions with low Z_i/A_i such as O^{5+} are heated more strongly than ions with higher Z_i/A_i ratio such as protons. Therefore, heavier elements are sufficiently heated to leave the solar surface without help of the Coulomb drag. In the steady flow of the fast solar wind not much fractionation of any kind in the total flux of elemental or isotopic species is expected or observed, e.g. in form of the FIP effect [27], so that wave-heating mainly has influence on the temperatures of ion species. At 1 AU the temperatures are observed to be proportional to the ion mass [28]. The effect of the ion temperatures on the fractionation due to variable Coulomb friction has been evaluated [25]. Isothermal ion species experience the fractionation summarized above, whereas ion species with temperatures proportional to the ion mass experience stronger fractionation in the 'slow' solar wind and still fairly weak fractionation in the 'fast' wind. In the source regions of the streamer-belt ('slow') solar wind the species appear to be rather isothermal [29, 30].

Gravitational settling in closed coronal magnetic loops, however, may cause quite strong depletions or enrichments of heavy isotopes in CMEs and in 'blobs', if statically stratified closed loop material is fed to the solar wind. In hydrostatic and thermal equilibrium in a coronal loop, the pressure of a species with atomic mass M scales as [31]

$$p_M(r) = p_M(r_0)\exp(-\int_{r_0}^{r_1} \frac{r_\odot^2}{\Lambda_{0,M} r^2} dr)$$

with the scale height $\Lambda_{0,M} = 50T/M$ in meters, T the temperature in K, $r_\odot = 6.96 \times 10^8$ m the solar radius, r_0 the closest distance from the center of the Sun where hydrostatic and thermal equilibrium is reached, and r_1 the typical height of the loop. This can be rewritten as

$$p_M(\rho) = p_M(\rho_0)\exp(-\frac{14M}{T_6}\frac{\rho_1 - \rho_0}{\rho_1 \rho_0})$$

with T_6 the temperature in MK and $\rho = r/r_\odot$. For example, in the core of a helmet streamer observed with the UVCS spectrometer on board SOHO, at $\rho_1 = 1.7$ and at a measured temperature $T_6 \approx 1.3$, the elemental abundance of O was reduced by one order of magnitude compared to its typical 'slow' solar wind abundance [32]. This means $\rho_1 - \rho_0 \approx 0.04$. A fairly strong depletion of the heavier isotopes of Ne, Mg, and Si of up to 20% per amu may be associated with the largest coronal loops. However, the effect on the mean isotopic abundance ratio in the 'slow' solar wind is weaker because not all of the 'slow' solar wind is fed by closed loop material. Furthermore, the elements Ne, Mg, and Si are presumably also depleted in these 'blobs' so that their contribution to the long-time averages of the 'slow' solar wind isotopic abundances remains small. Assuming an upper limit of 20% of 'slow' solar wind flux originating from closed coronal magnetic field structures an upper limit of 0.5% per amu is estimated here for the depletion of the heavier isotopes of Ne, Mg, and Si in the 'slow' solar wind. Coronal mass ejections may carry material, which either is depleted, enriched, or not fractionated in heavy isotopes, depending on the altitude of the material in the pre-CME loop, and on the temperature profile and other physical properties of the loop. In any case, it is wise to exclude the CME-related solar wind from the analysis during the recently launched GENESIS mission.

DISCUSSION

The data in Table I demonstrate that the solar isotopic composition inferred from the in-situ solar wind measurements, averaged over long-time periods but excluding CME events, agrees well, except for He, with the solar system isotopic composition inferred from meteorites and planetary samples [15]. The values for the Ne isotopic composition in Ref. [15] already include the Apollo SWC result [23], confirmed by MTOF data [6], that the Ne isotopes in Earth's atmosphere are strongly fractionated – most likely due to hydrodynamic escape – with respect to solar system (solar wind) Ne isotopes. The differences between the solar wind ^{4}He/^{3}He ratios and the solar system ^{4}He/^{3}He ratio [15] are well explained by processes in the interior of the Sun and by fractionation through variable Coulomb friction in the source regions of the solar wind [16].

However, the solar wind ^{15}N/^{14}N ratio of this work, although rather uncertain, needs to be discussed in more detail. The value derived from MTOF data is compatible with measurements on young lunar soils [33] reflecting the solar N/Ar elemental abundance ratio and with the ^{15}N/^{14}N ratio in Earth's atmosphere. It is also compatible with the N isotopic composition in very pristine solar system material, namely in chondrites [34, 35], where 'heavy' N is associated with Ne of solar isotopic composition. It deviates from the values measured in lunar ilmenites [36] and in the Jovian troposphere [37]. A recent publication [38] suggests a protosolar isotopic ratio ^{15}N/^{14}N = $3.1^{+0.5}_{-0.4} \times 10^{-3}$, inferred from HCN in comet Hale-Bopp [39]. Although the latter value and the value of this work, ^{15}N/^{14}N = $(5 \pm 1.4) \times 10^{-3}$, are quite different, their 1σ-margins still overlap. In the past, it was discussed that recent solar wind N may not reflect the N in the early solar nebula because ^{15}N may be enriched due to spallation of ^{16}O by energetic flare protons in the solar atmosphere [40]. The solar wind ^{7}Li/^{6}Li ratio inferred from measurements on lunar samples [41] indicates a sufficient energetic proton flux to produce significant amounts of spallation ^{15}N over the Sun's lifetime. On the other hand, this appears to be incompatible with observed amounts of ^{11}B from spallation [42]. It should also be mentioned that ^{15}N is depleted in the Sun's interior because of the nuclear CNO cycle, but it is assumed that this depletion is not observable at the Sun's surface due to insufficient diffusion [42]. As predicted in the SOHO/CELIAS proposal, the present-day solar wind ^{15}N/^{14}N ratio cannot be determined unambigously from MTOF data, so that the discussion may remain open until a more precise value is available from the GENESIS mission.

CONCLUSIONS

The solar wind, in particular the coronal-hole associated solar wind, is an authentic witness of the isotopic composition of the early solar nebula for elements not influenced by processes in the Sun's interior. The interpretation of in-situ measurements of solar wind abundances is free of the pecularities to interprete abundances derived from planetary and meteoritic samples. Present in-situ solar wind measurements are limited by instrumental uncertainties. To identify Galactic nuclear processes, the precision of in-situ measurements should be increased to the permil level. This is planned for the GENESIS mission. However, pre-cautions must be taken to eliminate biases due to isotopic fractionation in the solar wind by recording its parameters.

ACKNOWLEDGMENTS

The experimental work was supported by the Swiss National Science Foundation, by the PRODEX programme of ESA, by NASA grant NAG5-2754, and by DARA, Germany, with grants 50 OC 89056 and 50 OC 9605. The MTOF sensor was developed in the CELIAS consortium [5], in a project led by the University of Maryland space physics group. The WAVE entrance system was built under the guidance of the Physics Institute of the University of Bern at INTEC, Bern. The integrated sensor and the charge exchange of solar wind ions in thin carbon foils were calibrated at the University of Bern, at the Strahlenzentrum of the University of Giessen, Germany, and at the Centre d'Etudes Nucléaires de Grenoble, France. The Technical University of Braunschweig, Germany, contributed the Data Processing Unit.

REFERENCES

1. Benz, W., Kallenbach, R., and Lugmair, G.W., *From Dust to Terrestrial Planets*, *Space Sci. Rev.*, *92*, Kluwer Acad. Publ., Dordrecht, The Netherlands, 2000.
2. Sheeley, N.R., Jr., Wang, Y.-M., Hawley, S.H., Brueckner, G.E., Dere, K.P., Howard, R.A., Koomen, M.J., Korendyke, C.M., Michels, D.J., Paswaters, S.E., Socker, D.G., St. Cyr, O.C., Wang, D., Lamy, P.L., Llebaria, A., Schwenn, R., Simnett, G.M., Plunkett, S., and Biesecker, D.A., Measurements of flow speeds in the corona between 2 and 30 $R_{\odot}$, *Astrophys. J.*, *484*, 472, 1997.
3. Kallenbach, R., *Isotopic Composition of the Solar Wind*, Habilitationsschrift (Thesis in order to receive the venia docendi), University of Bern, 2000.
4. Gonin, M., Kallenbach, R., Bochsler, P., and Bürgi, A., 1995, Charge exchange of low energy particles passing through thin carbon foils: Dependence on foil thickness and charge yields of Mg, Ca, Ti, Co and Ni, *Nucl. Instr. and Meth. B*, *101*, 313–320, 1995.
5. Hovestadt, D., Hilchenbach, M., Bürgi, A., Klecker, B., Laeverenz, P., Scholer, M., Grünwaldt, H., Axford, W.I., Livi, S., Marsch, E., Wilken, B., Winterhoff, P., Ipavich, F.M., Bedini, P., Coplan, M.A., Galvin, A.B., Gloeckler, G., Bochsler, P., Balsiger, H., Fischer, J., Geiss, J., Kallenbach, R., Wurz, P., Reiche, K.U., Gliem, F., Judge, D.L., Hsieh, K.H., Möbius, E., Lee, M.A., Managadze, G.G., Verigin, M.I., and Neugebauer, M., Charge, element, and isotope analysis system onboard SOHO, *Sol. Phys.*, *162*, 441, 1995.
6. Kallenbach, R., Ipavich, F.M., Bochsler, P., Hefti, S., Hovestadt, D., Grünwaldt, H., Hilchenbach, M., Axford, W.I., Balsiger, H., Bürgi, A., Coplan, M.A., Galvin, A.B., Geiss, J., Gliem, F., Gloeckler, G., Hsieh, K.C., Klecker, B., Lee, M.A., Livi, S., Managadze, G.G., Marsch, E., Möbius, E., Neugebauer, M., Reiche, K.-U., Scholer, M., Verigin, M.I., Wilken, B., and Wurz, P., Isotopic composition of solar wind neon measured by CELIAS/MTOF on board SOHO, *J. Geophys. Res.*, *102*, 26,895, 1997.
7. Kallenbach, R., Bochsler, P., Ipavich, F.M., Galvin, A.B., Bodmer, R., Hefti, S., Kucharek, H., Gliem, F., Grünwaldt, H., Hilchenbach, M., Klecker, B., Hovestadt, D., and the CELIAS Team, Limits on the fractionation of isotopes in the solar wind as observed with SOHO/CELIAS/MTOF, *Proc. 31st ESLAB Symp., 22-25 Sept. 1997, ESA SP, 415*, 33, 1998.
8. Kallenbach, R., Geiss, J., Ipavich, F.M., Gloeckler, G., Bochsler, P., Gliem, F., Hilchenbach, M., and Hovestadt, D., Isotopic composition of solar wind nitrogen: First in-situ determination by CELIAS/MTOF on board SOHO, *Astrophys. J.*, *507*, L185, 1998.
9. Kallenbach, R., Ipavich, F.M., Kucharek, H., Bochsler, P., Galvin, A.B., Geiss, J., Gliem, F., Gloeckler, G., Grünwaldt, H., Hefti, S., Hovestadt, D., and Hilchenbach, M., Fractionation of Si, Ne, and Mg isotopes in the solar wind as measured by SOHO/CELIAS/MTOF, *Space Sci. Rev.*, *85*, 357, 1998.
10. von Steiger, R., and Geiss, J., Supply of fractionated gases to the corona, *Astron. Astrophys.*, *225*, 222, 1989.
11. von Steiger, R., Wimmer-Schweingruber, R.F., Geiss, J., and Gloeckler, G., Abundance variations in the solar wind, *Adv. Space Res.*, *15*, (7)3, 1995.
12. Kallenbach, R., Ipavich, F.M., Kucharek, H., Bochsler, P., Galvin, A.B., Geiss, J., Gliem, F., Gloeckler, G., Grünwaldt, H., Hilchenbach, M., and Hovestadt, D., Solar wind isotopic abundance ratios of Ne, Mg, and Si measured by SOHO/CELIAS/MTOF as diagnostic tool for the inner corona, *Phys. Chem. Earth*, *24*, 415, 1998.
13. Aellig, M.R., Hefti, S., Grünwaldt, H., Bochsler, P., Wurz, P., Ipavich, F.M., and Hovestadt, D., The Fe/O elemental abundance ratio in the solar wind as observed with SOHO/CELIAS/CTOF, *J. Geophys. Res.*, *104*, 24,769, 1999.
14. Kucharek, H., Ipavich, F.M., Kallenbach, R., Klecker, B., Grünwaldt, H., Aellig, M.R., and Bochsler, P., Isotopic fractionation in slow and coronal hole associated solar wind, *Proc. 203rd IAUS Symp.*, in press, 2001.
15. Anders, E., and Grevesse, N., Abundances of the elements – Meteoritic and solar, *Geochim. Cosmochim. Acta*, *53*, 197, 1989.
16. Gloeckler, G., and Geiss, J., Measurement of the abundance of Helium-3 in the Sun and in the Local Interstellar Cloud with SWICS on ULYSSES, *Space Sci. Rev.*, *84*, 275, 1998.
17. Wimmer-Schweingruber, R.F., Bochsler, P., and Wurz, P., Isotopes in the solar wind: New results from ACE, SOHO, and WIND, *Solar Wind Nine*, edited by S.R. Habbal, R. Esser, J.V. Hollweg, and P.A. Isenberg, pp. 147–152, AIP, 1999.
18. Collier, M.R., Hamilton, D.C., Gloeckler, G., Ho, G., Bochsler, P., Bodmer, R., and Sheldon, R., Oxygen 16 to oxygen 18 abundance ratio in the solar wind observed by Wind/MASS, *J. Geophys. Res.*, *103*, 7, 1998.
19. Weygand, J.M., Ipavich, F.M., Wurz, P., Paquette, J.A., and Bochsler, P., *ESA SP, 446*, 701, 1999.

20. Oetliker, M., Hovestadt, D., Klecker, B., Collier, M.R., Gloeckler, G., Hamilton, D.C., Ipavich, F.M., Bochsler, P., and Managadze, G.G., The isotopic composition of iron in the solar wind: First measurements with the MASS sensor on the WIND spacecraft, *Astrophys. J., 474*, L69, 1997.
21. Leske, R.A., Cohen, C.M.S., Cummings, A.C., Mewaldt, R.A., Stone, E.C., Dougherty, B.L., Wiedenbeck, M.E., Christian, E.R., and von Rosenvinge, T.T., Unusual isotopic composition of solar energetic particles observed in the November 6, 1997, event, *Geophys. Res. Lett., 26*, 153, 1999.
22. Wimmer-Schweingruber, R.F., Bochsler, P., Gloeckler, G., Ipavich, F.M., Geiss, J., Kallenbach, R., Fisk, L.A., Hefti, S., and Zurbuchen, T.H., On the bulk isotopic composition of magnesium and silicon during the May 1998 CME: ACE/SWIMS, *Geophys. Res. Lett., 26*, 165, 1999.
23. Geiss, J., Bühler, F., Cerutti, H., Eberhardt, P., and Filleux, C., Solar wind composition experiment, *NASA SP, 315*, 14.1, 1972.
24. Geiss, J., Hirt, P., and Leutwyler, On acceleration and motion of ions in corona and solar wind, *Sol. Phys., 12*, 458, 1970.
25. Bodmer, R., and Bochsler, P., Influence of Coulomb collisions on isotopic and elemental fractionation in the solar wind acceleration process, *J. Geophys. Res., 105*, 47, 2000.
26. Cranmer, S.R., Ion cyclotron wave dissipation in the solar corona: The summed effect of more than 2000 ion species, *Astrophys. J., 532*, 1197, 2000, and references therein.
27. von Steiger, R., Schwadron, N.A., Fisk, L.A., Geiss, J., Gloeckler, G., Hefti, S., Wilken, B., Wimmer-Schweingruber, R.F., and Zurbuchen, T.H., Composition of quasi-stationary solar wind flows from SWICS/ULYSSES, *J. Geophys. Res., 105*, 27,217, 2000.
28. Bochsler, P., Geiss, J., and Joos, R., Kinetic temperatures of heavy ions in the solar wind, *J. Geophys. Res., 90* 10,779, 1985.
29. Borrini, G., Gosling, J.T., Bame, S.J., Feldman, W.C., and Wilcox, J.M., Solar wind helium and hydrogen structure near the heliospheric current sheet: A signal of coronal streamers at 1 AU, *J. Geophys. Res., 86*, 4565, 1981.
30. Hefti, S., Grünwaldt, H., Ipavich, F.M., Bochsler, P., Hovestadt, D., Aellig, M.R., Hilchenbach, M., Kallenbach, R., Galvin, A.B., Geiss, J., Gliem, F., Gloeckler, G., Klecker, B., Marsch, E., Möbius, E., Neugebauer, M., and Wurz, P., Kinetic properties of solar wind minor ions and protons measured with SOHO/CELIAS, *J. Geophys. Res., 103*, 29,697, 1998.
31. Priest, E.R., *Solar Magneto-Hydromagnetics, Geophys. Astrophys. Monogr., 21*, 118, 1982.
32. Raymond, J.C., Kohl, J.L., Noci, G., Antonucci, E., Tondello, G., Huber, M.C.E., Gardner, L.D., Nicolosi, P., Fineschi, S., Romoli, M., Spadaro, D., Siegmund, O.H.W., Benna, C., Ciaravella, A., Cranmer, S., Giordano, S., Karovska, M., Martin, R., Michels, J., Modigliani, A., Naletto, G., Panasyuk, A., Pernechele, C., Poletto, G., Smith, P.L., Suleiman, R.M., and Strachan, L., Composition of coronal streamers from the SOHO ultraviolet coronagraph spectrometer, *Sol. Phys., 175*, 645, 1997.
33. Kim, J.S., Kim, Y., Marti, K., and Kerridge, J.F., Nitrogen isotope abundances in the recent solar wind, *Nature, 375*, 383, 1995.
34. Kung, C.C., and Clayton, R.N., Nitrogen abundances and isotopic compositions in stony meteorites, *Earth Planet. Sci. Lett., 38*, 421, 1978.
35. Sugiura, N., Kiyota, K., and Hashizume, K., Nitrogen components in primitive ordinary chondrites, *Met. Planet. Sci., 33*, 463, 1998, and references therein.
36. Hashizume, K., Chaussidon, M., and Marty, B., Nitrogen isotope analyses of lunar regolith using an ion microprobe; in search of the solar wind component, 31st Lunar Planet. Sci. Conf., #1565, 2000.
37. Fouchet, T., Lellouch, E., Bézard, B., Encrenaz, T., Drossart, P., Feuchtgruber, H., and de Graauw, T., ISO-SWS observations of Jupiter: Measurement of the ammonia tropospheric profile and of the $^{15}N/^{14}N$ isotopic ratio, *Icarus, 143*, 223, 2000.
38. Owen, T., Mahaffy, P.R., Niemann, H.B., Atreya, S., and Wong, M., Protosolar nitrogen, *Astrophys. J., 553*, L77, 2001.
39. Jewitt, D.C., Matthews, H.E., Owen, T., and Meier, R., Measurements of $^{12}C/^{13}C$, $^{14}N/^{15}N$, and $^{32}S/^{34}S$ ratios in comet Hale-Bopp (C/1995 O1), *Science, 278*, 90, 1997.
40. Kerridge, J.F., Solar nitrogen – Evidence for a secular increase in the ratio of nitrogen-15 to nitrogen-14, *Science, 188*, 162, 1975.
41. Chaussidon, M., and Robert, F., Lithium nucleosynthesis in the Sun inferred from the solar-wind $^{7}Li/^{6}Li$ ratio, *Nature, 402*, 270, 1999.
42. Geiss, J., and Bochsler, P., Nitrogen isotopes in the solar system, *Geochim. Cosmochim. Acta, 46*, 529, 1982.

SOLAR WIND IRON ISOTOPIC ABUNDANCES: RESULTS FROM SOHO/CELIAS/MTOF

F.M. Ipavich*, J.A. Paquette*, P. Bochsler†, S.E. Lasley* and P. Wurz†

*Department of Physics, University of Maryland, College Park, MD, USA 20742
†Physikalisches Institut der Universität Bern, Switzerland, CH3012

Abstract. The MTOF sensor uses time of flight measurements in a harmonic potential region to identify elements and isotopes in the solar wind with excellent mass resolution. The combination of MTOF's large bandwidth electrostatic deflection system and the 3-axis stabilized orientation of SOHO results in excellent counting statistics. We report relative abundances of the iron isotopes with mass 54, 56 and 57 amu. Since these isotopes are chemically identical, we expect little fractionation either in the solar wind or in the instrument, resulting in relatively small estimated uncertainties. Our results agree, within the measurement uncertainties, with terrestrial values.

INTRODUCTION

Observations of rare elements and isotopes in the solar wind are important because they help determine how neutral atoms in the relatively cool photosphere become the highly ionized atoms in the corona and then ultimately the supersonic plasma we call the solar wind. The results will also help establish the isotopic composition of the primordial solar nebula.

The good agreement of photospheric and meteoritic abundances [1] for a wide range of refractory and moderately volatile elements are consistent with a common origin of solar and planetary matter. The Sun is considered to contain a largely unfractionated sample of matter from the protosolar nebula; hence knowledge of solar composition can reveal the composition of the local interstellar medium some 4.6 Gy ago, and as well shed light on the chemical and isotopic evolution of planetary matter [2]. The solar wind composition is expected to be that of the solar outer convective zone, differing only as a result of possible fractionation processes. Elemental composition is subject to the well-known fractionation process ordered by the first ionization potential [3], reflecting the separation of neutrals and ions in the upper chromosphere; the magnitude of this fractionation is of order 3 or so in the slow solar wind. Isotopic composition is, however, not subject to this fractionation process; typical isotopic fractionation is expected to be at the percent level.

The Mass Time-of-Flight Spectrometer (MTOF) is one of four sensors (CTOF, MTOF, STOF, SEM) comprising the Charge, Element, Isotope Analysis System (CELIAS) Experiment [4] on the Solar Heliospheric Observatory (SOHO) spacecraft. CELIAS uses three time-of-flight (TOF) sensors to make composition measurements. MTOF has unprecedented mass resolution and collection power for solar wind composition studies, and can identify rare elements and isotopes that were previously not resolvable from more abundant neighboring species, or were not previously observable at all.

INSTRUMENTATION

The Mass Time-Of-Flight (MTOF) sensor of the CELIAS investigation is actually two subsensors. The MTOF/Main is the primary unit, providing elemental and isotopic abundance measurements of heavy ions in the solar wind. The MTOF/Proton Monitor is an auxiliary unit designed to measure the solar wind proton parameters. Both units are housed within a common structure which also contains the low voltage power converter, the high voltage power supplies, the analog electronics, and the digital electronics.

The MTOF Main Sensor

The MTOF Main Sensor (illustrated in Figure 1) is a high-mass ($M/\Delta M \approx 100$) resolution system that provides unprecedented solar wind composition data over a wide range of solar wind conditions. The high sensitivity of MTOF allows for the first time an accurate determi-

CP598, *Solar and Galactic Composition*, edited by R. F. Wimmer-Schweingruber

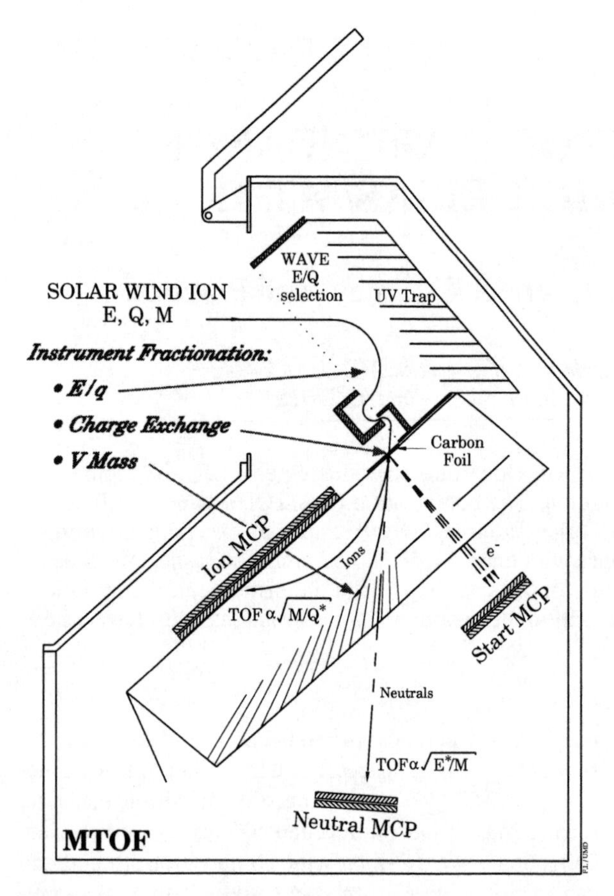

FIGURE 1. Schematic of the MTOF (Mass Time of Flight) Instrument. Solar wind ions enter through an energy/charge filter (the WAVE) and pass through a carbon foil, leaving them with a charge state q^*. They then enter a region where a special electric field bends those with $q^* > 0$ down to a microchannel plate. For an ion of mass M, the time of flight in this region is $\propto \sqrt{M/q^*}$.

nation of the abundances of many of the elements and isotopes in the solar wind. The MTOF sensor consists of the Wide-Angle, Variable Energy/charge (WAVE) passband deflection system and the time-of-flight High-Mass Resolution Spectrometer (HMRS).

MTOF owes its excellent mass resolution capability to a specially designed electric field configuration in its time-of-flight region. An ion's time-of-flight through this region is *independent* of its initial energy or angle. Sub-nanosecond TOF measurements translate into mass resolutions of a fraction of an AMU.

A prototype of the MTOF sensor, known as MASS, was flown on the WIND spacecraft (launched about a year before SOHO), and was able to identify a number of elements and isotopes for the first time. The MTOF sensor incorporates a number of enhancements over its prototype, including position-sensing and signal amplitude measurements, and improved background-rejection techniques. MTOF also generates a 2 to 3 order of magnitude improvement in counting statistics, thanks to: (a) the fact that SOHO always looks in the solar direction (unlike the spinning WIND spacecraft); and (b) a novel entrance deflection system with a very wide bandwidth (400% vs. the more typical 5% in other solar wind instruments), requiring only a few voltage steps to cover the entire energy-per-charge distribution of the solar wind. As a consequence of the very wide deflection system passband, MTOF cannot determine the charge state distribution of solar wind ions.

The MASS and MTOF sensors were designed and fabricated by the University of Maryland; their entrance deflection systems were built by the University of Bern, which also provided the primary calibration facility for these instruments.

THE MTOF TOF SYSTEM

The heart of the MTOF sensor is the time-of-flight High Mass Resolution Spectrometer (HMRS). The principal of operation of the HMRS is based on the fact that the time of flight t of an ion of mass M is proportional to $\sqrt{M/q^*}$ in the presence of an electric field that increases linearly with distance. Here q^* is the ion's charge after penetrating a thin carbon foil. Hence, the measurement of t gives unambiguous values of M/q^* for individual ions. The required electric field is produced by a combination of a hyperbolic plate set at a large positive voltage V_H (typically 20 to 25 kV) and a V-shaped plate at ground potential. The high mass resolution of the MTOF sensor is due to the fact that t is independent of the ion energy and the angle at which the ion enters the HMRS electric field. The value of V_H determines the maximum ion energy that can be deflected by the electric field, but does not affect the mass resolution. Since t can be measured with the precision of a fraction of a nanosecond, high mass resolution is achieved. Figure 2 displays a spectrum, where the TOF has been converted into mass units, near the mass range of iron.

The practical implementation of the HMRS requires an accurate measurement of t and the conversion of multiply-charged solar wind ions into singly-charged ions. This is accomplished with a combination of microchannel plate (MCP) detectors and a thin carbon foil at the entrance to the HMRS electric-field region. Ions passing through the foil undergo a large number of collisions with the carbon atoms, resulting in some energy loss, moderate scattering, and charge exchange. As a result of the charge exchange, ions with initial charge state q emerge from the foil with charge state q^*, where q^* is typically 0 or +1. As the ions leave the foil they also produce secondary electrons that are deflected to a "Start" MCP assembly that generates a start signal for the time-

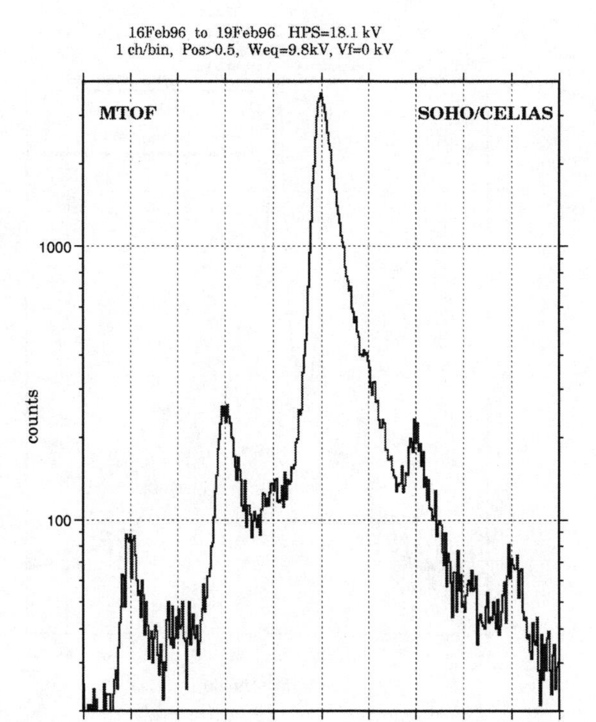

FIGURE 2. Observed solar wind mass spectrum accumulated over a 3 day period in 1996

of-flight analysis. The "stop" signal for particles with q^* > 0 is generated at a second large area "Ion" MCP assembly located behind the ground surface of the electric field region; for particles with q^* = 0, the "stop" signal is generated by the "Neutral" MCP.

The anodes for all three MCPs provide not only the required timing signals, but also amplitude signals that are pulse height-analyzed by the electronics. These amplitude signals provide some information about particle mass and energy. In addition, the Neutral and Ion MCP anodes provide 1-dimensional position information. The Neutral position is a measure of the solar wind ion's flow direction in the solar ecliptic plane, useful for interpretation of the data. The Ion position allows us to determine the E/q of an ion, and is also useful for rejecting background events.

The entire HMRS is maintained at a 'floating' voltage, V_F, in the range -5 to +5 kV. For high solar wind speeds, a positive V_F decelerates heavy ions so that they can be deflected by the hyperbola voltage and detected by the Ion MCP. For low solar wind speeds a negative V_F will accelerate the ions before they reach the carbon foil, and thereby improve the efficiency of their detection in the HMRS.

The MTOF Proton Monitor SubSensor

The MTOF Proton Monitor (PM) is designed to measure the solar wind proton bulk speed, density and thermal speed. The bulk speed (and, to a lesser extent, the thermal speed) is necessary to the evaluation of data from the main MTOF sensor. The PM is described in detail elsewhere [5].

MAXIMUM LIKELIHOOD TECHNIQUE

The counting rates in a series of Time of Flight Channels were simulated using a maximum likelihood technique. In this technique, the simulated count rate in a particular channel was taken as the mean, μ_i, of a Poisson distribution. Assuming that this Poisson distribution is the parent distribution from which the measurements were derived, the probability that any given number of counts N_i was observed in that channel is

$$P_i = \frac{\mu_i^{N_i}}{N_i!} e^{-\mu_i}$$

The probability that the measured TOF spectrum was observed, given the assumptions in the simulation, is simply the product of the probabilities for each channel:

$$P_{Tot} = \prod P_i$$

The goal of is to maximize this probability by altering the parameters of the model function, thus altering the μ_i. In practice, it is easier and equivalent to deal with the natural log of this expression:

$$\ln(P_{Tot}) = \sum_i N_i ln(\mu_i) - \mu_i - ln(N_i!)$$

A constrained minimization routine (E04KDF from the NAG Library; see [6]) was used to minimize the negative logarithm of the probability (equivalent to maximizing the positive logarithm.)

The model function used had 24 parameters. Fourteen of these were the heights of the peaks corresponding to masses 48, 49, and 52-63. Of the remaining 10, two described a linear background, two were needed to convert from mass to time of flight channel, two were associated with peaks caused by electronic "ringing" and the remaining 4 described the shape and width of the peaks. The form of the model function was a modified asymmetric Lorentzian:

$$\frac{A}{1 + ((X - X_0)/\Gamma)^n}$$

where A is the amplitude (in counts) for each of the 14 mass values, and X_0 is the peak location (in time of flight channel number) that is derived from the mass values. The exponent, n, and the half width at half maximum, Γ, are the same for all species, but are allowed to be different on either side of the peak.

OBSERVATIONS

The mass spectrum displayed in Figure 2 was accumulated over a 3 day time period in February 1996 during which time the solar wind was of a nondescript interstream type. The maximum likelihood technique was applied to derive the abundance ratios of the Fe isotopes. These abundance ratios are then corrected by the ratios of instrument efficiencies for these isotopes. The efficiency corrections depend weakly on the solar wind speed and temperature, the deflection system and hyperbolic plate voltage settings, and the incident charge state distributions of the Fe ions, and were in the range of 1% to 4% for the time intervals selected for this paper.

The solar wind behavior during another interstream time period is shown in Figure 3. The mass spectrum accumulated during this time interval is displayed in Figure 4.

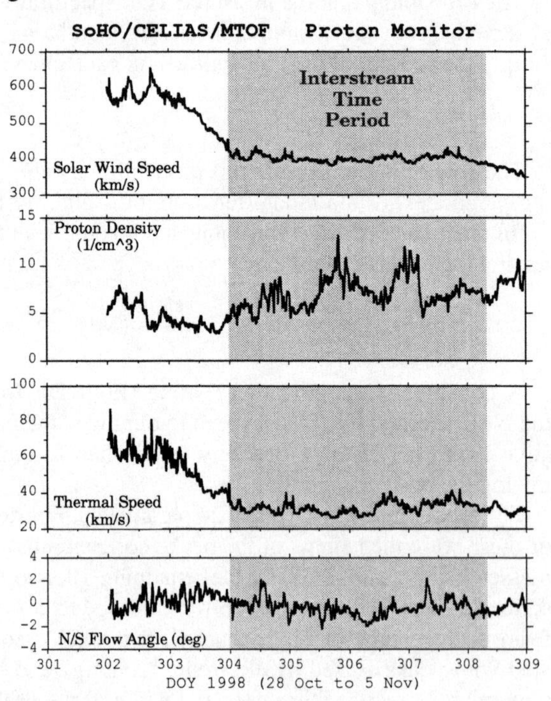

FIGURE 3. Solar wind behavior during a typical interstream time period.

In addition to the two individual time periods discussed above, mass spectra have also been obtained for two 12-month time periods (April 1996 to March 1997 and March 1997 to March 1998). For both 12-month periods, only those 5-minute intervals in which the solar wind speed was between 380 and 400 km/s were included. This was done in order to minimize the variation of the relative efficiencies.

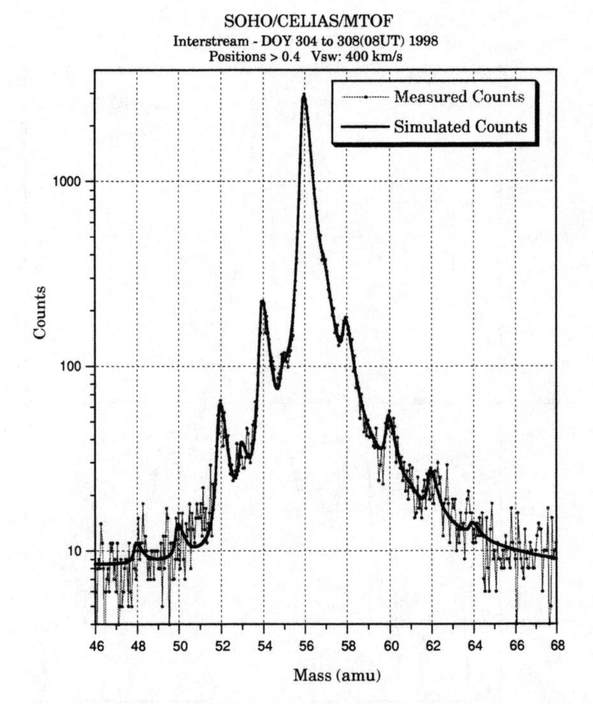

FIGURE 4. Mass spectrum accumulated during the shaded time period indicated in Figure 3. The fit from the maximum likelihood technique is indicated by the solid line.

RESULTS

The efficiency-corrected ratios derived for Fe^{54}/Fe^{56} and Fe^{57}/Fe^{56} in the 4 intervals discussed above are presented in Figure 5 and Figure 6, respectively. In each figure the terrestrial value from [7] is indicated by the solid horizontal line.

The results averaged over all 4 time periods are shown in Table 1, along with the accepted terrestrial values from [7]. Oetliker et al. [8] derived solar wind Fe isotopic ratios using the Wind/MASS sensor. Their results have larger uncertainties but are in agreement with the present work. They obtained a value (in percent) of 8.5 (+0.5, -2.2) for Fe^{54}/Fe^{56} and an upper limit of 5% for Fe^{57}/Fe^{56}.

The Fe isotope ratios in Table 1 are, within the measurement uncertainties, consistent with no fractionation of Fe isotopes in the solar wind. They are however also consistent with a modest fractionation effect. A mass dependent isotopic fractionation has been reported for Ne, Mg and Si [9], with the heavier isotope depleted in the slow solar wind. The average magnitude of this effect

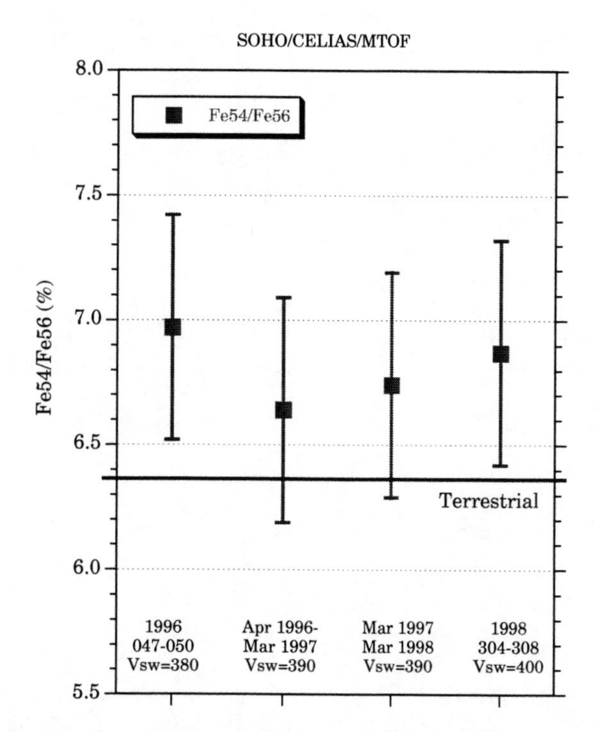

FIGURE 5. The derived ratios of Fe^{54} to Fe^{56} obtained during 4 distinct time intervals. The error bars denote 1-σ uncertainties and are dominated by systematic effects. The terrestrial value is indicated by the solid horizontal line.

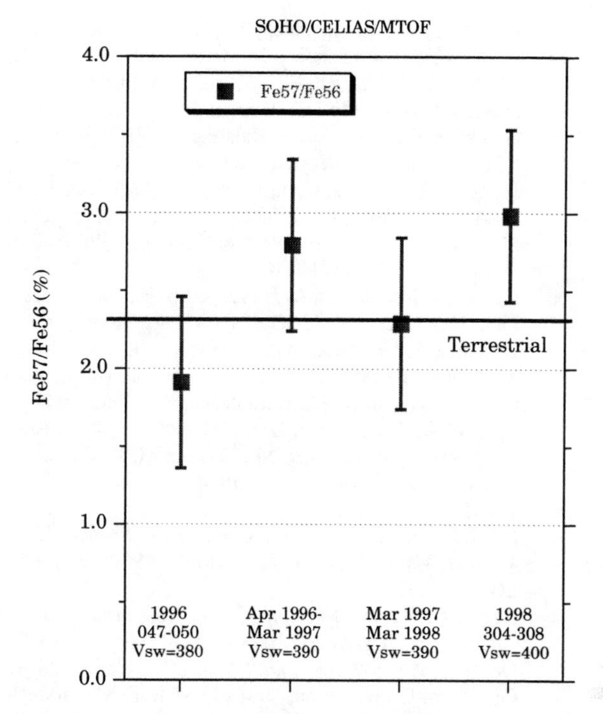

FIGURE 6. The derived ratios of Fe^{57} to Fe^{56} obtained during 4 distinct time intervals. The error bars denote 1-σ uncertainties. The terrestrial value is indicated by the solid horizontal line.

TABLE 1. Iron Isotopic Abundance Ratios

	Fe Isotopic Ratio (in percent)	
	This Work	Terrestrial*
Fe^{54}/Fe^{56}	6.8 ± 0.4	6.37
Fe^{57}/Fe^{56}	2.5 ± 0.5	2.31

* Beard and Johnson, 1999

was reported to be approximately 1.4% per amu. The Fe isotopic measurements reported here were all obtained from such slow solar wind flows. The fractionation effect from [9] would predict an enhancement of some 2.8% for Fe^{54}/Fe^{56} relative to the unfractionated solar composition, a result also consistent with our observed enhancement of about $(7 \pm 6)\%$ above the terrestrial value.

The current uncertainties in the isotope ratios presented in this paper are dominated by systematic effects. One such effect is the lack of knowledge of the precise shape of the TOF spectrum of Fe^{56} at TOF values that are far from the most probable value. We are attempting to improve our knowledge of the response in the wings of the TOF distribution by calibrating the MTOF spare instrument at an accelerator facility at the University of Bern [10]. The derived mass distribution can also be improved by carefully correcting for the effects caused by "electronic walk", in which different amplitude signals create slightly different times of flight. There is also a very small but statistically significant variation in the TOF with the residual ion energy (which can be determined from the MCP position measurement) and also with instrument temperature. We estimate that correcting for all of the above effects will decrease our uncertainty in the iron isotope ratios by perhaps a factor of two. It is also planned to examine the Fe isotope ratios in higher speed solar wind flows.

ACKNOWLEDGMENTS

We are grateful to all the individuals who participated in the design, the construction, and the calibration of CELIAS/MTOF. This research was supported by NASA grants NAG5-7678 and NAG5-9282.

REFERENCES

1. Anders, E., and Grevesse, N., *Geochimica et Cosmochimica Acta*, **53**, 197–214 (1989).
2. Bochsler, P., *Reviews of Geophysics*, **38**, 247–266 (1999).
3. Geiss, J., *Space Science reviews*, **85**, 241–252 (1998).

4. Hovestadt, D., Hilchenbach, M., Bürgi, A., Klecker, B., Laeverenz, P., Scholer, M., Grünwaldt, H., Axford, W. I., Livi, S., Marsch, E., Wilken, B., Winterhoff, H. P., Ipavich, F. M., Bedini, P., Coplan, M. A., Galvin, A. B., Gloeckler, G., Bochsler, P., Balsiger, H., Fischer, J., Geiss, J., Kallenbach, R., Wurz, P., Reiche, K.-U., Gliem, F., Judge, D. L., Ogawa, H. S., Hsieh, K. C., Möbius, E., Lee, M. A., Managadze, G. G., Verigin, M. I., and Neugebauer, M., *The SOHO Mission*, Kluwer Academic Publishers, Dordrecht, 1995, pp. 441–481.
5. Ipavich, F. M., Galvin, A. B., Lasley, S. E., Paquette, J. A., Hefti, S., Reiche, K.-U., Coplan, M. A., Gloeckler, G., Bochsler, P., Hovestadt, D., Grünwaldt, H., Hilchenbach, M., Gliem, F., Axford, W. I., Balsiger, A., Bürgi, A., Geiss, J., Hsieh, K. C., Kallenbach, R., Klecker, B., Lee, M. A., Managadze, G. G., Marsch, E., Möbius, E., Neugebauer, M., Scholer, M., Verigin, M. I., Wilken, B., and Wurz, P., *Journal of Geophysical Reasearch*, **103**, 17205–17213 (1998).
6. NAG, Numerical Algorithm Group, *NAG Fortran Library Manual, Mark 18*, NAG Ltd, Oxford, 1997, vol. 4, chap. E04.
7. Beard, B. L., and Johnson, C. M., *Geochimica et Cosmochimica Acta*, **63**, 1653–1660 (1999).
8. Oetliker, M., Hovestadt, D., Klecker, B., Collier, M. R., Gloeckler, G., Hamilton, D. C., Ipavich, F. M., Bochsler, P., and Managadze, G. G., *The Astrophysical Journal Letters*, **474**, L69–L72 (1997).
9. Kallenbach, R., Ipavich, F. M., Kucharek, H., Bochsler, P., Galvin, A. B., Geiss, J., Gliem, F., Gloeckler, G., Grünwaldt, H., Hefti, S., Hilchenbach, M., and Hovestadt, D., *Space Science Reviews*, **85**, 357–370 (1998).
10. Marti, A., Schletti, R., Wurz, P., and Bochsler, P., *Reviews of Scientific Instruments*, **72**, 1354–1360 (2001).

Isotopic Abundances in the Solar Corona as Inferred from ACE Measurements of Solar Energetic Particles

R. A. Leske*, R. A. Mewaldt*, C. M. S. Cohen*, E. R. Christian†, A. C. Cummings*, P. L. Slocum**, E. C. Stone*, T. T. von Rosenvinge† and M. E. Wiedenbeck**

*California Institute of Technology, Pasadena, CA 91125 USA
†NASA/Goddard Space Flight Center, Greenbelt, MD 20771 USA
**Jet Propulsion Laboratory, Pasadena, CA 91109 USA

Abstract. The isotopic composition of solar energetic particles (SEPs) has been measured using the Solar Isotope Spectrometer on the Advanced Composition Explorer. The measurements include up to 12 isotope abundance ratios for ten elements from C through Ni at energies of tens of MeV/nucleon in 18 large SEP events that have occurred since November 1997. These measurements clearly establish that SEP isotopic composition can vary widely (by factors of > 3) from event to event, presumably due to mass fractionation processes during particle acceleration and/or transport. Elemental and isotopic abundance ratios are strongly correlated, suggesting that elemental and isotopic fractionation relative to the coronal source are largely governed by the same processes. Using empirical correlations to correct for the fractionation and obtain the coronal isotopic composition yields preliminary abundance values in good agreement with those found in the solar wind, with comparable accuracy.

INTRODUCTION

A primary goal of the Advanced Composition Explorer (ACE) mission is to better establish the elemental and isotopic composition of the Sun. Solar energetic particles (SEPs) provide a sample of solar material that may be used for such studies, but particle acceleration and transport processes can affect the arriving composition. Two distinct categories of SEP events, impulsive and gradual, are generally recognized [1]. Particles in gradual events are thought to originate as solar wind or coronal material accelerated by large shocks driven by coronal mass ejections. Elements in gradual SEP events have been measured for many years [e.g. 2]. Their abundances have been found to be highly variable from event to event but correlated with the ionic charge to mass ratio, Q/M [3]. When corrected for this fractionation [3, 4] or averaged over many events [5], SEP abundances can be used to determine the coronal elemental composition more accurately than is possible from spectroscopic measurements for some elements such as noble gases. In principle, the coronal isotopic composition can be similarly obtained from SEPs [6, 7], which has not been possible spectroscopically for more than a few isotopes.

Unlike the case for SEP elemental measurements, before the launch of ACE there were only a few SEP heavy isotope measurements, and these included only elements up to Si [see, e.g. 7, and references therein]. To obtain adequate statistical accuracy, the earlier measurements sometimes required sums over several SEP events [8, 9, 10] and the resulting values usually agreed with terrestrial abundances but with rather large uncertainties. Isolated differences were noted for some gradual events [6, 7], and significant enrichments of ^{22}Ne were found in ^{3}He-rich periods [11].

In recent studies using ACE data, enhancements by up to a factor of ~2 were reported in the 6 November 1997 SEP event for many heavy/light isotope abundance ratios from ^{13}C/^{12}C to ^{60}Ni/^{58}Ni [12]. Using the Solar Isotope Spectrometer (SIS) on ACE, the ^{22}Ne/^{20}Ne ratio was observed to vary by a factor of > 3 from event to event at energies of 24–72 MeV/nucleon [13], and similar or greater variability was found for Ne in other events at lower energies [14]. In the present work, we extend the previous ACE/SIS studies and present isotopic abundance measurements for C, O, Ne, Mg, Si, S, Ar, Ca, Fe, and Ni in as many as 18 individual SEP events. The isotopic composition is highly variable, but using the abundance correlations between different species we empirically correct for the variation and obtain preliminary coronal isotopic abundances from SEPs. Observations in a subset of the events over a more limited energy interval are also reported in [15, 16].

CP598, *Solar and Galactic Composition,* edited by R. F. Wimmer-Schweingruber

OBSERVATIONS AND ANALYSIS

Using the dE/dx versus residual energy technique in a pair of silicon solid-state detector telescopes, the SIS instrument allows the nuclear charge, Z, mass, M, and total kinetic energy, E, to be determined for particles with energies of $\sim$10 to $\sim$100 MeV/nucleon [17]. For this study, we selected SEP events with sufficient fluxes of high energy heavy ions ($E \gtrsim 15$ MeV/nucleon, where mass resolution is best) to obtain statistically meaningful isotope abundances. Time profiles of the 18 selected events are shown in Figure 1. Their peak intensities vary by more than three orders of magnitude at these energies, with the events in the year 2000 (bottom panel) containing most of the smallest events considered for this study as well as the two largest SEP events in this solar cycle. The very high counting rates and correspondingly high rate of chance coincidences at the peaks of these two largest events severely degraded mass resolution in SIS (although the resolution is still adequate to separate elements), and therefore we restricted the isotopic analysis to the decay phases of these two events.

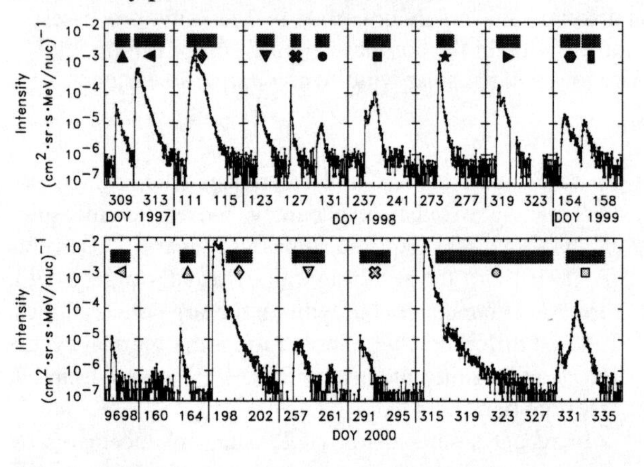

FIGURE 1. Time profiles of the 18 SEP events examined here, using hourly-averaged intensities of 21-64 MeV/nucleon oxygen from SIS on ACE. Shaded bars indicate time periods used for the isotope analysis; symbols represent these periods in Figures 3–5.

Isotopes of elements up through Ni are measured by SIS with a mass resolution which varies with Z and E; for the species and energies studied here it typically ranges from $\sim$0.15 to $\sim$0.3 amu. Details of the analysis required to obtain isotope abundance ratios and examples of mass histograms are given elsewhere [12, 13], but in most cases the good mass resolution makes the determination of abundances straightforward.

Obtaining coronal abundances from these data is not so simple, however, due to the fact that the SEP isotopic abundances may vary significantly from event to event [13], as shown for the ^{22}Ne/^{20}Ne ratio in Figure 2. Surprisingly, it appears that the composition variability was greatly reduced and nearly absent in the 1999–2000 time frame compared to that seen in 1997–1998. The reduced χ^2 obtained from fitting a constant to the first 9 points in Figure 2 is 83, while for the last 9 points it is a much more reasonable 1.3, indicating a statistically significant change in the variability. Tracking the variability of the composition in future events may help to determine whether this is merely a statistical aberration or an unexplained feature of the solar cycle.

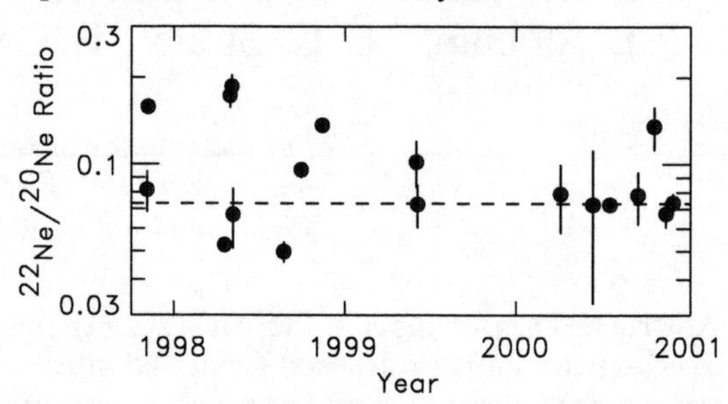

FIGURE 2. The SEP ^{22}Ne/^{20}Ne ratio measured by ACE/SIS at $E > 15$ MeV/nucleon plotted versus the date of the event.

In obtaining coronal abundances from the highly variable SEP isotope measurements, we are guided by the experience gained in coping with a similar situation encountered for elemental abundances. The variations of heavy ion elemental abundances in individual gradual events relative to coronal values have been found to scale reasonably well as a power law in the ionic charge to mass ratio, Q/M [3], with a different power law index for each SEP event. If the Q/M ratio is indeed the relevant organizing parameter, then the same physical mechanism responsible for the elemental fractionation, whatever it is, should also produce variations in the isotopic abundances, since Q/M will differ for two isotopes of the same element through the mass number. This implies there should be a predictable correlation between the abundances of various species, in particular between elemental and isotopic abundances. Following [6], if we base the power law fractionation index on the abundance ratio of any two reference species, such as Fe/O, Na/Mg, or, in general terms, R_1/R_2, (and remembering that $x^{\ln y} = y^{\ln x}$) it readily follows that we would expect the enhancement or depletion of the SEP abundance ratio for isotopes a and b of element X to be:

$$\frac{(^a\mathrm{X}/^b\mathrm{X})_{\mathrm{SEP}}}{(^a\mathrm{X}/^b\mathrm{X})_{\mathrm{corona}}} = \left(\frac{(\mathrm{R}_1/\mathrm{R}_2)_{\mathrm{SEP}}}{(\mathrm{R}_1/\mathrm{R}_2)_{\mathrm{corona}}}\right)^{\frac{\ln(b/a)}{\ln[(Q/M)_{\mathrm{R}_1}/(Q/M)_{\mathrm{R}_2}]}} \quad (1)$$

using the fact that Q should be the same for two isotopes of the same element.

To evaluate the expected enhancement from equation (1), the ionic charge state Q must be known for the reference species. Ionic charges are not often measured at SIS energies of tens of MeV/nucleon, and the measurements that exist show considerable variability for elements such as Fe [see, e.g. 18, and references therein]. Although measurements of Q at lower energies might be used [3], they may not apply to SIS data since several events have been clearly shown to exhibit energy-dependent charge states for heavy elements [19, 20, 21]. In addition, the relevant value of Q in equation (1) is that which the particles have when the elemental and isotopic fractionation takes place, which may be quite different from the value at 1 AU if fractionation happens early and if, for example, stripping occurs in acceleration through the corona [22, 23]. Another complication not addressed in equation (1) is the fact that in gradual SEP events the abundances of elements with low first ionization potential (FIP), such as Fe, are generally enhanced over those with high FIP, such as O, by an amount which also varies from event to event [24, 25]. In spite of the above considerations, a reasonable correlation has been shown between isotopic abundances and the Fe/O ratio [12, 13, 15], but uncertainties in the value of Q(Fe) at SIS energies make it difficult to directly compare the correlation with predictions.

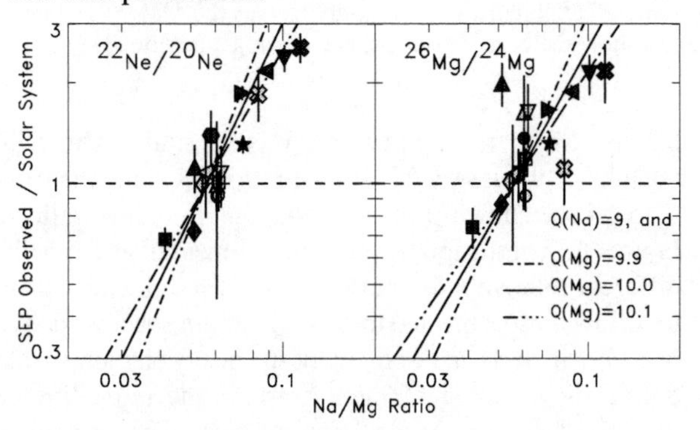

FIGURE 3. The ^{22}Ne/^{20}Ne (left) and ^{26}Mg/^{24}Mg (right) isotopic ratios versus the Na/Mg elemental abundance ratio in each of the SEP events of Figure 1. Both isotope ratios have been normalized to their respective "solar system" value [26]. Diagonal lines show the correlations expected from equation (1), assuming the charge states of Na and Mg as indicated.

With the appropriate choice of reference species, such as Na/Mg, we do find that the isotopic and elemental abundances tend to be correlated approximately as expected from equation (1), as shown in Figure 3. Both Na and Mg are low-FIP elements so this ratio is unaffected by variable FIP fractionation. As pointed out by Cohen et al. [27], both elements are theoretically expected to have ~2 electrons attached over a broad range of coronal temperatures [28], and since ^{23}Na is neutron-rich with respect to ^{24}Mg, there is a significant difference in Q/M. The solid lines in Figure 3 show the correlations expected from equation (1), assuming Q(Na)=9 and Q(Mg)=10. While this very simple model provides a good first order fit to the data, both isotope correlations appear to be shallower than expected.

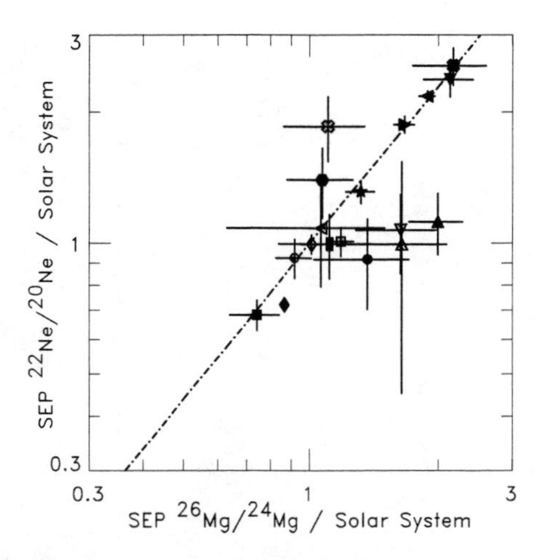

FIGURE 4. The ^{22}Ne/^{20}Ne versus ^{26}Mg/^{24}Mg isotopic ratios in each of the SEP events shown in Figure 1, normalized to standard solar system values [26]. The diagonal line shows the correlation expected using equation (1).

The predicted correlations are very sensitive to Q, and a small change of only 1% in the Q(Na)/Q(Mg) ratio changes the expected slope considerably, as also shown in Figure 3. Detailed equilibrium calculations [28] show that at a constant temperature Q(Na)/Q(Mg) is > 0.9 for all temperatures from 0.5–10 MK, that is, the expected correlation should be steeper than the solid line in Figure 3 and more discrepant with the data. However, the assumption of a constant temperature for all elements is probably unwarranted. In fact, Mg in particular is often found to have a mean Q typical of higher temperatures than most other elements [29, 30], which would result in a lower value of Q(Na)/Q(Mg) more consistent with the observed correlation. In any case, the relatively small amount of scatter in Figure 3 suggests that at the time of fractionation, Q(Na)/Q(Mg) does not differ by more than a few percent from event to event. This is considerably less than the event-to-event variability in individual charge states observed at 1 AU [31] and may provide a clue as to how and when mass fractionation takes place.

If the residual scatter in Figure 3 is indeed due to variability in the Q(Na)/Q(Mg) value and if Q is the same for 2 isotopes of the same element, then a better correlation should be possible using an isotope ratio as the reference value. This is illustrated by the correlation between the ^{22}Ne/^{20}Ne and ^{26}Mg/^{24}Mg ratios in Figure 4 which, for most of the events with the better-determined values, agrees very well with the expected correlation. At least no systematic deviation from the expected trend is evident. Although the outliers are small events with large uncertainties and most are not seriously discrepant statistically, it is interesting to note that they tend to lie near a value of unity on one of the two axes, as if only one isotope ratio is fractionated while the other is unaf-

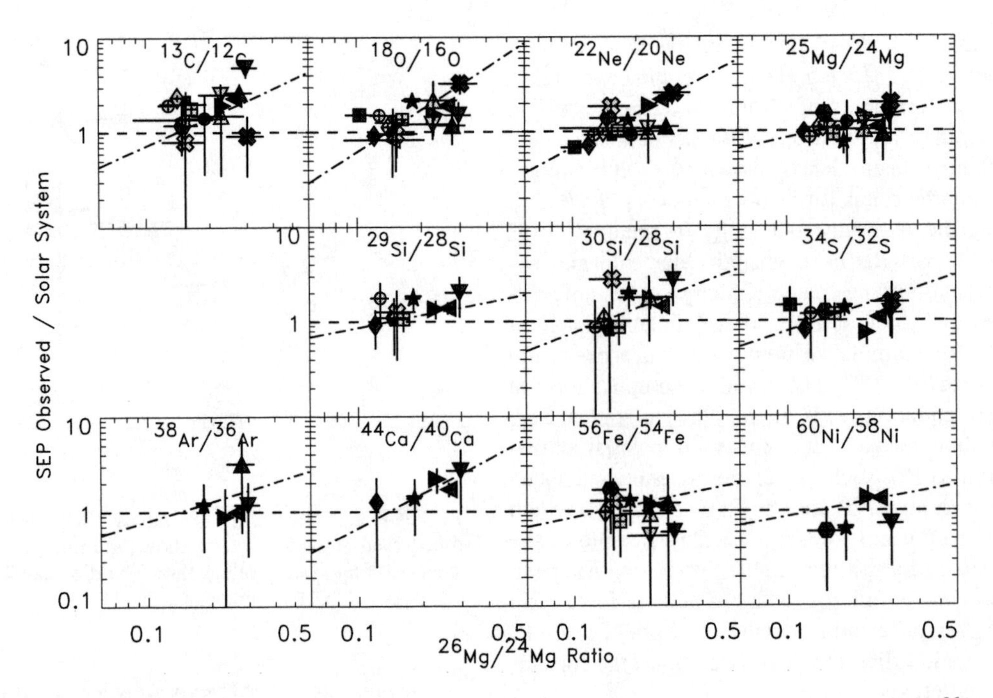

FIGURE 5. Eleven SEP isotope abundance ratios (normalized to standard abundances [26]) plotted versus the $^{26}Mg/^{24}Mg$ ratio. Symbols indicate the SEP events shown in Figure 1. The diagonal lines show the correlations expected using equation (1).

fected. Since these outlying events are among the smallest in our sample, even a small amount of contamination from impulsive events might significantly alter their composition. Of the ~6 outliers, 5 exhibit modest Fe enhancements, with Fe/O~0.5. Most of the events in our study contain significant enhancements of ^{3}He [16, 32], which could be due to residual material from impulsive flares resident in the interplanetary medium which is later accelerated by a shock [33]. The ion cyclotron wave resonances [34, 35] or cascading Alvén waves [36] responsible for the ^{3}He enrichment of impulsive events might also selectively enhance other species with discrete values of Q/M at higher harmonics of the ^{3}He cyclotron frequency [37], and in fact significant enrichments of both ^{22}Ne and ^{26}Mg have been reported in ^{3}He-rich periods [11, 38]. If it were possible for the resonance to affect a narrow enough frequency range, perhaps only one of these two species might be enhanced, resulting in a pattern such as appears to be present in Figure 4. It will be interesting to see if this pattern persists as more events accumulate during this solar cycle.

RESULTS

The isotopic composition of Ne is interesting since it differs in various solar system materials. Therefore, of the 2 isotopic ratios in Figure 4, we chose the $^{26}Mg/^{24}Mg$ ratio as our abundance standard so that we can solve for the SEP Ne composition in this study. The SEP abundance values for 11 isotope ratios for elements from C to Ni are shown plotted versus this reference ratio in Figure 5. The different expected slopes arise from the different relative mass number ratios, and the data seem to follow these expected trends for certain species such as Ne, Mg, Si, and Ca. For many of the heavy elements from S and above, the agreement between the more limited data and the expectations is not as clear. The agreement might break down with increasing distance in Q/M from Mg if the actual dependence on Q/M is not a simple power law as we assumed. For ^{13}C, 12 of the 15 data points fall above the expected correlation, including most of those for which the $^{26}Mg/^{24}Mg$ and other ratios show little or no fractionation. The reason for this is not at all clear, but if this preliminary result holds up, it suggests that ^{13}C routinely is enhanced or ^{12}C is depleted in SEP events relative to terrestrial abundances. In future work we plan to extend the isotope measurements to include ^{15}N to see if it is similarly affected.

To correct for the Q/M-dependent mass fractionation, we solve equation (1) for the coronal isotope ratios. For example, using the $^{26}Mg/^{24}Mg$ ratio as the reference ratio R_1/R_2, the $^{22}Ne/^{20}Ne$ coronal value obtained from any SEP event would be:

$$\left(\frac{^{22}\mathrm{Ne}}{^{20}\mathrm{Ne}}\right)_{\mathrm{cor.}} = \left(\frac{^{22}\mathrm{Ne}}{^{20}\mathrm{Ne}}\right)_{\mathrm{SEP}} \cdot \left(\frac{(^{26}\mathrm{Mg}/^{24}\mathrm{Mg})_{\mathrm{cor.}}}{(^{26}\mathrm{Mg}/^{24}\mathrm{Mg})_{\mathrm{SEP}}}\right)^{\frac{\ln(20/22)}{\ln(24/26)}} \tag{2}$$

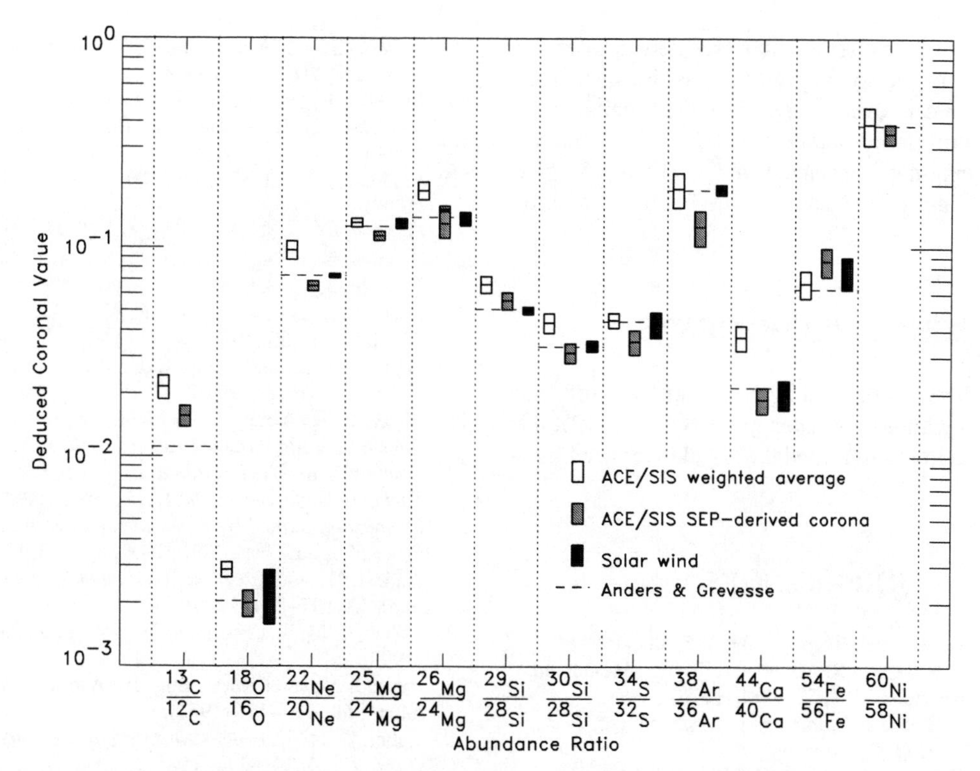

FIGURE 6. Deduced coronal source isotopic abundance ratio averages from SIS SEP measurements without correcting for fractionation (open boxes) and after correction (light grey boxes) as in equation (2). For comparison, standard solar system values (dashed lines; [26]) and measured solar wind values (dark grey boxes; [see 39, and references therein]) are shown. The ^{26}Mg/^{24}Mg ratio served as the reference value for the fractionation corrections for everything other than the ^{26}Mg/^{24}Mg ratio, for which Na/Mg at the ionic charge ratios considered in Figure 3 was used.

Note that the effect of uncertainties in the selected reference value is easy to determine from this expression, the propagation of errors is straightforward, and the exponents are simple constants and do not depend on the measured values.

Preliminary solar coronal isotopic abundances obtained following the example of equation (2) and averaging over all the SIS measurements are shown in Figure 6. For comparison, we have also calculated the weighted average without correcting for the fractionation, but including in the weighting the width of the parent population distribution added in quadrature to the statistical uncertainties. This may be a more representative value for cases such as S or Fe where the data may not follow the expected fractionation correlations in Figure 5, and with a large enough data set (if unbiased by selection effects) may even average out to the coronal value as seems to be the case for elemental abundances [5]. Also, the uncorrected average is the appropriate one to consider for assessing the average arriving solar particle composition at 1 AU. For example, the uncorrected average ^{22}Ne/^{20}Ne ratio we find here is consistent with the value of $\sim$0.09 of the so-called SEP component detected in lunar soils [40]. Both our corrected and uncorrected SEP values are also compared with the standard solar system values [26] and existing solar wind values [39] in Figure 6.

Although the results are preliminary, it is encouraging that this early attempt to obtain coronal abundances from the fractionated SEPs seems to yield reasonable values. With the exception of ^{13}C, all of the isotope abundances are within 2.5σ of the Anders and Grevesse [26] "solar system" values. Both the corrected noble gas isotopes ^{22}Ne and ^{38}Ar appear a bit low compared to Anders and Grevesse, but for these two species Anders and Grevesse adopted the solar wind values as their standard without accounting for mass fractionation in the solar wind of perhaps several percent [41]. So far, Ni isotope abundances have not been reported from solar wind data, so the SEP value given here is the first determination of the ^{60}Ni/^{58}Ni ratio in the corona.

In many cases the uncertainties on the SEP-derived coronal isotope values are comparable to those obtained from solar wind measurements. Accumulating additional SEP events will help reduce the uncertainties for the heaviest species such as Ar to Ni where there are still only a few measurements, but for most of the others a better theoretical understanding of the mass fractionation process is required to make much further progress.

Tracking some of the puzzles uncovered here through the declining part of the solar cycle, such as the decreasing variability in recent events (Figure 2), the possible fractionation of only some isotope ratios (Figure 4), and the apparent common enhancement of ^{13}C (Figure 5) may help to shed light on the nature of the fractionation process.

ACKNOWLEDGMENTS

This research was supported by NASA at the California Institute of Technology (under grant NAG5-6912), the Jet Propulsion Laboratory, and the Goddard Space Flight Center.

REFERENCES

1. Reames, D. V., *Revs. Geophys.*, **33**, 585–589 (1995).
2. Teegarden, B. J., von Rosenvinge, T. T., and McDonald, F. B., *Astrophys. J.*, **180**, 571–581 (1973).
3. Breneman, H. H., and Stone, E. C., *Astrophys. J. Lett.*, **299**, L57–L61 (1985).
4. Garrard, T. L., and Stone, E. C., *Proc. 23rd Internat. Cosmic Ray Conf. (Calgary)*, **3**, 384–387 (1993).
5. Reames, D. V., *Adv. Space Res.*, **15**, (7)41–(7)51 (1995).
6. Mewaldt, R. A., and Stone, E. C., *Astrophys. J.*, **337**, 959–963 (1989).
7. Williams, D. L., Leske, R. A., Mewaldt, R. A., and Stone, E. C., *Space Sci. Rev.*, **85**, 379–386 (1998).
8. Dietrich, W. F., and Simpson, J. A., *Astrophys. J. Lett.*, **231**, L91–L95 (1979).
9. Dietrich, W. F., and Simpson, J. A., *Astrophys. J. Lett.*, **245**, L41–L44 (1981).
10. Simpson, J. A., Wefel, J. P., and Zamow, R., *Proc. 18th Internat. Cosmic Ray Conf. (Bangalore)*, **10**, 322–325 (1983).
11. Mason, G. M., Mazur, J. E., and Hamilton, D. C., *Astrophys. J.*, **425**, 843–848 (1994).
12. Leske, R. A., et al., *Geophys. Res. Lett.*, **26**, 153–156 (1999).
13. Leske, R. A., et al., *Geophys. Res. Lett.*, **26**, 2693–2696 (1999).
14. Dwyer, J. R., Mason, G. M., Mazur, J. E., Gold, R. E., and Krimigis, S. M., *Proc. 26th Internat. Cosmic Ray Conf. (Salt Lake City)*, **6**, 147–150 (1999).
15. Leske, R. A., et al., *Proc. 26th Internat. Cosmic Ray Conf. (Salt Lake City)*, **6**, 139–142 (1999).
16. Leske, R. A., et al., "Measurements of the Heavy-Ion Elemental and Isotopic Composition in Large Solar Energetic Particle Events from ACE", in *High Energy Solar Physics: Anticipating HESSI*, edited by R. Ramaty and N. Mandzhavidze, ASP Conf. Ser. 206, Astronomical Society of the Pacific, San Francisco, 2000, pp. 118–123.
17. Stone, E. C., et al., *Space Sci. Rev.*, **86**, 357–408 (1998).
18. Leske, R. A., Mewaldt, R. A., Cummings, A. C., Stone, E. C., and von Rosenvinge, T. T., "The Ionic Charge State Composition at High Energies in Large Solar Energetic Particle Events in Solar Cycle 23", in *Proc. Joint SOHO–ACE Workshop 2001*, AIP Conf. Proc., AIP, New York, 2001, this volume.
19. Mazur, J. E., Mason, G. M., Looper, M. D., Leske, R. A., and Mewaldt, R. A., *Geophys. Res. Lett.*, **26**, 173–176 (1999).
20. Möbius, E., et al., *Geophys. Res. Lett.*, **26**, 145–148 (1999).
21. Oetliker, M., et al., *Astrophys. J.*, **477**, 495–501 (1997).
22. Reames, D. V., Ng, C. K., and Tylka, A. J., *Geophys. Res. Lett.*, **26**, 3585–3588 (1999).
23. Barghouty, A. F., and Mewaldt, R. A., *Astrophys. J. Lett.*, **520**, L127–L130 (1999).
24. Garrard, T. L., and Stone, E. C., *Adv. Space Res.*, **14**, (10)589–(10)598 (1994).
25. Mewaldt, R. A., et al., "Variable Fractionation of Solar Energetic Particles According to First Ionization Potential", in *Acceleration and Transport of Energetic Particles Observed in the Heliosphere: ACE 2000 Symposium*, edited by R. A. Mewaldt et al., AIP Conf. Proc. 528, AIP, New York, 2000, pp. 123–126.
26. Anders, E., and Grevesse, N., *Geochim. Cosmochim. Acta*, **53**, 197–214 (1989).
27. Cohen, C. M. S., et al., *Geophys. Res. Lett.*, **26**, 2697–2700 (1999).
28. Arnaud, M., and Rothenflug, R., *Astron. Astrophys. Suppl.*, **60**, 425–457 (1985).
29. Luhn, A., et al., *Proc. 19th Internat. Cosmic Ray Conf. (La Jolla)*, **4**, 241–244 (1985).
30. Leske, R. A., Cummings, J. R., Mewaldt, R. A., Stone, E. C., and von Rosenvinge, T. T., *Astrophys. J. Lett.*, **452**, L149–L152 (1995).
31. Möbius, E., et al., "Survey of Ionic Charge States of Solar Energetic Particle Events During the First Year of ACE", in *Acceleration and Transport of Energetic Particles Observed in the Heliosphere: ACE 2000 Symposium*, edited by R. A. Mewaldt et al., AIP Conf. Proc. 528, AIP, New York, 2000, pp. 131–134.
32. Wiedenbeck, M. E., et al., "Enhanced Abundances of ^{3}He in Large Solar Energetic Particle Events", in *Acceleration and Transport of Energetic Particles Observed in the Heliosphere: ACE 2000 Symposium*, edited by R. A. Mewaldt et al., AIP Conf. Proc. 528, AIP, New York, 2000, pp. 107–110.
33. Mason, G. M., Mazur, J. E., and Dwyer, J. E., *Astrophys. J. Lett.*, **525**, L133–L136 (1999).
34. Fisk, L. A., *Astrophys. J.*, **224**, 1048–1055 (1978).
35. Temerin, M., and Roth, I., *Astrophys. J. Lett.*, **391**, L105–L108 (1992).
36. Miller, J. A., *Space Sci. Rev.*, **86**, 79–105 (1998).
37. Bochsler, P., and Kallenbach, R., *Meteoritics*, **29**, 653–658 (1994).
38. Slocum, P. L., et al., "Measurements of Heavy Elements and Isotopes in Small Solar Energetic Particle Events", in *Proc. Joint SOHO–ACE Workshop 2001*, AIP Conf. Proc., AIP, New York, 2001, this volume.
39. Wimmer-Schweingruber, R. F., Bochsler, P., and Wurz, P., "Isotopes in the Solar Wind: New Results from ACE, SOHO, and WIND", in *Solar Wind Nine*, edited by S. R. Habbal et al., AIP Conf. Proc. 471, AIP, New York, 1999, pp. 147–152.
40. Wieler, R., *Space Sci. Rev.*, **85**, 303–314 (1998).
41. Kallenbach, R., et al., *Space Sci. Rev.*, **85**, 357–370 (1998).

A Solar Wind Coronal Origin Study from SOHO/UVCS and ACE/SWICS Joint Analysis

Y.-K. Ko*, T. Zurbuchen†, L. Strachan*, P. Riley** and J. C. Raymond*

*Harvard-Smithsonian Center for Astrophysics, Cambridge, MA USA
†Dept of Atmospheric, Oceanic and Space Sciences, University of Michigan, Ann Arbor, MI USA
**Science Applications International Corporation, San Diego, CA USA

Abstract. The solar wind ionic charge composition is a powerful tool to distinguish between the slow wind and the coronal-hole associated fast wind. The solar wind heavy ions are believed to be 'frozen-in' within 5 solar radii of the Sun which falls right in the range of SOHO/UVCS coronal observations. We present a joint analysis from SOHO/UVCS and ACE/SWICS which attempts to establish observational evidence of the coronal origin of the solar wind. To connect the solar wind with its coronal origin, we adopt a 3-D MHD model as a guide to link the solar wind at 1 AU to structures in the inner corona. We relate in-situ measured properties of the solar wind (elemental abundances and charge state distributions) with remotely sensed signatures in the corona, namely outflow velocity, electron temperature and elemental abundance.

INTRODUCTION

Solar wind heavy ion composition and elemental abundances are recognized to be a powerful tool to distinguish different types of the solar wind [1]. The solar wind ions measured in-situ are generally 'frozen-in' [2] within 5 solar radii, thus contain the information of the electron temperature (thermal or non-thermal), electron density and ion outflow velocities in the inner corona [3,4]. The mechanisms determining the solar wind elemental abundance are believed to operate at the chromospheric level and the upper transition region [e.g. 5,6]. Therefore the inner corona is the place that is closely related to these solar wind parameters. The UVCS instrument [7] aboard the SOHO spacecraft is designed to observe the solar corona from 1.4 $R_\odot$ to 10 $R_\odot$. The spectrocopic data taken by UVCS can be used to derive the ion kinetics, electron temperature, and elemental abundances [8,9]. This provides an ideal opportunity to incorporate both the coronal and the solar wind data to investigate the coronal origin of the solar wind. In this paper, we make such attempts by a joint analysis from SOHO/UVCS and ACE/SWICS. Note that the solar wind plasma measured by ACE should come from near the central meridian. On the other hand, UVCS observes the solar corona above the source region when it rotates to the west limb a few days later. To make comparison between UVCS and ACE data, we need to assume that the physical properties of the source region/corona do not change significantly within a few days.

UVCS OBSERVATIONS

UVCS observations (for details of the UVCS instrument, see [7]) were made from Oct. 18-24, 1999 at the west limb of the Sun. The observations were composed of two parts. The first part (denoted as 'intensity data') is a mirror scan at 1.5, 1.7, 2.0, 2.5, 3.5, and 4.5 $R_\odot$ with the objectives of obtaining radial dependences of the coronal plasma properties. The second part (denoted as 'abundance data') is a grating scan at 1.7 $R_\odot$ with the objectives of obtaining electron temperature and elemental abundances. Every three days (on Oct. 20 and 23) we did a modified plan with the mirror scan at 1.7, 2.0, and 2.5 $R_\odot$, then we took the same mirror scan but at another grating position. Figure 1 shows the UVCS pointings on Oct. 18, 20, 22, and 24. The slit width was 100 μm. The field-of-view is thus 28" $\times$ 40′.

In October 1999, the Sun was on its way toward solar maximum with the polarity reversal process toward its most active stage (as can be seen from the coronal source surface field maps, http://quake.stanford.edu/~wso/coronal.html). The

CP598, *Solar and Galactic Composition*, edited by R. F. Wimmer-Schweingruber

FIGURE 1. UVCS pointings on Oct. 18, 20, 22, and 24 of 1999. Plotted are the OVI channel slit positions on the composite image of EIT 195, UVCS Lyα synoptic (except Oct. 20), and LASCO C2 on those days.

FIGURE 2. SOHO/EIT 195 image at 21:48 UT and the NSO/KP He 10830Å coronal hole map on Oct. 13.

solar magnetic field is far from dipolar. Active regions exist at both low and high latitudes. Equatorial and mid-latitude coronal holes appear replacing the the two polar coronal holes during solar minimum. Correspondingly, the large scale morphology of the corona can change within a day. Around October 5, 1999, a large equatorial coronal hole appeared from the east limb and it was at the west limb when UVCS observation started on Oct. 18. Fig.2 shows the SOHO/EIT 195 image and the NSO/KP He 10830Å coronal hole map on Oct. 13 when this coronal hole was near the disk center. Note that there was a group of active regions at the east side of this coronal hole. Therefore, we can see from Fig.1 that the corona at the west limb during our observation started dark above this coronal hole, then gradually turned bright as those active regions rotated to the limb.

ACE OBSERVATIONS

We use the ion composition and elemental abundance data from ACE/SWICS (for a description of the ACE/SWICS instrument, see [10]) to compare with the coronal plasma properties. Since it takes 2-4 days for the wind plasma of 400-700 km/s to reach ACE, and the source region of the wind will take a few more days to rotate to the west limb, the solar wind plasma would be measured by ACE a few days before its source region would be observed by UVCS. Fig. 3 shows the solar wind He velocity, the O^{+7}/O^{+6} freezing-in temperature, the abundance ratios for $(Mg/Mg_{ph})/(O/O_{ph})$ (relative to their photospheric values), and $(Fe/Fe_{ph})/(O/O_{ph})$ from DOY (day-of-year) 286 (Oct. 14) to DOY 296. Note that the abundance for Mg, O, Fe shown here are measured from Mg^{+10}, O^{+6}, and Fe^{+7-+12}, respectively; therefore they are modulated by the electron temperature where these ions are frozen-in. This is important to realize especially for Mg since Mg^{+10} ionic fraction increases by 50% if the freezing-in T_e increasing from 6.10 to 6.25 (in log10 scale). We group the data into several time periods according to the solar wind speed and they are represented by different symbols in Fig.3. It is obvious that the fast wind originates from the equatorial coronal hole and it smoothly changes to slow wind in 6 days. Note that since it takes longer for the slow wind to reach ACE, the actual change in wind speed near the Sun should take a much shorter time (see the section below). However, the smooth change from fast to slow wind indicates the crossing of the coronal hole boundary toward the the active regions. It unambiguously implies that this is the period that should relate to the UVCS observations. There was a CME starting around DOY 294 (Oct. 21) which shows features of the magnetic cloud and bi-direction electrons. We thus concentrate on the ACE data before DOY 294.

TRACING THE SOLAR WIND BACK TO THE CORONA

In order to connect the solar wind measured by ACE with its source region, i.e. to know where in longitude and latitude the wind plasma measured at a given time originates from, some kind of trace-back model

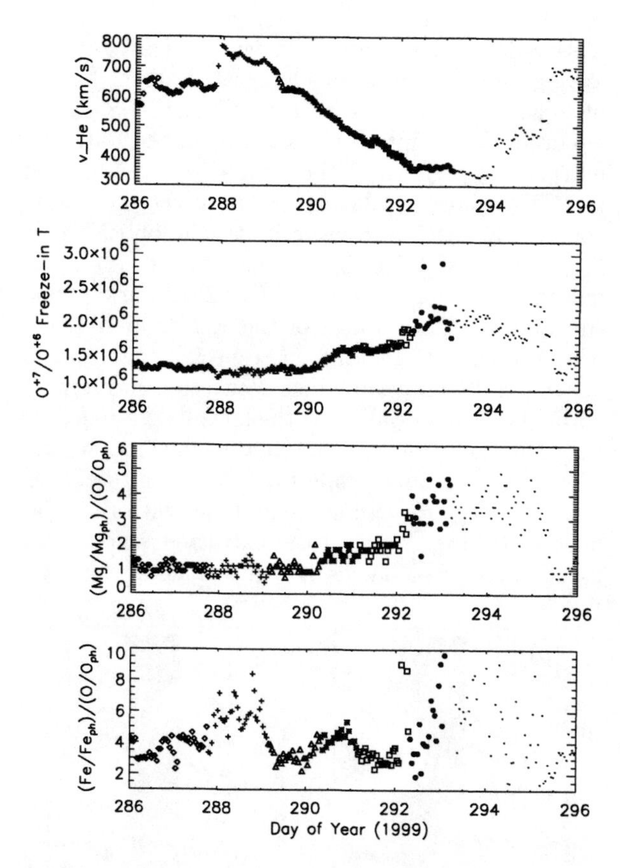

FIGURE 3. He velocity, O^{+7}/O^{+6} freeze-in temperature (in K), Mg/O and Fe/O abundance ratios relative to their photospheric values from ACE/SWICS.

is needed. A common one is the 'ballistic model' (i.e. constant velocity model) which assumes the solar wind trajectory is radial and the speed is constant once it leaves the Sun. Therefore the longitude of the source region mainly depends on the wind speed and the latitude is the same as that of ACE. A more sophisticated approach is to use a 3-D MHD simulation. To facilitate the mapping of ACE solar wind data back to the Sun, we first constructed an equilibrium MHD solution of the solar corona appropriate for the Whole Sun Month observations [e.g. 11,12]. The details of the algorithm used to advance the MHD equations are explained elsewhere [e.g. 13] and here we make a few brief remarks. We restrict ourselves to polytropic solutions and set γ=1.05 in the corona. This reduced polytropic index provides a very rough approximation of the effects of parallel thermal conduction in the corona. The inner radial boundary condition (at 1 $R_\odot$) is derived from the observed line-of-sight photospheric magnetic field and the upper radial boundary is set to 30 $R_\odot$. Reasonable initial values for the magnetofluid parameters are specified and the model is run forward in time until an equilibrium solution is achieved. We traced field lines from the location of ACE ballistically back to 30 $R_\odot$, at which point we traced the appropriate field line all the way to the solar surface by this MHD model. The ballistic approximation appears to be a reasonable approximation between 30 $R_\odot$ and 1 AU where dynamic interactions are only beginning to alter the flow pattern of the solar wind, particularly during solar maximum, and the flow is essentially radial. Within 30 $R_\odot$, we account for the super-radial expansion of the field lines that can result in a significant shift in the source latitude of the plasma by tracing field lines within the model solution.

Fig.4 plots the source position of the solar wind at 1.0, 1.7, 2.5, 3.5, 4.5 $R_\odot$ traced back from the ACE position. For comparison, the ballistic result is also plotted. Note the various symbols indicating respective time period and speed range (cf. Fig.3). One characteristic of the MHD model is that the slow solar wind always originates from the open field region near the coronal hole boundaries and the fast solar wind originates deeper in the coronal hole [11]. The 3-D MHD trace-back results differ from the ballistic one in that the solar wind trajectory is super-radial in the inner corona and the source region can be at different latitude from where ACE is (dotted line. Also compare the latitude of the coronal hole in Fig.2). The MHD model implies that the change-over of the fast wind to the slow wind occurs within 2 days below 2.5 $R_\odot$, while it is longer in the ballistic model.

CONNECTING THE CORONA AND THE SOLAR WIND

Fig.4 implies that the solar wind source region traced back from the ACE position lies within 15 degrees in latitude north of the equator. The question is if we can see coronal properties that are correlated with those of the solar wind. The task is not easy. The reason is that the observed line intensity is the emissivity integrated along the line of sight. During the early UVCS observations, the coronal hole was very dark and presumably it dominated the line of sight, at least at high heights (cf. Fig.1). Later on, the bright corona (presumably hot and dense materials of the active region loops) gradually rotated into the line of sight. According to the MHD model, the source of the solar wind plasma comes from the region between the coronal hole and the active regions. Some assumptions for the geometry and physical parameters for both structures are needed in order to separate the contributions to the

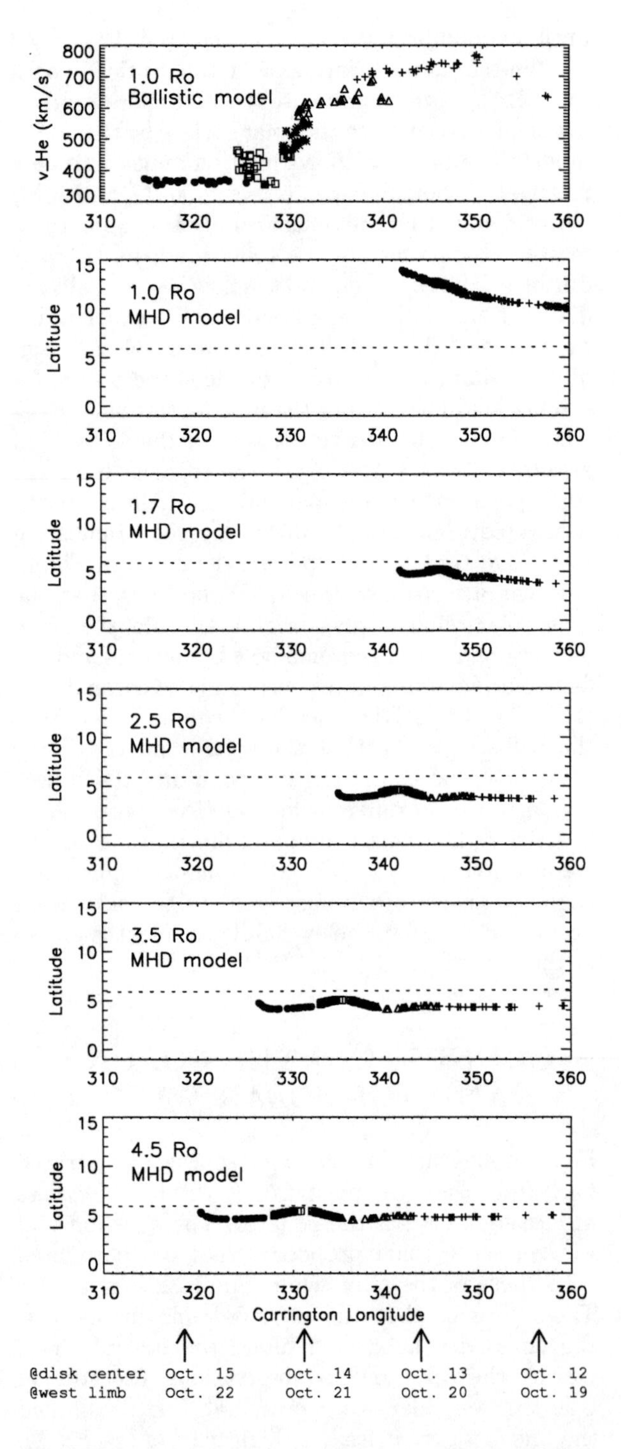

FIGURE 4. The trace-back of the solar wind from the ballistic and 3D MHD models. The approximate dates of the Carrington longitude when it passes the central meridian and the west limb are marked at the bottom of the plots. Different symbols represent different time periods grouped by the solar wind speed (cf. Fig.3). Note that most of the wind data observed by ACE during DOY 286-287 (open diamonds in Fig.3) are mapped back to the previous Carrington rotation, and not shown here.

observed line intensities. For now, we investigate the average coronal properties integrated along the line of sight and compare with the ACE/SWICS data.

For both the intensity and abundance data, we extract two regions: 1) the 'center' region (denoted as 'CT') which is 732 arcsec above the west limb covering 27 to 10 degrees in latititude for 1.5 to 4.5 $R_\odot$, respectively. This is the region that contains the traced-back solar wind from ACE. 2) A 'dark' region (denoted as 'CH') south of the 'CT' region either near the edge of the slit, or, especially at 3.5 and 4.5 $R_\odot$, a dark section which seems to contain more coronal hole material along the line of sight (e.g., see Fig.1, at 4.5 $R_\odot$ on Oct. 22). In the following, we will discuss the coronal properties within and between these two regions, and compare them with the solar wind data. Fig. 5 shows the extracted sections for two dates plotted on the Lyα fluxes along the slit.

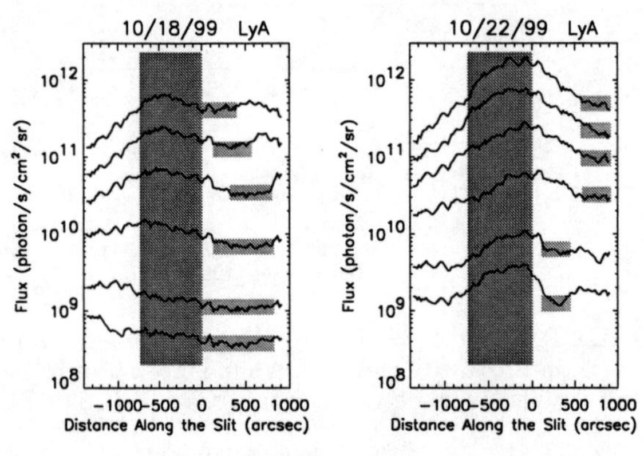

FIGURE 5. Extracted 'CT' (dark shaded areas) and 'CH' (light shaded areas) regions for Oct. 18 and 22. Curves from top to bottom are for 1.5, 1.7, 2.0, 2.5, 3.5, and 4.5 $R_\odot$, respectively. South is toward the right.

Fig.6 plots the Lyα and OVI 1032 kinetic temperatures from the line widths corrected for instrumental broadening, and the OVI 1032 to 1037 intensity ratios versus the observation day at each height. The kinetic temperature is defined from the 1/e width $\delta\lambda$:

$$T_{kin} \equiv \frac{mc^2}{2k}(\frac{\delta\lambda}{\lambda})^2 = T_{thermal} + T_{non-thermal}$$

which, without further knowledge of the non-thermal contribution, can be taken as the upper limit of the electron temperature, assuming ions are in thermal equilibrium with the electrons. The OVI 1032/1037 ratio is an indication of the outflow velocity of the O^{+5} ions [e.g. 14] for a given configuration of the electron temperature and density along the line of sight. These parameters are plotted versus time in order to compare with the solar wind data (Fig.3). We can see that: 1) The Lyα T_{kin} increases with time for the first 5 days at every

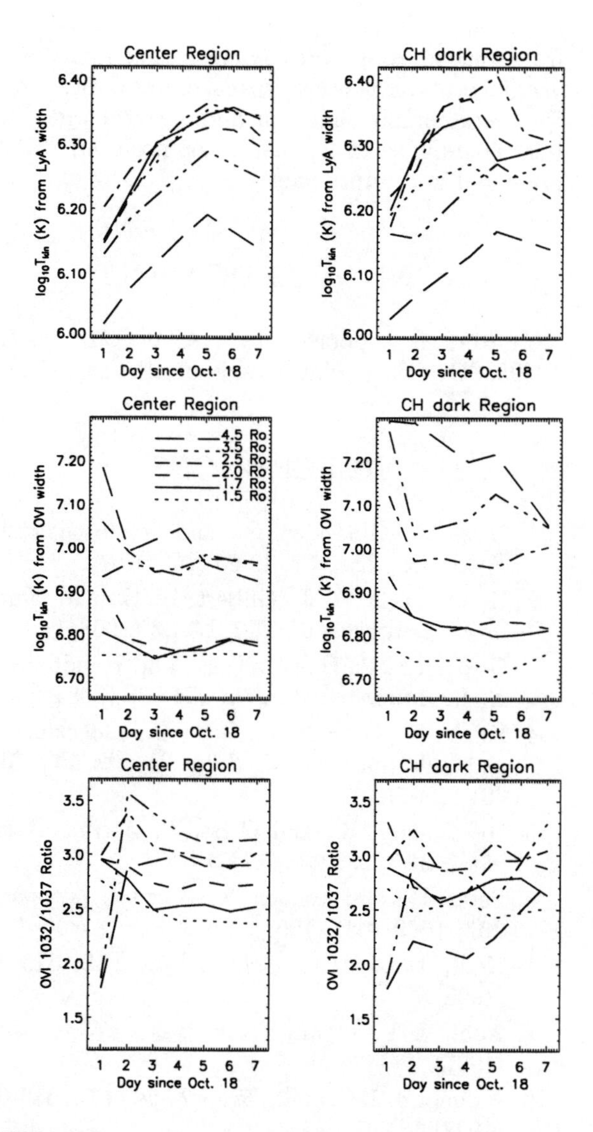

FIGURE 6. Lyα and OVI 1032 kinetic temperatures and the OVI 1032 to 1037 ratio versus time at each height. The line symbols for the heights are indicated on the left-center plot.

height. Those at low heights can be compared with the O^{+7}/O^{+6} freezing-in temperature of the solar wind. For the CT region T_{kin} is approximately the same below 2.5 $R_{\odot}$ then decreases at higher heights. 2) The OVI 1032 T_{kin} decreases with time but increases with height. This implies stronger nonthermal heating toward high heights [e.g. 15], especially for the CH region. This dependence of the line widths (i.e. T_{kin}) with height is similar to that of the equatorial streamer [8]. 3) The OVI ratios indicate that substantial outflow starts beyond 3.5 $R_{\odot}$. The OVI ratio is expected to increase with height for low outflow speeds ($\ll$ 100 km/s) since the density decreases (see Fig.6 below 2.5 $R_{\odot}$). Thus, in general, when the ratios decrease with height, it implies that the outflow speed increases. The effect is stronger for the CH region. At small heights, the ratio decreases with time which is likely due to the increase in the line-of-sight electron density as the active regions rotate toward the limb. At large heights, the ratio increases with time implying decreasing outflow speed and it seems to happen mainly at the first two days. This is consistent with the fast to slow wind transition in the ACE data. Note that on Oct. 18 at 3.5 and 4.5 $R_{\odot}$, we need to combine the CT and CH data to get good statistics for OVI 1037, so the OVI ratio is the same for both regions.

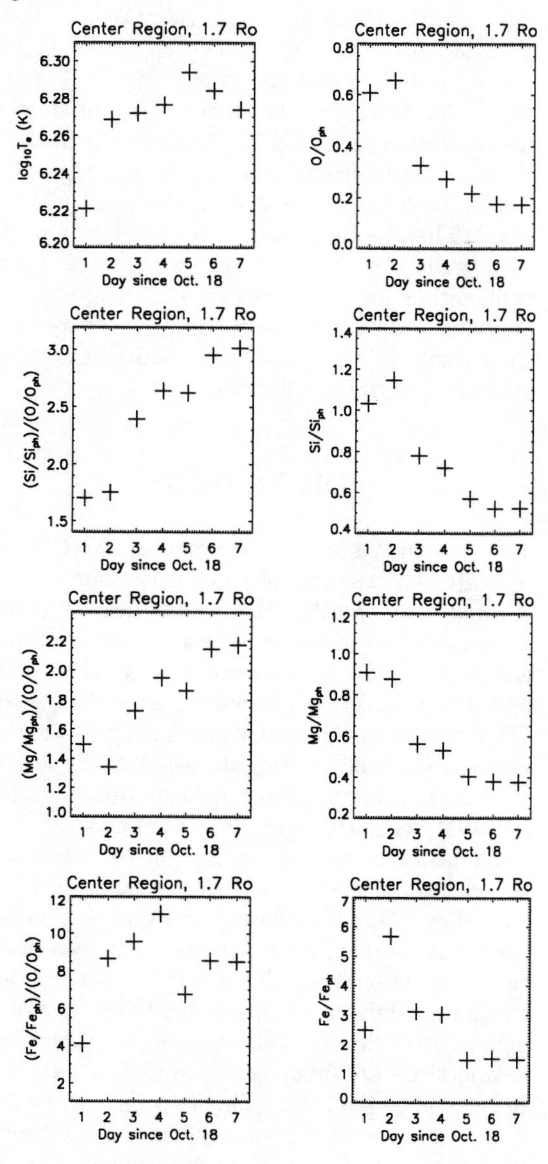

FIGURE 7. The electron temperature and elemental abundance from the CT region at 1.7 $R_{\odot}$ plotted versus time.

At 1.7 $R_\odot$, the abundance data allow us to obtain the electron temperature from the line ratios of Fe X 1028, [Fe XII] 1242, and Fe XIII] 510 assuming a single temperature along the line of sight. We can then obtain estimates for the elemental abundance from line ratios relative to O or H [9]. A more confident result is obtained in the CT region where these lines are brighter. Fig. 7 plots the resulting T_e and X/X_{ph} (absolute abundance which is relative to H, right panels in Fig.7) and $(X/X_{ph})/(O/O_{ph})$. They are from O: OVI 1032 abundance data, Si: the average of Si XII 520 abundance data and Si XII 499 intensity data; Mg: Mg X 609 intensity data; Fe: [Fe XII] 1242 abundance data. We can see that $(Si/Si_{ph})/(O/O_{ph})$ and $(Mg/Mg_{ph})/(O/O_{ph})$ increase with time, while $(Fe/Fe_{ph})/(O/O_{ph})$ does not have any systematic trend. Both T_e and the abundance relative to O seem to be consistent with the solar wind data (Fig.3). However, the systematic change of these parameters in the CT region has a time scale longer than that in the solar wind ($\sim$ 2 days). The absolute abundances all show a decrease with time. At this moment, we do not have an explanation for this phenomenon. The solar wind abundance data relative to hydrogen may help to shed some light on if this dependence is actually associated with the solar wind.

DISCUSSION

The line widths and OVI ratios in the CH region indicate that the line of sight is dominated by the coronal hole. On the other hand, the CT region is dominated by the hot and dense materials above the active region loops. Nevertheless, we find a likely correlation in T_e and elemental abundances in the CT region with the solar wind. This may have some implication on associating the origin of the slow wind with mixing of the closed field materials, although our analysis at this point is not conclusive.

Elemental abundance is thought to be a clear distinguisher for the fast and slow solar wind by their FIP effect. Our preliminery analysis here indicates that it is not straigtforward to make these calculations for the corona. Elemental abundances in the corona are difficult to obtain due to the line-of-sight effects. Also the methods to determine the electron temperature and abundances are subject to various uncertainties [16]. Some ions (e.g. Si XII, Fe X) are more sensitive to the temperature in the temperature range of concern than the others (e.g. O VI, Mg X, Fe XII), thus a small uncertainlty in T_e results in large uncertainty in the abundances derived from some lines. A more complete and detailed analysis, incorporating other aspects of this data set with Doppler dimming modelling and the solar wind data (e.g. abundance to H, charge composition) will be presented in a future paper.

ACKNOWLEDGMENTS

This work is supported by NASA grant NAG5-10093. SOHO is a joint mission of ESA and NASA.

REFERENCES

1. Geiss, J., Gloeckler, G., and von Steiger, R., *Space Sci. Rev.*, **72**, 49-60 (1994).
2. Hundhausen, A. J., Gilbert, H. E., and Bame, S. J., *Astrophys. J.*, **152**, L3-L5 (1968).
3. Owocki, S. P., Holzer, T. E., and Hundhausen, A. J., *Astrophys. J.*, **275**, 354-366 (1984).
4. Ko, Y.-K., G., Fisk, L., Geiss, J., Gloeckler, G., and Guhathakurta, M., *Solar Physics*, **171**, 345-361 (1997).
5. von Steiger, R., and Geiss, J., *Astron. Astrophys.*, **225**, 222-238 (1989).
6. Peter, H., and Marsch, E., *Astron. Astrophys.*, **333**, 1069-1081 (1998).
7. Kohl, J. L. et al., *Solar Phys.*, **162**, 313-356 (1995).
8. Kohl, J. L., et al., *Solar Phys.*, **175**, 613-644 (1997).
9. Raymond, J. C. et al., *Solar Phys.*, **175**, 645-665 (1997).
10. Gloeckler, G. et al., *Space Sci. Rev.*, **86**, 495 (1998).
11. Linker, J. A. et al., *J. Geophys. Res.*, **104**, 9809-9830 (1999).
12. Riley, P., Linker, J. A., and Mikic, Z., 'Solar cycle variations and the structure of the heliosphere: MHD simulations', submitted to Proceedings of Chapman Conference on Space Weather, April 2000.
13. Mikic, Z., Linker, J. A., Schnack, D. D., Lionello, R., and Tarditi, A., *Phys. Plasmas*, **6**, 2217 (1999).
14. Noci, G., Kohl, J. L., and Withbroe, G. L., *Astrophys. J.*, **315**, 706-715 (1987).
15. Cranmer, S. R., Field, G. B., and Kohl, J. L., *Astrophys. J.*, **518**, 937-947 (1999).
16. Raymond, J. C. et al., this volume (2001).

Comparison between Average Charge States and Abundances of Ions in CMEs and the Slow Solar Wind

A. A. Reinard*, T. H. Zurbuchen*, L. A. Fisk*, S. T. Lepri*, R. M. Skoug† and G. Gloeckler**

*Department of Atmospheric, Oceanic, and Space Science, University of Michigan, Ann Arbor, MI, 48109, USA
†Space and Atmospheric Sciences Group, Los Alamos National Lab,Los Alamos, NM 87545
**Department of Physics, University of Maryland, College Park, MD, 20742, Department of Atmospheric, Oceanic, and Space Science, University of Michigan, Ann Arbor, MI, 48109, USA

Abstract.
We present results from a comparison of CME and slow solar wind ejecta detected at the ACE spacecraft in 1998 and 1999. CME events were identified based on the observation of counterstreaming halo electrons from SWEPAM data. We discuss the compositional signatures in the framework of a recent model of the coronal magnetic field by Fisk and Schwadron [1]. We conclude that slow solar wind and CMEs have a common source in the corona, presumably coronal loops. The largest amount of fractionation is found in helium and in charge state composition. The former is related to collisional effects in the corona and the latter is attributed to the anomalous heating and propagation properties of some CMEs.

INTRODUCTION

Coronal mass ejections (CMEs) are eruptions of coronal magnetic field and plasma into the solar wind. The first continuous space-borne observations of CME ejections were from coronagraphs on OSO-7 and Skylab in the early 1970s [2, 3]. The mechanisms responsible for these eruptions are not well understood [4]. CMEs in the vicinity of the Earth are identified based on a subset of specific characteristics including, but not limited to, the following: descending velocity profiles, enhanced and smoothly rotating magnetic field, low in situ kinetic temperatures, and enhanced α/proton ratios. In addition, the closed nature of the magnetic field lines of a CME allows identification based on counterstreaming halo electrons [5].

For this investigation we use data from the SWICS (Solar Wind Ion Composition Spectrometer), MAG (Magnetometer instrument) and SWEPAM (Solar Wind Electron, Proton, and Alpha Monitor) instruments on board the ACE (Advanced Composition Explorer) spacecraft. The SWICS instrument measures the composition of the solar wind by use of an electrostatic deflection system, a linear time of flight with post-acceleration, and a solid state detector. This process allows unambiguous determination of the mass and mass per charge of most ions. With this information we can distinguish different ion charge states and determine elemental abundances [6]. The MAG instrument consists of two magnetometers that measure the amplitude and direction of the interplanetary magnetic field [7]. The SWEPAM instrument consists of an ion spectrometer and an electron spectrometer which measure the solar wind proton, helium and electron distributions [8]. Our study utilizes counterstreaming halo electron events identified in the SWEPAM data to determine the periods of CME events.

Composition measurements provide a direct measure of the source conditions in the coronal environment. Compositional patterns observed in situ "freeze-in" during the expansion of the CME from the corona to 1 AU and thus provides an important link between remote solar observations and in situ plasma observations. Composition signatures typically fall in two parts: ionic composition of given elements and elemental abundances.

Ionic compositional anomalies observed in some CMEs [9, 10, 11] are caused by unusual thermal conditions of the CME plasma in the low corona or chromosphere. These signatures are typically found in 50% of all ejecta (identified by plasma properties) [12, 13]. The details of the freeze-in process are generally rather complex [14, 15] and strongly dependent on the electron temperature profile. Non-thermal properties of near solar electron distribution functions can also be reflected in the observed charge states [16, 17].

CP598, *Solar and Galactic Composition*, edited by R. F. Wimmer-Schweingruber

The elemental composition reflects the coronal source of the CME prior to the ejection into space. The slow solar wind and most CMEs observed in the ecliptic plane are fractionated relative to the photospheric composition. Typically, elements of low First Ionization Potential (FIP), such as Fe, Mg, Si, are enhanced over high-FIP elements (Ne, He) [18]. This fractionation presumably occurs in closed magnetic structures [19, 20]. This phenomenon has recently been observed using spectroscopic composition measurements on SOHO [21].

Using these composition parameters, we now address the following important questions:

1. Is there a difference between the source and expansion properties of slow wind and CME plasma?
2. What are the consequences of such differences in the appearance of CMEs at 1 AU?

These questions will be studied using the average properties of CMEs and slow solar wind periods. We will then discuss these data in the framework of a solar coronal model by Fisk and Schwadron [1].

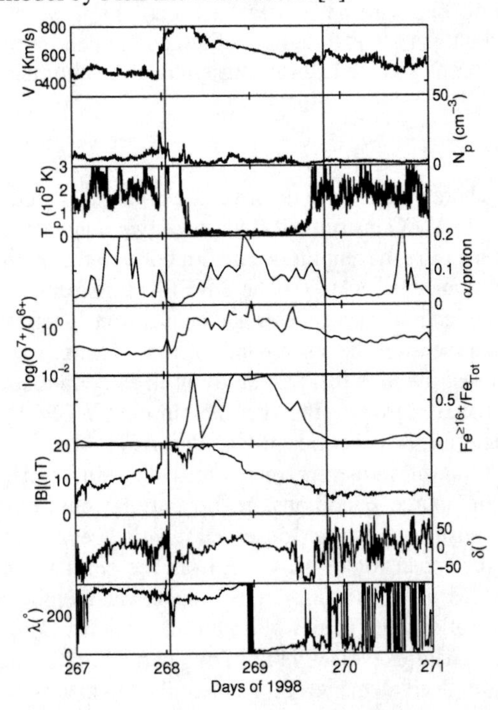

FIGURE 1. CME event identified by counterstreaming halo electrons (solid vertical lines). This event has increased charge states and α/proton ratio and decreased kinetic temperature.

GENERAL CHARACTERISTICS OF CME COMPOSITION

In this section we compare the compositional patterns of CMEs and slow solar wind. CMEs are identified based on the observation of counterstreaming halo electrons by the SWEPAM instrument. Figure 1 shows an example of one CME in this study. The period of counterstreaming electrons occurred between the dark vertical lines. The first panel shows the proton velocity (V_p) which has a typical expansion profile consisting of a sharp increase followed by a slow decrease. The second and third panels show proton density (N_p) and kinetic temperature (T_p). The kinetic temperature clearly decreases during this CME event. The fourth through sixth panels show composition parameters: α/proton ratio, O^{+7}/O^{+6} ratio, and the abundance of $Fe^{\geq+16}$ divided by total iron abundance. These compositional parameters are enhanced during this CME. The seventh through ninth panels show total magnetic field magnitude and field components latitude (δ) and longitude (λ).

Our study included a total of 56 CME events during 1998 and early 1999. For each CME the composition parameters shown in Figure 1 (in addition to He/O and Fe/O) were averaged over the entire CME as defined by the presence of counterstreaming electrons. This procedure allows us to do a statistical survey of CME composition. It should be emphasized that the data presented in this paper are, in some cases, averages over a long period of time and so effects such as high freeze-in temperatures and high He/H, among others, are muted if the enhancements only occurred over a fraction of the period of counterstreaming electrons.

For each CME time period we selected a "slow solar wind" time period by subtracting six days from the beginning time of the CME. This delay was corrected if there was any suspected interference with a previous CME or a high speed stream. The average speed of all "slow solar wind" periods is 415±100 km/s. Average values of several parameters were calculated for these slow solar wind periods over time intervals similar to the CME durations. We note that there are sometimes significant variations in the duration of CME associated times [22], which may indicate that an ejecta observed at 1 AU may be composed of a number of individual CMEs.

We now present survey plots that compare average CME composition with slow solar wind composition. The O^{+7}/O^{+6} ratio is used to order these data. When possible, we include average coronal hole composition from Ulysses during the south coronal hole passage between September 8-18, 1994. Each of the survey plots has the same format as described in Figure 2.

We first show a comparison of O^{+7}/O^{+6} and average iron charge states (Figure 2). Both quantities can be directly related to the electron temperature at their respective freeze-in points. O^{+7}/O^{+6} typically freezes-in close to the solar surface ($< 2\ R_\odot$) and iron freezes-in further out (3–4 $R_\odot$). The different freeze-in radii are a consequence of the different recombination rates of the two elements ([18] and references therein). Assuming a stan-

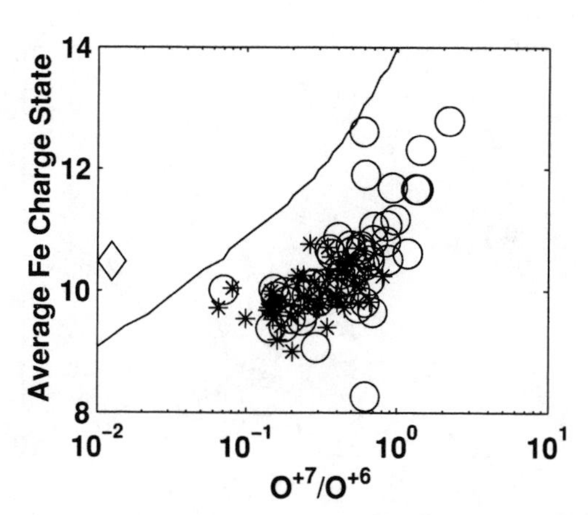

FIGURE 2. O^{+7}/O^{+6} ratio versus the average iron charge state for CMEs identified by counterstreaming halo electrons (circles), slow solar wind periods (stars) in 1998 and 1999, and an average coronal hole value (diamond). The line indicates values of collisional equilibrium of iron and oxygen with thermal electrons. O^{+7}/O^{+6} shows a positive correlation with iron charge state in both the slow solar wind and CME periods.

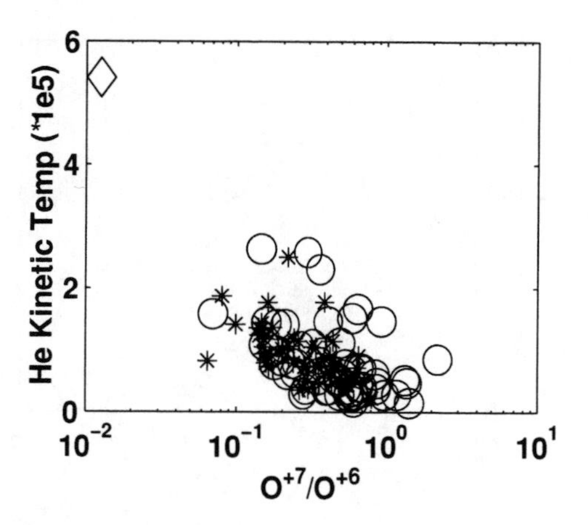

FIGURE 3. Same as Figure 2 but for O^{+7}/O^{+6} versus in situ helium kinetic temperature. O^{+7}/O^{+6} and kinetic temperature are inversely related.

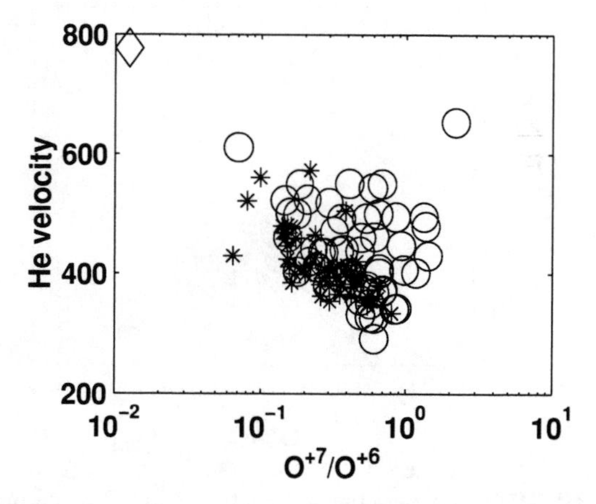

FIGURE 4. Same as Figure 2 but for O^{+7}/O^{+6} versus in situ helium velocity. O^{+7}/O^{+6} and velocity are inversely correlated for slow solar wind, while CME events have a broader range in O^{+7}/O^{+6} for a given velocity. The average CME speed is clearly larger than average slow solar wind speed.

dard electron temperature profile [23] in which a temperature peak occurs around 2 $R_\odot$, oxygen freezes in at a positive temperature gradient and iron freezes in at a negative temperature gradient. Small changes in velocity, temperature, or density would thus tend to cause an anti-correlation in oxygen and iron freeze-in temperatures [32]. Figure 2 clearly shows a positive correlation between O^{+7}/O^{+6} and the average iron charge state in both the slow solar wind and CME associated plasma. Note also that only CMEs are associated with the hottest freeze-in temperatures [9, 10, 24]. For comparison, local thermodynamic equilibrium values [25, 26] have been plotted in a solid line. All CME and slow solar wind points lie below this line, indicating a lower freeze-in temperature for iron than oxygen, which is consistent with the effects of a negative temperature gradient in the corona and contrary to the standard electron temperature profile.

Figure 3 compares the ion kinetic temperature with O^{+7}/O^{+6}. We use helium rather than electron kinetic temperature due to the smaller dependence on spacecraft charging effects. The anti-correlation between O^{+7}/O^{+6} and the kinetic temperature may be a signature of thermally-driven expansion close to the Sun. Note that some CMEs have a relatively hot in situ kinetic temperature, in contrast with the typical picture of a CME event. These events bear closer investigation. Figure 4 shows an equivalent plot for the average speed. Helium velocity is used because it is a very good approximation for velocities of the heavier elements and because statistics allow a more accurate measurement. Typically, CME speeds exceed that of slow solar wind events and show a larger variability. There is a possible anti-correlation between O^{+7}/O^{+6} and speed in CME events, while for the slow solar wind the anti-correlation is clear.

In Figure 5 we see no apparent correlation between O^{+7}/O^{+6} and Fe/O. Fe/O is generally ~0.1 for FIP associated plasma and ~0.04 in the photosphere [27]. Our values in both CME and slow solar wind events fall between those values. CME events, once again, greatly overlap the slow solar wind events, and account for the hotter events and those events with the highest Fe/O. This observation is probably due to CME events being a more extreme examples of slow wind phenomenon.

He/O and He/H give another measure of fractionation as helium is a very-high FIP element. Our results (Figures 6 and 7) indicate that there is no ordering due to frac-

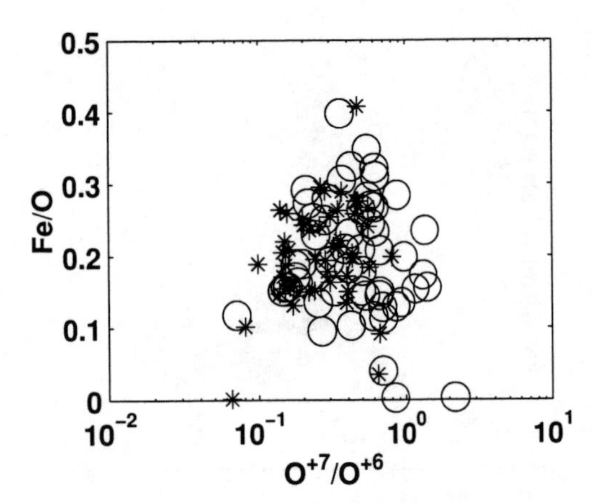

FIGURE 5. Same as Figure 2 but for O^{+7}/O^{+6} versus the iron to oxygen ratio. No relation is apparent.

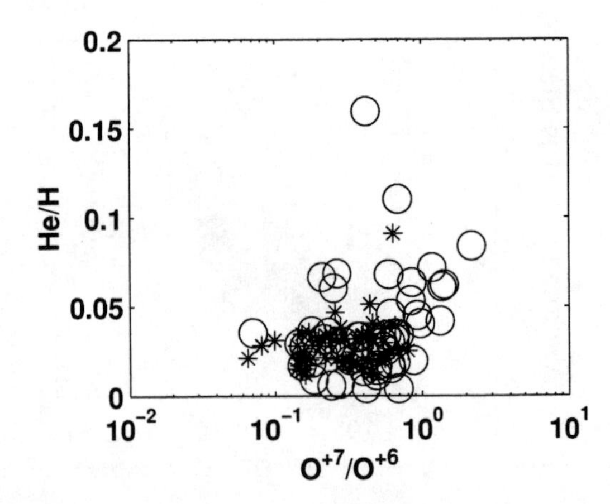

FIGURE 7. Same as Figure 2 but for O^{+7}/O^{+6} versus the helium to hydrogen ratio. CMEs with high values of O^{+7}/O^{+6} can have very high He/H

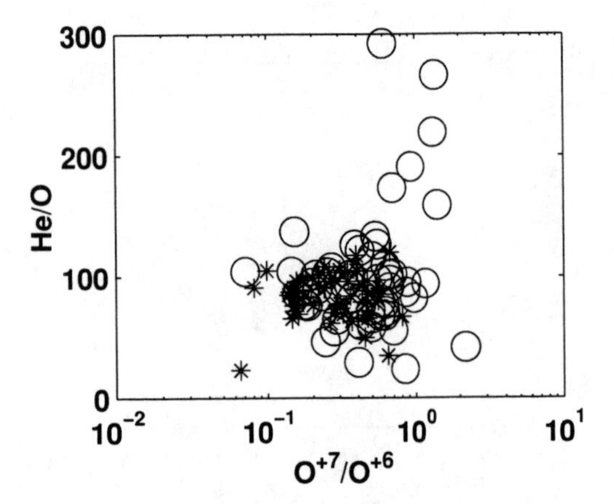

FIGURE 6. Same as Figure 2 but for O^{+7}/O^{+6} versus the helium to oxygen ratio. CMEs with high values of O^{+7}/O^{+6} can show enhanced He/O.

tionation based on He/O or He/H except in very hot CME events, where a large amount of variability occurs. This variability is much larger than in any other ionic species studied here. He^{2+} has the smallest collisional cross-section of all solar wind ions [28] and is therefore less coupled to the solar wind protons than any other heavy ion species. The consequent fractionation may therefore be of different nature than the FIP fractionation process, as indicated by comparisons of Figures 5 and 6.

DISCUSSION

A recent series of papers by Fisk et al. [1, 29, 30] have argued that there is large-scale and relatively rapid transport of open magnetic flux in the solar corona. Transport occurs by a diffusive process in which open field lines reconnect with closed magnetic loops and execute a series of random jumps in location. This process provides an explanation for the origin of the slow solar wind. The reconnection process and subsequent displacements of open field lines release energy into the solar corona that is sufficient to accelerate the solar wind [1, 30]. Material stored on the closed field lines is also released to form the mass flux in the solar wind.

The closed magnetic loops referred to in this theory have been identified by Feldman et al. [21]. The loops have heights of order 100,000 km, and electron freeze-in temperatures $\sim$1.4 x 10^6 K. Of more significance, they are FIP-enhanced in their composition [31], strongly suggesting that they are the origin of the slow solar wind.

CMEs also originate in coronal loops. It is reasonable to expect, then, that the characteristics of the slow solar wind and CMEs should be similar, with perhaps CMEs being simply a more extreme version, i.e. originating from larger and hotter loops [22]. This expectation is consistent with the observations presented here, though a more in-depth study is needed to confirm this hypothesis.

Low latitude corona has monotonically decreasing temperature

In Figure 2 we see that in both the slow solar wind and in most CME events the average iron charge state varies between 9 and 11 and the O^{+7}/O^{+6} ratio varies between 0.1 and 1. Both data sets follow a similar trend which is linear. This result is inconsistent with both the slow solar wind and CME events arising from a corona with an electron temperature peak. If the peaked profile were valid in these periods we would expect oxygen and iron based electron temperatures to be anti-correlated as described in Ko et al., [32]. These results provide evidence that the slow solar wind and CME events both originate in a corona with a monotonically decreasing electron temperature profile.

The electron temperature of solar loops, $\sim 1.4 \times 10^6$ K [21], are consistent with charge states for O^{+7}/O^{+6} that are observed in the heliosphere, i.e. the data are consistent with the freeze-in of O^{+7} and O^{+6} occurring in the loop. As the material expands out from the Sun after reconnection releases it from the loop, any subsequent heating is insufficient to overcome a temperature that we find decreases monotonically with distance. This expansion of the slow solar wind should be contrasted with the fast solar wind from the polar coronal hole, which Geiss et al. [18] determined had a temperature maximum that lies between the freeze-in points of O^{+7}/O^{+6} and iron.

CME expansion dominates the in situ kinetic temperatures

Figures 3 and 4 show a decrease in kinetic temperature and velocity with increased O^{+7}/O^{+6}. This finding is a signature of expansion and subsequent cooling of the solar wind. The expansion process is driven by the internal pressure of the CME. We can indirectly measure the internal pressure through the O^{+7}/O^{+6} ratio. Unless there is a fortuitous drop in the magnetic field or density, a sharp increase in electron temperature is indicative of a stronger internal CME pressure. The larger the initial pressure the faster the expansion in the low corona. This expansion results in adiabatic cooling of the plasma.

Elemental composition of CMEs and slow solar wind is similar, but not well ordered by coronal freeze-in temperature

Figure 5 indicates that CMEs and the slow solar wind have a similar elemental composition. However, we see no relationship between the Fe/O and O^{+7}/O^{+6} ratios, indicating that compositional parameters are not well-ordered by the coronal temperature in the structures, presumably loops, that gave rise to the CMEs and the slow solar wind. To some extent this result is surprising. The mechanisms by which the low FIP elements are enhanced in loops, described by Schwadron et al. [20], involve wave heating on the loops. Such heating will take time, and should also increase the electron freeze-in temperature in the loops and thus the O^{+7}/O^{+6} ratio. We might have expected, therefore, a more direct correlation between the Fe/O and the O^{+7}/O^{+6} ratios. We note that average solar wind measurements from Aellig et al. and references therein [35] showed an increase of log(Fe/O) with $\log(O^{+7}/O^{+6})$ during a study of 80 days in 1996. The differing results may be related to the observations occurring at different points in the solar cycle.

Helium is enhanced in "hot" CME events

Figures 6 and 7 show that in CMEs with very large O^{+7}/O^{+6} helium can be enhanced (averages values for He/O are ~ 75 in the slow solar wind and ~ 115 in the photosphere [27]). The primordial helium abundance relative to hydrogen is $\sim 8\%$ (see [34] and references therein). In comparison, the average solar wind helium abundance is only ~ 4 %. The depletion of helium presumably occurs in the chromosphere or the low corona [33]. The details of the fractionation process are currently not well understood. It was first suggested by Geiss et al. [28], that the fractionation may be related to insufficient Coulomb drag in the low corona that would deplete the corona of helium. If this is the case some energetic CMEs might pick up these low coronal helium enhancements and move them into the solar wind. Our results support this possibility.

CONCLUSIONS

We have presented a series of observations of the elemental and charge-state composition in CMEs and the slow solar wind, and of correlations between the composition. The most striking conclusion are the similarities between the CMEs and the slow wind. They share a common elemental composition, monotonically decreasing kinetic temperature profiles, and an apparently similar expansion history. These observations are consistent with CMEs and the slow solar wind having a common origin of closed magnetic loops in the corona, with perhaps the only difference being the mechanism by which the material is released from the loops. In Fisk and Schwadron [1] a mechanism is provided for the release of material from loops to form the slow solar wind as open field lines reconnect with the closed loops. In CMEs the loops themselves are released into the solar wind.

ACKNOWLEDGMENTS

This work was supported in part by NASA grant S02-GSRP-121 and by NASA contracts NAG5-2810 and NAG5-711. Work at Los Alamos was performed under the auspices of the US Department of Energy with financial support from the NASA ACE program.

REFERENCES

1. Fisk, L. A., N. A. Schwadron, The behavior of the open magnetic flux of the sun, *Astrophys. J.*, in press.
2. Tousey, R., The solar corona, *Space Res.*, **13**, 713, 1973.
3. Gosling, J. T., E. Hildner, R. M. MacQueen, R. H. Munro, A. I. Poland, and C. L. Ross, Mass ejections from the sun: a view from skylab, *J. Geophys. Res.*, **79**, 4581, 1974.
4. Klimchuk, J. A., Theory of coronal mass ejections, *J. Geophys. Res.*, in press.
5. Gosling, J.T., Coronal mass ejections and magnetic flux ropes in interplanetary space, in *Physics of Magnetic Flux Ropes*, Washington D. C., 343, 1990.
6. Gloeckler, G, J Cain, F. M. Ipavich, E. O. Tums, P. Bedini, L. A. Fisk, T. H. Zurbuchen, P. Bochsler, J. Fischer, R. F. Wimmer-schweingruber, J. Geiss, and R. Kallenbach, Investigation of the composition of solar and interstellar matter using solar wind and pickup ion measurements with SWICS and SWIMS on the ACE spacecraft, *Space Sci. Rev.*, **86**, 497, 1998.
7. Smith, C. W., J. L'Heureux, N. Ness, F., M. H. Acuna, L. F. Burlaga, and J. Scheifele, The Ace magnetic fields experiment, *Space Sci. Rev.*, **86**, 613, 1998.
8. McComas, D. J., S. J. Bame, P. Barker, W. C. Feldman, J. L. Phillips, P. Riley, J. W. Griffee, Solar Wind Electron Proton Alpha Monitor (SWEPAM) for the Advanced Composition Explorer, *Space Sci. Rev.*, **86**, 563, 1998.
9. Fenimore, E. E., Solar wind flows associated with hot heavy ions, *Astrophys. J.*, **235**, 245, 1980.
10. Henke, T., J. Woch, U. Mall, S. Livi, B. Wilken, R. Schwenn, G. Gloecker, R. von Steiger, R. J. Forsyth, and A. Balogh, Differences in the O^{+7}/O^{+6} ratio in magnetic cloud and non-cloud coronal mass ejections, *Geophys. Res. Lett.*, **25**, 3465, 1998.
11. Gloeckler, G., L. A. Fisk, S. Hefti, N. A. Schwadron, T. H. Zurbuchen, F. M. Ipavich, J. Geiss, P. Bochsler, R. F. Wimmer-Schweingruber, Unusual composition of the solar wind in the 2-3 May 1998 CME observed with SWICS on ACE, *Geophys. Res. Lett.*, **26**, 157, 1999.
12. Burlaga, L. F., R. M. Skoug, C. W. Smith, D. Webb, T. H. Zurbuchen, A. A. Reinard, Fast Ejecta during Ascending Phase of Solar Cycle 23: ACE observations, 1998-1999, *J. Geophys. Res.*, in press.
13. Lepri, S. T., T. H. Zurbuchen, L. A. Fisk, I. G. Richardson, H. V. Cane, G. Gloeckler, Iron charge distribution as an identifier of interplanetary coronal mass ejections, *J. Geophys. Res.*, in press.
14. Buergi, A. and J. Geiss, Helium and minor ions in the corona and solar wind: Dynamics and charge states, *Sol. Phys.*, **103**, 347, 1986.
15. Ko, Y-K, G. Gloeckler, C. Cohen, A. B. Galvin, Solar wind ionic charge states during the Ulysses pole-to-pole pass, *J. Geophys. Res.*, **104**, 17005, 1999.
16. Burgi, A., Effects of non-Maxwellian electron velocity distribution functions and nonspherical geometry on minor ions in the solar wind, *J. Geophys. Res.*, **92**, 1057, 1987.
17. Owocki, S. P., A. J. Hundhausen, The effect of a coronal shock wave on the solar wind ionization state, *Astrophys. J.*, **274**, 414, 1983.
18. Geiss, J., G. Gloeckler, R. von-Steiger, H. Balsiger, L. A. Fisk, A. B. Galvin, F. M. Ipavich., S. Livi., J. F. McKenzie, K. W. Ogilvie, B. Wilken, The southern high-speed stream: results from the SWICS instrument on Ulysses, *Science*, **268**, 1033, 1995.
19. Zurbuchen, T. H., L. A. Fisk, G. Gloeckler, and N. A. Schwadron, Elemental and isotopic fractionation in closed magnetic structures, *Space Sci. Rev.*, **85**, 397, 1998.
20. Schwadron, N. A., L. A. Fisk, and T. H. Zurbuchen, Elemental fractionation in the slow solar wind, *Astrophys. J.*, **521**, 859, 1999.
21. Feldman, U., K. G. Widing, and H. P. Warren, Morphology of the quiet solar upper atmosphere in the $4E^4 < T_e < 1.4E^6$K temperature regime, *Astrophys. J.*, **522**, 1133, 1999.
22. Zurbuchen, T. H., S. Hefti, L. A. Fisk, G. Gloeckler, and N. A. Schwadron, Magnetic structure of the slow solar wind: Constraints from composition data, *J. Geophys. Res.*, **105**, 18327, 2000.
23. Vernazza, J. E., E. H. Avrett, R. Loeser, Structure of the solar chromosphere. III - Models of the EUV brightness components of the quiet-sun, *Astrophys. J. Supplement Series*, **45**, 635, 1981.
24. Galvin, A. B., F. M. Ipavich, G. Gloeckler, D. Hovestadt, S. J. Bame, B. Klecker, M. Scholer, and B. T. Tsurutani, Solar wind iron charge states preceding a driver plasma, *J. Geophys. Res.*, **92**, 12069, 1987.
25. Arnaud,M, R. Rothenflug, An updated evaluation of recombination and ionization rates, *Astronomy and Astrophysics Supplement Series*, **60**, 425, 1985.
26. Arnaud, M, J. Raymond, Iron ionization and recombination rates and ionization equilibrium, *Astrophys. J.*, **398**, 394, 1992.
27. von Steiger, R., J. Geiss, G. Gloeckler, Composition of the solar wind, in *Cosmic Winds and the Heliosphere*, University of Arizona Press, 581, 1997.
28. Geiss, J., P. Hirt, H. Leutwyler, On the acceleration and motion of ions in the corona and solar wind, *Solar Phys.*, **12**, 458, 1970.
29. Fisk, L. A., Motion of the footpoints of heliospheric magnetic field lines at the Sun: Implications for recurrent energetic particle events at high heliographic latitudes, *J. Geophys. Res.*, **101**, 15547, 1996.
30. Fisk, L. A., T. H. Zurbuchen, N. A. Schwadron, On the coronal magnetic field: consequences of large-scale motion, *Astrophys. J.*, **521**, 868, 1999.
31. Feldman, U., U. Schühle, K. G., Widing, J. M. Laming, Coronal composition above the solar equator and the north pole as determined from spectra acquired by the SUMER instrument on SOHO, *Astrophys. J.*, **505**, 999, 1998.
32. Ko, Y-K., L. A. Fisk, J. Geiss, G. Gloeckler,and M. Guhathakurta, An empirical study of the electron temperature and heavy ion velocities in the south polar coronal hole, *Solar Phys.*, **171**, 345, 1997.
33. Laming, J. M. and U. Feldman, The solar helium abundance in the outer corona determined from observations with SUMER/SOHO, *Astrophys. J.*, **546**, 552, 2001.
34. Coplan, M. A., K. W. Ogilvie, P. Bochsler, J. Geiss, Interpretation of He-3 abundance variations in the solar wind, *Solar Phys.*, **93**, 415, 1984.
35. Aellig, M. R., S. Hefti, H. Grünwaldt, P. Bochsler, P. Wurz, F. M. Ipavich, D. Hovestadt, The Fe/O elemental abundance ratio in the solar wind as observed with SOHO CELIAS CTOF, *J. Geophys. Res.*, **104**, 24769, 1999.

Composition of magnetic cloud plasmas during 1997 and 1998

P. Wurz*, R.F. Wimmer-Schweingruber*, K. Issautier*, P. Bochsler*, A. B. Galvin†, J. A. Paquette** and F. M. Ipavich**

*Physics Institute, University of Bern, Sidlerstrasse 5, CH-3012 Bern, Switzerland
†EOS Space Sciences, University of New Hampshire, Morse Hall, Durham, NH 03824, USA
**Dept. Physics and Astronomy, University of Maryland, College Park, MD 20742, USA

Abstract. We present a study of the elemental composition of a sub-set of coronal mass ejections, namely events which have been identified of being of the magnetic cloud type (MC). We used plasma data from the MTOF sensor of the CELIAS instrument of the SOHO mission. So far we have investigated MCs of 1997 and 1998. The study covers the proton, alpha, and heavy ion elemental abundances. Considerable variations from event to event exist with regard to the density of the individual species with respect to regular "slow" solar wind preceding the MC plasma. However, two general features are observed. First, for the heavy elements (carbon through iron), which can be regarded as tracers in the solar wind plasma, a mass-dependent enrichment of ions monotonically increasing with mass is observed. The enrichment can be explained by a previously published theoretical model assuming coronal plasma loops on the solar surface being the precursor structure of the MC. Second, when comparing the MC plasma to regular solar wind composition, a net depletion of the lighter ions, helium through oxygen, is always observed. Proton and alpha particle abundances have to be regarded separately since they represent the main plasma.

INTRODUCTION

In a recent study it was reported that the plasma of the coronal mass ejection (CME) of 6 January 1997 was strongly mass-fractionated favoring heavier elements with respect to lighter ones [1, 2]. From the magnetic field measurements on WIND it has been concluded that the 6 January 1997 CME falls into the group of magnetic cloud (MC) events [3]. An overview of the 6 January 1997 CME event has been given by Fox et al. [4], which covers the launch of the CME, its propagation through interplanetary space, and its effect on the Earth's magnetosphere. Following this observation, the measured mass fractionation was successfully modelled by assuming large coronal loops, being the precursors of the magnetic cloud (MC) plasma, where the mass fractionation is established by diffusion across magnetic field lines [5].

A review covering the present understanding of CMEs has been given recently by Gosling [6], for the compositional aspects of CMEs see the review by Galvin [7]. CMEs with magnetic cloud topology generally exhibit somewhat higher freeze-in temperatures, i.e., the charge-state distribution of a particular element is shifted toward higher charge states than for the ambient undisturbed solar wind [8, 9]. Observational signatures of MCs consist of an enhanced magnetic field strength, a smooth rotation of the magnetic field direction as the cloud passes the spacecraft, and a low proton temperature. It has been found earlier that near 1 AU about one third of all CMEs in the ecliptic plane are magnetic cloud events [10].

In this paper we present further experimental evidence for the elemental fractionation of heavy ions in MCs using data from the CELIAS/MTOF instrument on the SOHO mission. There were two MCs during 1997 and five MCs during 1998, which passed the SOHO spacecraft located close to the Earth at first Lagrangian point, L1. These events are listed in Table 1. We present the analysis for five of these MCs. All these CME events are of the magnetic cloud type. In addition, the events have been selected such that the charge-state distributions are similar to what is observed in regular solar wind using ACE/SWICS quick-look data. Note that this selection includes the shift of charge-states in MCs mentioned in the previous paragraph. The sequence of three CME events from 2–3 May 1998 of which the second one is of MC nature, had very unusual charge state distributions [11] which lead to the exclusion of this event from the present analysis. A more sophisticated analysis due to these complications will be necessary and will be presented in the

CP598, *Solar and Galactic Composition*, edited by R. F. Wimmer-Schweingruber

TABLE 1. List of magnetic clouds during 1997 and 1998 that could be observed with solar wind particle instrumentation near Earth. Exact times of analyzed intervals for the reference periods and the MCs cloud are given as day-of-year (DOY).

MC Event*	Ref. start time [DOY]	Ref. end time [DOY]	MC start time [DOY]	MC end time [DOY]	Solar wind speed† [km/s]	Remarks
10–11 Jan 1997	8.00	9.00	10.27	10.98	440	MC followed by filament
7 Nov 1997	301.00	302.20	311.30	312.50	420	CME preceding MC
2–3 May 1998	—	—	—	—	510	Multiple CMEs
2 June 1998	151.72	153.00	153.43	153.65	430	
24 June 1998	173.00	174.00	175.42	176.00	390	
25 Sep 1998	—	—	—	—	—	SOHO not operational
8 Nov 1998	310.00	311.25	312.18	313.73	520	

* Date the event was observed at 1 ~ AU
† Average speed during MC duration

future. For the 25 September 1998 MC event the SOHO spacecraft was not operational.

DATA ANALYSIS

In this study we evaluated the elements C, N, O, Ne, Na, Mg, Al, Si, S, Ar, Ca, and Fe. From the ions recorded with the MTOF sensor, the CELIAS data processing unit accumulates time-of-flight (TOF) spectra for 5 minutes, which then are transmitted to ground. The raw counts for each mass peak of the different elements were extracted from each of the transmitted TOF spectra by fitting a model function of the peak shape and the background [12]. Subsequently, the overall efficiency of the MTOF sensor was calculated for each element and for each accumulation interval. To obtain particle fluxes for the chosen elements, the instrument response of the MTOF sensor comprising the transmission of the entrance system and the response of the isochronous TOF mass spectrometer, was taken into account in great detail [12].

The actual solar wind plasma parameters, which were measured by the Proton Monitor (PM) a sub-sensor of the MTOF sensor, are needed as input parameters for the instrument response of the MTOF sensor. The quality of the determination of the solar wind plasma parameters with the PM is quite good [13] and better than required for the determination of densities with the MTOF sensor. Another input parameter needed for the determination of the MTOF instrument response, in particular for the determination of the transmission of the entrance system, is the charge-state distribution of each element for each accumulation interval. The MTOF sensor determines only the mass of the incoming ion (with high resolution, however), but not its charge. The CTOF sensor was supposed to provide this information for a few key elements, but since mid August of 1996 problems in the CTOF sensor electronics prohibit these measurements. Thus, we had to resort to a model for the charge-state distributions of the elements. We derive the so-called freeze-in temperature from a semi-empirical model using the solar wind velocity as input parameter [12]. This model also accounts for ionization resulting from non-maxwellian electron distributions. From the freeze-in temperature we obtained charge distributions for each element by assuming an ionization equilibrium in the corona and by applying ionization and recombination rates for electronic collisions from Arnaud and co-workers [14, 15]. The application of the instrument function to the measured count rates yielded densities for the different elements. We extensively checked if the instrument function introduces a mass bias, but so far we did not find such an effect in the data analysis.

The MTOF sensor settings are cycled in a sequence consisting of four to six steps, which were optimized to cover a broad range of solar wind conditions. The stepping sequence includes two voltage settings for the entrance system and up to three values for the potential difference between the entrance system and the TOF mass spectrometer (negative, zero, and positive potential difference). For the present analysis only the steps with negative or zero potential difference have been used. In principle a time resolution of five minutes, the dwell time for each step, can be obtained if the sensitivity of the MTOF sensor is high enough for the particular element considered. For typical solar wind conditions and for the more abundant heavy elements in the solar wind, it is indeed possible to derive densities with such a high time resolution, as has been demonstrated earlier [2].

RESULTS

We compare the densities for the different elements during the MC with the corresponding densities during a reference solar wind period. To account for the variability of the solar wind with time, or with location of origin or with solar activity, we used for reference a day of

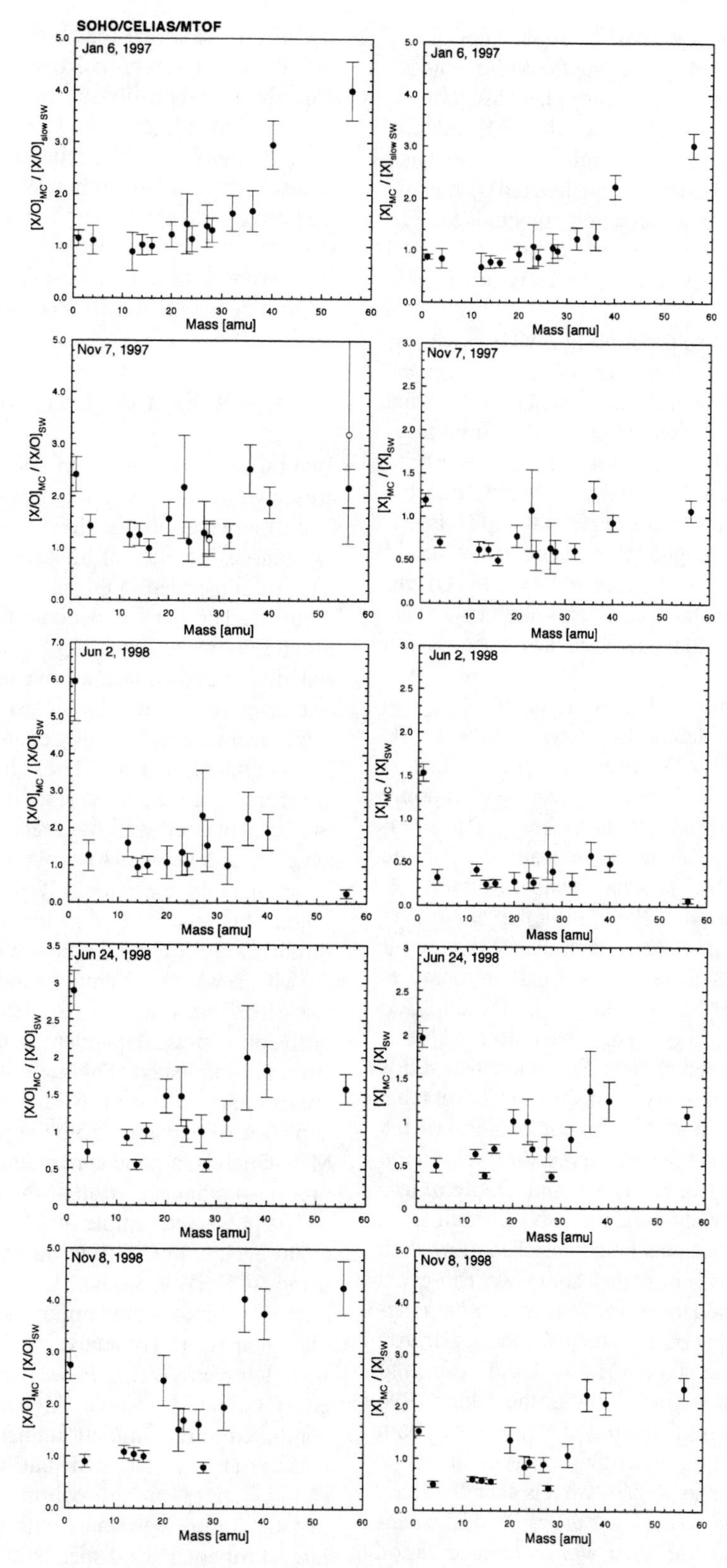

FIGURE 1. Results of the analysis of five MC events. The densities in the MC are compared to the respective densities in the preceding reference period of slow solar wind (see Table 1). Left column shows a comparison of abundance ratios with respect to oxygen; right column shows the density ratios of MC plasma and preceding SW plasma. The open symbols in the panel for the 7 November 1997 event were taken from [16].

slow solar wind preceding the MC by about a day. Since we chose reference periods preceding the MC events we can safely assume that these reference plasmas are unaffected by the disruption which caused the CME release. Note that the slow solar wind, e.g. solar wind associated with the streamer belt, is already fractionated by mechanisms governed by the first ionization potential (the FIP effect) [17, 18]. The exact time periods for the MC and the reference solar wind used in the analysis are given in Table 1.

Figure 1 shows the results for all five MC in two formats, one where data are given with reference to oxygen and one where direct comparison of MC and solar wind plasma is plotted. We derived the proton data from measurements with the PM, a sub-sensor of the MTOF sensor. The helium data as well as all the heavy ion data were derived from mass spectra recorded with the MTOF sensor. Since MTOF was designed to measure heavy ions in the solar wind He is largely suppressed by the MTOF entrance system by design; thus the determination of the He density is problematic and the He data have to be viewed with caution.

The left column in the Figure 1 shows the ratio of abundances with reference to oxygen in the MC cloud versus the solar wind reference period (that is $[X/O]_{MC}/[X/O]_{ref.\ SW}$ is plotted), which is a commonly used format to display variations in heavy ion abundances. The top panel shows a re-evaluation of the 10–11 January 1997 MC plasma considering more elements than Wurz et al. [2]. Within the error bars the initial results [2] have been reproduced. For the 7 November 1997 event there are two earlier measurements from WIND/MASS [16], $[Si/O]_{MC}/[Si/O]_{slowSW}$ and $[Fe/O]_{MC}/[Fe/O]_{slowSW}$, which have been added to the Figure. The reported Si and Fe abundance ratios agree with the present analysis within the error bars. For all five events we find that the composition of the MC is markedly different from the preceding solar wind plasma. In four out of five events we find a more or less organized mass fractionation for the heavy elements, He through Fe, with heavier ions being enriched more than the lighter ones. There is of course some event-to-event variability in the abundance of the ions and in the magnitude of the enrichment of the heavy elements. For iron the enrichment is in the range of 1.5 to 4 with respect to the preceding slow solar wind. Only for the 2 June 1998 event we find that the iron abundance is lower, by about a factor of five, than in the preceding solar wind.

To get a better understanding of what is actually going on in the MCs we have to consider the ratios of the densities in the MC versus the solar wind reference period (e.g. $[X]_{MC}/[X]_{ref.\ SW}$). These data are shown in the right column of Figure 1. At first glance the data looks qualitatively the same as the abundance data (left column) revealing again the mass fractionation. However, the striking difference is that for all events the lighter elements, with the exception of hydrogen, are actually depleted to about half to their density in the preceding solar wind plasma. Depending on the strength of the mass fractionation the densities of heavier elements reach solar wind values (events 7 November 1997 and 24 June 1998) or are even enriched compared to the solar wind (events 10 January 1997 and 8 November 1998). Again, the 2 June 1998 event does not match this pattern and we find that the iron density is only 0.06 of its solar wind value.

DISCUSSION AND CONCLUSIONS

Two things appear to be in common for the MC events presented in this study: the mass-dependent fractionation (with the exception of the 2 June 1998 event) and the substantial depletion of lighter elements. These two findings are illustrated in the summary of the data shown in Figure 2. The mass-dependent fractionation is observed for all minor ions including He. Protons, being the major constituent of the plasma, have their own life and have to be considered separately. Since the protons constitute the main plasma and the heavy elements are only tracer particles in the plasma, a different behavior of the protons is not surprising. The mass fractionation of the heavy ions can be explained well by a recent theoretical model [5], which was developed to explain the observed mass fractionation in the plasma of MC of the 10–11 January 1997 event. This model explains the mass fractionation by assuming large coronal loops as the precursor structure for the MC in which different elements are depleted as a result of diffusion across magnetic field lines. Since this diffusion is mass dependent a mass-dependent fractionation is established. This model can also reproduce the present data very well. Note that the model is based on depletion of elements from the precursor structure of the MC, which is in good agreement with the present finding of a substantial depletion of the lighter elements.

The presented sample of CME events are all magnetic cloud events. In addition, the events have been selected using ACE/SWICS quick-look data such that the charge state distributions are similar to what is observed in regular solar wind. By analyzing 56 CME events recorded with Ulysses/SWICS it was observed that MCs generally exhibit somewhat higher freeze-in temperatures compared to the ambient undisturbed solar wind [8, 9], i.e., the charge-state distributions are shifted to higher charges states. In the ecliptic the increase in freeze-in temperature correlates with solar wind speed and is largest for solar wind speeds exceeding 700 km/s [8, 9]. The MC events we analyzed are at moderate solar wind speeds (see Table 1) and therefore the charge-state distributions are similar to regular solar wind, which was ver-

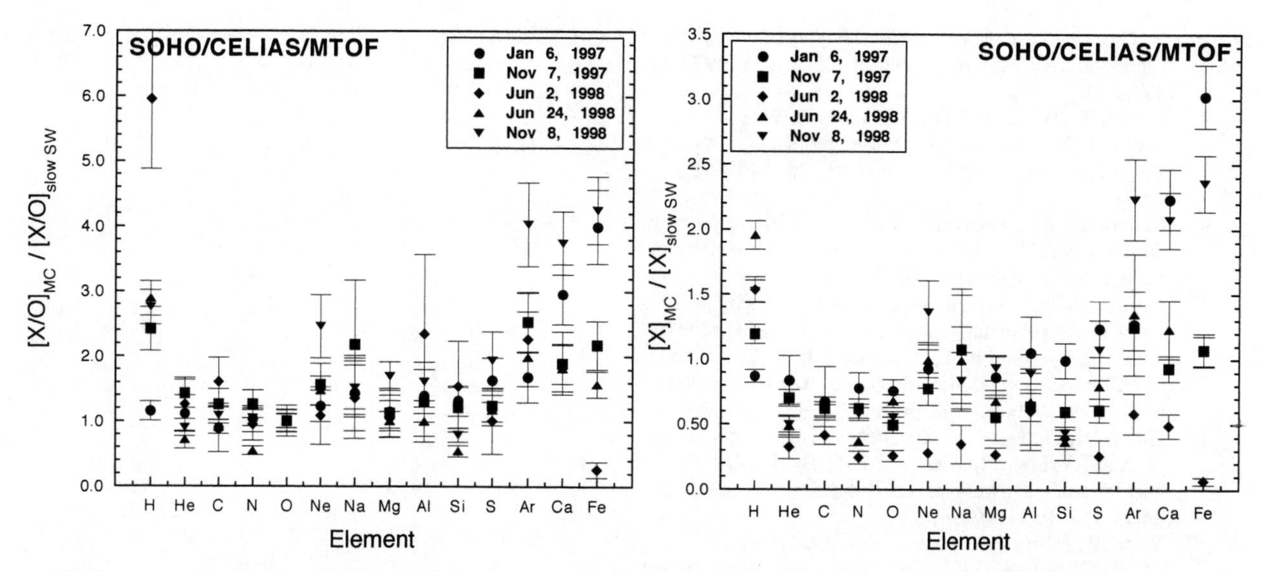

FIGURE 2. Summary of the analysis of five observed MC events showing the range of mass fractionation. Data and plotting format are the same as in Figure 1.

ified by checking the ACE/SWICS quick-look data. Our selection criteria lead to the exclusion of the 2–3 May 1998 event. Such a selection of events might introduce a bias in the result, in the sense that MCs with charge-state distributions significantly different from regular solar wind might also show different mass fractionations, if at all. On the other hand, we considered 5 out of a total of 7 events from the 1997–1998 time period in the present analysis. In addition, for the 2–3 May 1998 event an Fe/O ratio of 0.28 ± 0.10 was found [11], which is an increase of the Fe/O ratio by a factor of 2 compared to regular solar wind. This observation fits well into the general pattern of mass-dependent fractionation we observed. Thus we feel that the findings are quite representative for MCs in the ecliptic.

In the future we will analyze also the MC events from 1999 until present and extend our analysis also to events with unusual charge state distribution, like the 2–3 May 1998 event, the latter by using the actual charge-state distributions measured by ACE/SWICS. We have to await these analyses to see if the common features we found for MCs so far, the mass-dependent fractionation and the substantial depletion of lighter elements, will be observed there as well.

ACKNOWLEDGMENTS

CELIAS is a joint effort of five hardware institutions under the direction of the Max-Planck Institut für Extraterrestrische Physik (pre-launch) and the University of Bern (post-launch). The University of Maryland was the prime hardware institution for MTOF, the University of Bern provided the entrance system, and the Technical University of Braunschweig provided the DPU. This work is supported by the Swiss National Science Foundation.

REFERENCES

1. Wurz, P., Ipavich, F. M., Galvin, A. B., Bochsler, P., Aellig, M. R., Kallenbach, R., Hovestadt, D., Grünwaldt, H., Hilchenbach, M., Axford, W. I., Balsiger, H., Bürgi, A., Coplan, M. A., Geiss, J., Gliem, F., Gloeckler, G., Hefti, S., Hsieh, H. C., Klecker, B., Lee, M. A., Livi, S., Managadze, G. G., Marsch, E., Möbius, E., Neugebauer, M., Reiche, K.-U., Scholer, M., Verigin, M. I., and Wilken, B., *ESA*, **SP-415**, 395–400 (1997).
2. Wurz, P., Ipavich, F., Galvin, A., Bochsler, P., Aellig, M., Kallenbach, R., Hovestadt, D., Grünwaldt, H., Hilchenbach, M., Axford, W., Balsiger, H., Bürgi, A., Coplan, M., Geiss, J., Gliem, F., Gloeckler, G., Hefti, S., Hsieh, H. C., Klecker, B., Lee, M. A., Managadze, G. G., Marsch, E., Möbius, E., Neugebauer, M., Reiche, K.-U., Scholer, M., Verigin, M. I., and Wilken, B., *Geophys. Res. Lett.*, **25(14)**, 2557–2560 (1998).
3. Burlaga, L., Fritzenreiter, R., Lepping, R., Ogilvie, K., Szabo, A., Lazarus, A., Steinberg, J., Gloeckler, G., Howard, R., Michels, D., Farrugia, C., Lin, R. P., and Larson, D. E., *J. Geophys. Res.*, **103**, 277–286 (1998).
4. Fox, N. J., Peredo, M., and Thompson, B. J., *Geophys. Res. Lett.*, **25(14)**, 2461–2464 (1998).
5. Wurz, P., Bochsler, P., and Lee, M. A., *J. Geophys. Res.*, **105(A12)**, 27239–27249 (2000).
6. Gosling, J. T., "Coronal mass ejections: An overview", in *Coronal Mass Ejections*, edited by N. Crooker, J. A. Joselyn, and J. Feynman, Geophys. Monograph 99, American Institute of Physics, 1997, pp. 9–16.
7. Galvin, A. B., "Minor ion composition in CME-related solar wind", in *Coronal Mass Ejections*, edited by

N. Crooker, J. A. Joselyn, and J. Feynman, Geophys. Monograph 99, American Institute of Physics, 1997, pp. 253–260.
8. Henke, T., Woch, J., Mall, U., Livi, S., Wilken, B., Schwenn, R., Gloeckler, G., v. Steiger, R., Forsyth, R. J., and Balogh, A., *Geophys. Res. Lett.*, **25**, 3465–3468 (1998).
9. Neukomm, R., *Composition of Coronal Mass Ejections Derived with SWICS/Ulysses*, Phd thesis, University of Bern, Bern, Switzerland (1998).
10. Gosling, J. T., "Coronal mass ejections and magnetic flux ropes in interplanetary space", in *Physics of Magnetic Flux Ropes*, edited by C. T. Russell, E. R. Priest, and L. C. Lee, Geophys. Monograph 58, American Institute of Physics, 1990, pp. 343–364.
11. Gloeckler, G., Hefti, S., Zurbuchen, T. H., Schwadron, N. A., Fisk, L. A., Ipavich, F. M., Geiss, J., Bochsler, P., and Wimmer-Schweingruber, R. F., *Geophys. Res. Lett.*, **26(2)**, 157–160 (1999).
12. Wurz, P., *Heavy ions in the solar wind: Results from SOHO/CELIAS/MTOF*, Habilitation thesis, University of Bern, Bern, Switzerland (1999).
13. Ipavich, F. M., Galvin, A. B., Lasley, S. E., Paquette, J. A., Hefti, S., Reiche, K.-U., Coplan, M. A., Gloeckler, G., Bochsler, P., Hovestadt, D., Grünwaldt, H., Hilchenbach, M., Gliem, F., Axford, W. I., Balsiger, H., Bürgi, A., Geiss, J., Hsieh, K. C., Kallenbach, R., Klecker, B., Lee, M. A., Mangadze, G. G., Marsch, E., Möbius, E., Neugebauer, M., Scholer, M., Verigin, M. I., Wilken, B., and Wurz, P., *J. Geophys. Res.*, **103(A8)**, 17205–17214 (1997).
14. Arnaud, M., and Rothenflug, R., *Astron. Astrophys. Suppl. Ser.*, **60**, 425–457 (1985).
15. Arnaud, M., and Raymond, J., *Astrophys. J.*, **398**, 394–406 (1992).
16. Wimmer-Schweingruber, R. F., Kern, O., and Hamilton, D. C., *Geophys. Res. Lett.*, **26(23)**, 3541–3544 (1999).
17. Geiss, J., *Space Sci. Rev.*, **33**, 201–217 (1982).
18. von Steiger, R., "Solar wind composition and charge states", in *Solar Wind Eight*, edited by D. Winterhalter, J. T. Gosling, S. R. Habbal, W. S. Kurth, and M. Neugebauer, AIP Press, 1995, pp. 193–198.

Energetic Particle Composition

Energetic Particle Composition

Donald V. Reames

NASA Goddard Space Flight Center
Greenbelt, MD, USA

Abstract. Abundances of elements and isotopes have been essential for identifying and measuring the sources of the energetic ions and for studying the physical processes of acceleration and transport for each particle population in the heliosphere. Many of the sources are surprising, in a few cases the acceleration bias is extreme, but an understanding of the fundamental physics allows us to use energetic ions to determine abundances for the average solar corona, the high-speed solar wind, and the local interstellar medium.

INTRODUCTION

Energetic particles are accelerated by a variety of physical mechanisms at many sites throughout the heliosphere that may be listed as follows (see reviews by Lee (19) and Reames (36)):

Heliospheric Sources and Energetic Particles

1) Solar flares – ^{3}He/^{4}He and (Z>50)/O enhanced

2) CME-driven shock waves – large SEP events

3) Planetary magnetospheres
  a) Radiation belts – neutron albedo
  b) Io belt of S and O
  c) Trapped ACRs
  d) Ion conics

4) Planetary bow shocks
  – 'Upstream' events

5) Co-rotating interaction regions (CIRs)

6) Heliospheric termination shock
  – Anomalous Cosmic Rays (ACRs)
    – Interstellar pickup ions

7) Galactic Cosmic Rays (GCRs)

In recent years, we have learned to divide solar energetic particle (SEP) events into the small 'impulsive' events, accelerated in solar flares, and large, long-duration, 'gradual' events where acceleration occurs at shock waves driven out from the Sun by coronal mass ejections (CMEs) (9, 13, 14, 15, 17, 18, 36, 37). Resonant wave-particle interactions in flares can produce 1000-fold enhancements in ^{3}He/^{4}He and (Z>50)/O. On the other hand, abundances averaged over many large gradual SEP events allow the study of fractionation of the solar coronal material relative to the photosphere – the solar FIP (first ionization potential) effect. We are beginning to understand and model the dynamic physical processes of wave-particle interactions near shocks that explain abundance variations with time during one SEP event and from one event to another. We now see why averaging works. However, these abundances are complicated by shock re-acceleration of residual ions from impulsive flares, and by the exponential rollovers in the high-energy spectra, called spectral 'knees.'

Several magnetospheric populations have extremely interesting abundances. The main radiation belts of Earth, Jupiter, and Saturn consist almost entirely of the element H. This incredibly simple abundance pattern reveals the source of these belts as the decay of energetic neutrons that are produced by interactions of GCRs with the atmospheres, rings, and moons of these planets. Another interesting population, seen in the Jovian magnetosphere, has comparable abundances of S and O at several MeV/amu, without accompanying Ne, Mg, or Si. The S and O are accelerated from disassociated gasses such as SO_2 emitted from the volcanoes of the Jovian moon Io. Surely, this is the 'smoking gun' of abundances.

The anomalous cosmic rays (ACRs) are an example of an extreme FIP-based, ion-neutral separation of

CP598, *Solar and Galactic Composition,* edited by R. F. Wimmer-Schweingruber
2001 American Institute of Physics 0-7354-0042-3

a source population. Low-FIP ions are ionized in the local interstellar medium, but high-FIP ions, such as H, He, N, O, Ne and Ar, are mostly neutral. The neutrals easily cross magnetic fields to enter the heliosphere; if they are ionized by solar ultraviolet or by charge exchange with solar-wind protons, they are 'picked up' by the solar wind and carried out to the heliospheric termination shock where they are accelerated to produce ACRs. The distribution function of the interstellar pickup ions remains flat out to twice the solar wind speed where they greatly outnumber normal solar wind ions, providing preferential injection. The existence of the interstellar pickup ions was predicted by Fisk, Kozlovsky, and Ramaty (7) to explain ACRs well before these ions were actually observed in the solar wind, although the pickup of interstellar He had been suggested previously to explain He^+ in the solar wind (11). Abundances of the local interstellar medium can be determined from ACRs and pickup ions given suitable models of photo-ionization and charge exchange. Modeling is not required to determine isotopic ratios such as $^{22}Ne/^{20}Ne$ ~0.1 (20)

Corotating interaction regions are produced where high-speed solar wind streams overtake slower solar wind emitted earlier by the rotating Sun. From this interaction, a forward shock propagates out into the slow wind and a reverse shock propagates sunward into the high-speed stream. Particles are accelerated at both shocks but are more intense at the reverse shock. Particle intensities at the shocks increase with distance out to 5-10 AU. Energetic ions streaming sunward from the reverse shock, measured near Earth, appeared to represent the abundances of the high-speed wind (45). Later, however, these ions were thought to originate from singly ionized 'inner-source' pickup ions from solar wind that is absorbed, neutralized, and re-emitted at interstellar dust grains passing near the Sun (8). Ionization-state measurements of these energetic ions from CIRs, reported by Möbius *et al.* (30) at this workshop, indicate that the ions (with the exception of ~25% of He and ~8% of Ne) are multiply ionized and hence they are accelerated from the high-speed solar wind after all. We have come full circle.

Nearly all of the heliospheric sources are 'invisible,' in that ion acceleration produces no measurable photons, so we must derive the physics from the energetic particles themselves. This requires that we distinguish the influences of injection, acceleration, and transport. Improving observations and models of this rich variety of events and sources have begun to make this possible. Many of the properties of the energetic-particle populations have been discussed in previous review articles (19, 36) and will not be repeated here. This review will focus on those recent observations and new theories and models that have extended our understanding or revised our perspective of energetic particles and their underlying source abundances.

IMPULSIVE SOLAR FLARES

Energetic particles from impulsive solar flares are characterized by extreme abundance enhancements resulting from acceleration by resonant wave-particle interactions in the flare plasma (*e.g.* 36, 47, 48). All of the elements through Si are fully ionized, and Fe of charge ~20 is observed (22, 29), indicating flare-heated plasma at a temperature of ~10 MK.

Enhanced abundances similar to those seen in energetic particles are also deduced from the Doppler-broadened γ-ray lines emitted from the energetic ions in solar flares (23, 31). Narrow γ-ray lines emitted from the ambient flare plasma show normal coronal abundances with no enhancements. Thus, the enhancements arise during acceleration.

Recent observations on the Wind spacecraft have yielded abundances for the dominant element groups in the $34 \le Z \le 82$ region (39). These abundances, along with those for $Z \le 26$ (36, 39, 41), are shown in Figure 1 as enhancements relative to coronal abundances. Abundance enhancements at high Z have been confirmed by Mazur *et al.* (25) at this workshop.

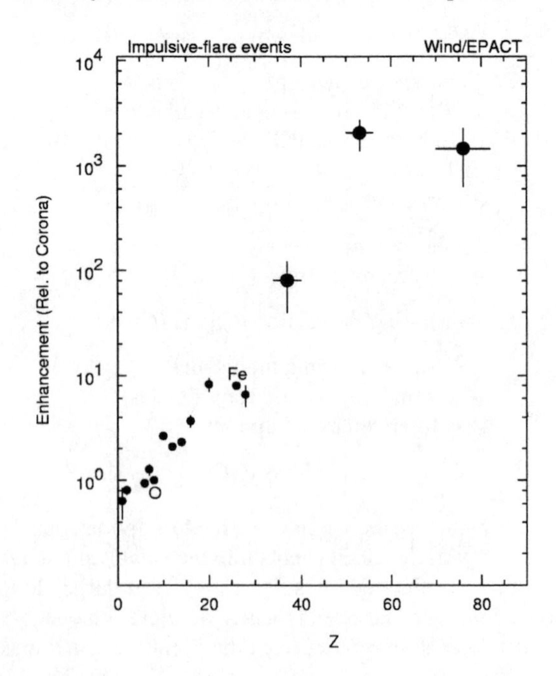

FIGURE 1. Abundance enhancements of heavy ions from impulsive solar flares.

Progressive enhancements of heavy ions may result from the physics of cascading waves (28) in which turbulent energy is generated at a large spatial scale and small wave number k, by magnetic reconnection above a flare. This energy Kolmogorov-cascades toward higher k and is first absorbed by ions with the lowest gyrofrequency and charge-to-mass ratio Q/A. Energy not absorbed by the heaviest ions continues to cascade toward lighter and lighter ions and is eventually absorbed by He or H.

Cascading waves may explain the abundance pattern in Figure 1, but, unfortunately, cannot explain the enhancements in $^3He/^4He$. This enhancement is believed to result from electromagnetic ion cyclotron (EMIC) waves produced between the gyrofrequencies of H and ^4He by electrons streaming down the magnetic field lines (47, 48). ^3He is the only species whose gyrofrequency lies in this region so it can efficiently absorb these waves. A similar mechanism produces 'ion conics' in the Earth's auroral region where electrons, ions, and waves can all be observed *in situ*.

The necessity for two acceleration mechanisms, with unknown spatial and temporal relationships, makes it difficult to understand ion acceleration in flares. The necessity for acceleration before further ionization and the lack of correlations in the abundance variations complicate the picture (38, 41).

GRADUAL SEP EVENTS

In essentially all of the large SEP events particles are accelerated at CME-driven shock waves (9, 13, 14, 15, 18, 19, 36, 37). Peak particle intensities are correlated with CME speed and only 1-2% of CMEs drive shocks that are fast enough to accelerate ions (37). Shocks from the largest, fastest CMEs span more than half the heliosphere. These shocks expand across magnetic field lines accelerating energetic particles as they go.

Owing to the spiral pattern of the interplanetary magnetic field lines, which the ions follow, the particle time profiles depend in a systematic way upon the longitude of the observer relative to the source, as shown in Figure 2 (4, 36, 40). An observer on the east flank of the shock sees a source at a western longitude on the Sun. As a function of time, this observer's connection point swings from the intense nose of the shock, near the Sun, to the weaker flank when the shock arrives at 1 AU. An observer on the west flank of the shock may see maximum intensity only after crossing through the shock into the region where field lines connect to the shock nose from behind. Acceleration may also weaken, especially at high energies, as the shock moves outward, but the relentless eastward swing of the observers connection point to the shock is always a major factor.

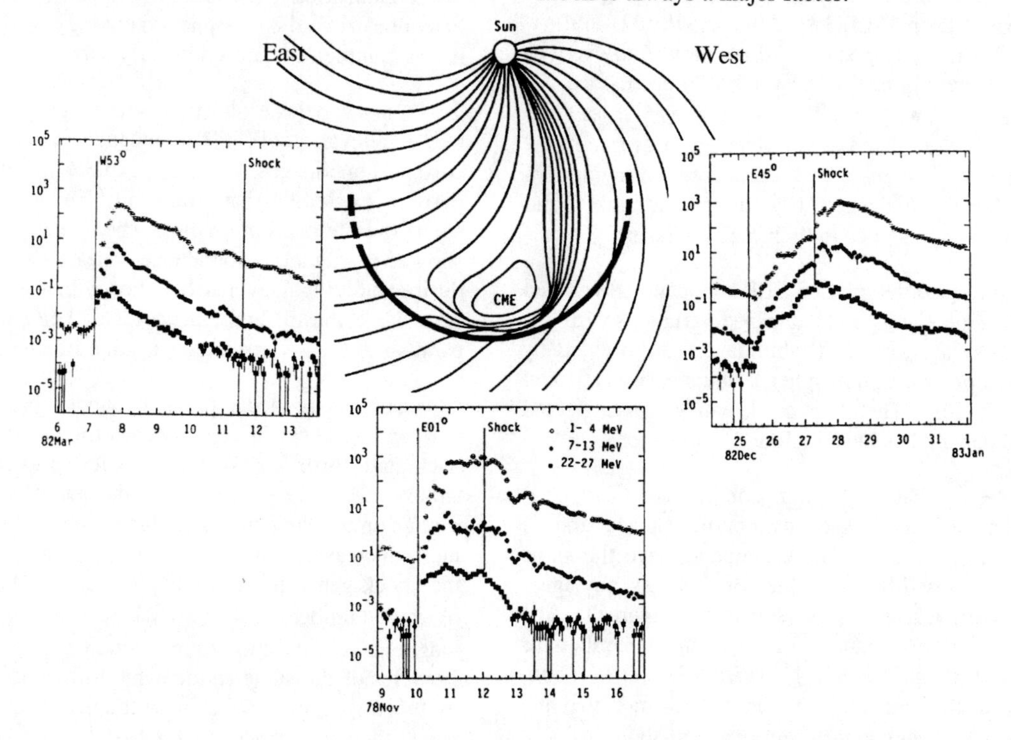

FIGURE 2. Intensity-time profiles for protons are shown for observers viewing a CME from three different longitudes.

A sufficiently fast CME near central meridian will produce an intensity peak at the time of shock passage (not seen in Figure 2), followed by a sharp decrease in intensity when the observer enters the CME or magnetic cloud, as is clearly seen in Figure 2. The reduced intensity inside the CME shows that little or no acceleration occurs at reconnection regions or other shock waves that might be *behind* the CME. Occasionally, however, new events at the Sun do fill this region behind the CME with energetic particles.

Average Abundances in Gradual Events

Historically, gradual SEP events have been used as a proxy for average coronal abundances. In his classic review, Meyer (26) realized that two different processes control SEP abundances when compared with photospheric abundances: 1) systematic event-to-event variations that depended upon Q/A of the ion (his 'mass bias') and 2) an overall dependence of the coronal source abundances on the first ionization potential (FIP) of the ions. Meyer also recognized impulsive ^{3}He-rich events as a separate population that he discussed in an appendix.

Breneman and Stone (2) significantly increased the number of elements measured and studied the Q/A dependence for 10 large SEP events observed on the Voyager spacecraft. They used the average Q/A values measured on ISEE-3 by Luhn *et al.* (22) and assumed that ionization states did not vary from event to event. They found a power-law dependence of enhancement *vs.* Q/A for ions with $Z \geq 6$ in several events, and they listed a complete set of average SEP abundances. If they had compared with modern photospheric abundances (10), they would have found no net Q/A dependence in their SEP averages.

Reames (35) determined SEP abundances averaged over 49 large events and examined variations. In Figure 3 these averaged SEP abundances are divided by the corresponding photospheric abundances (10) and plotted *vs.* FIP. The element H, which was neglected in early papers, has been included in this plot.

Evidence that averaging compensates for Q/A-dependent effects is seen by comparing Mg and Si with Fe in Figure 3. These elements have the same FIP but greatly different values of Q/A, yet they agree within statistical errors as seen in the Figure 3. The averaged SEP abundances for dominant elements have changed little in the last 15 years. However, abundances for the rarer elements have been improved and extended by Cohen *et al.* (5) at this workshop.

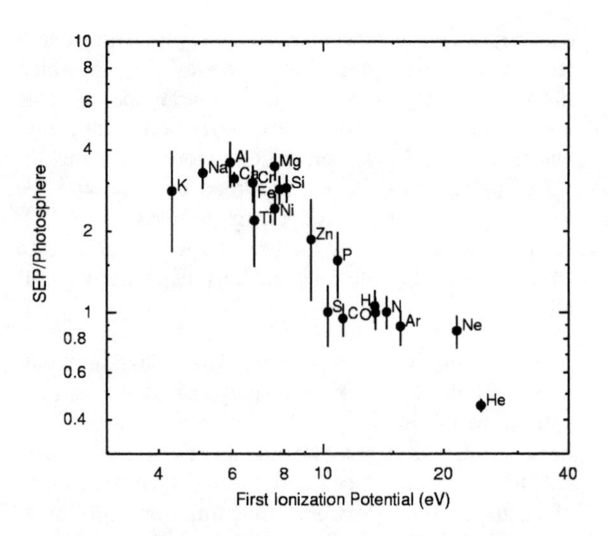

FIGURE 3. Averaged SEP coronal abundances relative to photospheric abundances are shown as a function of FIP.

Understanding SEP Abundance Variations

Unfortunately, the early treatment of SEP abundances and their variation from event to event was highly phenomenological. It left nagging questions. Why does averaging work so well? What actually causes the variations with Q/A. Why should a power-law organization exist and why does it break down? Why does H fit the Q/A phenomenology so poorly that it was completely ignored in early papers?

To answer these questions and to gain confidence in the relevance of SEP abundances as a proxy for coronal abundances, we must explore the physics of particle acceleration and transport. This necessity became even more compelling when new instruments showed systematic abundance variations with time during individual events, as shown in Figure 4 (36, 52). Note the different behavior of Fe/O in the two events and the uncorrelated variation in H/He.

Particles are accelerated at shocks because they gain an increment in velocity each time as they scatter back and forth across the velocity gradient of the shock (12, 17). At injection, the particles begin to scatter on ambient magnetic turbulence. As their velocity increases, those that begin to stream away from the shock generate or amplify resonant Alfvén waves of wave number, $k_{res}=B/\mu P$, where P is the particle's magnetic rigidity and μ the cosine of its pitch angle. Particles of the same rigidity that follow are scattered by the waves and increasingly trapped near the shock where they are further accelerated.

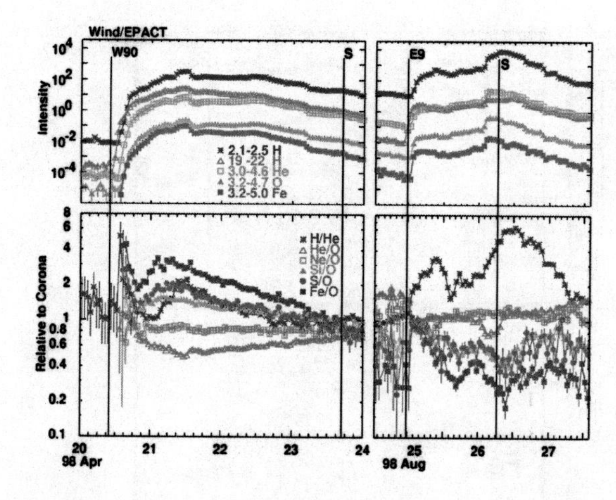

FIGURE 4. Selected intensities and relative abundances are shown *vs*. time for SEP events at two different longitudes.

At higher and higher energy, streaming particles grow new resonant waves. Eventually, at some high energy, the number of surviving particles becomes inadequate for wave growth to produce sufficient scattering, so these particles simply leak away from the shock. This leakage produces an exponential 'knee' in the particle spectrum that is otherwise a power-law in energy (3). We will discuss this knee in the next section.

Protons play a special role in wave generation at the shock. Since they are the most numerous species, they generate most of the waves, while the heavier ions act as test particles that probe the wave spectrum. SEP abundances are accumulated at the same velocity (*i.e.* energy/nucleon). However, ions with the same velocity will resonate with different regions of the wave spectrum because they have different values of Q/A, hence different rigidities. For wave spectra flatter than k^{-2}, for example, O will be scattered and trapped more efficiently than Fe of the same velocity, so that Fe/O is enhanced far away from the shock and suppressed nearer the shock.

Thus, Fe and O are merely redistributed in space along a magnetic flux tube by differential wave scattering. If we could integrate over space at a fixed time, we would obtain the coronal source abundances. However, since this is impractical, we can achieve a similar effect by averaging over SEP events at differing solar longitudes, since, as seen in Figure 2, events with western sources preferentially sample far ahead of the shock, and events with eastern sources preferentially sample near and behind the shock. *Abundance averaging over a large sample of events compensates for the spatial fractionation of elements by proton-generated Alfvén waves.* Of course, the abundances vary strongly with wave intensity even for events at a given longitude.

Detailed numerical calculations of Ng *et al.* (32, 33) follow the complete evolution of both particles and waves in space and time. These calculations can follow much of the complex behavior of abundances, as shown in Figure 5. The detailed time evolution depends upon the rate that the shock weakens; this is assumed to be linear in the simulation.

A critical feature of the wave-particle model has been the understanding it gives of the abundance of H, that was omitted in earlier studies. He/H and Fe/O are both ratios of high- to low-rigidity species at the same velocity. A power-law wave spectrum, like the Kolmogorov $k^{-5/3}$ spectrum, will produce power-law enhancements as a function of Q/A, so that He/H and Fe/O will behave similarly. Since the first particles to arrive propagate through a background Kolmogorov wave spectrum that is largely unmodified by wave growth, one might expect these ratios to begin at high values and decline with time. However, as shown in Figure 6, the ratios behave as expected in the 2000 April 4 event, but He/H behaves anomalously in the 1998 September 30 event (see also 43).

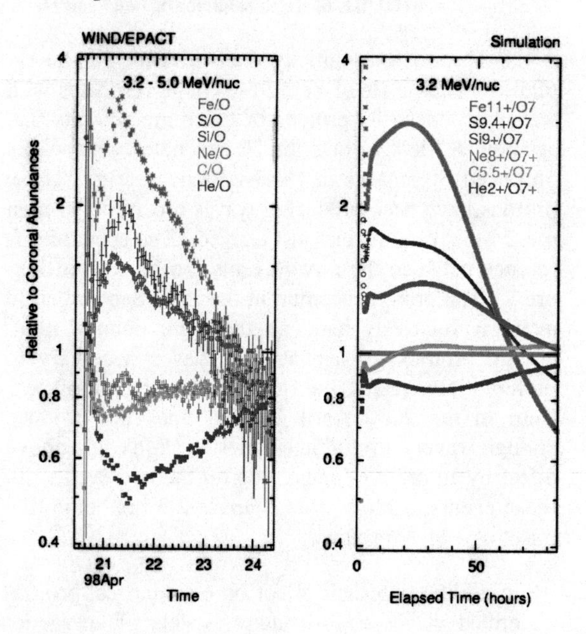

FIGURE 5. Comparison of observed and simulated abundance variations with time in the 1998 April 20 SEP event.

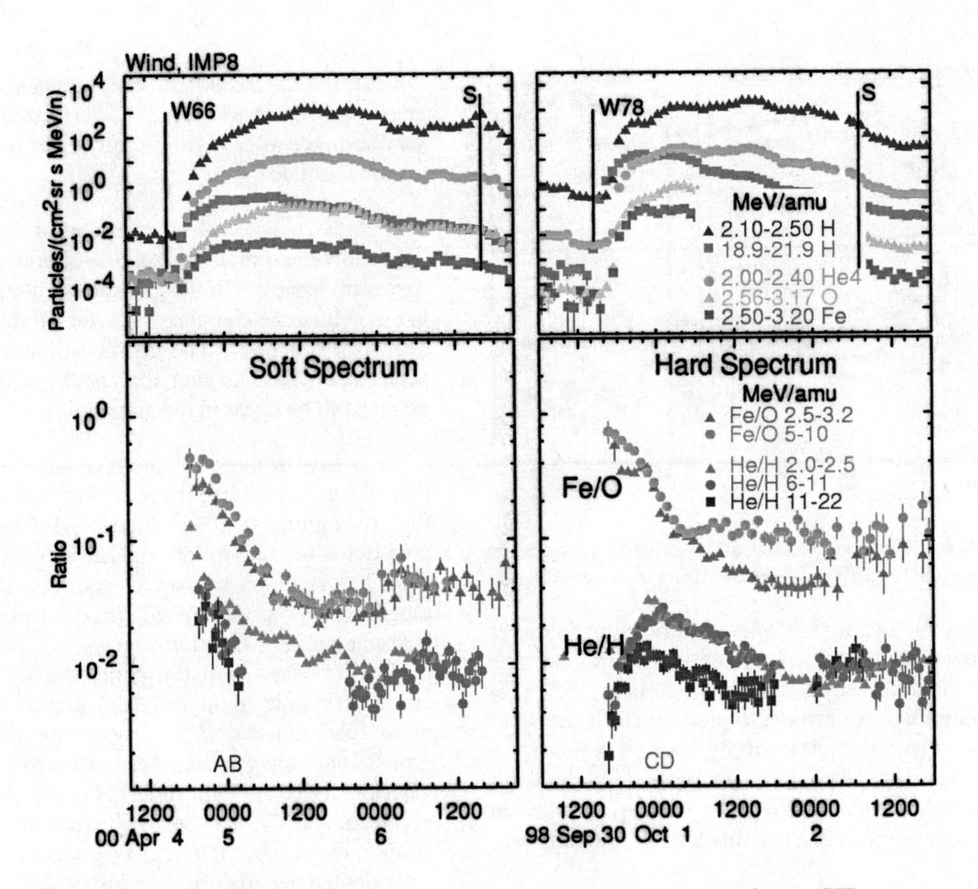

FIGURE 6. Time behavior of Fe/O and He/H are compared in the lower panels for two SEP events.

He/H ratios that initially rise with time can be explained because He of, say, 2 MeV/amu resonates with waves produced by protons of twice the velocity, *i.e.* at about 8 MeV. Thus, the He resonates with waves produced by protons that arrived much earlier. These protons have been producing waves much longer than the 2 MeV protons that just arrived. The spectral difference between these two events can be seen in Figure 7. The proton spectrum in the 1998 September 30 event is relatively hard, so there are enough high-energy protons to generate the waves necessary to preferentially scatter the He. The softer proton spectrum in the 2000 April 4 event does not produce enough waves. Proton intensities at 8 MeV and above differ by an order of magnitude for the two events. In these events, 8 MeV protons arrive 2-3 hrs before the onset of 2 MeV protons.

In addition to their effect on abundances, proton-generated waves control many aspects of energetic-particle behavior (37). They limit intensities early in events, flatten low-energy spectra (as seen in Figure 7), and rapidly reduce the streaming anisotropies in large SEP events. Even though the energy in proton-generated waves is limited to only a few percent of the energy in the protons themselves, the scattering that the waves produce greatly increases the acceleration efficiency of the shock, increasing attainable energies by factors of ~100 or more.

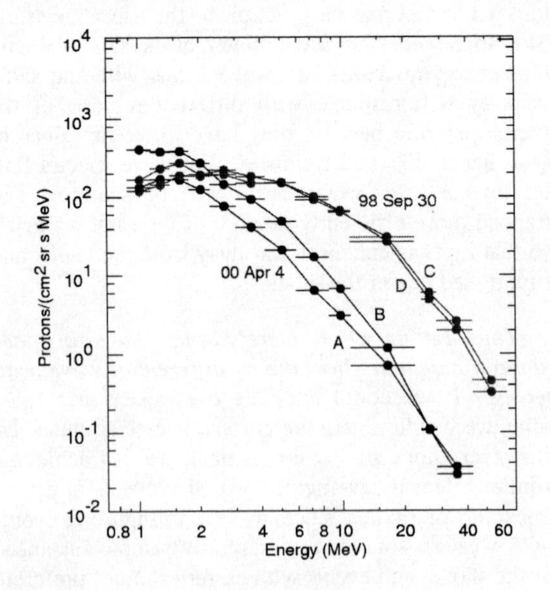

FIGURE 7. Proton energy spectra at times labeled A-D early in the two events shown in Figure 6.

Finally, it is important to realize that the transport of particles from SEP events may be complicated by the presence of CMEs and shocks that exist in interplanetary space prior to the onset of a new event. The recent event on 2000 July 14 is shown in Figure 8 (44). Intensities of protons below 100 MeV suddenly increase at an intervening shock that arrives at Earth about 5 hours after the SEP event onset. Particle intensities remain relatively flat between this early shock and the source shock that arrives on July 15, suggesting that particles are partially trapped between the two shocks. This trapping most likely affects abundance ratios like Fe/O, which rises and remains elevated until the shock passage on July 15. This behavior contrasts with that of the 1998 August 24 event (Figure 4), which also comes from a source near central meridian, but has no intervening shock.

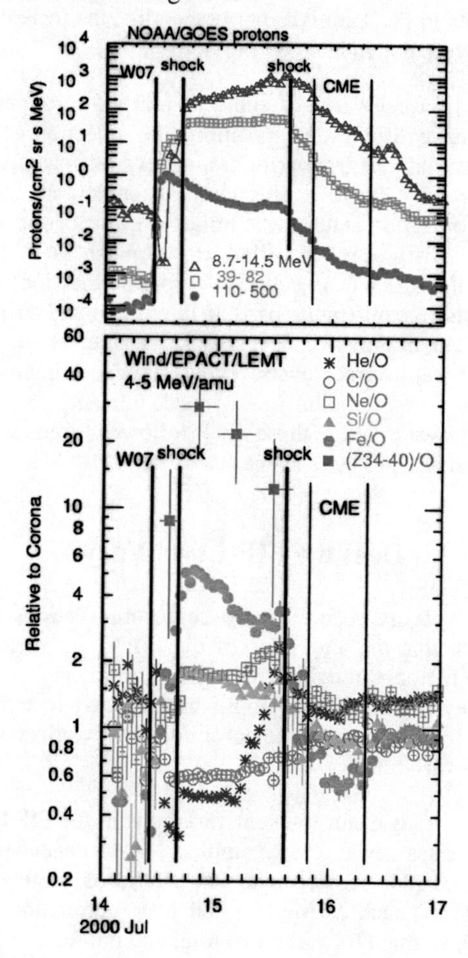

FIGURE 8. Proton intensities and relative abundances are shown *vs.* time for the 2000 July 14 event (44).

Spectral Knees

At sufficiently high energy, intensities of particles and resonant waves decrease, acceleration times increase, and particles leak away from the shock. The power-law spectrum of equilibrium shock acceleration is modified by the leakage (3) to a form such as $E^{-\gamma}\exp(-E/E_o)$ where we define the e-folding energy as the spectral 'knee' energy.

Tylka *et al.* (49) found that spectra in the 1998 April 20 event fit this form with $E_o = (Q/A)\,E_{oH}$, where E_{oH} is the knee energy for protons (see Figure 9). E_{oH} decreased slowly with time during the event from ~15 MeV to 10 MeV. Other events have a stronger or weaker dependence on Q/A and have larger variations of E_{oH} with time. Lovell *et al.* (21) derived the energy spectrum of the 1989 September 30 event using data from the ground-level neutron monitor network. For that event they found $E_{oH} \approx 1$ GeV. There are no instruments available to measure knee energies for ions above ~200 MeV/amu in SEP events.

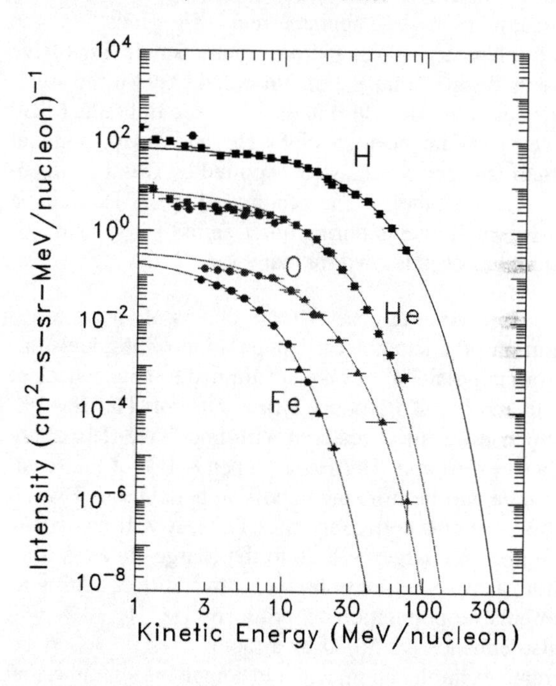

FIGURE 9. Ion spectra early in the 1998 April 20 event are fit to the form $E^{-\gamma}\exp(-E/E_o)$ by Tylka *et al.*(49) using data from IMP8, Wind, and ACE. E_o scales as Q/A in this event.

Spectral knees are a property of the acceleration, not a redistribution of particles in space. Therefore, averaging over events at energies above the knee will *not* recover coronal abundances.

The Seed Population

Shock waves accelerate ions from the high-energy tail of the thermal distribution. In the case of CME-driven shocks, ions from the corona and solar wind are sampled as the 'seed population' for acceleration. Ionization states for the accelerated ions are typical of the solar wind or the 1-2 MK coronal plasma. Charge states of Fe from 10 to 15 are usually seen, even up to energies of 200-600 MeV/amu (50). In a few events the shock begins sufficiently low in the coronal plasma that energetic ions are further stripped, producing ionization states that increase with particle energy (42). The transport of Fe with charge 20 is different from that for Fe of charge 10. Q/A-dependent acceleration and transport affect the relative abundances of different ionization states of a single element just as they affect the relative abundances of elements.

Fast shocks will accelerate *any* ions that they encounter at suprathermal velocities, such as the pickup ions in the case of the heliospheric termination shock. Mason *et al.* (24) have suggested that an accumulation of suprathermal ^{3}He ions in the interplanetary plasma from many small impulsive flares during solar maximum could explain the small increases in ^{3}He/^{4}He~1% that they see in gradual SEP events. Enhancements of ^{3}He and heavy ions at interplanetary shocks were reported by Desai *et al.* (6) at this workshop. The accumulation of ^{3}He and Fe from small events during quiet periods at solar maximum has been known for many years (46).

However, the observational effects of shock acceleration of a suprathermal population of residual ions from impulsive flares are not limited to abundances of ^{3}He and Fe. To produce a final ratio of ^{3}He/^{4}He~1%, suppose we inject material with impulsive-flare abundances (36) and ^{3}He/^{4}He~1. Then ~10% of the resultant Fe will be from the impulsive population. Even if this does not noticeably alter Fe/O, it will contribute Fe ions of charges ~18-20 to the charge-state distribution, as is sometimes observed. In addition, adding an impulsive population to produce ^{3}He/^{4}He ~1% will also enhance (Z>50)/O by a factor of ~10. As an extreme example, an injection of impulsive suprathermal ions to contribute 10% of ^{4}He will contribute half of the final Fe, with Q_{Fe}~20, add 25% of the final Ne, and enhance (Z>50)/O by a factor of ~100. Enhancements in ($34 \leq Z \leq 40$)/O by a factor of ~30 are actually seen in the 2000 July 14 event, and are shown in Figure 8.

Abundances in impulsive SEP events are not well correlated among themselves, so it is difficult to establish correlated enhancements in ^{3}He/^{4}He and Fe/O. However, correlations between Fe/O, Ne/O, and Q_{Fe} have already been reported (29). Re-acceleration of suprathermal ions from prior impulsive flares can explain most of the events that are not 'pure' gradual events as judged by abundances or ionization states. However, injection of this flare population might alter the SEP average abundances somewhat.

A new chapter in the rapidly evolving story of 'remnant impulsive suprathermals' has just been written by Tylka *et al.* (51). Those authors assume a small 5% injection of impulsive suprathermals into the CME-driven shock, and they use observed charge distributions for the injected ions. Since the spectral knee energy varies as Q/A in many events, the high-Q suprathermals persist to higher energies than the accelerated lower-Q solar-wind ions. This simulation quantitatively fits the observed increase in Q_{Fe} at high energies in the 2000 July 14 and the well-measured 1992 November 1 events. This model also explains increases in Fe/O at high energy that, like the increase in Q_{Fe}, were not understood previously.

If re-acceleration of suprathermal ions from small impulsive SEP events is important, injection of suprathermal ions from prior gradual events must also be important. However, because the latter abundances and ionization states are similar to those of the solar wind, this process is difficult to establish. Nevertheless, the efficient injection of suprathermal ions may increase the maximum particle intensities and energies that can be attained at a shock. Kahler *et al.* (16) found that 'overachievers,' events with peak intensities above the correlation line of peak intensity *vs.* CME speed, were often those that followed immediately behind another large event.

Does the FIP-Level Vary?

For many years, spectroscopic observations have shown that the amplitude of the FIP effect varies by large factors throughout the solar atmosphere (*e.g.* 53). However, SEP events might be expected to average over large regions of the corona so they would smooth these variations.

To study event-to-event variations in the FIP level, we must overcome complex Q/A dependences. Reames (34) showed that that a ratio of neighboring elements, such as Mg/Ne, had little correlation with Fe/O, so that Q/A variations might be minimal. Figure 10 shows Mg/Ne for 43 events as a function of time over a solar cycle. For this sample, the weighted mean separation of the FIP levels is 4.06±0.03. The variance of a single event from this mean is 18%. Presumably these variations come from uncorrectable (nonlinear) dependence upon Q/A, complicated by time dependence in the abundances like those shown in Figures 4, 5, 6, and 8.

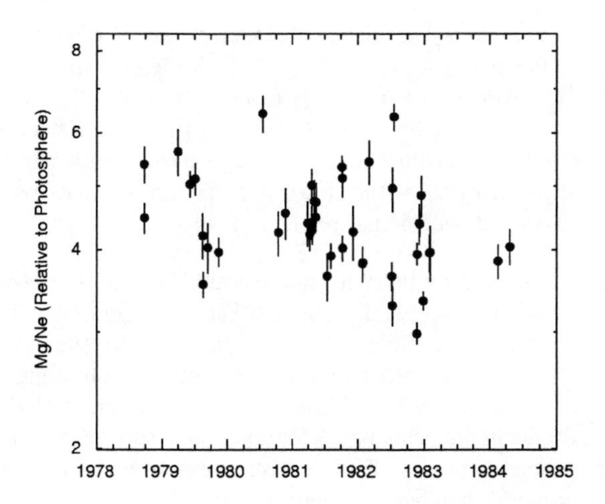

FIGURE 10. Mg/Ne is shown *vs.* time for 43 large SEP events (34).

More recently, Mewaldt *et al.* (27) fit the abundance enhancements to a power law in *Q/A*, treating the FIP level as an adjustable constant. They found somewhat larger variations. One wonders if part of the variation in the adjustable FIP level merely provided a partial compensation for the nonlinear departures from a power law in *Q/A* that are known to exist (see Figure 3.8 in reference 36). These authors also found a mean FIP amplitude of 4.0.

COROTATING INTERACTION REGIONS

As mentioned in the Introduction, ions accelerated from the high-speed solar wind at the reverse shock of CIRs were once believed to represent the abundances of the high-speed wind and the region above solar coronal holes (45). The abundances, averaged over 25 CIR events are shown, relative to photospheric abundances (10), as a function of FIP in Figure 11. While the statistics are somewhat poorer here than in SEP events, a smaller enhancement of low-FIP ions is seen in Figure 11 in comparison with Figure 3. It was known that interstellar He pickup ions could contribute to the energetic He from CIRs, but other pickup ions, such as O are rare in the inner heliosphere.

One of the historic problems with the abundances of energetic ions from CIRs is that the observed ratio of C/O= 0.89±0.05 is substantially larger than that of the SEP corona (0.465±0.013), photosphere (0.49± 0.10), or solar wind (0.71±0.07). Worse, the ratio seems to increase with the solar wind speed (36). No comparable variations are seen for C/O in SEP events although the range of corresponding shock speeds in SEP events is much greater than in reverse CIR shocks. However, differences between the solar wind and the photosphere and corona are also not understood.

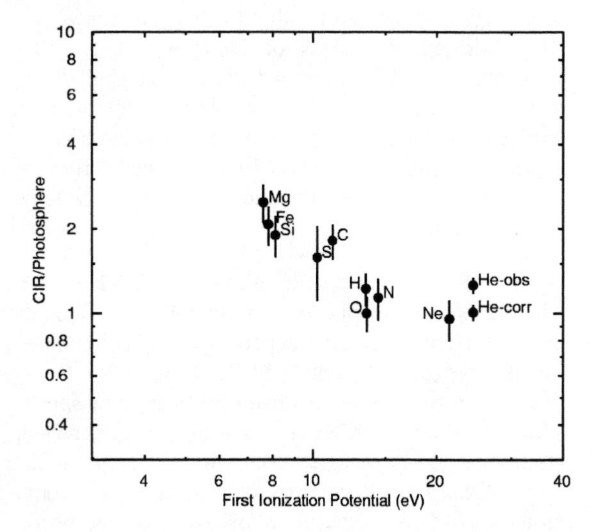

FIGURE 11. Average abundances of energetic ions from the reverse shock at CIRs, relative to the photosphere (10), are shown as a function of FIP. The element He is shown as observed and as corrected to remove interstellar pickup ions (see text).

The high value of C/O could not be explained by the presence of interstellar pickup ions, since C is suppressed in this population. A possible explanation advanced for the excess C was the 'inner source' of interstellar grains (8). Solar wind that is stopped by the grains and neutralized is subsequently 'recycled' and evaporated as neutrals that are photoionized and picked up by the solar wind. These singly charged ions C^+, O^+, and Ne^+ observed in the solar wind are attributed to the inner source (8). However, the distribution functions for these ions are well below those of the solar wind at all speeds. While the inner-source ions do have $C^+/O^+ \geq 1$, it is not clear why they would be preferentially accelerated.

Recent measurement of the ionization states of the energetic ions at 1 AU (30) show that most of the ions have charge states like those of the solar wind. The exceptions are that ~8% of Ne and ~25% of He are singly ionized, probably coming from the *interstellar* pickup-ion source. Thus, with these corrections, the abundances measured at 1 AU and shown in Figure 11 do indeed correspond to abundances of the fast solar wind. The correction for Ne is within errors, but the observed and corrected abundances for He are both shown in the figure. We have come full circle. However, the problem of explaining the high C/O has also returned.

SUMMARY

Energetic particle populations in the heliosphere come with a rich variety of abundances. They range from nearly pure H in radiation belts to flares with 1000-fold enhancements of heavy elements. The abundances reflect those of the underlying source plasma, as modified, in many cases, by fractionation processes that occur during injection, acceleration, and transport. Our challenge is to unravel these processes by distinguishing their dependence upon species, energy, and time.

Outside of regions of high magnetic fields such as planetary magnetospheres and solar flares, *all* of the sources seem to involve acceleration by collisionless shock waves. Gradual SEPs from CME-driven shocks, upstream particles from planetary bow shocks, CIRs, ACRs, and GCRs all allow us to probe the subtleties of shock acceleration with different source populations, shock parameters, and transport conditions. *All* of these sources show some degree of ion-neutral fractionation based upon FIP or a related variable. Gradual SEP events reflect abundances of the average corona and slow solar wind. CIRs and planetary bow shocks primarily reflect the fast solar wind that creates the highest shock speed. ACRs reflect the local interstellar medium as processed through the interstellar pickup ions, and GCRs reflect distant interstellar regions.

The self-consistent treatment of shock acceleration, based upon particle scattering by self-generated waves, was first applied to GCR acceleration (1), but has now been extended to other sources (12, 17). However, the time-equilibrium solutions that work well for slowly evolving shocks are ill suited to the unusually dynamic evolution of 'gradual' SEP events. With the aid of time-dependent models of abundance variations that have been developed recently, however, we are beginning to replace phenomenology with physical understanding.

For 16 years, the abundances averaged over many gradual SEP events at energies of a few MeV/amu, have served as a proxy for the average coronal abundance (26). We now understand this to be a natural consequence of using SEP events at different solar longitudes to sample the spatial redistribution of particles whose overall abundances are conserved to first order. Comparison of event-to-event spreads of Mg/Ne with those of Si/Mg or C/O, show that the FIP level varies less than 5-10% for events over a decade.

Abundance enhancements that exhibit a power-law dependence on Q/A (2) are produced when the spectra of the waves scattering the particles is a power-law flatter than k^{-2}, such as the $k^{-5/3}$ Kolmogorov spectrum. However, this behavior is usually seen only in small events or at extreme longitudes on the weak flanks of the CME-driven shock. In large events with strong wave growth, neither the wave spectra nor the abundance enhancements are power laws.

The erratic behavior in the abundance of H relative to other elements discouraged early workers from including H in SEP abundance tables. As the most abundant species, H dominates the production of particle-generates waves. Much of the behavior of He/H, for example, can be understood in terms of proton-generated waves. The element H can now be included with reasonable confidence.

Energy spectral knees, with their own species dependence, can distort the measure of coronal abundances, especially at high energy. Injection of a seed population of residual suprathermal ions from impulsive SEP events into the CME-driven shock can contribute enhancements in ^{3}He/^{4}He or Fe/O and elevated Q_{Fe} in gradual events. This explains the existence of 'mixed' or 'impure' gradual events, but it also suggests the need for a correction to SEP coronal abundances for some events.

SEP abundances from impulsive flares are acceleration dominated. They tell an interesting and complex story about resonant stochastic acceleration, but provide little information on coronal abundances. However, narrow γ-ray lines from flares do measure coronal abundances, while broad lines suggest the same enhancements seen in impulsive SEP events.

Energetic ions from the reverse shock in CIRs are a measure of abundances in high-speed solar wind streams that emerge from coronal holes. ACRs are an indirect measure of abundances in the local interstellar medium, although the origin of the rare low-FIP ions remains uncertain.

Energetic ions are a rich source of information about a variety of fundamental processes that take place in the heliosphere.

ACKNOWLEDGMENTS

I gratefully acknowledge the contribution made by Chee Ng, to this paper and to my personal education during the last few years, and I thank Allan Tylka for many helpful discussions. I also thank two unnamed referees for helpful comments.

REFERENCES

1. Bell, A. R.: 1978, *Mon. Not. Roy. Astron. Soc.*, **182**, 147.

2. Breneman, H. H., and Stone, E. C., *Astrophys. J. (Letters)* **299**, L57 (1985).

3. Ellison, D., and Ramaty, R., *Astrophys. J.* **298**, 400 (1995).

4. Cane, H. V., Reames, D. V., and von Rosenvinge, T. T., *J. Geophys. Res.* **93**, 9555 (1988).

5. Cohen, C. M. S. *et al.*, this volume (2001).

6. Desai, M. I. *et al.*, this volume (2001)

7. Fisk, L.A., Kozlovsky, B., and Ramaty, R., *Astrophys. J. (Letters)* **190**, L35 (1974).

8. Gloeckler, G., Fisk, L. A., Zurbuchen, T. H., and Schwadron, N. A., in *Acceleration and Transport of Energetic Particles Observed in the Heliosphere*, eds. R. A. Mewaldt, J. R. Jokipii, M. A. Lee, E. Möbius, and T. Zurbuchen, AIP Conf, Proc. **528**, 221 (2000).

9. Gosling, J. T., *J. Geophys. Res.* **98**, 18949 (1993).

10. Grevesse, N., and Sauval, A. J., *Space Science Revs.* **85**, 161 (1998).

11. Holzer, T. E. and Axford, W. I., *J. Geophys. Res.* **76**, 6965 (1971)

12. Jones, F. C., and Ellison, D. E., *Space Sci. Revs.* **58**, 259 (1991).

13. Kahler, S. W., *Ann. Rev. Astron. Astrophys.* **30**, 113 (1992).

14. Kahler, S. W., *Astrophys. J.* **428**, 837 (1994).

15. Kahler, S. W., *et al.*, *J. Geophys. Res.* **89**, 9683 (1984).

16. Kahler, S. W., Burkepile, J. T., and Reames, D. V., *Proc. 26th ICRC* (Salt Lake City) **6**, 248 (1999).

17. Lee, M. A., *J. Geophys. Res.* **88**, 6109 (1983).

18. Lee, M. A., in *Coronal Mass Ejections*, eds. N. Crooker, J. A. Jocelyn, J. Feynman, Geophys. Monograph **99**, (AGU press) p. 227 (1997).

19. Lee, M. A., in *Acceleration and Transport of Energetic Particles Observed in the Heliosphere*, eds. R. A. Mewaldt, J. R. Jokipii, M. A. Lee, E. Möbius, and T. Zurbuchen, AIP Conf. Proc. **528**, 3 (2000).

20. Leske, R. A. in *26th Int. Cosmic Ray Conf.* (Salt Lake City), eds. B. L. Dingus, D. B. Kieda, and M. H. Salamon AIP Conf. Proc. **516**, 274 (1999).

21. Lovell, J. L., Duldig, M. L., Humble, J. E., *J. Geophys. Res.* **103**, 23,733 (1998).

22. Luhn, A., Klecker, B., Hovestadt, D., and Möbius, E., *Astrophys. J.* **317**, 951 (1987).

23. Mandzhavidze, N., Ramaty, R., and Kozlovsky, B., *Astrophys. J.* **518**, 918 (1999).

24. Mason, G. M., Mazur, J. E., and Dwyer, J. R., *Astrophys. J. (Letters)* **525**, L133 (1999).

25. Mazur, J.E., Mason, G.M., Dwyer, J. R., Gold, R. E and Krimigis, S. M., this volume (2001)

26. Meyer, J. P., *Astrophys. J. Suppl.* **57**, 151 (1985).

27. Mewaldt, R. A., Cohen, C. M. S., Leske, R. A., Christial, E. R., Cummings, A. C., Slocum, P.L., Stone, E. C., von Rosenvinge, T. T., and Wiedenbeck, M. E., in *Acceleration and Transport of Energetic Particles Observed in the Heliosphere*, eds. R. A. Mewaldt, J. R. Jokipii, M. A. Lee, E. Möbius, and T. Zurbuchen, AIP Conf, Proc. **528**, 123 (2000).

28. Miller, J. A., and Reames, D. V., in *High Energy Solar Physics*, eds. R. Ramaty, N. Mandzhavidze, X.-M. Hua, AIP Conf. Proc. **374**, 450 (1996).

29. Möbius, E., *et al.*, in *Acceleration and Transport of Energetic Particles Observed in the Heliosphere*, eds. R. A. Mewaldt, J. R. Jokipii, M. A. Lee, E. Möbius, and T. Zurbuchen, AIP Conf, Proc. **528**, 131 (2000).

30. Möbius, E., *et al.*, this volume (2001).

31. Murphy, R J., Ramaty, R., Kozlovsky, B., and.Reames, D. V., *Astrophys. J.* **371**, 793 (1991).

32. Ng, C. K., Reames, D. V., and Tylka, A. J., *Geophys. Res. Lett.* **26**, 2145 (1999).

33. Ng, C. K., Reames, D. V., and Tylka, A. J., *Proc. 26th ICRC* (Salt Lake City) **6**, 151 (1999).

34. Reames, D. V., *Proc. First SOHO Workshop* (ESA SP-348), 315 (1992).

35. Reames, D. V., *Adv. Space Res.* **15** (7), 41 (1995).

36. Reames, D. V., *Space Science Revs.* **90**, 413 (1999).

37. Reames, D. V. in *26th Int. Cosmic Ray Conf.* (Salt Lake City), eds. B. L. Dingus, D. B. Kieda, and M. H. Salamon AIP Conf. Proc. 516, 289 (1999).

38. Reames, D. V., in *High Energy Solar Physics: Anticipating HESSI*, eds. R. Ramaty and N. Mandzhavidze, ASP Conf. Series 206, 102 (2000).

39. Reames, D. V., *Astrophys. J. (Letters)*, **540**, L111 (2000).

40. Reames, D. V., Kahler, S. W., and Ng, C. K., *Astrophys. J.* **491**, 414 (1997).

41. Reames, D. V., Meyer, J. P., and von Rosenvinge, T. T., *Astrophys. J. Suppl.*, **90**, 649 (1994).

42. Reames, D. V., Ng, C. K., and Tylka, A. J., *Geophys. Res. Lett.*, **26**, 3585 (1999).

43. Reames, D. V., Ng, C. K., and Tylka, A. J., *Astrophys. J. (Letters)* **531**, L83 (2000).

44. Reames, D. V., Ng, C. K., and Tylka, A. J., *Astrophys. J. (Letters)* **548**, L233 (2001).

45. Reames, D. V., Richardson, I. G., and Barbier, L. M., *Astrophys. J. (Letters)*, 382, L43 (1991).

46. Richardson, I. G., Reames, D. V., Wenzel, K. P. and Rodriguez-Pacheco, J., *Astrophys. J. (Letters)*, 363, L9 (1990).

47. Roth, I., and Temerin, M., *Astrophys. J.* **477**, 940 (1997).

48. Temerin, M., and Roth, I., *Astrophys. J. (Letters)* **391**, L105 (1992).

49. Tylka, A. J., Boberg, P. R., McGuire, R. E., Ng, C. K., and Reames, D. V., in *Acceleration and Transport of Energetic Particles Observed in the Heliosphere*, eds. R.A. Mewaldt, J.R. Jokipii, M.A. Lee, E. Moebius, and T.H. Zurbuchen, AIP Conf. Proc. 528, p 147 (2000).

50. Tylka, A. J., Boberg, P. R., Adams, J. H., Jr., Beahm, L. P., Dietrich, W. F., and Kleis, T., *Astrophys. J.* (*Letters*) **444**, L109 (1995).

51. Tylka, A. J., Cohen, C. M. S., Deitrich, W. F., Maclennan, C. G., McGuire, R. E., Ng, C. K., and Reames, D. V., *Astrophys. J.* (*Letters*), in press (2001).

52. Tylka, A. J., Reames, D. V., and Ng, C. K., *Geophys. Res. Lett.* **26**, 145 (1999).

53. Widing, K. G., and Feldman, U., *Astrophys. J.* **344**, 1046 (1989).

Long-Term Fluences of Energetic Particles in the Heliosphere

R. A. Mewaldt[1], G. M. Mason[2], G. Gloeckler[2], E. R. Christian[3], C. M. S. Cohen[1], A. C. Cummings[1], A. J. Davis[1], J. R. Dwyer[4], R. E. Gold[5], S. M. Krimigis[5], R. A. Leske[1], J. E. Mazur[6], E. C. Stone[1], T. T. von Rosenvinge[3], M. E. Wiedenbeck[7], and T. H. Zurbuchen[8]

[1]California Institute of Technology, Pasadena, CA 91125 USA
[2]University of Maryland, College Park, MD 20742 USA
[3]NASA/Goddard Space Flight Center, Greenbelt, MD 20771 USA
[4]Florida Institute of Technology, Melbourne, FL 32901 USA
[5]Johns Hopkins University/Applied Physics Laboratory, Laurel, MD 20723 USA
[6]The Aerospace Corporation, El Segundo, CA, 90009 USA
[7]Jet Propulsion Laboratory, Pasadena, CA 91009 USA
[8]University of Michigan, Ann Arbor, MI 48109 USA

Abstract: We report energy spectra of He, O, and Fe nuclei, extending from ~0.3 keV/nucleon to ~300 MeV/nucleon, integrated over the period from the Fall of 1997 to mid-2000. These fluence measurements were made at 1 AU using data from the SWICS, ULEIS, SIS, and CRIS instruments on ACE, and include contributions from fast and slow solar wind, coronal mass ejections, pickup ions, impulsive and gradual solar particle events, acceleration in corotating interaction regions and other interplanetary shocks, and anomalous and galactic cosmic rays. Fluence measurements of six additional species are presented in the energy region from ~0.04 to ~100 MeV/nucleon. We discuss the relative contributions of the various particle components, and comment on the shape and time dependence of the measured energy spectra.

INTRODUCTION

The various components of energetic particles observed in the heliosphere include rather steady sources such as the solar wind, pickup ions, and anomalous and galactic cosmic rays, as well as transient sources such as solar energetic particle events and interplanetary shocks. At the lowest energies, the typical scale of variations in the velocity, intensity, and composition of the solar wind is a factor of ~2, on time scales that range from hours to days to years. At the highest energies, anomalous and galactic cosmic rays vary in intensity over the solar cycle, but are essentially constant in composition. At intermediate energies ranging from the ~30 keV/nucleon to ~30 MeV/nucleon, the intensity, spectra, and composition of heliospheric particles are all observed to be highly variable (particularly at solar maximum), and apparently originate from a number of separate sources and acceleration processes. It is therefore of interest to integrate these highly variable components over an extended time period to obtain a direct measure of the longer-term fluence of energetic particles that originate on the Sun, in the heliosphere, and in the Galaxy.

The experiments on ACE are ideal for this kind of measurement, because they measure composition and energy spectra continuously from solar wind energies of ~1 keV/nucleon to cosmic ray energies of a few hundred MeV/nucleon [1]. In particular, ACE is able to explore the composition and energy spectra of ions in the relatively unknown suprathermal region from a few keV/nucleon to a few hundred keV/nucleon.

In this paper we report preliminary measurements of the long-term fluence of ions over six decades in energy/nucleon during a 33-month time period that includes both solar-minimum and solar-maximum conditions. The results of these measurements are of interest to understanding processes of particle acceleration and transport on the Sun and in the heliosphere, and to understanding the long-term contributions of energetic-particles to the lunar soil, meteorites, and planetary atmospheres. In a separate paper [2], we compare the observed fluences of solar wind and suprathermal nuclei with observations of energetic ions implanted in the lunar soil.

CP598, *Solar and Galactic Composition,* edited by R. F. Wimmer-Schweingruber

INSTRUMENTATION AND DATA ANALYSIS

The measurements reported here were obtained from the SWICS [3], ULEIS [4], SIS [5], and CRIS [6] instruments on ACE (see Table 1). Energetic particle fluxes from ULEIS, SIS and CRIS were accumulated over a 33-month time period extending from 1997:280 to 2000:184. Fluence data were obtained by summing hourly-average fluxes within a large number of separate energy intervals, taking into account the measured instrument live times and geometry factors. (Such hourly-average fluxes are a routine product of ACE Level-2 data processing, and are available at http://www.srl.caltech.edu/ACE/ASC/). Note that this period starts with solar-minimum conditions in 1997 and ends with solar-maximum conditions in 2000.

Table 1: Sources of Fluence Data

SWICS	0.0005 to 0.030 MeV/nuc
ULEIS	0.040 to ~5 MeV/nuc
SIS	~8 to ~100 MeV/nuc
CRIS	~80 to ~300 MeV/nuc

The SWICS data for this preliminary report were summed over an 11-month period extending from January through November in 1999. These 11-month fluences were multiplied by a factor of 3 to correspond to the 33-month period used by the higher-energy instruments. Because the solar wind is less variable than higher-energy solar and interplanetary components, this 1999 period should provide a reasonable representation of the longer time period.

As expected, this first comparison of absolute fluence measurements from four instruments led to re-normalization of some of the measurements. The fluence spectra from SIS and CRIS were found to be in agreement to within 10% where they overlapped, and were left unchanged. During the largest SEP events the ULEIS fluxes were somewhat low due to reduced detection efficiency in periods with intense fluxes. The correction factors for ULEIS during these intense periods were obtained by comparison with fluxes measured using the STEP sensor on Wind [7] and the ACE/SIS instrument; these latter two sensors are less susceptible to saturation than ULEIS. The correction to ULEIS fluences for He and CNO was a factor of 1.5 below 1.5 MeV/nucleon, and a factor of 2.2 at higher energies, while no corrections were made for species from Ne to Fe.

After applying small energy-dependent efficiency corrections, the ACE/SWICS oxygen abundance (based on previously un-calibrated matrix-rate data [3]) was normalized to the solar-wind composition measured by Ulysses/SWICS [8], which also brings it into reasonable agreement with the composition measured by ULEIS. These issues of absolute normalization will be examined in more detail in the future.

FLUENCE MEASUREMENTS

Measured fluences of He, O, and Fe extending from solar wind to cosmic ray energies are shown in Figure 1. All three species clearly have a common spectral shape. The peak at ~0.8 keV/nucleon in all three spectra corresponds to the slow-speed solar wind with a mean velocity of ~400 km/sec. The contribution of occasional higher-speed streams, sometimes including velocities as high as 1000 km/sec can also be seen. Including the intrinsic thermal speed of the solar wind, it appears from Figure 1 that the solar wind distribution extends to between 5 and 10 keV/nucleon before there is a change in slope. Beyond this is a long, suprathermal tail extending with a power- law slope of –2 to ~10 MeV/nucleon. Near ~10 MeV/nucleon all spectra exhibit a gradual "knee" and briefly steepen. Above ~100 MeV/nucleon, the modulated fluence of galactic cosmic rays (GCRs) begins to dominate, continuing on for many more decades in energy (see, e.g., [9]).

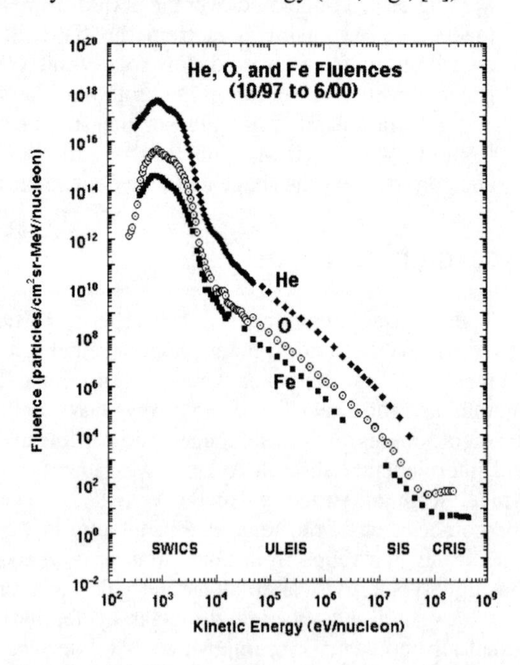

Figure 1: Fluences of He, O, and Fe nuclei measured over the period from 9/1997 to 6/2000 by the SWICS, ULEIS, SIS, and CRIS instruments on ACE. The SWICS fluences were measured over the first eleven months of 1999 and then multiplied by x3. The He fluence includes both doubly and singly-charged He.

In the intermediate region from ~30 keV/nucleon to ~30 MeV/nucleon the fluence spectra are a superposition of many separate "events". Some examples of these events are illustrated for oxygen in Figure 2. At energies from ~3 to 30 MeV/nucleon most of the fluence comes from the largest solar particle events that occur a few times a year during solar maximum (e.g., the 11/97 and 4/98 events). A time-intensity plot of high-energy SEP events observed by ACE appears in Figure 1 of Cohen et al. [10]; Figure 3 illustrates how these events contribute to the 10 MeV/nucleon fluence as a function of time. Note that over this 3-year period anomalous cosmic rays make only a very small contribution to the fluence at 1 AU – the contributions of large SEP events are considerably greater.

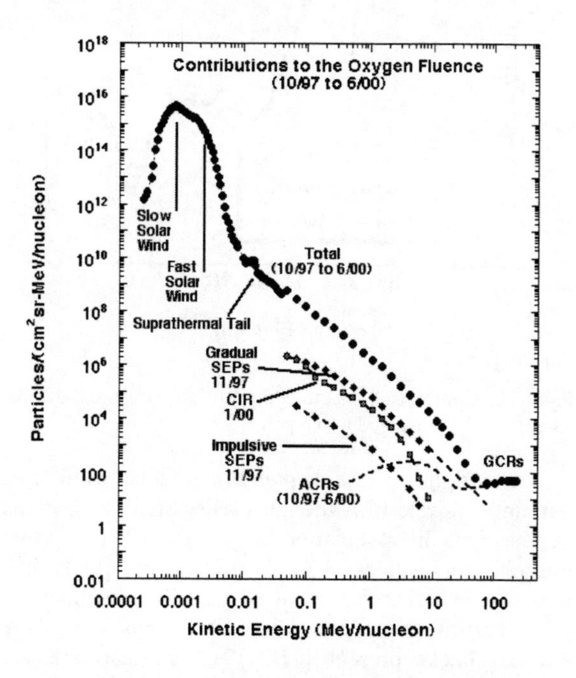

Figure 2: Illustration of some of the contributions to the oxygen fluence, including examples of impulsive and gradual SEP events, CIR-accelerated particles, and anomalous and galactic cosmic rays.

In the energy range from ~0.1 to 1 MeV/nucleon, there are important contributions from impulsive solar flares and from particles accelerated in corotating interaction regions (CIRs). Although these events are generally much smaller in size than gradual SEP events, the impulsive events, in particular, occur much more frequently. At ~0.1 MeV/nucleon there are no large individual events that dominate – rather, there appear to be similar contributions from as many as ~100 separate events of various kinds (see Figure 4).

At even lower energies (~10 keV/nucleon to ~50 keV/nucleon, measurements with SWICS/ACE have shown that suprathermal tails on the solar wind are continuously present [11], but the origin of these tails is presently a subject of investigation (see discussion below). At ~1 keV/nucleon the intensity of singly-charged interstellar and inner-source pickup ions is

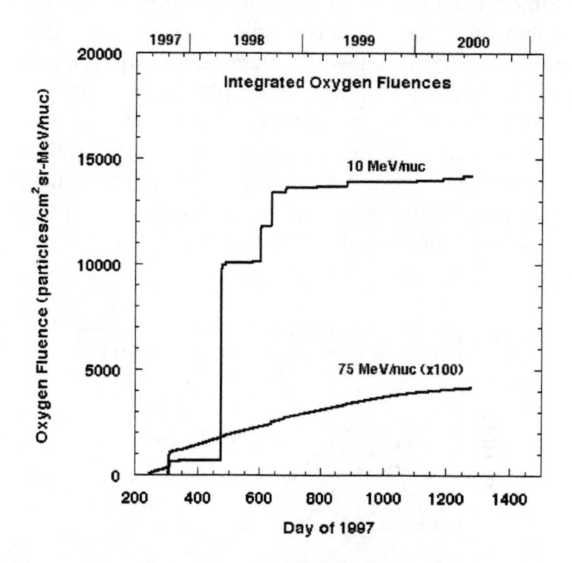

Figure 3: Time history of the integrated oxygen fluences at 10 and at 75 MeV/nucleon measured by SIS. In this case 27-day average fluxes were integrated. At 10 MeV/nucleon a few large SEP events contribute more than the continuous anomalous cosmic ray intensity. At 75 MeV/nucleon the steady, contribution of galactic cosmic rays is apparent.

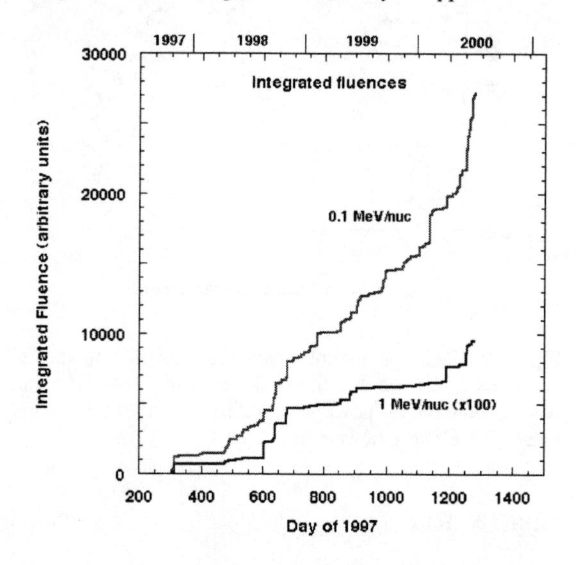

Figure 4: Time history of integrated daily oxygen fluences at 0.1 and 1 MeV/nucleon measured by ULEIS. Note that a large number of individual events contribute.

estimated to be several orders of magnitude lower than the solar-wind intensity (see Figure 9 of [11]). Using ULEIS and SIS data, fluence spectra have been constructed for a total of 9 species, extending from ~0.04 to ~100 MeV/nucleon. It is rather remarkable that all of these species show the same E^{-2} power-law spectral shapes from ~0.04 MeV/nucleon to ~10 MeV/nucleon, as demonstrated in Figure 5. Figure 6 compares the relative abundances at three representative energies. The three compositions appear to be very similar; there is no obvious evidence for contributions from sources with an unusual composition. Statistical uncertainties are negligible in Figure 6, but because data from two of the instruments was re-normalized, it is possible that there could be systematic uncertainties of as much as a factor of ~2.

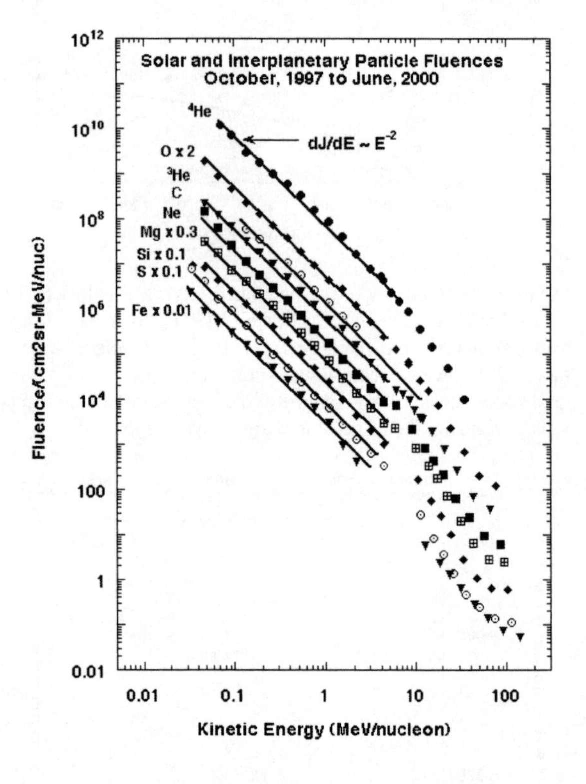

Figure 5: Solar and interplanetary fluences of nine species measured by ULEIS and SIS. It is surprising that all species have a common, E^{-2} power-law spectrum over a large energy range. The E^{-2} spectra were fit by eye.

DISCUSSION

It is somewhat surprising that the integrated fluences of ions from such a variable mix of sources show such a high degree of organization. All species are found to have a common spectral shape. In particular, it is remarkable that a common power-law slope applies to the region extending from ~10 keV/nucleon, where the suprathermal tails on the solar wind dominate, to ~10 MeV/nucleon, where large, gradual SEP events contribute most of the fluence.

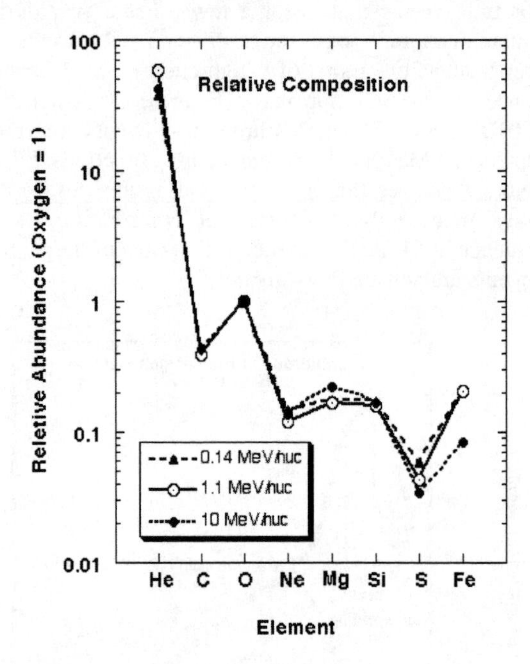

Figure 6: Comparison of the compositions measured at three energies, all normalized to oxygen = 1.

The origin of the suprathermal solar-wind tails remains a mystery. Although earlier studies observed similar tails in association with CIRs [12] or other interplanetary shocks, ACE studies of quiet time periods have shown that these tails are continuously present in the in-ecliptic solar wind, even when there are no shocks present [11]. The fact that tails are observed on the distributions of interstellar pickup ions as well as on solar-wind species argues that they must originate in interplanetary space and not on the Sun [11]. Fisk et al. [13] suggest that they are due to statistical acceleration by transit-time damping, but there is not as yet a consensus on their origin.

There are also questions as to the extent of the tails. In Figure 1 it is seen that the power-law portion of the spectrum extends continuously from ~10 keV/nucleon to ~10 MeV/nucleon, suggesting that all of these ions have a common origin. However, from Figures 3 and 4 it is appears that while a continuous process (or processes) might explain the bulk of the particles observed at ~0.1 MeV/nucleon, at 1 MeV/nucleon and at 10 MeV/nucleon isolated, individual events contribute most of the integrated fluence. It is clear that the origin of these tails will be a source of lively debate over the coming years.

Composition measurements provide some clues to the origin of the particles. The measured ^{3}He/^{4}He ratio above 100 keV/nucleon is ~0.01-0.02, a factor of more than 20 greater than the corresponding ratio in the solar wind. While this suggests that impulsive ^{3}He-rich solar flare events make an appreciable contribution to this energy region, ^{3}He is also overabundant in many gradual SEP events [14, 15], and it is likely that the ^{3}He fluence is due in part to the re-acceleration of remnant impulsive-flare material by interplanetary shocks [15, 16].

The time period covered by this study ended just before the "Bastille Day" event on July 14, 2000, the largest SEP event during this solar cycle. It is reasonable to ask to what extent inclusion of this event would modify the fluences in Figure 1. Using spectra compiled by Tylka et al. (personal communication, 2001) from a number of spacecraft including ACE and Wind, we find that the Bastille Day event contributes only a small fraction (<10%) of the 33-month fluences below 0.1 MeV/nucleon. However, over the energy range from ~2-3 to ~20-30 MeV/nucleon (depending somewhat on species) this single event contributes more than the 33 months combined. The greatest relative contribution is to He, since this event was enriched in He relative to heavier ions. The addition of this contribution does not significantly alter the E^{-2} character of the O and Fe fluences in Figures 1 and 5, but it does flatten the He spectrum somewhat. There are also other large events after July 2000 that would make significant contributions.

There are several similarities between the heliospheric spectra and the spectra of higher-energy galactic cosmic rays. Common spectral shapes are observed from ~10 keV/nucleon to ~10 MeV/nucleon, even though the individual events contributing to these spectra often have non-power-law spectra, with abundance ratios such as Fe/O that vary by a factor ranging from 0.1 to 10.

Similarly, it is possible that galactic cosmic rays also include contributions from several different sources that may have differing spectra and composition, yet the abundant GCR species from ~10^9 to ~10^{13} eV/nucleon appear to be consistent with a single common source spectrum, in this case somewhat steeper than –2. Note that in both of these cases the power-law regions end with a "knee", followed at higher energy by an "ankle". Although it is believed that shock-acceleration processes make the dominant contributions to the GCR spectra from 10^9 to 10^{14} eV/nucleon, the situation is presently less clear for the heliospheric spectra from 10^4 to 10^7 eV/nucleon.

SUMMARY

Continuous solar-wind, solar-particle, heliospheric, and cosmic-ray fluences have been measured from ~0.3 keV/nucleon to ~300 MeV/nucleon. Both solar minimum and solar maximum conditions were included, and we suggest that these spectra provide a good first approximation to the spectral shapes that would be obtained if such measurements were extended over an entire solar cycle. All species are found to have a common spectral shape. The solar wind (including both slow and high-speed wind) dominates the energy range up to ~8 keV/nucleon. In the energy range from ~10 keV/nucleon to ~10 MeV/nucleon as many as 100 or more separate events somehow combine to produce E^{-2} power-law spectra that are common to all of the nine species measured, including ^{3}He.

In the decade from ~5 to ~50 MeV/nucleon intense, gradual SEP events make the largest contribution, while at higher energies galactic cosmic rays take over. It is not yet known what processes make the largest contributions to the intermediate region from ~10 keV/nucleon to ~5 MeV/nucleon. The possibilities include impulsive SEP events, CIR events, or other, as yet unidentified processes that occur continuously on the Sun or in the inner heliosphere. We intend to address this question in the near future.

These are the first spectral measurements to extend continuously from solar-wind to cosmic-ray energies. Given the highly variable composition and intensity of the contributing events, the overall similarity of these fluence spectra is surprising.

References

1. Stone, E. C., et al., *Space Science Reviews*, 86, 1, 1998a.
2. Mewaldt, R. A., R. J Ogliore, G. M. Mason, and G. Gloeckler, "A New Look at Neon-C and SEP-Neon", this volume.
3. Gloeckler, G., et al., *Space Science Reviews*, 86, 449, 1998.
4. Mason, G. M., et al., *Space Science Reviews*, 86, 409, 1998.
5. Stone, E. C., et al., *Space Science Reviews*, 86, 357, 1998b.
6. Stone, E. C., et al., *Space Science Reviews*, 86, 285, 1998c.
7. von Rosenvinge, T. T., et al., *Space Science Reviews*, 71, 155, 1995.
8. von Steiger, R., et al. *J. Geophys. Research*, 105, No. A12, 27217, 2000.
9. Simpson, J. A., *Annual. Reviews of Nuclear and Particle Science* 33, 706, 1983.
10. Cohen, C. M. S., et al., "Solar Coronal Abundances of Rare Elements Based on Solar Energetic Particles", this volume, 2001.

11. Gloeckler, G., Fisk, L. A., Zurbuchen, T. H., and Schwadron, N. A., in AIP Conf. Proc. 528, *Acceleration and Transport of Energetic Particles Observed in the Heliosphere*, ed. R. A. Mewaldt et al., (New York: AIP), 221, 2000.
12. Chotoo, K., et al., *J. Geophys. Res.*, 105, 23107, 2000.
13. Fisk, L. A., Gloeckler, G., Zurbuchen, T. H., and Schwadron, N. A., in AIP Conf. Proc. 528, *Acceleration and Transport of Energetic Particles Observed in the Heliosphere*, ed. R. A. Mewaldt et al., (New York: AIP), 229, 2000.
14. Cohen, C. M. S. et al., *Geophys. Res. Letters*, 26, 2697, 1999.
15. Mason, G. M., et al., *Ap. J.* 425, L843, 1999.
16. Desai, M. I., et al. "Acceleration of ^{3}He Nuclei at Interplanetary Shocks, *Ap.J.*, 553, L89, 2001.

The Ionic Charge State Composition at High Energies in Large Solar Energetic Particle Events in Solar Cycle 23

R. A. Leske*, R. A. Mewaldt*, A. C. Cummings*, E. C. Stone* and T. T. von Rosenvinge†

*California Institute of Technology, Pasadena, CA 91125 USA
†NASA/Goddard Space Flight Center, Greenbelt, MD 20771 USA

Abstract. The ionic charge states of solar energetic particles (SEPs) depend upon the temperature of the source material and on the environment encountered during acceleration and transport during which electron stripping may occur. Measurements of SEP charge states at relatively high energies ($\gtrsim$ 15 MeV/nucleon) are possible with the Mass Spectrometer Telescope (MAST) on the Solar, Anomalous, and Magnetospheric Particle Explorer satellite by using the Earth's magnetic field as a particle rigidity filter. Using MAST data, we have determined ionic charge states of Fe and other elements in several of the largest SEP events of solar cycle 23. The charge states appear to be correlated with elemental abundances, with high charge states (~20 for Fe) for all elements in large Fe-rich events. We review the geomagnetic filter technique and summarize the results from MAST to date, with particular emphasis on new measurements in the very large 14 July 2000 SEP event. We compare the charge states determined by MAST with other measurements and with those expected from equilibrium calculations.

INTRODUCTION

Solar energetic particle (SEP) events are usually classified into one of two distinct types, gradual and impulsive [1]. Earlier studies found ionic charge states, Q, in gradual events to be similar to those expected for a plasma of T_e ~2 MK, with Q(Fe) ~15 at ~1 MeV/nucleon and with little event-to-event variability [2]. Such temperatures are typical of the corona, lending support to the idea that particles in gradual SEP events originate as solar wind or coronal material accelerated by large shocks driven by coronal mass ejections. In impulsive events, however, significantly higher charge states of ~20 were found for Fe [3], suggesting an origin in hotter, ~10^7 K flare plasma or indicating that considerable stripping took place. Recent measurements from the Solar Energetic Particle Ionic Charge Analyzer (SEPICA) on the Advanced Composition Explorer (ACE) have revealed a broad continuum of charge states, varying from event to event but correlated with elemental abundances [4], suggesting a greater complexity than earlier measurements indicated. Also, acceleration or transport produces large elemental and isotopic fractionation in SEP events [5, 6] and Q may also be affected.

Previous studies of ionic charge states using instruments employing electrostatic deflection [e.g. 2, 7] were limited to ~1 MeV/nucleon. In recent years, these studies have been extended to higher energies by using other techniques. One approach makes use of the Earth's magnetic field as a particle rigidity filter and has been successfully demonstrated at energies as high as ~70 MeV/nucleon [8] using the polar-orbiting Solar, Anomalous, and Magnetospheric Particle Explorer (SAMPEX) satellite. This same approach can be used at lower energies [9, 10, 11], which allows for cross-calibration with direct measurements [7]. In addition to event-to-event variability, these measurements have shown that Q may depend on energy [10, 11], which was confirmed by direct measurements at lower energies [7].

In this report, we illustrate the geomagnetic filter technique using data collected during the 14 July 2000 ("Bastille Day") SEP event by the Mass Spectrometer Telescope (MAST) on SAMPEX. We compare the new Bastille Day measurements with other SEP charge state measurements found by MAST to date and with charge states obtained by other methods. Further details of the analysis approach may be found in [8, 9, 10, 11, 12].

OBSERVATIONS

SAMPEX was launched in July 1992 into a 520×670 km 82° inclination Earth orbit [13]. MAST employs a silicon solid-state detector telescope with a collecting power of

CP598, *Solar and Galactic Composition*, edited by R. F. Wimmer-Schweingruber

~11 cm^{2}sr and uses the dE/dx versus residual energy technique to measure the nuclear charge, Z, mass, M, and total kinetic energy, E, for particles with E of ~10 to ~100 MeV/nucleon [14]. The energy interval depends on Z and covers higher E at higher Z, as illustrated in Figure 1 using data from the Bastille Day SEP event. In spite of the very high particle fluxes during this large event, all elements up through Ni ($Z = 28$) are clearly resolved. MAST is not directly sensitive to the ionic charge of the particles, as incident ions at these energies are quickly stripped in traversing a few microns of material.

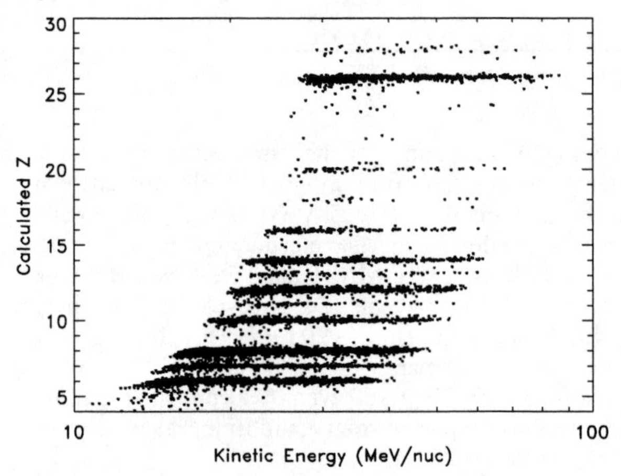

FIGURE 1. The calculated nuclear charge, Z, plotted versus kinetic energy for each heavy ion detected by SAMPEX/MAST from 14–21 July 2000 (DOY 196.0–203.0).

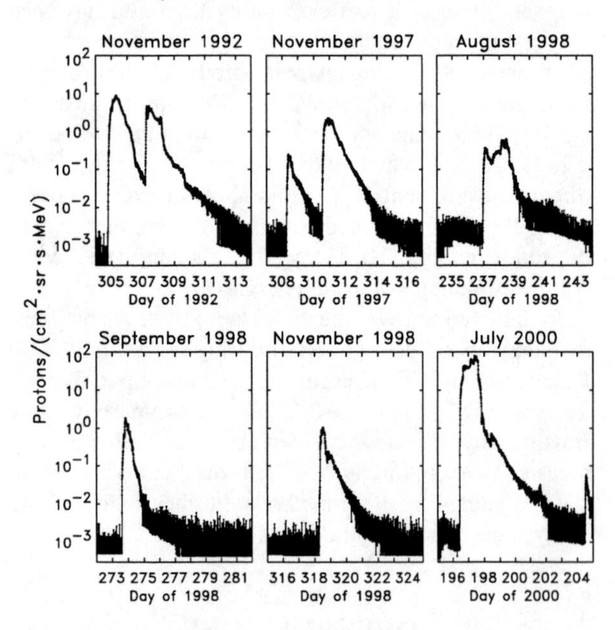

FIGURE 2. Time profiles of the SEP event periods examined here, using 5-min averaged intensities of 30–100 MeV protons from GOES 7 and 8.

Time profiles of the events examined in this study are shown in Figure 2, using 30–100 MeV protons from the GOES 7 and 8 spacecraft. The 1992 events occurred during the declining phase of solar cycle 22, while the remainder of the events shown were in solar cycle 23. The Bastille Day event was the largest SEP event so far in solar cycle 23, with particle intensities up to two orders of magnitude greater than in some of the other events in this study. Of the events shown in Figure 2, the 4 November 1997 event is excluded from this study because of unfavorable spacecraft pointing conditions.

Heavy ions at high energies often arrive earlier and disappear sooner than protons in SEP events. In the Bastille Day event the shortest time profiles were found for the heaviest ions, as shown in Figure 3. The heavy ions peaked near the very beginning of the event and had practically vanished at these energies by late on 15 July. For the charge state analysis we integrated over days 196.5 to 198.0 of 2000 and have not subdivided this interval to search for time variations in Q.

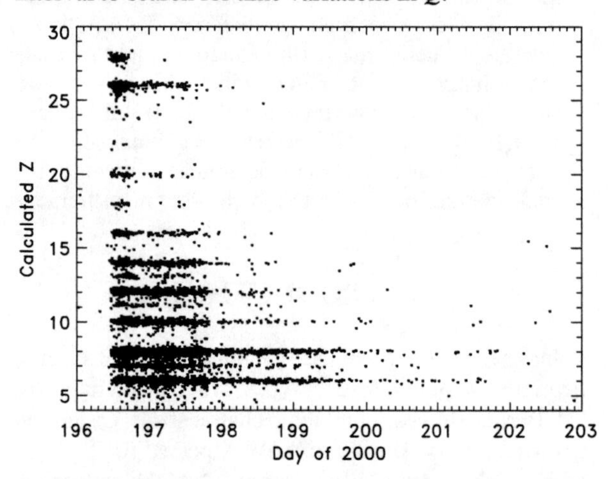

FIGURE 3. The calculated nuclear charge, Z, plotted versus arrival time for each heavy ion detected by SAMPEX/MAST during the Bastille Day event, illustrating the shorter time profiles for the heavier species.

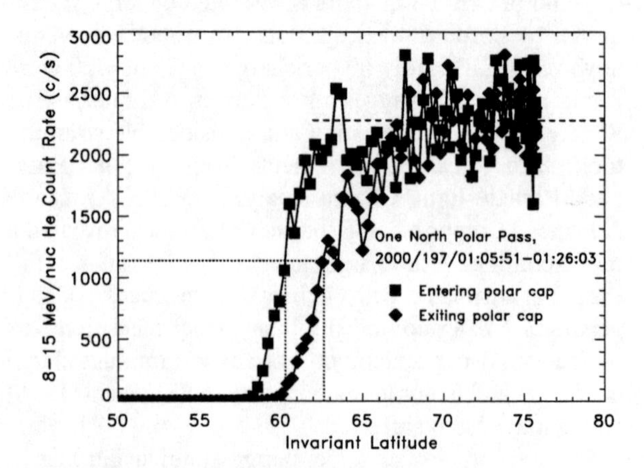

FIGURE 4. The count rate of 8–15 MeV/nucleon He measured every 6 s by MAST plotted versus invariant latitude for one crossing of the north polar cap on 15 July 2000.

ANALYSIS

From the near-polar Earth orbit of SAMPEX, MAST samples energetic particles at all geomagnetic latitudes, including both the polar latitudes where SEPs can reach low altitudes and the equatorial regions where they are excluded by the geomagnetic field. The transition between these two regimes is evident in Figure 4, which shows the count rate for 8–15 MeV/nucleon He plotted versus invariant latitude, Λ, for a single passage over the north polar cap. (For any point along a magnetic field line, Λ is the magnetic latitude at which the field line intersects the Earth's surface and is related to the magnetic L shell by $\cos^2\Lambda = 1/L$; see, e.g., [15]). We define the cutoff invariant latitude, Λ_C, to be that Λ at which the count rate drops to half its average value above 70°, as indicated by dotted lines in Figure 4. Since Λ_C depends on the cutoff rigidity (momentum per unit charge), R_C, of the type of particle being measured, it is sensitive to the particle's charge state. By measuring Λ_C for a given species and using the measurements of E from MAST, one can solve for Q if the relation between Λ_C and R_C is known. Calculations of Λ_C differ from measured values by up to a couple of degrees even in geomagnetically quiet periods [16], and the dynamic nature of the cutoff during the geomagnetic disturbances which often accompany SEP events is even more difficult to model accurately [17]. However, we have found it possible to empirically determine both the time variations in the cutoff and the relation between Λ_C and R_C.

When count rates are high, as in Figure 4, it is straightforward to measure Λ_C up to 4 times every orbit, at each crossing into and out of both the north and south polar caps. This allows us to directly determine how the cutoff varies on relatively fine timescales even during severe geomagnetic storms. The orbit-averaged Λ_C using the MAST 8–15 MeV/nucleon He rate during the Bastille Day event is shown in Figure 5. Cutoff variations in the earlier events and their relation to geomagnetic activity and local time (which accounts for the $\sim$2° difference in Λ_C from one side of the polar cap to the other in Figure 4) are discussed by Leske et al. [18]. The large cutoff suppression of $\gtrsim 10°$ during the Bastille Day event is greater than any of the others studied with MAST [18] and corresponds to a severe geomagnetic storm when the geomagnetic activity index, Dst, reached -300 nT.

The dependence of Λ_C on species and energy is needed to obtain the ionic charge states. To determine the cutoffs for various Z and E with sufficient statistical accuracy, we must sum over the duration of the entire SEP event. Since the time profiles differ for each Z and E (Figure 3), we must first correct for the time variability of the cutoff, otherwise the cutoffs for different species would be weighted to different times, and temporal variations in Λ_C would mask the dependence on Z and E we seek to measure. In the Bastille Day event, the particle intensity is low during the extreme cutoff suppression, so small errors in the time variability correction are less important. In other cases, however, especially those where two SEP events occur back-to-back as in November 1992 and November 1997 (Figure 2), the shock associated with the first event arrives near the peak of the second [18], and correcting for time variability is critical.

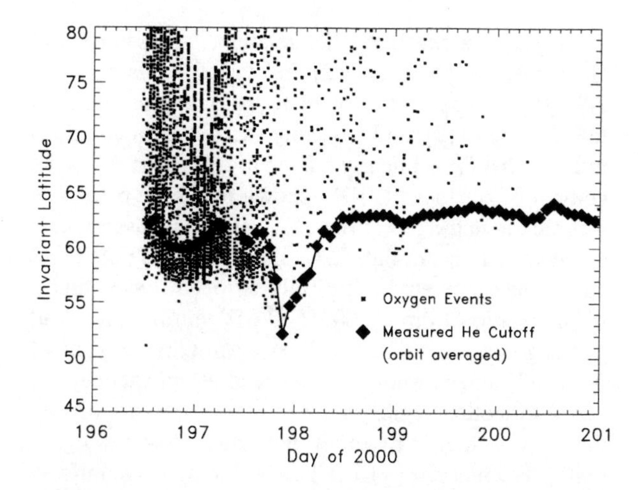

FIGURE 5. The invariant latitude at which each oxygen ion above $\sim$16 MeV/nucleon was detected by MAST during the Bastille Day event, plotted versus time, compared with the time dependence of the orbit-averaged cutoff latitude determined from the 8–15 MeV/nucleon He rate.

The invariant latitude at which each oxygen ion was detected by MAST during the Bastille Day event is also shown in Figure 5, and the abrupt lack of particles below $\sim$55° is due to the geomagnetic cutoff. The cutoff is several degrees lower for the higher energy (and higher rigidity) oxygen, but the variability in the oxygen cutoff follows that determined from the He rate rather well. Therefore, we take Λ_C measured using the rate of an abundant species such as He as a template to remove the cutoff variability from the higher Z particles where more limited statistics prevents our measuring Λ_C as accurately. For each detected particle, we find the difference between Λ_C for He at the nearest cutoff crossing and the average He cutoff value, and subtract this difference from the Λ at which the particle was detected.

After adjusting for the time variability, correcting for instrument livetime effects (which, for example, are responsible for the significant decrease in oxygen particles above the He cutoff in Figure 5), normalizing for the amount of time spent at each Λ, and summing over the duration of the event, the distributions in Λ for the heavy elements resemble the rate profile shown in Figure 4 [8, 12]. From linear fits to the low latitude edges of the distributions, we determine Λ_C for each abundant species in several energy intervals.

To derive a cutoff-rigidity relation, we assume that He is fully stripped ($Q = 2$) and that Q(C) is between +5.7 (the average value found in gradual events at much lower energies [2]) and +6 (fully stripped). Calculations indicate that R_C is linearly related to $\cos^4 \Lambda_C$ if Λ_C is the vertical cutoff [e.g., 17]. As shown in Figure 6 (and as was clearer in the 1992 events [8, 12]) our measurements agree with a linear relation. (It should also be noted that cutoff measurements using fully stripped nuclei find a linear relation from ~500–1600 MV during solar and geomagnetic quiet times [16]). A mean Q for oxygen of +7 [2, 8] agrees well with the trend established by He and C for both energies shown in Figure 6. Extrapolating the trend to higher rigidities, it is clear from the figure that Q(Fe) must be ~15 or less; it seems very unlikely that Q(Fe) could be, say, 20 or higher in this particular event. Linear fits to the combinations of the He and C data shown in the figure are used to establish the empirically-derived cutoff-rigidity relation for this event. Each Λ_C measured for the various Z and E is converted to a rigidity using each of the three relations, and the resulting uncertainty in the average rigidity encompasses the full spread in values obtained from the three different relations. From the deduced R_C and the measured E, the mean charge state and its uncertainty is obtained.

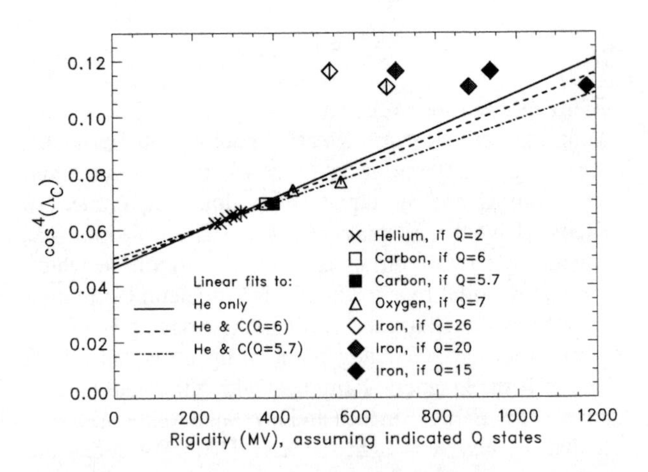

FIGURE 6. Measured $\cos^4 \Lambda_C$ during the Bastille Day event plotted versus rigidity calculated from the measured energy assuming the charge states indicated. For O and Fe, data at 2 energy points (at 3 different trial charge states each for Fe) are shown. Lines are fits to the subsets of the data indicated and serve as the relations between Λ_C and R_C used to obtain Q.

RESULTS

Charge states obtained by MAST in all of the SEP events studied to date, with their 1σ uncertainties, are shown in Figure 7. Uncertainties in the derived cutoff-rigidity relations dominate the overall uncertainties and might be reduced through further analysis. Except for a slightly low value for Q(Si), the charge states for the Bastille Day event are similar to those of the well-studied 1992 events.

A variety of other recent measurements of Q in gradual events obtained using other techniques (mostly at similar energies) are also shown Figure 7. Charge states from ACE/SIS [19] were deduced assuming the elemental and isotopic fractionation observed in the 6 November 1997 SEP event both scale as the same power law in Q/M. While the relatively high Q(Fe) value obtained from ACE/SIS agrees well with the value found by MAST in the same event, the Si charge states are very different. Other SAMPEX measurements in this event find Q(Si)~13 at ~2.5 MeV/nucleon and increasing with energy up to that point [11], apparently closer to the MAST value than to the value of 11.7 ± 0.2 deduced by Cohen et al. [19] at 12–60 MeV/nucleon. This may indicate systematic differences between the two approaches that are unaccounted for or additional stripping of Si after elemental and isotopic fractionation.

Values from the large gradual event of 20 April 1998 (not studied by MAST) shown from the work of Tylka et al. [24] were obtained from fits to the elemental energy spectra over a broad energy interval and agree well with the direct measurements from ACE/SEPICA [23] at much lower energies. For most elements, the measured charge states are often close to those expected from equilibrium calculations of collisional ionization in a 2 MK plasma [21, 22], a temperature typical of the solar corona. For the MAST 1997 and 1998 events shown, however, the high charge states of ~20 for Fe either require a higher temperature of ~10 MK or may suggest that additional stripping occurs during acceleration or transport [25, 26]. High energy tails on non-Maxwellian electron distributions might also generate somewhat higher charge states even in an otherwise relatively cold plasma, but this seems to be a larger effect for ions with a higher ionization potential such as O rather than Fe [e.g., 27, 28].

At lower energies of ~1 MeV/nucleon, a good correlation has been reported between Q(Fe) and the Fe/O ratio, with higher Q corresponding to enhanced Fe/O [4]. A similar trend is evident here, as shown in Table 1 using MAST Q measurements and the Fe/O ratio (for most of the events) obtained by ACE/SIS [5]. For the 25 August 1998 event, the Fe abundance above 30 MeV/nucleon was so low that it was not possible to accurately measure the cutoff location with the few particles detected by MAST, so no value for Q is given. Comparing Table 1 with Figure 2 of [4], it appears that for a given Fe/O value, Q(Fe) at MAST energies is generally higher than the value at lower energies. This may suggest that an increase in Q(Fe) with energy is fairly common, as has been measured in at least two events [7, 10, 11].

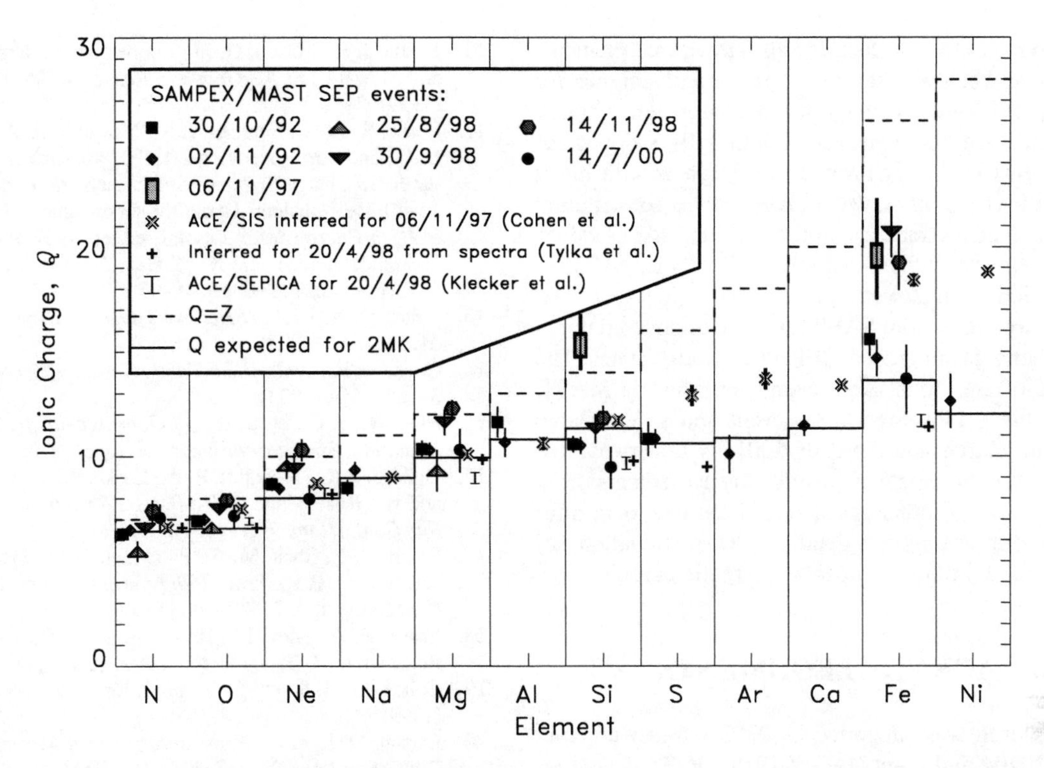

FIGURE 7. Charge states determined by MAST in SEP events in 1992 [8, 12], 1997 [11], 1998 [20], and 2000 (this work) compared with those expected in a 2 MK plasma (solid horizontal lines) [21, 22]. For comparison, charge states deduced in the 6 November 1997 event at similar energies are also shown [19], as well as charge states directly measured at low energies [23] and deduced from the spectral shape at high energies [24] in the 20 April 1998 event.

TABLE 1. Fe/O and Q(Fe) (at E ~28–65 MeV/nucleon) for MAST SEP events.

Event	Fe/O	Q(Fe)
30/10/92	$0.0702 \pm 0.0096^{*}$[29]	15.59 ± 0.81 [8]
02/11/92	$0.0932 \pm 0.0054^{*}$ [29]	14.69 ± 0.86 [8]
06/11/97	$0.900 \pm 0.006^{\dagger}$[5]	19.6 ± 2.4 [11]
25/08/98	$0.016 \pm 0.001^{\dagger}$ [5]	N/A
30/09/98	$0.299 \pm 0.003^{\dagger}$ [5]	20.7 ± 1.2 [20]
14/11/98	$0.761 \pm 0.006^{\dagger}$ [5]	19.2 ± 1.3 [20]
14/07/00	$0.0377 \pm 0.0007^{\dagger **}$	$13.7 \pm 1.7^{\ddagger}$

* $E = 30$–60 MeV/nucleon
† $E = 12$–60 MeV/nucleon
** From ACE/SIS [C. M. S. Cohen, private communication]
‡ This work

TABLE 2. Mean Q for Fe-poor and Fe-rich gradual events from MAST.

Element	avg Q, low Fe/O*	avg Q, high Fe/O†
N	6.36 ± 0.15	6.79 ± 0.27
O	6.90 ± 0.13	7.68 ± 0.21
Ne	8.63 ± 0.19	9.99 ± 0.42
Mg	10.26 ± 0.24	12.07 ± 0.33
Si	10.47 ± 0.27	(11.98 ± 0.42)**
Fe	15.02 ± 0.56	19.97 ± 0.82

* Average of events with Fe/O< 0.134
† Average of events with Fe/O> 0.134
** Average omitting the 06/11/97 event is 11.61 ± 0.44

It is unclear whether the correlation of composition and charge states is due to an admixture of impulsive SEP material into these otherwise gradual events [30], an origin of the material in hotter regions of the corona, or fractionation or charge-changing processes during acceleration or transport. The Q/M-dependent fractionation process that affects both elemental and isotopic abundances [e.g., 6] would not produce this correlation from material with a broad distribution of Fe charge states. If it did, then when Fe/O is enhanced, indicating enhancements of elements with lower Q/M, the lower Q(Fe) should also be enhanced, opposite to what is observed.

Using Table 1, if we divide the data into events where the high energy Fe/O ratio is depleted relative to the SEP-derived coronal value of 0.134 [31] or enhanced relative to the corona and calculate the average Q for each element in each of these two cases, we obtain the values given in Table 2. Because Q(Si) in the 6 November 1997 event is unusually high compared to that in the other two Fe-rich events, we list values in Table 2 both with and without this event included in the average. Both values are significantly higher than the average Q(Si) in the Fe-poor events. In fact, for all 6 measured elements from N to Fe, the mean Q is significantly higher for the Fe-rich events, being some 5 charge units higher at Fe and essentially consistent with fully stripped for all elements up through Mg. Table 2 agrees with lower energy observations showing charge states for different elements tend to track each other [4], and the Q values are very similar to those deduced indirectly by ACE/SIS in 4 Fe-rich gradual events [5].

Measurements from SAMPEX have uncovered unexpected behavior in gradual SEP event charge states. The analysis of additional large events detected by MAST, namely the 9 November 2000 event and several large events in March and April of 2001, is underway. Together with ongoing composition measurements from ACE, Wind, and other spacecraft, these new data offer the hope of revealing new details of the fractionation, acceleration, and transport of solar energetic particles.

ACKNOWLEDGMENTS

This research was supported by NASA under contract NAS5-30704 and grant NAGW-1919. GOES data (Figure 2) were obtained from the National Oceanic and Atmospheric Administration (NOAA), US Department of Commerce, through the Space Environment Center and the National Geophysical Data Center.

REFERENCES

1. Reames, D. V., *Revs. Geophys.*, **33**, 585–589 (1995).
2. Luhn, A., et al., *Proc. 19th Internat. Cosmic Ray Conf. (La Jolla)*, **4**, 241–244 (1985).
3. Luhn, A., Klecker, B., Hovestadt, D., and Möbius, E., *Astrophys. J.*, **317**, 951–955 (1987).
4. Möbius, E., et al., "Survey of Ionic Charge States of Solar Energetic Particle Events During the First Year of ACE", in *Acceleration and Transport of Energetic Particles Observed in the Heliosphere: ACE 2000 Symposium*, edited by R. A. Mewaldt et al., AIP Conf. Proc. 528, AIP, New York, 2000, pp. 131–134.
5. Cohen, C. M. S., et al., *Geophys. Res. Lett.*, **26**, 2697–2700 (1999).
6. Leske, R. A., et al., "Isotopic Abundances in the Solar Corona as Inferred from ACE Measurements of Solar Energetic Particles", in *Proc. Joint SOHO–ACE Workshop 2001*, AIP Conf. Proc., AIP, New York, 2001, this volume.
7. Möbius, E., et al., *Geophys. Res. Lett.*, **26**, 145–148 (1999).
8. Leske, R. A., Cummings, J. R., Mewaldt, R. A., Stone, E. C., and von Rosenvinge, T. T., *Astrophys. J. Lett.*, **452**, L149–L152 (1995).
9. Mason, G. M., Mazur, J. E., Looper, M. D., and Mewaldt, R. A., *Astrophys. J.*, **452**, 901–911 (1995).
10. Oetliker, M., et al., *Astrophys. J.*, **477**, 495–501 (1997).
11. Mazur, J. E., Mason, G. M., Looper, M. D., Leske, R. A., and Mewaldt, R. A., *Geophys. Res. Lett.*, **26**, 173–176 (1999).
12. Leske, R. A., Cummings, J. R., Mewaldt, R. A., Stone, E. C., and von Rosenvinge, T. T., "Measurements of the Ionic Charge States of Solar Energetic Particles at 15–70 MeV/nucleon Using the Geomagnetic Field", in *High Energy Solar Physics*, edited by R. Ramaty, N. Mandzhavidze, and X.-M. Hua, AIP Conf. Proc. 374, AIP, New York, 1996, pp. 86–95.
13. Baker, D. N., et al., *IEEE Trans. Geosci. Remote Sensing*, **31**, 531–541 (1993).
14. Cook, W. R., et al., *IEEE Trans. Geosci. Remote Sensing*, **31**, 557–564 (1993).
15. Roederer, J. G., *Dynamics of Geomagnetically Trapped Radiation*, Springer-Verlag, New York, 1970.
16. Ogliore, R. C., Mewaldt, R. A., Leske, R. A., Stone, E. C., and von Rosenvinge, T. T., *Proc. 27th Internat. Cosmic Ray Conf. (Hamburg)* (2001), in press.
17. Smart, D. F., Shea, M. A., Flückiger, E. O., Tylka, A. J., and Boberg, P. R., *Proc. 26th Internat. Cosmic Ray Conf. (Salt Lake City)*, **7**, 337–340 (1999).
18. Leske, R. A., Mewaldt, R. A., Stone, E. C., and von Rosenvinge, T. T., *J. Geophys. Res.* (2001), in press.
19. Cohen, C. M. S., et al., *Geophys. Res. Lett.*, **26**, 149–152 (1999).
20. Larson, D. J., et al., *Proc. 26th Internat. Cosmic Ray Conf. (Salt Lake City)*, **7**, 301–304 (1999).
21. Arnaud, M., and Rothenflug, R., *Astron. Astrophys. Suppl.*, **60**, 425–457 (1985).
22. Arnaud, M., and Raymond, J., *Astrophys. J.*, **398**, 394–406 (1992).
23. Klecker, B., et al., *Proc. 26th Internat. Cosmic Ray Conf. (Salt Lake City)*, **6**, 83–86 (1999).
24. Tylka, A. J., Boberg, P. R., McGuire, R. E., Ng, C. K., and Reames, D. V., "Temporal Evolution in the Spectra of Gradual Solar Energetic Particle Events", in *Acceleration and Transport of Energetic Particles Observed in the Heliosphere: ACE 2000 Symposium*, edited by R. A. Mewaldt et al., AIP Conf. Proc. 528, AIP, New York, 2000, pp. 147–152.
25. Barghouty, A. F., and Mewaldt, R. A., "Simulation of Charge-Equilibration and Acceleration of Solar Energetic Ions", in *Acceleration and Transport of Energetic Particles Observed in the Heliosphere: ACE 2000 Symposium*, edited by R. A. Mewaldt et al., AIP Conf. Proc. 528, AIP, New York, 2000, pp. 71–78.
26. Reames, D. V., Ng, C. K., and Tylka, A. J., *Geophys. Res. Lett.*, **26**, 3585–3588 (1999).
27. Owocki, S. P., and Scudder, J. D., *Astrophys. J.*, **270**, 758–768 (1983).
28. Ko, Y.-K., Fisk, L. A., Gloeckler, G., and Geiss, J., *Geophys. Res. Lett.*, **23**, 2785–2788 (1996).
29. Williams, D. L., *Measurements of the Isotopic Composition of Solar Energetic Particles with the MAST Instrument Aboard the SAMPEX Spacecraft*, Ph.D. thesis, California Institute of Technology, Pasadena, CA 91125, USA (1998).
30. Mason, G. M., Mazur, J. E., and Dwyer, J. E., *Astrophys. J. Lett.*, **525**, L133–L136 (1999).
31. Reames, D. V., *Adv. Space Res.*, **15**, (7)41–(7)51 (1995).

Measurements of Heavy Elements and Isotopes in Small Solar Energetic Particle Events

P.L. Slocum*, E.R. Christian†, C.M.S. Cohen**, A.C. Cummings**, R.A. Leske**, R.A. Mewaldt**, E.C. Stone**, T.T. von Rosenvinge† and M.E. Wiedenbeck*

*Jet Propulsion Laboratory, Pasadena, CA USA
†NASA/Goddard Space Flight Center, Greenbelt, MD USA
**California Institute of Technology, Pasadena, CA USA

Abstract. Using the Solar Isotope Spectrometer on the Advanced Composition Explorer, we have examined the $\sim$10-20 MeV/nucleon elemental and isotopic composition of heavy ($Z \geq 6$) energetic nuclei accelerated in 30 small solar energetic particle (SEP) events which occurred between 31 March 1998 and 2 January 2001. We have measured the average heavy element content, the $^{22}Ne/^{20}Ne$ ratio, and the $^{26}Mg/^{24}Mg$ ratio in these events, and find good agreement with past studies. We have categorized the events according to their $^{3}He/^{4}He$ ratios, and find significant enhancements in the neutron-rich heavy isotopes of Ne and Mg in the combined ^{3}He-rich data set: $^{22}Ne/^{20}Ne=0.17\pm0.05$ and $^{26}Mg/^{24}Mg=0.25\pm0.05$. We discuss the implications of these measurements for the acceleration of energetic nuclei in SEP events.

INTRODUCTION

Measurements of the elemental and isotopic composition of energetic nuclei from solar energetic particle (SEP) events can provide information about the acceleration mechanisms in these kinds of occurrences. SEP events have been classified into two main types: impulsive and gradual [1]. During gradual events, which have a duration on the order of days, the abundances of heavy SEPs have been shown to vary according to their charge-to-mass ratio [2]. The acceleration of the nuclei during these events is understood to occur at shock waves which are driven by coronal mass ejections (CMEs) [3, 4, 5, 6]. On average, the heavy element composition of SEP material from gradual events is expected to reflect that of the corona. Impulsive SEP events, however, commonly contain enhancements in $\sim$MeV Fe ($\sim 10\times$) and in $\sim$MeV Ne, Mg, and Si ($\sim 3\times$) relative to coronal abundances [7, 8]. Additionally, impulsive events frequently have $^{3}He/^{4}He$ ratios which are up to 3–4 orders of magnitude larger than the solar wind value of 0.0004 [9]. Impulsive SEP events typically last on the order of hours.

The observed differences between impulsive and gradual SEP events suggest that each has a distinct acceleration mechanism. As stated above, gradual events are known to be associated with CME-driven shocks. However, while the ^{3}He found in impulsive events may be selectively enhanced by ion cyclotron wave resonances [10, 11] or by cascading Alfven waves [12], the exact acceleration mechanism in impulsive events is not presently well understood. In this paper we have examined the heavy elemental and isotopic content of 30 small SEP events. We have classified the small events into two sub-groups according to their ^{3}He content, and have discussed the implications of their elemental and isotopic composition for possible acceleration models.

EVENT SELECTION

The Solar Isotope Spectrometer (SIS) consists of a pair of silicon solid state detector telescopes which measure the energy loss (E) and the rate of energy loss (dE/dx) of incident energetic nuclei with $10 \lesssim E \lesssim 100$ MeV/nucleon. Then, using the measured values of dE/dx and E, the charge and mass of the incident nuclei are derived through an iterative mathematical algorithm [13].

The 30 SEP events selected for this study were chosen according to three criteria. First, a set of 98 days between 31 March 1998 and 31 December 2000 were identified by requiring that their daily averaged SIS 11.0–26.5 MeV/nucleon Fe or 8.6–19.3 MeV/nucleon Mg fluxes be greater than a threshold of 5×10^{-7} particles/(s cm^{2}sr MeV/nuc), and that their daily averaged SIS 11.0–26.5 MeV/nucleon Fe fluxes be less than 2.5×10^{-6} particles/(s cm^{2}sr MeV/nuc). Second, the days which over-

CP598, *Solar and Galactic Composition*, edited by R. F. Wimmer-Schweingruber

lapped with any of the known large, gradual SEP events were excluded from the study. Finally, the SIS ~3–5 MeV/nucleon He fluxes on the selected days were examined subjectively for well–defined SEP event onset times and terminations. The 30 events with well–defined onset times and terminations became the final data set examined in this study, while the days which did not show well–defined time profiles in the ~3–5 MeV/nucleon He flux were discarded. Even so, due to the frequent occurrence of this type of small SEP event, it is still probable that some of the selected "good" events contain multiple injections from the same region on the Sun. This problem is not expected to generate any significant uncertainty in the results of this study, however, because we are only reporting abundances which have been averaged over many SEP events.

TABLE 1. Event times, ~4.5–5.5 MeV/nucleon ^{3}He/^{4}He ratios, and 11-22 MeV/nucleon Mg fluxes in (cm^{2}sr s MeV/nuc)$^{-1}$ for the 30 SEP events included in this study.

Event #	Time (Year:Day)	^{3}He/^{4}He	Mg Flux
0	1998: 93.1–104.8	0.035	1.7e-07
1	1998:119.7–121.3	0.045	1.9e-07
2	1998:121.3–122.2	0.041	5.7e-07
3	1998:147.4–148.6	0.066	7.6e-07
4	1998:167.7–170.6	0.042	7.3e-07
5	1998:249.1–251.2	0.100	2.5e-07
6	1998:251.6–254.2	0.177	4.3e-07
7	1998:270.2–272.1	0.110	1.0e-06
8	1998:294.1–297.0	0.037	1.2e-07
9	1998:309.7–311.7	0.045	1.6e-06
10	1998:311.7–316.4	0.062	1.5e-06
11	1998:327.5–334.9	0.041	2.3e-07
12	1999:147.3–148.4	0.039	3.9e-07
13	1999:172.7–176.0	0.064	1.1e-07
14	1999:320.7–321.8	0.062	3.5e-07
15	1999:321.7–327.2	0.043	3.4e-07
16	1999:362.0–364.2	0.053	1.2e-06
17	2000: 9.2– 13.5	0.037	3.9e-07
18	2000: 67.2– 68.8	0.090	4.2e-07
19	2000: 82.4– 87.2	0.041	1.1e-07
21	2000:125.3–126.7	0.071	4.0e-06
22	2000:136.7–138.8	0.038	6.4e-07
23	2000:138.8–140.7	0.039	7.7e-07
24	2000:144.7–147.9	0.196	1.9e-07
25	2000:169.2–172.9	0.054	1.6e-07
26	2000:175.3–177.2	0.051	1.4e-06
27	2000:177.2–180.6	0.032	3.9e-07
28	2000:224.0–225.4	0.038	1.0e-06
29	2000:225.4–226.6	0.045	2.2e-07
30	2000:226.6–228.6	0.044	7.1e-07

The final set of 30 events, listed in Table 1, were subdivided into two groups according to their total ~4.5–5.5 MeV/nucleon ^{3}He/^{4}He ratios. Of the 30 events, 13 had ^{3}He/^{4}He ratios greater than or equal to 0.065, while 17 had ^{3}He/^{4}He ratios which were less than 0.065. In general, the latter 17 events had ^{3}He/^{4}He ratios which were measured between 0.04 and 0.065. This apparent lower limit of 4% is probably due to spillover contamination of the ^{3}He peak by the ^{4}He peak. Although the nomenclature "^{3}He-rich" has generally been reserved for SEP events with ^{3}He/^{4}He$\geq$0.1, we have designated the first sub-group with ^{3}He/^{4}He$\geq$0.065 as "^{3}He-rich" and the second as "^{3}He-poor" in order to preserve adequate measurement statistics in the ^{3}He-rich sub-group.

While the unweighted average ~4.5–5.5 MeV/nucleon ^{3}He/^{4}He ratio in the ^{3}He-rich and ^{3}He-poor groups is ~12% and ~5% respectively, and impulsive events are expected to have higher ^{3}He/^{4}He ratios than gradual events, it is not clear that our two event sub-groups are cleanly divided as such. For example, the unweighted average duration of the events in the ^{3}He-rich (^{3}He-poor) subset is 2.4 (3.4) days, which is longer than expected for impulsive SEP events. It could be that while the ^{3}He-rich events contain a relatively large fraction of material which was originally accelerated in impulsive SEP events, they are not necessarily impulsive events. This idea stems from the theory by Mason et al. [15], which says that residual material from past impulsive SEP events may provide a source population for further acceleration by CME-driven shocks associated with gradual events. Experimental observations of SEP composition during large gradual events [14, 15, 16, 17] and during times of low solar activity [18, 19] have supported this theory.

LOW SOLAR ACTIVITY SPECTRA

The contributions to the small SEP event spectra from galactic cosmic rays (GCR) and anomalous cosmic rays (ACR) [20] have been estimated from SIS measurements during 241 days of low solar activity between 9 April, 1998 and 25 December, 2000. The 241 days were identified by examining the proton fluxes from the Electron, Proton, and Alpha Monitor (EPAM) instrument on board ACE, and requiring that the daily averaged 1.06-4.75 MeV proton flux be less than 0.16 (cm^2 sr s MeV/nuc)$^{-1}$. Next, the 241 days were grouped chronologically into three time periods of relatively constant solar modulation, based on inspection of SIS ~7-10 MeV/nucleon O fluxes. These time periods were 9 April, 1998–25 November, 1998, 26 November, 1998–14 January, 2000, and 14 January, 2000–25 December, 2000. For each of the three time periods, the average heavy element spectra were extracted from the daily measurements. These average heavy element spectra were weighted with the temporal fraction of the 30 small SEP events which occurred during each time period, and were subsequently summed for subtraction from the small SEP event spectra.

HEAVY ELEMENT MEASUREMENTS

The average energy spectra for the elements C, O, Ne, Na, Mg, Al, Si, and Fe were measured for the 30 small SEP events, as well as for the "^{3}He-rich" and "^{3}He-poor" subsets. These spectra were derived by taking the raw number of nuclei for each species detected during all of the SEP events combined, and dividing that number by the total amount of instrument livetime. Finally, the GCR and ACR components of the spectra were subtracted as described above to yield the SEP fluxes for each element. Figure 1 depicts the relative abundances of nine heavy element ratios in the energy range ~11–22 MeV/nucleon, normalized to coronal abundances [21], for both the ^{3}He-rich and ^{3}He-poor SEP event data sets.

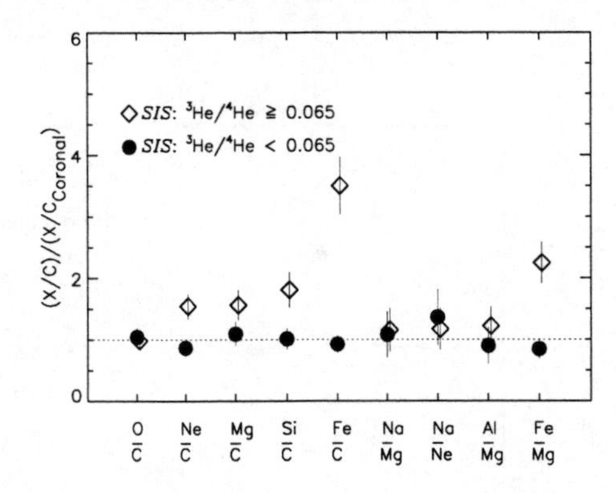

FIGURE 1. 11.0–21.8 MeV/nucleon abundances of O, Ne, Mg, Si, and Fe with respect to C, as well as the Na/Mg, Na/Ne, Al/Mg, and Fe/Mg ratios, normalized to coronal abundances [21], for the ^{3}He–rich (open diamonds) and ^{3}He–poor (filled circles) data sets.

From the figure, it is apparent that while the composition of the ^{3}He-poor data set generally reflects that of the corona, the ^{3}He-rich data set contains an average enhancement in Fe with respect to C of ~4×, as well as more modest enhancements in Ne, Mg, and Si of ~2×. The uncertainties shown in the plot are statistical in nature, and do not reflect any contribution from a possible non-statistical population spread in the events. This trend in the heavy element abundances is in agreement with past studies, which associate large ^{3}He/^{4}He ratios in SEP events with Fe, Ne, Mg, and Si enhancements at ~MeV energies [7, 8], although the magnitude of the enhancements are smaller in the above SIS data. Also of note is that since the coronal values chosen for normalization in this figure were derived from large gradual SEP events, it is likely that the ^{3}He-poor events consist primarily of small "gradual" events.

NEON AND MAGNESIUM ISOTOPE RATIOS

The isotope ratios of ~15-24 MeV/nucleon Ne and ~16-26 MeV/nucleon Mg have been extracted from the SIS data using a superposition of Gaussian fits to the derived charge histograms. Figure 2 shows the two charge histograms of Ne and Mg in these energy ranges. With the iterative charge calculation used, different isotopes of an element are assigned charges differing by ~0.1 charge units for each 1 amu mass difference. Thus the five peaks representing ^{20}Ne, ^{22}Ne, ^{24}Mg, ^{25}Mg, and ^{26}Mg are shown in the figure, and have been fit with a function composed of five Gaussian curves of identical width and uniformly spaced isotopes. Because the ^{25}Mg peak was not well-constrained by the data in the energy range ~16-26 MeV/nucleon, its height was fixed at 12.7% of the ^{24}Mg peak height in accordance with

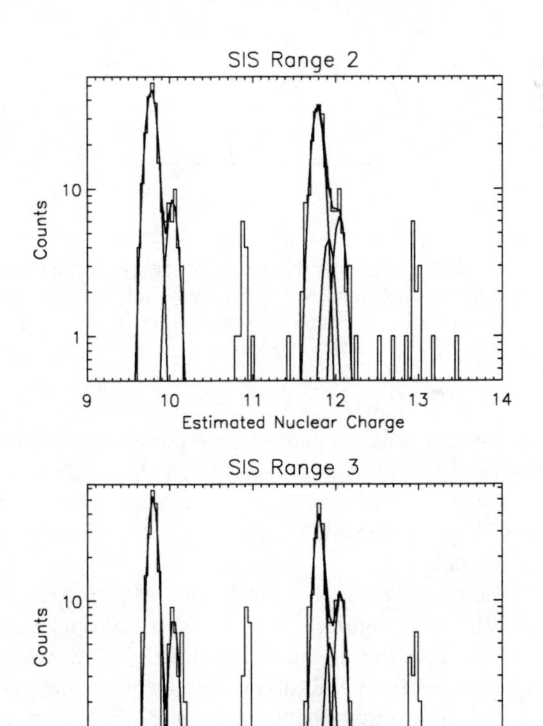

FIGURE 2. Estimated nuclear charge in the energy range 14.7–19.3 MeV/nucleon (top panel) and 17.6–24.1 MeV/nucleon (bottom panel) measured using SIS during the 30 small SEP events. The ^{20}Ne, ^{22}Ne, ^{24}Mg, ^{25}Mg, and ^{26}Mg peaks have been fit with a superposition of five Gaussian curves of identical width and uniformly spaced isotopes. The ^{25}Mg/^{24}Mg ratio of peak heights has been fixed at the average solar system value of 0.127 [22]

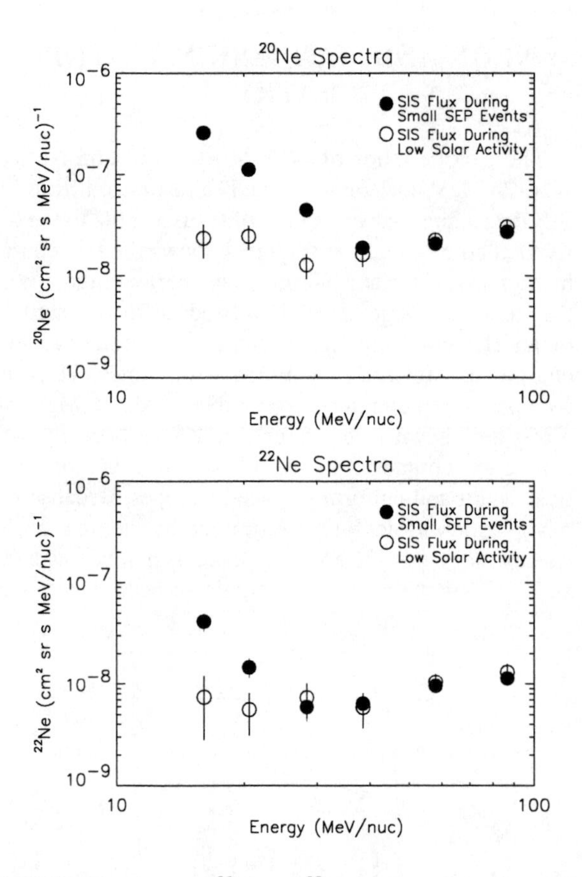

FIGURE 3. Average ^{20}Ne and ^{22}Ne energy spectra for the 30 small SEP events contained in this study (solid circles), and for the days of low solar activity over the same time period (open circles).

the average solar system value reported by Anders and Grevesse [22]. It is important to note that a factor of two variation about the assumed abundance of ^{25}Mg changes the ^{26}Mg/^{24}Mg result less than the statistical uncertainty in that ratio.

The measurements of the Ne and Mg isotope spectra are shown in Figures 3 and 4. The filled circles represent SIS data during the 30 small SEP events, while the open circles show the SIS measurements during periods of low solar activity between the small SEP events (see above). Because the ^{26}Mg quiet-time spectrum was difficult to extract below ~40 MeV/nucleon due to statistical limitations, the higher energy data points were fit with a power-law function which was extrapolated down to ~20 MeV/nucleon. Conversely, the Ne quiet-time spectrum contains a contribution from the ACRs which results in the beginning of a turn-up in the quiet-time spectrum at about 20-30 MeV/nucleon. All of the isotope spectra show clear low-energy (<30 MeV/nucleon) turn-ups in the SEP event spectra, which is evidence for the presence of solar particles in numbers well above the low solar activity background.

With the event spectra and low solar activity spectra shown in Figures 3–4, one can derive the small SEP event (with GCR and ACR components subtracted) ^{22}Ne/^{20}Ne and ^{26}Mg/^{24}Mg isotope ratios. This has been done for both the ^{3}He-rich and ^{3}He-poor event data sets. Figure 5 shows the results of the isotope ratio measurements below ~30 MeV/nucleon for each data set. On each plot, the average of the two data points, with one standard deviation of statistical uncertainty, is represented by the grey shaded area.

From the two top panels in Figure 5, it is apparent that the average SEP ^{22}Ne/^{20}Ne ratio is at least slightly enhanced over the solar wind value of 0.073 [23] for both the ^{3}He-rich and ^{3}He-poor event data sets. In the data set with ^{3}He/^{4}He$\geq$0.065, the ^{22}Ne/^{20}Ne ratio is found to be enhanced over the solar wind value by a factor of ~2.4×, with significance at the 2σ level, while in the ^{3}He-poor data set the ratio is only enhanced by 1σ. Similarly, in the bottom panels the ^{26}Mg/^{24}Mg ratio in the ^{3}He-rich

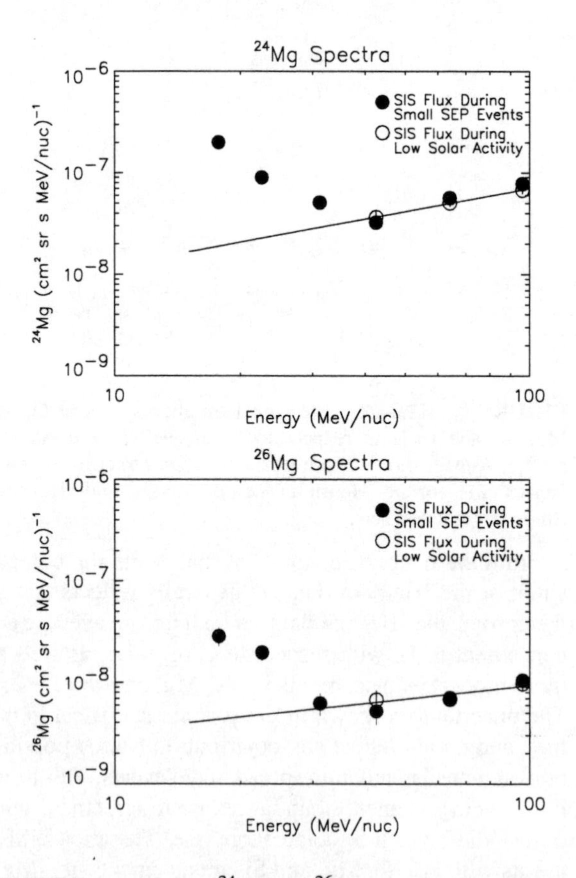

FIGURE 4. Average ^{24}Mg and ^{26}Mg energy spectra for the 30 small SEP events contained in this study (solid circles), and for the days of low solar activity over the same time period (open circles). The solid line represents the extrapolated galactic cosmic ray (GCR) spectrum from the fit with a power-law function to the three highest energy low solar activity data points on the plot.

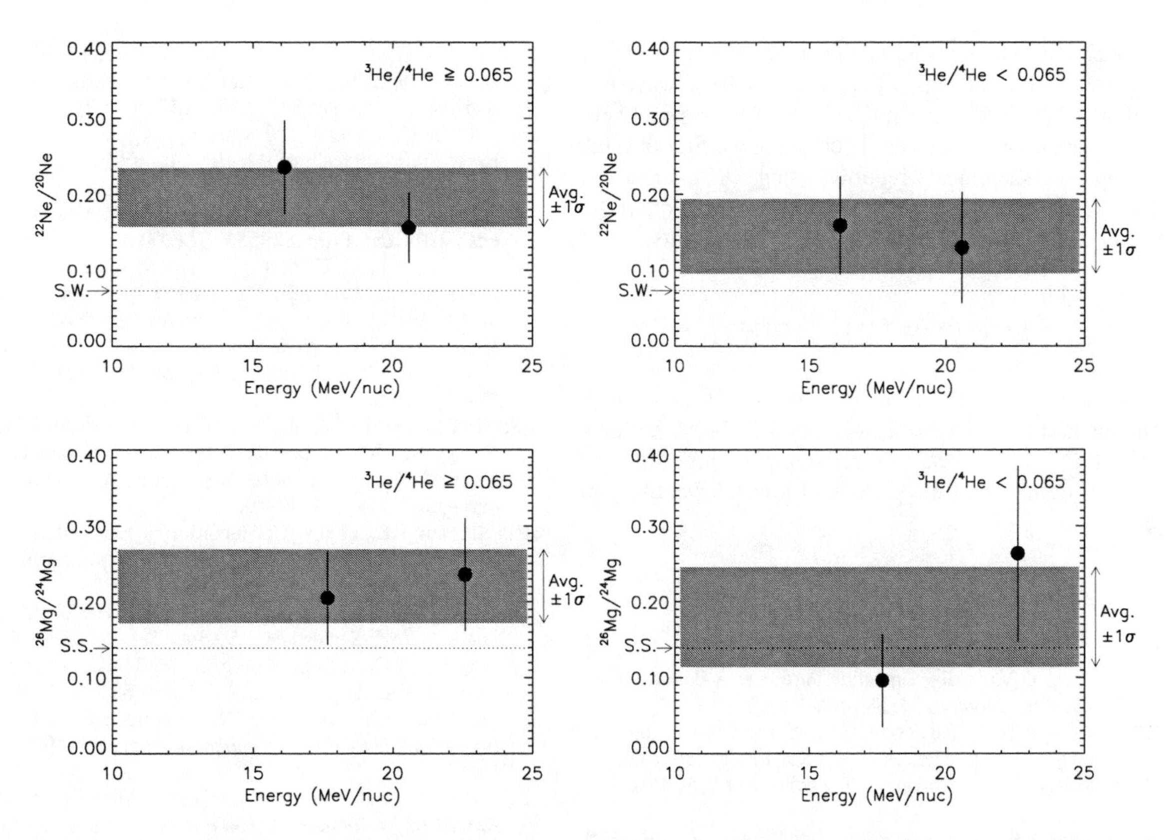

FIGURE 5. Top panels: ^{22}Ne/^{20}Ne ratios for the ^{3}He–rich and ^{3}He–poor data sets. The horizontal dotted line respresents the solar wind value of 0.073 [23]. Bottom panels: ^{26}Mg/^{24}Mg ratios for the ^{3}He–rich and ^{3}He–poor data sets. The horizontal dotted line respresents the average solar system value of 0.127 [22]. The shaded regions in each plot denote the average value of the two data points, with an uncertainty of one standard deviation.

SEP events shows an enhancement over the solar system value of 0.127 [22] by a factor of ~2×, at a significance level of 2.5σ. The data set with ^{3}He/^{4}He<0.065 yields a measurement of ^{26}Mg/^{24}Mg which is consistent with the solar system value.

While the ^{3}He–rich events have significant enhancements of neutron–rich heavy isotopes of Ne and Mg, it is important to note again that our definition of "^{3}He–rich" is less restrictive than the widely–accepted requirement that ^{3}He–rich SEP events have ^{3}He/^{4}He>0.1. For this reason, our "^{3}He–rich" subset of events may contain a smaller fraction of impulsive SEP material than it would have had with a more stringent cut on the ^{3}He content. It is possible that the heavy isotopic enhancements would have been even larger if statistics had allowed for such a study.

CONCLUSIONS

The average heavy element content of these 30 small SEP events has been shown to change significantly with the event ^{3}He/^{4}He ratio. Of the 30 events selected for the study, the 17 with ^{3}He/^{4}He<0.065 have an average ~11-22 MeV/nucleon heavy element composition which is very similar to the solar corona. The other 13 ^{3}He-rich events contain average enhancements in Fe/C of ~4, and enhancements in Ne/C, Mg/C, and Si/C of ~2. This trend is in qualitative agreement with past studies of ^{3}He-rich events at lower energies [7, 8].

The average ~15-25 MeV/nucleon Ne and Mg isotopic content of these SEP events also changes with the ^{3}He/^{4}He ratio. The ^{22}Ne/^{20}Ne ratio increases from 0.13±0.06 in the ^{3}He-poor events to 0.17±0.05 in the ^{3}He-rich data set, which is consistent with past measurements of "^{3}He-rich" events at ~MeV energies (^{22}Ne/^{20}Ne=0.29±0.10) [24]. Similarly, we find that the ^{26}Mg/^{24}Mg ratio increases to 0.25±0.05 in the ^{3}He-rich data set, consistent with previous studies (^{26}Mg/^{24}Mg=0.36±0.21) [24].

The observations of ^{3}He-rich SEP events with enhancements in neutron-rich heavy isotopes is consistent with the model of Miller [12]. This model predicts that in ^{3}He-rich events, isotopes with a lower charge-to-mass ratio should be enhanced relative to those with a higher ratio due to interactions with cascading Alfvén wave turbulence. While the exact magnitudes of the expected enhancements have not been determined, we find that the isotopic enhancements reported using SIS data are in qualitative agreement with this model and are quantitatively consistent with those of past measurements.

ACKNOWLEDGMENTS

This research was supported by NASA at Caltech (under grant NAG5-6912), JPL, and GSFC. We thank the EPAM science team for providing the proton fluxes used to define the time periods of low solar activity in this paper.

REFERENCES

1. Reames, D.V., "Solar Energetic Particles: A Paradigm Shift", *Rev. Geophys.* **33**, 585-589 (1995).
2. Breneman, H.H. and Stone, E.C., "Solar Coronal and Photospheric Abundances from Solar Energetic Particle Measurements", *Astrophys. J. Letters* **299**, L57-L61 (1985).
3. Gosling, J.T., "The Solar Flare Myth", *J. Geophys. Res.* **98**, 18949 (1993).
4. Kahler, S.W., "Solar Flares and Coronal Mass Ejections", *Ann. Rev. Astron. Astrophys.* **30**, 113 (1992).
5. Kahler, S.W., "Injection Profiles for Solar Energetic Particles as Functions of Coronal Mass Ejection Heights",*Astrophys. J.* **428**, 837 (1994).
6. Reames, D.V., "Particle Acceleration at the Sun and in the Heliosphere", *Space Sci. Revs.* **90**, 413 (1999).
7. Mason, G.M., et al., "The Heavy Ion Compositional Signature in 3He-rich Solar Particle Events", *Astrophys. J.* **303**, 849 (1986).
8. Reames, D.V., Meyer, J.P., and von Rosenvinge, T.T., "Energetic-Particle Abundances in Impulsive Solar Flare Events", *Astrophys. J.* **90**, 649–667 (1994).
9. Gloeckler, G., and Geiss, J., "Measurement of the Abundance of Helium-3 in the Sun and in the Local Interstellar Cloud with SWICS on Ulysses", *Space Sci. Rev.*, **84**, 275–284 (1998).
10. Fisk, L.A., "^{3}He-Rich Flares: A Possible Explanation", *Astrophys. J.* **224**, 1048–1055 (1978).
11. Temerin, M., and Roth, I., "The Production of ^{3}He and Heavy Ion Enrichments in ^{3}He-Rich Flares by Electromagnetic Hydrogen Cyclotron Waves", *Astrophys. J. Lett.* **391**, L105–L108 (1992).
12. Miller, J.A., "Particle Acceleration in Impulsive solar Flares", *Space Sci. Rev.* **86**, 79–105 (1998).
13. Stone, E.C. et al., "The Solar Isotope Spectrometer for the Advanced Composition Explorer", *Space Sci. Rev.* **86**, 357–408 (1998).
14. Cohen, C.M.S., et al., "New Observations of Heavy-Ion-Rich Solar Particle Events from ACE", *Geophys. Res. Lett.* **26:17**, 2697–2700 (1999).
15. Mason, G.M., Mazur, J.E., and Dwyer, J.R., "^{3}He Enhancements in Large Solar Energetic Particle Events", *Astrophys. J. Letters* **525**, L133–L136 (1999).
16. Cohen, C.M.S. et al., "The Isotopic Composition of Solar Energetic Particles", in *Acceleration and Transport of Energetic Particles Observed in the Heliosphere: ACE 2000 Symposium*, (AIP: New York), R. A. Mewaldt et al., eds., *AIP Conf. Proc.* **528**, 55–62 (2000).
17. Wiedenbeck, M. E., et al., "Enhanced Abundances of ^{3}He in Large Solar Energetic Particle Events", in *Acceleration and Transport of Energetic Particles Observed in the Heliosphere: ACE 2000 Symposium*, (AIP: New York), R. A. Mewaldt et al., eds., *AIP Conf. Proc.* **528**, 107–110 (2000).
18. Richardson, I.G. et al., "Quiet–Time Properties of Low–Energy (< 10 MeV/Nucleon) Interplanetary Ions During Solar Maximum and Solar Minimum", *Astrophys. J. Letters***363**, L9–L12 (1990).
19. Slocum, P.L., et al., "Composition of Energetic Particles at 1 AU During Periods of Moderate Interplanetary Particle Activity", *Adv. Space Res.*, in press (2001).
20. Fisk, L.A., et al., "Cosmic Rays in the Heliosphere", *Space Science Series of ISSI, Space Sci. Rev.*, **83** (1998).
21. Reames, D.V., "Solar Energetic Particles: Sampling Coronal Abundances", *Space Sci. Rev.* **85** 327–340 (1998).
22. Anders, E. and Grevesse, N., "Abundances of the Elements", *Geochim. Cosmochim. Acta.*, **53**, 197-214 (1989)
23. Geiss, J., et al., "Solar Wind Composition Experiment", *Apollo 16 Preliminary Science Report, NASA SP–315*, section 14 (1972).
24. Mason, G.M., Mazur, J.E., and Hamilton, D.C., "Heavy-Ion Isotopic Anomalies in ^{3}He-Rich Solar Particle Events", *Astrophys. J.* **425**, 843–848 (1994).

High-latitude Ulysses observations of the H/He intensity ratio under solar minimum and solar maximum conditions

D. Lario*, C.G. Maclennan†, E.C. Roelof*, J.T. Gosling**, G.C. Ho* and S.E. Hawkins III*

**The Johns Hopkins University Applied Physics Laboratory, Laurel, MD 20723-6099*
†Bell Laboratories, Lucent Technologies, Murray Hill, NJ 07974
***Los Alamos National Laboratory, Los Alamos, New Mexico, NW 87545*

Abstract. We analyze measurements of the 0.5-1.0 MeV/nucleon H/He intensity ratio from the Ulysses spacecraft during its first (1992-1994) and second (1999-2000) ascents to southern high latitude regions of the heliosphere. These cover a broad range of heliocentric distances (from 5.2 to 2.0 AU) and out-of-ecliptic latitudes (from 18°S to 80°S). During Ulysses' first southern pass, the HI-SCALE instrument measured a series of enhanced particle fluxes associated with the passage of a recurrent corotating interaction region (CIR). Low values (~6) of the H/He ratio were observed in these recurrent corotating events, with a clear minimum following the passage of the corotating reverse shock. When Ulysses reached high southern latitudes (>40°S), the H/He ratio always remained below ~10 except during two transient solar events that brought the ratio to high (>20) values. Ulysses' second southern pass was characterized by a higher average value of the H/He ratio. No recurrent pattern was observed in the energetic ion intensity, which was dominated by the occurrence of transient events of solar origin. Numerous stream interaction regions (SIRs) were observed in the solar wind and magnetic field data, many of which were bounded by forward and reverse shock pairs. The arrival of those SIRs at Ulysses usually produced an energetic particle flux enhancement, but they were not always characterized by a decrease of the H/He ratio; on the contrary, many SIRs showed a higher H/He ratio than some transient events. Within a SIR, however, the H/He ratio usually increased around the forward shock and decreased towards the reverse shock. Throughout the second ascent to southern heliolatitudes, the H/He ratio seldom decreased below ~10 even at high latitudes (>40°S). We interpret these higher values of the H/He ratio in terms of the increasing level of solar activity together with the poor definition and short life that corotating solar wind structures have under solar maximum conditions. The global filling of the heliosphere by transient solar events and the fact that in 1999-2000 Ulysses observed only intermediate (<650 km s^{-1}) solar wind speed (within which the pick-up He is less energetic than in the fast solar wind streams observed in 1992-1994) favored the protons with respect to alpha particles. Consequently, the average values of the H/He ratio observed by Ulysses during the rising phase of the solar cycle (1999-2000) were higher than those observed during the declining phase (1992-1994).

INTRODUCTION

The study of the suprathermal proton-to-helium abundance ratios during different phases of the solar cycle can be used as an indicator of how interplanetary acceleration mechanisms and "seed" particle populations change over the solar cycle. Previous studies [1, 2] have shown a fundamental solar cycle variation in the particle flux abundances. Shields et al. [1] showed that at the heliocentric distance of 1 AU solar minimum fluxes tend to be richer in helium than the fluxes measured during active periods. Richardson et al. [2] noted that the element abundances of corotating events (also observed at 1 AU) showed a clear transition from solar maximum to solar minimum. Solar minimum corotating events have element abundances that are well differentiated from the abundances measured during solar energetic particle (SEP) events, while at solar maximum the abundances in corotating events are more SEP-like [2].

Corotating energetic particle events are associated with the compression regions formed by the interaction between slow and fast solar wind streams. The rather simple configuration of the solar corona found at solar minimum, with large coronal holes over the polar caps, restricts the range of interaction between fast and slow solar wind to low heliographic latitudes ($\lesssim 40°$). The stability of the solar corona during solar minimum leads to the observation of recurrent compression regions on each

CP598, *Solar and Galactic Composition,* edited by R. F. Wimmer-Schweingruber

solar rotation, and hence the name of corotating interaction regions (CIRs). During solar maximum, the more complex arrangement of the solar corona, with small-scale coronal holes observed at all latitudes, increases the angular range where the interaction between slow and fast solar wind may take place. The dynamic character of the solar corona during active periods together with the small scale of the coronal holes leads to a less evolved and less periodic character of the solar wind stream interaction regions. Occasionally, these interaction regions may appear recurrently over a few consecutive rotations and thus called CIRs [3].

The out-of-ecliptic orbit of the Ulysses spacecraft with a period of 6.2 years allows us to study these solar cycle variations over a broad range of heliographic latitudes. At the end of 1992, the Ulysses spacecraft was at a distance of 5.2 AU from the Sun and moving south beyond 18°S. The spacecraft reached its most southern heliographic latitude of 80.2°S at a radial distance of 2.3 AU from the Sun in mid-September 1994. After a complete orbit over the poles of the Sun, in January 1999 Ulysses was again at a distance of 5.2 AU from the Sun and moving south beyond 18°S. During its second out-of-ecliptic orbit around the Sun, it reached a maximum latitude of 80.1°S at the end of November 2000. While the first excursion to southern latitudes (1992-1994) occurred during the declining phase of the solar cycle 22, the second ascent to high latitudes (1999-2000) coincided with the rising phase of the 23rd solar cycle.

In this paper we report measurements of the 0.5-1.0 MeV/nucl H/He ratio for the first and second Ulysses southern passes. While previous studies of the H/He ratio have limited their analysis to the ecliptic plane and at 1 AU from the Sun [1, 2, 4], or in the outer heliosphere but under solar minimum conditions [5, 6, 7, 8], we extend these studies to high heliographic latitudes and for solar maximum conditions. We describe how the observed H/He ratio is affected by the occurrence of transient events and the arrival of coronal mass ejections (CMEs) and stream interaction regions. We compare the evolution of the H/He ratio at high heliolatitudes during solar maximum and solar minimum and interpret its evolution in terms of the possible sources of energetic particles.

OBSERVATIONS

The measurements presented here were made with the CA (Composition Aperture) telescope of the HI-SCALE (Heliosphere Instrument for Spectra, Composition and Anisotropy at Low Energies) on the Ulysses spacecraft [9]. The CA telescope consists of a three element Si solid state detector telescope with a 5μ-thick first detector followed by two 200μ-thick second and third detectors. It uses a $\Delta E \times E$ detection scheme to measure particles of kinetic energy E satisfying the coincidence requirement in the first two detectors without triggering the third (anticoincidence) detector. The detected particles are analyzed using a four level priority-controlled pulse height analyzer (PHA). This analysis provides an unambiguous determination of the energetic proton and helium nuclei [9]. The small (but finite) backgrounds from the H and He PHA rates have not been subtracted; therefore, the H/He ratio is not computed if any of the rates are near background. We also use measurements from the Ulysses solar wind plasma experiment [10] and magnetometer [11], in order to identify the arrival of CMEs and stream interaction regions at the spacecraft. The occurrence of counterstreaming electron events has been used to determine the passage of CMEs over the spacecraft [12] and also those time intervals when Ulysses remains connected to shocks associated with stream interaction regions and CIRs [13].

Figure 1 shows the spin-averaged, three-day averaged fluxes of 0.5 to 1 MeV/nucl protons and helium together with the H/He ratio and the solar wind speed from 24 October 1992 to 24 October 1994 (top panel) and from 1 January 1999 to 1 January 2001 (bottom panel). During those two periods, Ulysses covered the heliocentric distances from 5.20 AU to 2.02 AU, and heliolatitudes from 19°S to 75°S, reaching a maximum latitude of 80.2 degrees in mid-September 1994 for the first orbit and at the end of November 2000 for the second orbit.

Solar Minimum: First Southern Pass

A recurrent pattern of particle flux increases with ~26-day periodicity was observed during the first southern pass. These enhancements were closely associated with the encounter of a recurrent high-speed (>750 km s^{-1}) solar wind stream. A compression interaction region, mostly bounded by forward-reverse shock (FS-RS) pairs, was formed at the leading edge of these high-speed streams (indicated by gray rectangles in the upper part of panel D in Figure 1). Rotation 15 at a latitude of ~34°S was the last rotation where a corotating FS-RS pair was observed. From then on, the spacecraft was immersed in the flow from the polar coronal hole and only reverse shocks were regularly observed. Above ~42°S, however, CIR-reverse shocks were observed only sporadically [14, 15]. Before the disappearance of recurrent CIR shocks, peaks in the ion intensity were observed close to both the forward and the reverse shocks. Even after no corotating shocks were seen (poleward of ~42°S), the recurrent peaks in the particle intensities were still observed, although the intensity decreased with increasing latitude. Superimposed on this regular feature were

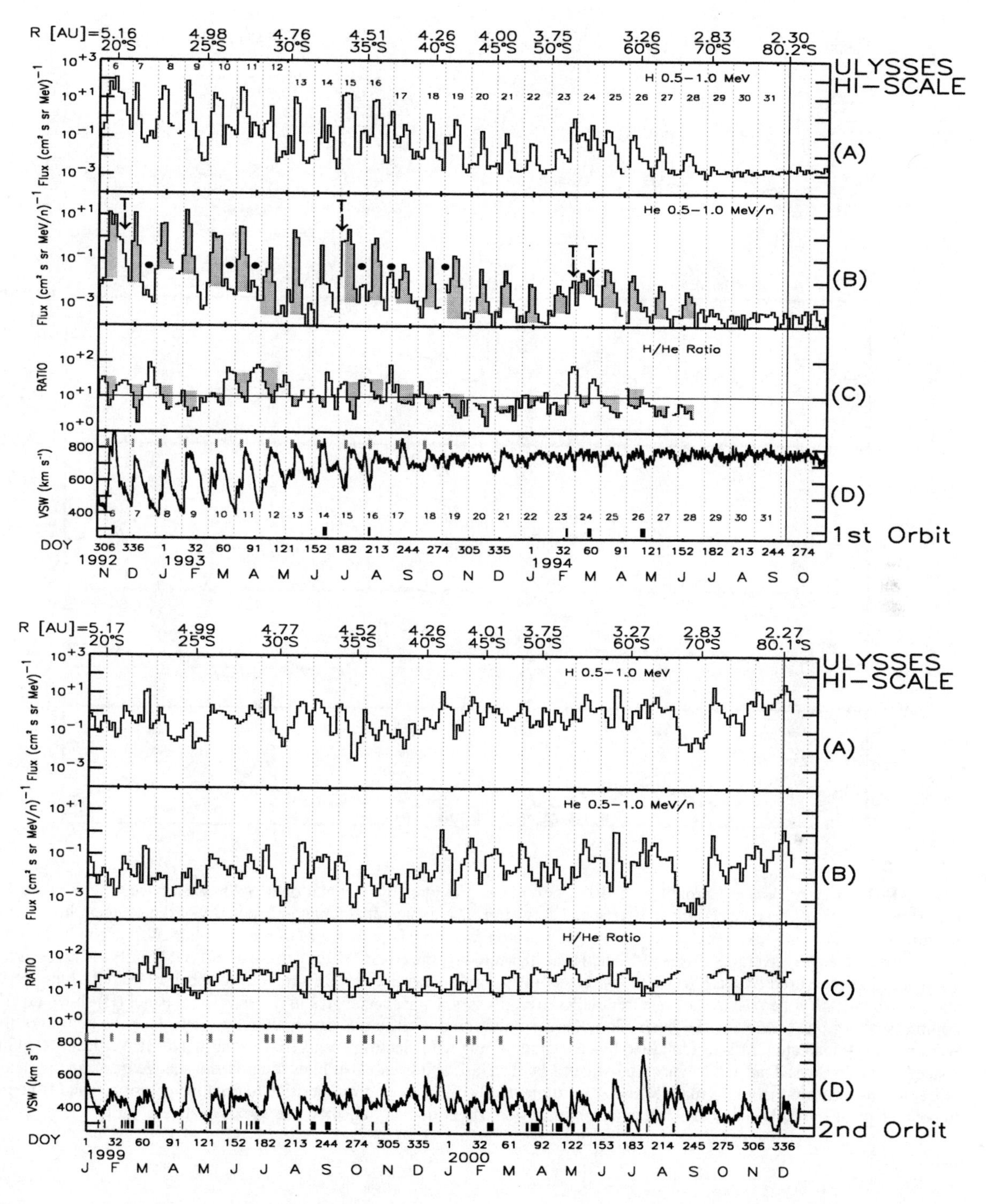

FIGURE 1. 3-day averages of the (A) 0.5-1.0 MeV proton flux; (B) 0.5-1.0 MeV/nucleon helium flux; and (C) H/He ratio as measured by the HI-SCALE instrument [9]. (D) Hourly averages of the solar wind speed as measured by the SWOOPS instrument [10]. The top set of four panels extends from 24 October 1992 to 24 October 1994 and the bottom set from 1 January 1999 to 1 January 2001. Shaded areas in the top panels identify the energetic particle events associated with CIRs. Upper rectangles in the solar wind panels (D) show the arrival of compression regions formed by the interaction between solar wind of different speeds; and black rectangles at the bottom of the solar wind panels (D) indicate the intervals associated with the passage of CMEs as identified by counterstreaming halo electron events in the solar wind [3]. Note that this identification has only been done up to a latitude of 75°S for the second orbit. The occurrence of transient particle events and "inter-events" are indicated by **T**'s and black dots in the top panel [16]. The vertical solid line marks the time when Ulysses reaches its maximum southern heliolatitude. Vertical dotted lines are spaced every 26 days. Rotation numbering scheme in the first orbit has been adapted from [14].

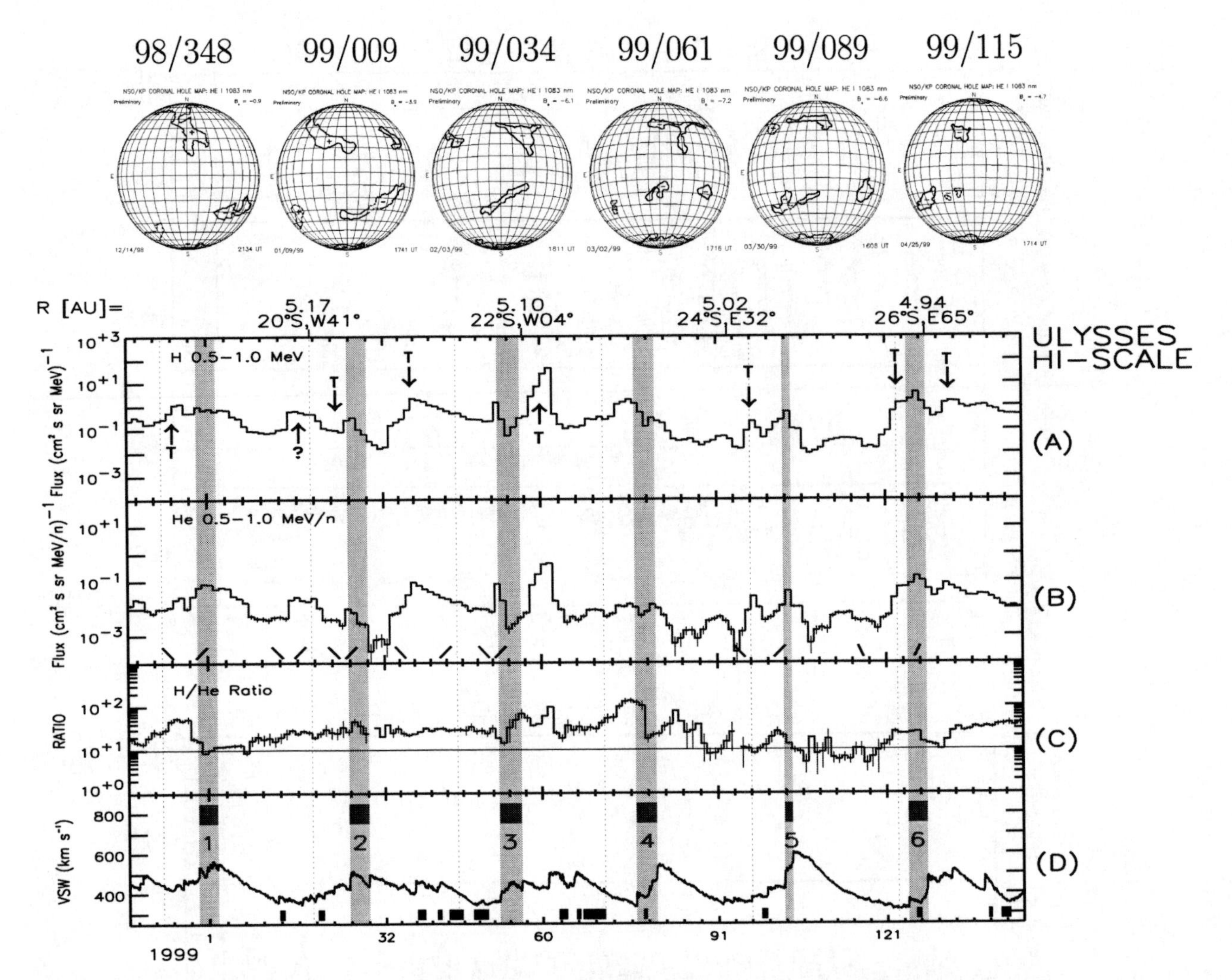

FIGURE 2. *Upper panel:* Coronal hole 1083 nm maps provided by the National Solar Observatory at Kitt Peak (solar Geophysical Data, 1999) taken on the day indicated at the top. The photospheric magnetic polarity of the southern hole is negative. *Bottom panel:* (A) Daily averages of the 0.5-1.0 MeV proton flux; (B) daily averages of the 0.5-1.0 MeV/nucleon helium flux; (C) ratio between the daily averages fluxes of H and He as measured by the HI-SCALE instrument [9]. (D) Hourly averages of the solar wind speed as measured by the SWOOPS instrument [10]. Time interval extends from 18 December 1998 to 21 June 1999. Gray shadow bars identify the passages of recurrent CIRs over Ulysses. Upper rectangles in the solar wind panel (D) show the arrival of compression regions formed by the interaction between solar wind of different speeds. Black rectangles at the bottom of the solar wind panel (D) indicate CMEs as identified by counterstreaming halo electron events in the solar wind [3]. Vertical dotted lines are spaced every 26 days. Transient events are indicated by **T**. Slash and back-slash lines identify the change of magnetic field sector at Ulysses as measured from the magnetometer instrument [11] on Ulysses; from positive to negative polarity (/) and from negative to positive polarity (\). The six CIRs that appear with a $\sim$26 day periodicity all show the same negative polarity.

several CME-related particle events (as in rotations 6, 14, 23, 24 and 26 indicated by black rectangles in the bottom part of the panel D in Figure 1), as well as particles associated with transient events of solar origin (as in rotations 6, 15, 23 and 24). Between rotations 7-8, 10-11, 11-12, 15-16, 16-17, and 18-19, Roelof et al. [16] noticed the occurrence of "inter-events" (indicated by black dots in panel B of Figure 1). Some of the characteristics of all these particle increases have been presented elsewhere [e.g., 5, 16] and the reader is referred to these papers for the salient details of these events.

Simnett et al. [17] performed a detailed study of the the evolution of the H/He ratio rotation by rotation. The main results of their analysis, apparent in the top panel of Figure 1, are that (1) the H/He ratio shows low ($\lesssim$10) values at the corotating events (indicated by shaded areas in panels B and C of the Figure 1), and high ($\gtrsim$50) values at the events of solar origin; (2) the H/He ratio decreases within the CIR, reaching a minimum value at the time of the reverse shock or during the crossing of the high-speed solar wind stream; (3) as Ulysses becomes immersed in the high-speed polar solar wind, the H/He ratio always remains below $\sim$10, with the exception of the two transient events in rotations 23 and 24. We note that the occurrence of an event of solar origin (as in rotations 23 and 24, or between rotations 10-11 and 11-12) may produce an increase of the H/He ratio which remains high for the subsequent solar rotations. We refer to [17] for further details on the evolution of the H/He ratio during the first southern pass.

Solar Maximum: Second Southern Pass

The second southern pass showed a completely different pattern. Ulysses observed both slow ($\sim$350 km s^{-1}) and an irregularly structured intermediate-speed ($\lesssim$650 km s^{-1}) solar wind [3]. The highest solar wind speed for the whole second southern pass ($\sim$750 km s^{-1}) was observed on day 193 of 2000 when Ulysses was at 62°S. Very low-speed ($\sim$300 km s^{-1}) solar wind was observed even when Ulysses was over the south pole. Although the speed difference between the slow and intermediate-speed solar wind was small, numerous stream interactions were well developed, many of which were bounded by FS-RS pairs [3]. While several of these interaction regions reappeared at roughly the solar rotation period over a few consecutive rotations (see for example the time interval shown in Figure 2), these structures were much less periodic than those observed in 1992-1993. A higher number of CME disturbances was also observed in the solar wind, with a larger rate at lower heliolatitudes, but they were still present at high latitudes [3].

In contrast to the first southern pass, energetic ion intensities fluctuated without any consistent pattern. Whereas at solar minimum the lowest ion fluxes occurred between two consecutive corotating events (unless there was an inter-event of solar origin), proton and helium fluxes at solar maximum were elevated throughout the second orbit. There was only one period between days 230 and 256 of 2000 when the proton fluxes were as low as one order of magnitude above background levels and helium fluxes were close to background levels (in that period we do not compute the H/He ratio). While solar minimum particle intensities were low or at background level at high heliolatitudes ($\gtrsim$70°S), at solar maximum several transient events were still observed even at 80°S. The elevated fluxes observed throughout the second southern pass indicate that the heliosphere was essentially completely populated at all heliolatitudes and heliolongitudes; the inner heliosphere was acting as a "reservoir" of low-energy ions [18]. The complicated temporal structures of the energetic particle intensities during the second southern pass make the classification of the fluxes in corotating or transient events difficult.

The H/He ratio usually maintained high ($\gtrsim$ 20) values throughout the second southern pass, with only occasional decreases below $\sim$10. There was no latitude dependence of the H/He ratio, because it remained high even at high heliolatitudes. Low values of the H/He ratio do not seem to be always correlated with high-speed streams. In fact, examination of the H/He ratio on a expanded time scale shows that the arrival of an interaction region or of an intermediate-speed stream does not always produce a decrease of the H/He ratio. Figure 2 shows in detail the first 4 months of 1999 when Ulysses observed a recurrence of six CIRs bounded by FS-RS pairs (gray shaded bars). The top panel of Figure 2 shows the small-scale southern coronal hole origin of the intermediate ($\sim$550 km s^{-1}) solar wind streams generating this sequence of CIRs. Ballistic projection of the solar wind and the negative magnetic field polarity observed during these streams support our identification. The arrival of CMEs after CIRs number 2 and 3 disturbed the structure of the solar wind streams observed at Ulysses. Numerous transient events (indicated by T's in panel A) occurred during that time interval. Near-relativistic electron flux increases observed on the ACE and Ulysses spacecraft together with solar observations were used to identify these events as being of solar origin [19].

Throughout the first four months of 1999, the H/He ratio never decreased below 10 (with the exception of the interval between CIR 5 and 6). However, the arrival of the CIRs number 1, 4, 5 and 6 did produce a decrease of the H/He ratio, with a clear minimum towards the reverse shock or during the crossing of the fast stream. So at least qualitatively, the behavior of the H/He ratio in CIRs during early 1999 was similar to that in 1992-1994. Exceptions to this behavior are CIRs number 2 and 3. The onset of a transient event within the CIR number 3 produced an increase of the H/He ratio at the reverse shock. The decaying phase of a transient event superimposed on CIR number 2 resulted in a high ($\sim$30) H/He ratio throughout the passage of this structure. The highest value of the H/He ratio during this sequence is found at the forward shock of the CIR 4, with a value higher than the one observed during most of the transient events. The same happened for example during the CIR on day 120 of 2000 when Ulysses was at 53°S. This CIR showed the highest (>100) H/He ratio observed at high latitudes, and clearly contrasted with the subsequent transient event on day 126 of 2000 which decreased the H/He ratio to values around $\sim$20.

DISCUSSION AND CONCLUSIONS

The main results of the comparison between the 0.5-1.0 MeV/nucleon H/He ratio measured during the first Ulysses' southern pass and the second southern pass are:

- particle fluxes at solar minimum are, on average, richer in helium than during solar maximum,
- whereas in solar minimum the H/He ratio allows us to distinguish between transient events and corotating events, during solar maximum such a distinction is not so clear,
- transient events of solar origin may be superimposed on CIR structures modifying the "normal" behavior of the H/He ratio, namely an increase around the forward shock and a decrease towards the reverse shock,
- the high-latitude heliosphere during solar maximum has characteristics similar to the heliosphere observed at the ecliptic plane. In particular, the H/He ratio does not show any latitudinal or radial dependence.

The different value of the H/He ratio observed for corotating events in solar maximum and solar minimum raises some questions about the seed particle population accelerated in CIRs. There are several heliospheric ion populations that are candidates for acceleration in CIRs, namely the thermal solar wind ions, a background population of solar energetic particles, and the interstellar and inner-source pick-up ions [20]. The low value of the H/He ratio observed in corotating events during solar minimum has been interpreted as a consequence of the pick-up He^+ acceleration in CIRs [21, 5]. It is well-known that pick-up ions are favored over solar wind ions for injection into acceleration mechanisms because of their higher energies in the solar wind frame [20]. However, since corotating streams are generally slower around solar maximum, and the maximum pick-up ion velocity scales directly with solar wind speed, their injection is less favored during solar maximum than in those cases when the wind speed is higher (i.e., during 1992-1994). Therefore, the lack of helium relative to protons during the second southern pass is due to a combination of the following factors: (1) the larger number of transient events, richer in protons, leads to a global filling of the heliosphere over a large range of heliolatitudes, and (2) the reverse shocks formed by the interaction between slow solar wind and intermediate solar wind are weaker and less efficient in accelerating the lower energy pick-up helium existing in this intermediate-speed solar wind. The combination of these two factors fits in the new picture of particle sources sketched by Mason [20] where several particle populations contribute to an interplanetary reservoir [18] of suprathermal ions. This reservoir has time-dependent inputs; during solar minimum the particles accelerated by CIRs are dominant, whereas during activity maximum the transient events of solar origin increase their contribution.

REFERENCES

1. Shields, J.C., et al., *J. Geophys. Res.* **90**, 9439-9453 (1985).
2. Richardson, I.G., et al., *J. Geophys. Res.* **98**, 13-32 (1993).
3. McComas, D.J., Gosling, J.T., and Skoug, R.M., *Geophys. Res. Lett.* **27**, 2437-2440 (2000).
4. Scholer, M., et al., *Astrophys. J.* **227**, 323-328 (1979).
5. Simnett, G.M., Sayle, K.A., and Roelof, E.C., *Geophys. Res. Lett.* **22**, 3365-3368 (1995).
6. Barnes, C.W., and Simpson, J.A., *Astrophys. J.* **210**, L91-L96 (1976).
7. Fränz, M., et al., *Geophys. Res. Lett.* **26**, 17-20 (1999).
8. Maclennan, C.G., and Lanzerotti, L.J., *Space Sci. Rev.* **72**, 297-302 (1995).
9. Lanzerotti, L.J., et al., *Astron. Astrophys.* **92**, 349-363 (1992).
10. Bame, S.J., et al., *Astron. Astrophys.* **92**, 237-266 (1992).
11. Balogh, A., et al., *Astron. Astrophys.* **92**, 221-236 (1992).
12. Gosling, J.T., et al., *J. Geophys. Res.* **92**, 8519-8535 (1987).
13. Gosling, J.T., et al., *Geophys. Res. Lett.* **20**, 2335-2338 (1993).
14. Bame, S.J., et al., *Geophys. Res. Lett.* **20**, 2323-2326 (1993).
15. Gosling, J.T., et al., *Space Sci. Rev.* **72**, 99-104 (1995).
16. Roelof, E.C., Simnett, G.M., and Armstrong, T.P., *Space Sci. Rev.* **72**, 309-314 (1995).
17. Simnett, G.M., et al., *Space Sci. Rev.* **72**, 327-330 (1995).
18. Roelof, E.C., et al., *Geophys. Res. Lett.* **19**, 1243-1246 (1992).
19. Lario, D., Roelof, E.C., Forsyth, R.J., and Gosling, J.T., *Space Sci. Rev.*, in press (2001).
20. Mason, G.M., in *Acceleration and transport of energetic particles observed in the heliosphere: ACE 2000 Symp.*, Eds. R.A. Mewaldt et al. (AIP Conf. Proc. 528) 234 (2000).
21. Gloeckler, G., et al., *J. Geophys. Res.* **99**, 17637-17643 (1994).

Energetic Particle Composition Measurements at High Heliographic Latitudes Around the Solar Activity Maximum

M. Y. Hofer*, R. G. Marsden*, T. R. Sanderson* and C. Tranquille*

*Space Science Dept. of ESA, ESTEC, P.O. Box 299, 2200 AG Noordwijk, The Netherlands.

Abstract. From 1999 until the end of 2000 the Ulysses spacecraft was traveling back towards the southern solar polar region from about 5 AU to less than 2.5 AU. In this time period the solar activity was close to the maximum. The results of a preliminary analysis of energetic particle composition data covering the energy range 4-20 MeV/n recorded by the COSPIN/LET instrument are shown. We attempt to use the composition signatures to identify the sources of energetic particles observed at high latitudes.

INTRODUCTION

Following aphelion passage in 1998 the Ulysses spacecraft began its second climb to high southern heliolatitudes, reaching its maximum latitude of 80.2 ° in November 2000. During this time period the spacecraft's heliocentric distance decreased from about 5 AU to less than 2.5 AU. In contrast to the first south polar passage in 1994 (e.g. [1], [2]), which took place as the level of solar activity was approaching the minimum of cycle 22, the recent high-latitude observations correspond to near-maximum activity conditions with large transient phenomena. In 1994, the 1-10 MeV/n particle fluxes measured at Ulysses consisted mainly of corotating interaction region (CIR)-related events and of the anomalous cosmic ray (ACR) [3] component, what was obtained based on the elemental abundance ratios. A decrease of the proton to alpha ratio towards around 10 is often used for the identification of particles accelerated at CIRs (e.g. [4], [5]). At the beginning of solar energetic particle (SEP) events the $^3He/^4He$ and the Fe/O ratios are often found to increase as a result from resonant wave-particle interaction in the flare plasma (see [6], [7]).

In this preliminary analysis the low energy fluxes recorded in 1999 and 2000 by the COSPIN/LET instrument on the Ulysses spacecraft are plotted in order to select two major events for a detailed analysis. For these selected time intervals the elemental abundance ratios are calculated to gain insight into the possible origin of the particles and the likely acceleration history. In particular, we are interested in identifying time intervals dominated by SEP events or CIR passages.

DESCRIPTION OF THE DATA

The low energy particle data used in this study is from the Low Energy Telescope (LET), one of the five telescopes in the Cosmic Ray and Solar Particle Investigation (COSPIN) onboard the Ulysses spacecraft [8]. The LET instrument records the fluxes and the composition of SEPs and of low energy cosmic ray nuclei from hydrogen up to iron over a range of energies from $\sim$1 MeV/n to 75 MeV/n (For details see [8], [3]).

RESULTS

In Figure 1 the proton and alpha intensities (upper panel) recorded by COSPIN/LET on Ulysses and the proton/alpha ratio in the in the $\sim$1 MeV/n range (proton: 1.2-3.0 MeV; alpha: 1.5-5.0 MeV/n) are plotted for the years 1999 and 2000. In addition, a panel showing the latitude of the spacecraft's trajectory (for the distance see [3]) is added to the figure. In 1999 and 2000 the spacecraft is traveling for the second time towards the highest southern heliolatitude of the trajectory reached in November 1999. The horizontal bars in the first panel in June 1999 and in July 2000 mark the time intervals selected for the current analysis.

The maximum proton intensity (1.2-3.0 MeV) in 1999 is about 60 protons/cm^2/s/sr/MeV at the beginning of July. In 2000 there are more than five events observed with maximum intensity above this level. Furthermore, the minimum value in the proton intensity recorded at the end of September in 1999 is never reached in 2000. At the end of 2000 the spacecraft is located above the southern solar pole where major particle events are observed.

CP598, *Solar and Galactic Composition,* edited by R. F. Wimmer-Schweingruber

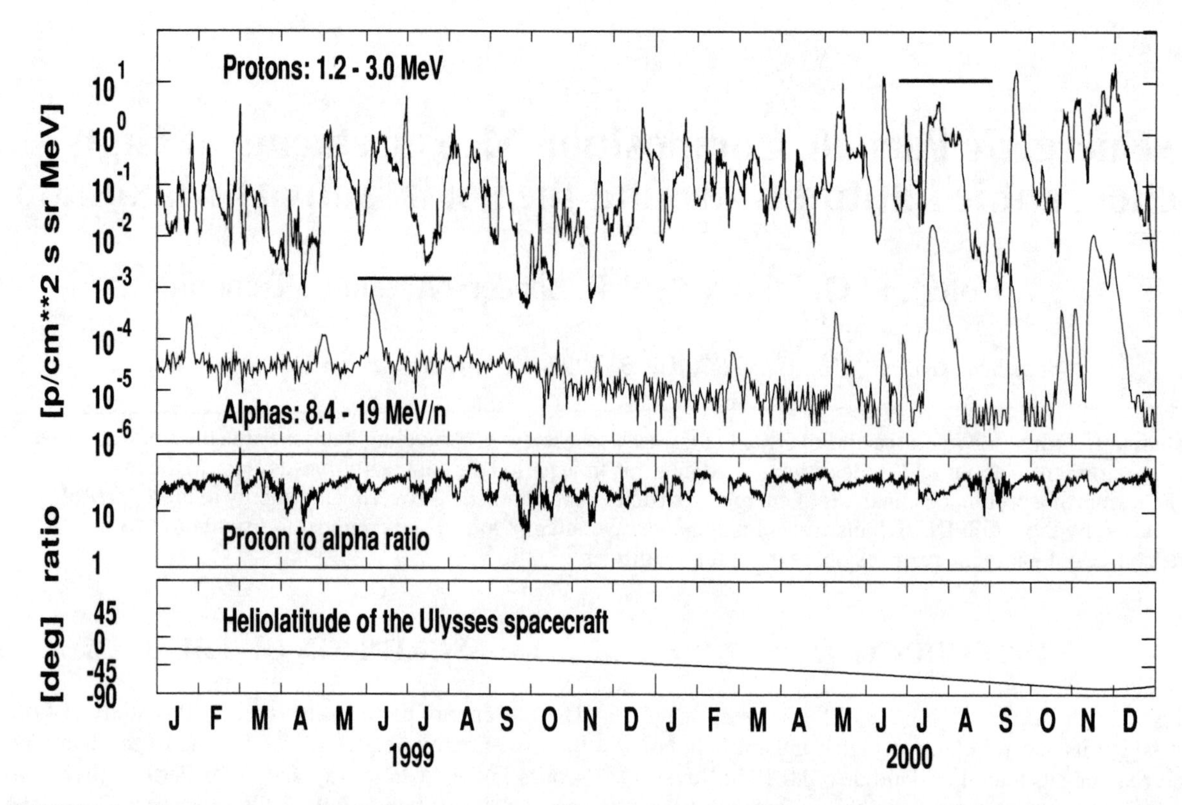

FIGURE 1. The proton (1.2-3.0 MeV) and the alpha (8.4-19 MeV/n) intensities, and the proton(1.2-3.0 MeV) / alpha(1.0-5.0 MeV/n) ratios recorded by COSPIN/LET on the Ulysses. The heliolatitudes of the trajectory of the Ulysses spacecraft are plotted in the third panel. The horizontal bars in June 1999 and in July 2000 mark the selected time intervals.

In 1999 the smallest proton/alpha ratios are lower than those in 2000, and the proton/alpha decreases are more frequent than in 2000. Based on these parameters there seems to exist a tendency for an increase of the activity from the beginning of 1999 towards the end of 2000.

The high variability in the particle data is accompanied by frequent changes in the magnetic field, in the solar wind velocity (for details please consult the Ulysses data archive: *http://helio.estec.esa.nl/Ulysses/archive/*) and by a complex shape of the heliospheric current sheet (HCS). During the entire time interval the HCS is highly tilted resulting in low speed streams coming from streamer belts and a well defined sector structure observed even over the pole [9].

Two large events in June 1999 and July 2000 are selected for the further analysis as marked by the horizontal bars in the first panel of Figure 1.

In Figure 2, the the oxygen intensity and the three day averaged abundance ratios for helium, carbon, nitrogen, neon and iron with respect to oxygen are shown for the selected time intervals and energy ranges as indicated. The ratios are represented in both panels of Figure 2 by symbols and the corresponding errors by vertical lines. In case of a single count, (which may cause an error larger than the value), the lower part of the error bar is not drawn. The black arrows at the bottom of the figure mark the time of the shock occurrence as determined from magnetic field and solar wind parameters ([10], [11]).

The upper panels in Figure 2 show the oxygen intensity and the elemental abundance ratios for the days 140-210 in 1999. Only three significant Fe/O values are obtained for this event.

The lower panels of Figure 2 show the same results for the selected (mainly gradual) event from day 180 until 250 in 2000 having less variations in the abundances than the event in June 1999. From day 220 on in 2000 the particle flux decreased rapidly, and therefore no statistically significant abundances can be derived.

Generally, the abundance ratios of He/O evolve in a mirrored way compared with those of the heavier ions what is visible in both events.

In Table 1 the averaged values of the statistically significant elemental abundances within a common regime in the selected time interval for helium, carbon,

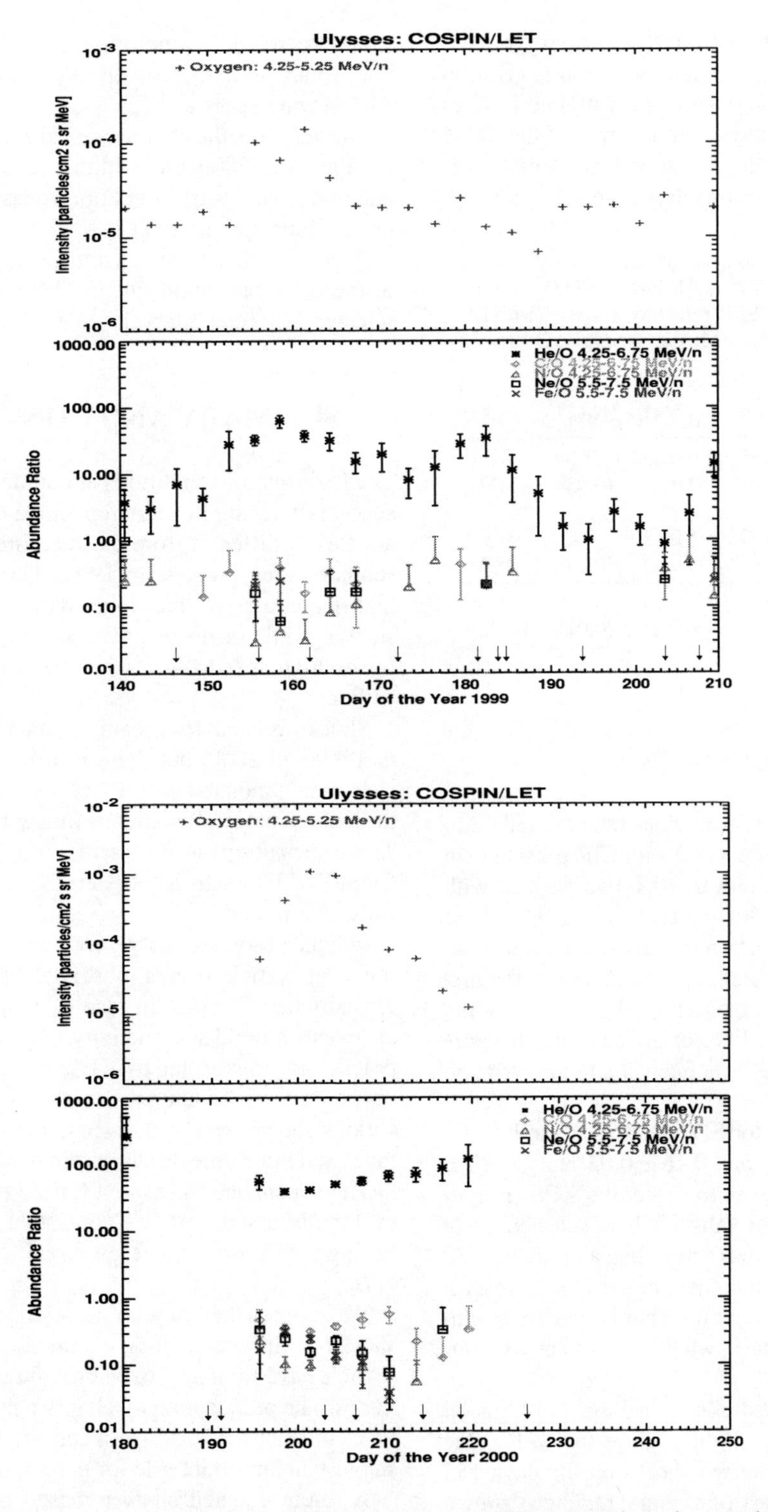

FIGURE 2. Three day averaged elemental abundances and corresponding oxygen intensity recorded during the selected events in 1999 and 2000. The values are represented by the corresponding symbols and the errors by vertical lines. The black downward arrows mark the time of the shock occurrence. Upper panels: Three day averaged values of the abundances of helium, carbon, nitrogen, neon, and iron with respect to oxygen from day 140 to 210 in 1999. Lower panels: Three day averaged values of the abundances of helium, carbon, nitrogen, neon, and iron with respect to oxygen from day 180 to 250 in 2000.

nitrogen, neon and iron with respect to oxygen are listed. The corresponding standard deviation is given in parentheses. The time intervals of each regime in days are given in the first column. The last row of the Table contains SEP values from Mason and Sanderson [12] derived in 1991, before Ulysses data were available.

TABLE 1. Averaged elemental abundance ratios during the selected time intervals in 1999 and in 2000 recorded by COSPIN/LET and the SEP reference values from [12]. (He/O, C/O, N/O: 4.25 - 6.75 MeV/n; Ne/O, Fe/O: 5.5-7.5 MeV/n)

	He/O	C/O	N/O	Ne/O	Fe/O
151-170, 1999	32 (6)	0.25 (0.05)	0.06 (0.015)	0.14 (0.02)	0.22 (0.02)
194-209, 2000	48 (3.8)	0.39 (0.04)	0.12 (0.02)	0.2 (0.04)	0.17 (0.04)
SEP [12]	55.2 (3)	0.48 (0.02)	0.13 (0.01)	0.15 (0.01)	0.16 (0.02)

DISCUSSION

Marsden et al. [4] used a proton/alpha ratio around 10 to identify particle fluxes associated with CIR passages. In 1999 there are about six and in 2000 two periods with values around this level. Gosling et al. [13] refer to these structures as stream interaction regions (SIR). Lario et al. [14] found a tendency for a recurrent structure in the first half of 1999. Nevertheless, based on the corresponding shock occurrence there is little or no recurrent structure present for the rest of the time interval and therefore no CIR could be identified.

The Fe/O abundances for SEP and CIR given by Mason and Sanderson [12] are 0.16 ± 0.02 and 0.097 ± 0.011, respectively, and even lower for the ACR component. The average of Fe/O values in both selected event are above the 10 percent level, indicating a probable SEP source. Cohen et al. [15] claim that most of the large SEP events in 1999 and 2000 are rather not Fe-enriched with respect to the coronal values, what is consistent with our findings.

The first of the selected events in June 1999 has an average Fe/O value only slightly above the SEP value as visible in Table 1. The two shocks on the days 155 and 161 mark the boundary of a central regime as shown in the upper panel Figure 2. The peak in the He/O ratio around day 181 occurs in vicinity of three shocks on the days 181, 183, and 184.

In 2000, in particular, around day 200 two high Fe/O numbers are derived which could be associated with the shock on day 203. During the following 10 days the values decrease. Reames et al. [16] interpret at least the initial part as a sign of an evolving non-Kolmogorov Alfven wave spectra.

Furthermore, the abundance values found in this study for the year 2000 are within the errors equal to the maximum values of in-ecliptic measurements recorded by the Wind spacecraft [17].

A possible small ACR contribution might be present in the data, but could not be distinguished during the selected SEP time intervals.

SUMMARY AND CONCLUSIONS

The heliospheric conditions encountered by the Ulysses spacecraft during its second passage over the south pole are clearly different from those recorded during the first southern solar passage in 1994. Throughout the recent high-latitude pass, the solar wind is highly variable, showing little or no recurrent structure, an no evidence of the high speed flows from polar coronal holes seen earlier.

Shocks related to stream interactions are also seen in 1999 and 2000, but these did not have the recurrent character associated with CIRs. Gosling et al. [13] refer to these structures as stream interaction regions (SIR). The contributions to the particle fluxes recorded by the COSPIN/LET instrument from these shocks are not yet fully determined.

Enhanced solar activity gave rise to a large number of transient particle events, observed up to highest southern latitudes. The occurrence of major energetic particle events around solar activity maximum over the solar poles is not unexpected based on the general understanding of the solar evolution. But, the corresponding observations shown here with even surprisingly large amplitudes and large time durations are made for the first time.

The abundances measured for the two mainly gradual events discussed here are consistent with the particles having a SEP origin and are equal to in-ecliptic values [17].

We suggest that they are accelerated by CME driven shocks. If the acceleration occurs at lower latitudes, the particle have somehow to be brought up to the higher latitudes. The propagation mechanism that allowed the particles to reach Ulysses is not certain, but the rapid onset suggest an important role for perpendicular diffusion.

A future detailed analysis based on this preliminary identification of the SIR and SEP dominated time intervals might give an even deeper understanding of the ongoing acceleration processes for particles in the MeV energy range.

ACKNOWLEDGMENTS

We acknowledge the use of the Ulysses Data System in the preparation of this paper. We thank Marek Szumlas for the help in the preparation of the figures.

REFERENCES

1. Simpson, J.A., et al., *Science*, **268**, 1019-1023, (1995).
2. Sanderson, T.R., et al., *Geophys. Res. Lett.*, **22(23)**, 3357-3361 (1995).
3. Tranquille, C., Marsden, R.G., and Sanderson, T.R., in *Solar and Galactic Composition*, edited by R.F. Wimmer-Schweinegruber, (AIP Conference Proceedings), in press (2001).
4. Marsden, R.G., et al., *Adv. Space Res.*, **13(6)**, 95-98 (1993).
5. Simnett, G.M., Sayle, K.A., and Roelof, E.C., *Geophys. Res. Lett.*, **22(23)**, 3365-3369 (1995).
6. Reames, D.V., et al., *Ap. J.*, **491**, 414-420 (1997).
7. Reames, D.V., *Space Science Rev.*, **90**, 413-491 (1999).
8. Simpson, J.A., et al., *Astron.&Astrophys. Suppl. Series*, **92(2)**, 365-399 (1992).
9. Sanderson, T.R., et al., The influence of the Sun's magnetic field on energetic particles at high heliospheric latitudes, submitted to *Geophys Res. Lett.* (2001).
10. Balogh, A., et al., *Astron.&Astrophys. Suppl.*, **92(2)**, 221-236 (1992).
11. Bame, S.J., et al., *Astron.&Astrophys. Suppl. Series*, **92(2)**, 237-265 (1992).
12. Mason, G.M. and Sanderson, T.R. *Space Science Rev.*, **89**, 77-90 (1999).
13. Gosling, J.T., et al., *Space Science Rev.*, in press (2001).
14. Lario, D., et al., High Latitude Ulysses Observations of the H/He intensity ratio under solar minimum and solar maximum conditions, in *Solar and Galactic Composition*, edited by R.F. Wimmer-Schweinegruber, (AIP Conference Proceedings), in press (2001).
15. Cohen, C.M., et al., *AGU Spring Meeting*, Abstract Booklet, (2001).
16. Reames, D.V., et al., *Ap. J.*, **531**, L83-L86 (2000).
17. Reames, D.V., Ng, C.K., and Tykla, A.J., Heavy ion abundances and spectra and the large gradual solar energetic particle event of 2000 July 14, accepted for publication in *Ap. J. Lett.*, (2000).

Ulysses Measurements of the Solar Cycle Variation of ~2-40 MeV/n Ions in the Inner Heliosphere

C. Tranquille*, R.G. Marsden* and T.R. Sanderson*

*Space Science Department of ESA, ESTEC, Noordwijk, The Netherlands

Abstract. We present measurements of energetic (~2 to 40 MeV/n) ion spectra and abundance ratios using the COSPIN/LET instrument flown onboard Ulysses. These measurements span almost a complete solar cycle in time and cover an extensive region of the inner heliosphere (heliocentric distance from 1 to 5 AU and heliographic latitude from the ecliptic to ±80°). We are able to characterise ions associated with solar energetic particle events and those that belong to the anomalous cosmic ray (ACR) component. Abundance ratios measured in the distinct particle regimes encountered throughout the Ulysses mission are collated and compared to similar measurements made by other spacecraft. We investigate in detail the evolution of the ACR oxygen energy spectrum during solar minimum and relate our observations to models of ACR transport.

INTRODUCTION

Ulysses, launched in October 1990, has provided continuous measurements of particles and fields in the inner heliosphere for almost a complete solar cycle. The mission began during the declining phase of the last (#22) solar cycle, with the in-ecliptic transit to Jupiter lasting approximately 1.5 years. During this leg of the mission, solar activity was high resulting in the most intense levels of charged particle fluxes seen at Ulysses to date [1]. In contrast, the first traversals of the southern [2] and northern solar poles [3] took place near solar minimum, with Ulysses immersed in steady, fast solar wind [4] originating from well formed coronal holes located at high heliolatitudes on the sun. Currently, the spacecraft is climbing to the northern polar regions of the sun for the second time, but on this occasion during the active phase of the new (#23) solar cycle [5]. When Ulysses returns to the northern solar pole in late 2001, it will have accumulated in situ measurements of the inner heliosphere to heliolatitudes of ±80°, over a complete eleven year solar cycle.

A major goal of the Ulysses mission is to measure the latitudinal variation of charged particles in the inner heliosphere, and to test models of their production and transport. The three main constituents of energetic ions in the inner heliosphere are 1) solar energetic particles (SEP), 2) low energy particles accelerated in the interplanetary medium (for example by travelling shocks [6] or by corotating interaction regions [7]) and 3) cosmic rays (composed primarily of galactic cosmic rays (GCR) and the anomalous cosmic ray (ACR) component).

Ion intensities during periods of maximum solar activity are dominated by solar energetic particles. It is well known that ACR and GCR ions are subject to solar modulation in the heliosphere, leading to larger intensities at solar minimum than at solar maximum [8].

The origin of ACR ions has been shown to be the neutral interstellar gas [9], which contains atoms having a high first ionization potential, such as helium, nitrogen, oxygen, neon and argon. These atoms, having zero electric charge, are able to penetrate into the inner heliosphere with relative ease. During their transport, they become singly ionized by solar ultraviolet radiation or by charge exchange with the solar wind. Once charged, they are picked up by the solar wind and convected out into the distant heliosphere [10], where they become energised by interactions with the termination shock [11]. The ions then diffuse inwards and experience the effects of convection and drift. The direction of the gradient drift is determined by the solar magnetic dipole polarity [12]. For the previous solar cycle (A>0), ACR ion propagation was characterised by positive radial and latitudinal gradients, resulting from preferential drift inwards and downwards from the heliospheric poles to the equatorial regions [13, 14]. During the current (since mid 2000) magnetic solar cycle (A<0), this configuration is reversed. Low intensities of ACR protons and carbon have also been measured in the intermediate heliosphere [15].

Measurements of the latitudinal gradient of ACR ions, including oxygen, have been reported extensively for the first orbit of Ulysses [16, 17, 18], when solar minimum conditions prevailed. A modest latitudinal gradient of a few percent per degree was found for ACR ion species

CP598, *Solar and Galactic Composition*, edited by R. F. Wimmer-Schweingruber

over the relevant energy range. The latitudinal gradient was found to be strongest at low latitudes and weakest at high latitudes. This is in contrast to GCRs which show small latitudinal gradients in the streamer belt (between 0° and 30°) [19]. The results for ACR ions are consistent with an independent analysis made using measurements from the SAMPEX, Pioneer and Voyager spacecraft [20]. A radial gradient of approximately 15%/AU, derived from a previous study using IMP and Pioneer data out to a distance of 15 AU [21], has been shown to be consistent with Ulysses measurements [22].

The Ulysses results also point to an asymmetry between the level of ACR intensities measured for the two hemispherical regions of the inner heliosphere [17], with greater fluxes seen in the northern hemisphere. A similar asymmetry was also found for galactic cosmic rays [23, 24]. A possible explanation for such an asymmetry may be a southward displacement of the heliospheric current sheet [25] resulting in a weaker magnetic field strength above the sun's magnetic equator.

In this paper we investigate the energy spectra and abundance ratios of ~2-40 MeV/n ions measured by the COSPIN/LET instrument throughout the complete mission, and compare the values to previous measurements made by other spacecraft. We examine in detail the profile of the ACR oxygen ion energy spectra, and infer some properties of the transport of these ions in the inner heliosphere.

INSTRUMENTATION

The data used in this analysis were obtained from the Low Energy Telescope (LET) of the Cosmic Ray and Solar Particle Investigation (COSPIN) flown onboard the Ulysses spacecraft [26]. The LET instrument measures the flux, energy spectra and ion composition of solar energetic particles and low energy cosmic ray nuclei in the energy range of ~1 MeV/n to ~50 MeV/n. The instrument uses the standard dE/dx versus E technique to provide pulse height analysis (PHA) information allowing the identification of the chemical species and energy of individual particles. The telescope is also able to provide counting rate information of protons, alpha particles and groups of particle species in fixed energy channels. The present study will focus on PHA data of ~2-40 MeV/n ions.

Figure 1 is a pulse height matrix for the complete mission constructed from a combination of two of the LET detectors (D1 and D2). The inset shows the matrix for protons and alpha particles (having significantly greater statistics) separately from the remaining ions in the main diagram. The box-like boundary of the helium track is an artefact of the PHA analysis software. This figure shows the ability of the instrument to differentiate ion species up to iron, with negligible shift in the detector response throughout the ten year lifetime of the mission. Analysis of the PHA charge histograms also confirms the expected performance of the LET instrument.

OBSERVATIONS AND DISCUSSION

Figure 2 is a colour spectrogram showing 10 day averages of oxygen intensities measured in 40 energy channels (of equal logarithmic width) from 1 to 100 MeV/n, throughout the complete Ulysses mission. Only channels between ~4 MeV/n and ~40 MeV/n provide valid oxygen intensities, and this is a restriction of the LET design and geometry. Intensities within non-zero energy bins have been interpolated with a third order polynomial to smooth the spectrogram. In addition to the time axis, a horizontal axis has been included to show the radial and latitudinal motion of the spacecraft, and important milestones, such as perihelion, aphelion and maximum southern and northern heliographic latitude, have been marked. The colour coding of the intensities has been chosen to accentuate a selected portion of the full dynamic range of the instrument. This was done to enhance details in the energy spectra of oxygen ions at solar minimum, at the expense of the high flux transient events seen during the first two years of the mission and during the Jupiter flyby.

The spectrogram clearly shows two distinct oxygen populations. At solar maximum, immediately after the launch of Ulysses, the oxygen population is characterised by discrete, high intensity events. These are oxygen ions of solar origin, which have been possibly accelerated close to the sun in the corona or alternatively in interplanetary space by travelling shocks or by CIRs. As solar activity decreases after the Jupiter flyby, the frequency of the transient events becomes significantly reduced. With the exception of a few isolated events during solar minimum, oxygen ions of solar origin do not return until after 1998, when the current solar activity cycle begins its active phase. The second population of oxygen ions seen in Figure 2 is that belonging to the ACR component. This is characterised in the spectrogram by a continuous distribution of intensity over the energy range of the LET instrument, lasting from mid 1993 for a five year period coincident with the last solar minimum. It is evident that ACR oxygen ions are subject to latitudinal gradients, with peak intensities seen when Ulysses was at maximum heliolatitudes, as expected for the configuration of the solar magnetic dipole for the previous magnetic cycle. There is also evidence for higher intensities in the northern hemisphere, over the complete energy range of the instrument. Radial gradients are also present in the

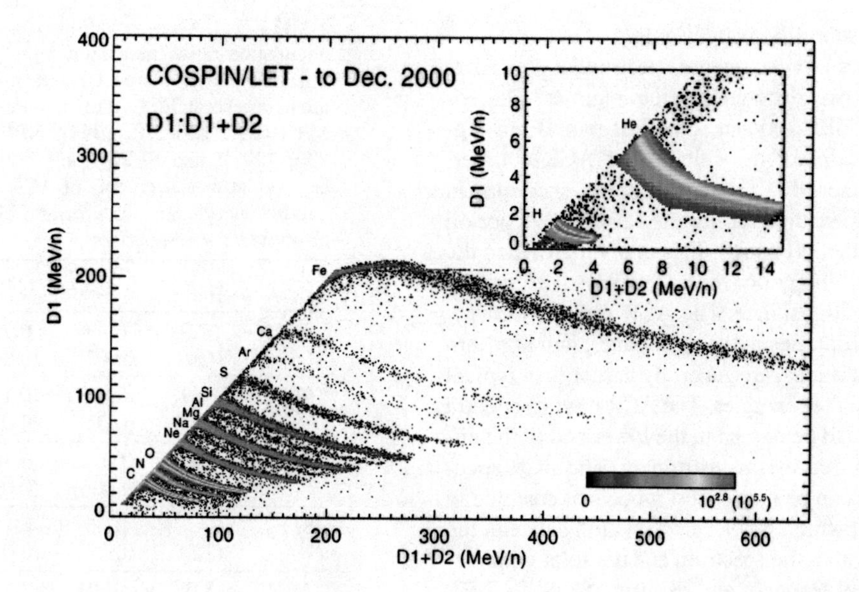

FIGURE 1.

The D1 versus D1+D2 pulse height matrix for the COSPIN/LET instrument measured from launch to December 2000.

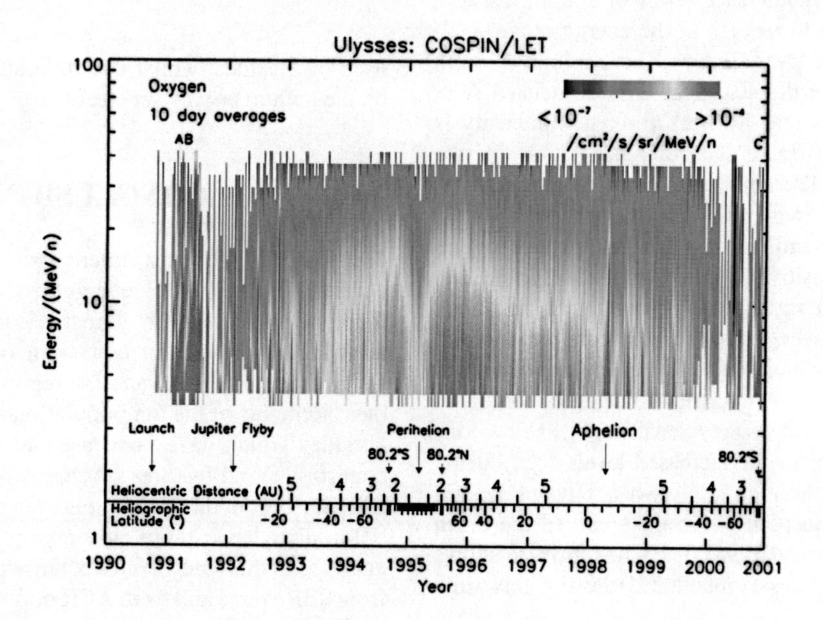

FIGURE 2. Colour spectrogram showing 10 day averages of oxygen intensities measured throughout the complete Ulysses mission. The SEP events (A, B and C) referred to in Figure 3 are marked.

data but are not so easily recognisable.

Figure 3 shows oxygen energy spectra that are representative of the two different particle regimes. The energy spectra of SEP oxygen ions (left panel) are significantly higher in intensity than their ACR counterparts at lower energies. The most intense spectrum in this figure was measured during the March 1991 period (plus symbol in the left panel). This interval provided the highest fluxes of energetic (MeV) particles seen during the Ulysses mission to date. The ACR spectra at maximum southern (triangles in the right panel) and northern (squares) heliolatitudes, are generally harder than typical SEP spectra at higher energies. The ACR oxygen spectra also exhibit a broad peak seen in the lower end of the energy range measured by the instrument. The ACR spectrum for oxygen can be used to test models of cosmic ray propagation [27] which predict a correlation between the energy of the peak in the spectrum and the solar wind velocity. A study by Heber et al. [28] using COSPIN/LET and EPAC data from Ulysses provided evidence for the existence of such a correlation.

Figure 4 shows abundance ratios of helium, carbon, nitrogen and neon to oxygen in the energy range of 4-8 MeV/n, averaged over 40 days. Vertical lines through each point indicate the statistical error associated with the measurements. The 4-8 MeV/n oxygen intensity is also shown in this figure as a reference profile for the abundance ratios. The abundance ratios of nitrogen and neon to oxygen remain fairly constant throughout the complete mission, unlike those for helium and carbon which are very sensitive to solar activity and latitudinal effects. The Ne/O ratio however does increase above the constant level when transient fluxes are measured at Ulysses. The abundance ratios of helium and carbon to oxygen show distinct boundaries for the population of ACR ions. The He/O is very sensitive to helium ions of solar origin and to the increased levels seen during the so-called fast latitude scan (when Ulysses passed through the ecliptic plane from the south to the north pole between 1994 and 1995). The C/O ratio also shows an increase (albeit less pronounced) during this time interval.

Table 1 summarises the abundance ratios seen for the ions presented in Figure 4. Values are given for the ecliptic transfer (ECL) the first southern (SPP1) and northern (NPP1) polar passes (both during solar minimum conditions) and for the second southern (SPP2) polar pass (during solar maximum). Periods of SEP events are excluded from the average polar pass values, as are measurements taken during the fast latitude scan. The abundance ratios measured during the first pair of polar passes are consistent with ACR values documented by Reames [29], with the exception of the He/O ratio. A possible reason for the discrepancy in the He/O ratio is the mismatch between the LET energy range of 4-8 MeV/n and that used by Reames, which could be significant depending on the helium energy spectrum.

TABLE 1. Average values of 4-8 MeV/n ion abundance ratios measured by Ulysses during different phases of the Ulysses mission (errors are in brackets): ECL (launch to February 1992); SPP1 (days 230-270, 1994); NPP1 (days 190-230, 1995); and SPP2 (days 310-350, 2000). Equivalent measurements of ACR ions (at comparable energies) as documented by Reames are also given for comparison.

	He/O	C/O	N/O	Ne/O
ECL	67.9	0.47	0.15	0.18
	(6.0)	(0.07)	(0.04)	(0.04)
SPP1	1.69	0.02	0.13	0.10
	(0.17)	(0.01)	(0.03)	(0.03)
NPP1	0.96	0.013	0.14	0.08
	(0.08)	(0.006)	(0.02)	(0.02)
SPP2	33.2	0.26	0.12	0.11
	(2.8)	(0.04)	(0.04)	(0.03)
ACR	5.0	< 0.01	0.12	0.07
	(1.0)		(0.01)	(0.01)

CONCLUSIONS

The COSPIN/LET instrument provides a coherent and continuous data set of energetic ion intensities in the inner heliosphere over almost a complete solar cycle, allowing the long term behaviour of these particles to be investigated. The data also represent a unique set of measurements of the ion population at high heliographic latitudes which have been used in previous studies to confirm general features of charged particle production and transport in the global heliosphere.

Ion abundance ratios and energy spectra have been analysed in this study to characterise particles originating from SEP events and from ACR populations as a function of the solar activity cycle. The recent series of energetic solar events seen since mid 2000 have provided significant heavy ion intensities at high latitudes which need to be investigated in detail. As the current active phase in the solar cycle comes to an end, the ACR component is expected to reappear and will be measured by the LET instrument.

ACKNOWLEDGMENTS

We acknowledge the use of the Ulysses Data System in the preparation of this paper.

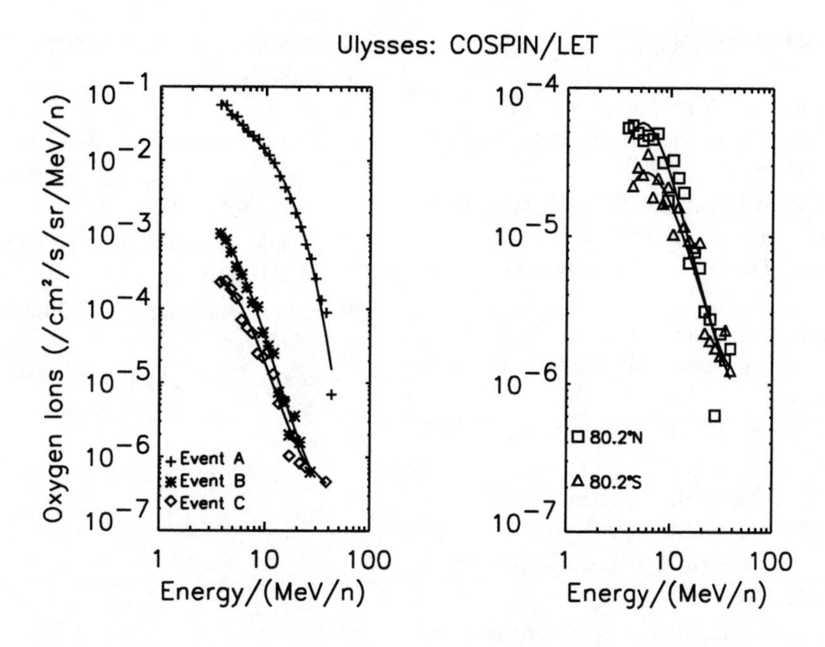

FIGURE 3. Oxygen ion energy spectra (10 day averages) measured by the COSPIN/LET instrument during SEP events (left panel) and at maximum southern and northern (right panel) heliolatitudes. The continuous lines represent polynomial fits to the spectra.

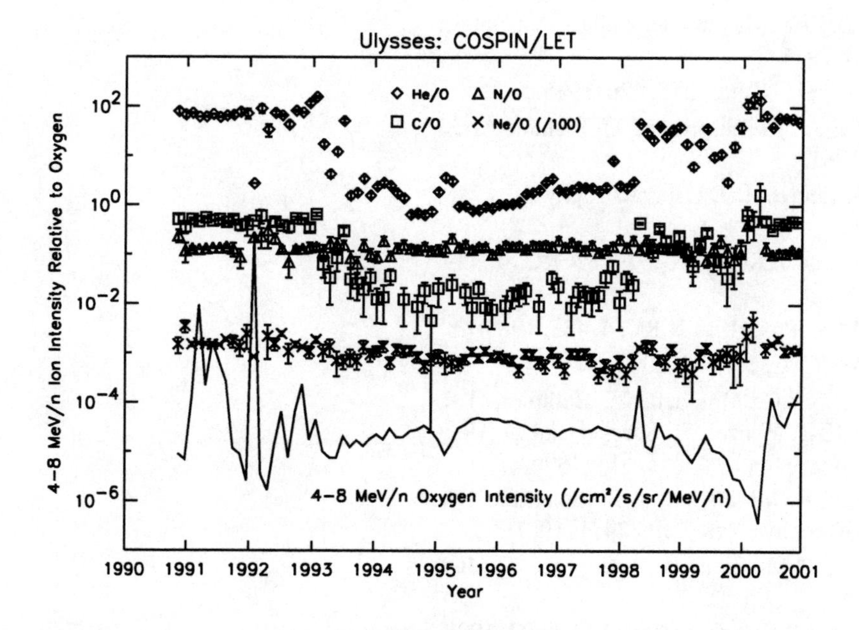

FIGURE 4. 40 day averages of abundance ratios of helium, carbon, nitrogen and neon (divided by 100) ions to oxygen, in the energy range of 4-8 MeV/n. Also shown is the 4-8 MeV/n oxygen intensity profile.

REFERENCES

1. Sanderson, T.R., R.G. Marsden, A.M. Heras, K.-P. Wenzel, J.D. Anglin, A. Balogh, R. Forsyth, Geophys. Res. Lett., **19**, 1263, 1992.
2. Sanderson, T.R., V. Bothmer, R.G. Marsden, K.J. Trattner, K.-P. Wenzel, A. Balogh, R.J. Forsyth, B.E. Goldstein, Geophys. Res. Lett., **22**, 3357, 1995.
3. Lanzerotti, L.J., C.G. Maclennan, R.E. Gold, S.M. Krimigis, T.P. Armstrong, M. Pick, Proc. 25th ICRC, **1**, 325, 1997.
4. McComas, D.J., et al., J. Geophys. Res., **105**, 10419, 2000.
5. Hofer, M.Y., R.G. Marsden, T.R. Sanderson, and C. Tranquille, these proceedings, 2001.
6. Tappin, S.J., E.C. Roelof, L.J. Lanzerotti, Astron. Astrophys., **292**, 311, 1994.
7. Desai, M.I., R.G. Marsden, T.R. Sanderson, A. Balogh, R.J. Forsyth, J.T. Gosling, J. Geophys. Res., **103**, 2003, 1998.
8. Potgieter, M.S., et al., Space Sci. Rev., in press, 2001.
9. Fisk, L.A., B. Kozlovsky and R. Ramaty, Astrophys. J., **190**, L35, 1974.
10. Mobius, E., et al., Nature, **318**, 426, 1985.
11. Pesses, M. E., J. R. Jokipii, and D. Eichler, Ap.J., **246**, L85, 1981.
12. Jokipii, J.R. and J.M. Davila, Astrophys. J., **248**, 1156, 1981.
13. Jokipii, J.R. and J. Kota, Proc. 19th ICRC, **4**, 449, 1985.
14. Potgieter, M.S., Proc. 19th ICRC, **4**, 425, 1985.
15. Leske, R.A., ICRC, **26**, 274, 2000.
16. Trattner K.J., R.G. Marsden, V. Bothmer, T.R. Sanderson, K.-P. Wenzel, B. Klecker and D. Hovestadt, Astron. Astrophys., **316**, 519, 1996.
17. Trattner, K.J., R.G. Marsden, V. Bothmer and T.R. Sanderson, Geophys. Res. Lett., **24**, 1719, 1997.
18. Tappin, S.J., G.M. Simnett, Astrophys. J., **469**, 33661, 1996.
19. Heber, B., et al.,J. Geophys. Res., **103**, 4809, 1998.
20. Cummings, A.C., J.B.Blake, J.R. Cummings, M. Franz, Proc. 24th ICRC, **4**, 800, 1995.
21. Webber, W.R. et al., Proc. 17th ICRC, **10**, 92, 1981.
22. Marsden, R.G., T.R. Sanderson, C. Tranquille, K.J. Trattner, A. Anttila, J. Torsti, Adv. Space Res., **23(3)**, 531, 1999.
23. Simpson, J.A., M. Zhang and S. Bame, Astrophys. J. Lett., **465**, L69, 1996.
24. Heber, B., et al., Geophys. Res. Lett., **23**, 1513, 1996.
25. Smith, E.J., J.R. Jokipii, J. Kota, R.P. Lepping, A. Szabo, Astrophys. J., **533**, 1084, 2000.
26. Simpson, J.A., et al., Astron. Astrophys. Suppl., **92(2)**, 365, 1992.
27. Moraal, H. and C.D. Steenberg, Proc. 26th ICRC, **7**, 543, 1999.
28. Heber, B., et al., Proc. 34th ESLAB Symposium, submitted, 2001.
29. Reames, D.V., Space Sci. Rev., **90**, 413, 1999.

Implications for Source Populations of Energetic Ions in Co-Rotating Interaction Regions from Ionic Charge States

D. Morris,[1] E. Möbius,[1] M. A. Lee,[1] M. A. Popecki,[1] B. Klecker,[2] L. M. Kistler,[1] A. B. Galvin,[1]

[1]*Space Science Center and Department of Physics, University of New Hampshire, Durham, NH, USA*
[2]*Max-Planck-Institut für extraterrestrische Physik, Garching, Germany*

Abstract. The ionic charge states of He have been observed in several co-rotating interaction regions (CIR) in 1999 and 2000 with ACE SEPICA. For all CIRs under study the He^+/He^{2+} ratio increases consistently from the start of the event towards the end, while the absolute flux of the energetic ions usually reaches a maximum close to the beginning of the event. With time, the spacecraft is magnetically connected to the compression region at increasing distance from the sun. Therefore, the increasing He^+/He^{2+} ratio can be interpreted as an increase in the relative importance of interstellar pickup ions over the solar wind as a source for the energetic ions. In addition, the observed He^+/He^{2+} ratio and its increase with the inferred distance from the sun provide an estimate of the injection and acceleration efficiencies for these species, which is found to be higher by about two orders of magnitude for pickup ions compared with the solar wind.

INTRODUCTION

A substantial contribution of He^+ to interplanetary energetic particle populations was first reported by Hovestadt et al. (1). Admixture of cold solar material was thought to be a possible source of these ions. Subsequently, the detection of interstellar pickup He^+ (2) in the inner heliosphere introduced another source of accelerated He^+ in interplanetary space.

Compression regions and associated shocks between adjacent fast and slow solar wind streams, called co-rotating interaction regions (CIRs), have long been known to accelerate particles efficiently. They are an important source of energetic particles in interplanetary space, generally during times of low solar activity (review (3)). The composition of these energetic ions is mostly similar to that of solar energetic particles and the solar wind. However, differences, notably for He and C, have been reported ((3) and references therein). For a CIR at 4.5 AU Gloeckler et al. (4) used Ulysses SWICS to identify interstellar pickup He^+ as the major contributor to He in CIRs for energies up to 60 keV. At this distance Fränz et al. (5) provided indirect evidence with Ulysses EPAC that He^+ must also be the main component at 0.6 - 2 MeV/n since He was found to be overabundant by a factor >2.5. He^+ was also observed as part of the suprathermal CIR population (up to 300 keV/Q) at 1 AU with SOHO STOF (6) and with Wind STICS (7).

These observations led to the suggestion that pickup ions may constitute an important source of suprathermal ions for further acceleration at interplanetary shocks (8). Based on the increased efficiency of pickup ions, with which they are injected into the acceleration processes to higher energies, Gloeckler et al. (8) have argued that inner source pickup ions may also contribute substantially to the energetic particle population in CIRs. As a consequence a substantial contribution of singly charged ions would be expected in the energetic heavy ion population of CIRs.

For a series of CIRs close to solar maximum in 1999 and 2000, Möbius et al. (10) have reported a substantial fraction (7 - 35%) of He^+ at 0.25 - 0.8 MeV/n, which they attributed to interstellar pickup ions. However, except for a 4.7% contribution of Ne^+ to Ne, most likely of interstellar origin, no significant singly charged component in the charge distributions of other heavy ions was found. In general the mean charge states resemble those of CME related solar energetic particle events and the solar wind. In this paper we extend our work on the He population in CIRs. We find a substantial variation of the He^+/He^{2+} ratio as time progresses from the beginning of each CIR, which is observed consistently in all of six events.

CP598, *Solar and Galactic Composition,* edited by R. F. Wimmer-Schweingruber

INSTRUMENT AND OBSERVATIONS

ACE was launched on August 25, 1997, and injected into a halo orbit around the Lagrangian point L1 on December 17, 1997 (11). Among a complement of high-resolution spectrometers to measure the composition of solar and local interstellar matter, as well as galactic cosmic rays, SEPICA provides the ionic charge state distribution of energetic particles.

To simultaneously determine the energy E, nuclear charge Z and ionic charge Q of incoming particles SEPICA combines electrostatic deflection in a collimator-analyzer assembly with an energy loss versus residual energy particle telescope. Z and E are determined in the latter, while Q is derived from the electrostatic deflection. A complete description of the SEPICA instrument and its data system may be found elsewhere (12). After problems with its pressure control valves for the proportional counters SEPICA has been operating for most of the time between early 1998 and 2000 with one of its two high geometric factor sensor units. Its ionic charge resolution at an energy of 1 MeV/Q is approximately $\Delta Q/Q = 0.3$ and its geometric factor is 0.09 cm^2sr.

Close to solar maximum a CIR caused by a low latitude coronal hole was observed with ACE. This CIR was tracked over six consecutive solar rotations in late 1999 and early 2000. First noticed during its appearance from DOY 337 to 340, 1999, was a distinct increase in the He^+/He^{2+} ratio over time, as shown in Fig. 1. The ratio has been taken over the energy range 0.4 – 0.8 MeV/n. The horizontal bars indicate the time interval, over which the ratio has been accumulated, and the vertical bars reflect the statistical error of the ratio. Roughly 1/2 day has been chosen as accumulation interval, but it usually is extended towards the end of the CIR, when the fluxes decrease and thus reduce the counting statistics. The vertical dashed line indicates the time when the spacecraft crossed the boundary between the fast and the fast compressed solar wind, later denoted F and F', respectively, in accordance with Chotoo et al. (7). The time periods of He observations for each recurrence are compiled in Table 1 along with the times of the traversal of the F – F' boundary, which evolves into a fast shock at larger distances from the sun.

A substantial increase of the He^+/He^{2+} ratio with time from values at the beginning of the CIR roughly between 0.1 and 0.2 to values at the end between about 0.5 and 1 is observed for all six recurrences. The ratio is between 0.2 and 0.3 at the crossing of the F – F' boundary, similar to the event averages (10).

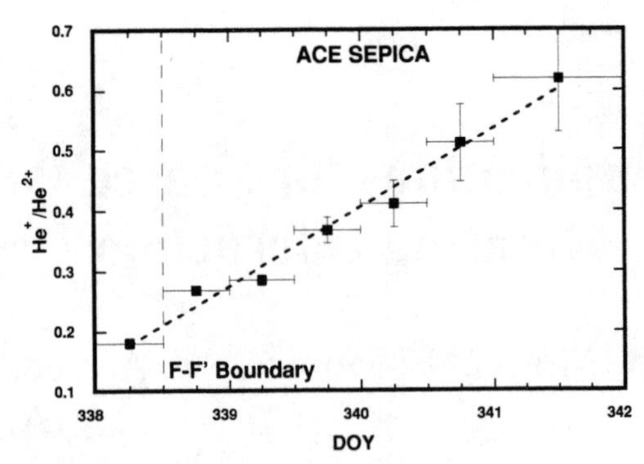

FIGURE 1: Variation of the He^+/He^{2+} ratio over the course of the CIR on DOY 338 – 342, 1999.

Table 1:

CIR	Dates	F – F' Boundary
1	284 - 288 1999	284 00 UT 1999
2	311 - 314 1999	312 17 UT 1999
3	338 - 342 1999	338 17 UT 1999
4	364 1999 - 004 2000	001 00 UT 2000
5	026 - 032 2000	028 12 UT 2000
6	054 - 058 2000	055 1430 UT 2000

DISCUSSION AND CONCLUSIONS

With ACE SEPICA we have observed a substantial increase in the He^+/He^{2+} ratio with time from the beginning of the CIR. This observation appears consistently in all six recurrences of the same CIR in 1999 through 2000. While He^{2+} represents the solar wind source, He^+ originates from interstellar pickup ions. As is shown in Fig. 2, a spacecraft at 1 AU is magnetically connected to the CIR at increasing distances r from the sun as time progresses from the start of the event. The integral pickup ion density n_{PU} varies as $1/r$ because of continuous production.

$$n_{PU}(r) = \frac{n_{Heo} \cdot \nu_i \cdot r_o^2}{V_{SW} \cdot r} \quad (1)$$

n_{Heo} is the interstellar He density, ν_i the ionization rate at $r_o = 1$ AU, and V_{sw} the solar wind speed. For simplicity we assume that n_{He} does not vary significantly with r, which is appropriate for He at $r > 1$ AU. The solar wind density decreases as

$$n_{SW}(r) = \frac{n_{SW}(r_o) \cdot r_o^2}{r^2} \quad (2)$$

because of flux conservation. Therefore, the ratio of He^+ pickup ions to solar wind He^{2+} increases linearly with r.

$$\frac{n_{PU}(r)}{n_{SW}(r)} = \frac{n_{Heo} \cdot \nu_i \cdot r}{n_{SW}(r_o) \cdot V_{SW}} \quad (3)$$

Most relevant for the production of energetic particles in a CIR at large distances, is the fast shock that travels into the fast solar wind, i.e. the F – F' boundary. Therefore, the time when the spacecraft crosses this boundary is the starting point with $r - 1$ AU = 0. We compute the radial distance from each point along the spacecraft path at 1 AU to the connection of the field line with the shock, assuming two Parker spirals that meet. One of them is typical for the local fast solar wind (dotted line through the S/C in Fig. 2) and the other one applies to the mean of the fast and slow solar wind speed (representative of the S' – F' interface in Fig. 2). This simplified picture ignores the progressive widening of the CIR with distance from the sun due to the shock motion into the fast wind, but corrects for the offset at 1 AU by starting at the F – F' boundary.

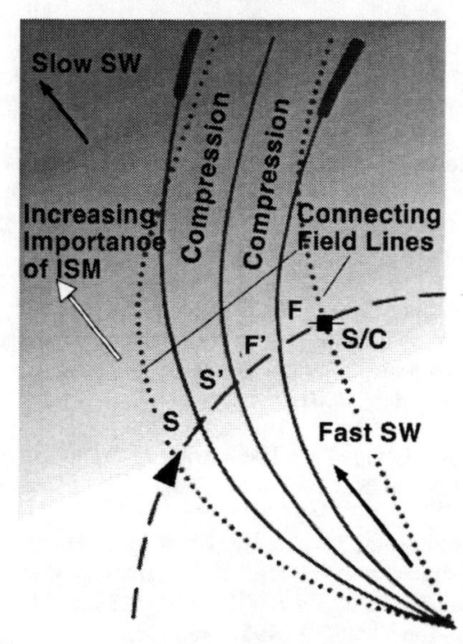

FIGURE 2: Schematic view of a CIR and apparent spacecraft path at 1 AU across the structure (adapted from (13)).

The He^{+}/He^{2+} ratios for events 2 - 6 are shown as a function of the deduced radial distance from the spacecraft in Fig. 3. Although event 1 shows a similar increase, it has been omitted from the compilation. Contrary to all other events it has a second flux maximum, which may originate from an interfering solar event, and both ratios and fluxes are much lower. The He^{+}/He^{2+} ratio in Fig. 3 increases approximately linearly with r with some scatter of the data points. This behavior is expected, if the increase of the interstellar source over the solar wind with distance from the sun is reflected in the energetic CIR population.

Using $n_{He} = 0.015$ cm^{-3} (14), an ionization rate of $1.25 \cdot 10^{-7}$ s^{-1} (15) for elevated solar activity, and V_{SW} = 600 km/s, a He^{+}/He^{2+} source ratio of $\approx 10^{-3}$ is obtained. This is substantially lower than the observed energetic particle ratios of 0.15 - 0.3 at 1 AU. This difference suggests strongly enhanced injection/acceleration efficiencies for interstellar pickup ions over the solar wind. Only suprathermal ions are efficiently accelerated at shocks. This provides a significant advantage for interstellar pickup ions, which populate a sphere in velocity space with $0 < V < 2V_{sw}$. Our values for the enhanced efficiency of He^{+} relative to He^{2+} (150 - 300 at 1 AU) are consistent with those at 4.5 AU (4).

As can be seen from eq. (3) the source ratio of He^{+} and He^{2+} at 1 AU and its slope with distance (in AU) from the sun should be approximately equal. As derived from the linear fit in Fig. 3, the slope of the ratio is substantially lower than the ratio at 1 AU, as if the interstellar source became less efficient with distance r. It should also be noted that our projected ratio for He^{+}/He^{2+} at 4.5 AU is lower by a factor of >10 compared with Ulysses observations at 4.5 AU. This seems to be compatible with the observation of a flatter slope. However, even if the slope were the same as the

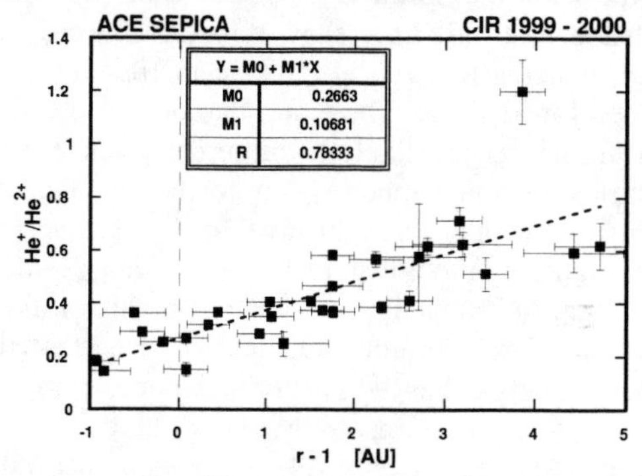

FIGURE 3: He^{+}/He^{2+} ratio versus inferred distance of connection to the CIR shock for events 2 – 6 in Table 1.

ratio at 1 AU, the projected He^{+}/He^{2+} ratio would reach ≈1.5, still lower by a factor of 6 than the value of ≈10 in Gloeckler et al. (4).

There are several possible explanations for such a behavior. Firstly, as pointed out the time-distance relation that we have used may be too simplistic, and a more thorough approach is necessary. A relation for the azimuthal shock motion and its connection to 1 AU has recently been evaluated analytically (16). Based on this relation the connection distances in Fig. 3 would be reduced by ≈30%. Secondly, particle transport effects along and perpendicular to the magnetic field may alter the ratios. He^{+} and He^{2+} have rigidities that are different by a factor of two. Thus transport along the field line from the CIR shock to the observer may alter the original source ratio (17). This effect will be tested in a separate investigation with different species that originate from the solar wind source, but have different rigidities because of their mass/charge ratios. In addition, Dwyer et al. (18) pointed out that transport perpendicular to the interplanetary magnetic field is important in CIRs. This may lead to mixing of populations from neighboring magnetic flux tubes and thus from different radial distances. Such a mixing would reduce the slope of the He^{+}/He^{2+} ratio. Finally, it is conceivable that Gloeckler et al. (4) have observed an exceptionally high He^+/He^{2+} ratio, as they have only studied one CIR.

In summary, the substantial contribution of He^+ in the energetic He population and its increase with the inferred distance from the sun in the CIRs under investigation suggest that interstellar He^+ pickup ions are a major contribution to these ions. Our result agrees with previous observations that $He^+/He^{2+} \approx 0.25$ in the suprathermal population of CIRs at 1 AU (6, 7) and $He^+/He^{2+} > 1$ in CIRs at 4.5 AU (4, 5). We confirm the earlier observations at 1 AU, extend them to higher energies, and demonstrate an increase of the He^+/He^{2+} ratio with the inferred distance from the sun. As Chotoo et al. (7) have shown, ions are already being accelerated effectively at 1 AU in a CIR, even without a developed shock. Therefore, the ions observed early in the events are likely to represent particle populations accelerated much closer to the sun than those observed with Ulysses. In any case the observed ratios are higher by more than two orders of magnitude than the ratio of interstellar pickup He^+ to solar wind He^{2+} at 1 AU. This enormous enhancement can be attributed to a strongly enhanced injection and acceleration efficiency for pickup ions over solar wind, as already argued by Gloeckler et al. (4). Only suprathermal ions can be effectively injected into acceleration. Pickup ions are essentially suprathermal in the frame of the solar wind, while the solar wind itself is rather cold.

ACKNOWLEDGMENTS

The authors are grateful to many unnamed individuals at UNH, at the MPE, and at TUB for their enthusiastic contributions to the completion of the ACE SEPICA instrument. The work was supported by NASA under NAS5-32626 and NAG 5-6912.

REFERENCES

1. Hovestadt, D., G. Gloeckler, B. Klecker and M. Scholer, Ionic charge state measurements during He^+-rich solar energetic particle events, Astrophys. J., 281, 463, 1984.

2. Möbius, E., et al., Direct observation of He^+ pick-up ions of interstellar origin in the solar wind, Nature, 318, 426 - 429, 1985.

3. Mason, G. M., and T. R. Sanderson, CIR associated energetic particles in the inner and middle heliosphere, Space Sci. Rev., 89, 77 - 90, 1999.

4. Gloeckler, G., et al., Acceleration of interstellar pickup ions in the disturbed solar wind observed on Ulysses, J. Geophys. Res., 99, 17637 - 17643, 1994.

5. Fränz, M., et al., Energetic particle abundances at CIR shocks, Geophys. Res. Lett., 26, 17 - 20, 1999.

6. Hilchenbach, M., et al., Observation of suprathermal helium at 1 AU: charge states in CIRs, in: Solar Wind Nine, S.R. Habbal, R. Esser, J.V. Hollweg and P.A. Isenberg, ed., 1999, American Inst. Physics: New York, pp. 605 - 608.

7. Chotoo, K., et al., The suprathermal seed population for corotating interaction region ions at 1 AU deduced from composition and spectra of H^+, He^{++}, and He^+ observed on Wind, J. Geophys. Res., 105, 23107 - 32122, 2000.

8. Gloeckler, G., Observation of injection and pre-acceleration processes in the slow solar wind, Space Sci. Rev., 89, 91 - 104, 1999.

9. Gloeckler, G., L. A. Fisk, J. Geiss, N. A. Schwadron, and T. H. Zurbuchen, Elemental composition of the inner source pickup ions, J. Geophys. Res. 105, 7459 - 7463, 2000.

10. Möbius, E., et al., Charge states of energetic (≈0.5 MeV/n) ions in corotating interaction regions at 1 AU and implications on source populations, Geophys. Res. Lett., subm., 2001.

11. Stone, E. C., et al., The Advanced Composition Explorer, Space Sci. Rev., 86, 1 - 22, 1998.

12. Möbius, E., et al., The Solar Energetic Particle Ionic Charge Analyzer (SEPICA) and the Data Processing Unit (S3DPU) for SWICS, SWIMS and SEPICA, Space Sci. Rev., 86, 449 - 495, 1998.

13. Richardson, I. G., L. M. Barbier, D. V. Reames, and T. T. von Rosenvinge, Co-rotating MeV/amu ion enhancements at ≤1 AU from 1978 to 1986, J. Geophys. Res., 98, 13 - 32, 1993.

14. Gloeckler, G., The abundance of atomic ^1H, ^4He and ^3He in the local interstellar cloud from pickup ion observations with SWICS on Ulysses, Space Sci. Rev., 78, 335 - 346, 1996.

15. Rucinski, D., et al., Ionization processes in the heliosphere - rates and methods of their determination, Space Sci. Rev., 78, 73 - 84, 1996.

16. Lee, M. A., An analytical theory of the morphology, flows, and shock compressions at co-rotating interaction regions in the solar wind, J. Geophys. Res. 105, 10491 - 10500, 2000.

17. Fisk, L. A. and M. A. Lee, Shock acceleration of energetic particles in corotating interaction regions in the solar wind, Astrophys. J., 237, 620, 1980.

18. Dwyer, J. R., et al., Perpendicular transport of low-energy corotating interaction region-associated nuclei, Astrophys. J. (Letters), 490, L115 - L118, 1997.

Galactic Composition

Galactic Abundances: Report of Working Group 3

B. Klecker*, V. Bothmer†, A. C. Cummings**, J. S. George**, J.W. Keller‡, E. Salerno§, U. J. Sofia¶, E. C. Stone**, F.-K. Thielemann||, M. E. Wiedenbeck††, F. Buclin§, E. R. Christian‡‡, E. O. Flückiger§, M. Y. Hofer§§, F. C. Jones‡‡, D. Kirilova¶¶, H. Kunow***, M. Laming†††, C. Tranquille§§ and K.-P. Wenzel§§

*Max-Planck-Institut für extraterrestrische Physik, 85740 Garching, Germany
†Max-Planck-Institut für Aeronomie, 37191, Katlenburg-Lindau, Germany
**Caltech, Pasadena, CA 91125, USA
‡Laboratory for Extraterrestrial Physics, GSFC, Greenbelt, MD 20 771, USA
§Physikalisches Institut, University Bern, CH-3012 Bern, Switze rland
¶Department of Astronomy, Whitman College, Walla Walla, WA 9936 2, USA
||Dept. of Physics & Astronomy, Univ. of Basel, Klingelbergstrasse 82, CH-4056 Basel, Switzerland
††JPL, Caltech, Pasadena, CA 91109, USA
‡‡NASA-GSFC, Greenbelt, MD 20771, USA
§§Space Science Department of ESA, ESTEC, 2200 AG Noordwijk, NL
¶¶Institute of Astronomy, Bulgarian Academy of Science, Sofia, B ulgaria
***IEAP, Universität Kiel, 24118 Kiel, Germany
†††Naval Research Laboratory, Washington DC 20375, USA

Abstract. We summarize the various methods and their limitations and strengths to derive galactic abundances from in-situ and remote-sensing measurements, both from ground-based observation s and from instruments in space. Because galactic abundances evolve in time and sp ace it is important to obtain information with a variety of different methods covering different regions from the Very Local Insterstellar M edium (VLISM) to the distant galaxy, and different times throughout the evolution of the galaxy. We discuss the study of the present-day VLISM with neutral gas, pickup ions, and Anomalous Cosmic Rays, the study of the local interstellar medium (ISM) at distances < 1.5 kpc utilizing absorption line me asurements in H I clouds, and the study of galactic cosmic rays, sampling contemporary (~ 15 Myr) s ources in the local ISM within a few kiloparsec of the solar system. Solar system abundances, derived from solar abundances and meteorite studies are discussed in several other chapters o f this volume. They provide samples of matter from the ISM from the time of solar system format ion, about 4.5 Gyr ago. The evolution of galactic abundances on longer time scales is discussed in the context of nuclear synthesis in the various contributing stellar objects.

INTRODUCTION

In Working Group 3 (WG3) of the SOHO-ACE workshop the various methods for obtaining galactic abundances have been discussed. Because galactic abundances are neither constant in time nor homogenous in space, it is important to obtain information on the galactic abundances with a variety of different methods covering different regions in space from the Very Local Interstellar Medium (VLISM) to the distant galaxy, and different times from the present to the early days of the galaxy, $\sim 10^{10}$ years ago. These various methods with their different measurement techniques have different limitations and strengths. In this report we will summarize these techniques or refer to other papers presented at this conference whereever possible to reduce overlap, provide references to recent results, discuss the limitations and strengths of the methods and give an outlook to future plans or further measurements needed.

The composition of the present-day VLISM can be studied by analyzing samples of the neutral interstellar gas that penetrates deep into the heliosphere. With presently available instrumentation it is possible to study this sample at various stages, starting with the neutral gas itself. A summary on neutral interstellar gas measurements available at present and in the near future is provided in chapter 1.

A more evolved sample of the VLISM are interstellar

CP598, *Solar and Galactic Composition*, edited by R. F. Wimmer-Schweingruber

pickup ions created by ionization by solar UV or charge exchange with solar wind protons. In the pickup process the ions are accelerated up to twice the solar wind velocity, i.e. to energies up to several keV/nucleon and can then be measured with ion composition mass spectrometers. For a summary of compositional measurements of the VLISM derived from the observation of pickup i ons see Gloeckler and Geiss, this volume [1].

Another, even more processed, sample of the VLISM are the Anomalous Cosmic Rays (ACR). In the 25 years since their discovery in low-energy quiet-time helium, nitrogen, and oxygen spectra, it has been well established that the origin of ACRs are pickup ions that are further accelerated in the outer heliosphere to energies of hundreds of MeV [2], presumably at the heliospheric termin ation shock [3], which is currently thought to be located at about 90 AU fro m the Sun. A summary of ACR measurements is provided in chapter 2 on Anomalous Cosmic Rays.

With relative speeds of interstellar neutrals of ~ 25 km/s and acceleration times of ACRs of a few years, the time delay between entering the heliosphere as neutral gas and the observation in the inner heliosphere as neutrals, pickup ions or ACRs is only of the order of a few years. Thus, all three samples provide information on the present-day abundances of the VLISM.

The present-day interstellar medium (ISM) at larger distances can be probed with

absorption line studies. H I clouds can provide compositional information on the present ISM at distances to about 1500 pc for a limited number of elements. These measur ements derive elemental abundances from gas-phase measurements complemented by grain models to determine a local galactic composition for N, O, and Kr. Results of this method are summarized in chapter 3 on H I Clouds.

Another sample of matter from beyond the solar system are Galactic Cosmic Rays (GCR). Galactic cosmic rays are believed to sample sources in the local interstellar me dium within a few kiloparsec of the solar system. In terms of galactic evolution, GCR s with their confinement time of ~ 15 Myr constitute also a contemporary sample of matter. Abundances of energetic nuclei near Earth reflect the source composition as well

as the acceleration and transport processes that have affected these particles. Chapter 4 on Galactic cosmic rays summarizes the present status and future persp ectives of GCR observations and their implications.

Solar system abundances derived from the measurement of solar abundances (see re ports of WG1 and WG2, this volume) and from the analysis of meteorites provide a sample o f matter from the solar system from the time of its formation about 4.5 Gyr ago. Elemental and isotopic abundances of meteorites can be studied in the laboratory in great detail. For a review of the elemental and isotopic abundances in meteorites see Hoppe, this volume [4], and r eferences therein.

Galactic abundances are neither constant in time nor are they universal on a sol ar level. Their evolution in space and time reflects the history of star formation and the lifetimes of the various contributing stellar objects (see [5], and references the rein). Thus, the evolution of galactic abundances can be investigated by the study of the surface composition of other stars of different age, i.e. different metallicity. Chapter 5 on Stella r Abundances provides a summary on stellar abundances as indicators of galactic evolution.

NEUTRALS

The solar system is moving through the VLISM approximately into the direction of Scorpio with a speed of about 25 km/s. As a result of this motion, an interstellar wind of neutral particles (mainly hydrogen with approximately 10 percent of helium) crosses the heliopause and enters the heliosphere with the same velocity. Neutrals from the VLISM and high energy GCRs with energies exceeding ~ 1 GeV are the only particles that can go beyond the heliopause. The solar wind plasma, being highly magnetized compared to the interstellar medium, acts as a shield for the low-energy ionized VLISM component.

The interstellar flux after entering the heliosphere is focused by the sun's gravity to form a focusing cone in a direction pointing away from the incoming local interstellar wind. The notable exception is the hydrogen for which the gravitational attraction is of about the same strength as the light pressure by solar photons. The Earth on its orbit around the Sun intercepts the stream of neutrals with collision angle and flux density that change according to the day of the year (Figure 1). In December the Earth, passing through the focusing cone, crosses the densest part of the interstellar flux. Nevertheless, during this period, the kinetic energy of the neutral atoms is not the highest.

Because of the vector addition of the Earth's motion, the energy of the interstellar atoms is highest during the period from January to May as seen from an Earth-bound observer (position A to B in Figure 1(a)). This condition enhances the velocity of the neutral atoms relative to the Earth from $\sim$25 km/s to $\sim$60-80 km/s.

Experimental Techniques

So far the dynamical properties of the flux of neutrals directly coming from the VLISM has been successfully used to make in situ measurements of the kinetic parameters of the interstellar gas flow as well as to study its

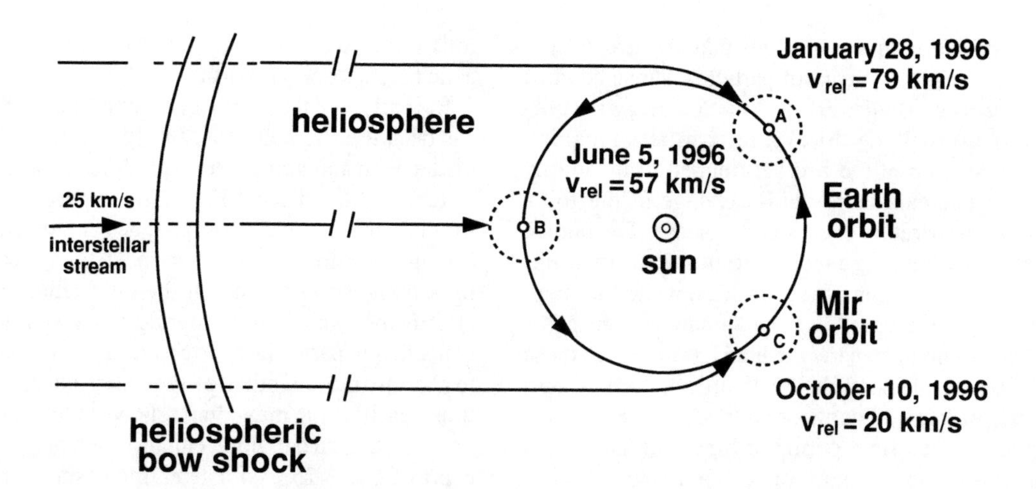

FIGURE 1. Seasonal variation of speed and direction of the interstellar neutral flux as consequence of the revolution of the Earth around the Sun. In December, the Earth is passing through the most dense part of the neutral gas , the focusing cone in the "downwind" direction (not shown in the picture), whil e from January to May, when the Earth is "upwind" of this pile-up, the helium flux density is lower but its velocity relative to Earth is higher.

elemental and isotopic composition. A complete review of the techniques for the detection of energetic neutral atoms currently in use and those developed for future applications in space research can be found in [6]. Working Group 3 discussed some of these techniques focusing the attention mainly on the pinhole camera principle used for the GAS experiment on Ulysses and the foil collection technique used for the COLLISA experiment on the Russian Space Station Mir whose recent results have been presented at this conference.

Pinhole camera: the GAS experiment on Ulysses

The GAS experiment on board the Ulysses spacecraft used, for the first time, an in situ technique to measure the properties of the local interstellar neutral gas. Up to now, the experiment allowed the determination of density, bulk velocity relative to the solar system and temperature of the interstellar flow by measuring the neutral helium penetrating the heliosphere [7]. In principle the GAS instrument (described in detail by Witte et al. [8]), detecting secondary particles emitted by a conversion plate when this is hit by the helium atoms, acts as a pinhole camera. Since the instrument is not equipped with any imaging detector, it is necessary to perform a two-dimensional scan over the hole sky in order to obtain a global picture of the neutral helium flux [9]. The efficiency of the instrument is highly energy dependent and neutral helium can only be detected when the particle's energy is larger than ~30 eV. Thus the detection of interstellar gas is only possible when the velocity vectors of the interstellar flux and Ulysses add up to a relative velocity high enough for detection. The energy-dependent efficiency also makes the sensor practically blind for the interstellar hydrogen. The detection probability for heavier elements would be high enough but the fluxes are so small that one can easily assume to measure exclusively interstellar neutral helium.

The advantages of the Ulysses GAS instrument are the rather simple design and the simple operation in space. Besides, the detection efficiency for energetic helium atoms is high (probably there is no other technique with comparable detection efficiency for helium). On the other hand, no mass and/or energy analysis of the detected particles can be achieved with this technique as well as no extraction of meaningful data is possible if the strict geometric and energetic instrumental requirements are not fulfilled.

So far, the GAS experiment onboard Ulysses has been the only instrument based o n the pinhole camera principle that has been used to study the neutral component of the interstellar inflow. However, future missions aiming to investigate the interstellar gas will very likely use an instrument of the Ulysses GAS type, probably extended to a full pinhole camera with a position-sensitive detector (alleviating, in this way, the need for a scan platform and for a spinning spacecraft).

The foil collection technique: the COLLISA experiment on Mir

The foil collection technique represents one of the earliest methods for the study of the elemental and isotopic composition of noble elements in space gas flows. The method is based on thin metal foils that are directly exposed in space to the stream of particles whose composition needs to be studied. If the kinetic energy is sufficiently large (usually ≥ 50 eV), the particles ramming against the surface of the foil get trapped in its atomic structure and remain there until a change in the foil's pressure or temperature causes their release. The choice of thin foils as trapping medium results from their extremely easy handling in space. Besides, while the small thickness assures a very low contamination from particles initially contained in the medium, compared to more massive objects, the possible use of large surfaces allows the capture of large amounts of particles. Once the foils have been exposed for a period of time long enough to allow a considerable capture of particles (from days to years, according to the particle flux density and to the experimental setup), they are recovered from space and delivered to Earth. Here, with a special procedure, the particles are extracted from the foil and sent to a noble gas mass spectrometer where their abundances can be measured. Due to the very low presence of noble gases both in the foils and in the laboratory mass spectrometers, the foil collection technique represents one of the most convenient methods for the analysis of this kind of particles. Other advantages offered by this method are the simple instrumentation and operation needed in space, plus the fact that the mass spectrometric analysis can be performed on Earth with resolutions and sensitivities much higher than is possible on a space-born instrument.

In 1965 Signer et al. [10] proposed for the first time an experiment aim ing to determine the elemental and isotopic composition of the solar wind, by means of the exposure of appropriate foils to the interplanetary plasma and subsequent extraction and analysis of the particles trapped in the foils. Such an experiment was later on carried out during the Apollo 11 and 12 missions, when aluminum foils (mounted on a pole stuck in the lunar soil) were exposed to flux of solar wind particles from the surface of the Moon [11]. In 1972 the foil sampling technique was used again to measure the noble gas fluxes in the auroral primary radiation [12, 13]. This time aluminum and platinum foils were mounted on two Nike-Tomahawk rockets that were launched directly into bright auroras. Some years later, during the Skylab missions, some foils were exposed to collect precipitating magnetospheric isotopes [14]. The foil collection technique was utilized for studying the neutral atoms coming from the Local Interstellar Cloud (LIC) for the first time in 1991 in the framework of the Interstellar Gas Experiment (IGE). On that occasion, several beryllium-copper foils were exposed to the flux of neutral interstellar particles from the outside of the Long Duration Exposure Facility (LDEF) satellite [15, 16]. Unfortunately, a malfunctions of the IGE electrical system during flight did not allow the exposure of all the foils, while the few foils exposed resulted highly contaminated with energetic magnetospheric particles.

Recently, for the COLLISA experiment, this method has been used to collect a sample of interstellar neutral matter with the aim to determine the helium and neon isotopic ratios in the LISM. The experiment was performed on board the Russian space station Mir. During Spring 1996 four beryllium-copper foils were exposed for $\sim$60 hours to the flux of the interstellar neutrals. As already mentioned, in this period of the year the particle collection is particularly effective since the Earth moves in the upwind direction of the interstellar flux of neutrals, reaching the maximum relative velocity ($\sim 60-80$ km/sec). Such a condition enhances the particles' kinetic energy up to $\sim$25 eV/nucleon. increasing the efficiency by which they get trapped into the foils. After exposure in space, the foils were recovered by the cosmonauts and brought back to Earth by the U.S. space shuttle Atlantis. The mass spectrometric analysis of the foils exposed to the interstellar flux has resulted in the clear detection of large amounts of ^{3}He and ^{4}He ($\sim 10^6$ and $\sim 10^{10}$ atoms respectively) with isotopic ratio ^{3}He/^{4}He equal to $\{1.71^{+0.50}_{-0.42}\} \times 10^{-4}$ [17].

New approaches for the study of the VLISM neutral component

Measurements of neutral atom density and composition require highly sensitive techniques because their fluxes in space are low. Detection and analysis of neutral atoms often require their ionization and this should be accomplished with efficiencies that approach unity. Long integration times are necessary for rare elements and large-aperture instruments should be considered when possible. Sample return techniques as discussed above can provide the necessary sensitivity but are not always applicable since low FIP elements from the VLISM do not arrive at 1 AU where sample return is most easily accomplished. Also, while the foil technique has been demonstrated to work well for the noble gases, it remains to be seen if the results for other elements will be of comparable quality. For in situ measurements new approaches and developments are required (see Wurz [6] for a review). There is reasonable hope, for example, that a measurement of the interstellar gas will be also possible with the IMAGE/LENA instrument. Besides, measurements of the interstellar neutral hydrogen distribution performed by the SWAN instrument on SOHO through observations of the scattered Lyman alpha (UV) radiation are already available [18].

At the conference, Keller et al. [19] have proposed neutral atom ionizat ion through charge exchange collisions with reactive metal vapors. These low ionization potential gases are mechanically confined to a collision cell through the spinning of rotor blades at turbomolecular speeds. Neutral atoms with finite electron affinity can pass through the cell and emerge as negative ions. The collisional cross section for charge exchange are large so that a relatively low density of vapor ($\sim 10^{13}$ cm^{-3}) is required for good ionization efficiency. The cell is opaque to the thermal metal vapor inside but partially transparent to the fast moving atoms that are to be detected. The product of the path length through the cell, the vapor density, and the collisional ionization cross section determines the ionization probability of an atom that passes through ([19]).

ANOMALOUS COSMIC RAYS

Anomalous cosmic rays (ACRs) are an accelerated sample of the local interstellar medium [2, 3]. They begin as interstellar neutral gas which drifts into the heliosphere at a speed of about 25 km s^{-1}. A portion of them become ionized by the processes of photoionization and/or charge exchange with the solar wind. These newly created ions are immediately picked up by the interplanetary magnetic field which is embedded in the expanding solar wind. The ions are then further accelerated in the outer heliosphere, gaining most of their energy at the solar wind termination shock, which is currently thought to be located at about 90 AU from the Sun [20].

Experimental Techniques

ACRs are observed in the multi-MeV/nucleon energy range principally by cosmic-ray solid-state detector telescopes on a variety of spacecraft in the heliosphere, including Voyager, ACE, SAMPEX, Wind, and Ulysses (see review in [21]). The most comprehensive and least fractionated sample of elemental abundances are obtained from the Voyager spacecraft in the outer heliosphere. Charge-state measurements are carried out at 1 AU by instruments on the SAMPEX spacecraft [22] in polar orbit around the Earth using the geomagnetic filtering technique [30]. SAMPEX also samples ACRs that have become trapped in the Earth's magnetosphere. The absolute intensity of this trapped population is >100 times that in the interplanetary medium [23]. Isotopic composition results for some elements are available from several spacecraft, including Voyager, SAMPEX, and ACE. The latter offers the statistically most accurate and comprehensive set of results.

Limitations and advantages of the method

In order to derive the composition of the local interstellar medium, account has to be taken of two additional fractionation steps beyond those affecting pickup ions. The two steps are the 1) injection and acceleration of the pickup ions at the termination shock and 2) the propagation inwards to the point of observation from the solar wind termination shock. By comparing observations from the two methods, it is possible to determine the nature of these fractionation steps.

ACRs offer a direct measurement of the isoptopic composition of the present-day VLISM [28]. For neutrals and pickup ions this is so far possible only for ^{3}He [17, 25]. In addition, the ACRs are critical in estimating the relative injection/acceleration efficiency at the termination shock. The pickup ion observations yield the flux of ions flowing into the shock and the ACR observations yield estimates of the resulting accelerated energy spectra. Comparing the two sets of observations yields the injection/acceleration efficiencies. These efficiencies can be determined for H, He, N, O, and Ne. In the case of Na and Ar, which are measured in the ACRs but not in the pickup ions, it is possible to estimate their interstellar neutral densities by extrapolating the injection/acceleration efficiency from that determined for N, O, and Ne. Another strength is that the charge state of the ACRs as a function of energy can be measured using the geomagnetic filtering technique with a polar orbiting satellite such as SAMPEX.

Results

- The elemental composition of H, He, N, O, Ne, Na, and Ar in the neutral VLISM, as deduced from ACR and pickup ion measurements, appears to be in good agreement with expectations from a recent model of the ionization state of the local interstellar medium [24, 26].
- Spectral features at low energy in other elements are seen which appear to require a source other than the VLISM [24].
- The isotopic composition of ACR O and Ne, and therefore the present-day VLISM, is consistent with that of the solar system within reasonably large uncertainties [27, 28, 29].
- Anomalous cosmic rays are primarily singly charged below about 350 MeV with a transition to higher charge states above that value [30, 31, 32].
- Heavier ions are preferentially injected into the ACR acceleration mechanism [1, 24, 33].

Outstanding problems and questions

- What are the limits on galactic evolution and what are the implications?
- What is the injection process at the solar wind termination shock?
- What is the origin of the minor ACR ions, e.g., Mg, Si, and S (see e.g. [21], and references therein)?

The Future

Greatly improved statistical accuracy and reduction of background due to galactic cosmic rays could be accomplished by an instrument with large collecting power in a polar orbit around Earth.

Measurements of the source spectra of ACRs would provide better estimates of the elemental and isotopic composition of the local interstellar medium, as well as better information on the injection and acceleration process. Voyager should provide exploratory measurements in the near future but a more advanced instrument on a mission to the termination shock is desired.

Because of the limitations due to the fractionation steps dicussed above, progress in 1) modeling the heliosheath region with respect to charge-exchange filtration effects on both the neutral and ionized component of the VLISM, and 2) modeling of the transport, pre-acceleration, injection, and acceleration of pickup ions to become ACRs will improve the derivation of the LISM composition from ACR measurements.

H I CLOUDS

H I clouds can provide reliable Galactic abundances for several elements. In fact, interstellar studies have used the gas-phase measurements of neutral interstellar clouds together with grain models to determine a local Galactic composition for N, O and Kr [34, 35, 36, 37, 38, 39]. These are the few elements for which we have gas-phase measurements and a good understanding of their dust abundances. Useful limits can be placed on the Galactic abundances of some species with less-well understood dust-phase characteristics, e.g. C and Sn. Although this method is quite limited in the elements that it can be used for, some very important species are among them.

Method

The method of determining the Galactic abundance from neutral (H I) clouds has an observational and a theoretical component. The ISM is composed of gas (the abundance of which can be measured) and dust (the abundance of which can be determined through grain modeling). If we add these two components together for a given element, we can easily obtain its Galactic abundance.

The gas-phase abundances are determined through absorption line studies. Most resonance lines of the dominant ions in neutral interstellar clouds occur in the ultraviolet region of the spectrum. Therefore one needs a high resolution UV spectrograph to make the measurements. The GHRS and STIS instruments aboard the Hubble Space Telescope (HST) are well designed for this purpose. The most reliable gas-phase abundances come from weak transitions of species. For extremely weak lines, the equivalent width or line strength alone gives an accurate abundance if the oscillator strength of the transition is well determined. These atomic constants have been updated and improved greatly since the HST was launched. This was necessitated by the high quality of the spectroscopic data. For absorption lines that are stronger than the weak limit, corrections must be made for unresolved saturation in the line cores. This is a straightforward procedure that is independent of the component structure of the absorbing regions [40, 41, 42]. Therefore, abundances in the gas can be well-determined without invoking any model dependent parameters. Gas-phase abundances determined in this way are usually acurate to 0.05 dex or $\sim$10%. Abundances from stronger absorption lines are much less reliable and therefore less useful for Galactic abundance studies.

The dust-phase abundances of the elements are necessarily model dependent because they cannot be directly observed. All of the cosmically abundant elements have *some* limits on their dust incorporation which are set by theories trying to reproduce extinction curves. However, theses grain models are not unique, so the limits placed on an elemental dust incorporation often has a large range. Only the elements which show the same levels of dust incorporation in all (or at least most) of the theories can be trusted for their dust-abundances. These elements, most of which are completely absent from dust [43], are the ones for which we can determine Galactic abundances.

Samples

The interstellar cloud samples for these studies are diffuse, $A_V \leq 1.0$ magnitudes, H I regions between the sun and O or B stars. The requirement of an early-type star results from the need for enough flux in the UV (above the Lyman limit) to observe very weak lines. These stars often have large values of $v \sin i$ which allow the stellar and interstellar absorption to be disentangled. The choice of diffuse neutral clouds facilitates the gas-phase abundance determinations because there is usually a well defined dominant ion for each element in these regions. Also, depletion onto dust is well understood in these clouds [44, 45, 34, 46] partly because no ice mantles can exist in them. Interstellar regions observed for these studies are all relatively local; the best current samples extend to only about 1500 pc from the sun [47, 48, 49, 50, 51]. This limit is imposed by current instrumentation and the flux requirements for high quality spectra. The sightlines do, however, sample a wide range of physical conditions present in diffuse clouds. This is judged by the fact that they have very different fractions of their hydrogen nuclei in the form of H_2 (as opposed to H I), $f(H_2)=2N(H_2)/[2N(H_2) + N(\text{H I})]$. These values, which range from about 10^{-6} to 0.6 in diffuse clouds, measure the balance between formation of H_2 molecules on grain surfaces and their photo- and/or chemical-destruction.

Limitations and advantages of the method

The neutral cloud method is limited in several ways. The most stringent of these is the need to understand the abundances of the elements in dust. Very few of these are known well enough for this method to work. A second limit is caused by the possibility of ionization in the gas. We are usually able to measure the dominant ion in the neutral clouds. If a substantial amount of an element is in a different ion state than expected, then we may underestimate its Galactic abundance. Previous studies have shown that this is not a problem for the species and regions that have so far been studied. Finally, unlike other methods, observations of interstellar lines reveal very little about isotope ratios because of limitations in data quality. The exception is boron for which isotope ratios have been determined from HST data [52, 53].

The greatest advantage of determining Galactic abundances from neutral interstellar clouds is that the method is so straight-forward. There is little-to-no model dependencies built into these abundances, especially if one is conservative about choosing elements with well determined dust-phase abundances. Therefore, the results from this method are extremely robust and reliable.

Results from neutral clouds

A discussion of the results based on this method are given in a paper by Sofia in these proceedings [43]. A summary of those results are given here. One of the most important findings from the interstellar studies is that the ISM within about 1500 pc of the sun in random directions is extremely uniform in composition. This indicates that a true local Galactic abundance does exist. The uniformity of the ISM measurements contrasts young star abundances which have significantly more scatter [39].

As far as specific ISM abundances are concerned, the local abundances of O [50], N [54] and probably C [47, 48] agree well with the solar photospheric measurements of Holweger [55]. The abundance of O in dust is relatively well known because of mineralogy, so its Galactic abundance can be well determined. The same is true for N which is probably not incorporated into dust at all. Carbon's abundance in dust is not well determined. Arguments based on the C/O ratio in young stars and the sun suggest, however, that it may be solar [39]. The fact that the local Galactic C, N and especially O abundances are solar is a very recently resurrected idea. Until recently the ISM community argued that the local Galactic abundance of O was subsolar based on gas-phase measurements of the ISM together with grain model considerations [34, 35, 36, 37, 50]. In the past several years, however, the reported solar abundances of O have been slowly dropping. The most recent photospheric abundance (reported in these proceedings [55]) is now in agreement with the abundances determined from the ISM years ago [47, 50]. This demonstrates the reliability of the neutral cloud method for determining the local Galactic composition.

Other elements in the local ISM do not appear to have solar abundances. For instance, the abundance of krypton in the ISM seems to be low compared to the solar system value [49]. Like N, Kr is not expected to be in dust at all so its Galactic abundance should be well determined. The apparent Kr underabundance may simply be the result of the difficulties in measuring Kr in the solar system (i.e. meteorites). If this is the case, then the ISM Kr/H value may be a good representation of the solar composition. Another element that appears not to agree with the sun is tin. Tin's incorporation into dust is not well understood so we are limited in what we can determine about its abundance. We can, however, confidently say that Sn is overabundant in the ISM with respect to the sun. This is because the measured gas-phase abundance alone is supersolar [51]. What we do not know, however, is to what extent it is incorporated into dust, and therefore how overabundant it is with respect to the sun.

The Future

As noted earlier, the major limit to this method is the uncertainty of the abundances of elements in dust; we can measure the gas-phase abundance for many more species than we have dust data for. Therefore, any improvements in the modeling of dust will help with the determination of Galactic abundances from the neutral

ISM. Particularly, the dust-phase abundance of C would be very important since we have a very good measurement of its gas-phase abundance in the local ISM [48]. Other important species include Si, Mg and Fe. However these species have variable depletion levels in diffuse clouds, possibly because they are able to grow on non-ice mantles. Finding Galactic abundances for these species, therefore, requires that their elemental incorporation into both grain cores and grain mantles be well understood.

In 2003 a new spectrograph, COS, will be placed on the Hubble Space Telescope. This instrument will be able to extend ISM studies of Galactic abundances to greater distances because it will have a sensitivity that is more than an order of magnitude greater than what is currently available on the HST. This will allow one to study how the Galactic abundances of elements change with distance from the sun. As noted earlier, the ISM abundance values are more tightly constrained than stellar values, so this will help to better define Galactic abundance gradients.

GALACTIC COSMIC RAYS

Galactic cosmic rays (GCRs) constitute another sample of matter from beyond the solar system that provides information about the composition of matter in the Galaxy. Abundances of energetic nuclei near Earth reflect the source composition as well as the acceleration mechanism and transport processes that have affected these particles. Measurements of interstellar dust provide information related to the pool of material available for acceleration.

Experimental technique

Energetic ions in the cosmic rays have been measured with balloon- and space-based instruments for many years. Recent satellite detectors such as the Cosmic-Ray Isotope Spectrometer (CRIS) [56] on NASA's Advanced Composition Explorer (ACE) and the High Energy Telescope (HET) [57] on the joint ESA/NASA Ulysses spacecraft, as well as earlier instruments on ISEE-3 [58, 59, 60], and Voyager 1 and 2 [61, 62, 63] have provided GCR composition measurements for ions with energies less than $\sim$500 MeV/nucleon. Typical spacecraft instruments use silicon detectors to measure the energy loss (dE/dx) and total energy for stopping nuclei. Trajectory information is provided by a scintillating optical-fiber hodoscope on ACE/CRIS, position-sensitive silicon wafers on Ulysses/HET, or gas-filled drift chambers on ISEE-3. At higher energies, the HEAO-3 Heavy Nuclei Experiment (HNE) [64, 65] and the French-Danish C2 [66, 67] detectors used ionization chambers, Cherenkov counters and flash tubes to identify incident particles. Magnetic spectrometers have been flown as balloon payloads [68, 69, 70].

In terms of galactic evolution, Galactic cosmic rays are a contemporary sample of matter; measurements of radioactive secondary isotopes such as ^{10}Be constrain the confinement time in the Galaxy to $\sim$15 Myr using a simple leaky-box model [72]. Galactic cosmic rays are believed to sample sources in the local interstellar medium (ISM) within a few kiloparsecs of the solar system. Diffusion models of GCR propagation in the galactic disk and halo, constrained by diffuse γ-ray emission from cosmic-ray interactions with ISM protons [73], limit the dimension perpendicular to the disk of the volume sampled by GCRs [74, 73, 75]. If the diffusion is isotropic, particles undergoing a random walk are not likely to move farther along the disk than they can in a perpendicular direction into the halo. This means that the cosmic rays we observe sample a comparatively small volume around the solar system.

The composition of the source material from which galactic cosmic rays are accelerated is thought to be similar to an average of the ISM composition with possible additional contributions from stellar winds or supernova ejecta. A known elemental fractionation that depends on atomic parameters such as the first ionization potential (FIP) or condensation temperature (volatility) may reflect a fractionation of the source material or a preferential acceleration of dust grains [76]. Once accelerated, either by a single shock or by multiple encounters with shocks or turbulent magnetic fields, the particles face on average a 15 Myr journey through a $\sim$5-10 g/cm^2 absorber, the interstellar medium. During this transport the relative abundances may be altered by spallation, radioactive decay, energy changing processes, and escape from the Galaxy. At the end of this the nuclei must still enter the heliosphere. The expanding solar wind, with its frozen-in magnetic field, tends to mix and lower the energies of the particles that are observed, modifying the spectra below a few GeV/nucleon.

Each of these processes must be modeled in order to understand the abundances at the source. Interstellar propagation calculations rely on measurements of the nuclear fragmentation cross sections, radioactive decay half-lives and the composition of the interstellar medium. For some species, secondary production by fragmentation overwhelms the primary contribution. The relative abundances of these isotopes tell us little about the source but do reflect the conditions during transport.

In spite of the many complex processes affecting the measured composition, galactic cosmic rays provide a rich variety of probes that constrain the parameters in the models. Elemental composition measurements have been made over a wide range in energy and mass. Now with ACE, we have a nearly complete set of precise isotopic GCR abundances from He to Zn below 500 MeV/nucleon. Data from HEAO-3, the Cosmic Ray Nucleus (CRN) experiment [71], and others at higher energies constrain the interstellar spectral indices which are essentially unaffected by heliospheric processes above ~10 GeV/nucleon.

Abundances for predominately secondary species depend on the details of transport through the Galaxy. Stable secondary isotopes constrain the total amount of matter the cosmic rays pass through. Radioactive β-decay secondaries probe the mean lifetime for escape from the Galaxy and the average density of the local interstellar medium [72]. Electron-capture secondaries reflect energy-changing processes in the interstellar medium and the possibility of multiple accelerations during propagation [77]. All of this information constrains the model predictions for the composition of the accelerated sample of material near the source.

Primary abundances allow the study of injection or acceleration fractionations imprinted on the source composition. Elements with similar FIP but different condensation temperatures can help distinguish between models of the GCR source. For example, low Na/Mg, Ge/Fe, Cu/Fe ratios, or a high P/S ratio in the GCR source compared to the solar-system abundances would favor models in which refractory cosmic rays are derived from ions sputtered from accelerated dust grains [78, 79]. Source ratios of these elements that are consistent with the solar-system values might reflect a FIP-based fractionation such as that observed in stellar coronae. Radioactive primaries such as ^{59}Ni probe the history of the source material before acceleration. Stable refractory (or low-FIP) primaries offer a direct look at the mix of material in the source pool.

Results

Perhaps the most remarkable result from decades of cosmic-ray composition measurements is that the relative composition of the refractory isotopes is very similar to that of the solar system. It is somewhat surprising that this should be the case, given the 4.5 Gyr time difference between the formation of the solar system and the cosmic rays that we observe today. Figure 2 compares ACE measurements of GCR source abundances for iron-group cosmic rays to the corresponding solar system abundances, normalized separately to ^{56}Fe=1 [80]. Further analysis indicates that the agreement also holds for ^{54}Fe, ^{40}Ca, and for the extremely rare ^{48}Ca at $\sim 10^{-4}$ times the ^{56}Fe abundance [81]. The largest differences are less than a factor of two over nearly four orders of magnitude in absolute abundance. This similarity imposes limits on models of galactic chemical evolution over the time since the solar system was formed. It is reasonable to hypothesize that the mix of sources from which the cosmic rays are accelerated is very similar to that which contributed the material from which the solar system was formed.

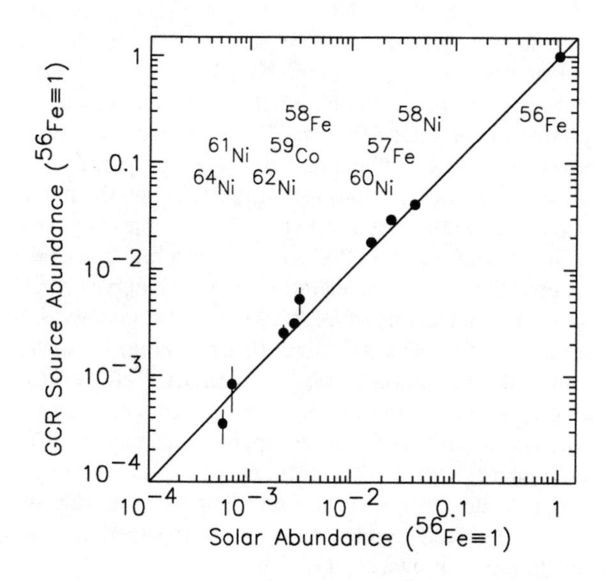

FIGURE 2. Comparison of cosmic-ray source abundances with corresponding solar-system values. Points along the diagonal line correspond to equal abundances, relative to ^{56}Fe, in the two samples of material [taken from 80].

Differences in the relative composition of the GCRs and the solar system provide important insight into the nature of the source. A well-documented excess of ^{22}Ne in the cosmic-ray source [82] may be due to a small admixture of material from the surface layers of massive Wolf-Rayet stars which are enriched in the end products of core helium burning. If this is the case, one might expect enhancements of the neutron-rich isotopes of Mg and Si, as well as heavy isotopes such as ^{58}Fe, ^{59}Co, and ^{61}Ni [84]. A small excess of ^{58}Fe in the ACE data (see Figure 2) may fit a pattern of isotope enhancements predicted by the Wolf-Rayet models employed to explain the ^{22}Ne anomaly [82, 83].

Galactic cosmic ray abundances of ultra-heavy elements (Z>30) show the imprints of stellar nucleosynthesis. The elements around Z ~ 39, 57, and 82 are dominated by contributions from isotopes produced in the (slow) "s-process" in which production chains stop at short-lived β-decay isotopes. Data from HEAO-3 HNE [65, 85] and Ariel-VI [86] show indications of enhancements for element groups which have contributions from nuclei formed predominately in the (rapid) "r-process" of explosive nucleosynthesis (e.g., in regions near Z~ 34,

53, 65, 77, and the actinides). Production of these isotopes requires an environment in which neutron capture rates dominate over nucleon decay. Observation of a high $^{78}Pt/^{82}Pb$ ratio by the TREK (track etch) experiment [87] also indicates an enhanced contribution of r-process material compared to the solar system. The actual site where the r-process occurs remains an unsolved mystery. It is clear, however, that both s- and r-process sources must contribute to account for the observed GCR abundances.

Primary isotopes such as ^{59}Ni, which decay by electron capture, probe the time delay between cosmic-ray production and acceleration. ^{59}Ni is produced in supernovae and decays with a half-life of $\sim 10^5$ years. Once accelerated, the electrons are stripped off and the nuclei become essentially stable. Figure 3 shows the calculated abundance at Earth of ^{59}Ni and ^{59}Co as a function of the elapsed time before acceleration. The solid curves correspond to the fraction of mass-59 material originally produced as ^{59}Ni. The ACE data (hatched area) is consistent with the complete decay of primary ^{59}Ni to ^{59}Co, implying that a time longer than 10^5 years elapsed before acceleration [88]. Since supernova shocks dissipate their energy over a 10^4 year time-scale (assuming a typical ISM density), this rules out the possibility that the observed galactic cosmic rays come from supernovae accelerating their own ejecta.

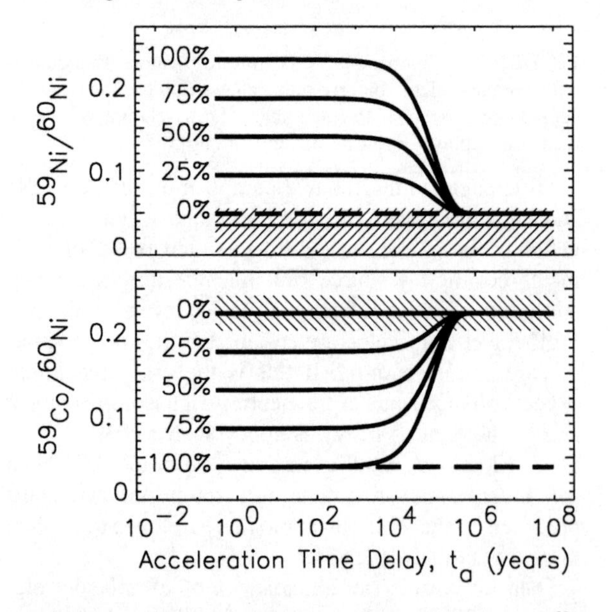

FIGURE 3. Abundances of mass-59 isotopes relative to ^{60}Ni observed at Earth as a function of the time delay between production and acceleration. The solid curves correspond to the fraction of mass-59 isotopes originally produced as ^{59}Ni in the source. Hatched areas represent the region allowed by the ACE data, consistent with a time delay greater than 10^5 years. [taken from 88]

The elemental fractionation inferred in the cosmic-ray source appears to be ordered by the first ionization potential or the generally correlated volatility. Determining which parameter controls the fractionation has strong implications for models of the source. A dependence on volatility would favor models involving a preferential acceleration of interstellar dust grains [76]. The source abundances of elements such as Na, P, Cu, and Ge which break the general correlation of FIP and volatility tend to favor volatility models [78], as do the Pb/Pt-group results from the TREK detector exposed on Mir [87].

Outstanding Problems

The outstanding problems pertaining to galactic cosmic rays at this time relate to questions presented above. What is the nature of the cosmic-ray sources? Are there identifiable components of the source composition which differ from the average ISM? What exactly is the material that is being accelerated? Is the observed elemental fractionation a FIP or volatility effect? What is the acceleration history of the cosmic rays, do they see one or multiple accelerations?

The Future

Improvements in the propagation model input parameters will be important for significant future progress. Data from ACE for most isotopes have reached a statistical precision such that uncertainties in the calculated source abundances are dominated by the precision of the nuclear fragmentation cross sections, and radioactive decay parameters.

New measurements of the fragmentation cross sections from ^{56}Fe and ^{60}Ni on a hydrogen target were made in October 2000 at the GSI heavy ion synchrotron in Darmstadt, Germany. Analysis of these data is underway and the results should greatly improve the accuracy of propagation models in the iron-nickel region and for the sub-Fe isotopes down through Ca where Fe fragmentation is the dominant source of secondary production. These data will allow better constraints on the Fe-Ni isotope excesses related to the acceleration time delay and Wolf-Rayet contributions. Future measurements of Mg, Si, and S fragmentation into Na and P would help distinguish between FIP and volatility while ultra-heavy production cross sections are needed to interpret the r-process signatures.

New electron-attachment cross sections and radioactive decay parameters are also needed. The ACE data clearly show the effects of electron capture decay during transport [77] but cannot yet determine whether cosmic rays get their full energy from a single shock or are accelerated gradually in multiple shocks. The confinement time in the Galaxy comes from analysis of radioactive decay secondaries such as ^{10}Be, ^{26}Al, ^{36}Cl, and ^{54}Mn which sample different effective volumes corresponding to their decay half-lives. Decays of ^{54}Mn in the laboratory are dominated by capture of the attached electrons. In the cosmic rays, where the electrons have been stripped away, the much rarer β^- decay is much more important. A precise measurement of half-life for this branch would make it possible to determine whether Fe and lighter elements reside in the Galaxy for similar times.

The ACE data provide a rich set of measurements for isotopes up to Zn. New measurements are needed for the ultra heavy species which include important signatures of r- and s- process nucleosynthesis as well as a number of important isotopes for distinguishing between FIP and volatility. The Advanced Cosmic-Ray Composition Experiment for the Space Station (ACCESS) has been proposed to measure elemental spectra up to Fe at energies approaching the "knee" ($\sim 3\times 10^{15}$ eV) to explore the limits of supernova shock acceleration. Abundances of elements with $Z>30$ may also be measured, looking for enhancements of r-process elements and depletions of elements for which FIP and volatility are poorly correlated. The Heavy Nuclei Experiment (HNX) is a mission presently undergoing a Phase A study that would expose silicon detector stacks and glass track-etch detectors for three years. The HNX experiment would provide elemental composition measurements from zinc up through the actinides.

Another important proposed mission, the Interstellar Probe, would directly measure the energetic particle composition of the interstellar medium outside the heliosphere, free from the effects of solar modulation. This information is presently inferred from astronomical observations but would be an important constraint on propagation models. A currently operating mission, Voyager I, is thought to be nearing the boundary of the heliosphere and may have a chance to make composition measurements outside the termination shock within the next 10-15 years.

Interstellar Dust

Interstellar dust is a repository of the refractory elements in all but the warmest interstellar clouds and as such contains within it some of the keys toward our understanding of the history and evolution of the galaxy. Dust comprises roughly 1% of the mass of the interstellar medium, which in turn makes up a third of the mass of the galactic disk [89].

The composition of dust can be inferred from the depletion of atoms from the gas phase in the interstellar clouds [90] assuming a standard composition (gas + dust). The gas abundances are determined by spectroscopic observation of absorption lines along various lines of sight, as described in the previous section. The inferred dust phase contribution constrains dust models and serves as a cross check of the assumed composition. The standard abundances against which depletions are measured are generally the solar abundances but it is recognized that deviations from the standard are possible if the observed cloud is not well mixed [91].

The characteristic broadening of spectral lines in solids prevents precise determination of dust composition through astronomical means, however, considerable information can be obtained from extinction, polarization and emissivity measurements. Solid state transitions are observed to some degree as well as broadened molecular transitions. For example, the absence of the OH stretch at 3 μm in absorption discounts the possibility of an ice-based composition of interstellar dust. Instead, the presence of this transition in cold dense interstellar clouds indicates that icy mantles have formed on the dust particles within them (see, e.g. discussion in [89]).

Fortunately the composition of local interstellar dust grains can be determined precisely though in situ measurements and sample return missions. The Ulysses and Galileo spacecraft have made direct measurements of the flux and size distribution of interstellar dust grains [92] within the solar system. However, grain sizes <0.1 μm are excluded from the inner heliosphere by radiation pressure and the interaction of the solar wind [93].

Future components of dust particle research include analysis of interstellar dust returned from the Stardust spacecraft. Although Stardust is primarily a comet sample return mission, some resources are devoted to interstellar dust collection. ESA is currently evaluating a proposed interstellar dust sample return mission (DUNE Galactic-Dust-Measurements-Near-Earth). Detailed analysis of smaller dust particles awaits a mission going outside the heliosphere such as the Interstellar Probe currently under study by NASA.

STELLAR ABUNDANCES AND GALACTIC EVOLUTION

Galactic abundances are neither constant in time and universal on a solar level [102] nor do they evolve in a simple fashion. Their evolution in space and time reflects the history of star formation and the lifetimes of the diverse contributing stellar objects like intermediate and low mass stars through planetary nebula ejection, mass loss from massive stars, as well as the ejecta of SNe II and SNe Ia. Planetary nebulae do contribute products of H- and He-burning and heavy s-process nuclei up to Pb and Bi, based on a sequence of neutron captures and beta-decays during (shell) He-burning. Galactic evolution for the elements in the range O through Ni is dominated by two alternative explosive stellar sources, i.e by the combined action of SNe II and Ia. The site for the production of the heaviest elements up to U and Th (the r-process based on neutron captures in environments with very high neutron densities) is still debated and most probably related to supernovae and/or neutron star ejecta (from binary mergers or jets).

Stars (with understood exceptions) do not change the surface composition during their evolution. Thus, surface abundances reflect the interstellar medium (out of which the stars formed) at the time of their formation. Therefore, observations of the surface composition (via spectra) of very low metallicity stars (i.e. very old stars) give a clue to gas abundances throughout the evolution of our Galaxy. This way they can serve as a test for all the contributing stellar yields. Stellar abundances are measured as a function of metallicity [Fe/H]=log$[(Fe/H)/(Fe/H)_\odot]$. Typical (Galactic chemical) evolution calculations suggest *roughly* the "metallicity-age" relation [Fe/H]=–1 at 10^9y, –2 at 10^8y, and –3 at 10^7y (see e.g., [115, 113, 101]). Observational data for [x/Fe] at low metallicities ($-2 <$[Fe/H]< -1), x standing for elements from O through Ca, show an enhancement of the alpha elements (O through Ca) by a factor of 2-3 (0.3 to 0.5 dex in [x/Fe]) in comparison to Fe [109, 108, 97, 104, 106, 107, 103, 99, 105, 96]. This is the clear fingerprint of the exclusive early contribution of fast evolving massive stars, i.e. SNe II. For [x/Fe]>-2 the scatter of this feature is very small. Below -2 it shows large variations of up to a factor of 10. As such very low metallicity stars were born when the Galaxy was only 10^7-10^8y old, approaching at $\approx 10^6$y the lifetime of individual massive stars, it is clear that one does not obserse well mixed average features, but the results of individual stellar explosions.

There have been groundbreaking r/Fe observations [110, 108, 111, 116, 100, 112] at low metallicities in recent years. The individual abundance features indicate a typical solar r-abundance pattern, at least above Ba. A very important result is also that Th and U could be observed. The r-abundances show a scatter of almost a factor of 1000 which still amounts to about a factor of 10 at [Fe/H]=-1, see Fig. 2 in [114]. This indicates on the one hand a large variation in r/Fe ratios in the individual production sites, and on the other hand a rarer occurance of r-process events as the asymptotic value is attained only later in galactic evolution. Nevertheless, the early occurrance of the r-process relates it to objects with massive stellar progenitors (either SNe II or neutron stars).

At about -2<[Fe/H]<-1.5 the onset of s-process nucleosynthesis from intermediate and low mass stars is observed, in agreement with the expected delay related to the stellar evolution timescales of such objects.

SNe Ia with their higher ratio of Fe-group elements to Si-Ca have to compensate for the respective overabundances in SNe II in order to obtain solar abundance ratios for the combined nucleosynthesis products. This effect is seen for metallicities in the range of -1<[Fe/H]<0. SNe II and SNe Ia, mixed in frequency ratios of $N_{\rm Ia}/N_{\rm II} = 0.15-0.27$ [98], require an Fe contribution from SNe Ia of 50-60%, if average ejected Fe masses of 0.1 $M_\odot$ and 0.6 $M_\odot$ are taken for SN II and SN Ia. An interesting feature is that also the Fe-group ejecta of both types of supernovae have to differ. For Ti and Mn one finds at low metallicities average SN II values [x/Fe] of 0.25 and -0.3, i.e. a clear signatures for a SN II behavior different from solar, which asks for the opposite effect in SN Ia.

ACKNOWLEDGEMENTS

We thank the organizers of the SOHO/ACE workshop on "Solar and Galactic Composition" for their hospitality.

REFERENCES

1. Gloeckler, G., and Geiss, J., "Composition of the Local Interstellar Cloud from Observations of Interstellar Pickup Ions" in *Solar and Galactic Composition*, edited by R. F. Wimmer-Schweingruber, AIP Conf. Proc., Woodbury, NY, 2001, this volume.
2. Fisk, L. A., Kozlovsky, B., and Ramaty, R., *Astrophys. J. Lett.*, **190**, L35–L38 (1974).
3. Pesses, M. E., Jokipii, J. R., and Eichler, D., *Astrophys. J. Lett.*, **246**, L85–L89 (1981).
4. Hoppe, P., "Elemental and Isotopic Abundances in Meteorites", in *Solar and Galactic Composition*, edited by R. F. Wimmer-Schweingruber, AIP Conf. Proc., Woodbury, NY, 2001, this volume.
5. Thielemann, F.-K., Argast, D., Brachwitz, F., and Martinez-Pinedo, G., "Stellar Nucleosynthesis and galactic Abundances" in *Solar and Galactic Composition*, edited by R. F. Wimmer-Schweingruber, AIP Conf. Proc., Woodbury, NY, 2001, this volume.
6. Wurz, P., "Detection of Energetic Neutral Atoms", in *The Outer Heliosphere: Beyond the Planets*, Copernicus Gesellschaft, 2000, chap. 11, pp. 251–288.
7. Witte, M., Banaszkiewicz, M., and Rosenbauer, H., *Space Sc. Rev.*, **78**, 289–296 (1996).
8. Witte, M., Rosenbauer, H., Keppler, E., Fahr, H., Hemmerich, P., Lauche, H., Loidl, A., and Zwick, R., *Astron. Astrophys. Suppl. Ser.*, **92**, 333–348 (1992).
9. Witte, M., Rosenbauer, H., and Banaszkiewicz, M., *Adv. Space Res.*, **13**, 121–130 (1993).
10. Signer, P., Eberhardt, P., and Geiss, J., *J. Geophys. Res.*, **70**, 2243–2244 (1965).
11. Geiss, J., Eberhardt, P., Bühler, F., Meister, J., and Signer, P., *J. Geophys. Res.*, **75**, 5972 (1970).
12. Axford, W., Bühler, F., Eberhardt, P., and Geiss, J., *J. Geophys. Res.*, **77**, 6724 (1972).

13. Bühler, F., Axford, W., Chivers, H., and Marti, K., *J. Geophys. Res.*, **81**, 111 (1976).
14. Lind, D., Geiss, J., and Stettler, W., *J. Geophys. Res.*, **84**, 6435 (1979).
15. Lind, D., Geiss, J., Bühler, F., and Eugster, O., "The Interstellar Gas Experiment, in: LDEF - 69 Months in Space", in *NASA Conf. Pub.*, 1991, vol. 3134, p. 585.
16. Bühler, F., Lind, D., Geiss, J., and Eugster, O., "The Interstellar Gas Experiment: Analysis in progress, in: LDEF 69 - Months in Space",in *NASA Conf. Pub.*, 1993, vol. 3194, p. 705.
17. Salerno, E., Bühler, F., Bochsler, P., Busemann, H., Eugster, O., Zastenker, N., Agafonov, Y., and Eismont, N., "Direct Measurement of $^3He/^4He$ in the LISM with the COLLISA Experiment" in *Solar and Galactic Composition*, edited by R. F. Wimmer-Schweingruber, AIP Conf. Proc., Woodbury, NY, 2001, this volume.
18. Brasken, M., and Kyrölä, E., *Astron. Astrophys.*, **332**, 732 (1998).
19. Keller, J., Ogilvie, K., and Coplan, M., "Measurement of Energetic Neutral Atom Composition in the Heliosphere and around Planetary Bodies" in *Solar and Galactic Composition*, edited by R. F. Wimmer-Schweingruber, AIP Conf. Proc., Woodbury, NY, 2001, this volume.
20. Stone, E. C., and Cummings, A. C., *Proc. 27th Intl. Cosmic Ray Conf. (Hamburg)* **10**, 4263–4266 (2001).
21. Klecker, B., Mewaldt, R. A., Bieber, J. W., Cummings, A. C., Drury, L., Giacalone, J., Jokipii, J. R., Jones, F. C., Krainev, M. B., Lee, M. A., Roux, J. A. L., Marsden, R. G., McDonald, F. B., McKibben, R. B., Steenberg, C. D., Baring, M. G., Ellison, D. C., Lanzerotti, L. J., Leske, R. A., Mazur, J. E., Moraal, H., Oetliker, M., Ptuskin, V. S., Selesnick, R. S., and Trattner, K. J., *Space Sci. Rev.*, **83**, 259–308 (1998).
22. Baker, D. N., Mason, G. M., Figueroa, O., Colon, G., Watzin, J. G. and Aleman, R. M., *IEEE Trans. Geosci. & Remote Sens.*, **31**, 531-541 (1993).
23. Selesnick, R. S., Cummings, A. C., Cummings, J. R., Mewaldt, R. A., Stone, E. C., and von Rosenvinge, T. T., *J. Geophys. Res.*, **100**, (A6), 9503-9518 (1995).
24. Cummings, A. C., Stone, E. C., and Steenberg, C. D., Composition of Anomalous Cosmic Rays and Other Heliospheric Ions (2001), in preparation for submittal to JGR.
25. Gloeckler, G., *Space Sci. Rev.*, **78**, 335–346 (1996).
26. Slavin, J. D., and Frisch, P. C., The Ionization of Nearby Interstellar Gas, submitted to ApJ (2001).
27. Leske, R. A., Mewaldt, R. A., Cummings, A. C., Cummings, J. R., Stone, E. C., and von Rosenvinge, T. T., *Space Sci. Rev.*, **78**, 149–154 (1996).
28. Leske, R. A., Mewaldt, R. A., Christian, E. R., Cohen, C. M. S., Cummings, A. C., Slocum, P. L., Stone, E. C., von Rosenvinge, T. T., and Wiedenbeck, M. E., "Observations of Anomalous Cosmic Rays at 1 AU", in *Acceleration and Transport of Energetic Particles Observed in the Heliosphere*, edited by R. A. Mewaldt, J. R. Jokipii, M. A. Lee, E. Moebius, and T. H. Zurbuchen, AIP Conf. Proc., **528**, Woodbury, 2000, pp. 293–300.
29. Leske, R. A., "Anomalous Cosmic Ray Composition from ACE", in *26th International Cosmic Ray Conference Invited, Rapporteur, and Highlight Papers*, edited by B. L. Dingus, D. B. Kieda, and M. H. Salamon, AIP **516**, Melville, 2000, pp. 274–282.
30. Klecker, B., McNab, M. C., Blake, J. B., Hamilton, D. C., Hovestadt, D., Kästle, H., Looper, M. D., Mason, G. M., Mazur, J. E., and Scholer, M., *Astrophys. J. Lett.*, **442**, L69–L72 (1995).
31. Mewaldt, R. A., Selesnick, R. S., Cummings, J. R., Stone, E. C., and von Rosenvinge, T. T., *Astrophys. J. Lett.*, **466**, L43–L46 (1996).
32. Barghouty, A. F., Jokipii, J. R., and Mewaldt, R. A., "The Transition from Singly to Multiply-Charged Anomalous Cosmic Rays: Simulation and Interpretation of SAMPEX Observations", in *Acceleration and Transport of Energetic Particles Observed in the Heliosphere*, edited by R. A. Mewaldt, J. R. Jokipii, M. A. Lee, E. Moebius, and T. H. Zurbuchen, AIP Conf. Proc., **528**, Woodbury, NY, 2000, pp. 337–340.
33. Cummings, A. C., and Stone, E. C., *Space Sci. Rev.*, **78**, 117–128 (1996).
34. Sofia, U. J., Cardelli, J. A., and Savage, B. D., *ApJ*, **430**, 650–666 (1994).
35. Snow, T. P., and Witt, A. N., *Science*, **270**, 1455 (1995).
36. Snow, T. P., and Witt, A. N., *ApJ*, **468**, L65–L68 (1996).
37. Mathis, J. S., *ApJ*, **472**, 643–665 (1996).
38. Snow, T. P., *Journal Geophys. Res.*, **105 A5**, 10239–10248 (2000).
39. Sofia, U. J., and Meyer, D. M., *ApJ*, submitted (2001).
40. Jenkins, E. B., *ApJ*, **471**, 292–301 (1996).
41. Savage, B. D., and Sembach, K. R., *ApJ*, **379**, 245–259 (1991).
42. Joseph, C. L., and Jenkins, E. B., *ApJ*, **368**, 201–214 (1991).
43. Sofia, U. J., "Limits to Galactic Abundances based on the Interstellar Medium", in *Solar and Galactic Composition*, edited by R. F. Wimmer-Schweingruber, AIP Conf. Proc., Woodbury, NY, 2001, this volume.
44. Jenkins, E. B., "Elemental Abuundances in the Interstellar Atomic Material", in *Interstellar Processes*, edited by D. J. Hallenbach and H. A. Thronson, Reidel, Dordrecht, 1987, pp. 533–559.
45. Savage, B. D., Cardelli, J. A., and Sofia, U. J., *ApJ*, **401**, 706–723 (1992).
46. Savage, B. D., and Sembach, K. R., *ARAA*, **34**, 229–330 (1996).
47. Cardelli, J. A., Meyer, D. M., Jura, M., and Savage, B. D., *ApJ*, **467**, 334–340 (1996).
48. Sofia, U. J., Cardelli, J. A., Guerin, K. P., and Meyer, D. M., *ApJ*, **482**, L105–L108 (1997).
49. Cardelli, J. A., and Meyer, D. M., *ApJ*, **477**, L57–L60 (1997).
50. Meyer, D. M., Jura, M., and Cardelli, J. A., *ApJ*, **493**, 222–229 (1998).
51. Sofia, U. J., Meyer, D. M., and Cardelli, J. A., *ApJ*, **522**, L137–L140 (1999).
52. Federman, S. R., Lambert, D. L., Cardelli, J. A., and Sheffer, Y., *Nature*, **381**, 764 (1996).
53. Lambert, D. L., Sheffer, Y., Federman, S. R., Cardelli, J. A., Sofia, U. J., and Knauth, D. C., *ApJ*, **496**, 614–622 (1998).
54. Meyer, D. M., Cardelli, J. A., and Sofia, U. J., *ApJ*, **490**, L103–L106 (1997).
55. Holweger, H., "Photospheric Abundances: Problems, Updates, Implications", in *Solar and Galactic Composition*, edited by R. F. Wimmer-Schweingruber, AIP Conf. Proc., Woodbury, NY, 2001, this volume.

56. Stone, E.C., et al., *Space Sci. Rev*, **96**, 285–356 (1998).
57. Simpson, J.A., et al., *A&AS*, **92**, 365 (1992).
58. Greiner, D.E., et al., *IEEE Trans. Geosci. Elec.*, **GE-16**, 163 (1978).
59. Wiedenbeck, M.E. and Greiner, D.E., *ApJ*, **247**, L119-L122 (1981).
60. Leske, R.A., *Ap. J.* **405**, 567-583 (1993).
61. Stone, E.C., et al., *Space Sci. Rev.*, **21**, 355 (1977).
62. Lukasiak, A., et al., *ApJ*, **488**, 454–461 (1997).
63. Webber, W.R., et al., *ApJ*, **457**, 435–439 (1996).
64. Binns, W.R., et al., *Nucl. Inst. Meth.*, **185**, 415 (1981).
65. Binns, W.R., et al., *ApJ* **346**, 997–1009 (1989).
66. Bouffard, M., et al., *Astrophys. Space Sci.*, **84**, 3 (1982).
67. Engelmann, J.J., et al., *Astron. Astrophys.*, **233**, 96–111 (1990).
68. Streitmatter, R.E., et al., *Proc. 23th Intl. Cosmic Ray Conf. (Calgary)* **6**, 623 (1993).
69. Moiseev, A., et al., *ApJ*, **474**, 489 (1997).
70. Hams, T., et al., *Proc. 26th Intl. Cosmic Ray Conf. (Salt Lake City)*, **3**, 121 (1999).
71. Grunsfeld, J.M. et al., *ApJ*, **327**, L31–L34 (1988).
72. Yanasak, N.E., et al., *ApJ*, submitted (2001).
73. Bloemen, J.B.G.M., *ARA&A*, **27**, 469 (1989).
74. Stecker, F.W. and Jones, F.C., *ApJ* **217**, 843 (1977).
75. Webber, W.R., Lee, M.A., and Gupta, M., *ApJ* **390**, 96 (1992).
76. Meyer, J.P, Drury, L., and Ellison, D.C., *ApJ* **487**, 182-196 (1997).
77. Niebur, S.M., et al., "Secondary Electron-Capture-Decay Isotopes and Implications for the Propagation of Galactic Cosmic Rays", in *Acceleration and Transport of Energetic Particles Observed in the Heliosphere*, AIP Conf. Proc., **528**, Woodbury, NY, 2000, pp. 406–409.
78. George, J. S., Wiedenbeck, M. E., Barghouty, A. F., et al., "Cosmic Ray Source Abundances and the Acceleration of Cosmic Rays", in *Acceleration and Transport of Energetic Particles Observed in the Heliosphere*, edited by R. A. Mewaldt, J. R. Jokipii, M. A. Lee, E. Moebius, and T. H. Zurbuchen, AIP Conf. Proc., **528**, Woodbury, NY, 2000, pp. 437–440.
79. George, J. S., Wiedenbeck, M. E., Binns, W. R., et al., "The Phosphorus/Sulfur Abundance Ratio as a Test of Galactic Cosmic-Ray Source Models", in *Solar and Galactic Composition*, edited by R. F. Wimmer-Schweingruber, AIP Conf. Proc., Woodbury, NY, 2001, this volume.
80. Wiedenbeck, M. E., et al., "Constraints on Cosmic-Ray Acceleration and Transport from Isotope Observations", in *Acceleration and Transport of Energetic Particles Observed in the Heliosphere*, edited by R. A. Mewaldt, J. R. Jokipii, M. A. Lee, E. Moebius, and T. H. Zurbuchen, AIP Conf. Proc, **528**, Woodbury, NY, 2000,pp. 363–370.
81. Wiedenbeck, M. E., et al., *Space Sci. Rev.*, accepted (2001).
82. Binns, W. R., Wiedenbeck, M. E., et al., "Galactic Cosmic Ray Neon Isotope Abundances Measured on ACE", in *Acceleration and Transport of Energetic Particles Observed in the Heliosphere*, edited by R. A. Mewaldt, J. R. Jokipii, M. A. Lee, E. Moebius, and T. H. Zurbuchen, AIP Conf. Proc., **528**, Woodbury, NY, 2000, pp. 413–416.
83. Binns, W. R., Wiedenbeck, M. E, Christian, E. R., et al., "GCR Neon Isotopic Abundances: Composition with Wolf-Rayet Star Models and Meteoritic Abundances", in *Solar and Galactic Composition*, edited by R. F. Wimmer-Schweingruber, AIP Conf. Proc., Woodbury, NY, 2001, this volume.
84. Prantzos, N. et al., *Proc. 19th Intl. Cosmic Ray Conf. (La Jolla)* **3**, 167–170 (1985).
85. Binns, W. R., et al., "The Abundances of the Heavier Elements in the Cosmic Radiation", in *Cosmic Abundances of Matter*, (ed. C.J. Waddington), AIP Conf. Proc., **183**, 147 (1989).
86. Fowler, P. H., et al., *ApJ*, **314**, 739 (1987).
87. Westphal, A. J., et al., *Nature*, **357**, 50 (1988).
88. Wiedenbeck, M. E., et al., *ApJ*, **523**, L61–L64 (1999).
89. Whittet, D. C. B., *Dust in the Galactic Environment*, Institute of Physics Publishing, New York (1992).
90. Sofia, U.J., *ASP Conf. Series*, **122**, 77 (1997).
91. Geis, D. R. and Lambert, D. L., *ApJ*, **427**, 232 (1992).
92. Frisch, P. C., et al., *ApJ*, **525**, 492 (1999).
93. Linde, T. J. and Gombosi, T. I., *J. Geophys. Res.*, **105** 10411–10417 (2000).
94. DuVernois, M. A. and Thayer, M. R., *ApJ*, **465**, 982-984 (1996).
95. Ormes, J.F. and Frier, P. *ApJ*, **222**, 471 (1978).
96. Argast, D., Samland, M., Gerhard, O.E., and Thielemann, F.-K. , *A&A*, **356**, 873 (2000).
97. Boesgaard A.M., King J.R., Deliyannis C.P., and Vogt S.S., *ApJ*, **117**, 492 (1999).
98. Cappellaro, E. et al. *A&A*, **322**, 431 (1997).
99. Carretta, E., Gratton, R.G., and Sneden, C., *A&A*, **356**, 238 (2000).
100. Cayrel, R. et al., *Nature*, **409**, 691 (2001).
101. Chiappini, C., Matteucci, F., Beers, T. C., Nomoto, K., *ApJ*, **515**, 226 (1999).
102. Grevesse, N. and Sauval, A. J., *Space Sci. Rev.*, **85**, 161 (1998).
103. Idiart T. and Thévenin F., *ApJ*, **541**, 207 (2000).
104. Israelian G., García López R.J., and Rebolo R., , *ApJ*, **507**, 805 (1998).
105. Israelian G., Rebolo R., García López R.J., et al., *ApJ*, **551**, 833 (2001).
106. Jehin E., Magain P., Neuforge C., et al., *A&A*, **341**, 241 (1999).
107. Matteucci, F., Romano, D., and Molaro, P., *A&A*, **341**, 458 (1999).
108. McWilliam, A., *Ann. Rev. Astron. Astrophys.*, **35**, 503, (1997).
109. Primas F., Molaro P., and Castelli F., *A&A*, **290**, 885 (1994).
110. Sneden, C., et al., *ApJ*, **467**, 819–840 (1996).
111. Sneden, C., et al., *ApJ*, **533**, L139–L142 (2000).
112. Sneden, C., et al., *ApJ*, **536**, L85–L88 (2000).
113. Thomas, D., Greggio, L., and Bender, R., *MNRAS*, **296**, 119, (1998).
114. Truran, J.W., Cowan, J.J., Sneden, C., Burris, D.L., and Pilacowski, C.A., in *The First Stars*, Springer, eds. A. Weiss et al. (2000).
115. Tsujimoto, T., et al. , *MNRAS*, **277**, 945 (1995).
116. Westin J., et al., *ApJ*, **530**, 783 (2000).

Limits to Galactic Abundances based on Gas-Phase Measurements in the Interstellar Medium

U. J. Sofia

Department of Astronomy, Whitman College, Walla Walla, WA 99362 USA

Abstract.
Gas-phase (including ions and molecules) interstellar measurements of some elements can place tight constraints on their Galactic, or at least their total (gas + dust) local interstellar abundances. The interstellar medium within 1500pc of the Sun seems to be well represented by a single standard reference abundance. Here we discuss the limits placed on that standard abundance by measurements of the elements O, C, N, Kr and Sn in the interstellar medium. These are the elements for which we have some idea of their dust-phase compositions. The abundance of Kr in the local (out to 1500 pc) interstellar medium is subsolar at about 60% of the meteoritic value as compared to H. Oxygen in the same regions appears to have solar abundances. Tin, a primarily s-process element, is supersolar in the local ISM, however we cannot determine to what extent. Nitrogen and carbon abundances are more difficult to ascertain, but they, like oxygen, appear to be solar within measurement errors.

1. INTRODUCTION

It is difficult to measure the total abundance of elements in the ISM because it is composed of both gas and solid-grain material. We can, however, measure the gas-phase abundances of many cosmically abundant elements in neutral (hydrogen) interstellar clouds if we have access to the UV where most resonance lines of dominant species occur. Since the advent of spectroscopy with the Hubble Space Telescope, particularly the GHRS and STIS instruments, the measurement of interstellar absorption features has become quite precise. This together with major improvements in the determination of transition oscillator stregths, e.g. Bergeson and Lawler [1], Cardelli and Savage [2], Fitzpatrick [3], Theodosiou and Federman [4], has allowed for interstellar gas-phase abundances to be determined often to an accuracy of 0.05 dex (about 10%) or less. These abundance determinations are very straight forward; they are not model dependent in any way so they are extremely reliable. This leaves the dust-phase abundance of the elements as the major uncertainty in the total (gas + dust) ISM composition.

The subject of neutral cloud abundances is usually addressed by assuming a Galactic abundance and measuring a gas-phase abundance in order to infer the dust abundance of an element. Some authors, however, have instead tried to determine the Galactic composition by combining gas-phase measurements with model constraints on dust abundances [5, 6, 7, 8, 9, 10, 11]. One can do this because if we *a priori* understand the role of an element in dust, and can measure its abundance in the gas, then we will have a good determination of the total ISM abundance.

In §2 we will discuss the elements for which we believe we can place some limit on in the local ISM composition (out to about 1500 pc from the Sun). We will discuss how the limits compare to the Sun in §3. Our conclusions are summarized in §4.

2. ELEMENTS

There is a very limited number of elements for which we have reliable information concerning dust-phase abundances. For these elements, a reliable local total-ISM (gas + dust) abundance can be well determined. Information (i.e. limits) for the total abundances of some other elements can be inferred from those abundances, or from their gas-phase abundances alone. The number of elements for which the total abundances or limits can be determined is quite low, but some very important species are among them.

Oxygen

Very good measurements of the gas-phase interstellar oxygen abundance exist for interstellar clouds. Meyer

CP598, *Solar and Galactic Composition*, edited by R. F. Wimmer-Schweingruber

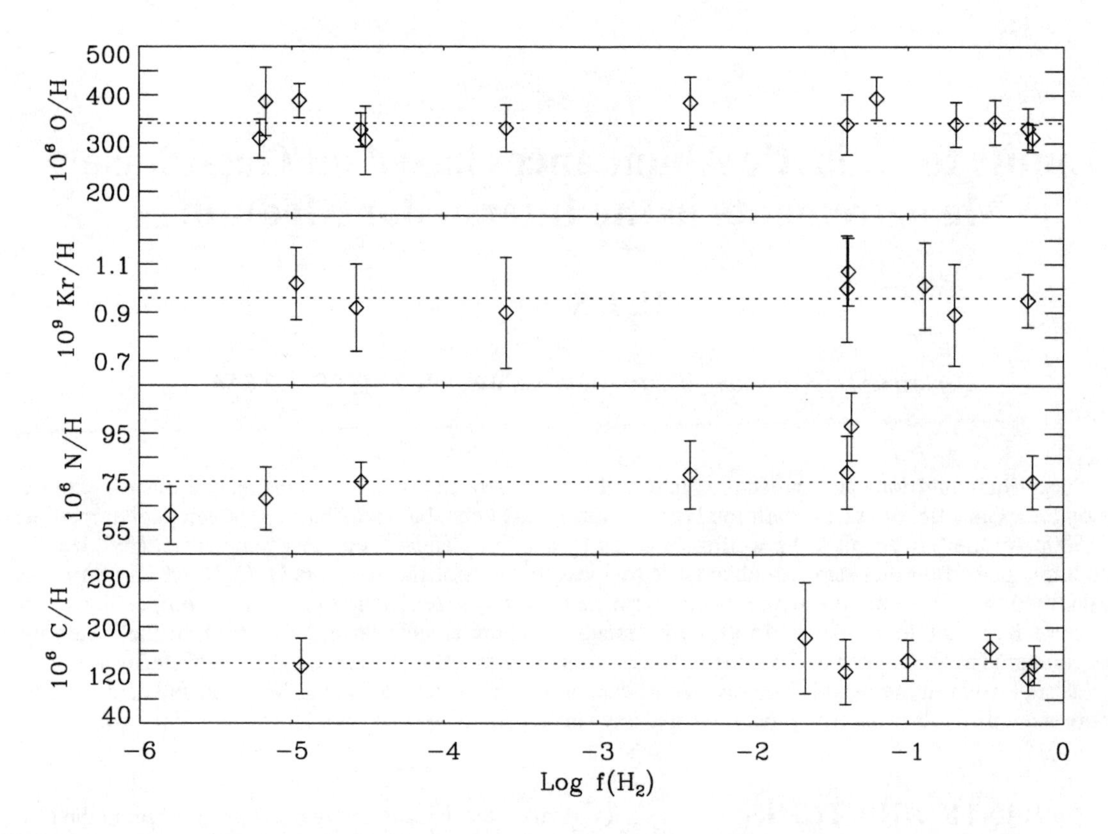

FIGURE 1. The interstellar gas-phase abundances of oxygen [10], krypton [12], nitrogen [13] and carbon [8, 14] with respect to hydrogen. The abundances are shown as number of atoms per 10^6 H nuclei except for Kr which is atoms per 10^9 H nuclei. The data with their 1 σ error bars are plotted as a function of the logarithmic fractional H_2 abundance. The dotted lines show the weighted average value for the measurements. Each of these elements apparently has a single abundance in the ISM out to about 1500 pc (the range over which the observations were made).

et al. [10] report the O/H number ratios for 13 diffuse cloud, $A_V \leq 1$, sightlines within 1500 pc of the Sun. They find that there are 319 ± 14 oxygen atoms in the gas per million H nuclei along these sightlines. The uniformity of the abundance with respect to hydrogen is amazing considering that the clouds have very different fractions of their H nuclei in the form of H_2, $f(H_2)=2N(H_2)/[N(HI) + 2N(H_2)]$; from about 60% toward ζ Oph to less than 1 part in 10^5 toward δ Ori [15] (see Figure 1). The large extremes in f(H_2) likely represent very different physical conditions in the clouds since this fraction indicates the equilibrium between the molecule's formation on grain surfaces and its chemical and/or photodestruction. The uniformity of the O/H values in diffuse clouds suggests that there is little exchange of O between the gas and dust, and that the local total-ISM O/H (gas + dust) abundance is probably constant. We should note that Cartledge et al. [16] have recently found that the constancy of the interstellar gas-phase O/H does not continue to denser regions such as translucent clouds. In these clouds the abundance of oxygen sometimes, but not always, shows an enhanced level of depletion.

O is likely the most abundant element in dust since it is so chemically reactive and abundant. The incorporation of O into dust, however, is limited in diffuse clouds since the grains do not have ice mantles, e.g. Savage et al. [17]. Therefore, O incorporation into grains in these sightlines is limited by the mineralogy of dust. Cardelli et al. [8] estimate that no more than 180 O atoms can be incorporated into dust per 10^6 H nuclei in a diffuse cloud sightline. If we assume that the maximum number of O atoms are in dust, then we conclude that the local total ISM abundance of oxygen is 499 ± 14 O atoms per million H nuclei. The uncertainty listed for this number only includes the uncertainty in the diffuse cloud gas-phase O/H; it does not include the error in the number of O atoms incorporated into the dust.

Krypton

Cardelli and Meyer [12] report 11 measurements of the gas-phase Kr/H in the diffuse neutral ISM and find that, like O/H, the values are uniform over a large range of $f(H_2)$ out to 1500 pc (see Figure 1). This again suggests a uniformity of the local interstellar abundances, and that little exchange of Kr occurs between the gas and dust phases of the ISM. This second point is not surprising because, as a noble gas, Kr is not likely to be incorporated into grains. Therefore, we believe that the dust-phase characteristics of Kr are well understood and that we can say that the Kr/H in the gas is a good representation of the total Kr/H abundance in the ISM. Cardelli and Meyer [12] find that there are 0.96 ± 0.05 Kr atoms per 10^9 H nuclei in the local ISM.

Kr is the only noble gas whose total interstellar abundance can be found through UV absorption studies. The only other candidate that has a measureable line of its dominant ion in the neutral ISM is argon. Its abundance, however, is greatly affected by ionization. Ionization is not an issue for the lines of sight with measured interstellar krypton because they sample substantially denser clouds (and are therefore better shielded from photoionization) than the sightlines with measured interstellar argon [18].

Nitrogen

Like krypton, nitrogen is not expected to be incorporated into dust. It is blocked from grain core formation because it is locked up as N_2 molecules in stellar atmospheres [19], and it should not form grain mantles in its atomic form [20]. Therefore, we can determine its total interstellar abundance by making the reasonable assumption that it is the same as the interstellar gas-phase abundance. Meyer et al. [10] measured the N/H abundance in 7 diffuse interstellar clouds and found that they all agreed within their errors at a value of 75 ± 4 N atoms per 10^6 H nuclei. This uniformity of the measurements again reinforces the notion that the total interstellar abundances are similar to a distance of at least 1500 pc from the Sun, the range over which the data are sampled (see Figure 1).

Carbon

The measured gas-phase C/H ratios in diffuse interstellar clouds with very different $f(H_2)$ values are constant at 140 ± 20 C atoms per 10^6 H nuclei (see Figure 1). This suggests little exchange between the gas and dust phases of C in the ISM, and once again supports the idea of a uniform ISM abundance out to at least 1500 pc. In order to use the gas-phase C abundance to find the Galactic abundance one needs to know how much of it is locked into grains. This is a difficult quantity to determine; current models vary greatly in the amount of C that they incorporate into grains.

Another way to find the Galactic C abundance is to assume that young stars are good representations of the interstellar composition. The abundances in these stars, however, do not agree and have C/O ratios that range from about 0.5 in B stars [11] to 0.7 in young F and G stars (and the Sun) [11, 21]. We therefore, cannot place a hard limit on the total Galactic C abundance. A Galactic C-to-O ratio of 0.5 (the B star ratio) is a likely lower limit to the possible C abundance. This would imply a total (gas+dust) ISM abundance of 250 C atoms per 10^6 H nuclei (based on the O abundance of 499 per 10^6 H nuclei given above). This is a practical lower limit because dust models have trouble reproducing extinction with less than this total C abundance [5, 6, 9]. B stars are bad representations of the ISM composition so their C/O is probably not applicable [11]. A local Galactic C abundance of 350 C atoms per 10^6 H nuclei (making the Galactic C/O the same as the Sun's) is probably a more realistic estimate [11].

Tin

Tin is an interesting element because it is formed almost entirely by the s-process in moderate-to-low mass AGB stars. Its total abundance in the ISM with respect to the Sun may therefore tell us something about stellar enrichment of the Galaxy in the past 4.5 Gyr. The gas-phase interstellar tin-to-hydrogen ration has been measured in 14 diffuse sightlines by Sofia et al. [22]. They found that a higher fraction of Sn was in the gas phase in low as compared to high $f(H_2)$ sightlines (see Figure 2). This is not a surprising characteristic for an element that is being exchanged between the gas and dust. In high $f(H_2)$ sightlines the grains are better protected, as are the H_2 molecules, against destruction [23]. The Sn/H ratio in the gas ranges from about 0.6 - 1.9 Sn atoms per 10^{10} H nuclei. We do not know enough about tin's incorporaton into dust to find a definitive total Sn/H for the diffuse ISM.

3. COMPARISON OF THE ISM ABUNBDANCES TO THE SUN

The argument has been made for years that the Sun has unusual abundances. The interstellar community was a great advocate for this because problems occurred when the dust-phase composition was inferred from the solar

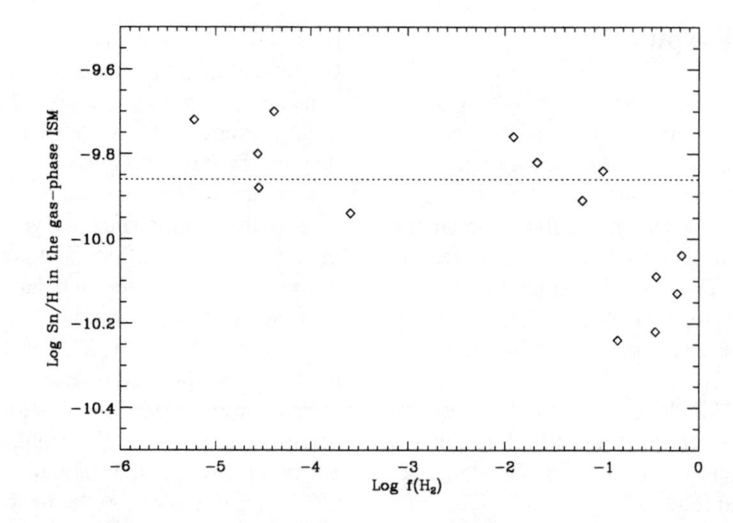

FIGURE 2. The logarithmic interstellar gas-phase abundance of tin with respect to hydrogen as a function of logarithmic fractional H_2 abundance. Unlike the elements in Figure 1, Sn shows enhanced depletions in denser diffuse-cloud regions. The lowest f(H_2) sightlines have an average abundance that is slightly supersolar. The solar abundance is shown by the dotted line.

standard [5]. A particular problem was the large dust-phase oxygen abundance. There was simply no place to chemically attach the number of oxygen atoms inferred to be in the dust. This led to the suggestion that the Sun is not a reasonable proxy to interstellar abundances and drove the search for a better representation of the interstellar standard. The reported solar abundance of oxygen, however, has been slowly dropping over the past several years [24, 25, 26, 27]. A recent photospheric measurement reported in these proceedings now places the oxygen abundance at log(O/H) = -3.264 ± 0.078 [21], significantly down from the Anders and Grevesse [24] value of -3.07 from 13 years earlier. This change brings the O/H number ratio in the Sun down to 545 ± 99 from 851 oxygen atoms per million H nuclei. Within the reported errors, this solar abundance now agrees with the interstellar abundance proposed by Cardelli et al. [8] and Meyer et al. [10].

The Kr/H value found for the ISM is approximately 60% below the meteoritic value [24]. Since the abundance of noble gases are so difficult to determine, the interstellar value may be a more reasonable approximation for the solar abundance.

The nitrogen abundance given above for the ISM is slightly lower than the solar photospheric value reported by Holweger [21]; 85 ± 22 N atoms per 10^6 H nuclei. This could mean that some N is incorporated into dust. The large uncertainty in the solar value, however, does not preclude it from agreeing with the total ISM abundance.

The C/H abundance is likely close to the solar value. The Sun and young F and G disk stars, which seem to well represent the local interstellar standard [11], have the same ratios of C/O. Since the total-interstellar O abundance now appears to agree with the solar value, this would imply that the total C/H ratio in the local ISM is also approximately solar.

Ususally an element such as tin with variable depletion levels tells us little about the total ISM abundance. Tin is different, though, because in the lowest f(H_2) sightlines, i.e. the sightlines with the most Sn in the gas, the measured gas-phase Sn/H is supersolar (by about 0.07 ± 0.04 dex). A characteristic of elements that show variable depletions is that even in the least depleted sightlines some of the element remains in dust. If this is the case for Sn, then we cannot say exactly how overabundant it is (because we do not know how much is in the dust in those regions). Sn is in the same column of the periodic table as Si, so they are chemically similar. We might expect, then, that they will be similarly incorporated into dust. In the lowest f(H_2) sightlines, Si is depleted by about 0.2 dex with respect to the Sun. This provides a possible limit of about 0.3 dex for the ISM overabundance of Sn/H. In any case, we can be confident that Sn is enriched in the ISM as compared to the Sun. This is the only element that has a measured gas-phase abundance that is supersolar.

4. CONCLUSIONS

Out to 1500pc, the interstellar gas-phase abundances of C/H, O/H and Kr/H are stable in diffuse clouds with different physical conditions as measured by the fraction of H in the form of H_2. The weighted average logarithmic gas-phase element-to-H abundances are -3.85 ± 0.06 [14], -3.50 ± 0.02 [10] and -9.02 ± 0.02 [12] for C,

O, and Kr. The RMS scatter values for these abundances are relatively low at 0.07, 0.05, and 0.06, respectively. This uniformity suggest that there *is* a standard local ISM abundance.

Interstellar measurements can be a useful tool for determining that standard abundance for those elements whose dust incorporation is well understood. From gas-phase measurements we find that the local total-interstellar abundances of O, N and probably C agree well with the solar photospheric measurements of Holweger [21]. The abundance of krypton appears to be low compared to the solar system value. This may simply be the result of the difficulty of measuring Kr in meteroites. Sn is overabundant with respect to the Sun, however its incorporation into dust grains does not allow us to determine by how much.

ACKNOWLEDGMENTS

This work was supported in part by a grant from Whitman College, and by the NASA grant NAG5-8249.

REFERENCES

1. Bergeson, S. D., and Lawler, J. E., *Astrophysical Journal*, **414**, L137–L140 (1993).
2. Cardelli, J. A., and Savage, B. D., *Astrophysical Journal*, **452**, 275—285 (1995).
3. Fitzpatrick, E. L., *Astrophysical Journal*, **482**, L199–L202 (1997).
4. Theodosiou, C. E., and Federman, S. R., *Astrophysical Journal*, **527**, 470–473 (1999).
5. Sofia, U. J., Cardelli, J. A., and Savage, B. D., *Astrophysical Journal*, **430**, 650–666 (1994).
6. Snow, T. P., and Witt, A. N., *Science*, **270**, 1455 (1995).
7. Snow, T. P., and Witt, A. N., *Astrophysical Journal*, **468**, L65–L68 (1996).
8. Cardelli, J. A., Meyer, D. M., Jura, M., and Savage, B. D., *Astrophysical Journal*, **467**, 334–340 (1996).
9. Mathis, J. S., *Astrophysical Journal*, **472**, 643–665 (1996).
10. Meyer, D. M., Jura, M., and Cardelli, J. A., *Astrophysical Journal*, **493**, 222–229 (1998).
11. Sofia, U. J., and Meyer, D. M., *Astrophysical Journal*, **554**, L221–L224 (2001).
12. Cardelli, J. A., and Meyer, D. M., *Astrophysical Journal*, **477**, L57–L60 (1997).
13. Meyer, D. M., Cardelli, J. A., and Sofia, U. J., *Astrophysical Journal*, **490**, L103–L106 (1997).
14. Sofia, U. J., Cardelli, J. A., Guerin, K. P., and Meyer, D. M., *Astrophysical Journal*, **482**, L105–L108 (1997).
15. Bohlin, R. C., Savage, B. D., and Drake, J. F., *Astrophysical Journal*, **224**, 132–142 (1978).
16. Cartledge, S. I. B., Meyer, D. M., Lauroesch, J. T., and Sofia, U. J., *Astrophysical Journal*, **submitted** (2001).
17. Savage, B. D., Cardelli, J. A., and Sofia, U. J., *Astrophysical Journal*, **401**, 706–723 (1992).
18. Sofia, U. J., and Jenkins, E. B., *Astrophysical Journal*, **499**, 951–965 (1998).
19. Gail, H.-P., and Sedlmayr, E., *Astron. and Astrophys.*, **166**, 225–236 (1986).
20. Barlow, M., *MNRAS*, **183**, 417–434 (1978).
21. Holweger, H., "Photospheric Abundances: Problems, Updates, Implications", in *Solar and Galactic Composition*, edited by R. F. Wimmer-Schweingruber, Springer-Verlag, Berlin, 2001, pp. XXX–XXX.
22. Sofia, U. J., Meyer, D. M., and Cardelli, J. A., *Astrophysical Journal*, **522**, L137–L140 (1999).
23. Cardelli, J. A., *Science*, **265**, 209–211 (1994).
24. Anders, E., and Grevesse, N., *Geochim. et Cosmochim. Acta*, **53**, 197–214 (1989).
25. Grevesse, N., and Noels, A., *Physica Scripta*, **T47**, 196–209 (1993).
26. Grevesse, N., Noels, A., and Sauval, A., "Standard Abundances", in *Cosmic Abundances*, edited by S. S. Holt and G. Sonneborn, Astronomical Society of the Pacific, San Francisco, 1996, pp. 117–124.
27. Grevesse, N., and Sauval, A., *Space Sci. Rev.*, **85**, 161–174 (1998).

Galactic Chemical Evolution

C. Chiappini* and F. Matteucci†

*Osservatorio Astronomico di Trieste - Via G.B. Tiepolo 11 - Trieste - TS - 34131 - Italy
†Università di Trieste - Via G. B. Tiepolo 11 - Trieste - TS - 34131 - Italy

Abstract. In this paper we review the current ideas about the formation of our Galaxy. In particular, the main ingredients necessary to build chemical evolution models (star formation, initial mass function and stellar yields) are described and discussed. A critical discussion about the main observational constraints available is also presented. Finally, our model predictions concerning the evolution of the abundances of several chemical elements (H, D, He, C, N, O, Ne, Mg, Si, Ca and Fe) are compared with observations relative to the solar neighborhood and the whole disk. We show that from this comparison we can constrain the history of the formation and evolution of the Milky Way as well as the nucleosynthesis theories concerning the Big Bang and the stars.

INTRODUCTION

In the past years a great deal of theoretical work has appeared concerning the chemical evolution of the Milky Way [59, 10, 9, 95, 99, 19, 18, 13, 71, 72, 8, 34, 17].

This is a consequence of the improvement of observations of chemical abundances in stars and gas, and the progress in our understanding of stellar evolution processes. Stellar evolution models are now converging to a more self-consistent description (see Thielemann this conference) and several research groups in the world are now able to compute one of the fundamental ingredients for chemical evolution models, the stellar yields. Calculations are now available for an extended range of stellar masses and for different initial metallicities [eg. 93, 50, 102].

Different approaches have been used so far for the description of the Milky Way formation and evolution. The most common ones are: *i)* Serial Formation in which the halo, thick and thin disk form in temporal sequence [eg. 59]; *ii)* Parallel Formation in which the various components start forming at the same time out of the same gas but evolve at different rates [68]; *iii)* The two-infall approach (see below).

As shown by Chiappini et al. [19], the most likely scenario, in the light of recent data, is the one where the Galaxy forms as a result of two main infall episodes. The serial approach predicts no overlapping in metallicities between the different stellar populations, against observational evidence [6]. Both approaches *i)* and *ii)* are at variance with the distribution of the angular momentum of stars in different components [104], indicating that the gas, out of which the stellar halo formed, did not participate in the formation of the disk. In the two-infall model, the first episode forms the halo and the gas lost by the halo rapidly (roughly 0.3-0.5 Gyr) accumulates at the center with the consequent formation of the bulge. During the second episode, a much slower infall of primordial gas gives rise to the disk with the gas accumulating faster in the inner than in the outer regions. In this scenario the formation of the halo and disk are almost completely disentangled although some halo gas eventually falls into the disk. This mechanism for disk formation is known as the "inside-out" scenario [49] and is quite successful in reproducing the main features of the Milky Way [19] as well as of external galaxies especially concerning abundance gradients [see 73]. In this paper we show the model predictions once the two-infall approach is adopted.

We believe that by improving the Milky Way model we will provide a solid basis for a more detailed understanding of the history of chemical enrichment not only of our Galaxy but also of other spirals. In particular, the slow timescale for the formation of the Milky Way disk implied by our chemical evolution model suggests that at high redshifts we should see smaller disks. An important aspect of this work will be to provide specific predictions for galaxy sizes of Milky Way-like galaxies as a function of redshift, which could in principle be tested against the growing body of observational data on distant galaxies. Although the interpretation of the observational data is still controversial [see 88], they clearly represent a powerful test for disk formation models.

CP598, *Solar and Galactic Composition*, edited by R. F. Wimmer-Schweingruber

OBSERVATIONAL CONSTRAINTS: A CRITICAL DISCUSSION

A good model of chemical evolution of the Galaxy should reproduce a number of constraints which is larger than the number of free parameters. Therefore, it is very important to choose a high quality set of observational data to be compared with model predictions. The set of observations that should be explained by the models are listed below. In what follows we comment on the most important ones.

Solar Vicinity

- Observational constraints at solar vicinity (Table 1)
- Solar Abundances (Table 2)
- Abundance ratios as function of [Fe/H]
- Age-Metallicity relationship
- Current supernovae rates (Table 1)
- Metallicity distribution of G dwarf disk stars

The solar abundances in principle should represent the chemical composition of the interstellar medium in the solar neighbourhood at the time of Sun formation (4.5 Gyrs ago). However, the abundance of oxygen in the Orion nebula is roughly solar when the new value reported by Holweger (this conference) is taken as a reference. This could be interpreted as an indication that the evolution of the solar vicinity in the last 4.5 Gyrs was very slow. In fact, when including the threshold in the process of star formation, we predict a small increase of the elements produced by massive stars (as is the case of oxygen) as a consequence of the threshold mechanism in the star formation rate. However, we have to consider also the possibility that the solar composition is not representative of the local ISM 4.5 Gyrs ago, an alternative being that the Sun was born in a region closer to the Galactic center and then moved to the present region (but see Holweger, this conference). The absolute solar values taken at a face value are not a very tight constraint to chemical evolution models as they are dependent on many model parameters.

The behaviour of the [α/Fe] ratio as a function of [Fe/H] is very useful to constrain chemical evolution models. The α-elements (O, Ne, Mg, etc..) are produced only by type II SNe (which have high mass progenitors with short life time), whereas most of the iron is produced by type Ia SNe, which are believed to be the result of the explosion of C-O white dwarfs in binary systems. Iron release from type Ia SNe begins not before several 10^7 years after the birth of a stellar generation, and the bulk of iron enrichment takes up to some Gyrs, depending on the assumptions on the binary system characteristics, explosion mechanism and star formation rate.

TABLE 1. Observed and predicted quantities at $R_\odot$ and $t = t_{Gal}$

	Model A [17]	Observed*
Fraction of halo stars	10 %	2—10 %
SNIa†	0.4	0.3 ± 0.2
SNII	0.8	1.2 ± 0.8
$SFR(R_\odot, t_{Gal})$**	2.6	2—10
$\Sigma_{gas}(R_\odot, t_{Gal})$‡	7.0	7—16
$\Sigma_{stars}(R_\odot, t_{Gal})$§	36.3	35 ± 5
$\Sigma_{gas}/\sigma_{tot}\ (R_\odot, t_{Gal})$	0.13	0.05—0.20
$\dot{\Sigma}_{infall}(R_\odot, t_{Gal})$¶	1.0	0.3—1.5
$\Delta Y/\Delta Z$	1.9	1—3
Nova outbursts (yr^{-1})	22	20—30
D_p/D_{now}	1.5	< 3

* References for the observed values can be found in [17, 79]
† SN rates are given in units of century^{-1}
** in units of $M_\odot$ pc^{-2} Gyr^{-1}
‡ in units of $M_\odot$ pc^{-2}
§ in units of $M_\odot$ pc^{-2}
¶ in units of $M_\odot$ pc^{-2} Gyr^{-1}

TABLE 2. Solar abundances by mass

	Model*	AG89†	GS98**	H2001‡
H	0.71	0.70		
D	3.3 (−5)	4.8 (−5)		
^{3}He	2.2 (−5)	2.9 (−5)		
^{4}He	2.7 (−1)	2.7 (−1)	2.7 (−1)	
^{12}C	3.5 (−3)	3.0 (−3)	2.8 (−3)§	3.3(−3)
^{16}O	7.1 (−3)	9.6 (−3)	7.7 (−3)	6.2(−3)
^{14}N	1.6 (−3)	1.1 (−3)	8.3 (−4)	8.5(−4)
^{13}C	4.7 (−5)	3.7 (−5)		
^{20}Ne	0.9 (−3)	1.6 (−3)		1.6(−3)
^{24}Mg	2.4 (−4)	5.1 (−4)		5.9(−3)
Si	6.9 (−4)	7.1 (−4)	7.1 (−4)	7.2(−4)
S	3.0 (−4)	4.2 (−4)	4.9 (−4)	
Ca	3.9 (−5)	6.2 (−5)	6.5 (−5)	
Fe	1.31 (−3)	1.3 (−3)	1.3 (−3)	1.1(−3)
Cu	7.7 (−7)	8.4 (−7)	7.3 (−7)	
Zn	2.3 (−6)	2.1 (−6)	1.9 (−6)	
Z	1.6 (−2)	1.9 (−2)	1.7 (−2)	

* Model A of [17] at 4.5 Gyr ago
† Anders and Grevesse [3] - meteoritic values
** Grevesse and Sauval [36] photospheric values. The meteoritic values reported by [36] for Si, S, Ca, Fe, Cu, and Zn agree with the photospheric ones except for S, Cu, and Zn, for which they report 3.6 (−4), 8.7 (−7), and 2.2 (−6), respectively. The value listed here for the abundance by mass of ^{4}He is that at the time of Sun formation
‡ Holweger, this conference
§ The observed abundances of GS98 and H2001 are elemental thus including different isotopes.

Therefore, the delayed arrival of the iron produced by type Ia SNe is responsible for the observed decrease in the [α/Fe] ratio as a function of the iron abundance in the solar vicinity [58, 59]. This fact makes the [α/Fe] ratio a very important tool to access the formation timescales of different galactic components and can be used as a cosmic clock.

In particular, recent data by Gratton et al. [35] suggest that there was a hiatus in the process of star formation during the halo-thin disk transition. Such a hiatus seems to be real since it is observed both in the plot of [Fe/O] versus [O/H] [19, 35] and in the plot of [Fe/Mg] versus [Mg/H] [32]. The evidence for this is shown by the steep increase of [Fe/O] and [Fe/Mg] at a particular value of [O/H] and [Mg/H], respectively, indicating that at a certain epoch (coinciding with the halo-disk transition) SNe II, responsible for the production of O and Mg, stopped exploding while Fe, produced by the long living SNe Ia, continued to be produced (see figure 4).

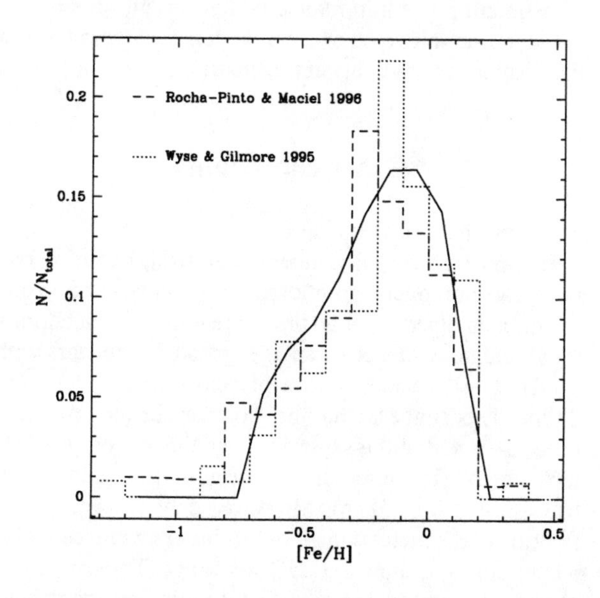

FIGURE 1. G-dwarf metallicity distribution. The curve shows the best model of Chiappini et al. [17].

Another fundamental constraint on chemical evolution models is the metallicity distribution of the G-dwarfs for the solar vicinity. The G-dwarf metallicity distribution is representative of the chemical enrichment of the Galaxy since these stars have lifetimes larger than or equal to the age of the Galaxy and hence can provide a complete record of its chemical history. Until 1995 the G-dwarf metallicity distribution which was adopted to compare with the models was the one published by Pagel and Patchett [64] and revised by Pagel [66] and Sommer-Larsen [91]. Later on, two different groups using new observations and up to date techniques, published new data on the G-dwarf distribution [77, 103]. The basic differences are in the new adopted catalog, namely, the Third Gliese Catalog, and in the calibration used to determine the metallicity. This calibration is based on Strömgren photometry which allows a more reliable estimation of the metallicity than does the one based on the UBV photometric system adopted in [64]. In particular, the new data show a well-defined peak in metallicity (between [Fe/H]=−0.3 and 0.), which was not evident in the previous data (see figure 1).

The age-metallicity relation [24] is not a strong constraint since it can be fitted by a variety of model assumptions and it shows a very large spread.

Radial Profiles - Disk

- Star Formation Rate profile
- Gas and Stars density distribution
- Radial Abundance Gradients

The observed ($HI + H_2$) distribution is taken from [21]. The surface density distribution of the total gas Σ_{gas} is obtained from the sum of the HI and H_2 distributions, $\Sigma_{HI} + \Sigma_{H_2}$, accounting for the helium and heavy elements fractions (thick line in the lower left panel of figure 2).

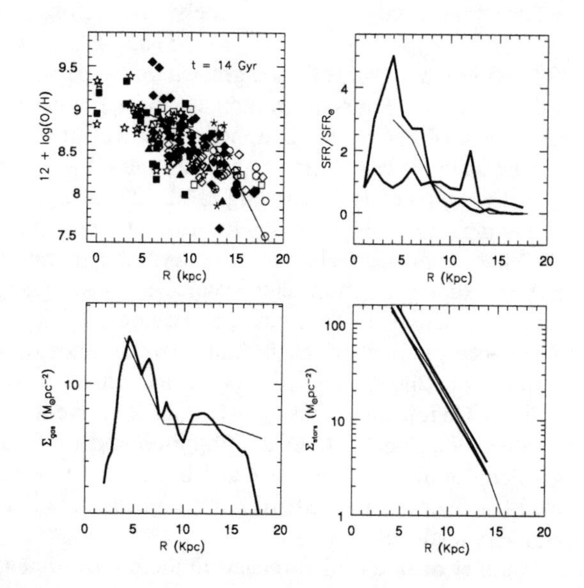

FIGURE 2. Predictions (thin lines) from the best model of [17]. In the upper left panel the oxygen gradient is shown. The upper right panel shows the radial profile of SFR (normalized by the star formation rate at the solar vicinity). The left and right lower panels show the gas and stellar surface densities respectively (the thick lines enclose the areas where the observations lie).

The stellar profile is exponentially decreasing outwards, with characteristic scale length $R_{stars} \sim 2.5-3$ kpc [83, 30]. Moreover, COBE observations suggest that the stellar disk has an outer edge of 4 kpc from the Sun [30]. To compare our model predictions on the stellar density profile along the Galactic disk to the observed one we consider $\Sigma_{stars}(R_{\odot}, t_{Gal}) = 35 \pm 5\ M_{\odot}\ pc^{-2}$ [33] and $R_{stars} = 2.5$ kpc (see lower right panel of figure 2).

The distributions of supernova remnants [37], of pulsars [51], of Lyman-continuum photons [40] and of molecular gas [74] all can be used to derive an estimate for the SFR along the Galactic disk. Since these observables cannot directly provide the absolute SFR without further assumptions — e.g. on the IMF and mass ranges for producing pulsars, supernovae, etc. [see 47], it is common practice to normalize them to their values at the solar radius, and then to trace the radial profile SFR(R)/SFR($R_\odot$) (the thick lines plotted in the upper right panel of figure 2 refer to the upper and lower limits obtained from the observational data listed above).

Among the radial constraints the most important and precise are the abundance gradients.

However, over the past decade, different authors using different observational tools often came to contradictory views on both the shape, the magnitude and the evolution of the abundance gradients along the disk. The controversial results originate from both theoretical and observational considerations.

Data from several sources, namely, HII regions [87, 29, 89, 1, 82]; PNe of type II [52, 53, 55] and B stars [90, 38] suggest a value for the gradient of oxygen of the order of $\simeq -0.07$ dex/kpc in the galactocentric distance range of 4-14 kpc (figure 2, upper left panel - gradients of other abundance elements can be seen in [17]).

However, recently Deharveng et al. [22] analyzed a new sample of 34 H II regions located between 6.6 and 17.7 kpc from the Galactic center and after a careful estimate of the electron temperatures in those objects they obtained, using their best observations, an oxygen abundance gradient which is flatter (by a factor of 2) than the one obtained in previous works based on H II regions. Their result seems to be in good agreement with results by Esteban et al.[26, 27, 28]. Proposals for flatter gradients or bimodal ones have also been made by works based on open clusters [31, 100] although the situation is still very unclear.

Another open question related to the observed abundance gradients concerns their variation with time: do the gradients steepen or flatten with time? This question cannot be answered properly by the presently available data [see 56]. However, PNe are the most promising objects to solve this problem. As it has been extensively discussed in the past few years, PNe Galactic distribution, kinematics, chemical composition and morphology clearly indicate that PNe comprise objects belonging to different populations [69, 54]. Previous work has shown that disk objects of type II are particularly useful in tracing the chemical enrichment of the interstellar medium at the time of the formation of the PN progenitors [53, 52].

In a recent work, Maciel and Quireza [55] obtained the gradients for a sample that includes the objects from Maciel and Köppen [53], Maciel and Chiappini [52] and Costa et al. [20] (the latter consists of a sample of PNe near the anticentre direction intended to derive a better estimate of the gradients at Galactocentric distances larger than the solar position). Their main conclusions were: *i)* there is an average gradient of -0.04 to -0.07 dex/kpc for what they call "inner" Galaxy (between 4 and 10 kpc — those authors assume $R_\odot$ = 7.6 kpc); *ii)* the gradients show a small variation for the different element ratios (see their table 2); *iii)* the PN gradients are generally slightly flatter than those derived from younger objects; *iv)* for larger galactocentric distances the PN gradients show some flattening. However, several uncertainties such as the small number of objects measured at galactocentric distances larger than 12 kpc, the problem of assigning precise ages to disk PNe, and the difficulties in estimating the importance of dynamical effects still prevent a definitive answer on the temporal variation of the abundance gradients in the galactic disk.

Other constraints

• Relative ages of Globular Clusters:

In a recent work, Rosenberg et al. [81], based on two new large homogeneous photometric databases of 35 and 15 globular clusters found that there is no age-metallicity trend and no evidence of an age spread for clusters with [Fe/H]< -1.2 and out to a galactocentric distance of 25kpc. This suggests that the globular cluster formation process started at the same zero age throughout the "internal halo" (by "internal" we mean up to a galactocentric distance of 25kpc). Moreover they showed that a fraction of the metal rich globular clusters were formed at a later time and show a $\simeq 25\%$ lower age. Those younger clusters located at larger galactocentric distances have typical halo kinematics. This is a very strong constraint and suggests that the timescale for the formation of the "inner-halo" was short, which is in agreement with the abundance ratios of metal-poor stars [18].

• Abundances and metallicity distribution of Bulge:

Abundance ratios are sensitive to details of galaxy evolution, and therefore represent a powerful tool for the study of the Galactic bulge formation. Recently, Barbuy [5] presented abundance ratios for bulge stars belonging to globular clusters. Those data show that many α-elements are overabundant suggesting a rapid bulge formation. Moreover, Matteucci et al. [60] suggested that in order to fit the observed metallicity distribution of giant bulge stars obtained by Rich [76] and McWilliam and Rich [62], the bulge should have been formed in a short timescale and probably with a top-heavy IMF. Those results are in agreement with the suggestion by Wyse and Gilmore [104], from angular momentum arguments, that the bulge formed out of the gas left from the halo formation [see also 25]. More data are needed, in particular to investigate the possible existence of abundance gradients in the bulge. In fact, the radial abundance gradients and the overall metallicity distribution are very useful in discriminating between the main bulge formation scenarios proposed so far, namely, monolithic or secular.

THE MODEL FOR THE MILKY WAY

In what follows the ingredients to build chemical evolution models and the specific choices that apply to our two-infall model are discussed briefly.

Ingredients

Star formation Rate

The process of star formation is still not understood and this is why we are forced to adopt a parametrization for this function. Many are the parametrizations adopted in the literature. A common approach is to use the so called "Schmidt-law" in which the star formation rate depends on a power between 1 and 2 of the volume or surface gas density. However, a known result is that to be able to reproduce the observed radial profiles (gas, stars, star formation rate and abundances) it is not enough to consider a radial variation of the thin disk formation timescale, but a radial dependence of the star formation itself is required [59].

Many are the parametrizations for such radial variation. One possibility is to assume that the star formation rate is not only a function of the gas density but also of the angular rotation speed of the gas. Another approach is to consider that the SFR has also a dependence on the total surface mass density (adopted in the present model). This last parametrization accounts for the feedback mechanism between star formation and heating of the interstellar medium, due to supernovae and stellar winds [eg. 92, 59, 11]. Observational evidence for such a law is provided by Dopita and Ryder [23]. Moreover, Kennicutt [46] suggested that either a SFR dependence on the total surface mass density or on the angular rotation speed of the gas leads to a good fit of the SFR measured from the Hα emission in other spiral galaxies. Kennicutt [45] (and more recently Martin and Kennicutt [57]) has also suggested the existence of a threshold gas density for the star formation of a few $M_{\odot}pc^{-2}$, below which the star formation stops.

Our prescription for the star formation rate (SFR) is [17]:

$$SFR \propto \Sigma_{tot}^{k_2} \Sigma_g^{k_1} \qquad (1)$$

where Σ_{tot} is the total surface mass density, Σ_g is the surface gas density, $k_1 = 1.5$ and $k_2 = 0.5$. A threshold in the surface gas density ($\sim 7\ M_{\odot}pc^{-2}$) is also assumed; when the gas density drops below this threshold the star formation stops.

The Initial Mass Function

There is at present no clear direct evidence that the IMF in the Galaxy has varied with time. A detailed discussion about possible observed variations in the IMF in different environments is given by Scalo [85], but such variations are comparable with the uncertainties still involved in the IMF determinations. The present uncertainties in the observational results prevent any conclusion concerning a universal IMF.

However, a variable IMF, which formed relatively more massive stars during the earlier phases of the evolution of the Galaxy compared to the one observed today in the solar vicinity, has often been suggested as being one of the possible solutions for the G-dwarf problem (namely the deficiency of metal-poor stars in the solar neighborhood when compared with the number of such stars predicted by the simple model). Such an IMF would also be physically plausible from the theoretical point of view if the IMF depends on a mass scale such as the Jeans mass [48]. Given the uncertainties in both theoretical and observational grounds, the proposed IMFs can in principle be tested only by means of a detailed chemical evolution model.

The effect of a variable IMF in a chemical evolution model was studied by Chiappini et al. [15]. In this work it was shown that a better agreeement with the observational constraints is obtained for a constant rather than variable IMF and this conclusion is mostly based on the abundance gradients and radial profiles of gas and SFR. Therefore a constant IMF should still be preferred when describing the evolution of the Galactic disk. In the present model we adopt the Scalo [86] IMF, constant in time.

Nucleosynthesis Prescriptions

One of the most important ingredients for chemical evolution models is the nucleosynthesis prescription and the computation of stellar yields. Below we describe the prescriptions adopted here:

• Light Elements produced during the Big Bang:

The elements formed during the Big Bang were H, D, ^3He, ^4He and ^7Li. Some of them were then consumed/produced by stars and hence what we observe today is a convolution of their primordial values with their evolution due to stellar processes. In the case of D, the primordial abundance value is an upper limit as this element is only consumed during stellar evolution. In fact the D evolution represents a very important constraint to chemical evolution models. As discussed in Tosi et al. [98] chemical evolution models which can reproduce the majority of observational constraints predict a depletion of D abundance of no more than a factor of 2-3. This of course can also be used to put limits on the D primordial abundance value. We assume primordial values for each of those elements and then trace their evolution in time. Our predictions for the evolution of D, ^{3}He, ^{4}He and ^{7}Li in the Galaxy can be found in the WG5 contribution [this conference and 16, 80]. In our model we have included the extra-mixing process for the ^{3}He production [14]. This "non-standard" convective mixing process would occur in RGB stars further consuming ^{3}He. In this way we are able to explain the ^{3}He observed in the interstellar medium and in the solar photosphere, overcoming the so called "helium-3 problem" (namely, the overproduction of this element by chemical evolution models adopting standard yields for low mass stars).

• Low and Intermediate mass stars ($0.8 \leq M/M_{\odot} \leq 8$):

Single stars in this mass range contribute to the Galactic enrichment through planetary nebula ejection and quiescent mass loss. They enrich the interstellar medium mainly in He, C and N. The adopted yields for the low and intermediate mass range stars are taken from van den Hoek and Groenewegen [101]. This new set of stellar yields allows a better agreement between the predicted C (and its isotopes) evolution with the observations (see next section). Moreover, we have included the explosive nucleosynthesis from nova outbursts (white dwarfs in binary systems giving rise to explosive nucleosynthesis and contributing mainly for the enrichment of ^{7}Li and of ^{13}C; [see 78].

• Type Ia Supernovae ($0.8 \leq M/M_{\odot} \leq 8$) :

Type Ia SNe are thought to originate from C-deflagration in C-O white dwarfs in binary systems. The type Ia SNe contribute to a substantial amount of iron ($\sim 0.6 M_{\odot}$ per event) and to non-negligible quantities of Si and S. They also contribute to other elements such as C, Ne, Ca, Mg and Ni, but in negligible amounts when compared with the masses of such elements ejected by type II SNe. The adopted nucleosynthesis prescriptions are from Thielemann et al. [94].

• Massive stars ($8 < M/M_{\odot} \leq 100$):

These stars are the progenitors of type II SNe. For this range of masses we adopt the yields computed by Woosley and Weaver [102] for the following elements: ^{4}He, ^{12}C, ^{13}C, ^{14}N, ^{16}O, ^{20}Ne, ^{24}Mg, ^{28}Si, ^{32}S, ^{40}Ca and ^{56}Fe. The major advantage of these calculations is that explosive nucleosynthesis is taken into account.

RESULTS

Figures from 1 to 6 and tables 1 and 2 show the predictions of our best model both for the solar vicinity and the whole disk.

The "two-infall" model allowed us to fit the observed metallicity distribution of the G-dwarfs by assuming a long timescale for the thin-disk formation. In particular, the fit of the G-dwarf [Fe/H] distribution requires that the local disk formed by infall of gas on a time scale of the order of 6-8 Gyr [see 19]. This long timescale for the thin-disk formation at the solar vicinity was then suggested also by more recent chemical evolution models [eg. 70, 13, 8, 42]. The same result is also suggested by chemodynamical models [84, 41].

A detailed discussion of the results can be found in Chiappini et al. [17, 19]. Here we call attention to some specific points.

What can we learn from the abundance ratios ?

Secondary/primary and s-process/primary ratios

Abundance ratios of a primary element over a secondary one are expected to decrease with time or metallicity [67] and to increase with the galactocentric distance when adopting the "inside-out" scenario. This fact can be useful to understand the origin of different elements and to give us information on the timescale of formation of galaxies. One example is the $^{12}C/^{13}C$ ratio. The temporal and spatial behaviour of the $^{12}C/^{13}C$ ratio predicted by models adopting the standard nucleosynthesis are flatter than observations [see 97]. A steeper gradient for $^{12}C/^{13}C$ can be achieved either by assuming that novae (white dwarfs in binary systems) are important producers of ^{13}C (restored into the ISM on longer timescales), or by adopting new C yields for low mass stars including deep extra mixing during the red giant phase associated with cool bottom processing [see 14].

For the $^{16}O/^{18}O$ ratio the problem resides in the fact that its predicted value in the ISM is higher by a factor 1.6 than that inferred from molecular cloud observations. Moreover, from the nucleosynthetic point of view ^{18}O is a neutron-rich element, namely an s-process element which should show, as do all s-process elements, a sort of secondary behaviour and, as a consequence, chemical evolution models predict that the $^{16}O/^{18}O$ ratio should decrease with time, being lower in the ISM than in the Sun, contrary to what is observed [see WG5 contribution and 97].

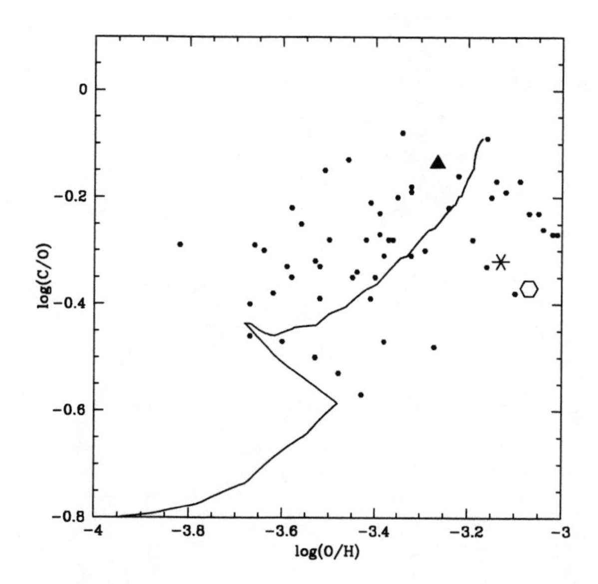

FIGURE 3. The line shows our model prediction for log(C/O) vs (O/H). The discontinuity seen in the model is the result of the gap in the star formation occurring at the end of the halo phase. The dots show the data by Gustafsson et al. [39]. The big simbols show the different location of the Sun in this diagram when adopting Anders and Grevesse [3] (hexagon), Grevesse and Sauval [36] (star) or Holweger (triangle - this conference)

Ratio between primary elements produced on different timescales

The abundance ratio of two primary elements that are restored into the interstellar medium by stars in different mass ranges, would show almost the same behavior discussed above. One example is the $^{16}O/^{12}C$ ratio. In this case both elements are primary but as ^{12}C is mainly restored into the interstellar medium by intermediate mass stars (and hence on larger timescales compared to the ^{16}O enrichment which comes mainly from massive stars), this ratio decreases as a function of metallicity. An important point about this abundance ratio is that models adopting the yields of Renzini and Voli [75] for low and intermediate mass stars could not reproduce the steep rise of C/O vs. O/H and the solar value for this ratio [19]. However, this problem is overcame once the yields of van den Hoek and Groenewegen [101] are adopted instead of the ones of Renzini and Voli [75], for low and intermediate mass stars (see figure 3) and [78].

The [O/Fe] vs [Fe/H] plot

It is worth noting that the two-infall model provides a good fit of the [α/Fe] versus [Fe/H] relation in the solar vicinity, as shown in Chiappini et al. [18]. Here we only show the plot of [Fe/O] versus [O/H] which indicates the existence of a gap in the SFR occurring at [O/H] ~ -0.2 dex (figure 4). In fact, if there is a gap in the SFR we should expect both a steep increase of [Fe/O] at a fixed [O/H] and a lack of stars corresponding to the gap period (the gap in Chiappini et al. [17] is more pronounced than in Chiappini et al. [18] given the lower value of the threshold adopted for the halo) [see discussion in 17, 25]. This gap, suggested also by the [Fe/Mg] data of Fuhrmann [32], in our models is due to the adoption of the threshold in the star formation process coupled with the assumption of a slow infall for the formation of the disk.

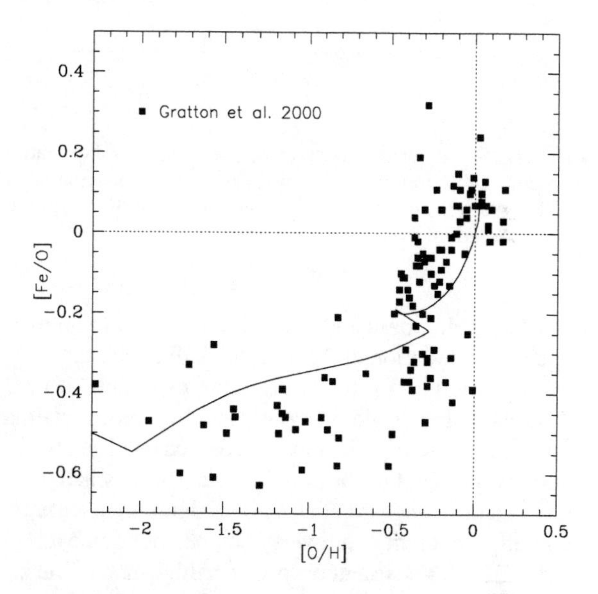

FIGURE 4. [Fe/O] vs [O/H] diagram. The data are from Gratton et al. [35] and the curve shows the prediction of the best model of Chiappini et al. [17].

Another interesting point about abundance ratios that can be understood using the particular case of [O/Fe] is discussed below. The data for oxygen from Gratton et al. [35] (figures 4 and 6) show a slight increase of the [O/Fe] ratio with decreasing [Fe/H] at variance with what happens for other α-elements which show a flatter plateau. This slight slope is well reproduced by theoretical models [18, 17] owing to the fact that the O/Fe production ratio from massive stars, in the adopted yields, is an increasing function of the initial stellar mass (figure 5). On the other hand, other elements such as Si and S are not predicted to have a large overabundance relative to Fe in metal poor stars, owing to the fact that they are also produced in a non-negligible way by type Ia SNe. The change in the slope, occurring at roughly [Fe/H] $=-1.0$ dex, is due to the bulk of iron produced by type Ia SNe which becomes important after a timescale of the order of 1 Gyr. This change in slope corresponds also to the

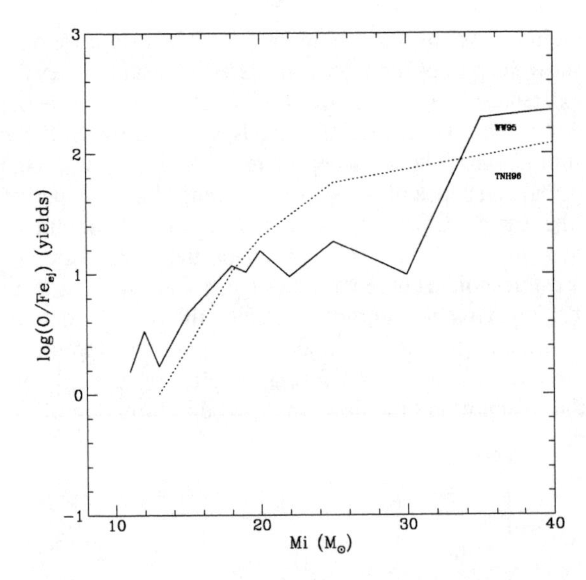

FIGURE 5. Ejected masses of oxygen and iron computed by stellar evolution models of massive stars Thielemann et al. [TNH96 93] and Woosley and Weaver [WW95 102] as a function of initial mass, M_i.

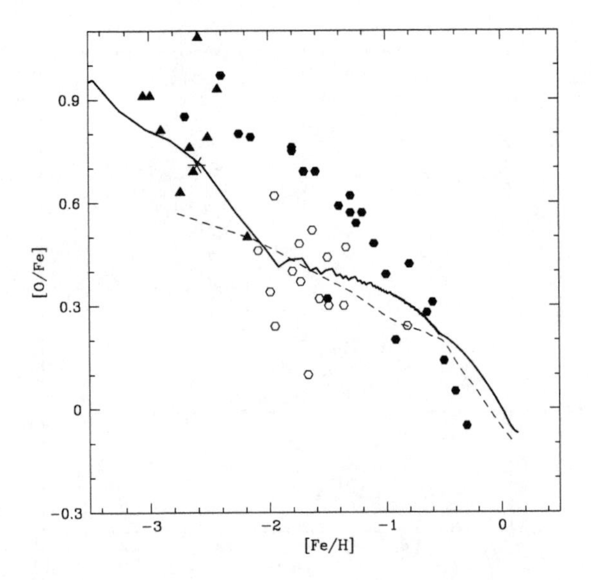

FIGURE 6. The solid line shows the predictions of Chiappini et al. [18] (adopting the yields of [93] instead of [102], for massive stars). The dashed line shows a fit to the data by Gratton et al. [35]. The open simbols are the observations reported by Meléndez et al. [63]. The filled hexagons represent the abundances obtained by Israelian et al. [44] from UV OH lines. The filled triangles show the data from Israelian et al. [43]. In particular, from [43] we ploted only the observations obtained from oxygen triplet lines with NLTE corrections. It can be seen that once the data of Israelian et al. [44] obtained from UV OH lines are not considered [see 4] the models provide a good fit to the most up to date observations.

end of the halo phase and thus allows us to have an estimate for the duration of the halo-thick disk phase.

Our prediction for the [O/Fe] ratio, especially at very low metallicities, is not in agreement with recent claims of a linear rising of this ratio with decreasing [Fe/H] obtained from UV OH lines (e.g. data from Israelian et al. [44] and Boesgaard et al. [7]). A detailed discussion about this apparent controversy can be found in Meléndez et al. [63]. Those authors obtained high-resolution infrared spectra in H-band in order to derive oxygen abundances from IR OH lines and found that for a sample of stars in the $-2.2 <$ [Fe/H] < -1.2, [O/Fe] $\simeq 0.4 \pm 0.2$ with no significant evidence for an increase of [O/Fe] with decreasing metallicity. Moreover, as shown by Asplund et al. [4], the traditional 1D LTE analyses of the UV OH lines can overestimate the O/Fe ratio while when adopting 3D analyses the results from UV OH lines are consistent with the one obtained from forbidden lines.

Figure 6 shows our model predictions (thick solid line) adopting the yields from Thielemann et al. [93] for massive stars, which predict the highest [O/Fe] at low metallicities, instead of Woosley and Weaver [102]. The recent data of Meléndez et al. [63] are also plotted (open simbols) and the star shows the very metal poor object recently measured by Cayrel et al. [12] from [OI] lines. The dashed line represents a mean fit to the oxygen data of Gratton et al. [35]. The triangles show the recent new results of Israelian et al. [43] (here we took only their measurements of the oxygen triplet lines with NLTE corrections in the iron abundance). As it can be seen, the model seems to fit the data quite well (even those of [43], whereas the previous data by the same authors - filled hexagons - based only on OH lines measurements cannot be explained by our models).

Solar Abundance Values

The solar abundances (by mass) predicted by our model are compared with the observed ones [3, 36, Holweger this conference] in Table 2. Since we assume a Galactic lifetime of 14 Gyrs the time of the Sun formation in our models corresponds to 9.5 Gyrs after the Big Bang. Given the uncertainties involved either in the observed determinations and in some of the chemical evolution parameters (namely, galaxy age, stellar yields, etc) we can consider that a model is in agreement with the observed values inside a factor 2 difference. From table 2 we can see that for most of the elements the observed and predicted values agree inside a factor of 2. We note that the values of Holweger and [36] for C, shown in table 2, include both ^{12}C and ^{13}C.

The radial profiles

As can be seen in figure 2, our model demonstrates a satisfactory fit to the elemental abundance gradients and it is also in good agreement with the observed radial profiles of the SFR, gas density and the number of stars in the disk. As shown in the previous sections, a decoupling between halo and disk phases is needed in order to best fit all the solar neighborhood observational constraints. However, the outer gradients are sensitive to the halo evolution, in particular to the amount of halo gas which ends up in the disk. This result is not surprising since the halo density is comparable to that of the outer disk, whereas it is negligible when compared to that of the inner disk. Therefore, the inner parts of the disk ($R < R_{\odot}$) evolve independently from the halo evolution in agreement with Chiappini et al. [19].

Moreover, we predict that the abundance gradients along the Galactic disk must have increased with time. Other authors find a flattening of the gradients ([see 97] for clear discussion on the possible scenarios for the evolution of the abundance gradients). In fact, different authors can fit the solar vicinity constraints and even the present time abundance gradients, but the papers seem to be divided into two groups concerning the evolution of the abundance gradients: in some of them the gradients steepen with time [96, 19, 84] while in others the abundance gradients flatten with time [2, 72, 42]. For the models that predict a flattening of the gradients with time, this can be explained as follows: in the inner parts of the disk those models assume a very high efficiency in the chemical enrichment process already in the earliest phases of the Galaxy evolution, thus soon reaching a maximum metallicity in the gas which then remains constant or decreases due to the gas recycled by dying stars. At the same time, in the outermost disk regions the lack of any pre-enrichment from the halo phase and the fact that those models do not include a threshold in the star formation process in the disk, produces a growth of metallicity larger than the one we found. This also explains an important difference between the results shown by Chiappini et al. [17] and those of Hou et al. [42]. As in Hou et al. [42] model the halo phase is completely decoupled from the disk evolution (even in the outer parts), their initial metallicities at larger Galactocentric distances are very small. Moreover, since they do not consider a threshold on the star formation process in the disk, their predicted abundances in the outermost parts of the disk keep increasing with time leading to a flattening of the abundance gradients.

The lack of good data for the *outer* Milky Way disk abundances still prevents us from testing the predictions for the evolution of abundance gradients and represents one of the main reasons for the non-uniqueness of the various chemical evolution models (a discussion about this specific problem can be found in Tosi [97]; and a discussion about the possible reasons for the above discrepancy can be found in Chiappini et al. [17]).

The gradients of different elements are predicted to be slightly different, owing to their different nucleosynthesis histories. In particular, Fe and N, which are produced on longer timescales than the α-elements, show steeper gradients. Unfortunately, the available observations cannot yet confirm or disprove this, because the predicted differences are below the limit of detectability.

DISCUSSION

We conclude this paper by discussing the scenario for the formation of the Milky Way that emerges once the best observables and the predictions of our detailed chemical evolution model are put together.

What do the observables in the Milky Way tell us ?

- The galactic disk formed inside-out and mainly out of extragalactic gas.
- The SFR should be a strongly varying function of radius.
- The solar vicinity region should have formed by slow infall of primordial (or metal-poor) gas over a time scale of the order of 7 Gyr.
- The bulge formed on a much shorter timescale than the disk and from the same gas which formed the halo.
- The inner halo formed on a shorter timescale than the outer halo.
- The SFR probably stopped for a certain period ($\leq$ 1Gyr) during the halo-thin disk transition.
- The majority of the α-elements should have been produced on short timescales relative to the age of the Galaxy ($\simeq$ 15 Gyr) (type II SNe) whereas the Fe-peak elements should have been restored with a large delay by type Ia SNe, in agreement with current ideas on nucleosynthesis and SN progenitors.
- The IMF should have been rather constant during the galactic lifetime.

Finally we would like to conclude this paper by calling attention to a forthcoming new important observational constraint on chemical evolution models, namely, the abundances in the interstellar medium (see WG3) which will help to constrain the last 4.5 Gyrs of evolution of our Galaxy. The results shown in this conference (e.g. Gloeckler, Wiedenbeck, WG3), together with the assumption that the sun represents the abundance of the ISM 4.5 Gyrs ago, seem to suggest a slow evolution for α-elements during this period. This leads to an interesting final question: is this another indication of the importance of the threshold in the process of star formation ?

ACKNOWLEDGMENTS

C. Chiappini would like to thank the organizers of the SOHO-ACE conference for the invitation and for the very stimulating environment created during the meeting. This work was partially supported by the Bern University and ESA space agency.

REFERENCES

1. Afflerbach, A., Churchwell, E., & Werner, M.W. 1997, ApJ, 478, 190
2. Allen, C., Carigi, L. & Peimbert, M. 1998, ApJ, 494, 247
3. Anders, E., & Grevesse, N. 1989, Geochim. Cosmochim. Acta, 53, 197
4. Asplund, M. & Garcia Pérez, A.E. 2001, astroph/0104071
5. Barbuy, B. 2000, in The Chemical Evolution of the Milky Way: Stars versus Clusters, eds. F. Matteucci & F. Giovannelli (Kluwer: Dordrecht), p.291
6. Beers, T.C., & Sommer-Larsen, J. 1995, ApJS, 96, 175
7. Boesgaard, A. N., King, J. R., Deliyannis, C. P., & Vogt, S. S. 1999, AJ, 117, 492
8. Boissier, S., & Prantzos, N. 1999, MNRAS, 307, 857
9. Carigi, L. 1996, Rev. Mex. Astron. & Astrofis., 4, 123
10. Carigi, L. 1994, ApJ, 424, 181
11. Carraro, G., Ng, Y.K., & Portinari, L. 1998, MNRAS, 296, 1045
12. Cayrel, R., Andersen, J., Barbuy, B., Beers, T.C., Bonifacio, P., François, P., Hill, V., Molaro, P., Nordström, B., Pletz, B., Primas, Spite, F. & Spite, M. 2001, astroph-0104357
13. Chang, R.X., Hou, J.L., Shu, C.G., & Fu, C.Q. 1999, A&A, 350, 38
14. Charbonnel, C. & do Nascimento, J.D.Jr. 1998, A&A 336, 915
15. Chiappini, C., Matteucci, F., & Padoan, P. 2000, ApJ, 528, 711
16. Chiappini, C., & Matteucci, F. 2000, in IAU Symp. 198: The Light Elements and their Evolution, ed. L. da Silva, M. Spite & J. R. de Medeiros, p.540
17. Chiappini, C., Matteucci, F., Romano, D. 2001, ApJ 554, 1044
18. Chiappini, C., Matteucci, F., Beers, T. & Nomoto, K. 1999, ApJ, 515, 226
19. Chiappini, C., Matteucci, F., & Gratton, R. 1997, ApJ, 477, 765
20. Costa, R.D.D., Chiappini, C., Maciel, W.J., & de Freitas Pacheco, J.A. 1997, in Advances in Stellar Evolution (Cambridge: Cambridge Univ. Press), 159
21. Dame, T.M. 1993, in Back to the Galaxy, ed. S. Holt & F. Verter, 267
22. Deharveng, L., Peña, M., Caplan, J., & Costero, R. 2000, MNRAS, 311, 329
23. Dopita, M. A., Ryder, S. D. 1994, ApJ 430, 163
24. Edvardsson, B., Andersen, J., Gustafsson, B., Lambert, D.L., Nissen, P. E. & Tomkin, J. 1993, A&A 275, 101
25. Elmegreen, B.G. 1999, ApJ 517, 103
26. Esteban, C., Peimbert, M., Torres-Peimbert, S., & Escalante, V. 1998, MNRAS, 295, 401
27. Esteban, C., Peimbert, M., Torres-Peimbert, S., & García-Rojas, J. 1999a, Rev. Mex. Astron. Astrof., 35, 65
28. Esteban, C., Peimbert, M., Torres-Peimbert, S., García-Rojas, J., & Rodríguez, M. 1999b, ApJS, 120, 113
29. Fich, M., & Silkey, M. 1991, ApJ, 366, 107
30. Freudenreich, H. 1998, ApJ, 492, 495
31. Friel, E.D. 1999, Ap&SS, 265, 271
32. Fuhrmann, K. 1998, A&A, 338, 161
33. Gilmore, G., Wyse, R.F.G., & Kuijken, K. 1989, ARA&A, 27, 555
34. Goswami, A. & Prantzos, N. 2000, A&A, 359, 191
35. Gratton, R.G., Carretta, E., Matteucci, F., & Sneden, C. 2000, A&A, 358, 671
36. Grevesse, N., & Sauval, A.J. 1998, Space Sci. Rev., 85, 161
37. Guibert, J., Lequeux, J., Viallefond, F. 1978, A&A, 68, 1
38. Gummersbach, C.A., Kaufer, A., Schäfer, D.R., Szeifert, T., & Wolf, B. 1998, A&A, 338, 881
39. Gustafsson, B., Karlsson, T., Olsson, E., Edvardsson, B. & Ryde, N. 1999, A&A 342, 426
40. Güsten, R., & Mezger, P.G. 1982, Vistas Astron., 26, 159
41. Hensler, G. 1999, Ap&SS, 265, 397
42. Hou, J.L., Prantzos, N., & Boissier, S. 2000, A&A, in press
43. Israelian, G., Rebolo, R., López, R.J.G., Bonifacio, P., Molaro, P., Basri, G. & Shchukina, N. 2001, astroph/0101032
44. Israelian, G., Garcia-Lopez, R. & Rebolo, R. 1998, ApJ, 507, 805
45. Kennicutt, R.C., Jr. 1989, ApJ, 344, 685
46. Kennicutt, R.C., Jr. 1998, ApJ, 498, 541
47. Lacey, C. G., Fall, S. M. 1985, ApJ 290, 154
48. Larson, R.B. 1998, MNRAS, 301, 569
49. Larson, R.B. 1976, MNRAS, 176, 31
50. Limongi, M., Chieffi, A., Straniero, O. 2000, in The Chemical Evolution of the Milky Way: Stars versus Clusters, eds. F. Matteucci & F. Giovannelli (Kluwer: Dordrecht), p.473
51. Lyne, A.G., Manchester, R.N., Taylor, J.H. 1985, MNRAS, 213, 613
52. Maciel, W.J., & Chiappini, C. 1994, Ap&SS, 219, 231
53. Maciel, W.J., & Köppen, J. 1994, A&A, 282, 436
54. Maciel, W.J. 1997, IAU Symp. 180, ed. H.J. Habing and J. Lamers, Dordrecht, p.397
55. Maciel, W.J., & Quireza, C. 1999, A&A, 345, 629
56. Maciel, W.J. 2000, in The Chemical Evolution of the Milky Way: Stars versus Clusters, eds. F. Matteucci & F. Giovannelli (Kluwer: Dordrecht), p. 81
57. Martin, C.L. & Kennicutt, R.C. 2001, astroph/0103181
58. Matteucci, F. Greggio, L. 1986, A&A, 154, 279
59. Matteucci, F., & François, P. 1989, MNRAS, 239, 885
60. Matteucci, F., Romano, D. & Molaro, P. 1999, A&A 341, 458
61. Mathews, G. J. & Schramm, D. N 1993, ApJ 404, 468
62. McWilliam, A. & Rich, R.M. 1994, ApJS 91, 749
63. Meléndez, J., Barbuy, B. & Spite, F. 2001, astroph-0104184
64. Pagel, B.E.J., & Patchett, B.E. 1975, MNRAS, 172, 13

65. Pagel, B.E.J., & Tautvaisiene, G. 1995, MNRAS, 276, 505
66. Pagel, B.E.J. 1989, in Evolutionary Phenomena in Galaxies, eds. J.E. Beckman & B.E.J. Pagel, Cambridge University Press, p. 201
67. Pagel, B.E.J. 1997, Nucleosynthesis and Chemical Evolution of Galaxies, Cambridge University Press
68. Pardi, C., Ferrini, F., Matteucci, F. 1995, ApJ 444, 207
69. Peimbert, M. 1978, in Planetary Nebulae (Kluwer: Dordrecht), 215
70. Portinari, L., Chiosi, C., & Bressan, A. 1998, A&A 334, 505
71. Portinari, L., & Chiosi, C. 1999, A&A, 350, 827
72. Portinari, L., & Chiosi, C. 2000, A&A, 355, 929
73. Prantzos, N., Boissier, S. 2000, MNRAS, 313, 338
74. Rana, N.C. 1991, ARA&A, 29, 129
75. Renzini, A., & Voli, M. 1981, A&A, 94, 175
76. Rich, R.M. 1988, ApJ 95, 828
77. Rocha-Pinto, H. J., Maciel, W. J. 1996, MNRAS 279, 447
78. Romano, D. & Matteucci, F. 2000, in The Chemical Evolution of the Milky Way: Stars versus Clusters, eds. F. Matteucci & F. Giovannelli (Kluwer: Dordrecht), p.547
79. Romano, D., Matteucci, F., Salucci, P., Chiappini, C. 2000, ApJ, 539, 235
80. Romano, D., Matteucci, F., Ventura, P. 2001, A&A (submitted)
81. Rosenberg, A., Saviane, I., Piotto, G., & Aparicio, A. 1999, AJ, 118, 2306
82. Rudolph, A.L., Simpson, J.P., Haas, M.R., Erickson, E.F., & Fich, M. 1997, ApJ, 489, 94
83. Sackett, P.D. 1997, ApJ, 483, 103
84. Samland, M., Hensler, G., & Theis, C. 1997, ApJ, 476, 544
85. Scalo, J.M. 1998, 38th Hertmonceux Conference, eds. G. Gilmore and D. Howell, ASP Conf. Ser., Vol. 142, p. 201
86. Scalo, J.M. 1986, Fundam. Cosmic Phys., 11, 1
87. Shaver, P.A., McGee, R.X., Newton, L.M., Danks, A.C., & Pottasch, S.R. 1983, MNRAS, 204, 53
88. Simard, L., Koo, D.C., Faber, S.M., Sarajedini, V.L., Vogt, N.P., Phillips, A.C., Gebhardt, K., Illingworth, G.D., & Wu, K.L. 1999, ApJ, 519, 563
89. Simpson, J.P., Colgan, S.W.J., Rubin, R.H., Erickson, E.F., & Haas, M.R. 1995, ApJ, 444, 721
90. Smartt, S.J., & Rolleston, W.R.J. 1997, ApJ, 481, L47
91. Sommer-Larsen, J. 1991, MNRAS 249, 368
92. Talbot, R.J. & Arnett, W.D. 1975, ApJ 197, 551
93. Thielemann, F. K., Nomoto, K. & Hashimoto, M. 1996, ApJ 460, 408
94. Thielemann, F.,K., Nomoto, K. & Hashimoto, M. 1993, in Origin and Evolution of the Elements, ed. N. Prantzos et al., Cambridge University Press, p. 297
95. Timmes, F. X., Woosley, S.E., & Weaver, T. A. 1995, ApJS, 98, 617
96. Tosi, M. 1988, A& A, 197, 47
97. Tosi, M. 2000, in The Chemical Evolution of the Milky Way: Stars versus Clusters, ed. F. Matteucci & F. Giovannelli, (Kluwer: Dordrecht), 505
98. Tosi, M., Steigman, G., Matteucci, F. & Chiappini, C. 1998, ApJ 498, 226
99. Tsujimoto, T., Yoshii, Y., Nomoto, K., Shigeyama, T. 1995 A&A, 302, 704
100. Twarog, B.A., Ashman, K.M., & Anthony-Twarog, B.J. 1997, ApJ, 114, 2556
101. van den Hoek, L.B., & Groenewegen, M.A.T. 1997, A&AS, 123, 305
102. Woosley, S. E., Weaver, T. A. 1995, ApJS 101, 181
103. Wyse, R. F. G., Gilmore, G. 1995, AJ 110, 2771
104. Wyse, R. F. G., Gilmore, G. 1992, AJ 104, 144

Stellar Nucleosynthesis and Galactic Abundances

F.-K. Thielemann, D. Argast, F. Brachwitz, G. Martinez-Pinedo

Dept. of Physics & Astronomy, Klingelbergstrasse 82, CH-4056 Basel, Switzerland

Abstract. Galactic abundances are neither constant in time nor do they evolve in a simple fashion, e.g. by an enrichment of heavy elements in constant relative proportions. Instead, their evolution in space and time reflects the history of star formation and the lifetimes of the diverse contributing stellar objects. Stellar winds from intermediate and massive stars, as well as supernovae of type Ia and type II/Ib/Ic are the main contributors to nucleosynthesis in galaxies. Despite many efforts, a full and self-consistent understanding of supernova explosion mechanisms is not existing, yet. However, they leave fingerprints, seen either in spectra, lightcurves, radioactivities/decay gamma-rays or in galactic evolution. The aim of the present paper is to highlight how model uncertainties can be constrained from abundance observations.

INTRODUCTION

The evolution of galaxies is dominated by its main contributors SNe II, SNe Ia, and planetary nebulae or, if expressed in terms of stellar progenitor masses, by massive stars with $M>8M_\odot$ and intermediate or low mass stars with $M<8M_\odot$, either as single stars or in binary systems.

Planetary nebulae do contribute to light elements and heavy s-process nuclei up to Pb and Bi. These are either products of H- and He-burning (mainly ^{4}He, ^{14}N, ^{12}C, [^{16}O, ^{22}Ne]) or based on a sequence of neutron captures and beta-decays, acting on pre-existing heavier nuclei. The required neutrons are provided by (α,n)-reactions in He-burning. Galactic evolution for the elements in the range O through Ni is dominated by alternative explosive stellar sources, i.e by the combined action of SNe II and Ia. The site for the production of the heaviest elements up to U and Th (the r-process based on neutron captures in environments with very high neutron densities) is still debated and most probably related to supernovae and/or neutron star ejecta (from binary mergers or jets).

The surface abundances of stars, indicating the interstellar medium abundances at the time of their birth, are a clear indicator of galactic evolution as a function of metallicity [Fe/H]=log$[(Fe/H)/(Fe/H)_\odot]$. Observational data for [x/Fe] at low metallicities ($-2<$[Fe/H]<-1), x standing for elements from O through Ca, show an enhancement of the alpha elements (O through Ca) by a factor of 2-3 (0.3 to 0.5 dex in [x/Fe]) in comparison to Fe [12, 37, 2]. This is the clear fingerprint of the exclusive early contribution of fast evolving massive stars, i.e. SNe II. The higher ratio of Fe-group elements to Si-Ca in SNe Ia has to compensate for these overabundances in SNe II in order to obtain solar abundance ratios for the combined nucleosynthesis products at solar metallicity [Fe/H]=0. Such global tests can constrain the average features of SNe II and SNe Ia mixed in frequency ratios of $N_{Ia}/N_{II}=0.15-0.27$ [58, 5] and require an Fe contribution from SNe Ia of 50-60%, if average ejected Fe masses of 0.1 $M_\odot$ and 0.6 $M_\odot$ are taken for SN II and SN Ia.

An interesting feature is that also the Fe-group ejecta of both types of supernovae have to differ. For Ti, Sc, Cr, Mn, Co, Ni one finds at low metallicities average SN II values [x/Fe] of 0.25, 0, -0.1, -0.3, -0.1, -0.1 [21]. Taken the typical uncertainty of 0.1 dex, this leaves Ti and Mn as elements with clear signatures for a SN II behavior different from solar, which asks for the opposite effect in SN Ia.

The features discussed so far apply to average SNe II and Ia compositions. A more complete understanding, however, would require observational clues for the range of individual SNe II and Ia which can directly constrain theoretical models. After presenting shortly the present status of SNe II and Ia nucleosynthesis predictions, we want to give examples how that might be done.

STELLAR EVOLUTION AND WIND EJECTA

H-burning converts ^{1}H into ^{4}He via pp-chains or the CNO-cycles. The simplest PPI chain is initiated by ^{1}H(p,$e^+\nu$)^{2}H(p,γ)^{3}He and completed by ^{3}He(^{3}He,2p)^{4}He. The dominant CNOI-cycle chain ^{12}C(p,γ)^{13}N($e^+\nu$)^{13}C(p,γ)^{14}N(p,γ)^{15}O($e^+\nu$)^{15}N(p,α)^{12}C

CP598, *Solar and Galactic Composition,* edited by R. F. Wimmer-Schweingruber

is controlled by the slowest reaction $^{14}N(p,\gamma)^{15}O$. Further burning stages are characterized by their major reactions, which are in He-burning $^{4}He(2\alpha,\gamma)^{12}C$ (triple-alpha) and $^{12}C(\alpha,\gamma)^{16}O$, in C-burning $^{12}C(^{12}C,\alpha)^{20}Ne$, and in O-burning $^{16}O(^{16}O,\alpha)^{28}Si$. The alternative to fusion reactions are photodisintegrations which start to play a role at sufficiently high temperatures when $30kT \approx Q$ (the Q-value or energy release of the inverse capture reaction). This ensures the existence of photons with energies $>Q$ in the Planck distribution and leads to Ne-Burning [$^{20}Ne(\gamma,\alpha)^{16}O$, $^{20}Ne(\alpha,\gamma)^{24}Mg$] at $T>1.5\times10^{9}$K (preceding O-burning) due to a small Q-value of $\approx$4 MeV and Si-burning at temperatures in excess of 3×10^{9}K (initiated like Ne-burning by photodisintegrations [of ^{28}Si]). The latter ends in a chemical equilibrium with an abundance distribution around Fe (nuclear statistical equilibrium, NSE). This is possible because at these temperatures fusion reactions are fast due to the permitted penetration of the Coulomb barriers at the corresponding energies while the typical Q-values of 8-10 MeV permit photodisintegrations as well.

The high densities in late phases of O- and Si-burning result in partially or fully degenerate electrons with increasing Fermi energies [38]. When these supercede the Q-value thresholds of electron capture reactions, this allows for electron capture on an increasing number of initially Si-group and later Fe-group (pf-shell) nuclei. Because sd-shell reactions were well understood in the past [11], O-burning predictions are quite reliable. The recent progress in calculating pf-shell rates [22] led to drastic changes in the late phases of Si-burning [15].

Stars with masses $M>8M_{\odot}$ develop an onion-like composition structure, after passing through all hydrostatic burning stages, and produce a collapsing core at the end of their evolution, which proceeds to nuclear densities [8, 57, 15]. The recent change in electron capture rates [22] sets new conditions for the subsequent Fe-core collapse after Si-burning, the size of the Fe-core and its electron fraction $Y_e=\langle Z/A\rangle$ [29]. It has been pointed out that stellar rotation adds important features to the evolution in general and composition in particular (via diverse mixing processes) [14, 32]. The metallicity plays an essential role in the amount of mass loss of massive stars [23, 27, 28]. Less massive stars experience core and shell H- and He-burning and end as C/O white dwarfs after strong episodes of mass loss [13]. Their ejected nucleosynthesis yields have been predicted by Renzini and Voli [47] and recently updated [59].

TYPE II SUPERNOVAE

The present situation is that self-consistent spherically-symmetric calculations (with the presently known microphysics) do not yield successful explosions [33, 24, 45] based on neutrino energy deposition from the hot collapsed central core (neutron star) in the adjacent layers. This seems to be the same for multi-D calculations, which however lack good neutrino transport schemes and do not yet consider the combined action of rotation and magnetic fields. The hope that the neutrino driven explosion mechanism could still succeed is based on uncertainties which affect neutrino luminosities (neutrino opacities with nucleons and nuclei and convection in the hot proto-neutron star) as well as the efficiency of neutrino energy deposition (convection in the adjacent layers).

Nevertheless, observations show typical kinetic energies of 10^{51} erg in supernova remnants. This permits one to perform light curve as well as explosive nucleosynthesis calculations by introducing a shock of appropriate energy in the pre-collapse stellar model [61, 51, 20, 34, 57, 46]. Such induced calculations lack self-consistency and cannot predict the ejected ^{56}Ni-masses from the innermost explosive Si-burning layers (powering supernova light curves by the decay chain ^{56}Ni-^{56}Co-^{56}Fe) due to missing knowledge about the mass cut between the neutron star and supernova ejecta.

A correct prediction of the amount of Fe-group nuclei ejected (which includes also one of the so-called alpha elements, i.e. Ti) and their relative composition depends directly on the explosion mechanism and the size of the collapsing Fe-core. Three types of uncertainties are inherent in the Fe-group ejecta, related to (i) the total amount of Fe(group) nuclei ejected and the mass cut between neutron star and ejecta, mostly measured by ^{56}Ni decaying to ^{56}Fe, (ii) the total explosion energy which influences the entropy of the ejecta and with it the amount of radioactive ^{44}Ti as well as ^{48}Cr, the latter decaying later to ^{48}Ti and being responsible for elemental Ti, and (iii) finally the neutron richness or $Y_e=\langle Z/A\rangle$ of the ejecta, dependent on stellar structure, electron captures and neutrino interactions. Y_e influences strongly the ratios of isotopes 57/56 in Ni(Co,Fe) and the overall elemental Ni/Fe ratio. The latter being dominated by ^{58}Ni and ^{56}Fe. The pending understanding of the explosion mechanism also affects possible r-process yields for SNe II [50, 63, 44, 9, 31].

The intermediate mass elements Si-Ca provide information about the explosion energy and the stellar structure of the progenitor star, while abundances for elements like O and Mg are essentially determined by the stellar progenitor evolution.

TYPE IA SUPERNOVAE

There are strong observational and theoretical indications that SNe Ia are thermonuclear explosions of accreting white dwarfs in binary stellar systems [18, 42, 41, 26] with carbon ignition and a thermonuclear runway causing a complete explosive disruption of the white dwarf [40, 60]. The mass accretion rates determine the ignition densities. A flame front then propagates at a subsonic speed as a deflagration wave due to heat transport across the front [16]. The averaged spherical flame speed depends on the development of instabilities of various scales at the flame front. Multi-dimensional hydro simulations suggest a speed $v_{\rm def}$ as slow as a few percent of the sound speed v_s in the central region of the white dwarf. Electron capture affects the central electron fraction Y_e and depends on (i) the electron capture rates of nuclei, (ii) $v_{\rm def}$, influencing the time duration of matter at high temperatures (and with it the availability of free protons for electron captures), and (iii) the central density of the white dwarf ρ_{ign} (increasing the electron chemical potential i.e. their Fermi energy) [21, 3, 22]. After an initial deflagration in the central layers, the deflagration might turn into a detonation (supersonic burning front) at lower densities [36].

Nucleosynthesis constraints can help to find the "average" SN Ia conditions responsible for their contribution to galactic evolution, i.e. especially the Fe-group composition. While ignition densities ρ_{ign} determine the very central amount of electron capture and thus Y_e, the deflagration speed $v_{\rm def}$ determines the resulting Y_e -gradient as a function of radius [21]. Y_e values of 0.47-0.485 lead to dominant abundances of ^{54}Fe and ^{58}Ni, values between 0.46 and 0.47 produce dominantly ^{56}Fe, values in the range of 0.45 and below are responsible for ^{58}Fe, ^{54}Cr, ^{50}Ti, ^{64}Ni, and values below 0.43-0.42 are responsible for ^{48}Ca. The intermediate Y_e-values 0.47-0.485 exist in all cases, but the masses encountered which experience these conditions depend on the Y_e-gradient and thus v_{def}. Whether the lower vales with $Y_e<0.45$ are attained, depends on the central ignition density ρ_{ign}. Therefore, ^{54}Fe and ^{58}Ni are indicators of v_{def} while ^{58}Fe, ^{54}Cr, ^{50}Ti, ^{64}Ni, and ^{48}Ca are a measure of ρ_{ign}. These are the (hydrodynamic) model parameters. An additional uncertainty is the central C/O ratio of the exploding white dwarf [17]. Nuclear uncertainties based on electron capture rates enter as well [3, 4]. Conclusions from these results are: (i) a v_{def} in the range 1.5–3% of the sound speed is preferred [21], and (ii) the change in electron capture rates [22] made it possible to have ignition densities as high as $\rho_{ign}=2\times10^9$ g cm^{-3} without destroying the agreement with solar abundances of very neutron-rich species [21, 3]. It seems, however, hard to produce amounts of ^{48}Ca sufficient to explain solar abundances from SNe Ia when applying more realistic electron capture rates and C/O ratios [62, 4].

OBSERVATIONAL CONSTRAINTS AND GALACTIC EVOLUTION

Galactic evolution can serve as a test for all contributing stellar yields, i.e. especially intermediate and low mass stars through planetary nebula ejection, mass loss from massive stars, as well as the ejecta of SNe II and SNe Ia. This is discussed in much more detail in a different contribution to this conference [7]. Here we want to focus on specific observational clues to SNe Ia and SNe II nucleosynthesis and indicate how very low metallicity stars might witness individual rather than only integrated SNe II yields.

There have been detailed discussions [52, 52, 21, 51] on observational constraints from supernova spectra, lightcurves, and X-ray and gamma-ray observations of remnants. They refer mainly to ^{56,57}Ni, and ^{44}Ti abundances and possibly stable Ni/Fe ratios in SNe II, giving insight into the details of the explosion mechanism with respect to the mass cut between the neutron star and the SN ejecta, the total energy of the explosion, the entropy and the Y_e in the innermost ejecta (see section 3). The intermediate mass elements Si-Ca provide information about the explosion energy and the stellar structure of the progenitor star, while elements like O and Mg are essentially determined by the stellar progenitor evolution.

In SNe Ia the ^{56}Ni production and the Si-Ca/Fe ratio are related to the total explosion energy and burning front speed in layers of the exploding white dwarf extending as far as $M(r)=1{\rm M}_\odot$ and beyond. Constraints on the ignition density and burning front speed in the central regions (as discussed in section 4) are reflected e.g. in minor isotopic abundances like ^{50}Ti and ^{54}Cr, where direct supernova or remnant observations cannot be used. Here one can only make use of global abundance constraints from galactic evolution, which therefore permit only statements about an "average" SN Ia. The application of present day electron capture rates makes it hard to account for our solar system ^{48}Ca [4]. On the other hand, observations of varying stable Ni/Fe ratios during the evolution of a supernova (by spectral means) related to ^{54}Fe and ^{58}Ni, might give clues to the metallicity of the exploding white dwarf [19].

Stars (with understood exceptions) do not change the surface composition during their evolution. Thus, surface abundances reflect the interstellar medium (out of which the stars formed) at the time of their formation. Therefore, observations of the surface composition (via spectra) of very low metallicity stars (i.e. very old stars) give a clue to gas abundances throughout the evolution of our Galaxy. Here we want to focus on recent promising trends in galactic evolution modeling, which might provide constraints on individual supernova models rather than only global properties of SNe II and SNe Ia. The reason for this possibility is the fact that there is no instantaneous mixing of ejecta with the interstellar medium, and therefore early phases of galactic evolution can present a connection between low metallicity star observations and a single supernova event. On average, each supernova pollutes a volume of the interstellar medium containing $\approx (3-5) \times 10^4 M_\odot$. (Each volume of the interstellar medium containing $\approx 3 \times 10^4 M_\odot$ needs to be enriched by $\approx 10^3$ SNe in order to obtain solar metallicities).

After a supernova polluted such a previously pristine environment mass it results in values for [x/Fe] and [Fe/H] in the remnant and a scatter in [x/Fe] is expected for the same [Fe/H]. The amount of the polluted volume depends on the explosion energy, and if there is a strong variation of explosion energies with progenitor mass this could affect the relation between [x/Fe] and the metallicity [Fe/H] [34]. The scatter in [x/Fe] expected for the same [Fe/H] is observed up to metallicities of [Fe/H]=-2 where it vanishes because overlapping contributions from many SNe II behave like a well mixed medium [1].

Some results from such an approach [1, 2] are displayed in Fig.1 which shows O/Fe, Mg/Fe, Si/Fe and Ca/Fe ratios. Three types of symbols are given, circles, squares and dots. The squares show observations of low metallicity stars with a scatter at very low metallicities, as expected if mixing of ejecta with the ISM is not instantaneous and contributions from individual supernovae with different progenitor masses give their different signatures. This scatter apparently becomes as small as the observational uncertainties at [Fe/H]=-2, where everywhere in the ISM the average signature of IMF integrated SNe II yields emerges. The open circles indicate the x/Fe and Fe/H ratios in a volume of $5 \times 10^4 M_\odot$ of an initially pristine (big bang) composition, polluted by a single supernova with our ejecta compositions [51, 39]. The numbers inside the circles refer to the progenitor mass of a supernova whose ejecta composition was mixed with $5 \times 10^4 M_\odot$ of prestine ISM. Such a remnant would therefore show directly the [Fe/H] metallicity caused by a single SN event. However, we compare with observations from low metallicity stars rather than remnants. The model stars are indicated by dots [1, 2]. They assume a stochastic star formation at random positions in the ISM and with random progenitor masses (however a statistical distribution according to a Salpeter IMF). One sees on the one hand that much smaller contributions to a stellar progenitor cloud can be possible than expected from a single remnant (leading to Fe/H ratios much smaller than given in the circles). On the other hand, successive enrichment of many (and finally overlapping) remnants leads to a metallicity evolution which approaches at [Fe/H]=-2 the IMF-averaged SNe II yields. These are apparently in agreement with the observations.

The features (at [Fe/H]=-2) agree nicely with previous galactic evolution calculations which applied our yields [56, 54, 30, 7], assuming instantaneous mixing of yields with the ISM. However, the observed and predicted scatter at very low metallicities bears information which was previously unavailable. (This approach relies on the assumption that these difficult low metallicity observations are correct.) The scatter for Si and Ca/Fe seems essentially correct, possibly only slightly too large, which could indicate that the lower mass SNe II (11 and 13 $M_\odot$) produce slightly too small ratios and the high mas end (50 $M_\odot$) slightly too large ones. This can be related to the uncertainty of the mass cut and the assumptions made in our previous papers on the Fe-yields as a function of progenitor mass. Thus, while the average x/Fe is correct, the progenitor mass dependence of the Fe-yields could be slightly too strong. This is, however, a small effect for the elements Si and Ca, produced in explosive burning. Much larger deviations can be seen for O and Mg, which are dominated by uncertainties in stellar evolution (not the explosion). Here the lower mass stars (11, 13, 15 $M_\odot$) predict clearly much too small ratios, far beyond the obervational scatter, while the 50 $M_\odot$ model predicts slightly too large ones. This indicates a problem in the stellar models which might improve with more recent calculations [8, 57, 14, 15] and has to be tested.

The main conclusion we can draw from these results is that such investigations can also test individual stellar yields rather than only IMF integrated samples. This is a large advantage over the very few data points we have from individual supernova observations. Such tests seem also very useful for other applications, where one is not certain about the stellar site of a nucleosynthesis product or possible contributions from objects with different evolution timescales. Hypernova contributions should be considered in a similar way [35].

The r-process is an example where the alternative site to supernovae, neutron star mergers, occur with a much smaller frequency. In that case, the mixed phase (orrurring for SNe II at about [Fe/H]=-2) should be delayed to larger metallicities. In addition, one expects with the large amounts of r-process ejecta from each occasionally occurring neutron star merger [10] a much larger scatter than for the smoothly changing supernova yields as a function of stellar mass. Both effects are seen in the r/Fe observations (a scatter of almost a factor of 1000 [49, 6] at low metallicities, which still amounts to about a factor of 10 at [Fe/H]=-1, see Fig. 2 in [55]. Combined with the suppression of abundances below A=130 in low-metallicity stars, these could be taken as supportive features for a fission cycling r-process from a low frequency source in galactic evolution. At least it argues for a very

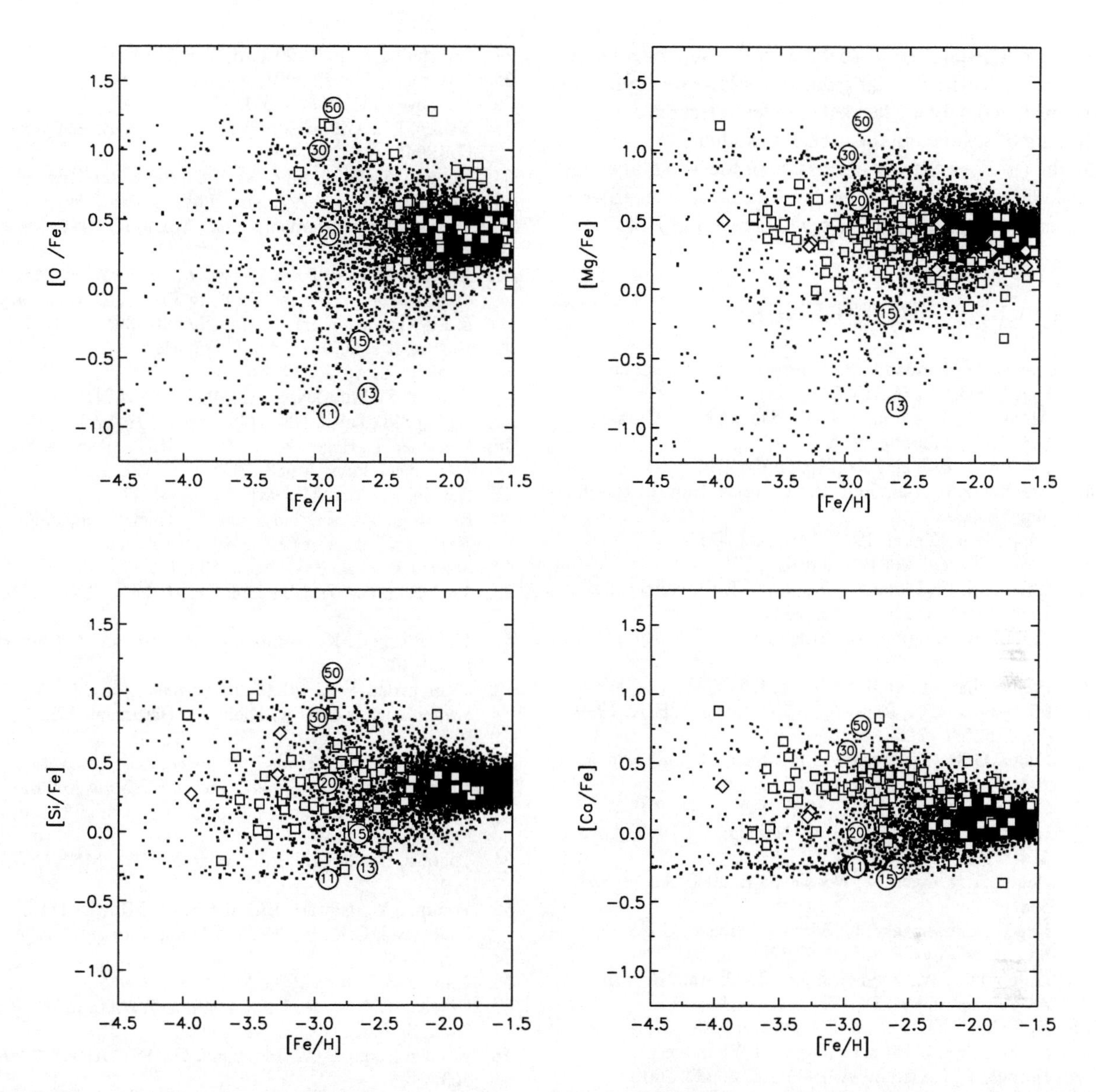

FIGURE 1. Comparison of low metallicity observations with SN II yields and a galactic evolution model. Three types of symbols are given, circles, squares and dots. The squares show observations of low metallicity stars with a scatter at very low metallicities. The open circles indicate the x/Fe and Fe/H ratios in a volume of $5 \times 10^4 M_\odot$ of an initially pristine (big bang) composition, polluted by a single supernova with our ejecta compositions. The numbers inside the circles refer to the progenitor mass of a supernova whose ejecta composition was mixed with $5 \times 10^4 M_\odot$ of prestine ISM. The (galactic evolution) model stars are indicated by dots (see the text for a detailed discussion).

rare event, even if some specific type of supernova is responsible for it. Similar galactic evolution calculations as presented here are needed in order to test the expected amount of scatter as a function of metallicity to give clues on the r-process site. A discussion of the advantages and disadvantages of both possible r-process sources (SNe II vs. neutron star mergers) is given in refs. [43, 48].

REFERENCES

1. Argast, D., Samland, M., Gerhard, O.E., Thielemann, F.-K. 2000, A & A 356, 873
2. Argast, D., Thielemann, F.-K., Samland, M., Gerhard, O.E. 2001, submitted to A & A
3. Brachwitz, F. et al. 2000, Ap. J., 536, 934
4. Brachwitz, F. et al. 2001, Ph. D. thesis, Univ. of Basel, unpublished
5. Cappellaro, E. et al. 1997, A&A 322, 431
6. Cayrel, R. et al. 2001, Nature 409, 691
7. Chiappini, C., Matteucci, F., Beers, T. C., Nomoto, K. 1999, ApJ 515, 226 and this volume
8. Chieffi, A., Limongi, M., Straniero, O. 1998, Ap. J. 502, 737
9. Freiburghaus, C. et al. 1999, Ap. J. 516, 381
10. Freiburghaus, C., Rosswog, S., Thielemann, F.-K. 1999, Ap. J. 525, L121
11. Fuller, G.M., Fowler, W.A., Newman, M. 1985, Ap. J. 293, 1
12. Gratton, R.G., Sneden, C. 1991, A & A 241, 501
13. Hashimoto, M., Iwamoto, K., Nomoto, K., 1993, Ap. J. 414, L105
14. Heger, A., Langer, N., Woosley, S.E. 2000, Ap. J., 528, 368
15. Heger, A., Langanke, K., Martinez-Pinedo, G., Woosley, S.E. 2001, Phys. Rev. Lett 86, 1768
16. Hillebrandt, W., Niemeyer, J.C. 2000, Ann. Rev. of Astron. Astrophys. 38, 191
17. Höflich, P. 2001, in 1st KIAS Astrophysics Workshop, Seoul/Corea, IAP-Publishing, ed. I. Yi, in press
18. Höflich, P., Khokhlov, A. 1996, Ap. J., 457, 500
19. Höflich, P., Wheeler, J.C., & Thielemann, F.-K. 1998, ApJ 495, 617
20. Hoffman, R.D. et al. 1999, Ap. J. 521, 735
21. Iwamoto, K. et al. 1999, Ap. J. Suppl. 125, 439
22. Langanke, K., Martinez-Pinedo, G. 2000, Nucl. Phys. A673, 481
23. Langer, N., Fliegner, J., Heger, A., Woosley, S.E. 1997, Nucl. Phys. A621, 457c
24. Liebendörfer, M. et al. 2001, Phys. Rev. D 6310, 3004
25. Liebendörfer, M. et al. 2001, Phys. Rev. D 6310, 4003
26. Livio, M. 2000 in *Type Ia Supernovae: Theory and Cosmology*, Cambridge Univ. Press, in press
27. Maeder, A., Meynet, G. 2000, A&A 361, 159
28. Maeder, A., Meynet, G. 2000, ARA&A 38, 143
29. Martinez-Pinedo, G. et al. 2000, ApJS, 126, 493
30. Matteucci, F., Romano, D., Molaro, P. 1999, A&A 341, 458
31. McLaughlin, G.C. et al. 1999, Phys. Rev. C59, 2873
32. Meynet, G., Maeder, A. 2000, A&A 361, 101
33. Mezzacappa, A. et al. 2001, PRL 86, 1935
34. Nakamura, T. et al. 1999, Ap. J. 517, 193
35. Nakamura, T. et al. 2001, Ap, J. 555, 880
36. Niemeyer, J.C. 1999, Ap. J. 523, L57
37. Nissen, P.E., Gustafsson, B., Edvardsson, B., & Gilmore, G. 1994, A&A 285, 440
38. Nomoto, K., Hashimoto, M. 1988, Phys. Rep. 163, 13
39. Nomoto, K. et al. 1997, Nucl. Phys., A161, 79c
40. Nomoto, K., Thielemann, F.-K., Yokoi, K. 1984, Ap. J. 286, 644
41. Nomoto, K. et al. 2000, in *Type Ia Supernovae: Theory and Cosmology*, Cambridge Univ. Press, eds. J. Niemeyer & J.W. Truran, in press (astro-ph/9907386)
42. Nugent, P. et al. 1997, Ap. J. 485, 812
43. Qian, Y.-Z., Ap. J. 534, L67
44. Qian, Y.-Z., Woosley, S. E. 1996, Ap. J., 471, 331
45. Rampp, M., Janka, H.T. 2000, Ap. J. 539, L33
46. Rauscher, T., Heger, A., Hoffman, R.D., Woosley, S.E. 2001, Nucl. Phys. A 688, 193c
47. Renzini, A., Voli, M. 1981, A & A 94, 175
48. Rosswog, S.K., Freiburghaus, C., Thielemann, F.-K. 2001, Nucl. Phys. A 688, 344c
49. Sneden, C. et al. 2000, Ap. J. 533, 139
50. Takahashi, K., Witti, J., Janka, H.-T. 1994, A&A, 286, 857
51. Thielemann, F.-K., Nomoto, K., Hashimoto, M. 1996, Ap. J. 460, 408
52. Thielemann, F.-K. 2000, in *Astronomy with Radioactivities*, eds. R. Diehl, D. Hartmann, MPE Report, p. 123
53. Thielemann, F.-K. et al. 2001, in *The Largest Explosions Since the Big Bang: Supernovae and Gamma-Ray Bursts*, eds. M. Livio, K. Sahu, N. Panagia, Cambridge University Press
54. Thomas, D., Greggio, L., Bender, R. 1998, MNRAS 296, 119
55. Truran, J.W., Cowan, J.J., Sneden, C., Burris, D.L., Pilacowski, C.A., in *The First Stars*, Springer, eds. A. Weiss et al.
56. Tsujimoto, T. et al. 1995, MNRAS 277, 945
57. Umeda, H., Nomoto, K., Nakamura, T. 2000, in *The First Stars*, Springer, eds. A. Weiss et al.
58. van den Bergh, S., & Tammann, G. 1991, ARA&A 29, 363
59. van den Hoek, L.B., Groenewegen, M.A.T. 1997, A & AS 123, 305
60. Woosley, S.E., Weaver, T.A. 1994, in *Les Houches, Session LIV, Supernovae*, eds. S.R. Bludman, R. Mochkovitch, J. Zinn-Justin, Elsevier Science Publ., p. 63
61. Woosley, S.E., Weaver, T.A. 1995, Ap. J. Suppl. 101, 181
62. Woosley, S.E. 1997, Ap. J. 476, 801
63. Woosley, S.E. et al. 1994, Ap. J. 433, 229

ACKNOWLEDGMENTS

This review (supported by the Swiss NSF) would not have been possible without presenting results from joint collaborations with J.J. Cowan, D. Dean, C. Freiburghaus, O.E. Gerhard, W.R. Hix, K. Iwamoto, K. Langanke, M. Liebendörfer, A. Mezzacappa, K. Nomoto, P. Höflich, T. Rauscher, S. Rosswog, M. Samland, M. Strayer and H. Umeda.

The Cosmic-Ray Contribution to Galactic Abundances of the Light Elements: Interpretation of GCR LiBeB Abundance Measurements from ACE/CRIS

N. E. Yanasak*, G. A. de Nolfo†, W. R. Binns**, E. R. Christian†, A. C. Cummings*, A. J. Davis*, J. S. George*, P. L. Hink**, M. H. Israel**, R. A. Leske*, M. Lijowski**, R. A. Mewaldt*, E. C. Stone*, T. T. von Rosenvinge† and M. E. Wiedenbeck‡

*Space Radiation Laboratory, California Institute of Technology, Pasadena, CA 91125 USA
†Laboratory for High Energy Astrophysics, NASA/Goddard Space Flight Center, Greenbelt, MD 20771 USA
**Dept. of Physics and McDonnell Center for the Space Sciences, Washington University, St. Louis, MO 63130 USA
‡Jet Propulsion Laboratory, California Institute of Technology, Pasadena, CA 91109 USA

Abstract. Inelastic collisions between the galactic cosmic rays (GCRs) and the interstellar medium (ISM) are responsible for producing essentially all of the light elements Li, Be, and B (LiBeB) observed in the cosmic rays. Previous calculations (e.g., [1]) have shown that GCR fragmentation can explain the bulk of the existing LiBeB abundance in the present day Galaxy. However, elemental abundances of LiBeB in old halo stars indicate inconsistencies with this explanation. We have used a simple leaky-box model to predict the cosmic-ray elemental and isotopic abundances of LiBeB in the present epoch. We conducted a survey of recent scientific literature on fragmentation cross sections and have calculated the amount of uncertainty they introduce into our model. The predicted particle intensities of this model were compared with high energy (E_{ISM}=200-500 MeV/nucleon) cosmic-ray data from the Cosmic Ray Isotope Spectrometer (CRIS), which indicates fairly good agreement with absolute fluxes for $Z \geq 5$ and relative isotopic abundances for all LiBeB species.

INTRODUCTION

It has been generally accepted for almost 30 years that production of the bulk of Li, Be, and B (LiBeB) in the present Galaxy can be attributed to inelastic collisions between cosmic rays and the interstellar medium (ISM). Specifically, large abundances of LiBeB arise from fragmentation of C, N, and O (CNO) in the ISM by cosmic ray H and He, fragmentation of cosmic-ray CNO species by the H and He in the ISM, and $\alpha - \alpha$ fusion in collisions between cosmic-ray and ISM nuclei [2]. Many previous studies have shown that these mechanisms can generally account for the all of the present-day local abundances of ^{6}Li, ^{9}Be, and ^{10}B [1]. Other sources of ^{7}Li and ^{11}B are required in addition to spallogenic mechanisms, leading to the suggestion that a significant component of the observable ^{7}Li in the ISM is a product of Big Bang nucleosynthesis [3]. Calculations by Woosley and Weaver suggest another small contribution to ^{7}Li and ^{11}B galactic abundances from neutrino-driven spallation of ^{12}C within Type II supernovae[4].

The observed elemental ratios Be/H, B/H, and Fe/H in low-metallicity halo stars formed in the early Galaxy tell a different story, indicating an overabundance of LiBeB in early epochs that cannot be accounted for by fragmentation of cosmic-ray CNO (e.g.,[5]). This interpretation assumes that the average ISM in any epoch serves as a source of material both for star formation and for cosmic rays in that epoch [6], and that these contribute LiBeB and other fragmentation products to the ISM at a somewhat later time. To explain both GCR spallogenic calculations and halo star abundances, possible solutions are that LiBeB species are created predominantly via fragmentation in the ISM of low-energy CO nuclei from SN II and Wolf-Rayet stars [7], or that cosmic rays are accelerated out of the metal-enriched supernova ejecta in superbubbles [8].

To address the origin of the LiBeB species, a precise calculation of the contribution from GCR fragmentation is needed. Our group has been using a simple cosmic-ray transport model to simulate cosmic-ray propagation in the galaxy for $4 \leq Z \leq 28$, based upon the formalism of Meneguzzi, Audouze, and Reeves [9], and we have recently begun work on improving this model to pre-

CP598, *Solar and Galactic Composition*, edited by R. F. Wimmer-Schweingruber

dict present epoch LiBeB GCR isotopic abundances and spectra observed at Earth. Because of their importance in any GCR transport model, we have re-examined fragmentation cross-section data in the scientific literature appearing both before and after the work of Read and Viola[10]. To test our model predictions, we used new abundance measurements made by the Cosmic Ray Isotope Spectrometer (CRIS) during the past three years [11]. The precision of the dataset from CRIS, $\sim$14% for statistical and systematic uncertainties in the LiBeB abundances, is high enough that cross-section uncertainties in the model must be considered. We have calculated the uncertainties that cross sections contribute to our model predictions. To determine in future studies whether we find an inconsistency between LiBeB halo star and spallogenic abundances, we wish to extend our predictions of the LiBeB production rate in the GCRs to lower energy. This includes using GCR H and He spectra from other experiments to predict LiBeB production rates from fragmentation of CNO atoms in the ISM and from $\alpha-\alpha$ fusion.

PROPAGATION MODEL

A steady-state, leaky-box model (e.g., [12]) was used for alculating the post-propagation GCR abundances observed by CRIS. Previously, abundance predictions from this model for cosmic-ray clock species led to the determination of the cosmic-ray confinement time [13]. A thorough review of the model input parameters (e.g., ionization fraction, ISM composition) was conducted by Yanasak et al. [13] to insure consistency with current literature. These parameters are similar to Davis, et al. [15], with a slight adjustment in the average value of the mean GCR pathlength through the ISM before escape to account for the lower ionization fraction used in this study. Abundance predictions for Z>4 are virtually identical for both models.

With the availability of precise cosmic-ray data from CRIS, uncertainties in the fragmentation cross sections have become a dominant limitation to the study of rare cosmic-ray species that are generated predominantly via spallation [13]. A re-examination of cross-section estimates and their uncertainties was undertaken for reactions involving ^{7}Be, Li isotopes, and products decaying to LiBeB species (e.g., ^{6}He, ^{10,11}C), to include recent cross-section measurements. Partial and total fragmentation cross sections for nuclei of mass A=9-56 were previously updated in Yanasak et al. [13]. The "excitation functions" of Read and Viola [10], based on previous cross-section measurements, provide an estimate of isobaric production cross sections as a function of energy, and these are useful for predicting thermal LiBeB abundances in the ISM. However, species such as ^{7}Be and ^{10}Be which ultimately decay to ^{7}Li and ^{10}B are significantly abundant in the GCRs, and their partial production cross sections are necessary for our model. Michel et al. [16] and Sisterson et al. [17] have precisely surveyed the partial Be production cross sections for fragmentation of CNO on hydrogen over a range of energies $E\sim 10-400$ MeV/nucleon. Higher energy cross-section measurements (E$\sim$365-600 MeV/nucleon, from [18, 19]) for p+CNO$\rightarrow$LiBeB reactions have also been made since the compilation of Read and Viola. In addition, measurements for He+CNO$\rightarrow$LiBeB reactions are now available [18, 20]).

In this study, three different methods for evaluating partial cross sections were used. For some reactions, enough cross-section data exist to define a function which can be interpolated at a particular energy to determine a cross-section value (Method 1). For other reactions which have a small number of measurements, the excitation functions of Read and Viola provided an energy dependence, and these were normalized where necessary for agreement with available cross-section data (Method 2). Finally, in cases where the parent nucleus is not CNO, the energy dependence of Silberberg, Tsao, and Barghouty [14] was used (Method 3a), and the cross sections for He-induced fragmentation in these cases were scaled using the parameterization of Hirzebruch, Winkel, and Heinrich [21]. For collisions with parent species A>16, the Silberberg et al. [14] energy dependence was also normalized to cross-section data where it exists (Method 3b). As a check, the isobaric cross-section data compiled in Read and Viola were compared to the sum of partial cross sections from the above methods, for reactions involving CNO parent nuclei.

Table 1 compares the contribution to LiBeB abundances from each of the three methods for calculating partial cross sections, typical uncertainties associated with cross-section measurements using each method, and the uncertainty contribution to the LiBeB abundances. The % secondary contributions to LiBeB species were determined by using our model to compute abundances after setting the cross sections for other methods equal to zero. Tertiary reactions, where B nuclei fragment to Li and Be, were not included in the calculation of the % contributions in Table 1, and their importance will be discussed later. Using the average cross-section uncertainties for each method, the amount of uncertainty in the total predicted LiBeB abundances was calculated assuming a steady-state solution of the leaky-box model and following the formalism of Wiedenbeck [22] adapted for use with secondary GCR species. The average cross-section uncertainties for each method were derived differently. For Methods 2 and 3b where few cross-section measurements exist, the average uncertainty was taken

TABLE 1. Cross-section contributions to uncertainties in predicted LiBeB abundances

Cross-sectional energy dependence function	Examples of reactions	(Li,Be,B)% secondary contributions *	Typical cross section uncertainty†	% Uncertainty in total predicted (Li,Be,B) abundance
Interpolated function	p+CNO→Be p+(^{12}C,^{16}O)→Li,B	(62.5,78.5,61.1)%	6% (p:5%, He:6%)	(2.6,2.7,1.9)%
Read and Viola	p+^{14}N→Li,B He+(^{12}C,^{16}O)→Li	(21.0,2.8,26.1)%	13% (p:17%, He:10%)	(1.5,0.4,2.4)%
Silberberg et al. (normalized to data)	p+^{22}Ne→B	(0.8,4.6,3.9)%	14%**	(0.0,<0.1,<0.1)%
(no data)	(He,p)+(A>16)→LiBeB	(15.7,14.1,8.9)%	20% [14] (p:20%, He:40%)	(0.3,0.3,0.1)%‡

* Excluding tertiary reactions (e.g., p+B→Li).
† For an individual reaction, between E=100-10000 MeV/nucleon
** No He cross-section measurements in this class
‡ Corresponds to 40% uncertainty in the He cross sections.

as the reduced standard deviation $\sigma = 1/\sqrt{\Sigma_i(1/\sigma_i^2)}$, where σ_i is an individual measurement uncertainty. For Method 1, an average percent deviation off the function weighted by measurement uncertainties was calculated as well as a reduced χ^2 comparing the measurements with the average. For most reactions, χ^2 was smaller than one, and the uncertainty was taken as the reduced standard deviation. For a few reactions, fluctuations not represented by the measurement uncertainties result in a large χ^2, and in these cases, the actual reduced standard deviation was estimated by making $\sigma^2 = \chi^2/\sqrt{\Sigma_i(1/\sigma_i^2)}$.

To estimate the uncertainty for reactions without any measurements (Method 3a), we compared the formulae of Silberberg et al. [14] for well-determined reactions to cross-section measurements. Silberberg et al. [14] estimate a general uncertainty in their formulae of 20%, and we found a similar standard deviation in the measurement distribution around their formulae. However, for reactions induced by ISM helium, the standard deviation is significantly larger (from $\sim 0.4-2.0$). The reason for this may be that the scaling of Hirzebruch et al. [21] was determined for $E_{ISM} \geq 700$ MeV/nucleon and may break down at lower energies. The function of Tsao et al. [23] that scales ISM hydrogen reactions to helium will be compared in the future to Hirzebruch et al. [21] to determine if there is better agreement.

As shown in Table 1, the efforts of many previous cross-section experiments have collectively resulted in measurements for reactions that contribute almost 85% to the total LiBeB GCR abundances. In addition to these reactions, the contribution from fragmentation of B isotopes to ^{6}He (which decays to ^{6}Li), Li, and Be may not be negligible in all cases because of the large relative abundance of B to Li and Be. Cross-section measurements are available for the p+B→Be reactions, and their uncertainties contribute 1% to the total Be abundance uncertainty. Unfortunately, we are not aware of the existence of high energy measurements of p+B→Li. Using the formulae of Silberberg et al. [14], the fragmentation of B boosts the abundance of Li from secondary reactions by an additional ~12%. Although the uncertainty for this reaction is unknown, a 20% cross-section uncertainty [14] results in a maximum 3% Li abundance uncertainty. Adding the abundance uncertainties in Table 1 in quadrature, and adding 1% to Be and 3% to Li for uncertainties from tertiary reactions, we find that cross-section uncertainties affect the total predicted LiBeB elemental abundances in any model by ~4.3%, 2.9%, and 2.5% for Li, Be, and B, respectively. It is important to note that all cross-section experimental measurements were assumed to be uncorrelated in this discussion. Multiple measurements for particular reactions exist that were made by individual experiments, and these data may suffer from correlated shifts in the cross-section normalization, resulting in somewhat larger abundance uncertainties than those derived above. However, the number of cross-section experiments surveyed by this study (>70) that measure dominant reactions producing LiBeB is greater than the number of those reactions ($\lesssim$24), so uncertainty correlations are suppressed in general.

The total fragmentation cross sections were also updated in this study, although the long interaction length for LiBeB in hydrogen compared to the mean transport pathlength through the ISM makes uncertainties for these less important. The energy dependence of total fragmentation formulae from Letaw, Silberberg, and Tsao [25]

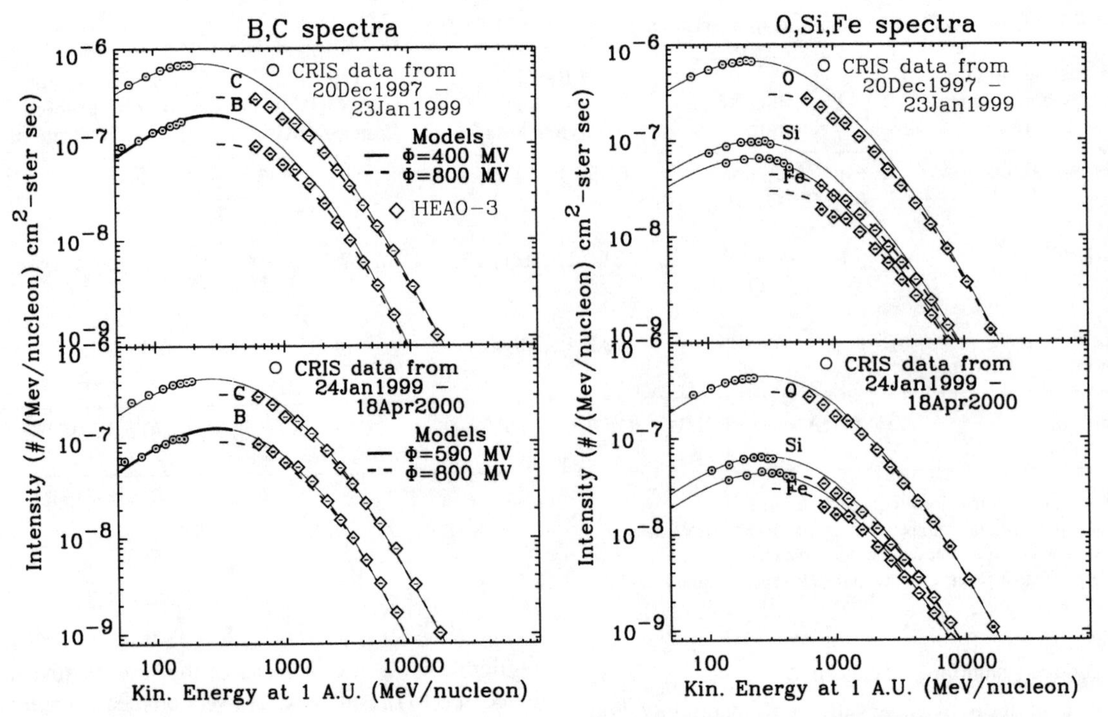

FIGURE 1. Flux measurements from CRIS for B and abundant primary GCR species. Data from the HEAO-3 spacecraft are shown for comparison [24]. Also shown are predictions for CRIS spectra from our model (solid line), uncertainties for B predictions (thick line below 400 MeV/nucleon), and model predictions at a higher level of modulation ϕ=800 MV comparable to HEAO-3 data (dashed line).

is consistent with LiBeB cross-section data at GCR energies in the ISM probed by CRIS (E_{ISM} ~200-500 MeV/nucleon), and at all energies for $Z \geq 8$. However, at lower energies relevant to galactic evolution studies, the formulae of Tripathi, Cucinotta, and Wilson [26] are in better agreement with LiBeB data. A total fragmentation cross-section energy dependence is chosen which matches Letaw et al. at the high energies and Tripathi et al. at lower energies.

Our steady-state model implicitly uses an exponential GCR pathlength distribution. The mean ISM pathlength was adjusted to match four secondary-to-primary ratios from CRIS (B/C, F/Ne, P/S, and (Sc+Ti+V)/Fe), using HEAO-3 data at higher energies as a measure of consistency [24]. Solar modulation of the GCR spectra was simulated using the spherically-symmetric model described by Fisk [27]. Levels of modulation were determined via choosing a source spectrum and matching post-propagation spectral shapes in our model to HEAO-3 and CRIS data [15].

DISCUSSION

Predictions for the absolute spectrum of B as well as other GCR dominant nuclei C, O, Si, and Fe are shown in Figure 1. The thicker curve shown for the predicted B spectrum at $E \lesssim 200$ MeV/nucleon indicates one standard deviation of uncertainty from cross sections. Data from CRIS and HEAO-3[24] match both the absolute intensities and energy dependences of the predicted spectra well for B, C, O, Si, and Fe during both time periods chosen. Statistical uncertainties for the CRIS are small compared with the size of the plotted circles, and an additional ~10% uncertainty in the instrument systematics should also be added in quadrature [11].

Figure 2 shows comparisons between isotopic ratios from model predictions and data from the first period in Figure 1. Small corrections ($\leq$ 10%) have been made to the data to account for differences in average measurement energy of the data for each species. Uncertainties for CRIS in this figure account for both statistical and systematic uncertainties. The hatched regions shown in Figure 2 are one standard deviation of uncertainty in the model prediction from cross sections, somewhat larger than those shown in Table 1 for elemental species because the uncertainties for individual isotopes are considered. The average value of the data and the model generally agree, although model uncertainties prevent a useful comparison to the energy dependences of the data. In particular, the ^{7}Be/^{9}Be ratio shows slightly more energy dependence than what is expected from the model. However, the uncertainty contribution from the

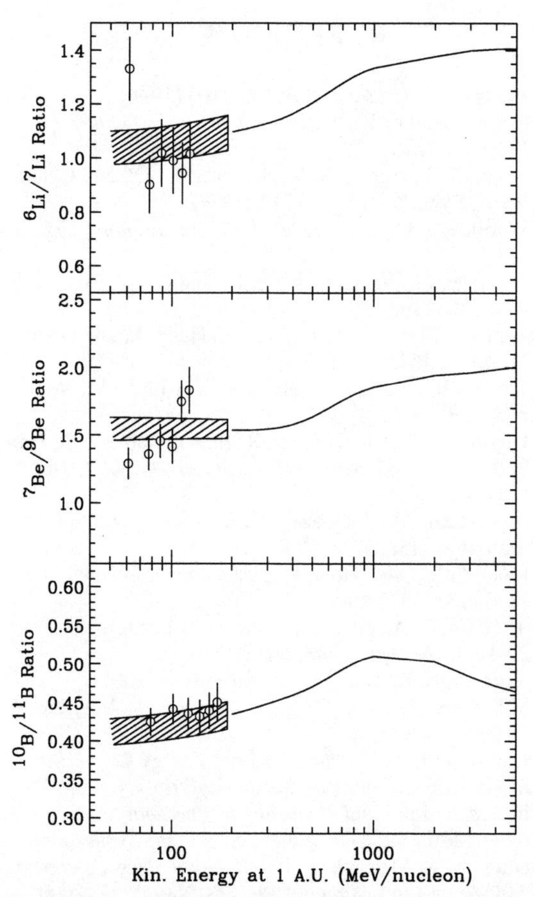

FIGURE 2. Relative isotopic abundances from CRIS (open circles) and model predictions (solid line). The hatched region indicates one standard deviation from cross-section uncertainties. Note that the y-axis scale for each ratio is different and has an offset.

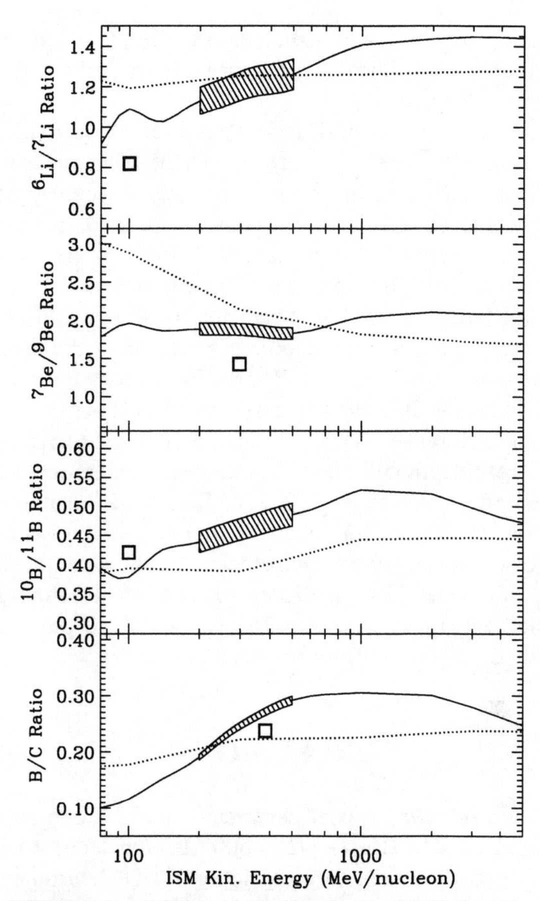

FIGURE 3. Model predictions in the ISM from this study (solid line). The size of the hatched region represents one standard deviation of model uncertainty (see Figure 2). The dotted line is a predictions from Meneguzzi et al. [9], and the squares are predictions from a model by Lukasiak, McDonald, and Webber [28].

(p,He)+B→Be,Li reactions is not included in the predicted uncertainties shown in Figure 2, which should improve the agreement.

We are achieving good agreement between CRIS data and our model for Be and B species, and our model predictions compare favorably with other computations of the GCR spallogenic contribution to LiBeB abundances. Propagation parameters from the model of Garcia-Munoz et al. [29] are used by a number of studies (e.g., [5]), which give comparable results to our model within the uncertainty of $\pm30\%$ suggested by the authors for their GCR mean escape pathlength. The ratios of stable secondary isotopic species with the exception of ^{10}B, which has a small contribution from the decay of ^{10}Be, should not be affected by differences in the choice of mean pathlength. Differences in the Li/C and Be/C ratios at 1 A.U. predicted by both models are $\lesssim$10% at 50 MeV/nucleon, increasing to $\sim$20% at 1000 MeV/nucleon. More importantly, the mean pathlength from [29] is not independent of the choices for production and fragmentation cross sections presented in the appendix of that study. One must take care that any study incorporating new cross-section measurements and using a previously-reported GCR mean pathlength is able to predict adequately GCR abundance data for species produced via fragmentation.

Figure 3 shows predictions from our model in the ISM for several elemental and isotopic abundance ratios as solid line and a hatched band, with the width of the band representing the typical model uncertainty from cross sections. Also shown are GCR abundance predictions from Meneguzzi et al. [9] (dotted line) and from a Monte Carlo diffusion model of Lukasiak et al. [28] (boxes). The energy assignment for the ^{7}Be/^{9}Be and B/C ratios from Lukasiak et al. [28] is somewhat uncertain because these data were measured within the solar system and extrapolated to the ISM. Agreement between the latter study and our model is good within uncertainties, with the possible exception of ^{7}Be/^{9}Be. One possible explanation for the lower abundance of ^{7}Be predicted by Lukasiak et al. [28] is that the cross sections used in each study are different.

Unlike the work of Garcia-Munoz et al. [29] and Lukasiak et al. [28], the early model of Meneguzzi et al. [9] assumes a mean pathlength that is indepenent of energy, and the effect of this choice appears as a generally flat secondary-to-primary ratio B/C. At energies probed by CRIS, the agreement between our model and that of Meneguzzi et al. is satisfactory or shows a difference from a lack of measured cross sections at the time of the earlier study (e.g., ^{10}B/^{11}B). Predictions by both models for the B/C ratio at lower energies shows a significant difference ($\sim$1.8x). Using an energy-dependent mean pathlength will effectively decrease the amount of LiBeB production from GCR CNO fragmentation at low energies relative to higher energies. Because the lower energy particles will thermalize more quickly than those with higher energies, the Meneguzzi et al. model should predict more production of LiBeB by GCR CNO parents with E_{ISM}>100 MeV/nucleon than our model.

CONCLUSIONS

We have surveyed model parameters at GCR energies probed by CRIS (E_{ISM} $\sim$200-500 MeV/nucleon). Our model gives a satisfactory prediction for GCR primary and secondary species with $Z \geq 4$, and it shows good agreement for relative isotopic abundances. In some cases (notably (p,He)+B$\rightarrow$Li), a lack of cross-section measurements limits our understanding of the model inputs. However, with additional measurements made since Read and Viola [10], uncertainties for the cross sections for some reactions have improved, and we have calculated estimates of the magnitude of our uncertainty. Comparisons with previous models show general agreement in the energy range relevant to CRIS data. Future work will include turning our model around to predict the amount of LiBeB produced via inelastic, GCR p and He interactions with CNO in the ISM, and we will also investigate the low-energy contribution from $\alpha - \alpha$ fusion to ^{6,7}Li.

ACKNOWLEDGMENTS

This research was supported by NASA at the California Institute of Technology, (grant NAG5-6912), the Jet Propulsion Laboratory, the NASA/Goddard Space Flight Center, and Washington University.

REFERENCES

1. Reeves, H., *Rev. Mod. Phys.*, **66**, 193 (1994).
2. Reeves, H., Fowler, W. A., and Hoyle, F., *Nature Phys. Sci.*, **226**, 727 (1970).
3. King, C. H., Austin, S. M., Rossner, H. H., and Chien, W. S., *Phys. Rev. C*, **16**, 1712 (1977).
4. Woosley, S. E., and Weaver, T. A., *Ap. J. Supp.*, **101**, 181 (1995).
5. Lemoine, M., Vangioni-Flam, E., and Casse, M., *Ap. J.*, **499**, 735 (1998).
6. Vangioni-Flam, E., Casse, M., Audouze, J., and Oberto, Y., *Ap. J.*, **364**, 568 (1990).
7. Casse, M., Lehoucq, R., and Vangioni-Flam, E., *Nature*, **373**, 318 (1995).
8. Higdon, H. C., Lingenfelter, R. E., and Ramaty, R., *Proc. 26th Int. Cosmic-Ray Conf. (Salt Lake City)*, **4**, 144 (1999).
9. Meneguzzi, M., Audouze, J., and Reeves, H., *Astron. Astrophys.*, **15**, 337 (1971).
10. Read, S. M., and Viola, V. E., *Atomic Data Nucl. Data Tables*, **31**, 359 (1984).
11. de Nolfo, G. A., et al., *this conference* (2001).
12. Leske, R. A., *Ap. J.*, **405**, 567 (1993).
13. Yanasak, N. E., et al., *Ap. J.*, **submitted** (2001).
14. Silberberg, R., Tsao, C. H., and Barghouty, A. F., *Ap. J.*, **501**, 911 (1998).
15. Davis, A. J., et al., "On the Low Energy Decrease in Galactic Cosmic Ray Secondary/Primary Ratios", in *Acceleration and Transport of Energetic Particles Observed in the Heliosphere: ACE 2000 Symposium*, edited by R. Mewaldt et al., AIP Conference Proceedings 528, American Institute of Physics, New York, 2000, p. 421.
16. Michel, R., et al., *Nucl. Instr. and Meth. B*, **103**, 183 (1995).
17. Sisterson, J. M., et al., *Nucl. Instr. and Meth. B*, **123**, 324 (1997).
18. Webber, W. R., Kish, J. C., and Schrier, D. A., *Phys. Rev. C*, **41**, 547 (1990).
19. Webber, W. R., et al., *Ap. J.*, **508**, 949 (1998).
20. Mercer, D. J., Austin, S. M., and Glagola, B. G., *Phys. Rev. C*, **55**, 946 (1997).
21. Hirzebruch, S. E., Winkel, E., and Heinrich, W., *Proc. 23rd Int. Cosmic-Ray Conf. (Calgary)*, **2**, 175 (1993).
22. Wiedenbeck, M. E., "The Effect of Cross-Section Uncertainties on the Derivation of Source Abundances From Cosmic-Ray Composition Observations", in *Composition and Origin of Cosmic Rays*, edited by M. M. Shapiro, D. Reidel, Dordrecht, 1983, p. 343.
23. Tsao, C. H., Silberberg, R., and Barghouty, A. F., *Ap. J.*, **501**, 920 (1998).
24. Englemann, J. J., et al., *Astron. Astrophys.*, **233**, 96 (1990).
25. Letaw, J. R., Silberberg, R., and Tsao, C. H., *Ap. J. Suppl.*, **51**, 271 (1983).
26. Tripathi, R. K., Cucinotta, F. A., and Wilson, J. W., *NASA Technical Report TP-1999-209726* (1999).
27. Fisk, L. A., *JGR*, **76**, 221 (1971).
28. Lukasiak, A., McDonald, F. B., and Webber, W. R., *Proc. 26rd Int. Cosmic-Ray Conf. (Salt Lake City)*, **3**, 41 (1999).
29. Garcia-Munoz, M., Simpson, J. A., Guzik, T. G., Wefel, J. P., and Margolis, S. H., *Ap. J. Suppl.*, **64**, 269 (1987).

Measurements of the Isotopes of Lithium, Beryllium, and Boron from ACE/CRIS

G. A. de Nolfo*, N.E. Yanasak†, W.R. Binns**, A.C. Cummings†, E.R. Christian*, J.S. George†, P.L. Hink**, M.H. Israel**, R.A. Leske†, M. Lijowski**, R.A. Mewaldt†, E.C. Stone†, T.T. von Rosenvinge* and M.E. Wiedenbeck‡

**Laboratory for High Energy Astrophysics, NASA/Goddard Space Flight Center, Greenbelt, MD, 20771 USA*
†Space Radiation Laboratory, California Institute of Technology, Pasadena, CA, 91125 USA
***Department of Physics and McDonnell Center for the Space Sciences, Washington University, St. Louis, MO, 63130 USA*
‡Jet Propulsion Laboratory, California Institute of Technology, Pasadena, CA, 91109 USA

Abstract. The isotopes of lithium, beryllium, and boron (LiBeB) are known in nature to be produced primarily by CNO spallation and α - α fusion from interactions between cosmic rays and interstellar nuclei. While the dominant source of LiBeB isotopes in the present epoch is cosmic-ray interactions, other sources are known to exist, including the production of ^{7}Li from big bang nucleosynthesis. Precise observations of galactic cosmic-ray LiBeB in addition to accurate modeling of cosmic-ray transport can help to constrain the relative importance among the different production mechanisms. The Cosmic Ray Isotope Spectrometer (CRIS) on the Advanced Composition Explorer (ACE) has measured nuclei with $2 \lesssim Z \lesssim 30$ in the energy range ~30-500 MeV/nucleon since 1997 with good statistical accuracy. We present measurements of the isotopic abundances of LiBeB and discuss these observations in the context of previous cosmic-ray measurements and spectroscopic observations.

INTRODUCTION

The elements lithium, beryllium, and boron (LiBeB) are rare in our galaxy and offer important constraints on our understanding of nuclear astrophysics. These fragile elements are not generated in appreciable amounts in stellar nucleosynthesis but are destroyed in stellar interiors, and this is reflected by their low abundance in nature (as measured in meteorites and stellar photospheres) [1]. Galactic cosmic-ray (GCR) LiBeB nuclei, on the other hand, are found to be in significant excess over their solar-system abundances. This excess is quite large (LiBeB/CNO ~0.25 for GCRs compared with $\sim 10^{-6}$ for solar-system values) and is attributed to the spallation of CNO GCRs on the interstellar medium (ISM) [2]. A few isotopic discrepancies between GCR abundances and solar-system abundances, specifically the ratios of ^{6}Li/^{7}Li and ^{11}B/^{10}B, are not consistent with spallation production, suggesting the need for other sources of the lithium and boron isotopes such as stellar nucleosynthesis [3] and neutrino spallation [4]. In addition, some ^{7}Li production is expected from primordial nucleosynthesis [5].

The evolution of LiBeB over the last 10 Gyr has been assessed through observations of low-metallicity halo stars [6], [7]. These observations indicate an approximately linear correlation between the elemental ratios of Be/H or B/H and Fe/H (or metallicity of the star), in contradiction with the expected quadratic behavior for the production of LiBeB nuclei from GCRs released into an ISM, which is itself increasing in metallicity with time from the contributions of evolving stars. A linear correlation suggests a primary origin of LiBeB nuclei not coupled to the ISM metallicity, such as the fragmentation products from freshly synthesized supernova ejecta. Several scenarios have been postulated to explain the observed correlation, including the origin of LiBeB species from within superbubble regions [8], [9], [10]. In this model, supernovae Type II (SN II), which are common in the early Galaxy and originate from stars in large associations, tend to sweep out a bubble filled with a warm, low density mixture of freshly synthesized supernovae ejecta and remaining interstellar material. Low-energy carbon and oxygen (<100 MeV/nucleon) are produced and accelerated within SN II, perhaps at the reverse shock [11],[12]. Once ejected, these low-energy carbon and oxygen nuclei are subsequently accelerated by massive stars or at the forward shocks of SN II in these

CP598, *Solar and Galactic Composition,* edited by R. F. Wimmer-Schweingruber

superbubble regions and interact with interstellar hydrogen and helium to produce LiBeB abundances via fragmentation in agreement with observations of old Population II stars [13], [14]. Other models have been proposed, including the origin of Be and B nuclei by GCRs accelerated as debris from grains formed in SN II ejecta [15], [16].

Observations of GCR LiBeB nuclei help to constrain models of the origin and evolution of present epoch abundances. The Cosmic Ray Isotope Spectrometer (CRIS) on ACE has been measuring the isotopic composition of $2 \lesssim Z \lesssim 30$ in the energy range ~30-500 MeV/nucleon since 1997, including the abundances of LiBeB between ~30-200 MeV/nucleon. CRIS observations of LiBeB nuclei originate almost exclusively from fast CNO GCRs interacting with the ISM to produce secondary LiBeB nuclei between 200-500 MeV/nucleon. The resulting GCR LiBeB nuclei depend on the fragmentation rate and the rate of escape from the Galaxy. Indeed, most of the GCR LiBeB escape from the Galaxy. As the escape rate is essentially the same for all LiBeB species of equal energy per nucleon [17], the relative abundances are a direct measure of the relative production rates. Approximately half of the LiBeB nuclei produced by GCRs, however, result from the fragmentation of ISM by GCR protons and helium whose energy spectra peak around 1 GeV. Provided the cross-sections are not too different between the interactions of ISM (p,He) + GCR (CNO) -> LiBeB around 1 GeV and the inverse interactions of ISM (CNO) + GCR (p,He) -> LiBeB in the energy range covered by CRIS, the observed abundances of LiBeB isotopic ratios from CRIS should reflect the actual isotopic ratios observed in the ISM. Indeed, based on a review of the cross-sections of Read and Viola [18] by Yanasak et al. (these proceedings), the cross-sections between 200-500 MeV/nucleon and those between 500-2000 MeV/nucleon for the inverse reactions are on average less than 20 % different. The ISM ratio ^{6}Li/^{7}Li, however, may not be similar to the GCR ratio since the cross-section for $\alpha + \alpha$ fusion to A=6,7 increases by more than 2 orders of magnitude between 10 and 100 MeV/nucleon, implying that a large fraction of ISM lithium may be produced at these low energies. Thus, GCR LiBeB observations over a wide range in energy can provide important constraints on cosmic-ray origin and evolution in addition to deciphering the dominant sources of LiBeB over time. In this paper, we present the measurements of GCR LiBeB from December 1997 through April 2000 and compare these observations with previously published results.

DATA ANALYSIS

CRIS identifies the charge and mass of incident particles using multiple dE/dx measurements and the total energy deposited within stacked silicon solid-state detectors [19]. The mass resolution is refined with the measurement of particle trajectory via a scintillating optical fiber hodoscope (SOFT) [19]. In the present study, CRIS observations of LiBeB are separated into two separate time intervals, corresponding to differing levels of solar modulation. The first time period ranges from December 20, 1997 to January 23, 1999, which corresponds to a total of 311 days, excluding periods of intense solar activity. The second period ranges from January 24, 1999 to April 18, 2000 corresponding to a total of 270 days, again excluding periods of intense solar activity. The effect of solar modulation experienced during these two time intervals is determined using a spherically symmetric model described by Fisk [20]. The solar modulation parameter, ϕ, is determined by fitting elemental spectra from CRIS at low energies and from HEAO-3 at high energies with the predictions of a propagation model [21]. The first interval corresponds to $\phi \sim 400$ MV and the second time interval corresponds to $\phi \sim 590$ MV.

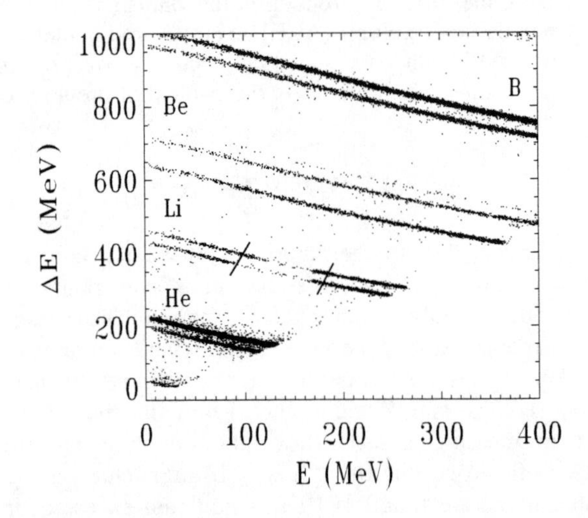

FIGURE 1. Energy loss (ΔE) versus the energy (E) deposited in a detector element midway through the stack for He, Li, Be, and B events. The solid lines indicate the region excluded from the analysis to exclude events within the low duty-cycle helium event buffer.

CRIS is designed to transmit only a small sample of the highly abundant helium events [19]. Since the onboard classification of an event as helium is rather coarse and based only on pulse heights, and since lithium is adjacent to helium, lithium events within certain energy intervals are treated as helium and sampled at a reduced duty cycle. Figure 1 shows an example of the energy loss versus the energy deposited in two detectors mid-

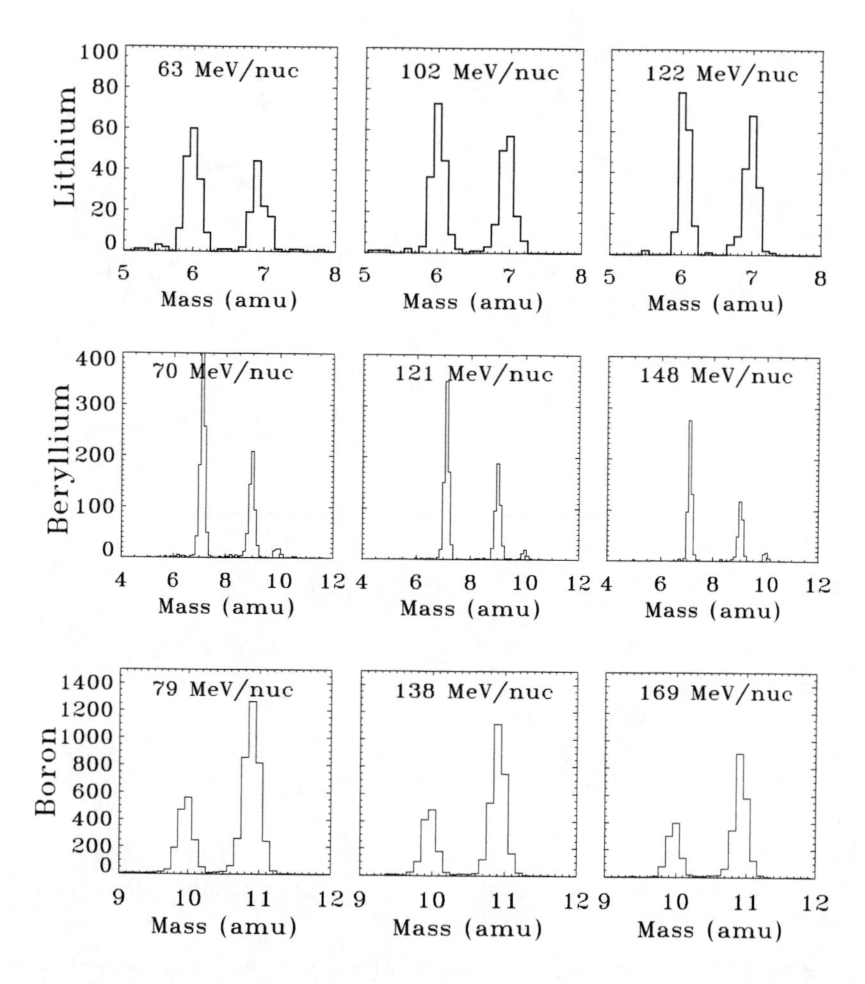

FIGURE 2. Mass histograms for the isotopes of LiBeB in three of the six energy ranges covered in this study. The labels refer to the median energy for each range.

way through the CRIS stack for events with $Z < 6$. From Figure 1, it is clear that CRIS obtains excellent mass resolution. The solid lines indicate a region where lithium events have been classified into a helium event buffer which is sampled at a lower rate, resulting in a lower density region in the track. In the present analysis, events located within an interval such as that shown in Figure 1 are excluded, ensuring that those events with pulse heights satisfying the helium event buffer are eliminated from the analysis.

Particles are selected on the basis of high-quality mass and charge identification. Valid events are required to trigger all three SOFT hodoscope planes. Furthermore, events that stop near the faces of the detector elements are rejected from the analysis. Additional background events are removed by demanding consistency between the measures of charge and mass from various combinations of the stack detector pulse heights [19].

The resulting mass histograms for LiBeB events in the first time period are shown in Figure 2 for three of the six energy bins analyzed in this study, ranging from ~30 MeV/nucleon to ~200 MeV/nucleon. The isotopes of LiBeB are clearly separated and the number of events for each isotope is simply determined by adding the number of events under each peak. Only events with an opening angle from the detector zenith of <25 degrees are included in the analysis. Corrections are applied to account for the probability of a particle surviving fragmentation within the instrument and for the charge dependent tracking efficiency with the SOFT hodoscope. The spallation correction accounts for the number of interaction lengths penetrated and is based on the cross-section formula of Westfall *et al.* [22].

RESULTS AND DISCUSSION

Figure 3 shows the elemental ratio of B/C and the isotopic ratios of ^{6}Li/^{7}Li, ^{7}Be/^{9}Be, and ^{11}B/^{10}B for the first time period along with previous measurements of these

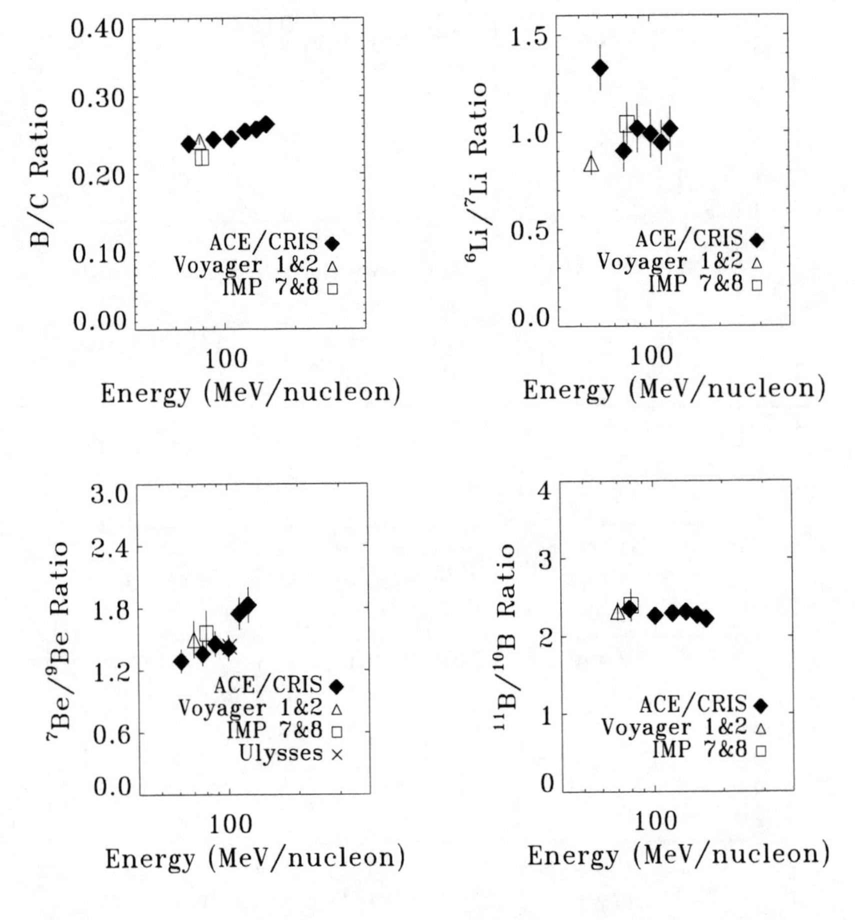

FIGURE 3. The ratios of ^{6}Li/^{7}Li, ^{6}Li/C, ^{7}Li/C and Li/C obtained from CRIS during the first time interval compared with previous measurements. The observations of IMP 7& 8 and Voyager 1& 2 were obtained during comparable periods of solar modulation.

ratios from Voyager 1 & 2 [23] and IMP 7 & 8 [24] and Ulysses [25]. Observations from Voyager 1 & 2 were made over 19 years at an average solar modulation level of $\phi \sim 450$ MV and IMP 7 & 8 observations were made at a solar modulation level of $\phi \sim 430$ MV. These measurements were obtained during periods of comparable solar modulation to that experienced by CRIS during the first time interval. CRIS observations of B/C and the isotopic ratios of ^{6}Li/^{7}Li, ^{7}Be/^{9}Be, and ^{11}B/^{10}B are in agreement with both Voyager and IMP to within the statistical accuracy of the measurements. Indeed, CRIS observations are unique in that there are now sufficient events to measure these ratios as a function of energy, providing further constraints on cosmic-ray propagation models.

Table 1 compares the isotopic measurements of LiBeB with previous measurements, where the results have been averaged over the entire energy range covered by CRIS. The uncertainties for CRIS in Table 1 represent only the statistical uncertainty. Typical systematic uncertainties are ~10% for beryllium and lithium due to the SOFT tracking efficiency. The predictions of two propagation models from Reeves [3] and Lukasiak et al. [23] are also shown in Table 1. While CRIS results agree with previous measurements and with the predictions of the propagation model by Lukasiak et al. [23], the measured isotopic ratios differ, not unexpectedly, with the calculations of Reeves [3] for GCRs in the several GeV/nucleon energy range. The effect of energy dependent cross-sections on the production of GCR LiBeB nuclei and the expected energy dependence of the LiBeB ratios compared with these observations will be explored in detail in a companion paper in these proceedings (Yanasak *et al.*) [26].

Comparing the relatively young GCRs ($\sim 1.5 \times 10^7$ years [27]) with presolar isotopic abundances can shed light on the evolution of LiBeB in general. The ratio of ^{7}Li/^{6}Li is well below the solar-system value of ~12 determined from meteorites [1]. The ^{11}B/^{10}B ratio is also well below the ratios determined from meteorites (4.05 ±0.2) [28] and from present abundance measurements

TABLE 1. LiBeB ratios compared with other observations and with the Leaky Box Model (LBM) prediction of Lukasiak et al. 1999 [23] and of Reeves et al. 1994 [3]. The solar modulation for CRIS in period I and II are $\phi \sim 400$ MV and 590 MV, respectively. Uncertainties are statistical only.

Ratios	CRIS I (40-130) (MeV/nucleon)	CRIS II (40-130) (MeV/nucleon)	IMP 7&8 (31-151) (MeV/nucleon)	Voyager 1&2 $\sim$80 (MeV/nucleon)	LMB (1999) $\sim$100 (MeV/nucleon)	Reeves (1994) $\sim$1 (GeV/nucleon)
^{7}Li/^{6}Li	0.97 ± 0.05	1.12 ± 0.08	0.97 ± 0.11	1.19 ± 0.09	1.26	1.4
^{7}Be/^{9}Be	1.54 ± 0.06	1.41 ± 0.07	1.56 ± 0.12	1.50 ± 0.18	1.43	–
^{11}B/^{10}B	2.29 ± 0.04	2.36 ± 0.04	2.40 ± 0.21	2.32 ± 0.12	2.49	2.5
B/C	0.250 ± 0.004	0.265 ± 0.004	0.220 ± 0.009	0.240 ± 0.005	0.237	–

of the ISM (3.4 ± 0.7) [29]. In fact, the meteoritic observations of ^{7}Li/^{6}Li and ^{11}B/^{10}B, representing protosolar measurements, are consistent with measurements of Population I stars (see compilation from [30] and [31]). The difference between the GCR ^{7}Li/^{6}Li and the ^{11}B/^{10}B ratios and the solar-system values has been attributed to excess ^{7}Li and ^{11}B, postulated to originate in part from neutrino spallation in core collapse supernovae [16]. The isotope ^{7}Li is also produced during primordial nucleosynthesis ($\sim$1/10 solar) [32]. Furthermore, halo stars with even lower metallicities [Fe/H] ~ -2.3 indicate a ^{6}Li/^{7}Li ratio of 0.05 [31]. While GCRs play a dominant role in the production of LiBeB nuclei, the origin and evolution of these species is complex as is reflected in the variety of observations from different epochs over the last 10 Gyr.

SUMMARY

The isotopic ratios of LiBeB and C are in agreement with previous measurements of GCRs. The observed ratios during both time periods are also in agreement suggesting that the effects of solar modulation are not significant, though this will be tested further with future observations from CRIS and the Solar Isotope Experiment (SIS) on-board ACE during periods of extreme solar modulation. While current propagation models are in general agreement with the LiBeB observations, CRIS observations over an extended energy range will provide further constraints on propagation models as will be discussed in the companion paper (Yanasak *et al.*) [26].

ACKNOWLEDGMENTS

This work was supported by NASA at NASA/Goddard Space Flight Center, the Jet Propulsion Laboratory, Washington University and the California Institute of Technology (under grant NAG5-6912).

REFERENCES

1. Anders, E., and Grevesse, N., *Geochim. Cosmochim. Acta*, **53**, 197 (1989).
2. Reeves, H., et al., *Nature*, **226**, 727 (1970).
3. Reeves, H., *Rev. Mod. Phys.*, **66**, 193 (1994).
4. Woosley, S., and Weaver, T., *ApJ Suppl.*, **101**, 181 (1995).
5. Schramm, D., *Origin and Evolution of the Elements*, N. Prantzos et al., Cambridge University, Cambridge, England, 1993, p. 112.
6. Ryan, S., et al., *ApJ*, **388**, 184 (1994).
7. Duncan, D., et al., *ApJ*, **488**, 338 (1997).
8. Bykov, A., "LiBeB", in *LiBeB Cosmic Rays and Gamma-Ray Line Astronomy*, edited by R. Ramaty et al., ASP Conference Series, 1999, vol. 171, p. 146.
9. Parizot, E., et al., *Astron. Astrophys.*, **328**, 107 (1997).
10. Higdon, H., et al., *Proc. 26th ICRC*, **4**, 144 (1999).
11. Parizot, E., and Drury, L., *Astron. Astrophys.*, **346**, 329 (1999).
12. Ramaty, R., et al., *ApJ*, **488**, 730 (1997).
13. Cásse, M., et al., *Nature*, **373**, 318 (1995).
14. Ellison, D., et al., *ApJ*, **487**, 197 (1997).
15. Ryan, S., et al., *ApJ*, **488**, 730 (1997).
16. Vangioni-Flam, E., et al., *ApJ*, **468**, 199 (1996).
17. Garcia-Munoz, M., et al., *ApJ Suppl.*, **64**, 269 (1987).
18. Read, S., and Viola, V., *Atomic Data Nucl. Data Tables*, **31**, 359 (1984).
19. Stone, E., et al., *Space Sci. Rev.*, **86**, 285 (1998).
20. Fisk, L., *JGR*, **76**, 221 (1971).
21. Davis, A., *Phys. Rev. C*, **528**, 421 (2000).
22. Westfall, G., *Phys. Rev. C*, **19**, 1309 (1979).
23. Lukasiak, A., et al., *Proc. 26th ICRC*, **3**, 41 (1999).
24. Garcia-Munoz, M., et al., *Proc. 17th ICRC*, **2**, 72 (1981).
25. Connell, J., *ApJ*, **501**, L59 (1998).
26. Yanasak, N., et al., *these proceedings* (2001).
27. Yanasak, N., et al., *Proc. 26th ICRC*, **3**, 9 (1999).
28. Chaussidon, M., and Robert, F., *Nature*, **374**, 337 (1995).
29. Lambert, D., et al., *ApJ*, **494**, 614 (1998).
30. Lemoine, M., et al., *ApJ*, **499**, 735 (1998).
31. Hobbs, L., *Physics Reports*, **333-334**, 449 (2000).
32. Spite, F., and Spite, M., *Origin and Evolution of the Elements*, N. Prantzos et al., Cambridge University, Cambridge, England, 1993, p. 201.

GCR Neon Isotopic Abundances: Comparison with Wolf-Rayet Star Models and Meteoritic Abundances

W.R. Binns*, M.E. Wiedenbeck†, E.R. Christian**, A.C. Cummings‡, J.S. George‡, M.H. Israel*, R.A. Leske‡, R.A. Mewaldt‡, E.C. Stone‡, T.T. von Rosenvinge**, and N.E. Yanasak‡

* *Washington University, St. Louis, MO 63130 USA*
†*Jet Propulsion Laboratory, California Institute of Technology, Pasadena, CA 91109 USA*
***NASA/Goddard Space Flight Center, Greenbelt, MD 20771 USA*
‡*California Institute of Technology, Pasadena, CA 91125 USA*

ABSTRACT

Measurements of the neon isotopic abundances from the ACE-CRIS experiment are presented. These abundances have been obtained in seven energy intervals over the energy range of ~80≤E≤280 MeV/nucleon. The ^{22}Ne/^{20}Ne source ratio is derived using the measured ^{21}Ne/^{20}Ne abundance as a "tracer" of secondary production of the neon isotopes. We find that the ^{22}Ne/^{20}Ne abundance ratio at the cosmic-ray source is a factor of 5.0±0.2 greater than in the solar wind. The GCR ^{22}Ne/^{20}Ne ratio is also shown to be considerably larger than that found in anomalous cosmic rays, solar energetic particles, most meteoritic samples of matter, and interplanetary dust particles. Recent two-component Wolf-Rayet models provide predictions for the ^{22}Ne/^{20}Ne ratio and other isotope ratios. Comparison of the CRIS neon, iron, and nickel isotopic source abundance ratios with predictions indicate possible enhanced abundances of some neutron-rich nuclides that are expected to accompany the ^{22}Ne excess.

1. Introduction

The abundance ratio of ^{22}Ne/^{20}Ne in the galactic cosmic rays (GCR) is well known to be high compared to the solar wind (SW) [1]. For neon, the solar wind (SW) is believed to best represent the composition of the pre-solar nebula [2]. The GCR neon isotopic abundances were measured first by a balloon-borne experiment [3]. Subsequent satellite experiments on IMP-7 [4], ISEE-3 [5,6], Voyager [7], Ulysses [8], and CRRES [9] verified this overabundance with more precise measurements.

Woosley and Weaver [10] suggested that this overabundance could be explained by a model in which the synthesis of neutron-rich isotopes in massive stars is directly proportional to their initial metallicity (fraction of elements heavier than He). Cosmic rays originating in a region of the Galaxy with metallicity that is higher than the Solar System could produce an overabundance of ^{22}Ne/^{20}Ne. However, this should also lead to an overabundance of the neutron-rich isotopes of Mg and Si, which has not been observed [8]. Galactic chemical evolution might also play a role in the ^{22}Ne/^{20}Ne overabundance, since cosmic rays may represent a more recent sample of matter than solar system abundances [11].

Olive and Schramm [12] suggested that an alternative to the conclusion that the GCR ^{22}Ne/^{20}Ne ratio is anomalously high is that the Solar System could instead be anomalously low. Supernovae in the near vicinity of the pre-solar nebula could have injected large amounts of ^{20}Ne and other α-particle nuclei shortly before the solar system formed, thus making the solar system anomalous rather than the GCRs.

Cassé and Paul [13] first suggested that the ^{22}Ne/^{20}Ne overabundance could be due to the injection and pre-acceleration of Wolf-Rayet (WR) star material that is mixed into interstellar medium to form the cosmic-ray source material. Soutoul and Legrain [14,15] have developed a diffusion model in which GCRs come preferentially from the inner Galaxy. The assumptions in this model are that 1) there is a gradient in the cosmic-ray density as a function of galacto-centric radius and that the density increases as one moves nearer the galactic center, 2) there is a larger gradient in the density of WR stars than for O stars as the galactocentric radius decreases [16], and 3) that these WR stars enrich the Galaxy locally in ^{22}Ne [17,18]. Mewaldt [19] reviews in greater detail the above models that predate his paper.

In this paper we present measurements of the neon isotopic abundances obtained by the Cosmic Ray Isotope Spectrometer (CRIS) instrument on the Advanced Composition Explorer (ACE) spacecraft [20]. The ^{22}Ne/^{20}Ne ratio has been measured as a function of energy over the energy range 77<E/M<278 MeV/nucleon. We derive the ^{22}Ne/^{20}Ne source ratio using our measurements of ^{21}Ne, ^{19}F, and ^{17}O abundances as "tracers" [21] of secondary production of the neon isotopes.

CP598, *Solar and Galactic Composition*, edited by R. F. Wimmer-Schweingruber

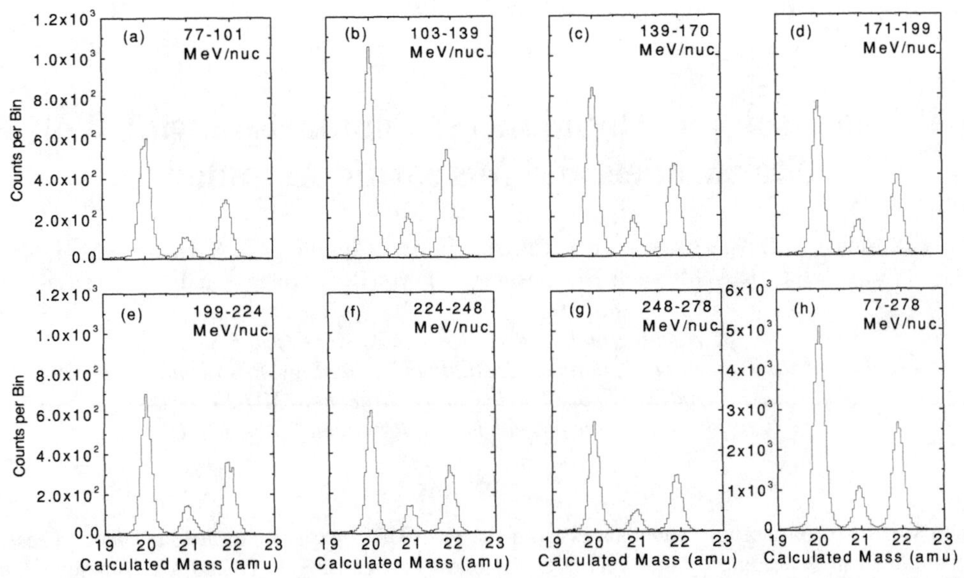

Figure 1-- Mass histograms of neon events in 7 energy bins (a-g). Figure 1h is the sum over all the energy intervals [22].

2. Measurements

The CRIS neon data are plotted in Figure 1a-g for events in 7 energy bins spanning the energy range of 77<E/M<278 MeV/nucleon. The sum of all events is shown in Figure 1h [22]. These events were collected from Dec. 4, 1997 through July 2, 2000 and are a selected, high-resolution data set. Events included in this analysis had trajectory angles ≤25° relative to the normal to the detector surfaces. Particles that stopped within 500 μm of the one surface of each silicon wafer having a dead layer were excluded from this analysis. The numbers of events selected corresponds to ~40% of the good events exclusive of these cuts. The rms mass resolution is 0.10 amu for events plotted in Figure 1, which is sufficiently good so that there is only a slight overlap of the particle distributions for adjacent masses. Mass cuts have therefore been taken at the minima between peaks and the events within the cuts counted to obtain abundances. In Figure 1, the data plotted consist of 6.0×10^4neon events. With the statistical accuracy of the CRIS data, it is possible, for the first time, to study the possible energy dependence of the ^{22}Ne/^{20}Ne ratio with high precision.

The ^{22}Ne/^{20}Ne and ^{21}Ne/^{20}Ne ratios measured by CRIS are plotted in Figure 2 as a function of energy. Corrections for energy spectra differences and nuclear interaction losses among the different isotopes have been made. The spectral dependence correction is ~7.5% for the ^{22}Ne/^{20}Ne ratio and ~3.7% for the ^{21}Ne/^{20}Ne ratio. The interaction correction is <1% for both ratios. As can be seen in Figure 2, the GCR ^{22}Ne/^{20}Ne ratio is approximately constant with a small decrease toward lower energies. A similar behavior is observed for ^{21}Ne/^{20}Ne, which also shows a roughly constant ratio with a slight decline for lower energies.

The ^{21}Ne/^{20}Ne ratio in the solar wind is low, 2.4 x 10^{-3} [1]. Because of the much higher abundance of ^{21}Ne in GCRs (higher by two orders of magnitude) it is believed to be almost entirely secondary (i.e., resulting from nuclear interactions of primary cosmic rays as they propagate through the interstellar medium). This makes ^{21}Ne suitable for use as a "tracer" isotope, with mass near that of ^{20}Ne and ^{22}Ne, for which the measured abundance can be scaled to estimate the fractions of ^{22}Ne and ^{20}Ne that are secondary [21].

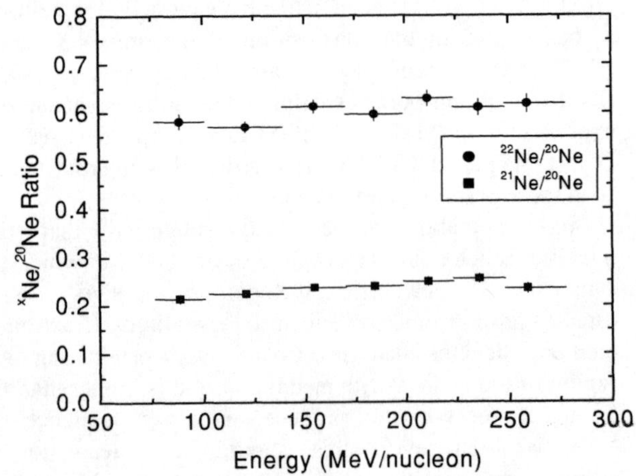

Figure 2—Plot of ^{22}Ne/^{20}Ne (circles) and ^{21}Ne/^{20}Ne (squares) ratios as a function of energy. In some cases the vertical error bars are smaller than the plotted point [22].

The CRIS Ne isotopic ratios averaged over the full energy range (taken from Fig 1h) have been compared with measurements made by other experiments [22]. The ratios reported from those experiments [5,7,8,9] have been adjusted to correspond to the approximate CRIS modulation level (~500 MV). The CRIS measurement has smaller statistical uncertainties than previous experiments, and the agreement is generally good, although the Voyager $^{22}Ne/^{20}Ne$ value is somewhat lower than our ACE-CRIS ratio.

The measured ratio of $^{22}Ne/^{20}Ne$ has been corrected for secondary production during propagation using the tracer method. Source abundances were also obtained using ^{19}F and ^{17}O secondary tracers. The $^{22}Ne/^{20}Ne$ galactic cosmic ray source ratios obtained using ^{21}Ne, ^{19}F, and ^{17}O as tracers were 0.350±0.004, 0.375±0.003, and 0.375±.004 respectively [22,23]. By averaging these three estimates, we obtain a "best" source abundance ratio of 0.366±0.006(stat.)±0.014(syst.) for $^{22}Ne/^{20}Ne$. The first quoted uncertainty in the abundance ratio is the statistical error, which takes into account both the measurement of $^{22}Ne/^{20}Ne$ and the uncertainty in the secondary correction associated with the statistics of the tracer isotope measurements. The second uncertainty is the systematic uncertainty estimate, calculated using the "sample standard deviation", obtained from the three tracer isotope estimates. In Figure 3 we plot the CRIS "best" value. The plotted error bar is the quadratic sum of the statistical and systematic uncertainties. It should be noted that the differing source ratio estimates from the three different tracer isotopes and the derived systematic uncertainties are comparable to those expected from nuclear cross-section uncertainties.

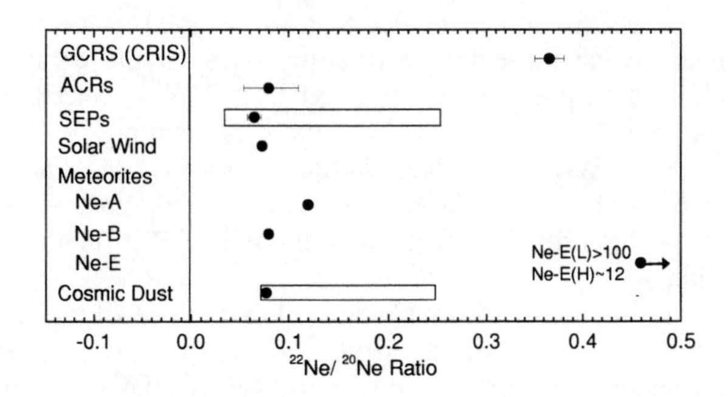

Fig. 3. The CRIS $^{22}Ne/^{20}Ne$ source abundance for GCRs are compared to Anomalous Cosmic Rays (ACRs) [24,25], solar wind [1,2], Solar Energetic Particles (SEPs) [26], meteoritic abundances [27], and cosmic dust (also called interplanetary dust particles or IDPs) [28]. The plotted error bar for CRIS is the quadratic sum of the statistical and systematic uncertainties (see text). The open bar for the SEP point represents the range of values measured in 11 solar events, while the data point and its uncertainty indicate the preliminary value deduced for the SEP source after accounting for the observed fractionation (Leske, private communication). Likewise the point plotted for Cosmic Dust is the average value obtained for 29 IDPs. The solid horizontal bar represents the spread in ratios for 27 of the particles measured, and the dotted horizontal bar the spread from the two remaining grains [28].

3. Discussion

Neon differs from most other elements in that it is observed to have a number of separate populations in "cosmic" matter with widely differing isotopic compositions.

Solar Wind--The solar wind is generally taken to give the best estimate of most isotopic ratios of noble gases including the Neon isotopes [2] in the presolar nebula. It is a sample of the solar corona and its isotopic composition is relatively stable on time scales comparable to the solar cycle. In Figure 3 the solar wind $^{22}Ne/^{20}Ne$ ratio is plotted [1,2].

Solar Energetic Particles (SEPs)--SEPs are believed to be a sample of the outer corona (gradual events) and the lower corona (impulsive events) [30]. A larger variation is observed from event-to-event than is the case for the solar wind. Although there are large event-to-event variations the average is close to the solar wind value. In Figure 3, the SEP neon for 11 different events is shown as a horizontal bar that gives the range of measured $^{22}Ne/^{20}Ne$ ratios [26]. The single data point included in the bar indicates the preliminary value deduced for the SEP source after accounting for the observed fractionation (Leske, private communication).

Anomalous Cosmic Rays (ACRs)—ACRs result from neutral atoms in the very local interstellar medium drifting into the heliosphere where they are ionized by solar UV or charge exchange with the solar wind. They are then swept out to the solar wind termination shock where they are accelerated to energies of typically tens of MeV [31]. The ACR $^{22}Ne/^{20}Ne$ abundance ratio [25,23] is plotted in Figure 3.

Meteorites—The Ne-A component in meteorites is found in carbonaceous chondrites and may contain presolar components [32]. Ne-B is believed to result from solar wind [1] and very low-energy SEP implantation in grains [33] since it has a $^{22}Ne/^{20}Ne$ ratio very similar to that of contemporary measurements of the solar wind. Ne-E, which is found in SiC and graphite grains, is mostly ^{22}Ne [27], and probably comes from He burning asymptotic giant branch (AGB) stars and the decay of ^{22}Na produced in supernovae or novae [29,35]. Ne-E is the only population found in meteorites that has a $^{22}Ne/^{20}Ne$ ratio greater than that found for GCRs. The abundance ratios of these meteoritic components are plotted in Figure 3.

Cosmic Dust—The sources of cosmic dust (or interplanetary dust particles; IDPs) are believed to be asteroid belt and cometary dust material. As such they represent primitive matter from both the inner and outer solar system. The world inventory of cosmic dust particles consists of several thousand particles (as of the year 2000) and Kehm [28] presents the $^{22}Ne/^{20}Ne$ ratio for 29 of these particles. In Figure 3, we plot the mean value as a single data point with the solid horizontal bar giving the range of ratios for 27 of the grains and the dotted horizontal bar giving the range of the remaining two grains [28].

The GCR source $^{22}Ne/^{20}Ne$ abundance ratio is 5.0±0.2 times greater than the SW value of 0.073. The large enhancement in $^{22}Ne/^{20}Ne$ for GCRs compared with ACRs, which show good agreement with solar wind abundances is striking. This enhancement is also seen in the ACE-SIS data [25] and in the SAMPEX results [24]. If the ACR ratio is representative of the general ISM, then there is apparently a large overabundance of high-energy ^{22}Ne cosmic rays relative to ^{20}Ne compared to low energy matter (i.e., gas). We note that it is not presently known to what extent the ACRs actually represent the general ISM. It should also be noted that the solar system may not have formed in the local ISM. It has been suggested that the Sun may have formed ~2 kpc closer to the galactic center than our present galactocentric radius of 8.5 kpc, based on the higher metallicity of the Sun relative to that of nearby stars [34].

Wolf-Rayet Stars--The overabundance of ^{22}Ne in GCRs may be a result of the acceleration of WR star material as first suggested by Cassé and Paul [13]. It has been estimated that $\sim 2\times10^{-5}$ solar masses of material per year is ejected from a typical WR star in high velocity winds [18, 37]. There are two dominant phases of WR stars, the WC and WN phases. Large quantities of He-burning material rich in ^{22}Ne are expelled from the stars when they are in the WC phase, resulting in greatly enhanced ^{22}Ne/^{20}Ne ratios. In the WN phase of WR stars, CNO cycling dominates with the resultant production of high ^{13}C/^{12}C and ^{14}N/^{16}O ratios, but no enhancement in the ^{22}Ne/^{20}Ne ratio [18,36]. It is thought that the pre-supernova WR winds might be swept up and accelerated by nearby supernovae or by the supernova shock from the WR star that ejected the material in the first place, without substantial mixing into the ambient ISM.

Recent modeling calculations by Meynet et al. [38] of the abundances of isotopes ejected from the surface of WR stars in the high velocity winds have incorporated the lower WR mass loss rates, updated nuclear reaction rates, and extended the reaction networks included in the models. Additionally, they have considered the effects of rotation, which has a strong effect on isotopic production. Their model is a two-component model that assumes that the GCRs are a mix of WR wind material and material with Solar System composition [2]. There are two free parameters in their models, the mixing fraction of WR and solar material (p), and the WR star metallicity. They have performed calculations for various effective metallicities of the WR stars for both non-rotating and rotating stars. However, their calculations for rotating stars at present [38] do not extend to elements heavier than Mg. In Figure 4 we show their WR model predictions for a WR star without rotation with an initial mass of 60 M_o and solar metallicity compared with the CRIS neon, iron, and nickel isotope ratios. The enhancement factors calculated by Meynet et al. [38] for individual isotopes have been used to calculate the mixing fraction required to obtain a ^{22}Ne/^{20}Ne ratio normalized to 5.0, which is our CRIS measured value. This normalization yields a mixing fraction of p=0.052. Comparing the predicted isotope ratios for iron and nickel with this mixing fraction to the derived CRIS abundance ratios results in generally good agreement, as shown in Figure 4, although the error bars for the rarest isotopes, e.g. ^{61}Ni/^{58}Ni, are rather large.

We have also calculated the required mixing fraction using the model results for WR stars with 60 M_o and solar metallicity rotating at an initial velocity of 300 km/s [38]. Rotation has a strong effect on the WR wind

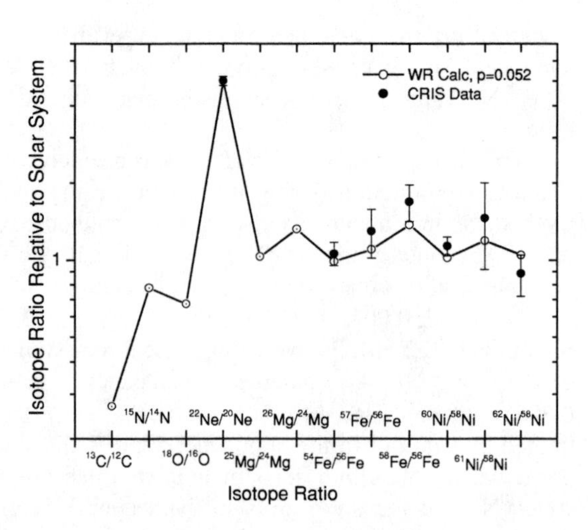

Figure 4—Isotopic ratios for iron and nickel isotopes are compared to a recent Wolf-Rayet modeling calculation (without rotation) [38]. The WR calculation is normalized to the ^{22}Ne/^{20}Ne ratio measured by CRIS.

composition and on the fraction of the stellar lifetime that the star spends in the WR and WC phases. Normalizing as before to the measured ^{22}Ne/^{20}Ne ratio of 5.0 we calculate a required mixing ratio p=0.215 which seems quite large. The predicted isotope ratios are not plotted in Figure 4 since, at present, the calculations have not been extended above ^{27}Al. Thus we see that a model that is intended to be more realistic than those with no rotation appears to make it more difficult to provide the observed ^{22}Ne/^{20}Ne enhancement. As noted by Meynet et al [38], these calculations are for a "typical" WR star with an initial mass of 60 M_o. Further theoretical work is needed to obtain predictions which include contributions from the full range of WR stars in the Galaxy.

Additionally, Soutoul and Legrain [14,15] have developed a two-component model that incorporates diffusion of cosmic rays, in which GCRs come preferentially from the inner Galaxy. In this model the overabundance at Earth of ^{22}Ne results from a ^{22}Ne galactocentric radial gradient that is steeper than for the "main" cosmic ray component. (This "main" component in [15] is represented by Oxygen. ^{20}Ne would also be part of this main component). They present calculations [15] for 2 GeV/nuc neon that can produce the observed ^{22}Ne excess for large ^{22}Ne gradients, provided that there is also a moderate gradient for the main component. Their modeling calculations in [14] show that the ^{22}Ne overabundance shows a weak energy dependence, decreasing slightly from 2 GeV/nuc to 1 GeV/nuc. Although their calculations did not extend down to the CRIS energy range, the weak energy dependence of their calculation suggests that their model can be made consistent with the observed ^{22}Ne excess in the cosmic rays.

4. Summary

In summary, our measurements have led to an improved value for the ^{22}Ne/^{20}Ne source abundance ratio that is a factor of 5.0±0.2 greater than for the solar wind. This ratio is significantly larger than any other available sample of "cosmic" matter with the exception of meteoritic Neon-E. For the two component WR model [38] (with no rotation) described above, a mixing fraction of about 5% is required to account for the $^{22}Ne/^{20}Ne$ source abundance ratio, and for that model the ratios of iron and nickel isotopes derived from CRIS data show good agreement within the measurement errors.

The diffusive two-component model [14,15] can also produce the observed $^{22}Ne/^{20}Ne$ overabundance.

Acknowledgments

This research was supported by the National Aeronautics and Space Administration at Washington University, the California Institute of Technology (under grant NAG5-6912), the Jet Propulsion Laboratory, and the Goddard Space Flight Center. It was also supported by the McDonnell Center for the Space Sciences at Washington University.

References

1. Geiss, J., Solar Wind Composition and Implications about the History of the Solar System, *Proc. of the 13th International Cosmic Ray Conference*, **5**, 3375, 1973.
2. Anders, E, and N. Grevesse, Abundances of the elements: Meteoritic and solar, *Geochim. et Cosmochim. Acta* **53**, 197, 1989.
3. Maehl, R., F.A. Hagen, A.J. Fisher, J.F. Ormes, Astrophysical Implications of the Isotopic Composition of Cosmic Rays, *Proc. of the 14th International Cosmic Ray Conference,* **1**, 367, 1975.
4. Garcia-Munoz, M., J.A. Simpson, and J.P. Wefel, The Isotopes of Neon in the GCRs, *ApJ*, **232**, L95, 1979.
5. Wiedenbeck, M.E., and D.E. Greiner, Isotopic Anomalies in the Galactic Cosmic Ray Source, *Phys. Rev. Lett.*, **46**, 682, 1981.
6. Mewaldt, R.A., J.D. Spalding, E.C. Stone, and R.E. Vogt, High Resolution Measurements of Galactic Cosmic-Ray Neon, Magnesium, and Silicon Isotopes, *ApJ* **235**, L95, 1980.
7. Lukasiak, A., P. Ferrando, F.B. McDonald, and W.R. Webber, Cosmic-Ray Isotopic Composition of C, N, O, Ne, Mg, Si Nuclei in the Energy Range 50-200 MeV per Nucleon Measured by the Voyager Spacecraft During the Solar Minimum Period, *ApJ*, **426**, 366, 1994.
8. Connell, J.J., and J.A. Simpson, High Resolution Measurements of the Isotopic Composition of GCR C, N, O, Ne, Mg, and Si from the Ulysses HET, *Proc. of the 25th International Cosmic Ray Conference*, **3**, 381, 1997.
9. DuVernois, M.A., M. Garcia-Munoz, K.R. Pyle, J.A. Simpson, and M.R. Thayer, The Isotopic Composition of Galactic Cosmic-Ray Elements from Carbon to Silicon: The Combined Release and Radiation Effects Satellite Investigation, *ApJ*, **466**, 457,1996.
10. Woosley, S.E., and T.A. Weaver, Anomalous Isotopic Composition of Cosmic Rays, *ApJ*, **243**, 651,1981.
11. Wiedenbeck, et al., Constraints on the Nucleosynthesis of Refractory Nuclides in Galactic Cosmic Rays, this conference.
12. Olive, K.A., and D.N. Schramm, OB Associations and the Nonuniversality of the Cosmic Abundances: Implications for Cosmic Rays and Meteorites, *ApJ*, **257**, 276, 1982.
13. Cassé, M., and J.A.Paul, On the Stellar Origin of the ^{22}Ne Excess in Cosmic Rays, *ApJ*, **258**, 860, 1982.
14. Soutoul, A., and Legrain, R., Galactic source distribution of heavy cosmic rays and the energy dependent overabundance of ^{22}Ne, *Proc. of the 26th International Cosmic Ray Conference*, **4**, 180, 1999.
15. Soutoul, A., and Legrain, R., ^{22}Ne excess in cosmic rays from the inner Galaxy, in Acceleration and Transport of Energetic Particles Observed in the Heliosphere: ACE 2000 Symposium, Ed Richard A. Mewaldt, et al., AIP Conference Proceedings **528**, 417, 2000.
16. Van der Hucht, K.A., B. Hidayat, A.G. Admiranto, K.R. Supelli, and C. Doom, The Galactic Distribution and Subtype Evolution of Wolf-Rayet Stars. III, *Astron. Astrophys.*, **199**, 217, 1988.
17. Meynet, G., and A. Maeder, WR Stars and Isotopic Anomalies in Cosmic Rays, *Adv. Space Res.*, **19**, 763, 1997.
18. Maeder, A. and G. Meynet, Isotopic anomalies in cosmic rays and the metallicity gradient in the Galaxy, *Astron. Astrophys.*, **278**, 406, 1993.
19. Mewaldt, R.A., The Abundances of Isotopes in the Cosmic Radiation, in Cosmic Abundances of Matter, Ed. By C.J. Waddington, AIP Conference Proceedings **183**, 124, 1989.
20. Stone, E.C., C.M.S. Cohen, W.R. Cook, A.C. Cummings, B. Gauld, et al., The Cosmic-Ray Isotope Spectrometer for the Advanced Composition Explorer, *Space Science Reviews,* **86**, 285, 1998.
21. Stone, E.C., and Wiedenbeck, M.E., A Secondary Tracer Approach to the Derivation of Galactic Cosmic-Ray Source Isotopic Abundances, *ApJ*, **231**, 606, 1979.
22. Binns, W.R., et al. (2001), Galactic Cosmic Ray Neon Isotopic Abundances Measured by the Cosmic Ray Isotope Spectrometer (CRIS) on ACE, *Advances in Space Research*, 27, 767, 2001.
23. Binns, W.R., M.E. Wiedenbeck, E.R. Christian, A.C. Cummings, J.S. George, et al., Galactic Cosmic Ray Neon Isotopic Abundances Measured on ACE: in Acceleration and Transport of Energetic Particles Observed in the Heliosphere, ACE 2000 Symposium, Ed. R.A. Mewaldt et al., AIP Conference Proceedings **528**, 413, 2000.
24. Leske, R.A, R.A. Mewaldt, A.C. Cummings, J.R. Cummings, E.C. Stone, et al., The Isotopic Composition of Anomalous Cosmic Rays from SAMPEX, *Space Sci. Revs.* **78**, 149, 1996.
25. Leske, R.A., R.A. Mewaldt, E.R. Christian, C.M.S. Cohen, A.C. Cummings, et al., Measurements of the Isotopic Composition of Anomalous Cosmic Ray N, O, and Ne from ACE, *Proc. of the 26th International Cosmic Ray Conference,* **7**, 539, 1999.
26. Leske, R.A., R.A. Mewaldt, C.M.S. Cohen, A.C. Cummings, E.C. Stone, et al., Event-to-event variations in the isotopic composition of neon in solar energetic particle events, *GRL*, **26**, 2693, 1999.
27. Ozima, M., and Podosek, F.A., Noble Gas Geochemistry, Cambridge: Cambridge University Press, 1983.
28. Kehm, K., Thesis, Washington University, 2000.
29. Amari, S., R.S. Lewis, and E. Anders, Interstellar Grains in Meteorites: III. Graphite and its Noble Gases, *Geochim. et Cosmochim. Acta*, **59**, 1411, 1995.
30. Cohen, C.M.S., R.A. Leske, E.R. Christian, A.C. Cummings, R.A. Mewaldt, P.L. Slocum, E.C. Stone, T.T. von Rosenvinge, and M.E. Wiedenbeck, in *Acceleration and Transport of Energetic Particles Observed in the Heliosphere: ACE 2000 Symposium*, Ed Richard A. Mewaldt, et al., AIP Conference Proceedings **528**, 55, 2000.
31. Leske. R.A., R.A. Mewaldt, E.R. Christian, C.M.S. Cohen, A.C. Cummings, P.L. Slocum, E.C. Stone, T.T. von

Rosenvinge, and M.E. Wiedenbeck, *in Acceleration and Transport of Energetic Particles Observed in the Heliosphere: ACE 2000 Symposium*, Ed Richard A. Mewaldt, et al., AIP Conference Proceedings **528**, 293, 2000.

32. Huss, G.R. and R.S. Lewis, Noble gases in presolar diamonds I: Three distinct components and their implications for diamond origins, *Meteoritics*, **29**, 791, 1994.
33. Weiler, R, H. Baur, and P. Signer, Noble gases from solar energetic particles revealed by closed system stepwise etching of lunar soil minerals, *Geochim. et Comochim. Acta*, **50**, 1997, 1986.
34. Wielen, R., Fuchs, B., and C. Dettbarn, On the Birth-place of the Sun and the Places of Formation of other Nearby Stars, *Astron. Astrophys.*, **314**, 438, 1996.
35. Lewis, R.S., S. Amari, and E. Anders, Interstellar grains in meteorites: II. SiC and its noble gases, *Geochim. et Cosmochim. Acta*, **58**, 471, 1994.
36. Chiosi, C, and A. Maeder, The Evolution of Massive Stars, *Ann. Rev. Astron. Astrophys*, **24**, 329, 1986.
37. Nugis, T., P.A. Crowther, and A.J. Willis, Clumping-corrected mass-loss rates of Wolf-Rayet stars, *Astron. Astrophys.* 333, 956, 1998.
38. Meynet, G., M. Arnould, G. Paulus, and A. Maeder, Wolf-Rayet Star Nucleosynthesis and the Isotopic Composition of the Galactic Cosmic Rays, *Space Sci. Rev.*, 2000 To be published, and Private Communications.
39. Wiedenbeck, M.E. et al., The Origin of Primary Cosmic Rays: Constraints from ACE Elemental and Isotopic Composition Observations, *Space Sci. Rev.*, 2001, Accepted for publication.

The Phosphorus/Sulfur Abundance Ratio as a Test of Galactic Cosmic-Ray Source Models

J.S. George*, M.E. Wiedenbeck†, W.R. Binns**, E.R. Christian‡, A.C. Cummings*, P.L. Hink**, R.A. Leske*, R.A. Mewaldt*, E.C. Stone*, T.T. von Rosenvinge‡ and N.E. Yanasak*

*California Institute of Technology, Pasadena, CA 91125 USA
†Jet Propulsion Laboratory, California Institute of Technology, Pasadena, CA 91109 USA
**Washington University, St. Louis, MO, 63130 USA
‡NASA/Goddard Space Flight Center, Greenbelt, MD 20771 USA

Abstract. Galactic cosmic-ray (GCR) elemental abundances display a fractionation compared to solar-system values that appears ordered by atomic properties such as the first ionization potential (FIP) or condensation temperature (volatility). Determining which parameter controls the observed fractionation is crucial to distinguish between GCR origin models. The Cosmic-Ray Isotope Spectrometer (CRIS) instrument on board NASA's Advanced Composition Explorer (ACE) spacecraft can measure the abundances of several elements that break the general correlation between FIP and volatility (e.g., Na, P, K, Cu, Zn, Ga, and Ge). Phosphorus is a particularly interesting case as it is a refractory (high condensation temperature) element with a FIP value nearly identical to that of its semi-volatile neighbor, sulfur. Using a leaky-box galactic propagation model we find that the P/S and Na/Mg ratios in the GCR source favor volatility as the controlling parameter.

INTRODUCTION

The origin of the source material for galactic cosmic rays (GCRs) remains one of the unsolved problems of particle astrophysics. Various views support origins as diverse as supernovae ejecta and interstellar dust [1]. The abundances of cosmic-ray elements measured near Earth contain information about the source composition and the effects of transport through the Galaxy. Careful modeling of these propagation effects can be used to determine the source composition and shed light on the sources that contribute to the cosmic rays [2]. The Cosmic-Ray Isotope Spectrometer (CRIS) instrument [3] on board the Advanced Composition Explorer (ACE) spacecraft has been making high-precision measurements of the GCR composition from He to Zn at typical energies of 50-500 MeV/nucleon.

Elemental GCR abundances display a long-observed elemental fractionation commonly parameterized by the first ionization potential (FIP) [4]. Specifically, elements with FIP values above ~ 10 eV are depleted in the GCR source by roughly a factor of five relative to solar-system abundances derived from meteoritic and photospheric observations [5]. A similar dependence on FIP, or possibly on first ionization time (FIT) [see, e. g. 6], has been observed in solar energetic particles, the solar corona, and the solar wind. This similarity led some to suggest that the GCR seed population was originally ejected at MeV/nucleon energies from the coronae of solar-like stars [4, 7]. These particles would have been further accelerated at a later time by some other mechanism, widely thought to be passing supernova shock waves [8].

Abundances of mass-59 isotopes measured by ACE essentially rule out the possibility that a supernova can accelerate its own ejecta [9]. The ^{59}Ni atoms in the source decay via electron capture to ^{59}Co with a half-life of $\sim 10^5$ years until they are accelerated, when the ions are fully stripped and become essentially stable. The amount of ^{59}Ni in the cosmic rays measures the elapsed time between production and acceleration. ACE data are consistent with the complete decay of all ^{59}Ni in the source. This implies an acceleration time delay greater than 10^5 years, far longer than the 10^4 years required to dissipate the energy from a supernova shock. Compositional arguments based on comparison of the GCR source abundances with those expected from supernova nucleosynthesis have also been used to support the idea of a long delay [10].

Current models envision large associations of OB stars

CP598, *Solar and Galactic Composition*, edited by R. F. Wimmer-Schweingruber

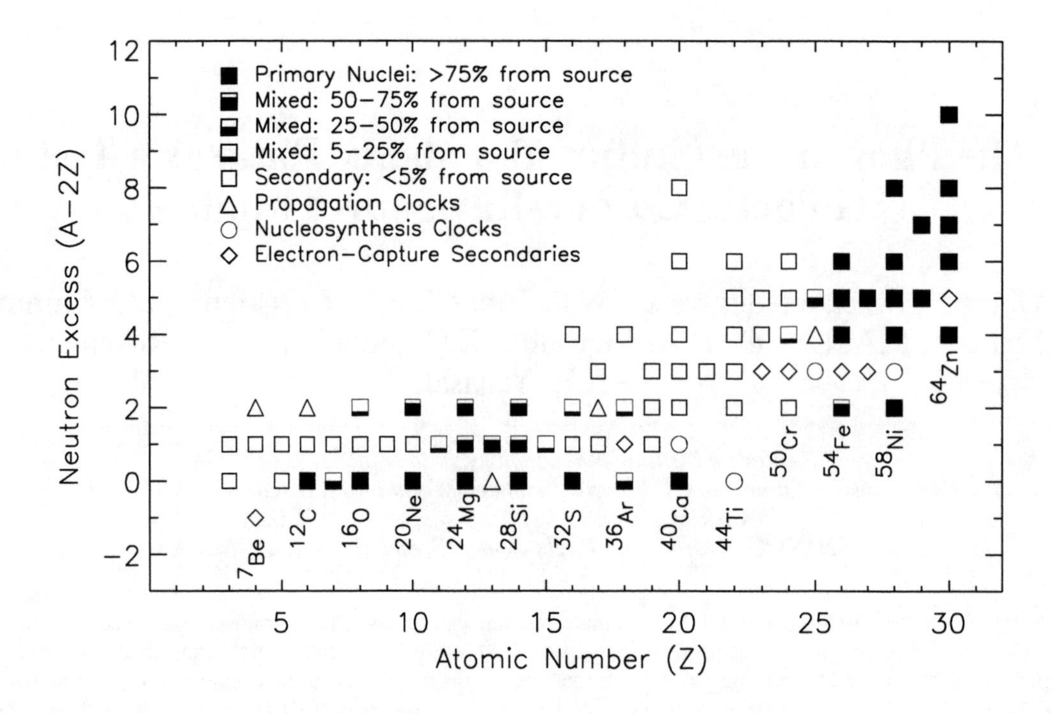

FIGURE 1. Cosmic-ray isotope chart coded according to the fraction of material observed at Earth that came directly from the source. Useful tracer isotopes are the stable nuclides with negligible primary contribution denoted by open squares.

where shock waves from consecutive supernovae sweep out a "superbubble", a region filled with a warm dilute mixture of fresh ejecta and interstellar medium (ISM) material stripped from the walls of the cavity [11, 12]. Later shocks accelerate material out of the bubble as cosmic rays.

An alternative model is based on the realization that the elemental fractionation may not be controlled by the first ionization potential, but rather by some correlated atomic property such as the condensation temperature, or volatility [13, 14, 1]. In this view, the refractory elements in the cosmic rays are thought to be sputtering products from interstellar dust grains. Weakly charged grains, with their very high mass-to-charge ratios, can easily be accelerated to energies of ~0.1 MeV/nucleon. Energetic ions sputtered off the grains through collisions with the interstellar gas are further accelerated by the same shock, thereby avoiding the need for a second separate acceleration mechanism. The GCR abundances of the refractory elements are set by the physical conditions, such as the temperature, under which the grains condensed. Ions in the gas phase would also be accelerated, accounting for the volatiles in the cosmic rays.

Determining which parameter controls the elemental fractionation is crucial for distinguishing between models of the source. The abundances of several elements that break the general correlation between FIP and volatility can indicate which is actually relevant. Seven of these elements (Na, P, K, Cu, Zn, Ga, and Ge) have had abundances measured by the CRIS instrument. The four heavy elements beyond nickel have relatively small corrections due to fragmentation of heavier nuclides. Abundances in this region fall rapidly with increasing nuclear charge so there is little material to contribute to secondary production by fragmentation during transport. These source abundances are reliably determined and limited mainly by statistics [15]. Of these four, Cu and Ge favor volatility models. The Ga abundance is consistent with FIP models with large statistical uncertainty. The volatile element Zn is depleted by a factor of 2.5 compared to the solar system value, however, as an intermediate-FIP element some depletion is expected in either model.

The source abundances of Na and P are limited by uncertainties in the propagation models used to correct for secondary production during transport. Nearly 80% of the Na and P observed at Earth is produced by fragmentation of heavier isotopes. Because of these large corrections, uncertainties in the propagation parameters, particularly in the nuclear fragmentation cross sections, are the dominant sources of error in the derived source abundances.

The Na source abundance has been previously shown to be depleted at the 2σ level relative to the solar-system abundance [15]. This depletion is consistent with that expected from volatility models. That result is updated here, and the analysis is extended to phosphorus, a somewhat refractory (T_c=1151K) intermediate-FIP element. The P/S ratio is particularly interesting because P and S have similar FIP values (10.49 and 10.36 eV, respectively) while S is more volatile (T_c=648K). A difference in the relative depletions of S and P compared to their solar-system values should be a good test of whether FIP or volatility controls the fractionation. The chief difficulty lies in making an accurate determination of the secondary contributions and in assessing the systematic uncertainties in the model.

One approach to the problem of extracting source abundances for elements such as Na and P is to use the abundances of purely secondary isotopes, i.e. those produced entirely by fragmentation, as tracers of the amount of secondary production [16]. These tracer isotopes constrain the GCR pathlength that controls the total amount of spallation. Reproducing the measured abundances of isotopes produced by spallation should provide a good estimate of the secondary contribution to ^{23}Na and ^{31}P, which have small primary components.

Figure 1 contains an isotope chart coded according to the fraction of material observed at Earth that comes directly from the source. The elements Na and P each consist of a single isotope in the cosmic rays. The chart can be used to identify appropriate tracer isotopes, those that are nearby in mass but have even larger secondary contributions than ^{23}Na and ^{31}P. We chose seven such isotopes, ^{17}O, ^{19}F, ^{21}Ne, ^{33}S, ^{36}S, ^{35}Cl, and ^{37}Cl, for use in this study.

THE ^{23}Na/^{24}Mg SOURCE RATIO

The points in Figure 2 show the result of a series of cosmic-ray propagation calculations. The steady-state "leaky-box" model gives equilibrium abundances in a homogeneous Galaxy, accounting for source injection, fragmentation, radioactive decay, energy loss, and particle escape [2]. The five points in each panel represent separate propagation calculations with values for the escape mean free path varying by up to a factor of two. The solid line is a linear fit. In each calculation, the source ratios of isotopes between Ne and Fe were adjusted relative to the stable primary isotope ^{24}Mg to reproduce the locally measured values. Data from the CRIS instrument used here were taken over a ~2.5 year period from August 1997 through April 2000. The shaded areas plot the percentage difference between the predicted and measured abundance of each tracer isotope compared with the source ratio required to account for the measured value of ^{23}Na/^{24}Mg=0.213±0.003 in each calculation. A model that correctly predicts the measured tracer abundance (0% difference) also provides an estimate of the amount of ^{23}Na in the source. The vertical line shows the solar-system ^{23}Na/^{24}Mg value of 0.068±0.005 [5].

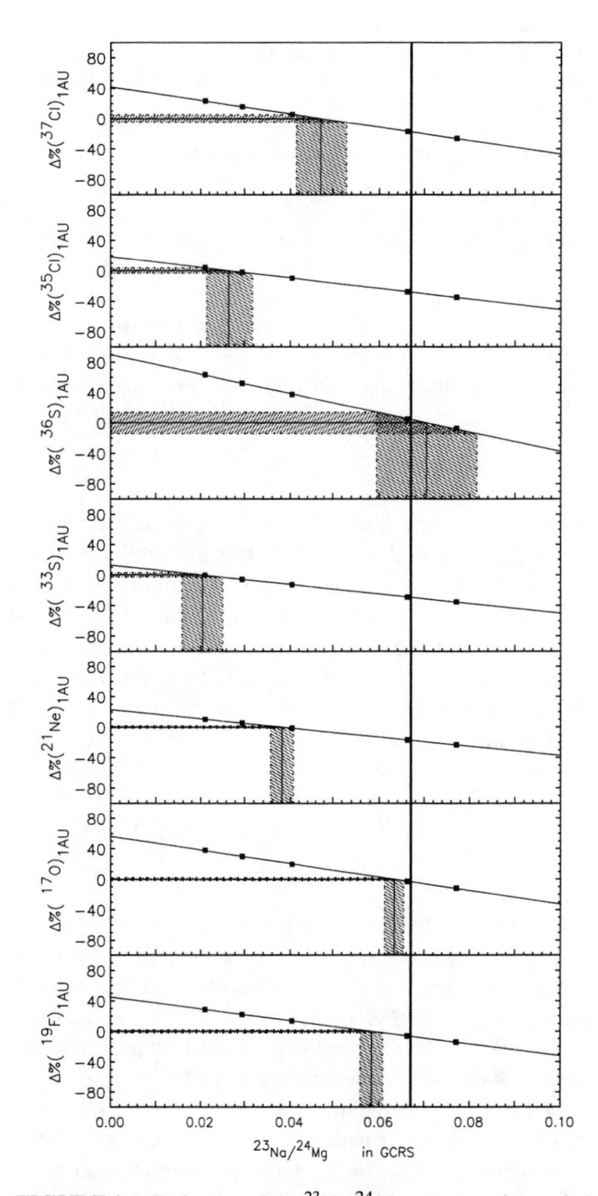

FIGURE 2. Estimates of the ^{23}Na/^{24}Mg source ratio needed to account for the observed ^{23}Na/^{24}Mg at Earth in various propagation calculations compared with the abundances of the secondary tracer isotopes predicted by those models. The ordinate in each case is the percent difference between the predicted local abundance and the measured value of the tracer. The vertical line indicates the solar-system ^{23}Na/^{24}Mg value [5]. Shaded areas indicate statistical uncertainty due to the measurement of the tracer isotope.

The propagation calculation used to estimate the secondary corrections is a steady-state leaky-box propagation model [17], based on the formalism of Meneguzzi,

TABLE 1. Relative uncertainties contributing to the estimates for the $^{23}Na/^{24}Mg$ source ratio from each of the tracer isotopes.

	Statistics		Cross sections		$\frac{^{23}Na}{^{24}Mg}$
	Tracer	$\frac{^{23}Na}{^{24}Mg}$	Tracer	^{23}Na	Source Ratio
^{17}O	4.3%	1.4%	8.8%	11.0%	0.063±0.009
^{19}F	5.1%	1.4%	9.9%	11.0%	0.058±0.009
^{21}Ne	8.8%	1.4%	6.4%	11.0%	0.038±0.006
^{33}S	24.6%	1.4%	6.5%	11.0%	0.021±0.006
^{36}S	15.7%	1.4%	13.7%	11.0%	0.070±0.017
^{35}Cl	20.4%	1.4%	10.0%	11.0%	0.027±0.007
^{37}Cl	12.6%	1.4%	7.9%	11.0%	0.048±0.009

et al. [2]. Source spectra were power laws in momentum with a spectral index of −2.3. The interstellar medium He/H ratio was assumed to be 0.11 by number with 16% of the hydrogen in an ionized state [18]. We used the Silberberg and Tsao semi-empirical forms of the nuclear cross sections [19, 20] to estimate the amount of fragmentation during transport. The escape mean free path was a function of the velocity β and rigidity R with a normalization Λ_0 which is varied in each calculation.

$$\begin{aligned} \Lambda_{esc} &= \Lambda_0\beta & R \leq 4.9\text{GV} \quad (1)\\ &= \Lambda_0\beta(R/4.9\text{GV})^{-0.7} & R > 4.9\text{GV} \end{aligned}$$

The effects of solar modulation were taken into account using a spherically symmetric Fisk model [21]. The average modulation level was determined by fitting the B/C and (Sc+Ti+V)/Fe secondary/primary ratios in the CRIS data [see 22]. The appropriate modulation parameter for this time period was determined to be ϕ=460 MV.

Table 1 lists the estimates of the $^{23}Na/^{24}Mg$ source ratio from each tracer isotope along with the main sources of uncertainty. The first column gives the spread in the estimate due to statistical uncertainty in the measurement of the tracer isotope as shown by the shaded areas in Figure 2. There is also a statistical contribution from the measurement of the $^{23}Na/^{24}Mg$ ratio itself, shown in the second column. The third and fourth columns represent uncertainties in the nuclear fragmentation cross sections for the formation of the tracer isotope and ^{23}Na itself. These values are determined by averaging the published uncertainties in the measured cross sections for producing each tracer from all known parents. Each uncertainty is weighted by the relative contribution that that parent makes to the daughter. At least one measurement at an appropriate energy is available for all relevant reactions.

The final column in Table 1 gives the individual estimates for the $^{23}Na/^{24}Mg$ source ratio including contributions from each source of uncertainty. Six of the seven estimates are lower than the solar-system values.

The best overall estimate of the $^{23}Na/^{24}Mg$ source ratio is taken as a weighted average μ and its uncertainty σ_μ of the individual estimates $x_i \pm \sigma_i$ given in Table 1.

$$\mu = \frac{\Sigma(x_i/\sigma_i^2)}{\Sigma(1/\sigma_i^2)} \quad \text{and} \quad \sigma_\mu^2 = \frac{1}{\Sigma(1/\sigma_i^2)} \qquad (2)$$

Some caution is warranted as this approach treats the individual uncertainties as being completely uncorrelated. In reality some contributions are highly correlated, such as that from the cross section for producing ^{23}Na. The combined uncertainty addresses the spread in the individual measurements but is not sensitive to a systematic shift.

The quality of the fit is estimated with a test statistic $\chi^2 = \Sigma(x_i - \mu)^2/\sigma_i^2$. Determining the mean from seven measurements leaves six remaining degrees of freedom, giving a reduced χ^2 value χ_ν^2 = 4.7. Such a large value points out that there are non-statistical fluctuations which are not represented by the stated uncertainties in Figure 2 and Table 1. These may suggest that the actual uncertainties in the cross-section measurements are larger than those ascribed to them in the literature. One way to characterize this additional uncertainty is to increase the uncertainty of the mean by an amount needed to make $\chi_\nu^2 = 1$, that is, $\sigma = \chi_\nu\sigma_\mu$.

With this approach we obtain an overall best value for $(^{23}Na/^{24}Mg)_{GCRS}$ of 0.038±0.006. Data from the *Ulysses* spacecraft [23] supports the use of solar-system isotopic abundances [5] to determine the elemental Na/Mg ratio in the source. We find the Na/Mg ratio in the GCR source relative to the solar system to be:

$$(\text{Na/Mg})_{\text{GCRS}}/(\text{Na/Mg})_{\text{SS}} = 0.57 \pm 0.11 \qquad (3)$$

This result is consistent with our previous result of 0.62±0.19 [15] using only the three lightest tracers taking the full spread of the individual estimates as a measure of the uncertainty. That calculation used the Webber et al. cross section parameterization [24] and a somewhat higher solar modulation level.

THE $^{31}P/^{32}S$ SOURCE RATIO

The same analysis has also been carried out for the source $^{31}P/^{32}S$ ratio. In this case, sulfur is known to be depleted relative to the solar-system value. The question being investigated is whether P and S are depleted by the same amount as would be expected from a FIP-controlled fractionation.

The results of the calculations are displayed in Figure 3. As with $^{23}Na/^{24}Mg$, only the statistical uncertainty due to measurement of the tracer isotope is shown. The measured $^{31}P/^{32}S$ value near Earth is 0.268±0.010. A

TABLE 2. Relative uncertainties contributing to the $^{31}P/^{32}S$ source ratio estimates from each of the tracer isotopes.

	Statistics		Cross sections		$\frac{^{31}P}{^{32}S}$
	Tracer	$\frac{^{31}P}{^{32}S}$	Tracer	^{31}P	Source Ratio
^{17}O	4.1%	3.6%	8.8%	5.0%	0.102±0.012
^{19}F	5.1%	3.6%	9.9%	5.0%	0.095±0.012
^{21}Ne	8.8%	3.6%	6.4%	5.0%	0.062±0.008
^{33}S	24.3%	3.6%	6.5%	5.0%	0.034±0.009
^{36}S	15.6%	3.6%	13.7%	5.0%	0.114±0.025
^{35}Cl	20.2%	3.6%	10.0%	5.0%	0.044±0.010
^{37}Cl	12.6%	3.6%	7.9%	5.0%	0.077±0.012

vertical line represents the solar-system $^{31}P/^{32}S$ ratio [5] of 0.021±0.003, expected if the depletion is the same. It is clear that all of the seven tracer isotopes predict a $^{31}P/^{32}S$ ratio higher than that of the solar system, i.e., that ^{31}P is less depleted than ^{32}S in the GCR source.

The individual estimates are given in Table 2 along with the corresponding contributions to their uncertainties. Using the same approach as before, we combine them into an overall best value of $(^{31}P/^{32}S)_{GCRS}$ = 0.065±0.01. The reduced χ^2 value folded into the uncertainty estimate was 6.3. This best estimate represents an enhancement of 3.1±0.7 over the solar-system value. Pending completion of the CRIS results for the S/Mg source ratio, we use the HEAO-3-C2 [25] value of 0.126±0.009 to find

$$(P/Mg)_{GCRS}/(P/Mg)_{SS} = 0.80 \pm 0.19 \quad (4)$$

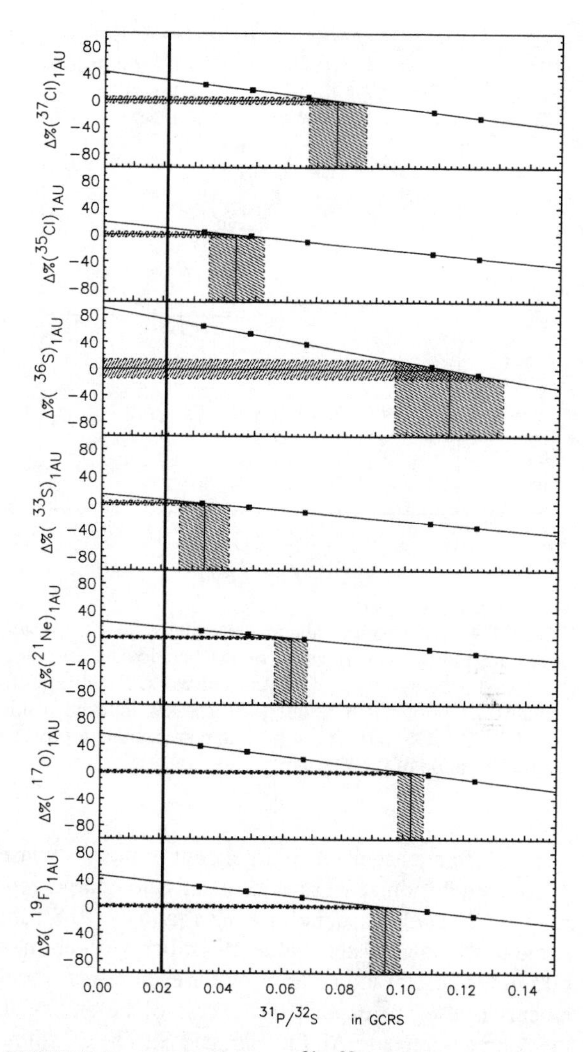

FIGURE 3. Estimates of the $^{31}P/^{32}S$ source ratio from various tracer isotopes as described in Figure 2 The vertical line indicates the solar-system $^{31}P/^{32}S$ value [5]. Shaded areas show statistical uncertainty from measurement of the tracer isotope.

DISCUSSION AND CONCLUSION

Using a secondary tracer isotope technique, we estimated the source abundances of ^{23}Na and ^{31}P based on new data from CRIS. An estimate based on seven tracer isotopes suggests that the semi-volatile element Na is depleted in the GCR source relative to the more refractory Mg, compared to solar-system abundances. This result is consistent with a prior analysis of Na based on three tracer isotopes [15]. Similarly, the less volatile P is enhanced with respect to the highly volatile S by a factor of three, even though each has a nearly identical FIP value. The P/Mg source ratio is consistent with the solar-system value or a small depletion due to the difference in condensation temperatures. It should be noted, however, that correlated errors such as incorrect cross sections for producing Na or P could systematically shift the individual estimates. This is a concern with the large corrections for secondary production that are used here.

Figure 4 summarizes the current results. The source abundance of each element relative to its solar-system value is plotted as a function of first ionization potential, normalized to Mg≡1. One possible FIP parameterization has been overlaid on the plot. The filled squares indicate the present results for Na and P while the right-pointing triangles represent previous results from CRIS data [15], and the crosses indicate data from the HEAO-3-C2 spacecraft [25] for context. The low values of Na/Mg, Cu/Fe, and Ge/Fe, and the enhancement of P/S are consistent with expectations of a volatility model. The overabundance of C may reflect additional sources of material unrelated to the present investigation [1].

Since Zn and S have nearly the same condensation temperature the apparent low S/Zn ratio seems at first to support FIP. Volatility models explain the low sulfur as a consequence of a rigidity dependence of the shock acceleration efficiency [1]. Volatile elements such

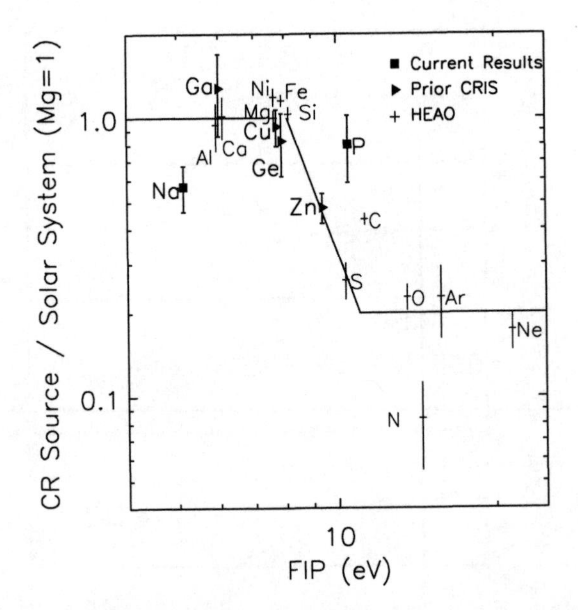

FIGURE 4. GCR source abundances relative to solar system values are plotted as a function of the first ionization potential. Filled squares are from the current work. Right-pointing triangles are prior CRIS results [15], crosses indicate results from HEAO-3-C2 [25]. The solid line represents one possible parameterization of the FIP step.

as S and Zn are accelerated largely out of the gas phase. A cool photo-ionized gas with a mass-to-charge ratio $A/Q \sim A^{0.7}$ could account for a S/Zn ratio of $\sim$0.6 compared to the solar system value. A similar, weaker, mass bias in the acceleration efficiency for even refractory elements is used to explain the excess of Fe and Ni in the source relative to Al, Ca, Mg, and Si. The Se source abundance would be an interesting test of a mass dependence. Se is heavier than Zn, but has a very similar FIP and condensation temperature. CRIS may be able to address this issue once the cosmic-ray fluxes return to their high solar minimum levels allowing for greater statistics.

A pattern seems to be emerging that supports volatility as the relevant parameter for the elemental fractionation of galactic cosmic rays. However, caution is warranted in assessing the significance of these results. Non-statistical fluctuations in the individual estimates are at least as large as the statistical uncertainties. The approach taken here is one quantitative method of characterizing the measurements but may still underestimate the actual spread. At the same time, these results do provide a stronger indication that some deficit of Na/Mg and some enhancement of P/S compared to the solar-system values is present in the galactic cosmic-ray source. Completion of the CRIS results for the primary elements will help place these results in context, but significant improvements for Na and P may require more precise measurements of the fragmentation cross sections.

ACKNOWLEDGMENTS

This work was supported by NASA at the California Institute of Technology (under grant NAG5-6912), the Jet Propulsion Laboratory, the Goddard Space Flight Center, and Washington University.

REFERENCES

1. Meyer, J., Drury, L., and Ellison, D., *ApJ*, **487**, 182–196 (1997).
2. Meneguzzi, M., Audouze, J., and Reeves, H., *A&A*, **15**, 337 (1971).
3. Stone, E., et al., *Space Sci. Rev.*, **96**, 285–356 (1998).
4. Cassé, M., and Goret, P., *ApJ*, **221**, 703 (1978).
5. Anders, E., and Grevesse, N., *Geochim. Cosmochim. Acta*, **53**, 197–214 (1989).
6. Geiss, J., *Space Sci. Rev.*, **85**, 241–252 (1998).
7. Meyer, J.-P., *ApJS*, **57**, 173–204 (1985).
8. Ginzburg, V., and Syrovatskii, S., *The Origin of Cosmic Rays*, Pergamon Press, Oxford, England, 1964.
9. Wiedenbeck, M., et al., *ApJL*, **523**, L61–L64 (1999).
10. Meyer, J., and Ellison, D., "The Origin of the Present Day Cosmic Rays (I)", in *LiBeB, Cosmic Rays, and Related X- and Gamma-Rays*, edited by R. Ramaty et al., ASP Conf. Series #171 (Astr. Soc. of the Pacific), San Francisco, 1999, pp. 187–206.
11. Higdon, J., Lingenfelter, R., and Ramaty, R., *ApJ*, **509**, L33–L36 (1998).
12. Parizot, E., and Drury, L., *A&A*, **349**, 673–684 (1999).
13. Bibring, J.-P., and Cesarsky, C., *Proc. 17th ICRC (Paris)*, **2**, 289 (1981).
14. Epstein, R., *MNRAS*, **193**, 723 (1980).
15. George, J., Wiedenbeck, M., Barghouty, A., et al., "Cosmic Ray Source Abundances and the Acceleration of Cosmic Rays", in [26], pp. 437–440.
16. Stone, E., and Wiedenbeck, M., *ApJ*, **231**, 606–623 (1979).
17. Leske, R., *ApJ*, **405**, 567–583 (1993).
18. Soutoul, A., Ferrando, P., and Webber, W., *Proc. 21st ICRC*, **3**, 337 (1990).
19. Silberberg, R., Tsao, C., and Barghouty, A., *ApJ*, **501**, 911 (1998).
20. Tsao, C., Silberberg, R., and Barghouty, A., *ApJ*, **501**, 920 (1997).
21. Fisk, L., *J. Geophys. Res.*, **76**, 221 (1971).
22. Davis, A., Mewaldt, R., et al., "On the Low Energy Decrease in Galactic Cosmic Ray Secondary/Primary Ratios", in [26], pp. 421–424.
23. Connell, J., and Simpson, J., *Proc. 23rd ICRC (Calgary)*, **1**, 559 (1993).
24. Webber, W., Kish, J., and Schrier, D., *Phys Rev C*, **41**, 566–571 (1990).
25. Engelmann, J., et al., *Astron. Astrophys.*, **233**, 96–111 (1990).
26. Mewaldt, R., et al., editors, *Acceleration and Transport of Energetic Particles Observed in the Heliosphere*, AIP Conference Series #528, New York, 2000.

Constraints on the Nucleosynthesis of Refractory Nuclides in Galactic Cosmic Rays

M. E. Wiedenbeck*, W. R. Binns†, E. R. Christian**, A. C. Cummings‡, A. J. Davis‡, J. S. George‡, P. L. Hink†, M. H. Israel†, R. A. Leske‡, R. A. Mewaldt‡, E. C. Stone‡, T. T. von Rosenvinge** and N. E. Yanasak‡

*Jet Propulsion Laboratory, California Institute of Technology, Pasadena, CA 91109 USA
†Washington University, St. Louis, MO 63130 USA
**NASA/Goddard Space Flight Center, Greenbelt, MD 20771 USA
‡California Institute of Technology, Pasadena, CA 91125 USA

Abstract. Abundances of the isotopes of the refractory elements Ca, Fe, Co, and Ni in the galactic cosmic-ray source are compared with corresponding abundances in solar-system matter. For the 12 nuclides considered, relative abundances agree to within a factor of 2, and typically within 20–30%. In addition, comparisons of cosmic-ray abundances with model calculations of supernova yields are used to argue that cosmic rays contain contributions from stars with a broad range of masses. Based on these and other results we suggest that cosmic rays probably represent a sample of contemporary interstellar matter, at least for refractory species.

INTRODUCTION

Over the past decade considerable progress has been made in precisely determining the isotopic composition of cosmic rays. Instruments on the Advanced Composition Explorer (ACE) and Ulysses missions have succeeded in resolving cosmic-ray isotopes up through the iron group, and the large geometrical acceptance of the Cosmic-Ray Isotope Spectrometer (CRIS) on ACE has made possible abundance measurements of very rare nuclides, including the first isotope measurements in the region $Z > 28$. Among the 80+ individual nuclides for which abundances have been measured in cosmic rays arriving near Earth, approximately one third are dominated by primary nuclei, which are accelerated in the Galaxy and have only relatively small contributions from secondary particles produced by fragmentation of heavier species during passage through the interstellar medium. For this subset of the observed nuclides it is possible to infer the abundances in the source material from which the particles were accelerated.

The increase in the number and precision of nuclidic abundance determinations in the cosmic-ray source is providing a new tool for studies of astrophysics in our Galaxy. These data are yielding information on the stellar sources in which the cosmic-ray material was synthesized. In addition, they are providing a detailed determination of the isotopic make-up of a sample of material which may be representative of the composition of contemporary interstellar matter.

OBSERVATIONS

The CRIS instrument [1] measures the charge, mass, and energy of detected nuclei using multiple measurements of dE/dx versus total energy made with stacks of silicon solid state detectors. In addition, particle trajectories are measured in a scintillating optical-fiber hodoscope to enable corrections for particle pathlengths through the silicon detectors. The geometrical factor of the instrument, ~ 250 cm^{2}sr, is large enough to obtain statistically significant measurements of even relatively rare nuclides. Figure 1 shows mass histograms obtained from CRIS for elements from Li through Cu. The events included in these histograms were selected to have ranges long enough to penetrate into at least the third silicon detector in the stack, yielding multiple mass measurements which are required to be consistent. In addition, to obtain a data set with particularly high resolution the events were required to have trajectories within 25° of the normal to the detector surfaces. In this data set, which contains approximately 30% of the particles collected, essentially all the stable and long-lived isotopes up through the iron group are resolved. For nuclides separated by 2 amu, including the rare isotopes ^{46}Ca, ^{48}Ca, ^{57}Co, ^{59}Co, ^{64}Ni, ^{63}Cu, and

CP598, *Solar and Galactic Composition*, edited by R. F. Wimmer-Schweingruber

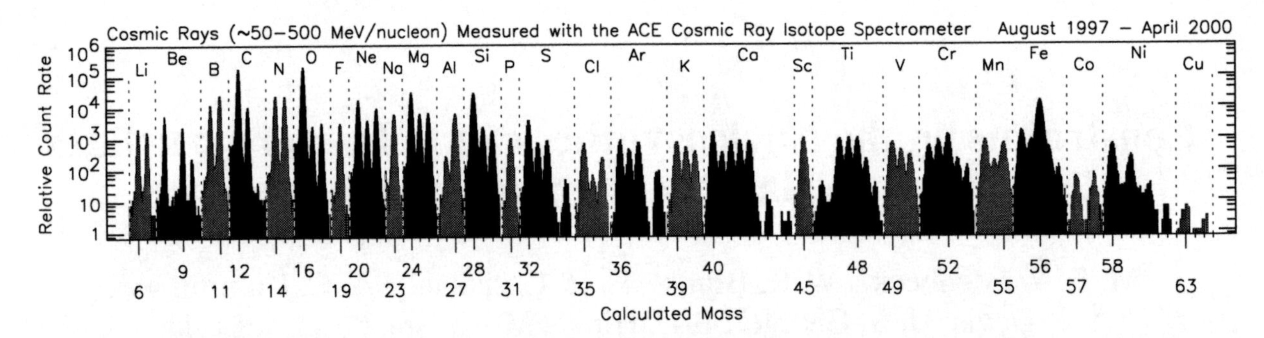

FIGURE 1. Cosmic-ray mass histograms from the Cosmic-Ray Isotope Spectrometer on ACE. Even and odd-Z elements are shaded different colors for clarity. Ticks along the abscissa indicate integral masses, with one tick labeled for each element. Note the logarithmic ordinate—the dynamic range in abundance shown is $> 10^5$.

^{65}Cu, particles with more extreme angles of incidence can also be used in the analysis, improving the statistical accuracy for these species.

The isotopic compositions of individual elements are derived from the relative areas of the isotope peaks in the mass histograms by applying relatively small corrections for differences in energy intervals and nuclear interaction probabilities in the instrument. A different analysis procedure is used to obtain the relative abundances of the elements, since energy intervals differ considerably for elements when charges are not closely spaced. The energy spectrum of each element is derived from the measured count rates and live times, combined with calculated energy intervals, geometrical factors, interaction loss probabilities, and detection efficiencies. Spectral shapes are generally similar, particularly for elements dominated by primary cosmic rays. Flux values are interpolated to a common energy per nucleon within the measurement intervals for the elements being compared. For the elements Ca, Fe, Co, and Ni considered here, an energy of 300 MeV/nucleon was used. The data analysis and the derived cosmic-ray composition near Earth are discussed more thoroughly in [2].

COSMIC-RAY SOURCE ABUNDANCES

In order to investigate the nucleosynthesis of cosmic-ray source material we have selected nuclides which 1) have relatively small contributions from secondary cosmic rays produced by spallation reactions in the interstellar medium and 2) are thought to have experienced minimal fractionation relative to one another during cosmic-ray injection and acceleration. The first requirement is satisfied by Fe and species slightly heavier than Fe, since abundances in this region are rapidly decreasing with increasing mass so that there is relatively little material that can fragment into a nuclide of interest. In addition, the extreme isotopes of calcium, ^{40}Ca and ^{48}Ca, have small secondary contributions. These doubly-magic nuclei have mass-to-charge ratios considerably different from that of ^{56}Fe, the most abundant nuclide that could fragment into them. The large change of A/Z required to produce these isotopes results in small production cross sections.

To avoid the complications of needing to correct for fractionation, our study focused on elements with low first ionization potential (FIP) and high condensation temperatures (low volatility). As has been extensively discussed (e.g, [3] and references therein), the ratio of cosmic-ray source elemental abundances to corresponding abundances in solar-system material are relatively constant for elements with FIP values less than about 9 eV, whereas higher-FIP elements are significantly depleted in cosmic rays. The physical mechanism for this fractionation is not conclusively established. Processes controlled by volatility rather than FIP have been proposed [3] and the general correlation of these two parameters makes it difficult to distinguish between these possibilities. Elements that are particularly volatile in spite of low FIP values have been avoided in the present study because of this uncertainty about how the elemental fractionation occurs. The elements Ca, Fe, Co, and Ni have both low FIP and low volatility, and thus should experience little, if any, fractionation relative to one another.

Source abundances can be reliably derived for the nuclides we are considering by taking advantage of the large number of isotopes with masses $40 < A < 56$ that are dominated by secondary fragmentation products. Using a simple leaky-box model of the interstellar propagation of cosmic rays and adjusting the escape length to reproduce the observed abundances of such secondaries we obtain estimates of the secondary contributions (generally small) to the nuclides for which source abundances are to be derived. These calculations are described in detail in [2].

The filled circles in Fig. 2 show the source composition derived from the CRIS observations, both as abundances relative to $^{56}\mathrm{Fe} \equiv 1$ and as ratios to solar system abundances [4]. In addition, source abundances re-

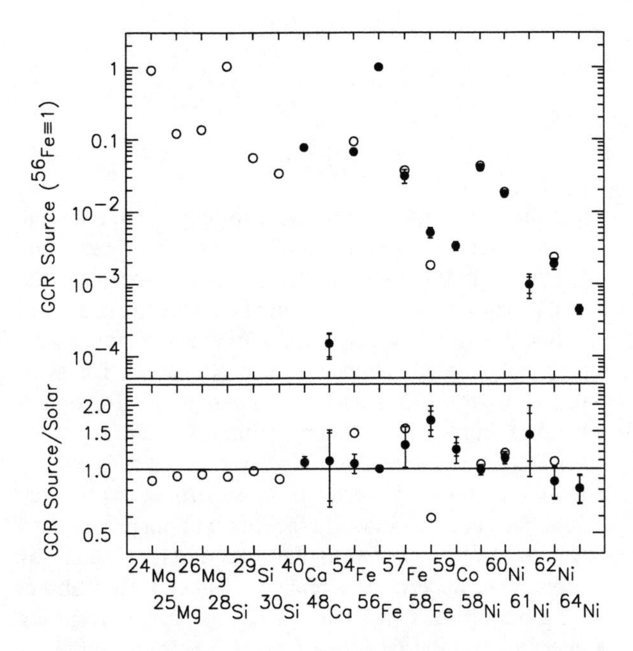

FIGURE 2. Cosmic-ray source composition from ACE (filled circles) and Ulysses (open circles). Cosmic-ray source abundances for the set of refractory nuclides shown range over nearly 4 orders of magnitude (upper panel), but in all cases are consistent with solar-system abundances [4] to within better than a factor of 2 (lower panel). For the ACE data the inner error bars represent statistical uncertainties while the outer error bars also incorporate estimates of systematic and propagation uncertainties [2]. Ulysses uncertainties (not shown) are discussed in [5, 6].

ported from Ulysses for the isotopes of Mg, Si, Fe, and Ni [5, 6] are plotted as the open circles. For normalizing the Ulysses abundances to ^{56}Fe we have used the source elemental ratios derived from HEAO-C2 observations [7]. In the cases where abundance measurements are available from both ACE and Ulysses, the agreement is generally quite good. Possible explanations of the most significant differences (for ^{54}Fe and ^{58}Fe) have been discussed in [2, 8]. CRIS also measures the isotopes of the refractory elements Mg and Si [2], but we have not yet performed propagation calculations that account for secondary-nuclide abundances over the full charge range $9 \leq Z \leq 28$ with the accuracy needed to take full advantage of the small statistical uncertainties in the CRIS observations.

The nuclides we are considering have absolute abundances ranging over a factor $\sim 10^4$. However, the abundance distribution is consistent with that found in solar system material to better than a factor of 2 for all of these nuclides, and typically within ~ 20–30%. This similarity is particularly striking in view of the fact that the various isotopes represented in Fig. 2 are thought to originate in a diverse range of stellar environments. Table 1 summarizes the nucleosynthesis processes and sites believed to be major contributors to the individual nuclides found in solar system material [9].

Table 1.
Major Nuclesynthetic Contributions[a]

Nucl.	Origin[b]	Nucl.	Origin[b]
^{24}Mg	C, Ne	^{56}Fe	xSi, Ia
^{25}Mg	C, Ne, He(s)	^{57}Fe	xSi, Ia
^{26}Mg	C, Ne, He(s)	^{58}Fe	He(s), nse-Ia-MCh
^{28}Si	xO, O	^{59}Co	He(s), α, Ia, ν
^{29}Si	Ne, xNe	^{58}Ni	α, Ia
^{30}Si	Ne, xNe	^{60}Ni	α, He(s)
^{40}Ca	xO, O	^{61}Ni	α, Ia-det, He(s)
^{48}Ca	nse-Ia-MCh	^{62}Ni	α, He(s)
^{54}Fe	Ia, xSi	^{64}Ni	He(s)

[a] from Woosley & Weaver (1995)
[b] Abbreviations for processes & sites are as follows: C, O, Ne: quiescent burning of these elements; xO, xNe, xSi: explosive burning of these elements; He(s): s-process during core/shell He burning; α: α-rich freeze out during SN II explosion; Ia: SN Ia igniting at Chandrasehkar mass; nse-Ia-MCh: inner n-rich regions of SN Ia igniting at Chandrasehkar mass; Ia-det: SN Ia from detonation at less than Chandrasehkar-mass; ν: neutrino process.

From the similarity between cosmic-ray source and solar-system compositions one expects the same processes to be important contributors to cosmic rays. However, the relative contributions of various nucleosynthetic processes to solar-system and galactic cosmic-ray (GCR) matter are not "universal". For example, in our neighboring galaxies, the Large and Small Magellanic Clouds, it has been found [10] that the contributions from Type Ia supernovae are significantly greater than in the Milky Way. Thus one can expect to obtain new insights into the origin of cosmic rays and/or solar system material by understanding the close correspondence of the compositions of these two populations.

SUPERNOVA CONTRIBUTIONS TO COSMIC RAYS

In massive stars, the production of most of the iron-group material occurs during the final stages of hydrostatic stellar evolution. The synthesis of these nuclides, for which the binding energy per nucleon is maximum, marks the exhaustion of a star's nuclear fuel. The ensuing collapse of the stellar core and its explosive rebound leads to further alteration of abundances. Detailed numerical calculations have been carried out by two groups [9, 11] to determine the composition of the material returned to the interstellar medium in such core-collapse supernovae (SN II) arising from a range of progenitor star masses.

For many nuclides very similar yields are reported

from these two investigations. However, for several nuclides of interest for this study there are important differences. We have chosen to adopt the results from [9], based on a detailed evaluation of the difference between the two calculational approaches [12]. It must also be recognized that massive-star models have not yet successfully accounted for the supernova core bounce that drives the explosive processing and ejection of material. Instead explosions are artificially put into the calculations, constrained by observed properties of SN II. Until the physics of the explosions is adequately understood, there remains the possibility of large errors in predicted yields, particularly in the iron-dominated stellar layers close to the "mass cut" separating the matter that is ejected from that which collapses into the remnant.

Important contributions of iron-group nuclides also come from Type Ia supernovae (SN Ia), which are thought to arise from the accretion of matter onto white dwarf stars in binary systems and the subsequent nuclear deflagration/detonation. Numerical calculations of these yields have also been reported [13]. However, the detailed physics of the SN Ia explosion are not yet firmly established, and various scenarios that have been considered lead to significantly different nucleosynthesis yields. For this study we have adopted the yields of the widely-used "W7" model (see [13] and references therein) as characteristic of the yields of low-mass ($M < 8M_\odot$) stars that evolve into white dwarves and ultimately explode as SN Ia.

To investigate the contributions of various supernova types to the cosmic-ray source, we have compared the cosmic-ray source abundances derived from the CRIS data with the available supernova yield calculations. These comparisons are shown in Fig. 3, where the cosmic-ray source abundances are plotted as the horizontal solid line with dotted lines indicating the $\pm 1\sigma$ uncertainty band. The SN II yields from various mass progenitor stars are plotted as the open symbols and the SN Ia yields as the filled symbol. For some ratios the calculations can reproduce the cosmic-ray abundances for a range of stellar masses. However, no single mass can account for most of the nuclides shown. Thus, for example, stars more massive than $\sim 15\ M_\odot$ can account for the observed ^{58}Ni/^{56}Fe, whereas to account for ^{64}Ni/^{56}Fe requires stars less massive than this value. For some species, most notably ^{48}Ca, none of these supernova models can adequately reproduce the observations.

To obtain quantitative limits on the amount of material that supernovae of various progenitor masses could have contributed to the cosmic-ray source, we have considered the effect of mixing material with the composition given by each of the supernova models into a pool of material from other sources. The ratio, by mass, of two nuclides X and Y in the mixed material is

$$(X/Y)_m = \frac{M_i(Y)\cdot(X/Y)_i + M_p(Y)\cdot(X/Y)_p}{M_i(Y)+M_p(Y)} \quad (1)$$

$$= f\cdot(X/Y)_i + (1-f)\cdot(X/Y)_p. \quad (2)$$

Here the subscripts refer to the resulting mixed material, m, the material from a particular supernova type, i, and material from the pool, p, into which the supernova material is being mixed. $M_i(Y)$ and $M_p(Y)$ are the masses of nuclide Y in the two original populations, while $(X/Y)_i$ and $(X/Y)_p$ are the abundance ratios by mass. The symbol $f \equiv M_i(Y)/\left(M_i(Y)+M_p(Y)\right)$ denotes the fraction of nuclide Y in the mix that came from sources of type i.

If $(X/Y)_i$ is larger than $(X/Y)_m$ then there is a maximum fraction of material from supernova type i that could have been included in the mix without producing a value of $(X/Y)_m$ exceeding that found in the cosmic-ray source. This maximum is realized if nuclide X is absent in the background material, $(X/Y)_p = 0$. The resulting upper limit on the fraction of nuclide Y from source i is given by $f_{\max} = (X/Y)_m/(X/Y)_i$.

It should be noted that this approach is providing an upper limit on the fractional contribution of a particular supernova type to the overall abundance of Y, the nuclide in the denominator of the ratio being considered. To convert this value into a limit on the fraction of the *total* mass contributed by the model (including all nuclides, not just X and Y), one would need to know the fraction by mass of nuclide Y in the material into which the supernova ejecta are being mixed—this would require an additional assumption about the nature of that material.

For illustration, we consider the ratio ^{58}Ni/^{56}Fe. Since the value of this abundance ratio expected from a supernova with a 11 $M_\odot$ progenitor is approximately 5 times the value in the cosmic-ray source (Fig. 3), this source could have contributed no more than $\sim$ 20% of the cosmic-ray ^{56}Fe. Figure 4 shows the upper limits derived in this way for each stellar model. The limits are most stringent for those models that produce ^{58}Ni/^{56}Fe with the largest excesses over the cosmic-ray value. When the cosmic-ray ^{58}Ni/^{56}Fe ratio is greater than that obtained from a given model, no constraint on the contribution of that model to the ^{56}Fe is obtained. (But one could obtain a limit on the fractional contribution to ^{58}Ni by considering the reciprocal ratio.)

Similar analyses can be done using the measured ratios of the other isotopes relative to ^{56}Fe. From each isotope one obtains a set of upper limits on the amounts of ^{56}Fe that could have been contributed by the various mass stars. Applying the limits from all the isotopes together one obtains a more stringent set of upper limits, with the amount of material from each stellar mass constrained by the most restrictive of the individual limits. Figure 5 shows (open circles) the upper limits obtained

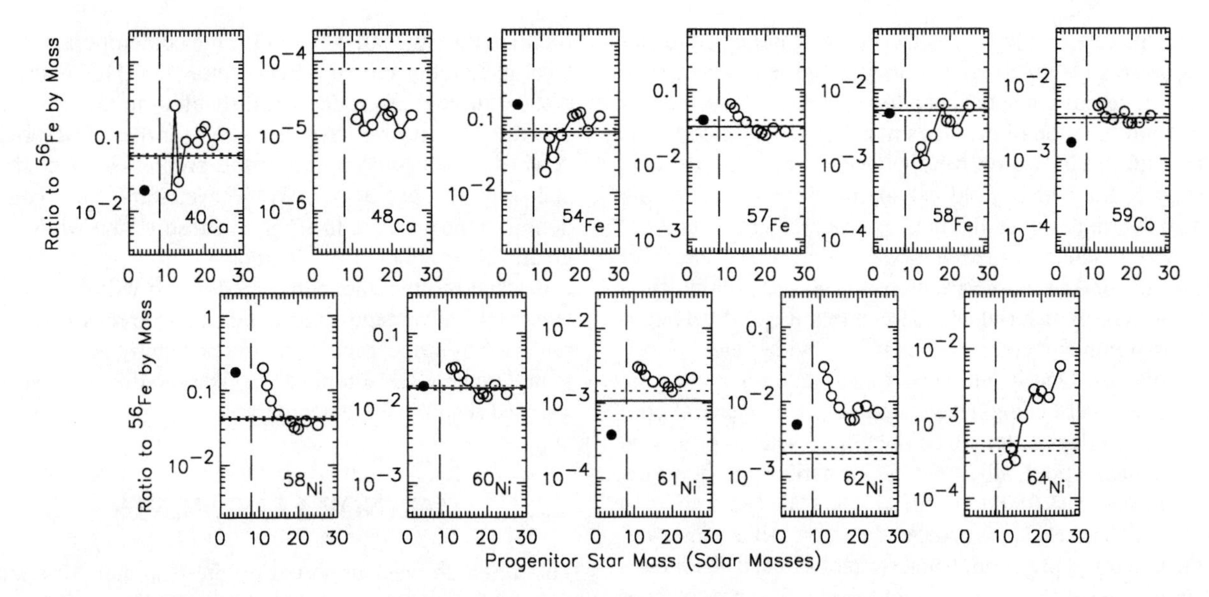

FIGURE 3. Comparison of GCR source abundance ratios with calculated values for yields of supernovae from various masses of progenitor star (filled circle: SN Ia [13]; open circles: SN II [9]). Cosmic-ray source values and their $\pm 1\sigma$ uncertainties are shown by the horizontal lines. The approximate mass boundary between low-mass stars that lead to SN Ia and high-mass stars that lead to SN II is shown by the dashed lines.

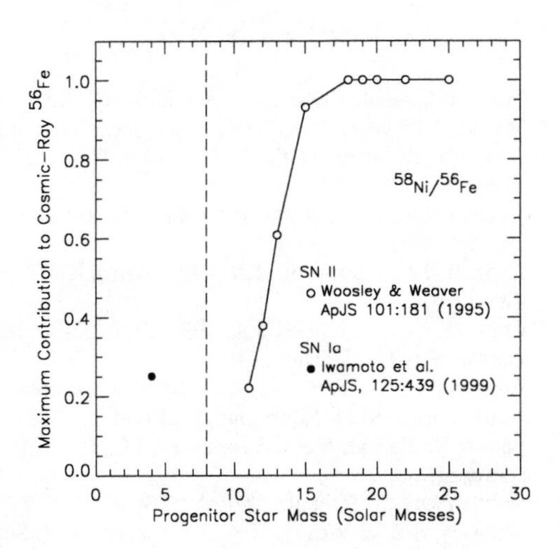

FIGURE 4. Limits on the maximum fraction of the GCR ^{56}Fe that could have originated from supernovae of various mass progenitor stars based on the observed ^{58}Ni/^{56}Fe ratio.

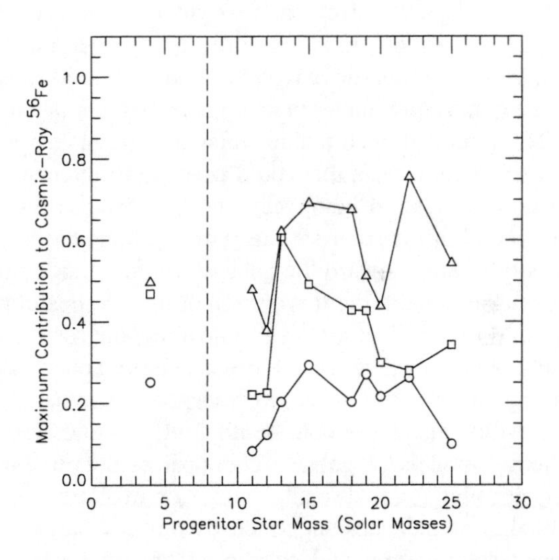

FIGURE 5. Limits on the maximum fraction of the GCR ^{56}Fe that could have originated from supernovae of various mass progenitor stars given the derived GCR source abundances for all the Ca, Fe, Co, and Ni isotopes studied. The calculated upper limits take into account the 1σ uncertainty in the cosmic-ray source value, but no uncertainties in the nucleosynthesis models. Circles, squares, and triangles indicate the limits obtained by disregarding the 0, 1, or 2 most restrictive constraints, respectively.

by requiring that none of the isotope ratios relative to ^{56}Fe exceed their cosmic-ray values in the overall mix of material. Of course, if there are significant errors in the model calculations for a particular isotope, any upper limit driven by this isotope would be invalidated. Figure 5 also shows the effect of allowing the violation of the most restrictive one or two individual isotope limits (open squares and triangles, respectively).

DISCUSSION

The most notable feature of the combined upper limits on the contributions of various mass stars to cosmic-ray source material (Fig. 5) is the fact that none of the star types considered could be dominating the production of this material. Since the abundance yields are a relatively smooth function of the star's mass (Fig. 3), we conclude that the limits we derive apply not just to the discrete masses for which yield calculations are available, but also for ranges of stellar masses around these values.

The essential constraints we have on the nature of the cosmic-ray source are its compositional similarity to solar-system material and the fact that it apparently contains significant contributions from a wide range of stellar masses. One would expect such features if the cosmic rays were being accelerated out of the general pool of interstellar matter, since the solar system condensed from such a pool, albeit ~ 4.5 Gyr earlier. Furthermore, the initial mass function (IMF) in the Galaxy is known to have a relatively smooth dependence on stellar mass, and the Galaxy is old enough that stellar birth and death rates should be similar provided that the IMF has not changed greatly over the past æon.

The suggestion that supernova material gets well mixed with interstellar matter before cosmic-ray acceleration occurs is supported by the observation that a time $\gtrsim 10^5$ yr elapses between nucleosynthesis and acceleration [14], since this time is significantly longer than that required for a supernova remnant to dissipate its energy as it expands into matter of average interstellar density.

There remain uncertainties related to possible spatial and temporal effects that could produce differences between the material being sampled by cosmic rays and that found in the solar system. For example, the higher density of stars toward the galactic center is thought to be causing a more rapid evolution of the composition of the ISM there. Whether this should affect the cosmic-ray composition observed near Earth depends on poorly constrained aspects of cosmic-ray transport in the Galaxy.

It would also be of considerable interest to determine whether models of galactic chemical evolution (GCE) can account for the small amount of evolution of interstellar abundances suggested by our comparison between solar-system and cosmic-ray material. Unfortunately, GCE studies generally have not been extended to cover the time between solar-system formation and the present, concentrating instead on evolution up to this time. Furthermore, reports of GCE results have often not included the abundances of individual nuclides that would be needed to take advantage of observations such as those shown in Fig. 2.

For reasons discussed above, the work on cosmic-ray source composition presented here has concentrated on refractory nuclides. A model accounting for the origin of cosmic-ray source material must ultimately also address the abundances of volatile species. The isotopic composition of the element neon remains a puzzle: the ratio ^{22}Ne/^{20}Ne in the cosmic-ray source exceeds the value found in the Sun and inferred for the contemporary very local interstellar medium by a factor ~ 5 ([15] and references therein), so it appears difficult to fit this element into the scenario we have discussed for the refractories. A widely-discussed solution to this problem invokes special sources (specifically, Wolf-Rayet stars) that would dominate the production of ^{22}Ne in cosmic rays while not greatly affecting the overall amount of this isotope being contributed to the interstellar medium. It will be important to take advantage of the detailed and precise cosmic-ray source composition data now becoming available to search for smaller isotopic differences predicted to be associated with the ^{22}Ne anomaly.

ACKNOWLEDGMENTS

This research was supported by the National Aeronautics and Space Administration at the California Institute of Technology (under grant NAG5-6912), the Jet Propulsion Laboratory, NASA's Goddard Space Flight Center, and Washington University.

REFERENCES

1. Stone, E.C. et al. 1998, Space Sci. Rev. 86, 285
2. Wiedenbeck, M.E. et al. 2001, Space Sci. Rev., in press
3. Meyer, J.-P., Drury, L. O'C., & Ellison, D. C. 1997, Astroph. J. 487, 182
4. Anders, E. & Grevesse, N. 1989, Geoch. Cosmoch. Acta 53, 197
5. Connell, J.J. & Simpson, J.A. 1997, Astroph. J. Letters 475, L61
6. Connell, J.J. & Simpson, J.A. 1997, Proc. 25th Internat. Cosmic Ray Conf. (Durban) 3, 381
7. Engelmann, J.J., Ferrando, P., Soutoul, A., Goret, P., Juliusson, E., Koch-Miramond, L., Lund, N., Masse, P., Peters, B., Petrou, N., & Rasmussen, I.L. 1990, Astron. Astroph. 233, 96
8. Wiedenbeck, M.E. et al. 2001, Adv. Space Res., 27, 773
9. Woosley, S.E. & Weaver, T.A. 1995, Astroph. J. Suppl. 101, 181
10. Tsujimoto, T., Nomoto, K., Yoshii, Y., Hashimoto, M., Yanagida, S., & Thielemann, F.-K. 1995, Mon. Not. Royal Astron. Soc. 277, 945
11. Thielemann, F.-K., Nomoto, K., & Hashimoto, M. 1996, Astroph. J. 460, 408
12. Hoffmann, R.D., Woosley, S.E., Weaver, T.A., Rauscher, T., & Thielemann, F.-K. 1999, Astroph. J. 521, 735
13. Iwamoto, K., Brachwitz, F., Nomoto, K., Kishimoto, N., Umeda., H., Hix, W.R., & Thielemann, F.-K. 1999, Astroph. J. Suppl. 125, 439
14. Wiedenbeck, M.E. 1999, Astroph. J. Letters 523, L61
15. Binns, W.R. et al. 2001, Joint SOHO–ACE Workshop, this volume (AIP: New York) submitted

Direct Measurement of $^3He/^4He$ in the LISM with the COLLISA experiment

E. Salerno*, F. Bühler*, P. Bochsler*, H. Busemann*, O. Eugster*, G. N. Zastenker†, Yu. N. Agafonov† and N. A. Eismont†

*Physikalisches Institut, University of Bern, Sidlerstrasse 5, CH-3012, Switzerland
†Space Research Institute (IKI), Russian Academy of Sciences, Profsoyuznaya ul. 84/32, 117997 Moscow, Russia

Abstract.
Results from direct measurements of the helium isotopic ratio in the closest regions of the Local Interstellar Medium (LISM) are presented. Neutral ^{3}He and ^{4}He atoms coming from the LISM were captured in space by means of the *foil collection technique*, a method already successfully used during the Apollo missions to determine the noble gas isotopic ratios in the solar wind. In the framework of the Swiss-Russian project COLLISA (COLLection of InterStellar Atoms), beryllium-copper foils were placed on the outer surface of the space station Mir and directly exposed to the flux of interstellar neutrals. The neutral particles of the LISM cross the heliopause and reach, almost unaltered, the Mir orbit at 400 km height above the Earth. Here, the kinetic energy of the interstellar flux ramming against the foils is sufficient to trap the particles into the atomic structure of the metal. After an exposure of ~60 hours, the foils were recovered by the cosmonauts and brought back to Earth by the American space shuttle Atlantis. The particles were then extracted with a stepwise heating procedure and their abundances were measured in the mass spectrometric laboratories of the University of Bern. The analysis performed so far allowed the detection of ^{3}He and ^{4}He atoms of interstellar origin. The measured interstellar ratio $^3\text{He}/^4\text{He} = \{1.70^{+0.50}_{-0.42}\} \times 10^{-4}$ is consistent with protosolar values obtained from meteorites and Jupiter's atmosphere. Such a result seems to confirm the hypothesis that no significant change of the ^{3}He abundance occurred in the LISM during the last 4.6 Gy.

INTRODUCTION

The measurement of the elemental and isotopic composition in different astrophysical sites (stellar interiors and winds, planetary rocks and atmospheres, neutral and ionized gas clouds, meteorites, etc.) gives direct information on the chemical structure of the galactic matter at different galactocentric distances and in different evolutionary epochs. As discussed by Chiappini (this volume) [1], the observed chemical abundances are extremely important since they can be used to test the predictions of Galactic Chemical Evolution models and to constrain their input parameters.

For instance, the accurate measurement of the helium isotopic abundances could help to solve one of the open issues of astrochemistry: the ^{3}He problem. Theoretical models predict that the ^{3}He produced during the primordial nucleosynthesis undergoes several astration processes which partially produce it (D is immediately burnt into ^{3}He in stars of all masses) and partially destroy it (^{3}He is significantly transformed into ^{4}He in the interiors of massive stars). The resulting ^{3}He net yield is a steeply decreasing function of the stellar initial mass [2]. This behavior leads to an overestimation of the ^{3}He solar abundance [3]. The ^{3}He abundance predicted by the models differs, in fact, by almost two orders of magnitude from the abundances observed in both presolar material [4] and LISM [5] ($^3\text{He/H} \sim 10^{-5}$), but is in agreement with observations of planetary nebulae ($^3\text{He/H} \sim 10^{-3}$) [6, 7, 8]. A solution to the problem was proposed in 1995 by Charbonnel [9] [see also 10, 11]. It consists in processing ^{3}He into heavier elements by an extra-mixing mechanism occurring below the convective zone of low-mass stars ($\lesssim 2\ M_\odot$) on the red giant branch. The values observed in planetary nebulae, however, indicate that some of these stars have to be net producers of ^{3}He . Therefore, it has been recently suggested by several authors that extra-mixing occurs only in a fraction of low mass stars. Galli et al. [12] showed that this fraction should be larger than 80%. Recently Chiappini and Matteucci [13], adopting a new version of their "two-infall" model, have predicted the evolution of ^{3}He for

CP598, *Solar and Galactic Composition*, edited by R. F. Wimmer-Schweingruber

different percentages of low mass stars in which extra-mixing should occur. They found that the best fit with observations is reached when this mechanism occurs in 93% of the stars. A similar result was found by Tosi [2]. According to the "Tosi-1" model, it is in fact possible to reproduce the abundances of ^{3}He observed in the Sun, in the LISM and in planetary nebulae if deep mixing is assumed to operate in $\sim$90% of the low mass stars. Although the extra-mixing mechanism seems to explain the apparent inconsistencies between the predicted abundances and those observed in different galactic objects, further investigations are necessary, both to find out its possible causes and to check its effects on later stellar evolution phases.

Since few decades, the observation of astrochemical "reservoirs" has been a fast growing area of research in both ground-based astronomy and space science. So far, many efforts have been made to improve the quality and the precision of such measurements. Amongst others, some experiments have been recently performed to determine the physical and chemical properties of the Local Interstellar Medium (LISM): the region of our Galaxy that extends within few hundreds of pc of the Sun. COLLISA is one of these experiments.

In this work we give a report on the experiment and present the results of the interstellar ^{3}He/^{4}He measurement. We then compare our ratio with similar values obtained from observations of meteorites, Jupiter's atmosphere and, in general, of all the astrophysical sites that could be representative of the present-day and the protosolar cloud. The consistency of these values is discussed. Finally, the agreement between the measured concentration of interstellar helium and the value predicted by the "hot gas" model is discussed.

THE EXPERIMENT

The COLLISA project [14, 15] is the result of a cooperation between the Group for Space Research and Planetary Sciences of the University of Bern and the Space Research Institute (IKI) of the Russian Academy of Sciences. The aim of the experiment is to collect and determine the helium isotopic composition of a sample of neutral atoms directly coming from the Local Interstellar Cloud (one of the several gas clouds wich compose the LISM and wherein the Sun is immersed at the moment). The experimental procedure is based on the *foil collection technique*. A detailed description of this method, as well as the astrophysical conditions necessary for its application to the collection of interstellar particles, is given by Klecker et al. (this volume). The foil collection technique was developed at the Physics Institute of the University of Bern and successfully used during the Apollo missions to collect solar wind ions on the surface of the Moon [16]. With this method some metal foils are directly exposed in space to a stream of particles. If the kinetic energy of the particles is sufficiently high, they get trapped in the foils. After the exposure, the foils are brought back to Earth and the amount of particles trapped within them is measured in mass spectrometers.

Studies on the scattering of interstellar helium in the Earth's atmosphere [15] have indicated that this process does not affect the collection of interstellar neutral atoms if the foil exposure is performed during minimum solar activity at altitudes higher than 200-300 Km. The Russian space station Mir, orbiting the Earth at a distance of $\sim$ 400 km, represents a perfect ground-base for the exposure of trapping foils.

In the framework of the COLLISA project four beryllium-copper foils (200 cm^2 wide and 15 μm thick) were exposed in Spring 1996, for approximately 60 hours, to the flux of the interstellar neutral atoms.

The foils, covered with a beryllium-oxide layer to further increase their trapping efficiency [17], were mounted on special cassettes plugged inside two collectors (see Figure 1). The collectors, named KOMZA I and II, were designed and constructed at IKI with the participation of the Space Physics Design Bureau. They were installed on the outside of the Spektr module of the Mir. The Spektr, already provided with KOMZAs, was launched and docked to the space station in 1995.

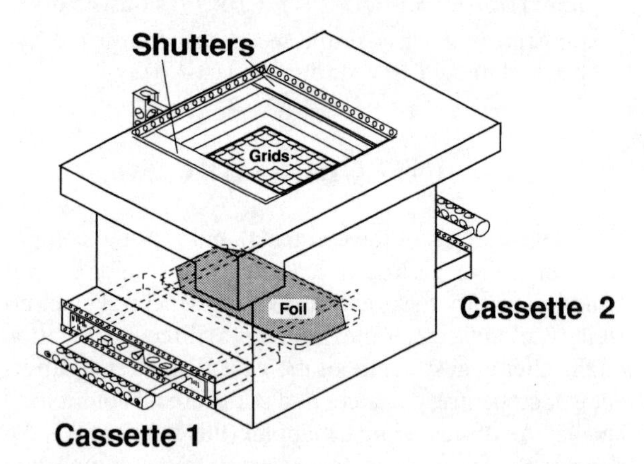

FIGURE 1. A KOMZA particle collector. Only one of the two cassettes (dashed lines) is shown.

The collectors were equipped with shutters that remained open only when the apertures were looking into the direction of the incoming interstellar atoms. The best conditions for the capture of the particles were reached during Spring time. As already mentioned in [18], in this period of the year the particle collection is particularly effective since the Earth moves in the upwind direction of the interstellar flux of neutrals. This condition enhances the velocity of the neutral atoms relative to Mir from

~25 km/s to ~60-80 km/s (~25 eV/AMU) increasing their trapping probability up to ~30% [17].

Care was taken to keep the shutters closed whenever a possible contamination of the foils with solar irradiation or with terrestrial atmospheric particles could have happened. To avoid foil contaminations, the shutters were also closed during the Mir working activities (docking, undocking, refuellings, switching on of cruise or altitude-control engines, etc.). Electrical grids, placed in the collectors above the foils, rejected < 100 eV electrons and positively charged ions with energies up to 5 keV. This precaution was taken in order to protect the foils from a possible contamination with terrestrial energetic magnestospheric ions. Heating plates were placed just below the foils to constantly keep them at 50°C during the exposure. The foils were heated to avoid the formation, on their surface, of condensation layers which could have reduced the trapping efficiency. After exposure, the foils were recovered by the cosmonauts of Mir and brought back to Earth by the U.S. space shuttle Atlantis.

THE MASS SPECTROMETRIC ANALYSIS

Once landed on Earth, the foils were delivered to the University of Bern for the measurement of the trapped particles. The first step of the analysis consisted in degassing the foils in a UHV high-temperature furnace. The extraction was performed in several temperature steps: 300°, 600°, 1100°, 1400° and 1700°C. Measurements, performed on foils that had previously been artificially bombarded with helium isotopes at different energies, showed in fact that particles implanted with typical interstellar energies (~25 eV/AMU) are released in a temperature range of 300°-1100°C. At temperatures below 300°C and above 1100°C, particles with lower and higher implantation energies are released, respectively. This is due to the fact that according to their velocity the atoms penetrate the foil to different depths: lower speeds lead to superficial trappings while high kinetic energies drive the particles deeper into the foil. As a consequence, the deeper the position of the particles in a foil, the higher the thermal energy necessary to extract them. In this way, the stepwise heating provides a further safety measure to separate the interstellar particles from the low energetic (<1 eV) atmospheric or high energetic (5000 eV) magnetospheric particles possibly captured by the foils. The measurement of the foils artificially bombarded also provided an estimate of the foil trapping efficiency "η" (i.e. the percentage of particles captured by the foil, compared to the total amount irradiated) [17]. Typical values of the trapping efficiency for the helium isotopes in the beryllium-oxide were: $\eta_3 = 0.18 \pm 0.04$ and $\eta_4 = 0.25 \pm 0.04$, while their mean ratio was $\eta_3/\eta_4 = 0.73 \pm 0.07$.

After the extraction, chemically active gases were trapped by getters. The helium was then transferred, for the measurement, into a Mass Analyzer Product 215-50 mass spectrometer with ion counting collector.

RESULTS

Figure 2 summarizes the results of the analysis performed on one (L641-2-1) of the four foils exposed on the Mir during Spring 1996.

The continuous lines show the accumulated amount of ^{3}He and ^{4}He (upper and lower panel respectively) released per mg by the foil, at each temperature step. For comparison, data derived from the analysis of the foils L461-3-2, L461-3-5, L460-3-1 never flown in space are also plotted (dashed lines). Such measurements were performed to estimate the background due to noble gases contained in the foil prior to the flight. Even though measures were taken during the foil preparation to avoid any possible contamination, small amounts of noble gases may be present in the atomic structure of the metal foils. These quantities (namely foil blanks) have to be accurately determined, to correct the yields of the exposed foils. The values of the foil blanks are obtained from analysis of foils that were not exposed on Mir but were treated identically as the exposed ones.

Given errors are mainly due to the background variation. The release profiles indicate that the detected ^{3}He and ^{4}He have a clear interstellar origin. The highest percentage of gas was in fact released in the temperature range of 300°-1100°C, which is the expected one for particles with interstellar energies. Above 1100°C the beryllium-copper reaches its melting point, releasing only terrestrial contaminations. Up to 1100°C, the ^{4}He extracted from the flown foil is $\{1.36 \pm 0.11\} \times 10^9$ atoms/cm^2. This quantity is more than one order larger than the amount of gas released by the blank foils ($\{1.16 \pm 0.20\} \times 10^8$ atoms/cm^2), clearly indicating that the particles detected in the flown foil could be only trapped in space. While the ^{4}He blank contributes about 8% to the total ^{4}He of the exposed foil, the ^{3}He blank corresponds to 39% of the total ^{3}He. This is probably due to the presence of residual tritium inside the metal. The tritium, maybe present in the atmosphere in higher levels in the past and already contained in the recycled copper used for the foil production, decays in ^{3}He with a half-life of 12.323 years. This time is short enough to produce tainting amounts of ^{3}He in the beryllium-copper foils of the COLLISA experiment.

The value of the helium isotopic ratio, determined

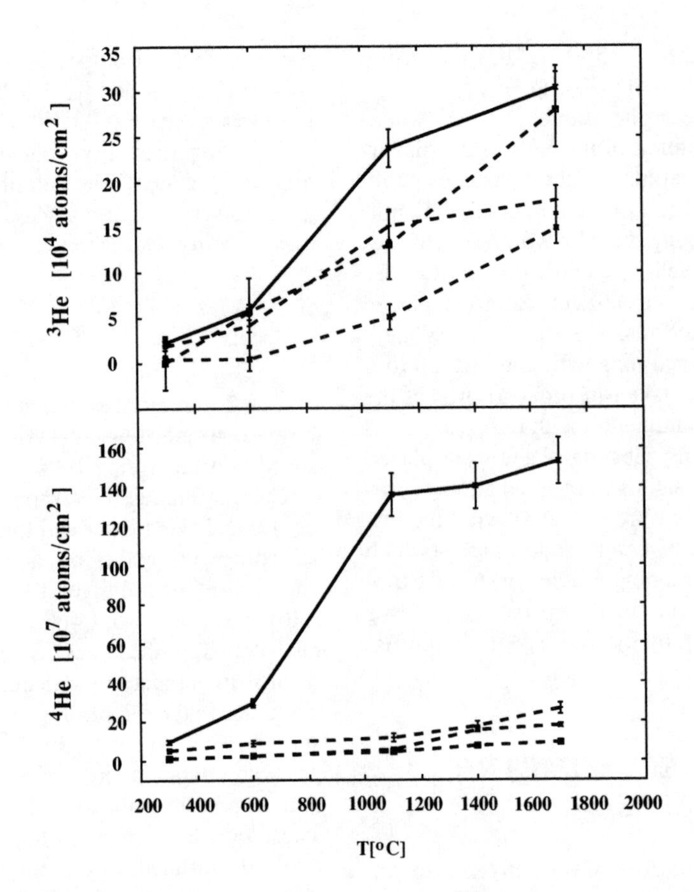

FIGURE 2. Helium released from the foils L461-2-1 (exposed on Mir to the interstellar flux), L461-3-2, L461-3-5 and L460-3-1 (blank foils)

after correcting the total ^{3}He and ^{4}He release for the foil blanks and the relative trapping efficiency η_3/η_4, is:

$$(^3\mathrm{He}/^4\mathrm{He})_{\mathrm{LISM}} = \{1.71^{+0.50}_{-0.42}\} \times 10^{-4}$$

For the determination of the error bars, a conservative range of ±25% has been adopted, due to the low number of foil blanks analyzed. The uncertainty of the upper limit was further increased to take into account possible systematic errors introduced in the determination of the relative trapping efficiency η_3/η_4 which would favor even more the lighter isotope.

TABLE 1. ^{3}He/^{4}He derived from observations of different astrophysical sites.

Reservoir	^{3}He/^{4}He	Source
LISM (neutrals)	$1.71^{+0.50}_{-0.42} \times 10^{-4}$	[This work]
LISM (pickup ions)	$2.48^{+0.68}_{-0.62} \times 10^{-4}$	[5]
Meteorites (Q-phase)	$1.3 \pm 0.02 \times 10^{-4}$	[19]
Meteorites	$1.5 \pm 0.3 \times 10^{-4}$	[4]
Jupiter atmosphere	$1.66 \pm 0.05 \times 10^{-4}$	[20]
Sun (OCZ)	$3.7 \pm 0.7 \times 10^{-4}$*	[21]

* consistent with Gloeckler et al. [22]

DISCUSSION

Comparison with Solar System and LISM Abundances

Protosolar and present-day LISM values of the ^{3}He/^{4}He ratio can be derived, directly and indirectly, from observations of various astrophysical reservoirs. Table 1 summarizes some of these values. Although consistent within the error limits, the value of the ^{3}He/^{4}He ratio in the Local Interstellar Medium observed with the COLLISA experiment is lower than that derived from the measurements of pickup ions [5]. However, the present-day LISM ratio, inferred from the measurement of neutral atoms is closer to the protosolar values observed in meteorites and in the Jupiter's atmosphere, suggesting that no substantial change in the LISM ratio, and therefore no significant increase of ^{3}He, occurred during the last 4.6 Gy.

The determination of the $^3He/^4He$ ratio in the solar Outer Convective Zone (OCZ), obtained from solar wind measurements, allows to calculate the protosolar (^3He + D)/H [23]. In the young Sun deuterium was in fact efficiently converted to ^3He. The helium has subsequently remained unprocessed in the material of the Outer Convective Zone, as is implied by the continuing presence in this region of the more reactive ^9Be. The $^3He/^4He$ ratio measured today in the outer convective zone can be therefore considered representative of the protosolar (^3He + D)/H [24]. However, due to the different settling of ^3He and ^4He out of the OCZ into deeper layers of the Sun [25], and to possible solar mixing processes [26], the present day $^3He/^4He$ in the OCZ could have increased of a few percent compared to the protosolar value. Geiss and Gloeckler [24], making an estimate of the contribution due to these two effects, found that this should not exceed $(5 \pm 3)\%$. Applying this correction to data derived by Bodmer and Bochsler [21] (in agreement with those observed by Gloeckler et al. [22]) and using the standard universal ratio He/H$\sim$ 0.1, one obtains:

$$\left[(^3\mathrm{He}+\mathrm{D})/\mathrm{H}\right]_{ps} = \{3.5 \pm 0.7\} \times 10^{-5}$$

This value leads to a protosolar $^3He/^4He$ ratio which agrees with those observed in meteorites, Jupiter and neutral LISM if a deuterium abundance of:

$$(\mathrm{D/H})_{ps} = \{1.9 \pm 0.7\} \times 10^{-5}$$

is assumed. Although consistent with the inferior limit of the error bars, this value is lower than the one found by Mahaffy et al. [20] in the Jovian atmosphere:

$$(\mathrm{D/H})_{Jupiter} = \{2.6 \pm 0.7\} \times 10^{-5}$$

suggesting a slight overestimation of the primordial deuterium abundance. Such an overestimation seems to be also confirmed by recent observations of Jupiter's and Saturn's atmospheres performed with the Short Wavelength Spectrometer onboard the Infrared Space Observatory [27].

Expected and Measured ^{4}He Concentrations

Model calculations of the expected concentrations of trapped interstellar ^4He have been performed in the framework of the COLLISA project [15]. The calculations were based on the "hot gas" model that describes the distribution of interstellar neutral ^4He atoms at the location of the Earth, taking into account the thermal velocity of the particles in the interstellar medium. The parameters for the interstellar helium used in the model are those given by Witte et al. [28] for the period November 1994 - June 1995 (Table 2).

TABLE 2. Interstellar neutral helium properties. Data from Witte et al. (1996)

	Nov. 94 - Jun. 95
Flow Speed	24.6 ± 1.1 km/s
Flow Direction (ecliptic longitude)	$74.7 \pm 1.3°$
Flow Direction (ecliptic latitude)	$-4.6 \pm 0.7°$
Temperature	5800 ± 700 K
Helium density	$1.4\ 10^{-2} cm^{-3}$
Photoionization	$>1.1\ 10^{-7} sec^{-1}$

The velocity distribution was assumed to be a shifted Maxwellian far upwind from the Sun. At Earth, it was modified by the solar attraction - differently for each day of the year, depending on the Earth's orbital velocity and location. The Mir orientation was known for each exposure, thus the shadowing by the KOMZA walls could reliably be assessed individually for each foil piece.

The model calculations of the expected concentrations of trapped interstellar ^4He on the foil L461-2-1 yield:

$$^4\mathrm{He} = 2.33 \times 10^9 \mathrm{atoms/cm^2}.$$

Such a value is approximately twice as high as the measured concentration:

$$^4\mathrm{He} = \{1.24 \pm 0.11\} \times 10^9 \mathrm{atoms/cm^2}.$$

(corresponding to an average accumulation rate of trapped interstellar atoms of $\{5.0 \pm 0.7\} \times 10^3$ atoms/cm^{2}s). A similar discrepancy factor has been found in previous measurements of COLLISA samples exposed to the interstellar flux [14]. The difference between predicted and measured concentrations could be due to the uncertainties in the determination of the flux of neutrals and, marginally, to that of the foil trapping efficiency.

CONCLUSION

In the framework of the COLLISA project we have determined the helium isotopic ratio in the closest regions of the Local Interstellar Medium using the foil collection technique. The value of the $^3He/^4He$ ratio was obtained from the analysis of one of the foils exposed on Mir during 1996. The present-day isotopic composition of neutral helium in the LISM is lower than that derived from the analysis of pickup ions, but it is consistent with the presolar cloud value, as derived from meteorites and Jupiter's atmosphere. The present-day ^3He abundance, derived from the COLLISA ratio, is therefore consistent with that observed in the presolar cloud, confirming the hypothesis that no substantial increase of ^{3}He occurred in the local interstellar medium during the last 4.6 Gy.

Measurements of more foils exposed to the interstellar flux are planned. They aim at confirming the previously found helium abundances and isotopic ratio and at detecting, for the first time, the interstellar ^{20}Ne/^{22}Ne ratio.

ACKNOWLEDGMENTS

The authors would like to thank all participants in the experiment COLLISA. We are specially grateful to the cosmonauts Sergey Avdeev, Thomas Reiter, Yury Onufrienko and Yury Usachev for exchanging the cassettes in space and to Armin Schaller for the technical support during the mass spectrometric measurements. This work was supported by the Swiss National Science Foundation.

REFERENCES

1. Chiappini, C., and Matteucci, F., Galactic Chemical Evolution (2001), this volume.
2. Tosi, M., "Evolution of D and ^{3}He in the Galaxy", in *The Light Elements and Their Evolution - IAU Symposum*, edited by L. da Silva, M. Spite, and J. de Medeiros, 2000, vol. 198 of *ASP Conference Series*.
3. Rood, R., Steigman, G., and Tinsley, B., *Astrophys. J.*, **207**, L57 (1976).
4. Geiss, J., "Primordial Abundances of Hydrogen and Helium Isotopes", in *Origin and Evolution of the Elements*, edited by N. Prantzos, E. Vangioni-Flam, and M. Cassé, Cambridge University press, Cambridge, 1993, p. 89.
5. Gloeckler, G., and Geiss, J., *Space Sci. Rev.*, **84**, 275 (1998).
6. Rood, R., Bania, T., and Wilson, T., *Nature*, **355**, 618 (1992).
7. Rood, R., Bania, T., Wilson, T., and Balser, D., "The Quest for the Cosmic Abundance of ^{3}He ", in *The Light Element Abundances, Proceedings of the ESO/EIPC Workshop*, edited by P. Crane, Springer, Berlin, 1995, p. 201.
8. Balser, D., Bania, T., Rood, R., and Wilson, T., *Astrophys. J.*, **483**, 320 (1997).
9. Charbonnel, C., *Astrophys. J.*, **453**, L41 (1995).
10. Hogan, C., *Astrophys. J.*, **441**, L17 (1995).
11. Boothroyd, A., and Malaney, R., *astro-ph/9512133* (1995).
12. Galli, D., Stanghellini, L., Tosi, M., and Palla, F., *Astrophys. J.*, **477**, 218 (1997).
13. Chiappini, C., and Matteucci, F., *astro-ph/0004030* (2000).
14. Bühler, F., Bassi, M., Bochsler, P., Eugster, O., Salerno, E., Zastenker, G., Agafonov, Y., Gevorkov, L., Eismont, N., Prudkoglyad, A., Khrapchenkov, V., and Shvets, N., *Astrophys. and Space Sci.*, **274**, 19 (2000).
15. Bassi, M., *COLLISA - An Experiment to Collect Interstellar Neutral Atoms in Metallic Foils on an Earth-Orbiting Satellite*, Ph.D. thesis, Universität Bern (1997).
16. Geiss, J., Bühler, F., Cerutti, H., Eberhardt, P., and Filleux, C., Apollo 16 preliminary science report, Tech. Rep. SP-315, Section 14, NASA (1972).
17. Filleux, C., Mörgeli, M., Stettler, W., Eberhardt, P., and Geiss, J., *Radiation Effects*, **46**, 1 (1980).
18. Klecker, B., and Bothmer, V. (2001), this volume.
19. Busemann, H., Baur, H., and Wieler, R., Protosolar and Circumstellar He Isotopic Ratios Deduced from "phase Q" in Carbonaceous Chondrites (2001), this volume.
20. Mahaffy, P., Donahue, T., Atreya, S., Owen, T., and Niemann, H., *Space Sci. Rev.*, **84**, 251 (1998).
21. Bodmer, R., and Bochsler, P., *Astron. & Astrophys.*, **337**, 921 (1998).
22. Gloeckler, G., Geiss, J., and Fisk, L., Composition of the Local Interstellar Cloud (2001), this volume.
23. Geiss, J., and Reeves, H., *Astron. & Astrophys.*, **18**, 126 (1972a).
24. Geiss, J., and Gloeckler, G., *Space Sci. Rev.*, **84**, 239 (1998).
25. Gautier, D., and Morel, P., *Astron. & Astrophys.*, **323**, L9 (1997).
26. Bochsler, P., Geiss, P., and Maeder, A., *Solar Phys.*, **128**, 203 (1990).
27. Lellouch, E., Bézard, B., Fouchet, T., Feuchtgruber, H., Encrenaz, T., and de Graauw, T., *Astron. & Astrophys.*, **370**, 610 (2001).
28. Witte, M., Banaszkiewicz, M., and Rosenbauer, H., *Space Sci. Rev.*, **78**, 289 (1996).

Composition of the Local Interstellar Cloud from Observations of Interstellar Pickup Ions

G. Gloeckler* and J. Geiss†

*Department of Physics and IPST, University of Maryland, College Park, Maryland 20742, USA
†International Space Science Institute, Hallerstrasse 6, CH-3012 Bern, Switzerland

Abstract. Observations of pickup ions give us a new method to probe remote regions in and beyond the heliosphere. Comprehensive and continuous measurements of H, He, N, O, and Ne especially with the Solar Wind Ion Composition Spectrometer on Ulysses, are providing a wealth of data that are used to infer the chemical and physical properties of the Local Interstellar Cloud (LIC) that surrounds our solar system. We present new results on the elemental composition and ionization state of the LIC. Comparing the abundance of He, N, O and Ne in the present day local interstellar medium to the corresponding solar system abundance of these elements we find that, contrary to expectations of overabundance of heavy elements in the LIC, these two compositions are about the same. This suggests that there has been little or no additional galactic evolution since the formation of the Sun 4.6 billion years ago, or that the materials in the solar system and in the LIC followed different evolutionary paths.

INTRODUCTION

The composition of the local interstellar medium can be studied using mass-spectrometric analysis of samples of the interstellar gas that penetrates deep into the heliosphere. One can analyze this sample at various stages in its evolution, starting with its simplest form, the interstellar gas itself, the atoms that drift with $\sim$25 km/s, a few eV, into the heliosphere. This has been achieved only for He atoms at the present time [35] whose density in the LIC has been determined from a sample of this interstellar gas. More evolved samples of interstellar gas are pickup ions created by ionization from the interstellar gas deep inside the heliosphere. The composition of these ions, immediately accelerated to several keV in the pickup process, can be measured using modern ion composition mass spectrometers such as the Solar Wind Ion Composition Spectrometer (SWICS) flown on Ulysses and ACE. The results presented here are based on analyses of pickup ions measured with SWICS on Ulysses.

The most evolved sample of the interstellar gas are the Anomalous Cosmic Rays (ACRs) which are the interstellar pickup ions further accelerated to hundreds of MeVs or more at the heliospheric termination shock at about 80–100 AU from the Sun. These energetic particles then diffuse back to fill the heliosphere where their composition can be measured [see 3]. The processes of acceleration and solar modulation introduce compositional biases, especially for H and He that must be corrected before interstellar abundances can be determined.

Because the samples of interstellar gas available to us inside the heliosphere are confined to the neutral component (the ionized component cannot enter the heliosphere), we are restricted to those elements that are not already predominantly ionized in the interstellar medium, namely, H, He, N, O, Ne and Ar.

Pickup ions are singly charged and have a characteristic velocity distribution with a sharp drop in density at twice the solar wind speed. Except for He and to some extent Ne, the flux of interstellar pickup ions decreases rapidly inside several AU from the Sun. Thus, in order to separate interstellar pickup ions from other sources of pickup ions one must do the measurements at distances larger than about 3 AU from the Sun. Furthermore, to identify pickup ions in the presence of the far more abundant solar wind, ion mass spectrometers with low background, such as SWICS [13], must be used.

MEASUREMENTS OF PICKUP ION VELOCITY DISTRIBUTIONS

To determine the density of a particular atomic species at the location of the termination shock (TS) from measurements of pickup ions observed in the inner heliosphere requires knowledge of both the total loss rate of these atoms due to ionization in the heliosphere and the pro-

CP598, *Solar and Galactic Composition,* edited by R. F. Wimmer-Schweingruber

duction rate of pickup ions from these atoms. The loss rate determines the shape of the measured velocity distribution of the corresponding pickup ion species [e. g. 10, 18]. The product of the pickup ion production rate and the density of the parent atoms are obtained from the measured pickup ion densities [18]. For data averaged over long time periods (about one year or more) the average value of the pickup ion production rate approaches that of the loss rate of the corresponding atoms.

Based on our analysis of a far more extensive data set than used in our previous publications, we present here improved measurements of the distribution functions of H^+, He^+, N^+, O^+ and Ne^+. Because the observations reported here were taken at large heliocentric distances near the ecliptic plane where the average magnetic field is nearly azimuthal, corrections for anisotropy due to weak pitch angle scattering [e. g. 15, 18, 10] were negligible. Furthermore, here we use ionization rates based on measured solar EUV fluxes [34] and include, for the first time, ionization due to electron impact [28, 29]. We then derive new values for the atomic densities of the neutral components of the interstellar gas at the termination shock and from these estimate elemental abundances in the LIC.

Hydrogen and Helium

In Figure 1 we show the 10-month time-averaged velocity distribution of pickup H^+ measured with SWICS when Ulysses was at low latitudes, near its aphelion at a heliocentric distance R of 5.3 AU. During the entire period Ulysses was in the slow solar wind with an average speed of 382 km/s. No specific time periods (e.g. shock passages) were excluded. This velocity spectrum, as well as those in the subsequent figures, shows the phase space density averaged over the view angles of the instrument [e. g. 14] as a function of W, the ion speed divided by the ambient solar wind speed. The dominant peak centered at $W = 1$ corresponds to solar wind protons. The flat portion of the spectrum between $W \approx 1.4$ and 2 is due to pickup hydrogen. The characteristic drop in density at $W \approx 2$ is less pronounced here compared to the H and He pickup ion distributions observed in the fast solar wind [9], because of the high velocity tails resulting from ions accelerated in the slow solar wind [17].

The loss rate of hydrogen atoms, β_{loss}, determines the shape of the pickup ion distribution, the flat portion of the spectrum [e. g. 18]. The bold curve which best matches the measured spectrum was calculated using β_{loss} of the form $(\beta_o + \beta_e R^{-p})$, where β_o is the sum of the charge-exchange and photoionization rates and β_e, the electron impact ionization rate, all normalized to 1 AU. The value of p was assumed to be 1.1 [28] except for He. Using

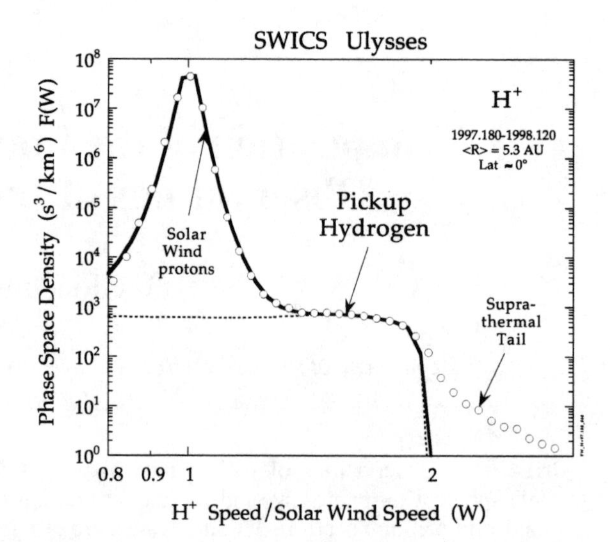

FIGURE 1. Time averaged phase space density of H^+ versus normalized speed W (ion speed divided by average solar wind speed) observed with SWICS on Ulysses in the slow solar wind at heliocentric distance R = 5.3 AU and low latitude. Interstellar pickup hydrogen is in the flat portion of the spectrum between $W \sim 1.4$ and 2. The average total loss rate of atomic hydrogen determines the shape of the pickup ion portion of the distribution. Solar wind protons dominate below $W \approx 1.4$ and accelerated protons form the suprathermal tail. The dotted curve is the model fit of pickup protons and the bold solid curve is the sum of solar wind and pickup protons. The best fit to the data, using the hot model [32], is for an average loss rate (at 1 AU) of atomic hydrogen, $\beta_{loss} = (5.5 + 2.4R^{-1.1}) \times 10^{-7}$ s^{-1} (see text) and μ (ratio of radiation pressure to gravitational force) equal to 1.1. Assuming that the average production rate of pickup hydrogen is equal to β_{loss} gives an atomic hydrogen density at the termination shock, $n_{H^o} = 0.097 \pm 0.015$ cm^{-3}.

the solar wind proton speed and density measured by SWICS, averaged over the time period given in Figure 1, we derive a hydrogen ionization rate from charge exchange with protons of 4.7×10^{-7} s^{-1}. The contribution from photoionization is estimated to be 0.8×10^{-7} s^{-1} [29], for a total $\beta_o = 5.5 \times 10^{-7}$ s^{-1}. To fit the spectral shape of the pickup H^+ distribution, however, required an additional ionization of H by electron impact [28, 29] of $\beta_e = 2.4 \times 10^{-7}$ s^{-1}. Because of the long averaging time (almost one year) used here, we make the reasonable assumption that the production rate of pickup ions is the same as the loss rate of the corresponding interstellar neutrals. From the best fit of the pickup ion model curve to the data of Figure 1 we obtain the neutral hydrogen density at the TS, n_{H_o}, to be 0.097 ± 0.015 cm^{-3}. The derived values of neutral densities at the termination shock, of β_o based on measured solar wind (from SWICS) and solar EUV [34] fluxes, and of β_e adjusted to fit the pickup ion spectral shapes are given in columns 2, 3 and 4 of Table 1.

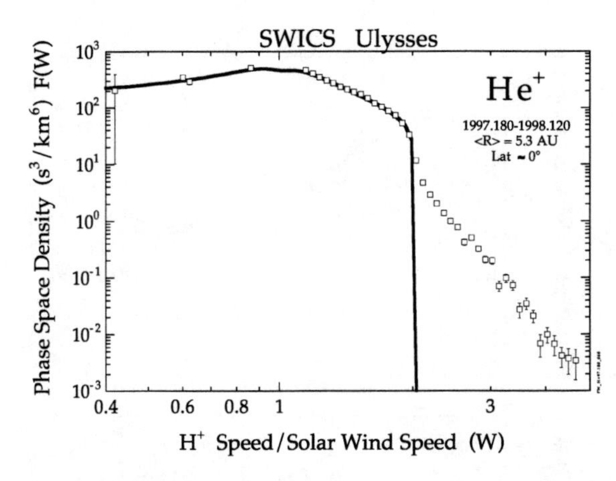

FIGURE 2. Same as Figure 1 but for He^+. The flat portion of the spectrum, below $W \approx 2$ is entirely pickup He^+. Above $W \approx 2$ is the suprathermal tail due to accelerated pickup He^+. The best fit to the data is for $\beta_{loss} = (1.0 + 0.6R^{-0.6}) \times 10^{-7}$ s^{-1}. The atomic He density at the termination shock is 0.016 ± 0.002 cm^{-3}.

The He^+ distribution function, shown in Figure 2, is purely interstellar since, in the solar wind, He is almost entirely doubly ionized. The characteristic density drop at $W \approx 2$ is clearly visible despite the presence of the pronounced suprathermal tail. We used the hot model [32] with a LIC gas temperature of 7000 K [35], inflow speed 25.7 km/s [35], and the measured value for $\beta_o = 1.0 \times 10^{-7}$ s^{-1}. The electron impact ionization rate required to fit the observed spectral shape was $\beta_e = 0.6 \times 10^{-7}$ s^{-1} and $p = 0.6$. With $\beta_{prod} = \beta_{loss}$ we find from the best fit of the model curve to the spectrum of Figure 2 an atomic helium density at the termination shock of 0.016 ± 0.002 cm^{-3}.

Unlike all other pickup ions, about 3% of interstellar pickup He is doubly ionized, predominantly by charge exchange with solar wind alpha particles. Previously, we have used pickup He^{++} measured in the fast solar wind to derive the interstellar neutral He density [16] because its production rate could be determined from the solar wind He flux. However, in the distant slow solar wind, where no anisotropy corrections are required, it is difficult to determine accurately the density of pickup He^{++} both because of prominent suprathermal tails and broader solar wind alpha particle distributions compared to those found in the high-latitude fast wind [e. g. 17].

Our revised value for the atomic H at the location of the termination shock is somewhat smaller than, but within errors of, our previously published density of 0.115 ± 0.025 cm^{-3} [16], which was derived from a different and far shorter data sample. In that case β_{prod} could not be assumed to be equal to β_{loss} and only the charge exchange portion of β_{prod} was measured. The present value of the atomic He density at the TS is nearly

TABLE 1. Densities of atoms at the termination shock (~80 to ~100 AU) and densities of atoms and ions in the local interstellar cloud.

Isotope	Density at Termination Shock (cm^{-3})	$\beta_{loss} = \beta_{prod}$ (10^{-7} s^{-1})*		Filtration Factor †	Neutral Fraction**	Densities in the Local Interstellar Cloud (cm^{-3})				LIC abundance to Protosolar Abundance‡
		β_o	β_e			atoms	ions	grains	total	
^1H	0.097 ± 0.015	5.5	2.4	0.54 ± 0.05	0.75 ± 0.07	0.180	0.059	0	0.239	1
^4He	0.016 ± 0.002	1.0	0.6	1	0.67 ± 0.07	0.016	0.0079	0	0.0239	1
^3He	$(4.0 \pm 1.0) \times 10^{-6}$	1.0	0.6	1	0.67 ± 0.07	4.0×10^{-6}	2.0×10^{-6}	0	6.0×10^{-6}	1.5 ± 0.5
^{14}N	$(7.8 \pm 1.5) \times 10^{-6}$	6.0	2.0	0.79 ± 0.17	0.81 ± 0.08	9.6×10^{-6}	2.2×10^{-6}	1.0×10^{-6}	1.3×10^{-5}	0.67 ± 0.26
^{16}O	$(5.3 \pm 0.8) \times 10^{-5}$	7.0	4.1	0.62 ± 0.06	0.84 ± 0.08	8.5×10^{-5}	1.7×10^{-5}	1.6×10^{-5}	1.2×10^{-4}	0.91 ± 0.29
^{20}Ne	$(7.6 \pm 1.5) \times 10^{-6}$	3.4	3.3	1	0.23 ± 0.02	7.6×10^{-6}	2.5×10^{-5}	0	3.3×10^{-5}	1.15 ± 0.39
^{36}Ar§	$(2.6 \pm 0.8) \times 10^{-7}$	9.0	8.0	1	0.32 ± 0.03	2.6×10^{-7}	5.5×10^{-7}	0	8.1×10^{-7}	1.35 ± 0.56

* Sum of the measured values for the average charge exchange and photoionization rates (β_o), and the electron impact ionization rate (β_e) at 1 AU deduced from spectral shapes of the respective functions.
† Derived using method of [16], with $n(H^o)/n(He^o) = 11.25$ and H/He = 10; see text for details.
** For H and He: derived using method of [16] (see text); for all the rest: based on results of model 19 of [30].
‡ ^1H and ^4He from [1] ^3He from [26] all others from [19], except for O for which the updated value of [20] was used.
§ For Ar the atomic density is derived from its ACR abundance [3]; see text for details.

identical to our previously published value [16]. The errors in the densities of atomic H and He are almost entirely systematic, resulting from uncertainties in the geometrical factor of the SWICS instrument, ionization rates and other model parameters.

Nitrogen, Oxygen and Neon

Other interstellar pickup ions that can be measured with SWICS are nitrogen, oxygen and neon [7]. These heavy interstellar pickup ions can only be measured when the solar wind speed is low, since otherwise a large portion of their spectra falls outside the energy range of SWICS. Argon, the only remaining interstellar pickup ion species expected to be present in the heliosphere is outside the mass per charge range of SWICS and has as yet not been detected. Because the densities of these heavy interstellar pickup ions are at least a factor of 10^2 lower than that of He^+, long accumulation times are required.

For N^+, O^+ and Ne^+ we averaged the data over a much longer (40 months) time period than for H^+ and He^+. This was done to improve the statistical accuracy of the distribution functions of the very rare heavy pickup ions and also to ensure that the average production rate, which must be used to derive atomic densities at the termination shock, was equal to the average loss rate which determines the spectral shapes of the velocity distributions. The velocity distributions of N^+, O^+ and Ne^+ are shown in Figure 3. The spectra of these ions are characterized by a flat portion at $\sim 1.3 < W <\sim 1.8$, a drop in density at $W \approx 2$ and a peak around $W \sim 1$. The flat portion results predominantly from interstellar pickup ions. The peak is due to another source of pickup ions that are believed to be produced close to the Sun and to be primarily "recycled" solar wind implanted in interplanetary dust and then released as slow moving neutrals from which these "inner source" pickup ions are then created [8, 9, 12, 17, 18]. The densities at the termination shock and the ionization rates (β_o and β_e) of interstellar N, O and Ne atoms are obtained from model fits to the flat portions of the corresponding velocity distributions and are listed in columns 2 and 4 of Table 1, respectively. Errors include both statistical and systematic uncertainties.

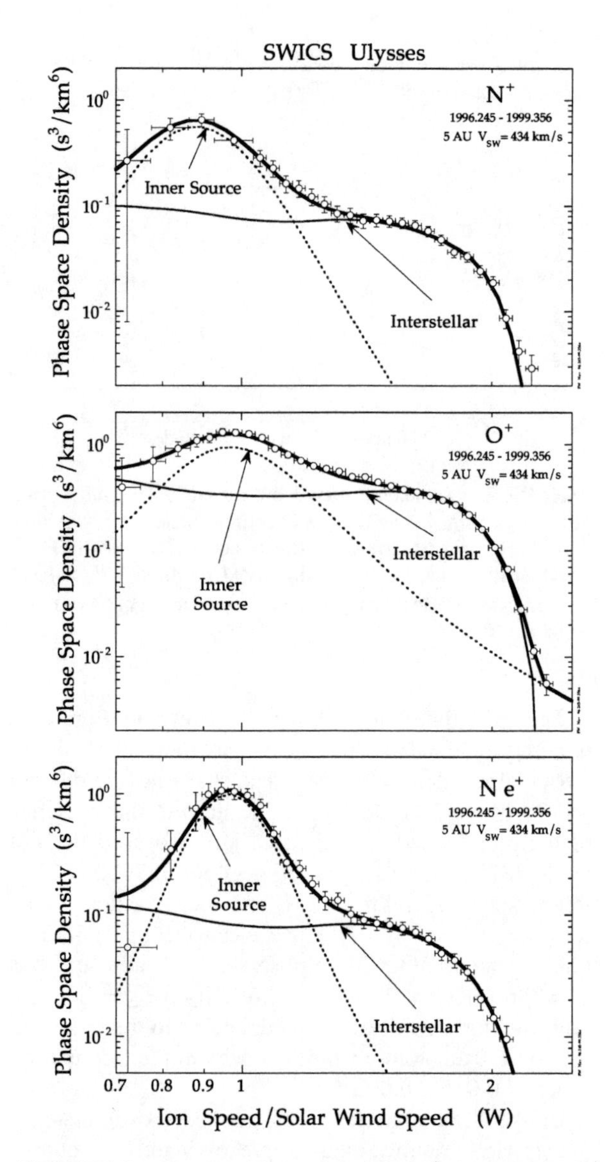

FIGURE 3. Same as Figure 1 but for N^+ (top panel), O^+ (middle panel) and Ne^+ (bottom panel). During the entire 40-month time period Ulysses was near 5 AU in the slow solar wind with an average speed of 434 km/s. The flat portion of the spectrum, between $W \approx 1.3$ and 1.8 is dominated by interstellar pickup ions. The peaks around $W \approx 1$ are due to inner source pickup ions produced close to the Sun. Suprathermal tails due to accelerated pickup ions are not measured because they fall outside the upper energy range of SWICS. The spectra have been smoothed. Model curves (labelled 'Interstellar') were fit to the interstellar pickup ion portions of the spectra to obtain average loss rates and atomic densities at the termination shock.

ACCELERATION EFFICIENCIES OF ANOMALOUS COSMIC RAYS

It is now well established that the source of anomalous cosmic rays (ACRs) are interstellar pickup ions [5] that have been accelerated to hundreds of MeV. It is believed that most of the acceleration takes place at the termination shock [27, 22], although some initial acceleration is likely to occur closer to the Sun [17, 18]. The composition of the ACRs is measured rather precisely [e. g. 3] and can be compared with extrapolated pickup ion densities near the location of the termination shock, as is done in Figure 4. The source of the most abundant elements

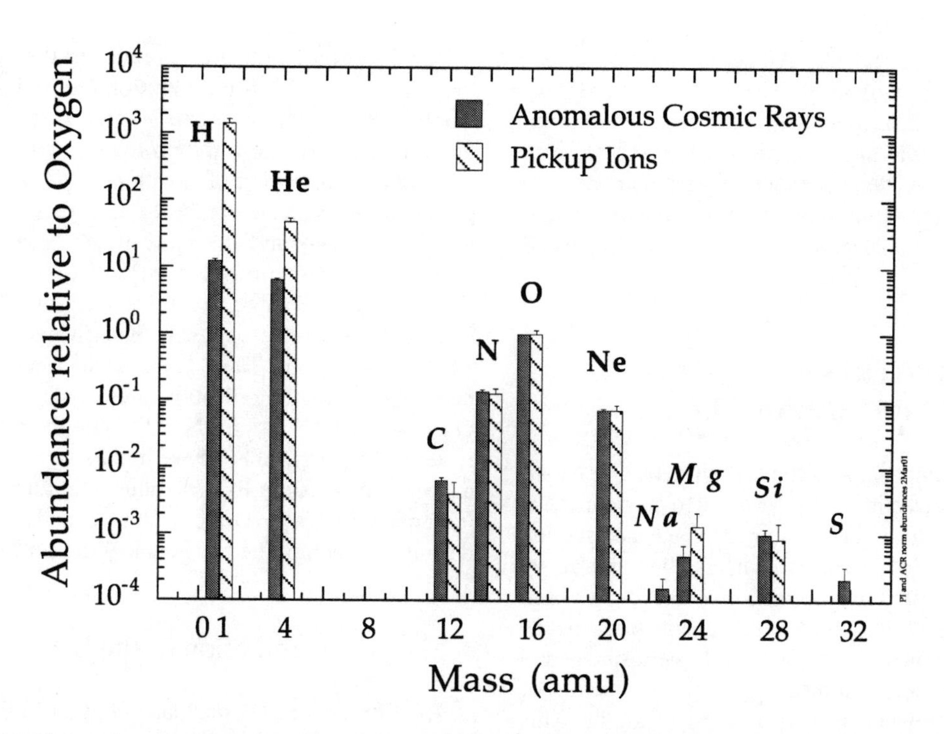

FIGURE 4. Comparison of abundance ratios relative to oxygen of Anomalous Cosmic Rays [e. g. 3] to corresponding ratios (also relative to O) of pickup ions. The abundance ratios for interstellar pickup ions were obtained from the density ratios of interstellar atoms at the termination shock (column 2 of Table 1). The densities of inner source pickup ions (C, Mg and Si) were extrapolated to the termination shock and divided by the interstellar plus inner source pickup ion abundance of O at the termination shock (cf. [12] for details).

in the ACRs (H, He, N, O, Ne, and Ar) is predominantly the interstellar gas. The source of the less abundant ACR elements, the minor ACRs (C, Na, Mg, Si and S) is not known.

Comparing the relative abundance of the ACR with that of interstellar pickup ions reveals nearly equal abundance ratios for N/O and Ne/O, and significantly lower ACR He/O and especially H/O ratios, as compared to corresponding interstellar pickup ion ratios. The relative abundances of the minor ACRs (C, Mg, Si) are similar to the respective ratios of inner source pickup ions extrapolated to the termination shock [12], suggesting that inner source pickup ions may be a significant source of the minor ACRs [17, 3].

The much lower abundance of H and He in the ACRs compared to their pickup ion abundances is believed to be due to the less efficient acceleration (and/or injection) of these light ions compared to the more massive particles. A measure of the relative acceleration efficiency is obtained by dividing the ACR abundance ratios (relative to oxygen) by the corresponding ratios of pickup ion fluxes at the termination shock. The pickup ion fluxes are obtained by model calculations using the densities of interstellar atoms at the termination shock (column 2 of Table 1) and the ion production rates listed in column 3 and 4 of Table 1. The results are shown in Figure 5.

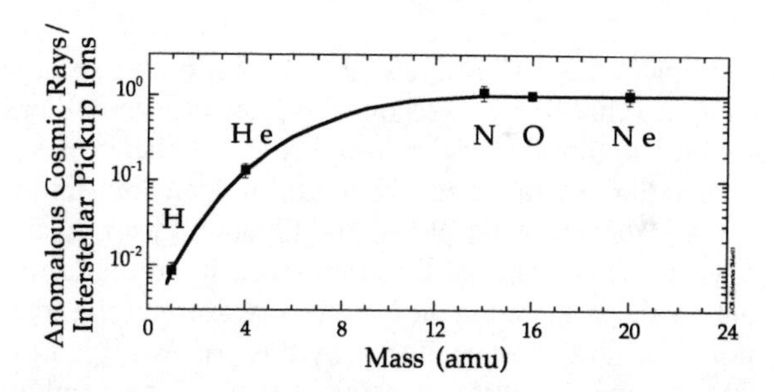

FIGURE 5. Ratios of abundances, normalized to oxygen, of ACR H, He, N and Ne to corresponding ratios of interstellar pickup ion abundances, extrapolated to the termination shock, versus particle mass. The curve drawn through the data points is a measure of the relative acceleration efficiency of the ACRs.

The curve drawn through the data points indicates the relative efficiency of accelerating particles to ACR energies. It appears that ions with mass/charge above ~ 10 are accelerated with about the same efficiency as O, while He and especially H have relatively lower efficiencies. Models for accelerating pickup ions to form the ACRs should be capable of reproducing the results shown in Figure 5.

Because observations of pickup argon are as yet not available, we use its ACR abundance to deduce the atomic density of interstellar Ar. Assuming that the acceleration efficiency of Ar is the same as of O, as suggested in Figure 5, it follows that the Ar/O pickup ion flux ratio is equal to the Ar/O ratio in the ACRs. From this we deduce the atomic density of Ar at the termination shock (column 2 of Table 1) using model calculations with values for its ionization rates (β_o and β_e) listed in columns 3 and 4 of Table 1.

COMPOSITION OF THE LOCAL INTERSTELLAR CLOUD

To derive the composition of the Local Interstellar Cloud (LIC) several corrections must be made to abundances at the termination shock. First, one must make a correction for the reduction in the density of interstellar neutrals by a process called filtration. Second, one must add to each element the density of ionized components that exist in the LIC. Interstellar ions cannot enter the heliosphere and thus cannot be measured directly. Finally, a fraction of some of the elements is likely to exist in grains. This fraction must be added to the ionized and neutral portion to estimate the total density of a given element in the LIC.

Filtration

A fraction of LIC neutrals that have a sufficiently large charge exchange cross section with protons becomes ionized in the filtration region, roughly located between the interstellar bowshock and the heliopause, and thus is excluded from the heliosphere. The filtration factor, F, defined to be the ratio of the atomic density at the termination shock to that in the LIC, is a measure of the fraction of neutrals lost or filtered by this process. It is possible to find the filtration factor of H and O by combining measurements of atomic densities of H, He and O at the termination shock (listed in column 2 of Table 1) with average column density ratios of the neutral components ([H^o/He^o] and [H^o/O^o]) obtained from relevant absorption lines from nearby stars. Since He is not filtered ($F = 1$ for He as well as for Ne and Ar because of their negligibly small charge exchange cross section with ~25 km/s protons [29]) its atomic density in the LIC is the same as its density at the termination shock, $n_{TS}(He^o) = n_{LIC}(He^o)$. The filtration factor for hydrogen, for example, is then $F_H = \{n_{TS}(H)/n_{TS}(He)\}/[H^o/He^o] = 0.54$ using corresponding values from column 2 of Table 1 and $[H^o/He^o] = 11.25$ [12]. The filtration factor of O ($F_O = 0.62$) is derived in a similar manner using $[H^o/O^o] = 2090$ [25].

The filtration factor is also related to the average proton density, $< p >$ in the filtration region [16]. For example, for oxygen $F_O \approx \exp(- < p > \sigma L)$, where L is the dimension of the filtration region and σ the charge exchange cross section of O with protons. With this simple equation we can obtain the nitrogen filtration factor, F_N from F_O and the ratio of the N and O charge exchange cross sections, $\log(F_N) = (\sigma_N/\sigma_O) \times \log(F_O)$. The charge exchange cross section for N is poorly known. We estimate (σ_N/σ_O), using values of β_o for N and O (column 3 of Table 1) from which we subtract the better known photo ionization rates of 4.3×10^{-7} s^{-1} for O and 4.7×10^{-1} s^{-1} for N [29]. With the resulting estimate for (σ_N/σ_O) of ~0.5 we find $F_N \approx 0.79$ and assign a liberal uncertainty to this value. The filtration factors (column 5 of Table 1) are used to calculate the atomic density for each of the 6 elements (column 7 of Table 1).

Ionization in the LIC

The filtration factor may also be used to find the proton density in the filtration region and then in the LIC (see [16] for details) since $< p >= -\log_e(F)/(\sigma L)$. Using the oxygen filtration factor we derived from measurements ($F_O = 0.62$), its charge exchange cross section with protons ($\sigma_N \approx 2.3 \times 10^{-15}$ cm^2 [29]) and an estimate of the size of the filtration region ($L = 155$ AU [21, 23]), gives $< p >= 0.088$ cm^{-3}. Assuming a weak bowshock with a compression ratio of $r \approx 1.5$ [16] we find the LIC proton density, $n_{LIC}(H^+) =< p > /r = 0.059$ cm^{-3}. The ionization fraction of H in the LIC, $\chi_H = n_{LIC}(H^+)/(n_{LIC}(H^+)+n_{LIC}(H^o)) = 0.25$. The ionization fraction of He ($\chi_{He} = 0.33$) is then obtained from the He atomic and the H total (atoms plus ions) densities assuming that the total density ratio of $^1H/^4He = 10$ [1]. The neutral fractions, $(1-\chi)$, of H and He are then 0.75 and 0.67 respectively.

The neutral fractions of all other LIC elements are taken from Slavin and Frisch [30], who calculate the photoionization of interstellar matter within ~5 pc of the Sun by observed radiation sources. We take results for their model #19 (listed here in column 6 of Table 1) because these match best the neutral fractions of H and He we derive. The neutral fractions are then combined with the corresponding atomic densities to give the densities of the ionized component for each element. The results are listed in column 8 of Table 1.

Dust

The amount of interstellar material contained in grains is difficult to estimate. Based on an upper limits given in

[2], we adopt $(13 \pm 13)\%$ as the fraction of oxygen contained in grains. Little nitrogen and negligible amounts of H, He, Ne and Ar are expected to be incorporated in dust. We assume that in the LIC $\sim(8 \pm 8)\%$ of total N and no H, He, Ne and Ar is in grains (see column 9 of Table 1).

COMPARISON OF INTERSTELLAR AND PROTOSOLAR ABUNDANCES

The total LIC densities, i.e. sums of the respective densities of neutrals, of ions and of the portions contained in dust for each of the isotopes are listed in column 10 of Table 1. The ratio of LIC to protosolar abundances is given in the last column of Table 1 and plotted in Figure 6. For the protosolar cloud (PSC) values of the helium isotopes we take $^1H/^4He$ to be 10 [1] and $^3He/^1H$ = $(1.66 \pm 0.05) \times 10^{-5}$ measured in the Jovian atmosphere by the Galileo Probe [26]). The abundance ratios of $^2H/^1H$ in the protosolar cloud and in the LIC are from [11] and [24] respectively. For the protosolar abundance of all other elements we use solar system abundances [19, 20] since these are believed to have remained unchanged over the lifetime of the Sun.

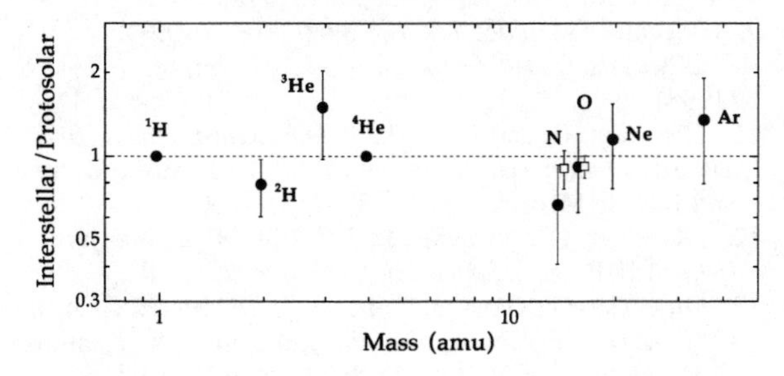

FIGURE 6. Interstellar abundances relative to protosolar abundances, both normalized to hydrogen, versus particle mass. *Filled circles*: the LIC (cf. column 11 of Table 1). *Open squares*: interstellar diffuse clouds (IDC) [31]. The average IDC gas phase abundances do not include possible contributions from dust, except for oxygen whose abundance was increased by 56%, an amount assumed to be incorporated in dust [31].

The data shown in Figure 6 indicate that, within estimated uncertainties, the compositions of the present day interstellar medium and the ~4.6 Gyrs older PSC are about the same. Contributions to the errors of the LIC abundance values for ^3He and heavy elements come from statistical and systematic errors listed in Table 1. The total uncertainty in the ratio of LIC abundance to protosolar abundance was obtained by the method of propagation of errors combining errors from Table 1 and those of [24] with errors given in [19, 20, 26]. For nitrogen both the filtration factor and the amount incorporated into dust is not well known, but each of the resulting corrections is relatively small and thus contributes little to the total error. The mean N/H in the LIC is about 67% of the PSC value. Oxygen is the most abundant of the heavy pickup ions and its value at the termination shock is determined with the highest accuracy. Its filtration and ionization fractions as well as the upper limit of the fraction of O contained in grains are all relatively small and reasonably well known. Thus, the LIC abundance of O is now fairly well established and is comparable to its PSC value. Ne and Ar are not filtered nor incorporated in grains. However, the large statistical errors and the high degree of ionization of these noble gases coupled with uncertainties in current estimates of their neutral fractions prevent an accurate determination of their total LIC abundance at the present time. C, Na, Mg, Al, Si, P, S, Ca and Fe are almost completely ionized in the LIC [30] and the small fraction that remains neutral is likely to be highly filtered. It would be extremely difficult if not impossible to determine the abundance of these elements from measurements of pickup ions or ACRs.

Comparing the interstellar diffuse cloud (IDC) abundances within ~1500 pc [31] with the PSC composition (open squares of Figure 6) one finds that IDC N/H and O/H are each slightly lower than the are in the PSC, and, within errors, the same as in the LIC. This suggests that the composition of the LIC may be representative of the gas composition in the interstellar medium within ~1500 pc.

INTERSTELLAR MAGNETIC FIELD AND TERMINATION SHOCK LOCATION

The density of LIC H and He ions and atoms determines the pressure that the LIC exerts on the heliosphere. Following the procedure outlined in [16] we obtain new estimates of the LIC magnetic field and the heliocentric distance of the termination shock at the latitude of Voyager 1 using values for the LIC H and He densities reported here and the one-year average solar wind speed and density obtained with SWICS on Ulysses during 2000. With values for H and He from Table 1 we find that the LIC magnetic field is $\sim 2\,\mu$G, the bow shock compression ratio is ~1.5, the electron density is ~ 0.067 cm^{-3}. Using pressure balance [16] we find the most probable distance to the termination shock at the stagnation point to be 85 AU in remarkable agreement with our previous estimate [16]. This would place the shock location at the 33.7° latitude of Voyager 1 about 7 AU further at ~92 AU [e. g. 36].

Voyager 1, now at 82 AU has not yet detected the shock. At the stagnation point the TS distance would then be >75 AU. This places upper limits on the hydrogen ionization fraction and on the magnetic field strength in the LIC of <0.31 and <3.1 μG respectively. Assuming that the LIC magnetic field is negligibly small ($< 0.3\,\mu$G) places an upper limit on the location of the termination shock of ~97 AU at the stagnation point or about ~104 AU at the position of Voyager 1.

DISCUSSION

Comparing the element abundances in the LIC with those inferred for the protosolar cloud (last column of Table 1 and Figure 6), we find that the LIC is not — as would be expected — enriched in the heavy elements with respect to the 4.6 billion years older protosolar cloud. This shows that at least in the regions where these clouds have been travelling, galactic evolution was slow during the last 30 percent in the life of the galaxy. On the other hand, the isotopes of hydrogen and helium show the general trend that is expected in our epoch: in the LIC, D/H is decreased and ^{3}He/^{4}He is increased relative to the values inferred for the PSC [e. g. 24, 6, 11].

The galactic evolution of element and isotope compositions is mainly determined by two processes. These are (a) nucleosynthesis in stars by which the heavier element content in the interstellar medium is increased with time, and (b) infall of relatively unprocessed material into the galactic disk. Both these processes are irregularly distributed in space and time. Thus, the evolution of nuclei cannot be expected to be monotonic and simultaneous in all places of the galactic disk, even if they are presently adjacent. The relative velocity between LIC and solar system is ~26 km/s [35], corresponding to 10 kps in 4.6×10^9 years. This underlines that in spite of their present proximity, there is no special genetic relationship between the LIC and the PSC, and it is perhaps not particularly odd that the solar system was somewhat less endowed than its present galactic neighborhood with material processed in stars. On the other hand, deuterium follows the expected trend. This is not necessarily a contradiction, because the D abundance in a present-day galactic sample depends strongly on the infall of relatively unprocessed material [33], and the infall rate is probably quite heterogenous.

ACKNOWLEDGMENTS

We gratefully acknowledge the numerous essential contributions of the many individuals at the Space Physics Group of the University of Maryland, the University of Bern, the Max-Planck-Institute für Aeronomie and the Technical University of Braunschweig which assured the success of the SWICS experiments on Ulysses. We thank C. Gloeckler for her help with data reduction, Ursula Pfander for preparing the camera-ready manuscript, Vladislav Izmodenov for stimulating discussions and Priscilla Frisch and Jonathan Slavin for providing us the neutral fractions. This work was supported in part by NASA/JPL contract 955460 and by NASA/Caltech grant NAG5-6912.

REFERENCES

1. Bahcall, J. N. and Pinsonneault, M. H., *Rev. Modern Phys.* **67**, 781, (1995).
2. Cardelli, J. A., Meyer, D. M., Jura, M., and Savage, B. D., *Astrophys. J.* **467**, 334–340, (1996).
3. Cummings, A. C., Stone, E. C. and Steenberg, C. D., *Proc. 26th Intl. Cosmic Ray Conf. (Utah)* **7**, 531–534, (1999).
4. Dupuis, J., Vennes, S., Bowyer, S., Pradhan, A. K., and Thejll, P., *Astrophys. J.* **455**, 574–589, (1995).
5. Fisk, L. A., Kozlovsky, B. and Ramaty, R., *Astrophys. J.* **190**, L35–L37, (1974).
6. Geiss, J. and Gloeckler, G., *Space Sci. Rev.* **84**, 239–250, (1998).
7. Geiss, J., Gloeckler, G., Mall, U., von Steiger, R., Galvin, A. B., and Ogilvie, K. W., *Astron. Astrophys.* **282**, 924–933, (1994).
8. Geiss, J., Gloeckler, G., Fisk L. A., and von Steiger, R., *J. Geophys. Res.* **100**, 23,373–23,377, (1995).
9. Gloeckler, G., *Space Sci. Rev.* **89**, 91–104, (1999).
10. Gloeckler, G. and Geiss, J., *Space Sci. Rev.* **86**, 127–159, (1998).
11. Gloeckler, G. and Geiss, J., in *The Light Elements and Their Evolution*, IUA Symposium 198 (L. da Silva, M. Spite and J. R. de Medeiros, eds.), 224–233, (2000).
12. Gloeckler, G, and Geiss, J., 34th ESLAB Symposium (Marsden, R., ed.), *Space Sci. Rev.*, in press (2001).
13. Gloeckler, G., Geiss, J., Balsiger, H., Bedini P., Cain, J. C., Fischer, J, Fisk, L. A., Galvin, A. B., Gliem, F., Hamilton, D. C., Hollweg, J. V., Ipavich, F. M., Joss, R., Livi, S., Lundgren, R., Mall, U., McKenzie, J. F., Ogilvie, K. W., Ottens, F., Rieck, W., Tums, E. O., von Steiger, R., Weiss, W. and Wilken, B., *Astron. Astrophys. Suppl. Ser.* **92**, 267–289, (1992).
14. Gloeckler, G., Geiss, J., Balsiger, H., Fisk, L. A., Galvin, A. B., Ipavich, F. M., Ogilvie, K. W., von Steiger, R. and Wilken, B., *Science* **261**, 70–73, (1993).
15. Gloeckler, G., Schwadron, N. A., Fisk L. A. and Geiss, J., *Geophys. Res. Lett.* **22**, 2665–2668, (1995).
16. Gloeckler, G., Fisk, L. A., and Geiss, J., *Nature* **386**, 374–377, (1997).
17. Gloeckler, G., Fisk, L. A., Zurbuchen, T. H. and Schwadron, N. A., in *Acceleration and Transport of Energetic Particles Observed in the Heliosphere, AIP Conference Proceedings* **#528**, (ACE-2000 Symposium, (R. A. Mewaldt, J. R. Jokipii, M. A. Lee, E. Möbius, and T. H. Zurbuchen, eds.), 221, (2000).
18. Gloeckler, G., Geiss, J., and Fisk, L. A., in *The Heliosphere near Solar Minimum: the Ulysses Perspectives*, (A. Balogh, R. G. Marsden, and E. J. Smith, eds.), Springer-Praxis, Berlin, 287–326, (2001).

19. Grevesse, N. and Sauval, A. J., *Space Sci Rev.* **85**, 161–174, (1998).
20. Holweger, H., in *Solar and Galactic Composition*, edited by R. F. Wimmer-Schweingruber, *AIP Conference Proceedings*, AIP, Woodbury, NY, in press (2001).
21. Izmodenov, V. V., Geiss, J., Lallement, R., Gloeckler, G., Baranov, V. B., and Malama, Y. G., *J. Geophys. Res.* **104**, 4731–4741, (1999).
22. Jokipii, J. R., *J. Geophys. Res.* **91**, 2929–2932, (1986).
23. Linde, T. J. and Gombosi, T. I., *J. Geophys. Res.* **105**, 10411–10418, (2000).
24. Linsky, J. L. and Wood, B. E., in *The Light Elements and Their Evolution*, IUA Symposium 198 (L. da Silva, M. Spite and J. R. de Medeiros, eds.), 141–150, (2000).
25. Linsky, J. L., Diplas, A., Wood, B. E., Brown, A., Ayres, T. R., and Savage, B. D., *Astrophys. J.* **451** 335–351, (1995).
26. Mahaffy, P. R., Donahue, T. M., Atreya, S. K., Owen, T. C., and Niemann, H. B., *Space Sci. Rev.* **84**, 251–263, (1998).
27. Pesses, M. E., Jokipii, J. R. and Eichler, D., *Astrophys. J.* **246**, L85-L88, (1981).
28. Rucinski, D., and Fahr, H. J., *Astron. Astrophys.* **224**, 290–298, (1989).
29. Rucinski, D., Cummings, A. C., Gloeckler, G., Lazarus, A. J., Möbius, E. and Witte, M., *Space Sci Rev.* **78**, 73–84, (1996).
30. Slavin, J. D. and Frisch, P. C., *Astrophys. J.*, in press (2001).
31. Sofia, U. J., *Proceedings of the Joint SOHO/ACE Workshop 2001 on Solar and Galactic Composition*, these proceedings, (2001).
32. Thomas, G. E., *Ann. Rev. Earth Planet. Sci.* **6**, 173–204, (1978).
33. Tosi, M., *Space Sciences Series of ISSI*, Vol. **4** and *Space Science Rev.* **84**, p. 207–218, (1998).
34. Viereck, R., McMullin, D. R., Puga, L., Tobiska, W. K., Webber, M., Judge, D. L., *Geophys. Res. Lett.* **28**, 1343–1346, (2001).
35. Witte, M., Banaszkiewicz, M. and Rosenbauer, H., *Space Sci Rev.* **78**, 289–296, (1996).
36. Zank, G. P. and Pauls, H. L., *Space Sci Rev* **78**, 95–106, (1996).

New Concept for the Measurement of Energetic Neutral Atom Composition and the Imaging of their Sources

J. W. Keller*, M. A. Coplan†, J. E. Lorenz** and K. W. Ogilve*

*Goddard Space Flight Center, Greenbelt, MD 20771
†IPST, University of Maryland, Collage Park, MD 20714
**Northrop Grumman - Litton Adv. Syst., College Park, MD 20740

Abstract. Enrgetic neutral atoms created by charge neutralization of ions convey information about the region from which they originate. Free of magnetic and electric fields the neutrals travel in straight lines and can be observed far from their source. The neutrals provide the means for imaging planetary magnetospheres and studying the interstellar gas and the neutral solar wind. Neutral atoms with energies above a few keV can be observed with instruments that are largely modifications of devices already designed to detect ions. For neutrals with lower energies, other techniques are required. We present a new concept for detecting and imaging neutral atoms below 1 keV energy that has the potential for improving detection efficiency and resolution by a factor of ten over existing methods. The proposed instrument will use an excited-atom electron-attachment cell for high neutral to ion conversion efficiency. A novel method to contain the gas but allow incident neutral atoms and converted ions to pass into and out of the cell with high efficiency is presented. Applications of the detector are discussed.

INTRODUCTION

Much of our knowledge of interplanetary space and the magnetospheres of planetary bodies comes from the measurement of charged particles and photons. To a great extent the neutral third component is missing. This is primarily due to technological difficulties associated with low flux and inefficient detectors for neutral particles. Recently, as new detection schemes have become available,[1][2] measurements of interplanetary and interstellar neutral atoms increased. The emphasis has been of fluxes of H and to a lesser extent He. These measurement techniques are largely modifications of methods for the detection of ions. For energies above 3 keV, neutrals can be ionized by passage through foils and for above 0.1 MeV, total energies can be measured using solid state detectors. For lower energy neutrals, other techniques are required. This energy range is important since the cross section for the creation of neutral atoms is largest in this range while it represents the energy regime of low-temperature geophysical plasmas.

We are presenting a new concept for detecting and imaging neutral atoms with energies between 10 eV to 10 keV. The proposed instrument will use a laser-excited alkali-vapor electron-attachment cell for high neutral to ion conversion. Neutral atoms are ionized by extracting loosely bound electrons from the metal atoms on collision. Containment of the conversion atoms in the cell is done with high speed rotating vanes. Incident neutral atoms and converted ions above a few eV in energy pass into and out of the cell with high efficiency. The measurement has the potential for improving detector efficiency and resolution over existing methods.

NEUTRAL ATOM CONVERSION

The proposed low energy neutral atom imaging technique is based on the conversion of neutral atoms to negative ions and subsequent analysis and detection of the negative ion. With the proposed technique, the neutral is converted to a negative ion by passing it through a vapor of excited alkali metal atoms contained in an electron attachment cell. As the atom passes through the cell it extracts an electron from one of the metal atoms and is converted to a negative ion with negligible change in energy or direction of travel. The energy and direction of travel of the negative ion after it leaves the cell are determined with conventional charged particle optics. The quality of the measurement depends on the number of particles collected by the detector while the time resolution depends on the period necessary to collect sufficient events for an image. For the method that is proposed the

CP598, *Solar and Galactic Composition*, edited by R. F. Wimmer-Schweingruber

number of imaging events per second is given by

$$I = I_0 \varepsilon \sigma n \ell \rho \qquad (1)$$

where, I_0 is the incident neutral atom rate, σ, is the cross section for the conversion of neutral atoms to negative ions in the conversion cell, n is the density of alkali metal atoms in the cell, ℓ is the length of the conversion cell, ρ is the duty cycle (~ 50% in our application), and ε is the efficiency with which the converted atoms can be detected. In order to achieve 1% overall detection efficiency, the product $n\ell\sigma\rho$ should exceed 0.01. Ion detection efficiency with current charged particle detectors is near unity and practical values of cell path length are from 5 to 10 cm. This means that σn must be in the range of 0.002 cm^{-1} so that a cross section of the order of 10^{-15} cm^2 and a target density of ~10^{12} atoms/cm^3 is required.

Alkali metal atom densities in the range 10^{12} atoms/cm^3 are not difficult to achieve in a closed cell, but in order to contain the atoms at these densities while allowing unimpeded passage into the cell for the incident neutral atoms and an exit for the converted negative ions requires new technologies. As discussed below, the requirements of an electron attachment cell that is opaque to alkali metal atoms and transparent to fast neutral atoms and negative ions can be met with what we call a turbotrap that uses the operating principles of a turbomolecular pump. Together, laser excitation and a turbotrap provide a system fully capable of imaging neutral atoms with high efficiency in the energy range from 10 eV to 10 keV.

A potential energy diagram illustrating the mechanism of electron attachment is shown in Figure 1. The potential curves for the reactant, neutral (N) and alkali metal atom (M), and products, negative ion (N^-) and ionized alkali (M^+) are shown. As N and M approach along the lower curve there is a finite probability that a transition is made to the upper curve in the region of the curve crossing at R_x. The cross section for negative ion formation is roughly proportional to R_x. If, through optical excitation, M is excited to M^*, the resulting reactant curve ($N + M^*$) is raised, increasing the value of the internuclear separation where crossing occures to R_x^*. From another point of view, the loosely bound outer electron of the alkali metal atom can be considered to be quasi-free and an incident neutral atom sees what amounts to a cold plasma from which electrons can attach themselves to the passing neutral atoms. The metal ion core provides the third body necessary for momentum conservation. Because electron transfer takes place at long range, scattering is minimized and the converted negative ion preserves the angle and energy information originally carried by the neutral.

Alkali metal vapor has been used both to increase the yield of negative ion production and as a tool for making

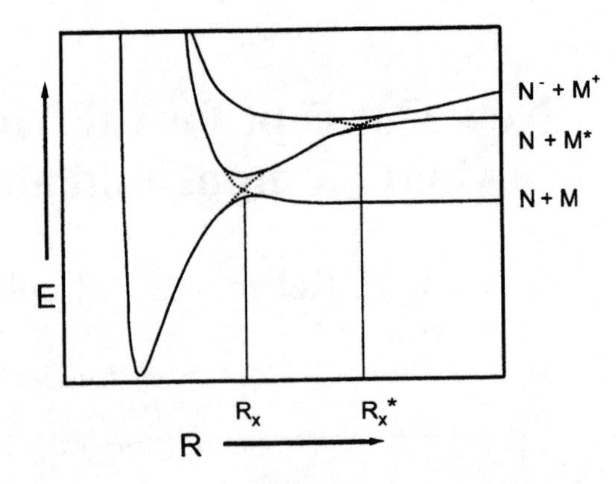

FIGURE 1. Potential energy curves for an alkali metal - neutral system. The curves represent the Coulombic ion-pair potential and the two neutral curves that arise from the interaction of neutral with both the ground and excited state metal atoms. Avoided crossing between static curves of the same symmetry occurs at R_x. and R_x^*.

polarized atoms and ions. In laboratory plasma physics, Brisson et al.[3] developed a low energy atom analyzer using a cesium heat pipe to measure the temperature of a laboratory plasma. In Brisson's example, the concerns addressed are similar to the space plasma measurements; negative ion formation in the plasma providing a tool for plasma temperature determinations. Evidence for electron attachment cross section enhancement by laser excitation is given by the calculations by Vora et al. [4] for the production of O^- in collisions between O and Cs. Cross section values of 10^{-15} cm^2 were obtained at 1 keV incident atom energy. When the Cs was excited, the calculated cross-section increased to 1.5 x 10^{-14} cm^2 at 10 eV, remaining above 10^{-15} cm^2 for energies up to 10 keV. Equivalent cross sections for the formation of hydrogen negative ions are smaller by an order of magnitude however for most observations we anticipate much higher fluxes of neutral hydrogen.

To fully take advantage of the increase in the cross-section with excitation it is necessary that a significant population of the metal atoms in the conversion cell be excited. This can be accomplished through the use of low-power diode laser. Off-the-shelf lasers are available that are capable of accessing the first excited states of cesium or rubidium. A single laser is insufficient to maintain a significant population excited atoms over the full dimensions of the electron-attachment cell (at least without the introduction of a complex optical system). However the large oscillator strengths of the $^2P_{3/2,1/2} - ^2S_{1/2}$ transitions result in the reabsorption of photons from fluorescing atoms by nearby ground state atoms.

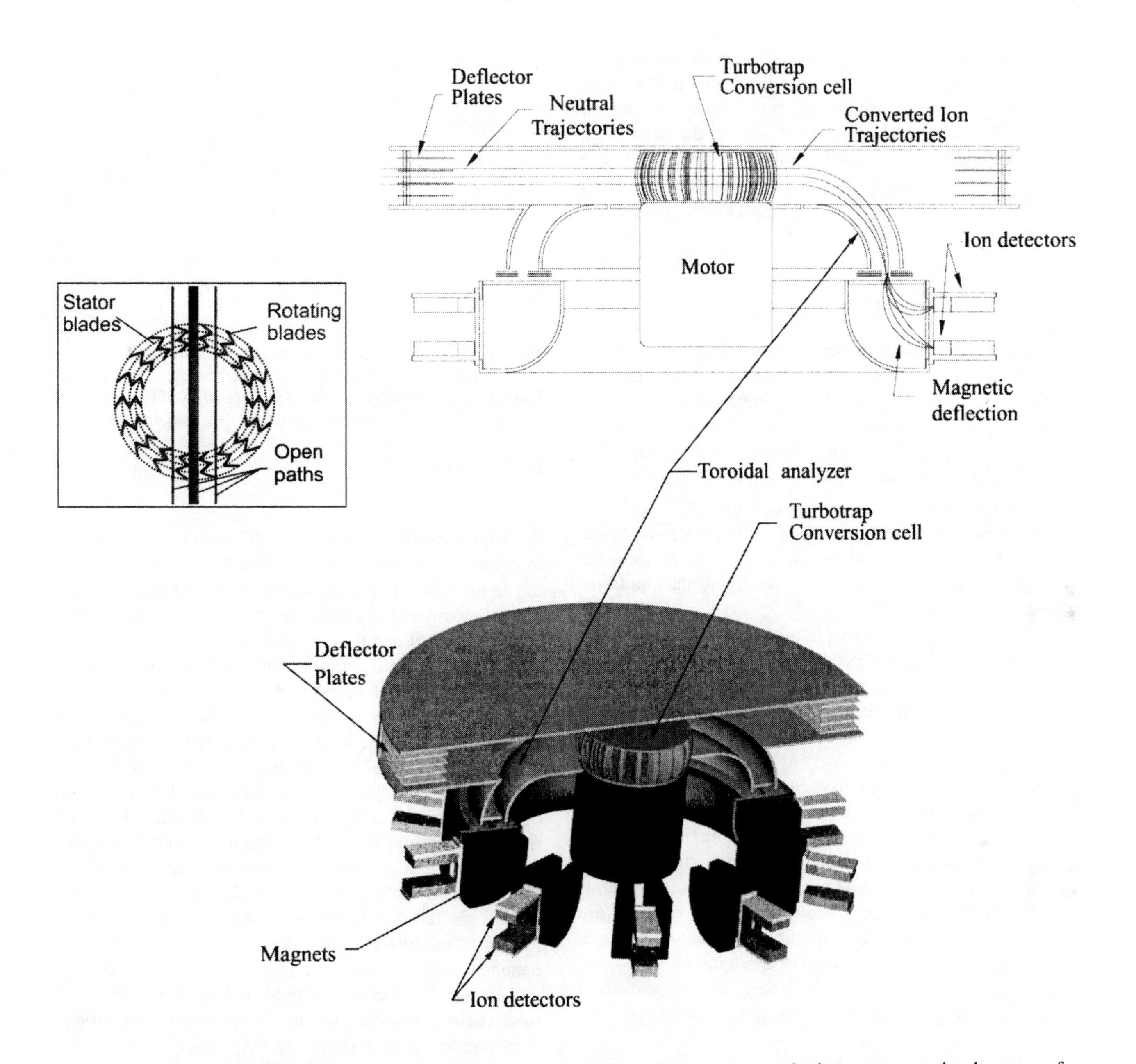

FIGURE 2. *Conceptual* design of a neutral atom detector system. Neutral atoms enter the instrument, passing by a set of electrostatic deflector plates that prevent charged particles from entering. Atoms continue through the turbotrap and emerge as negative ions. The ions are deflected into a electrostatic analyzer and are separated according to their energy. Subsequently magnetic deflection is used to determine the mass of the atom. The inset illustrates the turbotrap principle. Gaseous Cs atoms are confined to center of the cell by the action of rotating blades that sweep escaping atoms back into the trap. Nonmoving stator blades channel reflected atoms back into the trap. The blades are arranged in an open configuration that allows passage of energetic atoms.

In a phenomenon well know to the designers of resonance radiation lamps, the light is trapped in the vapor and only escapes through diffusion outward or when a particular atom fluoresces in the wing of the Lorentzian profile outside the Doppler width of the transition. The effective adsorption cross section for narrow bandwidth radiation tuned to the center of the Doppler broadened profile is given by, $\sigma_{optical}$ where

$$\sigma_{optical} = \frac{2}{\Delta\nu_d}\sqrt{\frac{\ln 2}{\pi}}\frac{\lambda_0^2}{8\pi}\frac{g_2}{g_1\tau} \tag{2}$$

where λ_0 is the wavelength of the transition, $\Delta\nu_d$ is the Doppler width, τ is natural lifetime of the excited state, and g_2 and g_1 are the degeneracies of the upper and lower levels respectively. Taking cesium, the room temperature Doppler width of the $^2P_{1/2} - ^2S_{1/2}$ transition at 8944 Å is 354 MHz with a lifetime of 3.8 x 10^{-8} s which results in a cross-section of 5.6 x 10^{-10} cm^2. For atom densities greater than 10^{11} cm^{-3} the mean free path of a photon will be much smaller than the dimensions of the electron-attachment cell, satisfying the condition necessary for radiation trapping.

The laser can also serve to monitor the density of the vapor, a requirement needed for calibration purposes. This can be accomplished by measuring the absorption of light through the vapor or by measuring the apparent fluorescence lifetime which has been extended by the radiation trapping process.[5][6]

Both hydrogen and oxygen have electron affinities high enough to assure significant conversion efficiency even without laser excitation. However for systems with low electron affinities, it is worth mentioning that this instrument concept could be inverted to observe neutral atoms with low ionization potentials. For example the Io torus could provide a significant source of energetic sodium atoms. By replacing cesium with a highly electronegative gas (i.e. I_2, UF_6) efficient positive ionization of sodium via electron transfer might be achieved.

THE TURBOTRAP

The electron attachment cell containing the alkali metal vapor must be opaque to the alkali metal atoms contained inside and transparent to the energetic neutrals and negative ions that must enter and leave. To confine the alkali metal atoms we exploit the large difference in velocity between the neutrals and negative ions and the thermal alkali metal atoms. To illustrate, at 60° C, 99% of cesium atoms in a Maxwell distribution have speeds below 0.5 km/s, while a 10 eV oxygen atom is moving considerably faster at 11 km/s. We propose a charge transfer cell that will be surrounded with rotating blades on a number of concentric circles. The rotational velocity (of order 30,000 rpm) will be sufficiently high to sweep alkali metal atoms moving toward the ends of the cell back into the center while at the same time slow enough that fast neutral atoms and negative ions can pass through unimpeded. This is illustrated in figure 2.

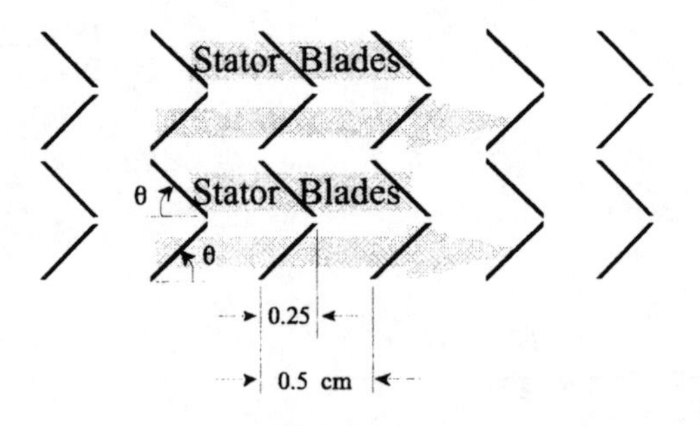

FIGURE 3. A model of a turbo trap was investigated using a Monte Carlo method to explore the trapping capabilities. Particles with a random trajectories were allowed to propagate through the model illustrated here and their eventual fate (escaped or trapped) was recorded. For simplicity, a linear geometry was investigated.

We have explored the turbotrap concept using a Monte Carlo simulation of particle trajectories. Our goal was explore the feasability of containing a thermal gas by the mechanical action of blades moving at turbomolecular speeds sweeping back escaping atoms. To simplify the calculation we adopted a linear geometry for the model of turbotrap illustrated in figure 3. Initial velocities were randomly chosen from a Maxwellian distribution and the particles were allowed to propagate through the trap. On collision with the blades the particles were either specularly reflected or reflected in a cosine distribution with respect to the blade normal depending on an arbitrarily defined sticking coefficient. This latter case corresponds to the possibility that the particle sticks to the surface and subsequently desorbs from it. The coding for the model is incomplete as it is currently two-dimensional but the preliminary results are encouraging. For an arrangement that includes two sets of rotating blades moving at 166 m/s and two sets of non-moving stator blades we find that over 98% of the particles are returned to the trap. Without the trap the flow of cesium vapor from the cell will be 40 g/cm^2/year and with it the flow can be reduced to 0.8 g/cm^2/year. The small amount of escaping cesium can be trapped on cold surfaces outside the trap to prevent it from contaminating the instrument electronics and the spacecraft. Surfaces inside the turbotrap will be coated with cesium, which can become a source of negative ions that may escape the trap. However background ions will be thermal (> 0.03 eV) below the pass energy of the subsequent electrostatic analyzer. In principle, higher detection efficiencies can be achieved using higher densities but this may limit the lifetime of the detector, more cesium would excape and the cell would have to run at higher temperatures.

For a 10 cm diameter set of blades 166 m/s corresponds to approximately 32,000 rpm. This rate is well within recent developments in spacecraft technology for reaction wheels and gyroscopes and the power required for the turbotrap is less than is required for reaction wheels because of the smaller moment of inertia.

APPLICATIONS

The turbotrap and gas-phase electron-attachment cell concept was developed to increase the sensitivity and resolution of neutral atom imaging detectors. If sucessfully developed, this next generation detector may have a number of specific applications, including those listed below.

- Planetary Magnetospheres - The Earth's magnetosphere is a source of energetic hydrogen and oxygen, both of which are electronegative and readily form negative ions on collision with alkali metal vapors. Recent observations include high energy neutral atoms by the Cassini spacecraft of Jupiter and of the Earth by the IMAGE spacecraft.[7]
- Local Interstellar Medium - Electronegative neutrals from the local interstellar medium which penetrate the heliosphere and survive transport to the detector without photoionization can be monitored directly with an electron-attachment cell based detector. Atoms with finite electronegativities such as carbon or oxygen are candidates for detection, although it may be necessary for the measurement to take place beyond 1 a.u to assure that the neutral flux will not be significantly diminished through photoionization. The interstellar medium moves at a velocity of 25 km/s with respect to the heliosphere which puts the velocity of neutral atoms within the regime of a low-energy neutral atom detector.
- Neutral Solar Wind - Neutral atoms in the solar wind occur almost exclusively as a result of charge exchange between solar wind ions and interstellar neutral atoms, and with material originating from outgassing of interstellar grains.[8][9] Another source of neutrals that can charge exchange with the solar wind is the geocorona, or analogous gas cloud around a gravitating body. Measurements of the NSW can be made from low Earth orbit since neutral atoms penetrate Earth's magnetic field. The flux of neutral H in the solar wind is approximately 10^{-4} of the flux of the wind itself, and has been observed recently using the IMAGE spacecraft.[10]
- Molecules - Many large molecules and small free radicals are electronegative. Although their velocity distribution in space can be expected to be thermal, the velocity necessary to pass though a turbotrap may arise from the relative motion of the spacecraft as it passes through the source. For example the Giotto spacecraft encountered the comet Halley at 70 km/s. Although restricted to electronegative molecules, the alkali-metal-vapor electron-attachment cell has the potential of improving dramatically the efficiency of a mass spectrometer that uses electron impact to ionize neutral molecules.

DISCUSSION

The turbotrap, if successfully developed, will serve as an efficient ionizer of energetic neutral atoms. As such it will be one component in a larger instrument system for detecting an analyzing these atoms. We illustrate one possible configuration in figure 2. This design concepts places the electron-attachment cell in the center of a toroidal electrostatic analyzer.[11] It would provide a field of view in the plane perpendicular to the axis of rotation of the turbotrap and would be suited for imaging planetary magnetospheres. Alternative designs might incorporate a linear geometry reminiscent of a jet engine which may be more suitable for directed neutral atom flows such as the neutral solar wind or the neutral interstellar medium.

A prototype of the turbotrap is currently under development by us. Our aim is to demonstrate the ability to confine a gas in the trap for extended periods. Engineering issues such as motor development will be addressed although an eventual flight instrument may require a magnetic bearing motor which is beyond the scope of our current effort. On a rotating spacecraft, the angular momentum arising from a high-speed rotating motor must be canceled. A duel motor design that replaces the stator with blades rotating in the opposite direction will serve to cancel the net angular momentum imparted to the spacecraft. This has the added advantage of reducing the required rotational speed of individual motors by a factor of two, helping to alleviate the increased complexity of a duel motor design.

REFERENCES

1. Gruntman, M., *Rev. Sci. Inst.*, **68**, 3617(1997).
2. Wurz, P., in *The Outer Heliosphere: Beyond the Planets*, K. Scherer, H. Fichtner, and E. Marsch, ed.s, Copernicus Gesellschaft e.V., Katlenburg-Lindau, Germany, 2000, pp. 251.

3. Brisson, D., Baity, F. W., Quon, B. H., Ray, J. A., and Barnett, C. F., *Rev. of Sci. Inst.* **51**, 511(1980).
4. Vora, R. B., Turner, J. E., and Compton, R. N.,*Phys. Rev. A*, **9**, 2532(1974).
5. Milne E., *J. London Math.* Soc. **1**, 1(1926).
6. Bonanno R., Boulmer J., Weiner J., *Comments At. Mol. Phys.*, **16**, 109(1985).
7. Burch, J. L., Mende, S. B., Mitchell, D. G., Moore, T. E., Pollock, C. J., Reinisch, B. W., Sandel, B. R., Fuselier, S. A., Gallagher, D. L., Green, J. L., Perez, J. D., Reiff, P. H., *Science*, **292**, 619(2001).
8. Gruntman, M. A. *J. Geo. Res.*, **99**, 19213(1994).
9. Moore, T. E. , Collier, M. R., Burch, J. L.,et al., *Geophys. Res. Lett.* **28**, 1143(2001).
10. Collier, M.R., Moore, T. E., K.W. Ogilvie, et al., *J. Geo. Res.*, in press.
11. Young D. T. , Bame, S. J., Thomesen, M. F., Martin, R. H., Burch, J. L., Marshall, J. A., and Reinhard, B., *Rev. Sci. Instr.*, **59**, 743(1988).

Fractionation, Acceleration, and Transport Processes

The Polar Coronal Holes and the Fast Solar Wind : Some Recent Results

S. Patsourakos*, S.-R. Habbal†, J.-C. Vial** and Y. Q. Hu‡

*Mullard Space Science Laboratory, University College London, Holmbury St Mary, Dorking, Surrey, RH5 6NT, UK
†University of Wales, Department of Physics, Penglais, Aberystwyth, Ceredigion, SW23, 3BZ, UK
**Institut d'Astrophysique Spatiale, Université Paris XI - CNRS, Bât. 121, F-91405 Orsay Cedex, FRANCE
‡University of Science and Technology of China, Hefei, Anhui, Peoples Republic of China

Abstract. We report on recent results on the source regions of the fast solar wind : the Polar Coronal Holes (PCH). They concern a comparison between the effective temperatures for a large set of different ions obtained from observations in the inner corona of PCH and from a fast wind numerical model based on the ion-cyclotron resonant dissipation of high-frequency Alfvén waves. We also report on some preliminary results from our modeling concerning the Fe/O ratio in the inner corona in PCH.

INTRODUCTION

A major breakthrough in our understanding of the fast solar wind resulted and still results from observations made by the Solar and Heliospheric Observatory (SOHO). Is seems that the fast solar wind emanates from the boundaries of the magnetic network in Polar Coronal Holes (1) and then continues through the interplume regions its journey into the outer corona (2) and (3). Electrons are rather cool in PCH and their temperature barely exceed 1 MK (4). On the other hand protons and minor ions are significantly hotter than electrons with minor ions being preferentially heated and accelerated with respect to protons (5).

The above results gave strong hints on the importance of the resonant interaction between ion-cyclotron (*ic*) waves with solar wind ions, for fast solar wind energetics and dynamics. The potential of the above mechanism for preferential heating and acceleration of the solar wind ions was recognized by a number of authors (e.g., (6), (7) and (8)). The essence of this mechanism is that high frequency ion-cyclotron waves resulting from either a turbulent cascade of low frequency Alfvén waves towards high frequencies (e.g., (9)) or from microflaring in the network (10) can dump their energy and momentum into plasma heating and acceleration when their frequency matches the local gyrofrequency of a given ion. Since the power spectrum of such waves is inversely proportional to the frequency of the waves this means that ions with a smaller gyrofrequency will be more strongly heated and accelerated than ions with a larger gyrofrequency. Moreover, this preferential heating and acceleration of low gyrofrequency ions, takes place closer to the Sun owing to the rapid decrease with distance of the magnetic field strength. When the dispersion of Alfvén waves by the minor ions is taken into account (e.g., (11) and (12)) preferential heating and acceleration of minor ions is even more enhanced : minor ions act as 'channels' for the wave heating to proceed from one gyrofrequency to a much faster one.

One of the most useful spectroscopic observables in the solar corona is the ion effective temperature. It essentially corresponds to a combination of thermal and nonthermal (waves, turbulence) motions along a given line of sight (LOS) in the corona. Since effective temperatures are directly associated to ion temperatures, they provide a powerful means of testing heating mechanisms. For example, in the frame of the models exposed in the previous paragraph, one can expect an ordering of the ion effective temperature with the ion gyrofrequency : the smaller the gyrofrequency the larger the effective temperature. This has been confirmed by (9) who confronted observations of the effective temperatures of Mg X and O VI made by UVCS in the outer corona with model calculations. SUMER observations of a number of different ions in the inner corona (14) have also demonstrated the existence of such a trend even though it is somehow weak. What is clearly needed is to confront effective temperatures of a large number of ions which scan a wide range of gyrofrequencies derived from observations and detailed model predictions. This is the main objective of the present paper. Moreover, we make an initial assessment on the pos-

CP598, *Solar and Galactic Composition*, edited by R. F. Wimmer-Schweingruber

TABLE 1. Initial conditions at 1 $R_\odot$ used for constructing the grid of our models described in the text. f_m, r1, σ refer to the formulation of flow tube cross-section given in (15). The ion densities are calculated by using the photospheric elemental abundances of (16) and ionization equilibrium calculations of (17). The Alfvén waves power spectrum is the same as in (9).

electron density	$2.5\times10^8 cm^{-3}$
electron(=proton, α-particle, ion) temperature	8.0×10^5 K
magnetic field strength	8 G
f_m	5
r1	1.31 $R_\odot$
σ	0.51 $R_\odot$
helium abundance	0.06
wave amplitude	30 kms^{-1}

sible impact that the physical mechanisms invoked by the models exposed above may have on elemental abundances in the inner corona of PCH.

The present paper is organized as follows. In the second section we outline the numerical model we used while in the third section we give some sample results of the model. In the fourth section we make a comparison between observed effective temperatures and model calculations concerning a number of different ions. The fifth section deals with the Fe/O abundance ratio in the inner corona of PCH as it is derived from the model. In the sixth section we discuss our results and present our conclusions.

DETAILS OF THE NUMERICAL MODEL

The numerical model used here is described in full detail in (9). It simultaneously solves the time-dependent conservation equations for mass, momentum, energy and wave-action for a four-species plasma (protons, electrons, α-particles and one minor ion) in a flow tube with the relevant physical quantities being functions of the radial distance only. The flow tube cross-section is described using the formalism of (15). A Kolmogorov power-spectrum describes the energy transfer from low-frequency left-hand-polarized Alfvén waves to high-frequency ion cyclotron waves, and with the resulting resonant interaction with the solar wind ions (solar wind bulk and minor species) is responsible for coronal heating and the acceleration of the solar wind. The distribution of wave energy to the ions is described by the quasi-linear theory of wave-particle interaction and by a cold-plasma dispersion relation. Grid points in space are distributed in such a way as to deal with steep gradients. The corresponding numerical code is run for a sufficient time ($\approx$ 35 days in physical time) inorder to reach a steady-state solution i.e. mass, momentum and energy to attain a constant value and it covers a distance range 1 $R_\odot$ - 1.2 AU. Minor ions could approximately be treated as test-particles in the ambient proton α-particles wind. For more details, the interested reader should reffer to (9) and the references therein.

In the present work we constructed a grid of solar wind solutions, employing the same initial conditions and comprising a number of different minor ions. As such we had at hand the radial profiles of the density, velocity and temperature for protons, α-particles and a number of minor ions. In Table 1 we give the initial conditions we used for each run made in the frame of the present work.

SAMPLE RESULTS

We present in Figures 1 and 2 some sample results from our modelling. In Figure 1 we show a number of properties of the background wind : protons and α-particles start to get appreciably heated from a distance of about 1.3 $R_\odot$, which corresponds to the distance from which most of the super-radial expansion of the flow tube starts taking place. To note here, that the α-particles to protons temperature ratio, is larger than their mass ratio which results from the inclusion of *dispersive* Alfvén waves. Contrary to protons and α-particles, electrons remain cool since on one hand they receive no direct heating and on the other hand Coulomb collisions are not sufficient enough as to sustain a common temperature between ions and electrons. Finally, after an initial lag α-particles start to flow faster than the protons.

In Figure 2 we note that the temperature and velocities of the selected iron ions, which correspond to the dominant ions in the inner corona of coronal holes, are clearly ordered in a decreasing charge state : the smaller the charge state (and thus the gyrofrequency) the more a given ion is heated and accelerated closer to the Sun. Moreover, iron ions are heated and accelerated more and closer to the Sun as compared to the solar wind bulk (protons and α-particles). The iron ions reach their terminal speed at a distance of about 3 $R_\odot$ which is signif-

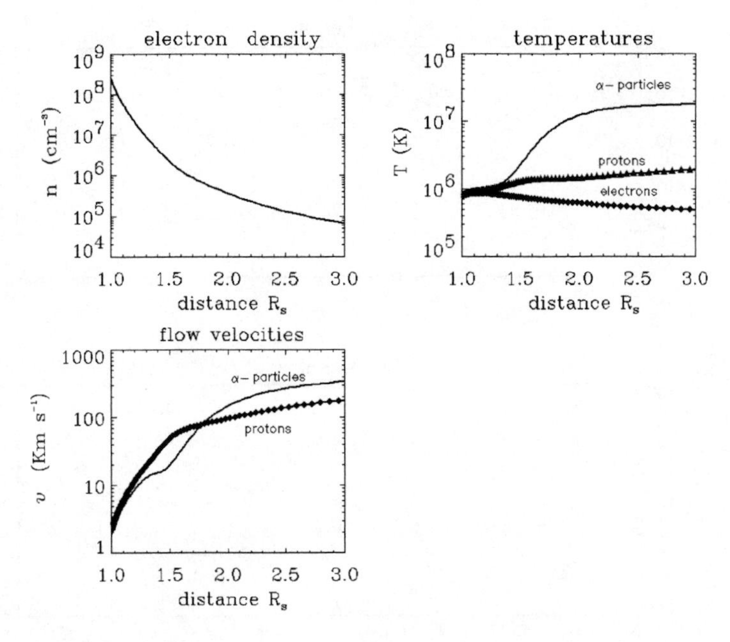

FIGURE 1. Radial profiles of (a) the electron density (upper left panel), (b) the electron, proton and α-particles temperatures (upper right panel) and (c) the proton and α-particles flow velocities (lower left panel).

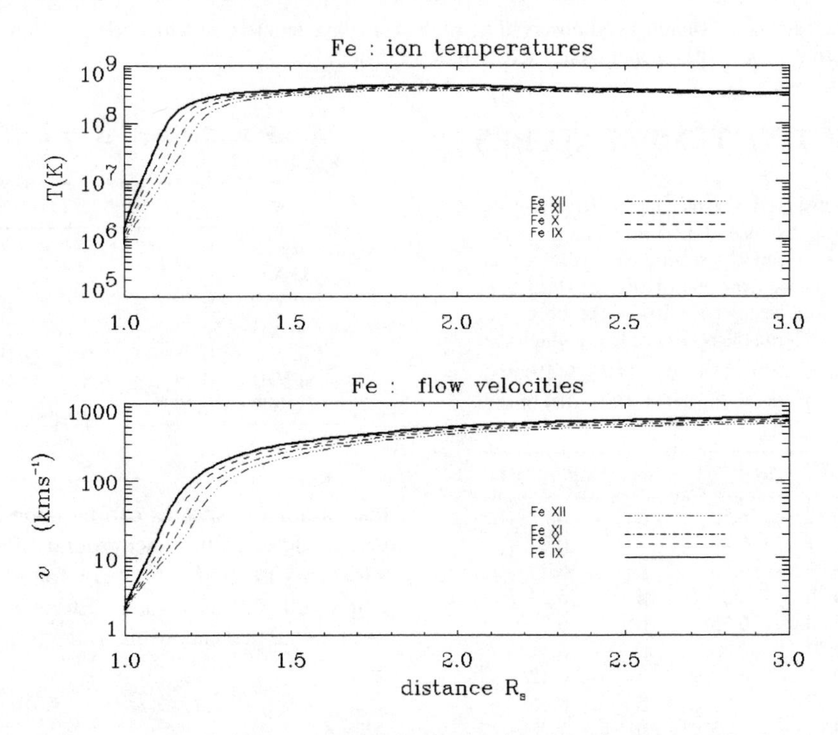

FIGURE 2. Calculated radial profiles of (a) the temperature (upper panel) and (b) the flow velocity (lower panel) for several iron ions.

icantly smaller than the corresponding distance for protons ($\approx 10 R_\odot$) and are extremely hot (temperatures of the order of 100 MK at 1.2 $R_\odot$). The significant difference in velocity between the different ion stages of iron in the inner corona (e.g., a factor ≈ 4 at 1.2 $R_\odot$ between Fe IX and Fe XII), may have a significant bearing on ionization balance calculations in the solar wind acceleration regions which normally use the assumption of a common flow velocity for all ionization stages of the same element.

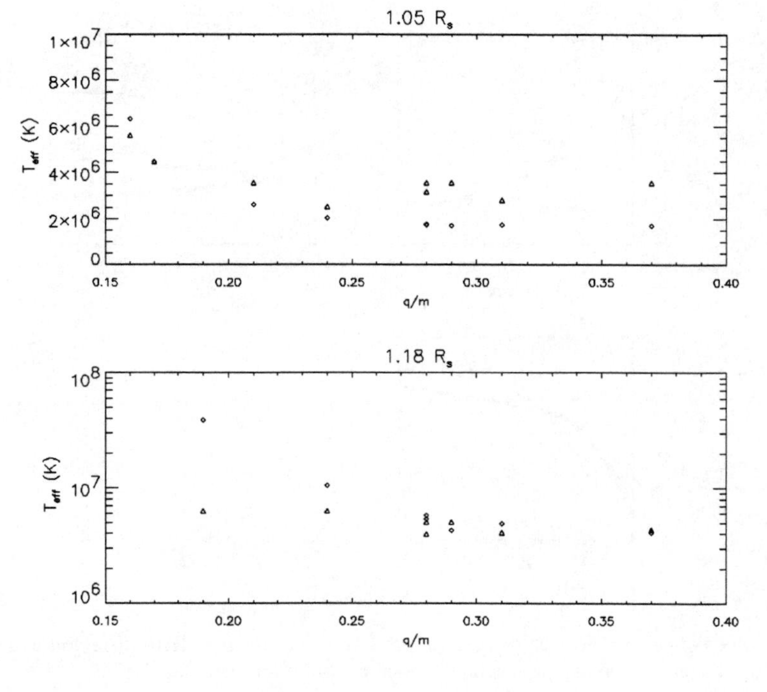

FIGURE 3. Model calculated (rhombs) and observed (triangles) effective temperatures for a series of ions as a function of their $\frac{q}{m}$ ratio. They concern distances 1.05 (upper panel) and 1.18 (lower panel) $R_\odot$.

ION EFFECTIVE TEMPERATURES

TABLE 2. Observed versus calculated effective temperatures at a distance of 1.05 $R_\odot$. The first column gives the used ion, the second column gives its corresponding charge over mass ratio, the third column gives the logarithm (base 10) of the observed T_{eff}, the fourth column the reference from which the T_{eff}^{obs} were extracted ; α : (18), β : (14), γ : (19) and the fifth column gives the logarithm (base 10) of the calculated T_{eff}.

ion	$\frac{q}{m}$	$\log(T_{eff}^{obs})$	ref	$\log(T_{eff}^{mod})$
Mg X	0.37	6.55	α	6.23
O VI	0.31	6.45	γ	6.24
Ne VIII	0.29	6.55	β	6.23
Ne VII	0.28	6.5	β	6.24
Mg VIII	0.28	6.55	β	6.25
Si VIII	0.24	6.4	β	6.31
Si VII	0.21	6.55	β	6.42
Fe XI	0.17	6.65	β	6.65
Fe X	0.16	6.75	β	6.8

From the model results we calculated the effective temperatures for a number of ions and compared them with published observations of effective temperatures in PCH. We note that we checked if the inclusion of a minor species in the calculations, considerably alters the properties of the background proton-α wind, in the same spirit as in (9). We found, as the above authors also did,

TABLE 3. The same as for Table 1, but now at at a distance 1.18 $R_\odot$.

ion	$\frac{q}{m}$	$\log(T_{eff}^{obs})$	ref	$\log(T_{eff}^{mod})$
Mg X	0.37	6.64	α	6.61
O VI	0.31	6.61	γ	6.69
Ne VIII	0.29	6.7	β	6.63
Ne VII	0.28	6.7	β	6.73
Mg VIII	0.28	6.6	β	6.76
Si VIII	0.24	6.8	β	7.02
Fe XII	0.19	6.8	β	7.59

that minor ion species can be approximately treated as test particles in the background solar wind (maximum variations of 10-15 % were found between solutions employing different ions). Effective temperatures T_{eff} were calculated according to the following relation :

$$T_{eff} = T_i + \frac{m_i}{2k} < \delta\upsilon^2 >, \tag{1}$$

where T_i is the ion temperature and $< \delta\upsilon^2 >$ is the velocity variance of unresolved motions along the LOS. For calculating $< \delta\upsilon^2 >$ we considered that it is associated (1) to the velocity amplitude of the Alfvén waves which propagate along the magnetic field lines and it writes as :

$$< \delta\upsilon^2 >= \frac{2p_w}{\rho}, \tag{2}$$

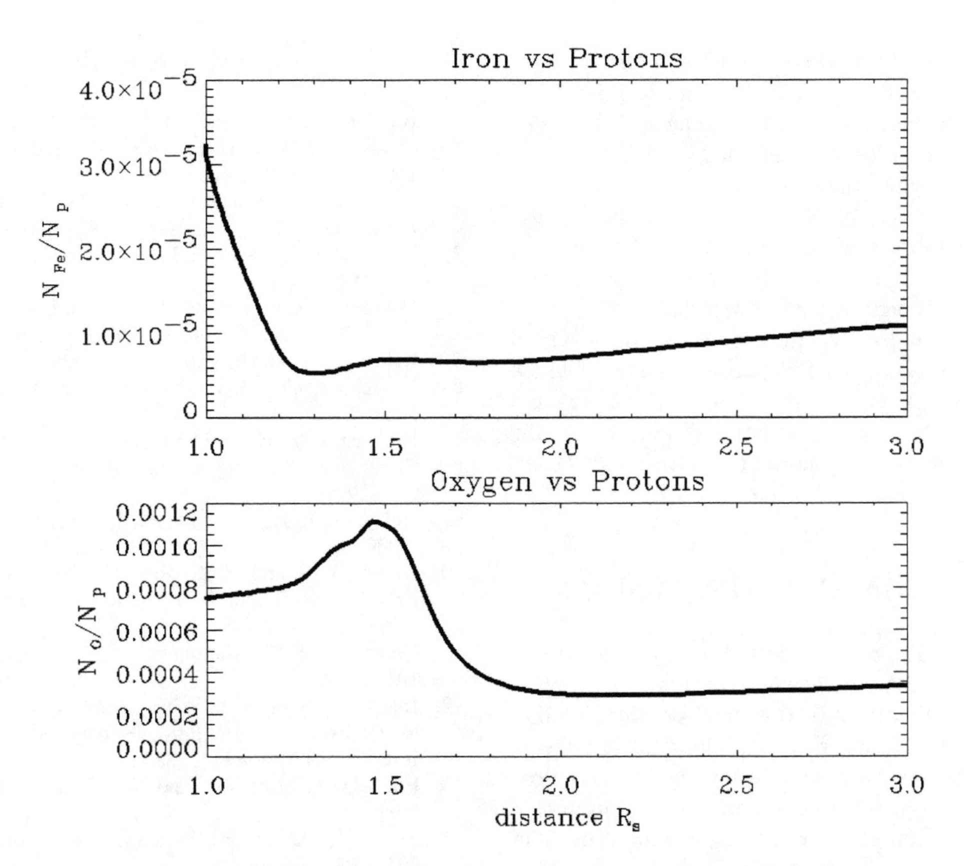

FIGURE 4. Radial variation of the the iron (upper panel) and oxygen (lower panel) abundances with respect to the Hydrogen as a function of the distance.

where p_w is the wave pressure and ρ the mass density and (2) to solar wind motions along the LOS that produce extra line broadening. For (2) we calculated the rms value of the ion flow velocities distribution along the LOS, by weighting each point taken into account in the calculation by the square of the corresponding electron density. T_{eff} were determined by the observations, by attributing the line width of the observed spectral lines of different ion species to an effective temperature.

The results of this comparison are tabulated in Tables 2 and 3 and shown in Figure 3. The calculated T_{eff} tend to decrease with increasing $\frac{q}{m}$ following the lines of the ion-cyclotron resonance mechanism. The plateau towards large $\frac{q}{m}$ can be explained by the existence of heavier ions (Fe, Si) in this range of $\frac{q}{m}$ than in the descending part of the T_{eff} versus $\frac{q}{m}$ curve which tends to compensate the decrease of T_{ion} with $\frac{q}{m}$). The observed T_{eff} are less 'well-ordered' with $\frac{q}{m}$ than the calculated but a trend is also evident. What we can conclude is that the calculated and observed T_{eff} are in a good agreement to a factor of maximum 2 in all but one cases. It can also be noted that while at $r = 1.18\,R_\odot$ we have $T_{eff}^{obs} > T_{eff}^{mod}$ this trend is reversed for $r = 1.05\,R_\odot$. The point of the maximum disagreement between the observations and the calculations (a factor ≈ 6) corresponds to observations made in the Fe XII line at 1.18 $R_\odot$. This line is particularly faint in coronal hole regions and observations may have underestimated its true width : the poor photon statistics in the line wings hamper to resolve the full line and thus the corresponding T_{eff}.

THE FE/O RATIO IN THE INNER CORONA OF POLAR CORONAL HOLES

Inorder to make a case on the possible implications the present model and the *ic* mechanism in general may have on elemental abundances in the inner corona, we show here some preliminary results on the Fe/O abundance ratio (Figure 4). *In-situ* observations (20) of this ratio in fast solar wind streams have shown that it takes values comparable to the photospheric ones. We approximated the iron and oxygen total densities by the sum over the densities of their dominant ionization stages (Fe IX, X, XI, XII and O VI, VII, VII correspondingly). It can be seen from Figure 4 that the iron abundance decreases with distance in the range 1.0-1.3 $R_\odot$ while for oxygen this decrease takes place in the range $\approx$ 1.5-1.8 $R_\odot$. The above behavior is a direct consequence of the different velocity profiles of oxygen and iron : iron is more massive than oxygen and it is therefore more efficiently accelerated closest to the Sun owing to the smallest gyrofrequencies of the iron ions as compared to the oxygen ions. Therefore the iron and oxygen start moving faster than the protons and thus become depleted at different radial distance regimes. From the above it follows that until a distance of $\approx$ 1.3 $R_\odot$ the Fe/O decreases a result which may be termed as an 'inverse FIP-effect' for iron and oxygen. We note here, that at larger distances from the Sun both elements retain more or less their abundances at 1 $R_\odot$.

DISCUSSION AND CONCLUSIONS

The rather good agreement between the observed and calculated T_{eff} for a significant number (10) of different ions with $\frac{q}{m}$ ratios from 0.16 to 0.37 provides another strong argument in favor of the importance of the *ic* mechanism for fast solar wind heating and acceleration. It is remarkable to note how well calculations fit the observations, given a number of simplifications made in the calculations - inclusion of one minor ion at any time and thus neglecting the effect of the other ions on the waves dispersion, use of a cold plasma relation which overestimates the effects of the waves. It could be that the above two effects compensate each other leading to a solution not too far from 'reality'. In any case, the four-fluid model we used seems capable of capturing the essential elements of fast solar wind energetics and dynamics - as it was also shown for the outer corona and 1 AU in (9) regardless its simplifications. An extension of such a comparison to other ions observed in the inner corona is foreseen.

In an *ic* heated and accelerated solar wind it seems that iron may be depleted with respect to oxygen in the inner corona. This kind of behavior may be termed as an inverse FIP-effect. Clearly, ionization balance calculations consistent with differential flows of ions of the *same* element inferred by modeling along with detailed observations of the Fe/O ratio in the inner corona of Polar Coronal Holes are needed to settle this issue.

ACKNOWLEDGMENTS

S. Patsourakos is financed by PPARC.

REFERENCES

1. Hassler, D. M., Dammasch, I. E., Lemaire, P., Brekke, P., Curdt, W., Mason, H. E., Vial, J.-C., and Wilhelm, K., 1999, *Science* **283**, 810 (1999).
2. Giordano, S., Antonucci, E., Noci, G., Romoli, M., and Kohl, J. L., Astron. & Astrophys. **531**, L79 (2000).
3. Patsourakos, S. and Vial, J.-C., Astron. & Astrophys. **359**, L1 (2000).
4. David, C., Gabriel, A.-H., Bely-Dubau, F., Lemaire, P., and Wilhelm, K., Astron. & Astrophys. **336**, L90 (1998).
5. Kohl, J. L., et al., ApJ **501**, L127 (1998).
6. Marsch, E., Goertz, C. K. and Richter, A. K., JGR **87**, 5030 (1982).
7. Hollweg, J. V., JGR **104** 506, (1999).
8. Cranmer, S. R., Field, G. B., and Kohl, J. L., ApJ **518**, 937 (1999).
9. Hu, Y. Q., Esser, R. and Habbal, S. R., JGR **105**, 5093 (2000).
10. Axford, W. I. and McKenzie, J.F., 'Solar Wind Seven' (eds. Marsch E. and Schwenn, R.), Pergamon Press, 1 (1992).
11. Gomberoff, L. F., Gratton, F. T., and Gravi, G., 1996, JGR **101**, 15661 (1996).
12. Hu, Y.Q., 'Solar Wind Nine', Habbal, S. R., Esser, R., Hollweg, J. V. and P. A. Isenberg (eds.), American Institute of Physics **471**, 285 (1999).
13. Hu, Y.Q., Habbal, S.R., and Xing, L., JGR **104**, 24819 (1999).
14. Tu, C.-Y., Marsch, E., Wilhelm, K. and Curdt, W., ApJ **503**, 475 (1998).
15. Kopp, R. A. and Holtzer, T. E., Sol. Phys. **49**, 43 (1976).
16. Meyer, J. P., ApJS **57**, 173 (1985).
17. Arnaud, M. and Rothenflug, R., Astron & Astrophys Supp **60**, 425 (1995).
18. Hassler, D. M., Rottman, G., Shoub, E. C. and Holtzer, T. E., ApJ **348**, L77 (1990).
19. Banerjee, D., Teriaca, L., Doyle, J. G. and Lemaire, P., Sol. Phys. **194**, 43 (2000).
20. Aellig, M.R., Hefti, S., Grünwaldt, H., et al., JGR **104**, 24769 (1999).

Ion Fractionation and Mixing Processes in the Turbosphere and the Solar Wind Formation Region: Scaling Approach

I. S. Veselovsky

Institute of Nuclear Physics, Moscow State University, Moscow 119899, Russia

Abstract. The kinetic theory of the ion distribution function formation in the plasma is briefly considered. Models are discussed of the individual particle and collective behavior. Governing dimensionless parameters are presented for numerous collisional and collisionless regimes of the ion generation, separation and mixing in the solar atmosphere. Global steady state and local non-stationary processes are shown to be interrelated in a complicated manner possibly leading to the observed diversity in composition variations of the solar wind. The strongest composition variations are produced at the shocks, but the mixing prevails in the sense that protons remain the dominant component even under these extreme conditions. The dynamical balance between the mixing and fractionation depends on many space-time scales simultaneously present in the plasma and electromagnetic fields, which explains the lack of simple universal classifications and poor understanding of the laminar or turbulent processes of the ion composition formation. For example, it is not clear if the persistent difference between the composition of the quasi-stationary slow and fast solar wind streams could be related mainly to the vertical stratification or to the horizontal inhomogeneity of the solar corona and dynamical processes in holes and streamers.

INTRODUCTION

Variations of the solar wind ion composition and their origins are poorly known and understood in many instances (see, e.g., reviews, [1, 2, 3, 4] and references therein), though the thermal diffusion, electric fields and Alfven waves are often involved in the separation processes.

The aim of this paper is to bring theoretical arguments helpful in attempts of a qualitative understanding of the observed diversity of the solar wind composition variations as a result of the multi-scale plasma evolution with a competition of different physical processes of the ion separation and mixing in the turbosphere and the solar wind formation region. The term turbosphere is used here to designate the region around the Sun where the velocity of the regular radial outflow v is less than the velocities δv of irregular vertical and horizontal motions: $v \ll \delta v$.

KINETIC REGIMES

In the deep regions of the solar atmosphere (chromosphere) where ionization and recombination processes happen with other atomic and radiative transitions and many-particle interactions are essential, one should consider correlation functions which could be in the equilibrium state or not. The macroscopic diffusion approximation can be used only when collisions are sufficiently frequent. Dimensionless parameters like ratios of the ionization, recombination, excitation and radiation times are very important physical quantities which are needed for the better understanding of the situation. In addition to this, several dimensionless parameters are useful indicators of the importance (unimportance) of the local/nonlocal conditions (the so called Trieste numbers which are ratios of internal and external energy, momentum and mass fluxes in the coronal and heliospheric structures under consideration). Trieste numbers characterize the openness degree and vary from zero in the open system to infinity in the isolated structures. They are of an order of one for many morphological entities in the solar wind formation region, which makes the analysis of the turbulent and "recuperating" systems very difficult and sensitive to the imposed initial and boundary conditions. It is very important, but sometimes neglected without sufficient grounds, that the plasma in the solar wind formation region is radiative. The useful dimensionless parameter in this respect, V_e, is the ratio of the kinetic plasma power to the emitted radiation power. The parameter V_e is small for active regions and large for coronal holes as a rule [5].

In the fully ionized coronal regions with the negligible radiation role (coronal holes) the one-particle distribu-

CP598, *Solar and Galactic Composition,* edited by R. F. Wimmer-Schweingruber

tion functions $F(\vec{r},\vec{v},t)$ obey the kinetic equations with the Landau collision term

$$\hat{D}f_i = \hat{S}t f_i,$$

where

$$\hat{D} = \frac{\partial}{\partial t} + \vec{v}\frac{\partial}{\partial \vec{r}} + \left\{ \frac{Z_i e}{m_i}(\vec{E} + \frac{1}{c}[\vec{v}\times\vec{B}]) - \frac{M_\odot G}{r^3}\vec{r} \right\} \frac{\partial f}{\partial \vec{v}}$$

is the transport operator in the phase space, $\hat{S}t$ is the Landau collision term [6]. The dimensionless Knudsen number $Kn = \lambda/l$, where λ is the Coulomb mean free path and l is the characteristic length, delimits collisional and collisionless regimes, $Kn \ll 1$ and $Kn \gg 1$, correspondingly.

Ion separation and mixing processes under frequent collisions when $|\hat{S}t| \gg |\hat{D}|$ or $Kn \ll 1$ can be described in the Coulomb diffusion approximation. The diffusion coefficient along the magnetic field is proportional to $D_{\parallel} \sim \lambda_i \sim \sigma^{-1}$, where λ_i is the mean free ion path, σ is the Coulomb cross section in the fully ionized plasma. The laminar diffusion across the field is suppressed accordingly: $D_{\perp} \approx D_{\parallel}(r_i/\lambda_i)^2 \sim \lambda_i^{-1}$, where r_i is the ion Larmor radius, $D_{\perp} \sim B^{-2}$. The laminar diffusion regimes [7] are violated by the multi-scale convection and waves in the turbosphere around the Sun where the plasma is out of equilibrium. In this case, some effective turbulent mixing length L should be used instead of λ in the corresponding transport estimates. The mixing length L in the chromosphere and in the corona is generally comparable to the standard heights. Morphologically, vertical motions in both directions along the loops and open magnetic structures take place here together with horizontal shuffling motions due to convective electric drifts, magnetic stresses and thermal pressure gradients in the intermingled low- and high-beta regions. Turbulent processes could enhance or depress the local and global transport of the bulk mass and its ion components depending on the situation and the geometry of the problem. It is very likely, that vertical and horizontal turbulent mixing efficiency is different in the sources of the slow and fast solar wind streams because of different plasma parameters and the magnetic field geometry. Nevertheless, the vertical mixing apparently prevails over the gravitational settling and other separation mechanisms. As a result, composition variations are observed to be rather moderate in the solar wind as rule.

GAS-KINETIC APPROXIMATION FOR THE SOLAR WIND IONS

The free energy of the solar wind plasma resides mainly in the inhomogeneous radial motion of ions, which is supermagnetosonic. Because of this, the cold gas of noninteracting particles can be a valuable zero order approximation when considering the kinetic equations for the ion distribution functions. In this approximation, the force-free radial motion is used in the method of characteristics neglecting transversal thermal velocities, electromagnetic and gravity fields [8].

The gas-kinetic theory of weak and strong inhomogeneities in the collisionless solar wind was developed using the small parameters M^{-1} and M_A^{-1}, where M and M_A are the Mach and Mach-Alfven numbers for ions. Distribution function corrections were analytically calculated in this way for ions in response to the nonstationary plasma density, bulk velocity and temperature perturbations at the given boundary near the Sun. The leading terms are of the order M^{-2} and M_A^{-2}. The details of calculations were described earlier [8, 9, 10]. In the fast streams from coronal holes $T_i \sim m_i$ and v_{Ti} are the thermal velocities of the ion species nearly equal to each other, hence partial Mach numbers $M_i \sim v/v_{Ti}$ are also approximately equal to each other for different ion species. As a result, the variations of the composition should be canceled in the linear approximation for this case. In the slow wind, $T_\alpha \sim T_p$ because of the sufficient collisions and $M_\alpha^2 \sim 4M_p^2$. Hence, composition fluctuations are mostly produced by the proton parameter variations which are four times stronger than alpha particle variations in the linear approximation. The linear approximation breaks down when $SM^{-2}\frac{\delta A}{A} \geq 1$, where $S = \frac{r-r_0}{vt}$ is the Strouhal number, $\delta A/A$ being the relative amplitude of parameter variations (density, velocity, temperature). It is clear that short time composition fluctuations are steepen and mixed at the corresponding distances because of the nonlinear evolution with the formation of interpenetrating streams or they are preserved when the nonlinearity is saturated. Saturation processes are difficult to evaluate a priory. For this purpose the a detailed information is needed about the dynamics taking into account the kinetic plasma and electromagnetic field characteristics in the real geometry situation. These considerations probably explain the lack of the universal composition behavior and the presence of the observed diversity with temporary mixed and fractionated regions. Nevertheless, the nonlinear mixing processes are obviously very important limitative factors during the ion separation. As a result, the fractionation is not sufficiently strong to eliminate protons from the ion composition or make them a secondary component in the known regions of the heliosphere. It appears that protons remain dominant ions everywhere and every time in the corona and in the solar wind in spite of many different ion separation mechanisms operating in the rarefied plasma.

The ballistic motion of particles in the force-free space and in the gravity field preserves the composition only when interpenetrating streams and the corresponding mixing processes are absent. Such laminar flows are locally and temporary possible under appropriate boundary and initial conditions in the solar corona. Due to the nonlinear kinematic steepening of the velocity profiles interpenetrating streams could appear in the kinetic approximation. Otherwise, the ballistic approximation breaks down and shock fronts develop. The mixed state with interpenetrating streams, once created, develop in different ways depending on the ratio of relaxation and dissipation times and lengths to the transit times and scales of the inhomogeneity given by the corresponding Reynolds numbers (ordinary and magnetic). In the one-scale approximation, only two possibilities exist in this respect: the homogenization of the mixture or its segregation after the transit stage. But the large free energy (kinetic and/or magnetic) of the interpenetrating ion streams could be a source of ion instabilities and the generation of new small-scale structures is possible [11]. Hence, the fragmentation, segregation and relaxation coexist with the merging, mixing and increase of structures. Both tendencies, the decay and growth of the structures are observed in the solar corona and the heliosphere permanently. They are documented in the composition measurements of quasistationary and transient solar wind flows, but the knowledge and understanding of the evolution is rather limited. Interpenetrating ion streams in the solar wind are sometimes observed, but their role in the overall plasma mixing is not clear and needs additional studies.

SELF-SIMILAR ION FRACTIONATION IN THE SOLAR WIND

The collisionless cold ion dynamics with the compensating electrons near the thermal equilibrium can be described by multifluid equations taking into account electric and magnetic fields [6]. The nonstationary self-similar variations in the ion composition of the solar wind have been analyzed based on this approach [12].

The governing equations for protons ($i = 1$) and alpha-particles ($i = 2$) in the case of motions along the external magnetic field are written as follows:

$$(u_i - \tau)\frac{dn_i}{d\tau} + n_i\frac{du_i}{d\tau} + \frac{kn_iu_i}{\tau} = 0 (i = 1,2),$$

$$(u_1 - \tau)\frac{du_1}{d\tau} + \frac{1}{2n}\frac{dn}{d\tau} = 0,$$

$$(u_2 - \tau)\frac{du_2}{d\tau} + \frac{1}{2n}\frac{m_1 Z_2}{m_2 Z_1}\frac{dn}{d\tau} = 0,$$

$$n = Z_1 n_1 + Z_2 n_2, (Z_1 = 1, Z_2 = 2),$$

where $u_1 = v_1\sqrt{\frac{m_1}{2T_e}}, u_2 = v_2\sqrt{\frac{m_1}{2T_e}}, \tau = \frac{r}{t}\sqrt{\frac{m_1}{2T_e}}$ are dimensionless ion velocities and the self-similar variable,

$$n_e = n_0 exp(e\varphi/T_e),$$

$k = 0, 1, 2$ for the planar, cylindrical or spherical geometry. The electric potential φ in the corona is of an order of $100\,V$.

There are two possible asymptotics at $\tau \gg 1$:
(I) $u_1 = C_1(1 - \frac{k}{u\tau^2}) + O(\tau^{-3})$,

$$n_1 = n_{10}(1 - \frac{3}{2}\frac{kC_1}{\tau}) +)(\tau^{-3}),$$

$$u_2 = C_2 - \frac{3}{8}\frac{m_1 Z_2}{m_2}\frac{kC_1}{\tau^2} + O(\tau^{-3}),$$

$$n_2 = n_{20} + O(\tau^{-1}),$$

where $C_{1,2}$ and $n_{10,20}$ are arbitrary constants;
(II) $u_1 = \tau + \sqrt{\frac{k+1}{2}} + O(\tau^{-1})$,

$$n_1 = n_{10} exp[-\tau\sqrt{2(k+1)}],$$

$$u_2 = \tau + \frac{m_1 Z_2}{m_2}\sqrt{\frac{k+1}{2}},$$

$$n_2 = n_{20} exp[-\tau\frac{m_2}{m_1 Z_2}\sqrt{2(k+1)}].$$

In the both cases (I and II) the condition $n_2 \ll n_1$ is assumed. The formulae are applicable also for other types of ions ($i = 2$), and not only for alpha-particles.

It is seen from these formulae that enhanced (depleted) n_2/n_1 regions could propagate faster or slower than unperturbed plasma flow depending on the boundary and initial conditions (which are beyond the scope of the self-similar assumptions). In the case (I) for weak perturbations, alpha-particles are moving faster than protons. For strong perturbations (II), alpha-particles are moving slower than protons. The potential electric fields and the electron distributions are controlling factors. The velocity difference $u_2 - u_1$ is positive in the case (I) when $C_1 \approx c_2$ and $u_2 - u_1 \sim \tau^{-2}$ decreases with increasing distance from the Sun, in a qualitative accordance with the Helios observations [13].

Theoretically, it is obvious that electric fields are involved in the ion dynamics not only along the magnetic field, but also in the nonstationary and quasistationary drifts across the magnetic field, which are strongly inhomogeneous and variable in the solar wind formation region. As a consequence, the characteristic velocities of the nonstationary ion mixing or separation along the magnetic field could be determined by the ion sound speed, and by the drift velocities across the magnetic fields [14, 15]. The drift velocities are often scaled as L/t, where L and t are characteristic lengths and times of the magnetic field variations. Contrary to the prejudice that magnetic fields always suppress the transport processes across them, observations in space and laboratory experiments, as well as theoretical arguments show many examples of the enhanced energy and mass transports perpendicular to magnetic fields in the presence of the inductive or potential electric fields due to the imposed initial or boundary conditions. Both external nonlocal fields and local fields generated in situ by the free energy available in the open, closed and partially open plasma systems of the solar corona and the heliosphere could be and are the drivers of the electromagnetic "separators" or "mixers" operating there.

The dimensionless Faraday number $F = \frac{j}{\rho c}\frac{l}{ct}$, where j and ρ are the electric current and the electric charge densities, could indicate on the dominance of the potential $F \ll 1$ or inductive $F \gg 1$ electric fields. The case $F \ll 1$ was considered here.

DIMENSIONLESS MULTIFLUID MHD SCALING

The demarkation of physically different regimes is possible with the use of the following dimensionless parameters and their combinations:

$Kn = \lambda_i/l$ is the Knudsen number,

$S = l/vt$ is the Strouhal number,

$M = v/c_s$ is the Mach number,

$M_A = v/V_A$ is the Mach-Alfven number,

$Fr = v^2 R(M_\odot G l)^{-1}$ is the Froude number,

where R is the heliospheric distance, $M_\odot$ is the mass of the Sun, G being the gravitation constant, and V_A, the Alfven speed. Additionally, the group of dissipative parameters appears:

$Re = vl/\nu$ is the (viscous) Reynolds number,

$Re_m = \sigma lv/c^2$ is the magnetic Reynolds number,

$Pr = \nu/\chi$ is the Prandtl number,

where ν is the kinematic viscosity, χ is the temperature conductivity and σ is the electric conductivity. All dissipative coefficients are tensor quantities in the strong magnetic field [7].

In the multifluid approximation, relative abundances of the species $\alpha_i j = n_i/n_j$ are important dimensionless parameters where n_i and n_j represent corresponding number densities. All parameters above are assumed to be partial quantities referred to the given type of ions (indexes i,j). For example, nonstationary $S > 1$, quasistationary $S < 1$ and intermediate $S \sim 1$ regimes are delimited by the corresponding Strouhal numbers $S = \frac{l}{vt}$.

The high (low) ion-electron temperature ratios $r = T_i/T_e$ could indicate on the dominant ion (electron) heating. The situation is rather simple in the low β regions: $r > 1$ usually means the viscous mechanical heating, for example due to the nonlinear acoustic wave damping. The opposite case $r < 1$ could mean the nonequilibrium Joule heating of electrons. The situation in the high β regions is much more complicated because of the anisotropy introduced by the magnetic field.

The dimensionless scaling approach is useful for a better understanding of the different physical situations in the binary cases (inertia and collisions, inertia and electromagnetic forces, inertia and gravity etc.) as well as the ternary and more complicated regimes when several physical factors are essential, for example, combined inertia, gravity and electromagnetic forces. In addition to the one-scale approximation and classification, multiscale states and mechanisms represent very interesting practical applications.

There are many tens of the physically different regimes of the dissipative macroscopic plasma behavior in the stratified solar atmosphere and the turbosphere around the Sun [5]. Electromagnetic and gas-kinetic drivers are known to be effective in the "mixers" and "separators" of the ion composition. Subsonic thermal pressure pulses in the low β regions, field aligned interpenetrating streams, secondary field-aligned flows driven by the nonstationary electric drifts and the continuity conditions steepen and decay with the formation of the $T_i \sim m_i$ states. The same states can be obtained as a result of the ion cyclotron In the situations far from the thermal and mechanical equilibrium in the solar corona and the heliosphere anomalous transport processes by Alfven waves and convective motions are ubiquitous. Collisions between particles tend to equalize the temperature differences. Both mechanisms, the convective mixing of different plasma parcels and wave-particle interactions are also capable of producing observed deviations from the relation $T_i \sim m_i$. Stronger or weaker dependences could result from different initial or boundary conditions in the case of the mixing. Instead of these nonlocal causes, locally steeper or shallower wave spectra can also lead to same qualitative results.

Subphotospheric convective motions generate electric currents and magnetic fields seen in the photospheric and chromospheric network elements, granules, spicules, loops and other nonstationary solar activity manifestations. These nonstationary magnetic fields produce inductive electric fields according to the equation $[\vec{\nabla} \times \vec{E}] = -\frac{1}{c}\frac{\partial \vec{B}}{\partial t}$. As a result, nonstationary electric drifts $\vec{V}(\vec{r},t) = c\frac{[\vec{E}\times\vec{B}]}{B^2}$ arise. Small-scale plasma elements are mixed across the magnetic field under the action of fluctuating transversal electric field.

The corresponding mixing lengths L for the mentioned motions could be less, comparable, or greater than the standard height H_i (0.1-100 Mm) for ions in the solar atmosphere at different levels with characteristic times 10^2 - 10^5 s and velocities of an order of ~ 0.1 - 100 km/s, but usually they are comparable to standard heights H_i.

Local vertical up- and down motions are nearly sonic or trans-sonic in spicules, jets, loops. Horizontal motions are sometimes faster, but sometimes slower than vertical ones. Convective flows and wave branches are involved in the motions in different proportions.

The turbulent mixing length L could also be less, comparable or greater than the Coulomb mean free path for ions λ_i.

Accordingly, different diffusion and mixing regimes are locally possible: molecular or turbulent ones under essential or negligible gravitational sedimentation convective or wave-like, with subsonic or supersonic motions in a magnetically dominated or unmagnetized conditions.

DISCUSSION

Ion-atom separation mechanisms in the partially ionized chromospheric plasma lead to the fractionation of elements according to their first ionization potentials, see e.g. the review of models [16].

Solar wind charge states are used in attempts of obtaining rough estimates of temperature profiles in coronal holes, when comparing calculated expansion times and ionization-recombination times assuming local collisional equilibrium with electrons. The situation in the slow wind and coronal mass ejections is much more uncertain, variable and complicated.

First ionization potential (FIP) fractionation takes place in the solar wind: elements Fe, Mg, Si with the low FIP ($\leq$ 10 eV) corresponding to the Lyman-alpha radiation contribution are more abundant than high FIP elements like inert gases. In the models of this phenomenon (VonSteiger and Geiss, 1989) the ratio of the ion trapping time in the magnetic field to the ionization time by UV radiation is important controlling factor and different types of the solar wind flows exist according to the role of the ion-atomic separation processes in the partially ionized solar atmosphere.

The difference between elemental compositions and charge states of the fast and slow solar wind streams was confirmed and investigated in detail in the recent study of the Ulysses data [17]. The charge state distributions are consistent with a single frozen-in temperature for each element only in the fast streams. The first ionization potential fractionation is clearly seen in the slow streams, it is much more weaker in fast streams.

Bürgi [18] have considered the steady-state three-fluid model with the cusp geometry and found the depleted alpha to proton flux ratio in the coronal streamer. The useful dimensionless parameter here is the geometry expansion factor $f = \frac{S_2}{S_1}$ for the corresponding flux tube cross sections with a possible focusing (defocusing) of flows and wave energy fluxes in the low β solar wind formation region [19].

Bochsler [20] have integrated numerically the energy balance equations for a buoyant expanding plasmoid taking into account its heating and radiative losses together with the ionization state changes which were shown to be a sensitive indicator of the physical processes in the inner corona. Useful dimensionless parameters here are represented among others by the α_i-ionization degrees, V_e - the ratio of the plasma kinetic power to the electromagnetic radiation (absorption) power [5].

The behavior of minor ions in the models is very complicated and depends on many factors: masses, charges, gravity, thermal pressure, magnetic stresses, thermal diffusion, expansion factors etc. [21]. The Coulomb friction plays different roles in the rapidly expanding magnetic structures near coronal streamers and in polar coronal holes. The solar wind composition and the composition of the source regions in the solar atmosphere needs additional comparative studies. In this respect, recent SOHO and ACE measurements of heavy ion composition variations can be used as a sensitive tool for the classification of the solar wind acceleration conditions in the transition region and in the solar corona [22, 23].

Ion composition measurements of coronal mass ejections bring some features which might be interpreted as signatures of non-equilibrium plasma mixtures of different temperature components [24]. Turbulent mixing processes could be much more effective than molecular or Coulomb diffusions in producing such mixtures.

CONCLUSIONS

Individual particle degrees of freedom (one-particle molecular kinetic approximation), few or many particle correlations (ionization and recombination processes, atomic and radiation phenomena) as well as collective (macroscopic radiative MHD description) behavior are possibly important ingredients to be taken into account when considering fractionation and mixing processes in the solar atmosphere.

Turbulent mixing makes the fractionation mechanisms less effective in many instances. Sufficiently strong quasistationary magnetic fields are sometimes stopping the transversal mixing and fractionation processes and could preserve the memory of the original composition formation in the solar wind.

The scaling approach is useful for a better qualitative understanding of the ion separation and mixing processes in the solar wind.

ACKNOWLEDGMENTS

The work is partially supported by the INTAS-ESA grant 99-00727, RFBR grants 00-15-96623, 01-02-16579, Federal Program "Astronomy" project 1.5.6.2 and the State Program "Universities of Russia" grant 99-0600. The author is grateful to the Organizing Committee and the Swiss National Science Foundation for the financial support enabling his attendance and the presentation of the paper at the First SOHO/ACE Workshop: "Solar and Galactic Composition".

REFERENCES

1. Hundhausen, A. J., *Coronal Expansion and Solar Wind*, Springer-Verlag, Heidelberg, 1972.
2. Hollweg, V., "Helium and Heavy Ions", in *Solar Wind Four*, Report No MPAE-W-100-81-31, 1981, pp. 414–424.
3. Neugebauer, M., "Observations of Solar-Wind Helium", in *Solar Wind Four*, Report No MPAE-W-100-81-31, 1981, pp. 425–433.
4. Ogilvie, K. W., and Coplan, M. A., *Rev. Geophys. Suppl.*, pp. 615–622 (1995).
5. Veselovsky, I. S., "Nearly Sonic and Trans-Sonic Convective Motions in the Solar Atmosphere Related to the Solar Wind Origin", in *Solar Wind Eight*, AIP 382, 1996, pp. 161–164.
6. Lifshits, E. M., and Pitaevsky, L. P., *Physical Kinetics*, Nauka (in Russian), Moscow, 1979.
7. Braginskii, S. I., "Transport Phenomena in Plasma", in *Questions of Plasma Theory, vol. 1.*, edited by M. A. Leontovich, Atomizdat (in Russian, also translated in English), Moscow, 1963.
8. Veselovsky, I. S., *Sov. Phys. JETP*, **50**, 681–684 (1979).
9. Veselovsky, I. S., *Geomagnetism and Aeronomy*, **20**, 769–776 (in Russian, also translated in English) (1980).
10. Veselovsky, I. S., *Geomagnetism and Aeronomy*, **21**, 968–972 (in Russian, also translated in English) (1981).
11. Veselovsky, I. S., and Shabansky, A. V., *Geomagnetism and Aeronomy*, **27**, 358–361 (in Russian, also translated in English) (1987).
12. Veselovsky, I. S., *Geomagnetism and Aeronomy*, **13**, 166–168 (in Russian, also translated in English) (1973).
13. Marsch, E., Mülhauser, K.-H., Pilipp, W., Schwenn, R., and Rosenbauer, H., "Initial Results on Solar Wind Alpha Particle Distributions as Measured by Helios Between 0.3 and 1 A.U.", in *Solar Wind Four*, Report No MPAE-W-100-81-31, 1981, pp. 443–449.
14. Veselovsky, I. S., "Nonstationary Electric Drifts in the Solar Atmosphere", in *New perspectives on Solar Prominences*, edited by D. Webb, D. Rust, and B. Schmieder, AIP 150, 1998, pp. 123–126.
15. Veselovsky, I. S., "Field-Aligned Electric Currents in the Heliosphere", in *Proc. 9th European Meeting on Solar Physics*, ESA SP-448, 1999, pp. 1217–1228.
16. von Steiger, R., "Solar Wind Composition and Charge States", in *Solar Wind Eight*, AIP 382, 1996, pp. 193–198.
17. von Steiger, R., Schwadron, N. A., Fisk, L. A., Geiss, J., Gloeckler, G., Hefti, S., Wilken, B., Wimmer-Schweingruber, R. F., and Zurbuchen, T., *J. Geophys. Res.*, **105**, 27217–27238 (2000).
18. Bürgi, A., "Dynamics of Alpha Particles in Coronal Streamer Type Geometry", in *Solar Wind Seven*, Pergamon Press, Oxford, 1991, pp. 333–336.
19. Veselovsky, I. S., *Geomagnetism and Aeronomy*, **36**, 1–7 (in Russian, also translated in English) (1996).
20. Bochsler, P., "Minor Ions - Tracers for Physical Processes in the Heliosphere", in *Solar Wind Seven*, Pergamon Press, Oxford, 1991, pp. 323–332.
21. Bodmer, R., and Bochsler, P., *J. Geophys. Res.*, **105**, 47–63 (2000).
22. von Steiger, R., and Zurbuchen, T. H., "Solar Wind Composition as a Diagnostic Tool", in *Solar and Galactic Composition*, Conference Booklet, Joint SOHO–ACE Workshop 2001, March 6–9, 2001, Bern, Switzerland, 2001, p. 45.
23. Ko, Y.-K., Zurbuchen, T., Strachan, L., Riley, P., and Raymond, J. C., "A Solar Wind Coronal Origin Study from SOHO/UVCS and ACE/SWICS Joint Analysis", in *Solar and Galactic Composition*, Conference Booklet, Joint SOHO–ACE Workshop 2001, March 6–9, 2001, Bern, Switzerland, 2001, p. 52.
24. Bochsler, P., "Mixed Solar Wind Originating From Coronal Regions of Different Temperature", in *Solar Wind Five*, NASA CP 2280, 1983, pp. 613–622.

Acceleration in a Current Sheet and Heavy Ion Abundances in Impulsive Solar Flares

Yuri E. Litvinenko

Institute for the Study of Earth, Oceans, and Space, University of New Hampshire, Durham, NH 03824-3525, USA

Abstract. The influence of collisional energy losses on stochastic particle acceleration in impulsive solar flares is considered in the context of preferential acceleration of heavy ions. It is shown that ion pre-acceleration in a reconnecting current sheet mitigates the effect of collisional energy losses, thus removing a strong sensitivity of the resulting anomalous abundances on the initial ion charge states. As an example, the expected Fe/O enhancement factors are computed and shown to be comparable with those observed with ACE SEPICA in a series of impulsive flares in 1998. One consequence of the model is that the preferential acceleration of heavy ions can occur only when the plasma gas pressure is large enough, $\beta \approx m_e/m_p$, which may explain the observed correlation between the heavy ion enrichment and selective ^{3}He acceleration in impulsive flares.

INTRODUCTION

Recent observations of impulsive solar flares suggest relatively low temperatures, $T \leq 3 \times 10^6$ K, in the region of the solar corona where charged particle acceleration takes place (see [19] for a review). This fact renewed the interest in two-stage particle acceleration models that attempt to explain the heavy ion enrichments observed in impulsive events [21]. Two-stage models postulate that the selective ion enrichment occurs at the first stage owing to an acceleration mechanism that depends on the initial mass-to-charge ratio A/Q of relatively weakly ionized heavy ions. At the second stage the ions are ionized, so that they are eventually observed to have almost identical A/Q ratios.

A specific model for the first-stage ion enrichment process is based on the effect of collisional energy losses experienced by the ions as they are being stochastically accelerated by plasma turbulence in the solar corona. The losses can play a significant role because of the above-mentioned low temperature at the flare site. It is theoretically established that the competition between stochastic energy gains and Coulomb energy losses leads to preferential acceleration of heavy ions in space plasma [6, 5]. Physically for an ion to enter the acceleration process, the energy gain rate has to exceed the loss rate due to collisions, giving an advantage $\sim Q^2/A$ to heavy ions.

Several variants of this preferential acceleration model have been suggested in application to impulsive solar flares. Particularly detailed calculations [18, 17] demonstrated that the model could indeed reproduce some typical features of heavy ion abundances in impulsive flares. The fraction of each ionic component accelerated, however, turned out to be very sensitive to the assumed plasma temperature that determines the ion charge state distribution in ionization equilibrium. The predicted ion enhancements were criticized for being much larger than the observed ones [11] unless additional averaging over temperature profile in a loop was introduced.

To agree better with observations, the mechanism for heavy ion enrichment based on Coulomb losses should depend only weakly on the ionic charge Q and mass A (defined here in units of the proton mass and electric charge). The simplest solution is that the ions should be pre-accelerated as a jet whose speed is independent of either Q or A [12]. Such high-speed jets are naturally produced through reconnection of magnetic field lines in a current sheet. Given that magnetic reconnection in the corona is the premier candidate for the mechanism of flare energy release, it appears useful to investigate the effect of ion acceleration in a flare current sheet on the formation of anomalous ion abundances in impulsive solar flares. This is the purpose of this paper.

Section 2 reviews the influence of energy losses on the composition of the accelerated particles. Application of the model to a data set from the SEPICA sensor on the ACE spacecraft confirms the unrealistically extreme sensitivity of the model to the plasma temperature. Section 3 shows that ion pre-acceleration in a reconnecting current

CP598, *Solar and Galactic Composition*, edited by R. F. Wimmer-Schweingruber

sheet is indeed important for the following ion enrichment processes in (post)flare loops, decreasing the sensitivity on temperature and improving the agreement with the SEPICA observations. Another result is a new connection between the conditions for the observed heavy ion enrichments and those for the selective ^{3}He acceleration. Conclusions are presented in Section 4.

ION SELECTION DUE TO COLLISIONAL ENERGY LOSSES

The starting point of the analysis is the assumption that ions in the flare loop are being accelerated stochastically by turbulence generated in the course of the flare:

$$d\mathcal{E}/dt = \alpha pc, \quad (1)$$

where p and $\mathcal{E}$ are the momentum and kinetic energy of an ion. The value of the parameter α is defined by a particular type of waves and is generally proportional to the turbulent energy density W. For instance $\alpha \sim (v_A^2/c)k_{\rm min}(W_A/U_B)$ in the case of the Alfvén turbulence, where v_A is the Alfvén speed, U_B is the magnetic field energy density, and $k_{\rm min}$ is the minimum wave vector magnitude of the wave spectrum [14].

Besides being accelerated, the ions also lose energy due to collisions with the ambient particles. The Coulomb loss rate for a particle of speed v, moving in a plasma with electron temperature $T_{\rm e}$ and density $n_{\rm e}$ is as follows:

$$P = \frac{8\pi^{1/2} n_{\rm e} e^4 Q^2}{(2kT_{\rm e} m_{\rm e})^{1/2}} \ln\Lambda F\left[\frac{v}{(2kT_{\rm e}/m_{\rm e})^{1/2}}\right] \quad (2)$$

with the function F reaching the maximum $F_{\rm max} \approx 0.5$ for $v \approx 1.5(2kT_{\rm e}/m_{\rm e})^{1/2}$ [2]. For a particle to be accelerated, the energy gain has to exceed the loss. Hence the condition $d\mathcal{E}/dt > P_{\rm max}$ determines the fraction of the ions entering the acceleration process [6, 18].

In order to stress the essential physical points made in this paper, a constant electron temperature $T_{\rm e}$ is assumed. Given a reliable model for the coronal temperature and density, the results for ion abundances can be integrated over a coronal loop and properly weighted according to the differential emission measure. Another simplifying assumption is that an initially cool flare loop heats up rapidly until the heating and loss processes are balanced for the dominant plasma component (protons):

$$(d\mathcal{E}/dt - P_{\rm max})_{\rm protons} \approx 0. \quad (3)$$

This condition leads to a simple expression [18] for the ion charge states, which can enter the acceleration process:

$$Q < A^{1/2}. \quad (4)$$

For example O and Ne initially can only have charge states $Q \leq Q_{\rm max} = 4$ whereas $Q \leq Q_{\rm max} = 7$ for Fe. It is this selection effect that can be responsible for the observed anomalous ion abundances. Recall that, in the framework of a two-stage model, the ions are fully stripped only after their initial acceleration in a low-temperature ($T \leq 3 \times 10^6$ K) plasma.

The predicted ion abundances are computed by summing over the charge states $Q \leq Q_{\rm max}$, making use of the standard tables of ionization equilibria [1]. In the simplest case the abundances are completely determined by the temperature. Enhancement factors, which are computed using the data on normal element abundances [19], can be directly compared with the ion enhancements observed in solar flares. By way of illustration, in this paper the model predictions are compared with the Fe/O enhancements detected with the SEPICA sensor on ACE [15, 16] in five well-observed impulsive flares in 1998 (days of year 136, 149, 249, 251, 252).

Comparison with the SEPICA data demonstrates that the collisional loss effect can indeed be responsible for large heavy ion enhancements in a million-degree plasma of the solar corona (Fig. 1). The results of the comparison, however, are not quite satisfactory because of the strong sensitivity of the predicted enrichments on the initial charge state distribution that is determined by the temperature. A notable characteristic feature of heavy ion enrichments in impulsive flares is that the relatively low enhancement factors do not change much from flare to flare. This is in contrast to the predictions of the simple model based on the action of collisions. One way to remove the discrepancy is to average the predicted abundances over the flare loop volume [18]. Alternatively, an additional mechanism may act to remove the strong sensitivity on the plasma parameters. One such mechanism is proposed below.

THE EFFECT OF ION ACCELERATION IN A CURRENT SHEET

Perhaps the simplest mechanism that can accelerate ions to suprathermal energies irrespective of their mass is through the production of fast jets, and the most likely way for a jet formation in solar flares is through magnetic reconnection in a current sheet (see [12] for a discussion).

One-dimensional magnetic field in such a sheet would have a zero plane and would not influence the particles that simply move along the electric field direction. On the contrary, studies of particle motion in realistic current sheets with two- and three-dimensional magnetic field show that the field controls the character of particle orbits and the escape speed. The escape speed is determined by

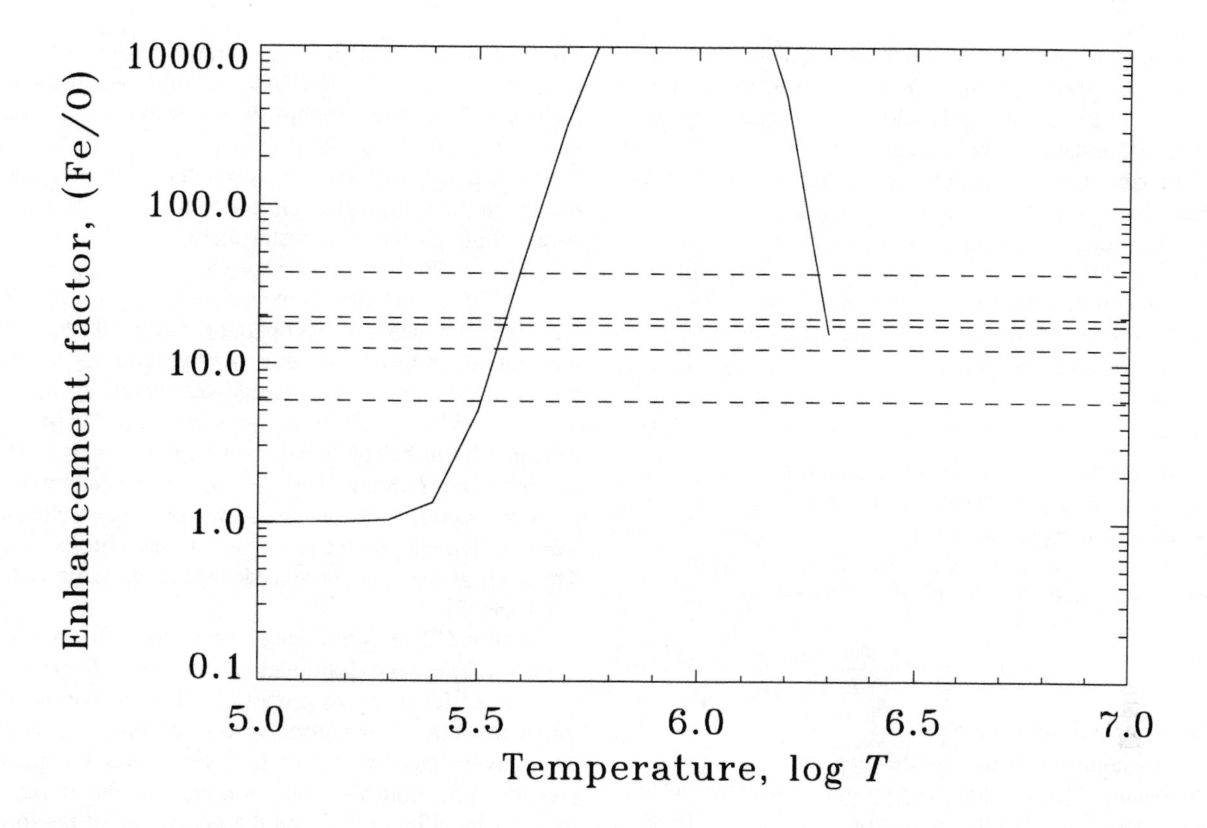

FIGURE 1. Predicted enhancement factors for the Fe/O ratio, compared with the data for five impulsive flares observed with SEPICA ACE in 1998 (Möbius et al., 1999): $\delta = 1.0$

both the electric field E in the sheet and the magnetic field component $B_\perp$ perpendicular to E:

$$v_{\rm esc} = 2cE/B_\perp. \tag{5}$$

This result is easily obtained in the reference frame where the electric field vanishes [22, 9], assuming that the speed remains nonrelativistic and the influence of the nonreconnecting magnetic field component directed along the electric field on the particle motion can be ignored for ions (for a review see [7], where the particle orbits are discussed for various parameter regimes). Thus the escape speed is independent of either A or Q. Self-consistent treatment [8] indicates that the speed is determined by the Alfvén speed based on the reconnecting component of the magnetic field:

$$v_{\rm esc} = v_{\rm A} = B/(4\pi n m_{\rm p})^{1/2}, \tag{6}$$

which is the same result as in the MHD analysis of reconnection.

An important aspect of the jet generation through magnetic reconnection is that the process is essentially collisonless [22, 9]. The acceleration time scales with the inverse gyro-frequency

$$t_{\rm acc} = \pi mc/(qB_\perp) \tag{7}$$

and is much shorter than the typical collisional energy loss time:

$$t_{\rm acc} \approx 10^{-3}(A/Q)\ {\rm s} \ll t_{\rm Coul} \approx 1\ {\rm s} \tag{8}$$

for ions with resulting energies of $0.1 - 1$ MeV per nucleon.

Magnetic reconnection is almost universally accepted as the mechanism of flare energy release. Hence it is useful to investigate quantitatively how the formation of anomalous ion abundances by the collisional energy loss effect is influenced by ion pre-acceleration in a flare current sheet.

It should be noted that the idea of particle pre-acceleration through the reconnection jet is different from the previously suggested model [18] in which the Coulomb losses modified the abundances in the reconnection inflow region before the ions entered the current sheet, whereas particle acceleration by the reconnection electric field in the sheet was treated as the second-stage process. On the contrary, this paper explores the possibility that pre-acceleration in the sheet is followed by stochastic turbulent acceleration in the flare loop (cf., [12, 13]) to the observed energies.

As before, stochastic ion acceleration is described by Equation (1), and the collisional loss rate is described by Equation (2). A significant difference, however, arises because of the rapid collisionless pre-acceleration of ions in the current sheet. Since the pre-acceleration timescale is so short, the Alfvén speed of the reconnection jet can be taken as an initial condition for the turbulent acceleration:

$$v(t=0) = v_A > 0. \tag{9}$$

If the electric field acceleration is strong enough, this initial speed can greatly exceed $1.5(2kT_e/m_e)^{1/2}$, and the ions will not experience the maximum collisional loss as they are being accelerated by the waves. For the purposes of an analytic treatment, it is useful to adopt

$$F(x \gg 1) \approx \frac{\pi^{1/2}}{2x} \tag{10}$$

as a simple limiting case [2].

Assuming again that the heating and loss processes are balanced for the dominant protons in the loop, which were not pre-accelerated, Equations (1), (2), and (10) are combined to give a modified condition for preferential ion acceleration:

$$Q < \delta A^{1/2} \tag{11}$$

if $\delta \geq 1$ (cf., Equation (4)). Here, ignoring factors of order unity,

$$\delta \approx \frac{v_A}{(kT_e/m_e)^{1/2}} \approx \left(\frac{m_p}{m_e}\beta\right)^{-1/2}, \tag{12}$$

and the plasma beta $\beta = 8\pi k n_e T_e/B^2$. For example $\delta \approx 1.3$ for the typical coronal parameters $B \approx 100$ G, $T \approx 10^6$ K, and $n \approx 10^9$ cm^{-3}.

Because the Coulomb loss rate falls off rapidly with increasing speed, pre-acceleration mitigates the selection effect of the losses, thus leading to a more modest enhancements in the heavy ion abundances. Comparison with the ACE SEPICA data, using the same methods and data set as in the previous section, indeed demonstrates a much better agreement of the model and the data (Fig. 2). Further tests of the model should include its application to the first comprehensive observation of the abundances of trans-iron elements in solar energetic particle events, in which the heavy-ion enhancements were measured to be as high as ≈ 1000 relative to O over their coronal values [20].

Collisional losses can be ignored altogether if $\delta \gg 1$. It follows that the selection effect disappears unless

$$\beta \geq m_e/m_p. \tag{13}$$

The same inequality defines a parameter regime in which the electromagnetic ion-cyclotron instability has the lowest threshold of any current-driven instability [4]. This may be more than a coincidence.

Recall that heavy-ion enhancements are well correlated with the spectacular ^{3}He enrichments in impulsive flares, although the relationship between the two phenomena is rather intricate. It appears that ^{3}He-rich flares are also heavy-ion rich but not *vice versa*, suggesting that the ^{3}He enrichment process operates at coronal sites that are enriched in heavy ions due to other reasons, but the process itself does not preferentially accelerate the heavy ions [10]. Whereas the heavy ions are likely to be accelerated by turbulent Alfvén waves, the selective ^{3}He acceleration is currently believed to be produced by resonant interaction with the electromagnetic ion-cyclotron waves generated by electron beams in the corona [23]. The electron beams are produced in the reconnecting current sheet [7].

Equation (13) may provide a new theoretical interpretation for the observed correlation between the heavy-ion and ^{3}He enrichments, as proposed below. It is in small impulsive flares that chromospheric evaporation leads to relatively large values of β. This creates favorable conditions for both the heavy ion enrichments through the collisional loss effect and the generation of the ion-cyclotron waves by electron beams that are generated at the site of magnetic reconnection. The waves in turn interact with the ^{3}He ions, resulting in the ^{3}He enrichments.

Suppose that initially β is small in a loop. As the flare progresses, MHD Alfvén waves are generated in the loop and start accelerating the heavy ions. If the processes of plasma heating and evaporation are strong enough, they eventually result in a larger β thus leading to larger ion enhancement factors and triggering the generation of ion-cyclotron waves responsible for the ^{3}He enrichment. A more detailed analysis is required to understand whether the model can explain the observed heavy-ion enhancements in the absence of the ^{3}He enrichment in some flares.

DISCUSSION

Collisionless magnetic reconnection in coronal current sheets is almost unanimously accepted as the flare energy release mechanism. Highly super-Dreicer electric fields of the order of a few V/cm are associated with rapid reconnection for typical parameters in the corona. Hence efficient particle acceleration in the current sheet is a signature of rapid magnetic reconnection in flares. This makes it necessary to introduce ion pre-acceleration into models for anomalous ion abundances in impulsive solar

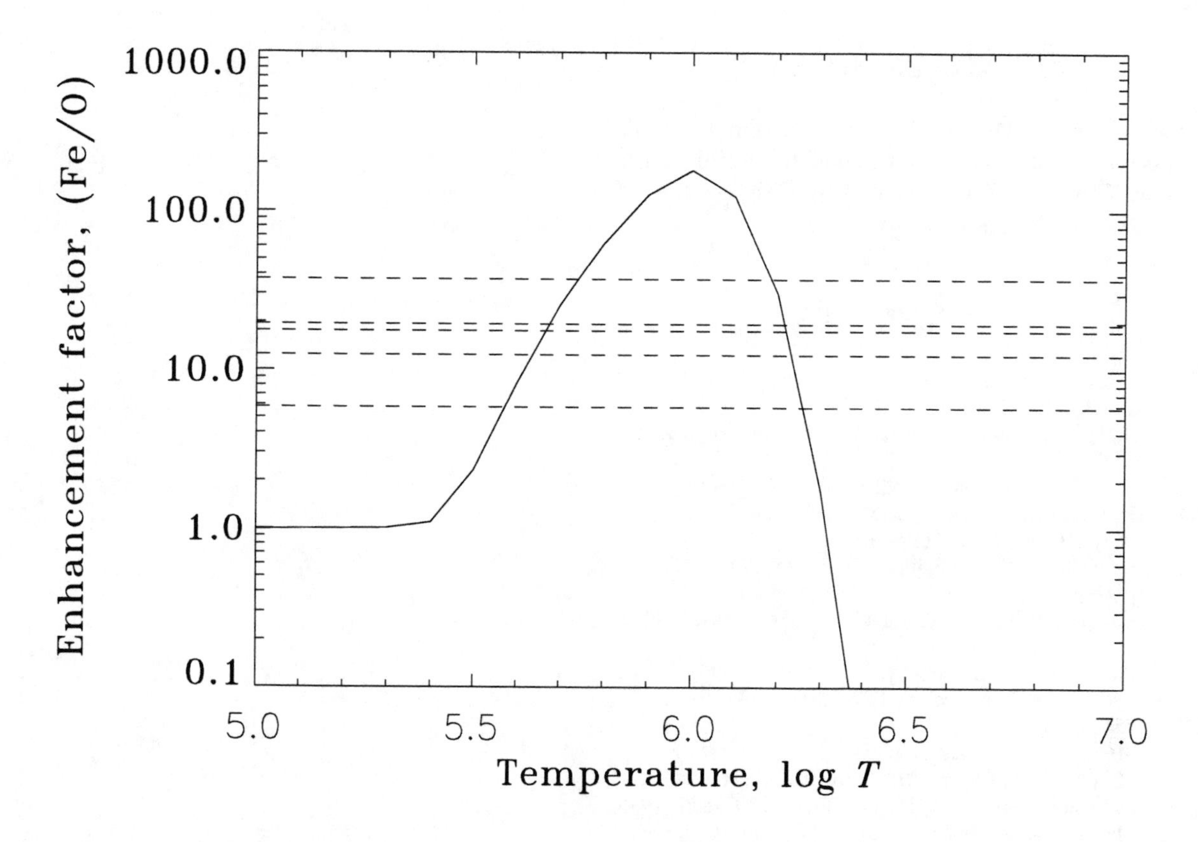

FIGURE 2. Predicted enhancement factors for the Fe/O ratio, compared with the data for five impulsive flares observed with SEPICA ACE in 1998 (Möbius et al., 1999): $\delta = 1.3$

flares. One such model, investigated in this paper, is based on an interplay of the processes of ion energy gain due to acceleration and energy loss due to collisions [6, 5, 18, 12].

It was demonstrated that pre-acceleration in a reconnecting current sheet mitigates the influence of the losses on stochastic particle acceleration, thus removing a very strong sensitivity of the resulting abundances on the initial distribution of ion charge states and leading to a better agreement with observations. This result is a simple consequence of the fact that the energy loss rate falls off rapidly with increasing energy. To explain why the eventually observed ions are almost fully stripped, the model can be developed further by relaxing the assumption of ionization equilibrium.

It was suggested some time ago [12] that initial energization through plasma jets is necessary to explain why the abundances of solar energetic particles are not too different from normal solar abundances. Application of this idea to the selective acceleration of heavy ions in impulsive flares leads to an interesting quantitative requirement, Equation (13), for the selection to occur. It is very interesting that the same requirement defines a parameter regime where the electromagnetic ion-cyclotron instability has the lowest threshold of any current-driven instability [4]. Resonant acceleration by the ion-cyclotron waves is the most likely mechanism for ^{3}He preferential acceleration in impulsive flares [23]. Thus the approach of this paper may provide a new theoretical connection between the processes of heavy-ion and ^{3}He enrichments, which is well established observationally [10].

Finally it should be noted that another model for the heavy ion enrichments invokes cyclotron damping of cascading turbulence. The damping is more effective for heavy ions for which the resonance condition is easier to satisfy than for protons [3]. Recently this approach has been shown to give good agreement with typical impulsive flare abundances [13]. It appears interesting to specify under what physical conditions either of the suggested mechanisms could be responsible for the observed ion abundances in flares.

ACKNOWLEDGMENTS

I am grateful to Dr. Grand Marnier for stimulating discussions and the anonymous referee for useful suggestions. This work was supported by NSF grant ATM-9813933 and NASA grant NAG5-7792.

REFERENCES

1. Arnaud, M., and Rothenflug, R., Astron. Astrophys. Suppl., 60, 425 (1985).
2. Butler, S. T., and Buckingham, M. J., Phys. Rev., 126, 1 (1962).
3. Eichler, D., Astrophys. J., 229, 413 (1979).
4. Forslund, D. W., Kindel, J. M., and Stroscio, M. A., J. Plasma Phys., 21, 127 (1979).
5. Korchak, A. A., and Filippov, B. P., Sov. Astron., 23, 323 (1979).
6. Korchak, A. A., and Syrovatskii, S. I., Sov. Phys. Doklady, 3, 983 (1958).
7. Litvinenko, Y. E., Astrophys. J., 462, 997 (1996).
8. Lyons, L. R., and Speiser, T. W., J. Geophys. Res., 90, 8543 (1985).
9. Martens, P. C. H., Astrophys. J., 330, L131 (1988).
10. Mason, G. M., Reames, D. V., Klecker, B., Hovestadt, D., and von Rosenvinge, T. T., Astrophys. J., 303, 849 (1986).
11. Meyer, J.-P., Astrophys. J. Suppl., 57, 151 (1985).
12. Melrose, D. B., Aust. J. Phys., 43, 703 (1990).
13. Miller, J. A., in High Energy Solar Physics–Anticipating HESSI, eds. R. Ramaty and N. Mandzhavidze (San Francisco: ASP), 145 (2000).
14. Miller, J. A., Guessoum, N., and Ramaty, R., Astrophys. J., 361, 701 (1990).
15. Möbius, E., Klecker, B., Popecki, M. A., et al., in 26th ICRC Proceedings, ed. D. Kieda et al., 6, 87 (1999).
16. Möbius, E., Klecker, B., Popecki, M. A., et al., in Acceleration and Transport of Energetic Particles Observed in the Heliosphere: ACE 2000 Symposium, ed. R. E. Mewaldt et al., 131 (2000).
17. Mullan, D. J., Astrophys. J., 268, 385 (1983).
18. Mullan, D. J., and Levine, R. H., Astrophys. J. Suppl., 47, 87 (1981).
19. Reames, D. V., Space Sci. Rev., 90, 413 (1999).
20. Reames, D. V., Astrophys. J., 540, L111 (2000).
21. Reames, D. V., Meyer, J. P., and Rosenvinge, T.T., Astrophys. J. Suppl., 90, 649 (1994).
22. Speiser, T. W., J. Geophys. Res., 70, 4219 (1965).
23. Temerin, M., and Roth, I., Astrophys. J., 391, L105 (1992).

On the Energy Dependence of Ionic Charge States in Solar Energetic Particle Events

B. Klecker[1], E. Möbius[2], M.A. Popecki[2], M.A. Lee [2], and A.T. Bogdanov[3]

(1) Max-Planck-Institut für extraterrestrische Physik, 85741 Garching, Germany
(2) Space Science Department and Department of Physics, University of New Hampshire, Durham, 03824, NH, USA
(3) Technische Universität Braunschweig, 38106 Braunschweig, Germany

ABSTRACT In large, gradual, interplanetary shock related solar energetic particle events an increase of the mean ionic charge of heavy ions with energy is often observed. There are large event to event variations. For Fe, for example, the observed variation in the energy range 0.1 – 1 Mev/nuc ranges from ~ 1 – 2 charge units to about 4 charge units. Large increases of the mean ionic charge with energy could be explained by acceleration at low coronal altitudes and charge stripping in this sufficiently dense environment. However, some of the events show evidence for local acceleration at interplanetary shocks. In this low–density environment the charge stripping mechanism would not work. We investigate rigidity dependent effects at quasi–parallel shocks, that could be due to mass per charge dependent effects of the acceleration, propagation, or loss processes, resulting in a mass per charge dependent exponential cutoff of the energy spectra. We show that a mass per charge dependent cutoff as observed recently in large, gradual solar energetic particle events will also result in an increase of the mean ionic charge with energy.

INTRODUCTION

The ionic charge composition of suprathermal ions is a sensitive indicator for the temperature of the source region. Besides that, the acceleration and transport processes depend significantly on velocity and rigidity, i.e. on the mass and ionic charge of the ions. With experiments onboard the SAMPEX, SOHO and ACE spacecraft earlier measurements with ISEE–3 (e.g. Hovestadt et al., 1981; Luhn et al., 1985) of the ionic charge composition of suprathermal ions accelerated at coronal or interplanetary shocks have been extended over a much larger energy range. With SAMPEX an increase of the mean ionic charge at energies > 10 MeV/nuc has first been observed for two gradual solar energetic particle events (SEP) in October / November 1992 (Mason et al., 1995; Leske et al., 1995; Oetliker et al., 1997, s. a. Fig. 1). New measurements with improved resolution and sensitivity with SOHO and ACE showed that the mean ionic charge in the energy range ~ 30 to 500 keV/nuc increases with energy in many events, with a large event–to–event variability (Möbius et al., 1999; Mazur et al., 1999, Bogdanov et al., 2000; Klecker et al., 2000). The increase of the mean ionic charge between solar wind energies and ~ 60 MeV/nuc is most noticeable for heavy ions (e.g. Si, Fe). Figure 1 shows as an example for the range of the energy dependence of the mean ionic charge of Fe the observations for 3 events in 1992, 1997, and 1998.

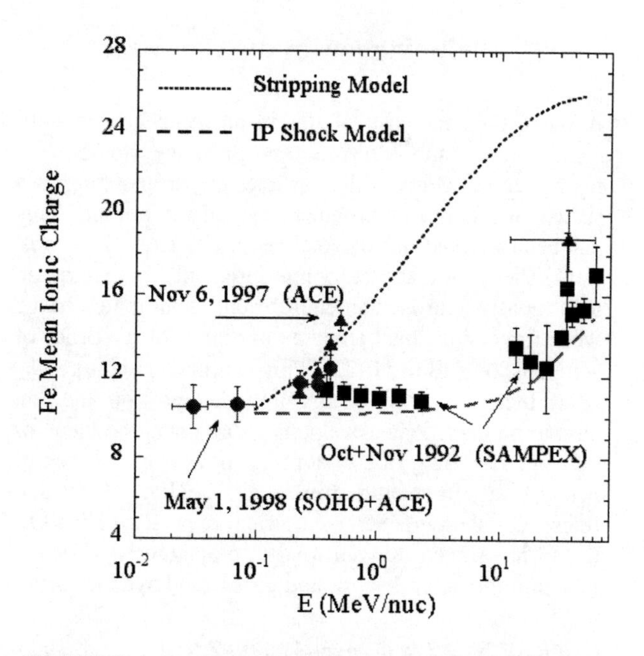

FIGURE 1. Typical cases of the energy dependence of the mean ionic charge of Fe as observed for 3 gradual SEP events in 1992, 1997 and 1998. The dotted line shows the energy dependence as computed for charge stripping low in the corona. The dashed line is taken from Fig. 3; for details see the text.

CP598, *Solar and Galactic Composition*, edited by R. F. Wimmer-Schweingruber

The November 1992 event shows an almost constant Fe mean ionic charge of ~ 11 at energies < 3 MeV/nuc with a large increase to ~ 17 at ~ 60 MeV/nuc. In the November 6, 1997 event, on the other hand, the mean ionic charge of Fe increases from ~ 10 to 14.5 at low energies, in the energy range 0.18 – 0.54 MeV/nuc, with a further increase up to ~ 18.5 in the energy range 12 – 60 MeV/nuc (Cohen, et al., 1999). However, it should be noted that in this case the mean ionic charge at high energies has been inferred from M/Q dependent composition arguments and has not been directly measured. The third example (May 2, 1998) shows the Fe mean ionic charge as determined for a particle intensity increase at energies < 1 MeV/nuc that was closely related to the passage of an interplanetary shock, and the acceleration most likely occurred in interplanetary space (Klecker al., 2000). In this case a moderate increase of the mean ionic charge from solar wind values (~ 10) at suprathermal energies (30 – 70 keV/nuc) by ~ 2 charge units to ~ 12 at ~ 400 keV/nuc was observed. For this event no ionic charge state measurements at higher energies are available.

INTERPRETATION OF AN ENERGY DEPENDENT IONIC CHARGE DISTRIBUTION

Stripping low in the Corona

A monotonic increase of the mean ionic charge with energy in the sub–MeV/nuc energy range would be a natural consequence if the particles are propagating in a sufficiently dense environment low in the corona. This has been pointed out by Reames et al., 1999. It is well established from laboratory measurements that energetic ions rapidly approach an equilibrium mean ionic charge when traversing small amounts of matter of the order of ~ $\mu g/cm^2$ (e.g. Betz, 1972). This equilibrium charge Q_{eq} (Z,v) increases with energy and depends on the ion atomic number, Z, and velocity, v and it is the basis of our understanding of the stopping power and ranges of heavy ions in matter (see e.g. Ziegler, 1980 and references therein). Semi–empirical expressions for Q_{eq} (Z, v) have been derived for the propagation of heavy ions through neutral solids and gases and have the form

$$Q_{eq}(Z,\beta) = Z * (1 - \exp(-125\,\beta / Z^{2/3}), \qquad (1)$$

where $\beta = v/c$, and c is the speed of light. The functional form of the exponent in (1) is based on the interactions of heavy ions with the electrons of the atoms of the target material. Note that the factor $Z^{2/3}/137$ is the electron velocity $\beta_{TF} = v_{TF}/c$ in the Thomas–Fermi atom model, in units of the speed of light (see e.g. Northcliff and Schilling, 1970, and references therein).

With the argument based on earlier work by Luhn and Hovestadt (1987) that in a plasma environment ionizing interactions will depend on the relative velocity β_{rel} of the ion and the thermal electrons, Reames et al. (1999) estimated the mean ionic charge of heavy ions for the limiting case of equilibrium mean charge states due to stripping by replacing β in equation (1) by $\beta_{rel} = \beta_{ion} + \beta_{e,th}$, where β_{ion} and $\beta_{e,th}$ are the ion speed and the thermal speed of the electrons, respectively, in units of the speed of light. The dotted line in Fig. 1 illustrates the resulting energy dependence of the equilibrium ionic charge of iron for a coronal temperature of 1.3 MK, using the approximation of Reames et al. (1999). Figure 1 shows that, both, the functional dependence of the increase of the mean ionic charge of Fe as observed in the Nov 6, 1997 SEP event and the absolute values in the energy range 0.2 – 0.5 MeV/nuc can be reasonable well reproduced. The requirements on electron density, N_e, and resident time, τ, to achieve this equilibrium ionic charge depend on energy and are at ~ 1 MeV/nuc $N_e\,\tau \geq 10^{10}$ $cm^{-3}s$ (Reames et al. 1999, Kocharov et al., 2000). This corresponds, for acceleration times of ~ 10 to 10000 s, to electron densities of 10^9 to 10^6 cm^{-3}, i.e. to radial distances at the Sun significantly below 2 solar radii. However, at higher energies the equilibrium ionic charge of Fe is significantly higher than the mean ionic charge of Fe inferred by Cohen et al. (1999) for the Nov 6, 1997 event. Whether this could be due to non–equilibrium effects as discussed below needs detailed model calculations and is beyond the scope of this paper.

The stripping process will also work for somewhat lower values of $N_e\,\tau$, with the difference that charge stripping equilibrium will not be reached and the mean ionic charge will be somewhat smaller than the equilibrium value and has to be computed by using the appropriate ionization and recombination cross sections. Models including the effects of shock acceleration, statistical acceleration and charge exchange reactions have been used to investigate the implications of charge exchange reactions on ionic charge states in gradual (Barghouty and Mewaldt, 1999, 2000; Kartavykh and Ostryakov, 1999) and impulsive solar energetic particle events (Ostryakov et al., 2000). The results showed that the mean ionic charge and the charge distribution in the energy range 0.20 – 0.50 MeV/nuc as observed in the Nov 6, 1997 event can be reproduced reasonably well (e.g. Stovpyuk and Ostryakov, 2001). However, these authors did not try to simultaneously fit the ionic charge of Fe at higher energies as inferred for this event by Cohen et al. (1999).

However, some events, in particular those showing large intensity enhancements at low energy at an interplanetary shock, consistent with local acceleration in interplanetary space, exhibit also small, but significant increases of the mean ionic charge of heavy ions with energy (Bogdanov et al., 2000, Klecker et al., 2000, see also Fig. 1). For acceleration in interplanetary space stripping will not work because of the small density, and other processes may play an important role.

Local Shock Acceleration and Energy Dependence of the Mean Ionic Charge

In the test particle limit of diffusive shock acceleration at a quasi–parallel, planar shock, steady state conditions, and no losses, the distribution function of ions with mass, M, and ionic charge, Q, is given by

$$f_{M,Q} = f_{0,M,Q}\; v/v_0^{-\gamma}, \qquad (2)$$

with particle velocity, v, and injection velocity, v_0, respectively. In this ideal case the spectral index γ is determined by the shock compression ratio (e.g. Axford et al., 1977; Blandford and Ostriker, 1978) and is independent of mass and ionic charge, i.e. no variations of the ionic charge (and elemental) composition with energy would be expected. However, if the energy spectra depend on both, velocity and mass per charge, the ionic charge composition and the elemental composition will vary with velocity, as shown below. Deviations from the simple power law energy dependence of the energy spectra can be expected if at least one of the above assumptions is violated. Non steady state conditions, particle losses, or finite shock size will generally result in a high–energy roll–over, with particle spectra falling off more steeply than described by a power law (e.g. Forman, 1981; Ellison and Ramaty, 1985, and references therein). Energetic particle spectra similar to steady state solutions of a planar shock model with exponential roll–over from losses and finite shock size have been described, for example, by Ellison and Ramaty (1985). They find for a diffusion coefficient κof the form $\kappa \sim \beta R$

$$j(E) = j_0\, E^{-\gamma} \exp(-\tilde{\square}\square), \text{ with} \qquad (3)$$

$$E_0\,(A/Q) \sim E_{0,p} * Q/A, \qquad (4)$$

where $E_{0,p}$ is the e–folding energy of protons. Spectra of this form have been reported recently for large, gradual events by Tylka et al. (2000). The observed spectra have been fitted using the spectral shape (3) and showed systematic differences of the e–folding energy E_0 for particles of different mass per charge ratio that could be approximated by $E_0\,(A/Q) \sim E_{0,p} * (Q/A)^{\delta}$, with δ in the range 0.8 – 2.3. In fact, Tylka et al. (2000) successfully used these systematic differences to infer mean ionic charge states of heavy ions for 2 large, gradual events in 1998.

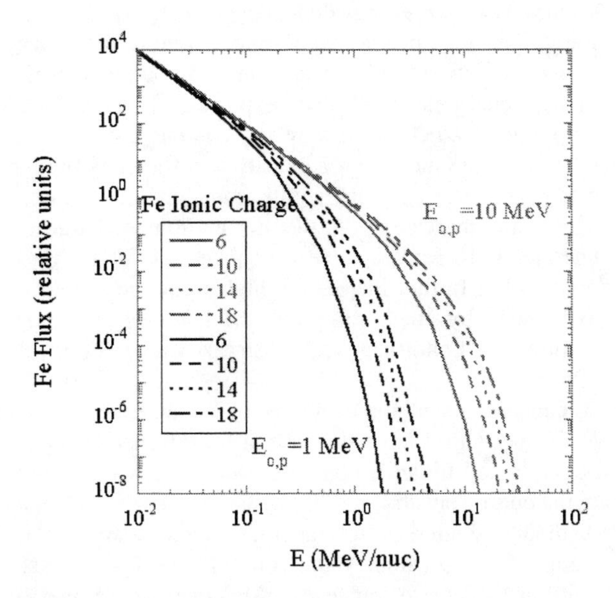

FIGURE 2. Model energy spectra of selected ionic charge states of Fe for power law spectra with an exponential cutoff of the form $j(E) = j_0\, E^{-\gamma} \exp(-E/E_0)$, with $E_0\,(A/Q) \sim E_{0,p} * Q/A$.

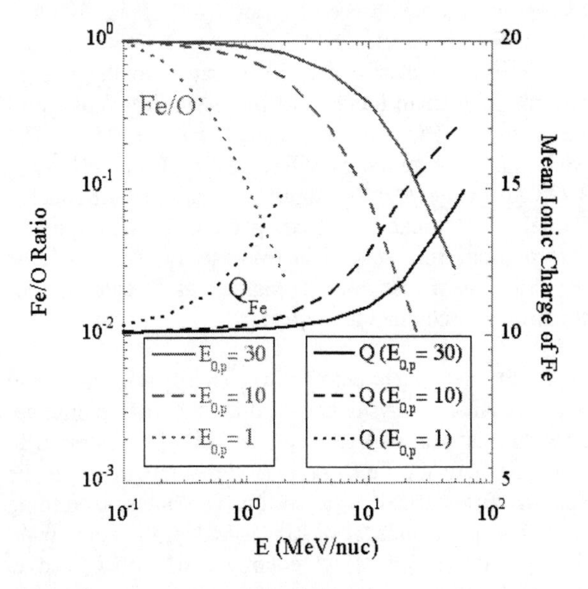

FIGURE 3. Mean ionic charge of Fe and Fe/O ratio as a function of energy, computed for 3 different values of the e–folding energy $E_{0,p}$.

If this spectral form holds for individual charge states, a moderate energy dependence of the mean ionic charge is a natural consequence. To illustrate this effect we computed in Fig. 2 the differential energy spectra of iron with ionic charge states of 6, 10, 14, and 18, using equations (3) and (4), $\delta = 1$, and $\gamma = 2$.
The effect of the rigidity dependent cutoff that varies systematically with ionic charge state is evident. A M/Q dependent cutoff of the form described above will result in a suppression of low ionic charge states, i.e. in an increase of the mean ionic charge with energy. At the same time, also the relative abundances of elements with different M/Q will vary with energy.

In Fig. 3 we computed the mean ionic charge of Fe for three values of the e–folding energy of protons (30 MeV, 10 MeV and 1 MeV), assuming solar wind abundances of the individual Fe charge states as reported for the May 1, 1998 Interplanetary Shock event (Klecker et al. 2000). Figure 3 demonstrates that values of $E_{0,p}$ in the range of ~ 10 – 30 MeV as observed by Tylka et al (2000) result, for this simple model, in an almost constant mean ionic charge at energies < 3 MeV/nuc, with an increase to Q_{Fe} ~ 15 – 17 at 50 MeV/nuc. For $E_{0,p}$ = 30 MeV, this functional dependence is very similar to the SAMPEX observations in the October 30 to November 6, 1992 event. This is illustrated by the dashed line in Fig. 1, where $E_{0,p}$ = 30 MeV and the same parameters have been used as in Fig. 3. Smaller values of E_0 will result in an increase of the mean ionic charge starting at lower energies, as demonstrated for $E_{0,p}$ = 1 MeV in Fig. 3. This example shows that small increases of the mean ionic charge of Fe by 1 – 2 charge units in the energy range 0.1 – 1 MeV/nuc could also be caused by a mass per charge dependent cutoff in the energy spectra. The small increase of the mean ionic charge of Fe from ~ 10 to 12.5 in the energy range 0.03 – 0.4 MeV/nuc as observed by SOHO and ACE in the May 1, 1998 SEP event (Fig. 1) would be consistent with such a model. For a quantitative comparison detailed spectral information from several instruments on ACE will be needed. This investigation is planned as an extension of the present study in a future paper.

A rigidity dependent cutoff of the energy spectra would also result in systematic variations of elemental abundances with particle velocity. This is illustrated in Fig. 3, where the Fe/O–ratio is computed from the Fe spectra for individual ionic charge states, assuming ionic charge abundances as measured in the solar wind during the May 1, 1998 event, as described above, summed over all charge states, and an oxygen spectrum computed also with equations (3) and (4). For oxygen a charge state of 6 and the same parameters have been used as for iron, and the Fe/O–ratio has been normalized to 1 at 0.01 MeV/nuc. Figure 3 illustrates that the increase of the mean ionic charge and the decrease of the Fe/O ratio are directly related, as expected. Values of $E_{0,p}$ in the range 10 – 30 will result in a decrease of the Fe/O–ratio in the energy range ~ 0.5 – 20 MeV/nuc by a factor of ~ 10. In case of the October 30 and Nov 2, 1992 events a decrease of the Fe/O–ratio from 0.41 ± 0.03 near 1 MeV/nuc (Mason et al., 1995) to 0.031 ± 0.007 and 0.071 ± 0.006 at ~ 28 – 65 MeV/nuc has been observed, indeed (Selesnick et al., 1993). Thus, the rigidity dependent cutoff effect seems to be a promising candidate for the interpretation of ionic charge variations with energy of the type observed in the Nov 2, 1992 event that are difficult to understand otherwise (s. a. discussion in Oetliker et al., 1997), or for only small increases of the mean ionic charge with energy at energies < 1 MeV/nuc.
In this mechanism, the variation of the mean ionic charge with energy is limited to the range of ionic charge states available in the source, i.e. there is the additional requirement that the initial ionic charge distributions consist of at least two ionic charge states. Therefore the largest effect can be expected for ions with a broad solar wind ionic charge distribution as observed for Fe, consistent with observations.

DISCUSSION AND SUMMARY

An increase of the mean ionic charge of heavy ions with energy has been observed in many solar energetic particle events. There is a large event–to–event variability of this energy dependence.
Large increases of the mean ionic charge in the energy range 0.03 to 0.5 MeV/nuc with a further increase at higher energies could be explained by additional stripping of electrons low in the corona (e.g. Q_m (Fe) ~ 15 at 1 MeV/nuc), if the density is sufficiently high to establish charge stripping equilibrium.
Moderate increases of the mean ionic charge at energies < 10 MeV/nuc (e.g. Q_m (Fe) ~ 13 at 10 MeV/nuc) with a further increase at higher energies could be explained by the stripping mechanism and non-equilibrium conditions (e.g. Ostryakov and Stovpyuk, 1999).
A constant mean ionic charge of Fe at energies < 1 MeV/nuc with a large increase at higher energies could be explained by an exponential, M/Q dependent, cutoff in the energy spectra.
A moderate increase of the mean ionic charge of Fe from ~ 9 – 10 (Solar Wind source) by ~ 1 – 3 charge units in the energy range < 1 MeV/nuc would also be consistent with a rigidity dependent exponential cutoff of the energy spectra, provided the e–folding energy E_0 is sufficiently small.

So far we discussed the two limiting cases, i.e. (1) stripping low in the corona, and (2) M/Q dependent acceleration effects in interplanetary space. In principle, a combination of the two effects could also occur. In both cases, the large variability of the energy dependence could be explained by the variability of the acceleration parameters and, in case of acceleration close to the Sun, by the variability of the electron density. In order to determine which of the mechanisms is important in individual SEP events, a precise determination of the ionic charge dependence on energy and the determination of the energy spectra over a wide energy range are essential. In the case of stripping, for a large range of coronal temperatures of ~ 10^6 to 10^7 K a relatively steep increase of the mean ionic charge in the energy range 0.1 – 1 MeV/nuc is expected (Kocharov et al., 2000), i.e. in the energy range where measurements from SAMPEX and ACE are available. For M/Q dependent cutoff effects and an exponential cutoff, on the other hand, at $E < 1$ MeV/nuc no energy dependence, or only a gradual increase, accompanied by compositional variations, would be expected. However, the approximation of an exponential cutoff may be too simplistic and not be a good fit for all cases. As has been shown by Lee (2000) in a model combining shock acceleration with interplanetary transport and upstream escape, the exponential cutoff could show a much more involved dependence on particle and shock parameters.

In any case, the precise determination of the ionic charge distribution in the energy range ~ 0.1 to 10 MeV/nuc provides a powerful tool for the investigation of source locations and acceleration processes of solar energetic particles in CME or interplanetary shock related events.

REFERENCES

1. Axford, W.I., Leer, E., and Skadron, G., Proc. 15th Internat. Cosmic Ray Conf. (Plovdov) **11**, 132 - 142 (1977).
2. Barghouty, A.F., and Mewaldt, R.A., Proceedings of the 26th ICRC **6**, 138 - 141 (1999).
3. Barghouty, A.F., and Mewaldt, R.A., ACE 2000 Symposium, AIP Conf. Proc. **528**, 71 - 78 (2000).
4. Betz, H.-D., Revs. Modern Phys. **44**, 465-539 (1972).
5. Blandford, R.D., and Ostriker J.P., Ap. J. (Letters) **221**, L29-L32 (1978).
6. Bogdanov, A.T., et al., ACE 2000 Symposium, AIP Conf. Proc. **528**, 143 - 146 (2000).
7. Cohen, C.M.S., et al., Geophys. Res. Lett. **26**, 2697 - 2700 (1999).
8. Ellison, D.C. and Ramaty, R. Astrophys. J.**298**, 400 - 408 (1985).
9. Forman, M.A., Adv. Space Res. **1**, 97 - 100 (1981).
10. Hovestadt, D., et al., Adv. Space Res. **1**, 61 - 64 (1981).
11. Hovestadt, D. et al., The SOHO Mission (editors: Fleck, B. et al.), Solar Physics **162**, 441 - 481 (1995).
12. Kartavykh, Y.Y. and V.M. Ostryakov, 26th ICRC **6**, 272 - 275 (1999).
13. Klecker, B. et al., ACE 2000 Symposium, AIP Conf. Proc. **528**, 135-138 (2000).
14. Kocharov, L. et al., Astron. & Astrophys. **357**, 716 - 724 (2000).
15. Lee, M. A., ACE 2000 Symposium, AIP Conf. Proc. **528**, 3 - 18 (2000).
16. Leske et al., Astrophys. J. Lett. **452**, L149 - L152 (1995).
17. Luhn, A. et al., Proc. 19th ICRC, **4**, 241 - 248 (1985).
18. Luhn, A., and D. Hovestadt, 1987, Astrophys. J. **317**, 852 - 857 (1987).
19. Mason, G.M., et al., Astrophys. J. **452**, 901 (1995).
20. Mazur, J.E., G.M. Mason, M.D. Looper, et al., Geophys. Res. Lett. **26**, 173 - 176 (1999).
21. Möbius, E., et al., Space Science Reviews **86**, 449 - 495 (1998).
22. Möbius, E., M. Popecki, B. Klecker, D. Hovestadt, et al., Geophys. Res. Lett. **26**, 145 - 148 (1999).
23. Northcliffe, L.C., and R.F. Schilling, Nuclear Data Tables, A7, 233 - 463 (1970).
24. Oetliker, M., B. Klecker, D. Hovestadt, et al., Astrophys. J. **477**, 495 - 501 (1997).
25. Ostryakov, V.M and Stovpyuk, M.F., Solar Physics **189**, 357-372 (1999).
26. Ostryakov, V.M., Kartavykh, Y.Y., Ruffolo, D., et al, J. Geophys. Res. **105**, A12, 27315 - 27322 (2000).
27. Reames, D.V., C.K. Ng, and A.J. Tylka, Geophys. Res. Lett. **26**, 3585 - 3588 (1999).
28. Selesnick, R.S., et al., Astrophys. J., **418**, L45 - L48 (1993).
29. Stovpyuk, M.F. and V.M. Ostryakov, Solar Physics **198**, 163 - 167 (2001).
30. Tylka, A.J. et al., ACE 2000 Symposium, AIP Conf. Proc. **528**, 147 - 152 (2000).
31. Ziegler, J.F., Handbook of stopping cross-sections for energetic loss in all elements, **Vol. 5**, Editor J.F. Ziegler, Pergamon Press, (1980).

Particle Acceleration in Interplanetary Shocks: Classification of Energetic Particle Events and Modeling

M. Den*, T. Yoshida† and K. Yamashita**

*4-2-1, Nukuikita, Koganei, Tokyo, 184-8795, Japan
†2-1-1, Bunkyo, Mito, Ibaraki, 310-8512, Japan
**Chiba, 243-8522, Japan

Abstract. Gradual solar energetic particle events using data observed by EPAM (Electron, Proton, and Alpha Monitor), SWEPAM (Solar Wind Electron, Proton and Alpha Monitor) and Magnetic Field Experiment (MAG) on the ACE are studied. The energetic particle events are classified in four types according to the variance in the flux, the characteristic duration time of the events and the maximum energy of the accelerated particles. We perform the modeling of typical events by using numerical simulations for two types. We apply the stochastic differential equation method coupled with the particle splitting to diffusive acceleration, and obtain the energy spectrum and the spatial distribution of the accelerated particles. The relation between the different classes of the events and injection model is discussed.

INTRODUCTION

Interplanetary shock waves play an important role as an "initiator" of geomagnetic disturbances and an acceleration of energetic particles. Shock accelerated particle intensities mostly begin to enhance in advance of shock passages with a duration ranging from several hours to more than 10 days, while only shock waves are observed without an enhancement of particle fluxes. There are many variations in particle-flux behavior and the maximum energy of accelerated particles depends on events. On the other hand, shock acceleration theories proposed to explain those evens are diffusive shock acceleration [1, 2] and shock drift acceleration [3]. (Stochastic acceleration, which is second-order Fermi acceleration, is not taken account here because it is inefficient mechanism comparing with other two theories.) Our questions are as follows: what causes variation in energetic particle events and how are they understood based on the acceleration theories?

Features of energetic proton events depend on several factors. The difference in an energy range of accelerated protons leads to different profile of a proton flux. van Nes et al.[4] studied proton energy spectra in the range of 35-1600 KeV by surveying 75 interplanetary shocks, sub-divided shock events into different four groups, and analyzed the relation between the effect of shock accelerations (diffusive or drift) and the shock strength. As for higher energy shock associated particle events, Kallenrode[5] examined the intensity profiles of 5-MeV protons for 351 interplanetary shocks, divided them into three groups, and compared the features of the particle events with those of the shocks. Cane et al.[6] claimed that the intensity-time profiles of solar proton events (SEPs) depended on longitude of the source region on the Sun by studying 235 solar proton events observed by multiple spacecrafts. Reames[7] examined the spatial distribution of high energy particles accelerated by coronal mass ejection(CME)-driven shocks by using multi-spacecraft data and concluded that the intensity profiles of the particles ahead of the shock could depend on the longitude of the point in the shock wave of the magnetic field connected to the observer. Using the assumption that the nose of the shock is the region where acceleration is the strongest, they explained the variation of the particle flux for the typical data.

In this paper, we study shock accelerated energetic particle events with energies ranging from 47 KeV to 4.75 MeV using particle data observed by the Electron, Proton and Alpha Monitor (EPAM) onboard the ACE spacecraft. This energy range covers energies with which van Nes carried out statistical study of particles and lower part of an energy range of particles which Cane et al., Reames and Kallenrode investigated. Energetic particles in this energy range showed various behavior which is due to multiple reasons such as shock strength, an acceleration mechanism and heliolongitude of source region of energetic particles. Shock associated particle events in this energy range are classified in four groups according to the variance in the flux, the characteristic

CP598, *Solar and Galactic Composition*, edited by R. F. Wimmer-Schweingruber

duration time of the events and the maximum energy of the accelerated particles. To gain important clue about the mechanism leading to different groups, shock parameters such as a compression ratio, a shock normal angle, a shock velocity and a maximum intensity of a particle flux are obtained for each group using solar wind data observed by SWEPAM, and MAG instruments onboard the ACE and CELIAS/MTOF Proton Monitor (PM) on the SOHO.

We try to combine acceleration theories and observations by modeling events in each type and to clarify the cause difference in particle events. At first, we perform modeling of typical events of types 1 and 2 by using numerical simulations. The well known acceleration mechanism of energetic particles associated with CME-driven shocks is first-order Fermi acceleration, however, there are few studies in which show how the simple shock diffusion acceleration theory could explain observations of energetic events. Lee and Ryan[8] solved analytically the time-dependent cosmic ray transport equation with some strong assumptions, e.g., the diffusion coefficient is self-similar and proportional to r^2 where r is radial distance, which is not widely accepted. Zank, Rice and Wu[9] also studied shock acceleration taking into account dynamical evolution of a spherical shock wave. Their approach is semi-analytical: hydrodynamical quantities such as a shock speed are calculated numerically, while they obtain the particle flux analytically with simplification of the transport equation. We solve the cosmic ray transport equation numerically without any assumptions concerned with the equation. Our calculation method is stochastic differential equation (SDE) method, a kind of Monte Calro simulation, coupled with the particle splitting. We consider that an injection of particles plays an important role for understanding of the mechanism leading to different types. We set different injection models for types 1 and 2, calculate the energetic particle flux and the spatial distribution, and compare them with observations.

In section 2, we show typical energetic particle events and describe classification. Section 3 gives modeling method and simulation results. Section 4 is devoted to our concluding remarks.

OBSERVATIONS AND CLASSIFICATION OF ENERGETIC PARTICLE EVENTS

Observations

A criterion of detection of interplanetary shock waves is that discontinuities in a solar wind density, a speed, a magnetic-field intensity, and a temperature all occur at the same time. We used the solar wind plasma data provided by SWEPAM teams, the magnetic field data observed by the MAG teams. CELIAS/MTOF PM on the SOHO also provides useful data for the solar wind data and we made use of the data in the case that SWEPAM observations were not available. As for a particle flux, ion data observed by EPAM were studied and we also used GOES particle data in case of large energetic events in which several 10-MeV protons were detected. We surveyed the data from November 6, 1997 to July 10, 2000 and found 68 interplanetary shock events. The number of events is much less than that, 112 events for the same period, listed by Smith[10]. One of reasons of this discrepancy is that the author adopted a different criterion for interplanetary shock events, namely, if there are discontinuities in either MAG or SWEPAM data, he regards the observations as shock events. Our criterion may be rather restricted one and detail comparison is needed, which is our future work. In this paper, we take into account only transient shocks, not corotaing shocks.

Classification

Our question is why there are various behaviors in diffusive region for the particle flux. As is well known, diffusive shock accelerated particles come in advance of the associated shock. However, difference between a start time of particle flux enhancement from the background level and a time of the shock passage, we call a "precursor time", distribute from about 0 hours to two days.

Thus, we classify energetic particle events in the following four types according to a variance in a particle flux, a characteristic duration time of particle events and a maximum energy of accelerated particles.

Type 1: Typical shock diffusive accelerated particle events. The precursor time is about one days and larger than 10-MeV protons are not accelerated (Fig. 1 upper left). This type corresponds to "PIP" events in [5].

Type 2: Solar protons and interplanetary shock accelerated protons are mixed (Fig. 1 upper right). Accordingly a range of the diffusion region cannot be identified. The associated CMEs in this type are accompanied by solar proton flares, so the events are almost associated with strong shocks. The maximum energy intensity ranges from 10^3 to 10^6. This type corresponds to "SIP" events in [5].

Type 3: Behavior of particle flux is similar as the one in type 1, but the precursor time and total duration of enhancement of the flux are very long (Fig. 1 lower left). Furthermore, larger than 10-MeV protons seem to be accelerated by CME-driven interplanetary shocks, not in the process of solar flares, because the peaks of the flux almost coincide with the shock passage time (Fig. 2). The maximum energy intensity ranges from 10^5 to 10^6.

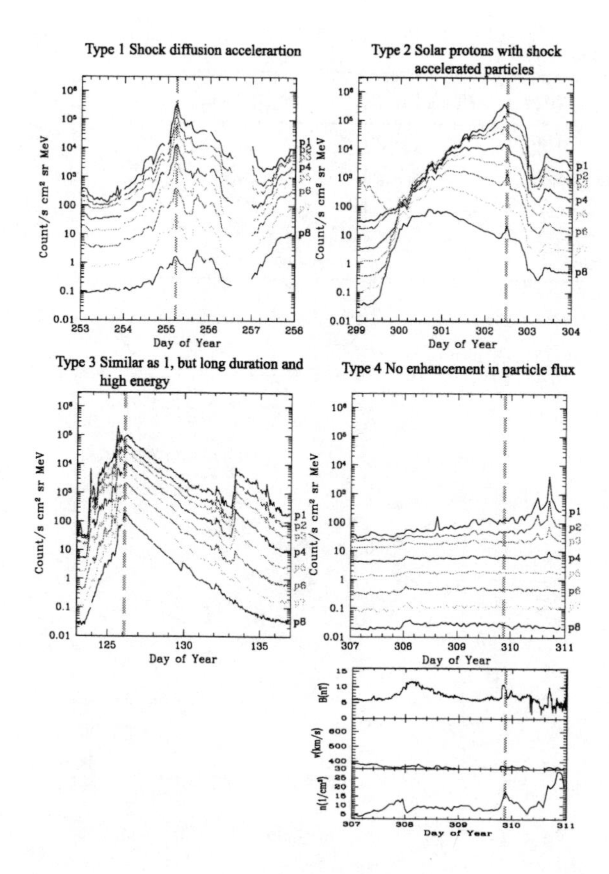

FIGURE 1. Intensity-time profile of particles with energy channels, p1 47-65 KeV, p2 65-112 KeV, 112-187 KeV, p4 187-310 KeV, p5 310-580 KeV, p6 580-1060 KeV, p7 1060-1910 KeV and p8 1910-4750 KeV, of typical events for four types. Shock passages were detected on 1999/255 day of year (DOY) (type 1), 2000/302 DOY (type 2), 1999/125 DOY (type 3) and 1999/309 (type 4). Vertical dotted lines mean shock passage time.

Type 4: There is no evident enhancement in a particle flux despite an interplanetary shock passage occurred (Fig. 1 two lower right figures). Weak shocks are related to particle events in this type and both [5] and [4] reported similar group.

Events belonging to van Nes et. al.'s group A are probably included in our types 2 and 3, and their groups B and C correspond to our type 1 ([4]), while we cannot find correspondence for our type 3 in both [5] and [4]. Events in type 3 suggest that interplanetary shocks can accelerate protons up to energies larger than 10 MeV. However, these events seldom occur comparing with events in other types, and this indicates some additional and strict conditions are needed for acceleration of MeV protons besides existence of shocks.

In Table 1 we list numbers of events, intensities of the peak flux and shock parameters, a compression ratio

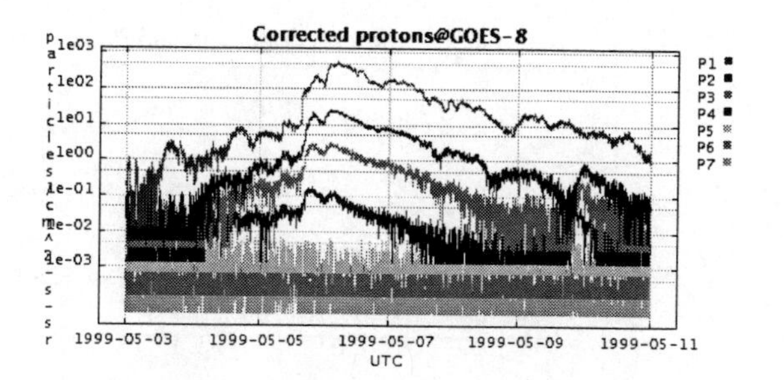

FIGURE 2. MeV proton flux for the period 1999 May 03 (123 DOY) to May 10 (130 DOY). Enhancement can be seen for channels p1 (0.6-4.2 MeV), p2 (4.2-8.7 MeV), p3 (8.7-14.5 MeV) and p4 (15.0-44.0 MeV). The remains are 39.0-82.0 MeV(p5), 84.0-200 MeV (p6) and 110.0-500.0 MeV (p7).

$r_n = \rho_2/\rho_1$, a magnetic compression ratio B_2/B_1 and an angle between the upstream magnetic field and the shock normal, θ_{B_n} where ρ and B are a solar wind plasma density and an intensity of an interplanetary magnetic field and suffix 1,2 mean upstream and downstream respectively. Systematic analysis such as correlations between these parameters and features of each type is needed for finding out some clues for the mechanism leading to different types and this is in preparation. In this paper, we pay an attention to types 1 and 2 and try to model each acceleration process by using numerical simulations. Modeling of events grouping in types 3 and 4 will be performed elsewhere.

MODELING: METHOD OF NUMERICAL SIMULATIONS AND MODELS

We apply the SDE method coupled with the particle splitting to diffusive shock acceleration. Numerical simulations by SDEs are much easier than solving the original the cosmic-ray transport equation, because the SDEs are ordinary differential equations.

Assuming the spherically symmetric geometry, the Fokker-Planck form of the the cosmic-ray transport equation is given by

$$\frac{\partial \phi}{\partial t} = -\frac{\partial}{\partial r}(v_r + \frac{2K}{r})\phi + \frac{\partial^2}{\partial r^2}K\phi + \frac{\partial}{\partial u}\{\frac{1}{3}(\frac{\partial v_r}{\partial r} + \frac{2v_r}{r})\}\phi. \tag{1}$$

Here we introduce following quantities $u = \ln(p/m_p c)$ and $\phi = 4\pi r^2 p^3 f$, where p is the proton momentum, f is the distribution function for protons, v_r is the radial velocity of the background flow, K is the spatial diffusion coefficient, m_p is the proton mass, and c is the speed of light. Equation (1) is equivalent to following SDEs,

$$dr = (v_r + 2K/r)dt + \sqrt{2K}dW_r \tag{2}$$

TABLE 1. Shock Parameters and Peak Flux Intensity

Type	Num.	r_n	B_2/B_1	θ_{B_n} (deg.)	Peak Flux(/cm^2/sec/str/MeV)
1	20	1.2-3.6	1.2-2.2	2.2-85.3	$3.9\times10^2 - 1.7\times10^6$
2	16	1.2-4.4	1.3-3.5	7.8-73.2	$3.1\times10^3 - 2.8\times10^6$
3	3	2.3-3.6	1.9-2.8	25.5-79.9	$3.3\times10^5 - 2.5\times10^6$
4	5	1.2-2.9	1.3-1.9	7.4-57.2	$2.1\times10^2 - 5.7\times10^3$
Not determined	20				

and

$$du = -\{\frac{1}{3}(\frac{\partial v_r}{\partial r} + \frac{2v_r}{r})\}dt, \quad (3)$$

where dW_r is a Wiener process given by the Gaussian distribution

$$P(dW_r) = (2\pi dt)^{-1/2}\exp(-dW_r{}^2/2dt). \quad (4)$$

The integration of the SDEs is performed by a simple Euler method.

A particle splitting is essential to achieve a wide dynamic range of the energy attained by accelerated particles. We set splitting surfaces u_i in energy space with an equal spacing in logarithmic scale. Each time an accelerated particle hits the surface u_i, the particle is split into w particles with the same energy and spatial position which the particle has attained. The statistical weight which is needed to calculate the final spectrum of the particles is decreased by a factor of w in each splitting. In our simulation we take $w = 2$ as a choice. This particle splitting method is permitted, because the process which we are considering is Markovian.

Two models are simulated as typical cases in types 1 and 2. In these models we adopt the following values, which are mainly obtained from observational data: For the model A as the typical case in type 1, the shock velocity $V_s = 396$ km/s, the compression ratio $r_n = 2.4$, and the the diffusion coefficient $K = 5.8\times10^{18}$ cm^2/s. For the model B as the typical case in type 2, $V_s = 852$ km/s, $r_n = 3.0$, and $K = 2.7\times10^{19}$ cm^2/s. The shocks start at $r_s = 0.1$AU where r_s is radial distance from the Sun and proceed with constant velocities. The diffusion coefficients are assumed to be constant and uniform. In the model A, particles are continuously injected with an energy of 0.05MeV at the shock front. In model B, twenty percent of particles are impulsively injected with an energy of 0.15MeV when the shock front arrives at 0.5AU and the remains are continuously injected with an energy of 0.05MeV.

Figures 3 and 4 are energy spectrum obtained from the particle data of typical events for types 1 and 2 respectively. Our simulation results are shown in Figs. 5 ~ 8. Simulated energy spectrum in all space are presented in Figures 5 and 6 and spatial distribution of the particles in Figs. 7 and 8

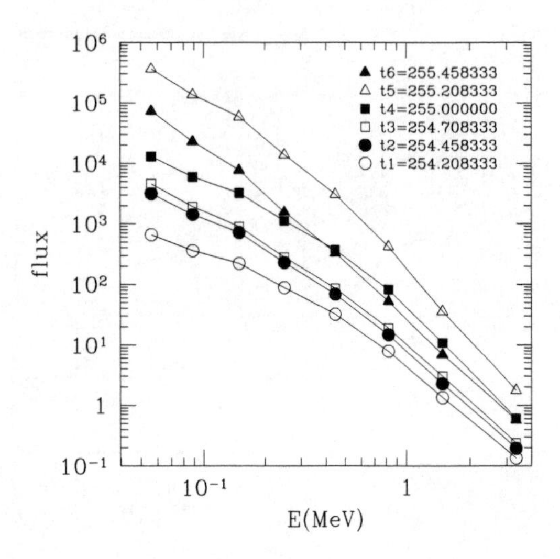

FIGURE 3. Evolution of energy spectrum for the event 1999/255 DOY.

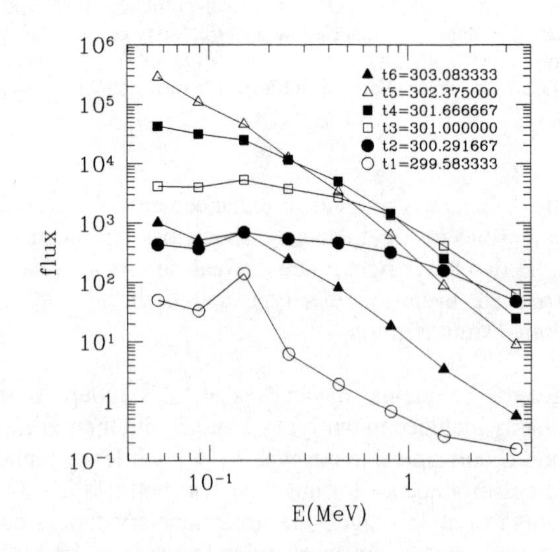

FIGURE 4. Same as Fig. 3 but for the event 2000/302 DOY.

The spectra in all space of the model A and B show that high energy end of the spectra increase with the time scale of particle acceleration by shock waves faster

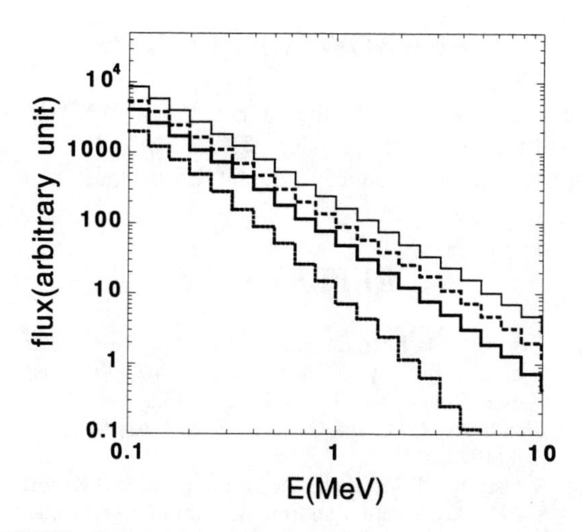

FIGURE 5. Evolution of energy spectrum in all space obtained by simulation for model A in which particles are injected continuously, and any particles are not injected impulsively. The upper lines show the later spectrum.

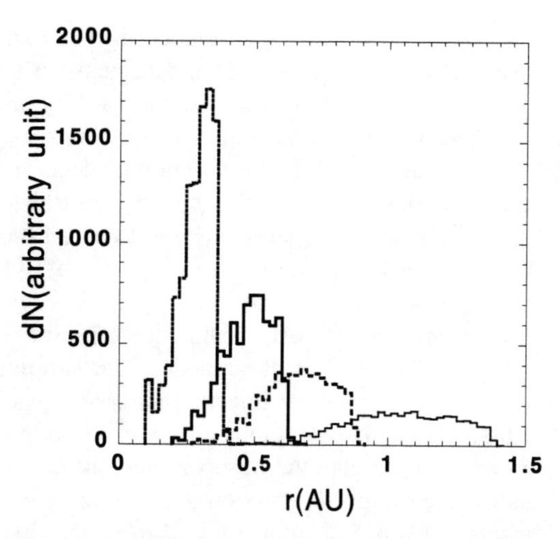

FIGURE 7. Evolution of spatial distribution of the particles for model A. The right hand side lines are the latter distribution.

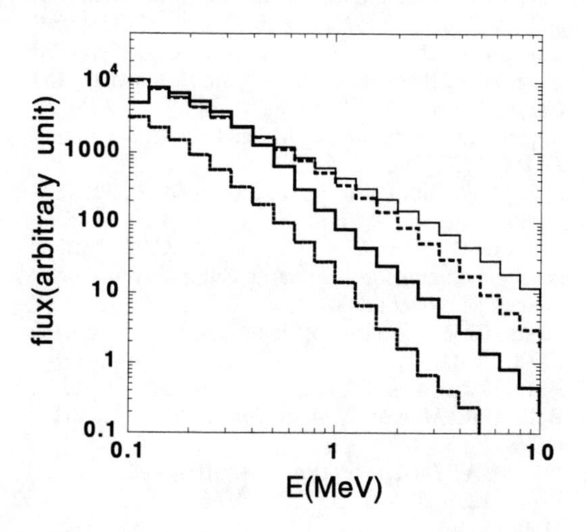

FIGURE 6. Evolution of energy spectrum in all space obtained by simulations for model B. Twenty percent of particles with energy of 0.15MeV are impulsively injected at 0.5AU and the remains are injected continuously. The upper lines show the later spectrum.

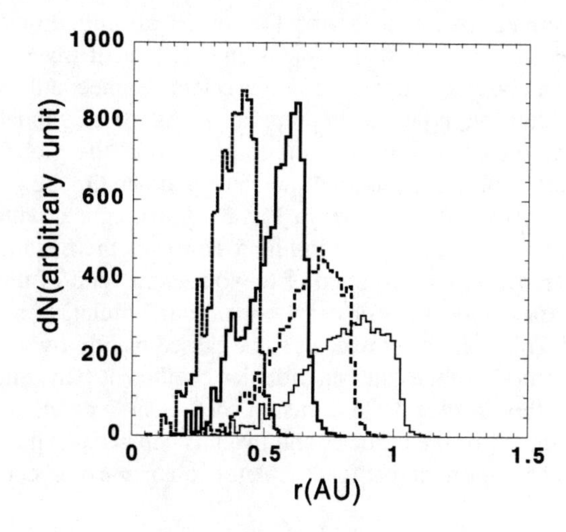

FIGURE 8. Same as Fig. 7 but for model B.

than the low energy end, while the observation seems to show the opposite behavior. However, the evolution of the spectra in the model B around 1 AU is similar to the observational spectra which belongs to type 2. In this case, the injected particles can be accelerated to higher energy range. As a result, the spectra become flatter than without injection. Although the spectra are steeper than the observational ones, our simulation results show that the impulse injection is important for type 2. The different behavior in the spectra between the observation for type 1 and model B is discussed in next section.

CONCLUDING REMARKS

We have studied the energetic particle events associated with the CME-driven interplanetary shock waves and classified them in four groups according to their intensity profiles of the flux, duration time for enhancement of the flux and maximum energy of accelerated particles. Our classes are (type 1) typical shock diffusion acceleration events, (type 2) larger than 10-MeV protons presumably accelerated in a flare process are also observed with the CME-driven shock accelerated particles, (type 3) larger than 10-MeV protons are observed, however, they seem to be accelerated by shock waves, not by energetic phenomena on the solar surface and precursor and total duration time are very long compared with events in type 1, (type 4) no or little peaks in the particle flux.

To gain some clues causing difference in types 1 and 2, we have performed modeling of typical events for types 1 and 2 by using numerical simulations and obtained the energy spectrum and the spatial distribution for each type. We have adopted different injection models for two types: continuous injection of 0.05 MeV particles for model A which is corresponding to type 1 and for model B, corresponding to type 2, impulsive injection of twenty percent of particles with an energy of 0.15MeV when the shock front arrives at 0.5AU continuous injection with an energy of 0.05MeV for the remains. Our simulation results show that impulsive injection makes the spectra flatter than with only continuous injection. Although the spectra are steeper than the observational ones, this indicates that the impulse injection is important for type 2. Various kinds of injection can lead to various evolutions of the spectrum of accelerated protons in the interplanetary shocks. The parameters of an injection is one of the important key to understand particle acceleration. Systematic parameter search for numerical simulations, statistical analysis of particle events and detail investigation about acceleration or no acceleration mechanism for other two types are in preparation. As for the model of the type 1 with only continuous injection, the global behavior of the spectrum of the energy at one time, e.g., t_5 (a shock passage time) in Fig. 3, almost agrees with the spectrum at later time in Fig. 5. However, there is the difference in the evolution of the low energy end of the spectra between the observation and our simulation result. This discrepancy seems to be caused mainly by our assumption of the constant diffusion coefficient. This implies that in type 1 the diffusion coefficient depends on the energy of the particles and that this dependence may play an important role for complete explanation of events for type 1.

The simulation method adopted here have been simple, i.e., our method is essentially one-dimensional, which indicates that shock drift acceleration cannot be included, and our assumptions of constant and uniform shock speeds and diffusion coefficients are not real. Furthermore, turbulence, or self-excited waves in a magnetic field caused by a wave-particle interaction is not taken into account. The existence of the turbulence is essential for shock acceleration and we consider that the self-excited waves are closely related to events in types 3 and 4. The self-consistent theories of shock acceleration (Bell(1978)[11] and Lee(1983))[12]) are presented and Ng and Reames(1994)[13] modeled the amplification and damping of waves numerically (also see Reames and Ng(1998)[14]). Thus use of full self-consistent multidimensional simulation method can clarify the acceleration mechanism and particle propagation process of events in each type. Our simulations can be regarded as one of the first step of the series of the subject of the shock accelerated high-energy particles.

ACKNOWLEDGMENTS

We would like to thank the ACE EPAM, SWEPAM and MAG teams and the SOHO CELIAS/MTOF PM team for providing their data and the GOES for their plots.

REFERENCES

1. Axford, W. I., Leer, E., and Skadron, G., "The acceleration of cosmic rays by shock waves", in *Proc. Int. Conf. Cosmic Rays 15th*, 1977, vol. 11, p. 132.
2. Blandford, R. D., and Ostriker, J. P., *Astrophys. J.*, **221**, L29 (1978).
3. Armstrong, T. P., Chne, G., Sarris, E. T., and Krimigis, S. M., "Acceleration and modulation of electrons and ions by propagating interplanetary shocks", in *Study of Travelling Interplanetary Phenomena*, edited by M. A. Shea, F. F. Smart, and S. T. Wu, D. Reidel, Hingham, Mass, 1977, p. 367.
4. van Nes, P., Reinhard, R., Sanderson, T. R., Wenzel, K.-P., and Zwickle, R. D., *J. Geophys. Res.*, **89**, 2122 (1984).
5. Kallenrode, M.-B., *J. Geophys. Res.*, **101**, 24393 (1996).
6. Cane, H. V., Reames, D. V., and von Rosenvinge, T. T., *J. Geophys. Res.*, **93**, 9555 (1988).
7. Reames, D. V., Barbier, L. M., and Ng, C. K., *Astrophys. J.*, **466**, 473 (1996).
8. Lee, M. A., and Ryan, J. M., *Astrophys. J.*, **303**, 829 (1986).
9. Zank, G. P., Rice, W. K. M., and Wu, C. C., Particle acceleration and coronal mass ejection-driven shocks: A theoretical model (2000).
10. Smith, C. W., Ace lists of disturbances and transients (2000), URL `http://www.bartol.udel.edu/~chuck/ace/ACElists/obs_list.html`.
11. Bell, A. R., *Monthly Notices Roy. Astron. Soc.*, **182**, 147 (1978).
12. Lee, M. A., *J. Geophys. Res.*, **88**, 6109 (1983).
13. Ng, C. K., and Reames, D. V., *Astrophys. J.*, **424**, 1032 (1994).
14. Reames, D. V., and Ng, C. K., *Astrophys. J.*, **504**, 1002 (1998).

Simulation of shock size asymmetry caused by charge exchange with pickup ions

H. Shimazu

Applied Research and Standards Division, Communications Research Laboratory, Koganei, Tokyo 184-8795 Japan

Abstract. This paper considers the interaction between the solar wind and unmagnetized planets using computer simulations. We used a three-dimensional hybrid code that treats kinetic ions and massless fluid electrons. The results showed that the shock shape and magnetic barrier intensity are asymmetrical in the direction of the electric field. The velocities of pickup ions and electrons differed near locations where charge exchange occurred. This velocity difference caused an electric current and it generated a strong magnetic barrier on the side of the planet to which the electric field was pointing. This magnetic barrier acted as an obstacle to the solar wind, and the shock was inflated on this side. Application of these results to the interaction between the solar wind and the interstellar medium near the heliopause is also discussed.

INTRODUCTION

Magnetohydrodynamics (MHD) simulations have been useful to investigators studying the interaction between the solar wind and planets with little or no magnetization (e. g., Venus) [1, 2, 3, 4, 5, 6]. Near Venus, ion reactions such as electron impact ionization, photo ionization and charge exchange are of significant interest. Electron impact ionization is a process in which neutral planetary particles are ionized by the impact of electrons. The electron impact ionization gives the highest ion production rate of the three processes [7]. Photo ionization is a process in which neutral planetary particles are ionized by the solar EUV flux. Charge exchange occurs between the solar wind and neutral planetary particles [8]. Many believe that these reactions have similar effects on macro-scale structures such as a shock because both reactions generate heavy ions (O^+ ions). These oxygen ions are picked up by the solar wind [9, 10] and then cause mass loading [11, 12]. This mass loading can change the macro-scale structures, and MHD simulations have shown that it can increase the shock height [13].

Unfortunately, MHD simulations cannot make allowance for changes in distribution functions caused by ion reactions. The hybrid code (kinetic ions and massless electron fluid) is one of the most useful simulation methods for studying ion distribution functions. In this study, the effects of photo ionization and charge exchange are compared by using simulations of three-dimensional global hybrid code. The electron impact ionization is also important, but it is difficult to include electron dynamics in the hybrid code. In this study the electron impact ionization is not considered. This study is aimed at understanding the physical processes rather than fitting simulation results to observations.

A hybrid code was used in a three-dimensional simulation of the interaction between the solar wind and the dayside portion of an unmagnetized planet [14, 15]. Later, a magnetotail region was included in the simulation box and it was shown that the resultant magnetic field configuration around Mars was consistent with observations [16].

There is asymmetry in the shock altitude around Venus in the direction of the $-v_{sw} \times B_{sw}$ convection electric field (v_{sw}: solar wind velocity observed at the planet; B_{sw}: interplanetary magnetic field). Observations obtained by the Pioneer Venus Orbiter (PVO) showed that the shock is further away from the planet on the side of the planet to which the convection electric field is pointing than on the other side [17, 18]. However simulations using hybrid code showed that the shock is further away from the planet on the opposite side [16, 19, 20, 21]. It was shown that this asymmetry is caused by downstream oxygen ions of planetary origin that reduce the downstream Alfvén and fast mode velocities [22]. This is because the shock cone angle depends on the ratio of the downstream fast mode velocity to the upstream flow velocity. Unfortunately, the direction of the asymmetry in the simulation did not agree with the observations near Venus. This discrepancy has been the subject of controversy. In this study, we found that when charge exchange is included, the direction of simulated shock altitude asym-

CP598, *Solar and Galactic Composition,* edited by R. F. Wimmer-Schweingruber

metry agrees with observations near Venus. The process in which charge exchange causes shock altitude asymmetry will also be described.

This paper is the summary of my paper [23] ("Effects of charge exchange and photo ionization on the interaction between the solar wind and unmagnetized planets", *J. Geophys. Res.*) but application of the results to the shock shape near the heliopause was added.

MODEL

The simulation code used was the same as that used by [22], except that the effects of photo ionization and charge exchange were included. The three-dimensional Cartesian coordinate system (x, y, and z) was used in the simulation. There were 64 equally spaced grid cells in each direction. The average number of particles in each cell was 16. The solar wind was emulated by using a super-Alfvénic plasma (velocity $v_{sw} = 4.0V_A$, where V_A is the upstream Alfvén velocity) continuously injected into the simulation system from the $x = 0$ plane. The solar wind ions were protons. The bulk velocity of the solar wind was parallel to the x axis, and the solar wind protons were removed from the simulation system when they reached the other boundary.

Planetary oxygen ions, distributed uniformly in a sphere of radius $R = L/8$, were placed at the center of the simulation box (L: box size). R was set to $6.4r_L = 25c/\omega_{pi}$, where $r_L = v_{sw}/\omega_{ci}$ is the Larmor radius of protons, ω_{ci} is the upstream proton cyclotron frequency, and ω_{pi} is the upstream proton plasma frequency. Under a typical upstream condition ($v_{sw} = 500$ km/s and $\omega_{ci} = 0.5$ Hz), R (6.4×10^3 km) corresponds to that of Venus. The proton inertial length c/ω_{pi} equals V_A/ω_{ci}, which is approximately the Larmor radius of shocked solar wind protons. This fact also means that a pickup oxygen ion has a Larmor radius of 16 c/ω_{pi}.

The sphere was also a source of planetary ions. This source corresponds to an ion generation process in Venusian ionosphere. The supply rate of these ions was assumed to be uniform and constant, and was the same as that used by [22]. This supply is needed to maintain the gaseous planet and is limited to the sphere. Ions added outside the planet by photo ionization or charge exchange will be discussed later. This uniform supply assumption was not realistic, but the resolution was not fine enough to include ionospheric processes such as plasma transport.

In this study, the ion flux from the ionosphere across the ionopause was low, and the ionosphere was not diffused on the upstream side.

The initial ambient magnetic field B was given by the potential field:

$$B = \begin{cases} \nabla\left\{B_0 y\left(1+\dfrac{R^3}{2r^3}\right)\right\} & (r > R) \\ 0 & (r < R), \end{cases} \tag{1}$$

where $B_0/(4\pi m_i n_0)^{1/2} \equiv V_A = 1.0 \times 10^{-4}c$, $r^2 = (x-L/2)^2+(y-L/2)^2+(z-L/2)^2$, m_i is the proton mass, n_0 is number density, and c is the speed of light. Electrical resistivity was not artificially introduced. In the upper ionosphere, gravity can be ignored because it is much weaker than the electromagnetic force exerted by the solar wind. In this study gravity was ignored because we only considered the upper part of the ionosphere.

We assumed that the temperature of the solar wind and that of the planetary plasma were the same and that the initial temperature of the ions was the same as that of the electron fluid. The electron temperature was constant in space and time. It was confirmed that the results did not depend so much on the electron temperature. We also assumed that the ratio of the ion thermal pressure to magnetic pressure of the solar wind (i.e., β_i) was 1.0. Thus β ($= \beta_i + \beta_e$; β_e is β for electrons) was 2.0. The sonic Mach number of the solar wind flow is 2.8. Under the initial conditions, the dynamic pressure of the solar wind was in balance with the thermal pressure of the planetary plasma at the subsolar point.

The ion production rate for photo ionization is given by

$$q_{photo} = \sigma\phi n_1, \tag{2}$$

where σ is the ionization cross section of atomic oxygen, ϕ is the ionization photon flux outside the atmosphere, $\sigma\phi = 13.5 \times 10^{-7}$/s at the heliocentric distance of Venus for solar maximum [7], and n_1 is the number density of the ionizable constituent. However, in this photo-ionization simulation, the ionization rate was assumed to be constant. Oxygen ions were added to the dayside portion of the 1.0 - 1.3R region at the same total rate as the planetary ions described earlier. The total production rate corresponds to 2.85×10^{24}/s and $q_{photo} = 2.17 \times 10^{-3}$ s^{-1} cm^{-3}. Thus n_1 is 1.61×10^4 cm^{-3}. The total ion production rate from photo ionization for Venus is 3.4×10^{24}/s [7]. This value is close to our value. The temperature of the newly added ions was the same as that of the planetary oxygen ions.

In the charge-exchange simulations, the reaction

$$H^+ + O \rightarrow H + O^+ \tag{3}$$

was assumed. Part of the solar wind protons that reached a distance of 1.3R from the center of the planet disappeared, and oxygen ions were added at this location. The temperature of the newly added ions was the same as that of the planetary oxygen ions. The charge exchange ionization rate is given by

$$q_{CE} = n_p v_p \sigma_c n_1, \quad (4)$$

where n_p is the solar wind (or magnetosheath) proton density, v_p is the solar wind (or magnetosheath) proton bulk velocity, and σ_c (8×10^{-16} cm^2) is the cross section for charge exchange with atomic oxygen. In this study, however, the charge-exchange rate was assumed to be constant. This rate corresponds to 2.36×10^{23}/s and q_{CE} is 1.80×10^{-4} s^{-1} cm^{-3}. When $n_p v_p$ is assumed to be 10^8 s^{-1} cm^{-2}, n_1 becomes the order of 10^3 cm^{-3}. The ion production rate from charge exchange for Venus is 1×10^{24}/s [7]. This is larger than the value we used. Because the charge exchange effect was more than that of photo ionization, as shown later, we used a lower charge exchange rate.

In the photo ionization model, heavy ions are added to the flow, whereas in the charge exchange model, they replace the solar wind protons. The momentum required for heavy ions must be supplied immediately when ions appear in the charge exchange model. Therefore the charge exchange process is more abrupt than the photo ionization process.

We used $1.3R$ for the ion reaction region, which is based on the estimates in [7]. Their estimates showed that both rates decrease at near exponential rates as the distance from the planet increases. They are the highest by far in the region of 1.0 - $1.3R$.

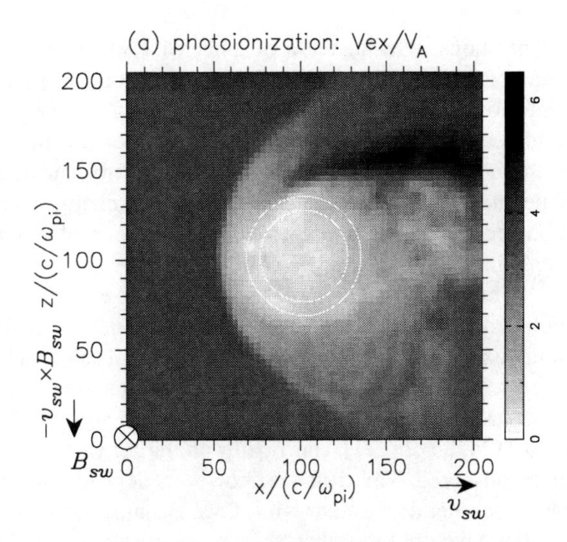

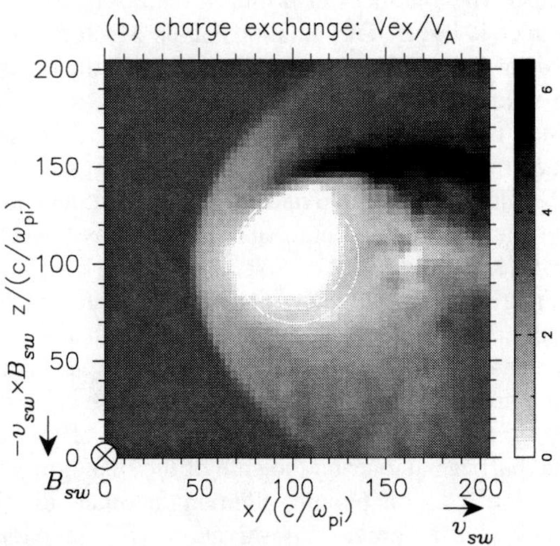

FIGURE 1. The x component of electron velocity for $y = L/2$ plane at $\omega_{ci}t = 37.5$ when (a) photo ionization was included and (b) charge exchange was included. Circles with the radii $1.0R$ and $1.3R$ are shown. The x axis is directed to upstream solar wind velocity v_{sw}, y axis to upstream magnetic field B_{sw}, and z axis to $v_{sw} \times B_{sw}$.

RESULTS

Figures 1a and 1b show the x components of the electron velocity for the $y = L/2$ plane when photo ionization and charge exchange were included. They are cross sections of the $y = L/2$ plane, cutting through the center of the planet. The simulation results shown here are for $t = 37.5/\omega_{ci}$, at which the solution is nearly stabilized. The period $37.5/\omega_{ci}$ corresponds to almost one gyro period of oxygen ions near the ionosheath/magnetosheath. Because the mass of an oxygen ion is 16 times that of a proton, and because the magnetic field is nearly three times that of the solar wind, the gyro frequency of an oxygen ion is $(3/16)\omega_{ci}$. Therefore the period is $2\pi/(3\omega_{ci}/16) = 34/\omega_{ci}$. In this period, a pickup oxygen ion, which has an average velocity of v_{sw}, moves $136c/\omega_{pi}$ ($0.66L$). Thus we can see the effects of the mass loading at the time $37.5/\omega_{ci}$.

The electron velocity v_e in the hybrid code is given by

$$v_e = (en_{H^+}v_{H^+} + en_{O^+}v_{O^+} - j)/e(n_{H^+} + n_{O^+}), \quad (5)$$

where j is electric current, e is the unit charge, v_{H^+} and v_{O^+} are bulk velocities of protons and oxygen ions, respectively, and n_{H^+} and n_{O^+} are densities of protons and oxygen ions, respectively.

As shown, the velocity decreased sharply upstream of the planet, and a bow shock was generated. Downstream of the planet, the shocked solar wind flow near $x = 150c/\omega_{pi}$ and $z = 140c/\omega_{pi}$ (upper side) is accelerated by the Lorentz force to a speed greater than that upstream. On the other side (lower side), the flow decelerated because of the mass loading of the planetary oxygen ions escaping from the planet. A magnetotail formed downstream of the planet, and the tail region was filled mainly with planetary ions. These are consistent with those of the hybrid simulation that had no photo ionization or charge exchange [22].

The shock is located at a distance of 1.93R in Figure 1a and 2.15R in Figure 1b at the subsolar point from the center of the planet. The position of the shock was determined as the distance from the center of the planet to the nose point where the velocity suddenly changes. When neither photo ionization nor charge exchange were included, the shock was located at a distance of 1.33R from the center (the other conditions were the same.) [22]. Thus, when photo ionization or charge exchange is included, the shock is inflated. This is because the additional oxygen ions play the role of a larger obstacle. The Venera 9 and 10, and the PVO observations showed that the distance to the nose of the bow shock of Venus is 1.1 - 1.6R [24, 25, 26, 27]. Our results are larger values than those observed. The distance, however, depends on the Mach number of the solar wind flow. Because we used a smaller Mach number than that in the observations, the distance is larger. This smaller Alfvén Mach number is used for numerical convenience. Although the height of the shock changed with the Mach number, it was confirmed that there were few differences in the physical structures when the Alfvén Mach numbers are 4 or 8. The different total ionization rates at Venus (that depends on the solar activity) and at this simulation may also cause the difference.

Figures 2a and 2b show the magnetic field intensity for the $x = L/2$ plane when photo ionization and charge exchange are included. These figures show asymmetry in the magnetic barrier in the direction of the convection electric field ($-z$ direction). The intensity of the magnetic barrier is larger on the side of the planet to which the electric field is pointing than on the other side. This agrees with the previous observations [28] and simulations [16, 22]. When these figures are compared, we find that the intensity of the magnetic barrier is larger when charge exchange is included (Figure 2b).

These figures show the penetration of the magnetic field into the obstacle especially under photo ionization. The penetration of the magnetic flux into the planet may result in small magnetic field compression in the ionosheath/magnetosheath. This compression is prominent when compared with that presented in Figure 2 of [5]. The penetration may change physical quantities near the planet less than those in reality. Thus a penetrable obstacle may decrease the asymmetries.

Figures 2a and 2b clearly show shock altitude asymmetry in the direction of the convection electric field ($-z$ direction). Figure 2a shows that the shock height is 7.98% smaller on the side of the planet to which the electric field is pointing than on the other side when photo ionization is included. The shock height was determined as the distance from the center of the planet to the point where the magnetic field intensity suddenly changes. The direction of the asymmetry agrees with that of previous simulations in which neither photo ionization nor charge exchange were included [16]. This asymmetry in shock altitude was explained by lower downstream Alfvén and fast mode velocities due to heavy ions on the side of the planet to which the electric field was pointing than on the other side [22]. The shock cone angle depends on the ratio of the downstream fast-mode velocity to the upstream flow velocity. However the direction of the asymmetry does not agree with observations near Venus [17, 18].

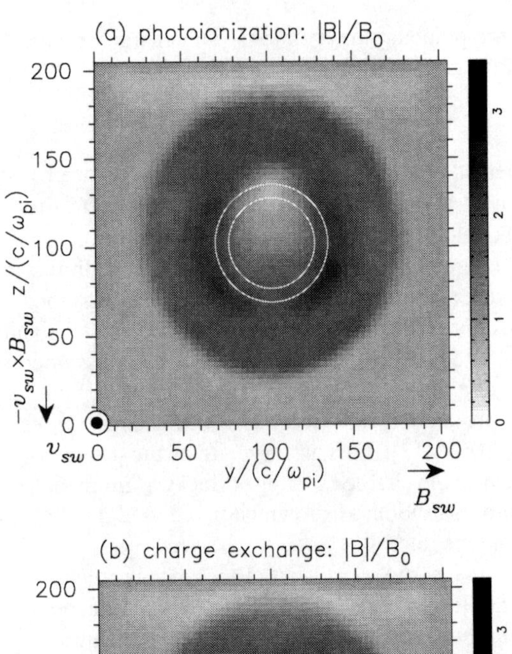

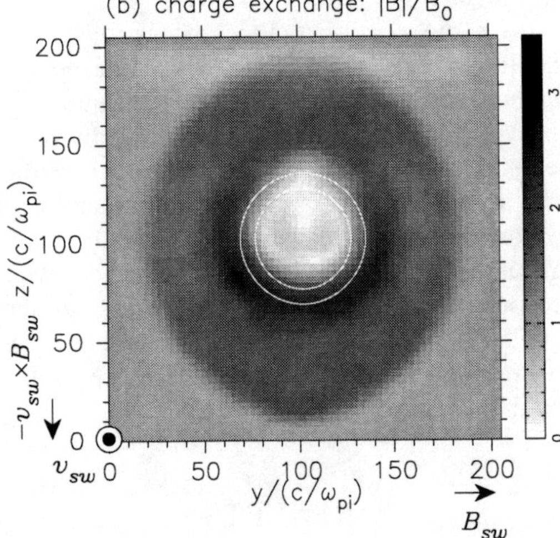

FIGURE 2. Magnetic field intensity for $x = L/2$ plane at $\omega_{ci}t = 37.5$ when (a) photo ionization was included and (b) charge exchange was included.

When charge exchange is included, the direction of the shock altitude asymmetry reverses (Figure 2b) and agrees with the observations. The shock height is 7.79% larger on the side of the planet to which the electric field is pointing than on the other side. These figures also show that when the charge exchange is included, the shock is further away from the planet than when the photo ionization is included. This is true in all directions.

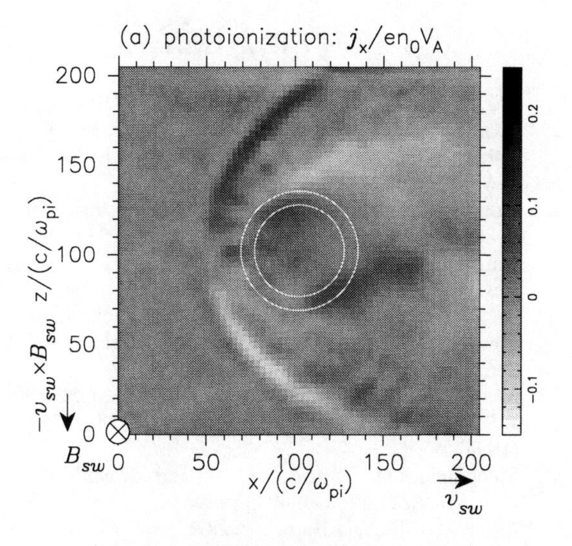

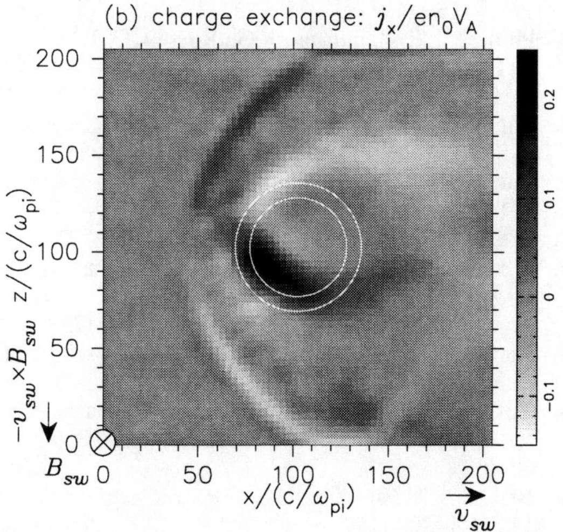

FIGURE 3. The x component of the electric currents for $y = L/2$ plane at $\omega_{ci}t = 37.5$ when (a) photo ionization was included and (b) charge exchange was included.

To investigate the cause of this difference in shock altitude, we consider the x component of the electric current j_x. Figures 3a and 3b show it. When we compare both figures, we find that the magnitude of j_x is larger when the charge exchange is included (Figure 3b). The current j_x was generated by the difference in the velocities between the ions and electrons when charge exchange occurred. One would see first a negative then a positive j_x current in the cycloidal motion of newly picked up ions on the lower side of Figure 3b. The newly picked up ions move at twice the solar wind speed at the maximum of their cycloidal motion and are at rest at the minimum. It is worth noting that the pickup ion Larmor radius is $(16/3)c/\omega_{pi}$ as shown earlier, and the ion travels 2π times the Larmor radius in one gyro-period. Thus the peaks of negative and positive current should be separated by approximately $30v/\omega_{ci}$, where v is the local flow speed. The separation between the peaks in Figure 3b is consistent with this value.

The cycloidal motion of pickup ions can generate both j_x and j_z currents. It was confirmed that the distributions of the j_x and j_z currents are coincident. Thus it is reasonable to consider that the currents responsible for the magnetic barrier are the j_z current on the side of the planet to which the electric field was pointing. The region where j_x is large in Figure 3b (which also corresponds to the region where j_z is large.) agrees with the magnetic barrier region. This magnetic barrier acted as an obstacle to the solar wind, and the shock was inflated on this side when the charge exchange was included.

We have examined the charge exchange caused by oxygen atoms. We also performed simulations that included the following charge exchange reaction:

$$H^+ + H \rightarrow H + H^+. \quad (6)$$

The simulation results showed that the direction of the shock altitude asymmetry was the same as that when oxygen was used. The charge exchange affected the shock altitude asymmetry in the same manner.

APPLICATION TO THE HELIOSPHERE

Shock shape asymmetry near the heliopause is considered in this section. The shock shape is important when we consider behavior of pickup ions near the heliopause.

Properties near Venus and those near the heliopause are compared. According to the table of properties near the heliopause [29], the magnetic field intensity near the heliopause is less than that near Venus by the order of 1, and the Larmor radius near the heliopause is larger than that near Venus. However the size of the heliosphere is much larger than that of the Venusian magnetosheath, and the ratio r_L/L_{system}, where L_{system} is the size of the system, is probably much smaller near the heliopause than that near Venus. Thus the finite Larmor radius effect would be less near the heliopause than that near Venus. No significant asymmetries associated with the finite Larmor radius effect would be found at the heliopause.

Since charge exchange changes the macro-scale electric current and also momentum near the shock as shown earlier, charge exchange may affect the macro-scale structure even when the ratio r_L/L_{system} is small. However charge exchange ionization rate near the heliopause is lower than that near Venus by the order of 8, because density near the heliopause is much less than that near Venus (see Eq. (4)) [29]. When we consider the lower charge exchange rate and smaller ratio of r_L/L_{system}, no significant asymmetries of the bow shock shape would be expected near the heliopause.

CONCLUSIONS

We compared the effects of photo ionization and charge exchange by using three-dimensional hybrid code simulations of the interaction between the solar wind and Venus.

When either photo ionization or charge exchange were included, the shock was inflated because of the presence of additional oxygen ions. The shock shape asymmetry in the direction of the convection electric field when the photo ionization was included was opposite to actual observations. However, when the charge exchange was included, the direction of the shock shape asymmetry agreed with that observed. This is because the current carried by picked up ions made the strong magnetic barrier.

The photo-ionization and charge-exchange rates were kept constant in the region between 1.0 and $1.3R$ in our simulations because this made it easy to work with and interpret the simulation results. This, however, is ideal. In a more realistic ionization rate model these ionization rates decrease in a nearly exponential fashion as the distance from the planet increases [7]. This is unlike our simulation, which used a step function.

To observationally confirm these simulation results, our model should be improved in more realistic way. Future work will include the electron impact ionization, the combined effects of the two ionization mechanisms that were analyzed separately, a spatial dependence of the ion production rates, and larger Mach numbers.

We applied the results to the shock near the heliopause. The shock shape is important when we consider behavior of pickup ions. Because the ratio r_L/L_{system} is probably much smaller near the heliopause than that near Venus, the finite Larmor radius effect would be less near the heliopause than that near Venus. Moreover since the charge exchange ionization rate near the heliopause is much lower than that near Venus by the order of 8, charge exchange may not affect asymmetry of the bow shock shape near the heliopause.

ACKNOWLEDGMENTS

The author thanks Dr. Shinobu Machida (Kyoto University), Dr. Motohiko Tanaka (National Institute for Fusion Science), Dr. Takashi Tanaka (Communications Research Laboratory), and Dr. Katsuhide Marubashi (Communications Research Laboratory) for their stimulating and insightful comments and suggestions during the course of this work. The author also appreciates the referee's valuable comments. Parts of this paper have appeared in [23] and were modified by permission of the American Geophysical Union.

REFERENCES

1. Wu, C. C., *Geophys. Res. Lett.*, **19**, 87–90 (1992).
2. Tanaka, T., *J. Geophys. Res.*, **98**, 17251–17262 (1993).
3. Cable, S., and Steinolfson, R. S., *J. Geophys. Res.*, **100**, 21645–21658 (1995).
4. Tanaka, T., and Murawski, K., *J. Geophys. Res.*, **102**, 19805–19821 (1997).
5. Kallio, E., Luhmann, J. G., and Lyon, J. G., *J. Geophys. Res.*, **103**, 4723–4737 (1998).
6. Bauske, R., Nagy, A. F., Gombosi, T. I., Zeeuw, D. L. D., Powell, K. G., and Luhmann, J. G., *J. Geophys. Res.*, **103**, 23625–23638 (1998).
7. Zhang, M. H. G., Luhmann, J. G., Nagy, A. F., Spreiter, J. R., and Stahara, S. S., *J. Geophys. Res.*, **98**, 3311–3318 (1993).
8. Stebbings, R. F., Smith, A. C. H., and Ehrhardt, H., *J. Geophys. Res.*, **69**, 2349–2355 (1964).
9. Mihalov, J. D., and Barnes, A., *Geophys. Res. Lett.*, **8**, 1277–1280 (1981).
10. Phillips, J. L., Luhmann, J. G., Russell, C. T., and Moore, K. R., *J. Geophys. Res.*, **92**, 9920–9930 (1987).
11. Intriligator, D. S., *Geophys. Res. Lett.*, **9**, 727–730 (1982).
12. Luhmann, J. G., Russell, C. T., Spreiter, J. R., and Stahara, S. S., *Adv. Space Res.*, **5**, (4)307–(4)311 (1985).
13. Murawski, K., and Steinolfson, R. S., *J. Geophys. Res.*, **101**, 2547–2560 (1996).
14. Brecht, S. H., *Geophys. Res. Lett.*, **17**, 1243–1246 (1990).
15. Moore, K. R., Thomas, V. A., and McComas, D. J., *J. Geophys. Res.*, **96**, 7779–7791 (1991).
16. Brecht, S. H., *J. Geophys. Res.*, **102**, 4743–4750 (1997).
17. Alexander, C. J., Luhmann, J. G., and Russell, C. T., *Geophys. Res. Lett.*, **13**, 917–920 (1986).
18. Russell, C. T., Chou, E., Luhmann, J. G., Gazis, P., Brace, L. H., and Hoegy, W. R., *J. Geophys. Res.*, **93**, 5461–5469 (1988).
19. Brecht, S. H., and Ferrante, J. R., *J. Geophys. Res.*, **96**, 11209–11220 (1991).
20. Brecht, S. H., Ferrante, J. R., and Luhmann, J. G., *J. Geophys. Res.*, **98**, 1345–1357 (1993).
21. Shimazu, H., *Earth Planets Space*, **51**, 383–393 (1999).
22. Shimazu, H., *J. Geophys. Res.*, **106**, 8333–8342 (2001).
23. Shimazu, H., *J. Geophys. Res.*, **106**, 18751–18761 (2001).
24. Verigin, M. I., Gringauz, K. I., Gombosi, T., Breus, T. K., Bezrukikh, V. V., Remizov, A. P., and Volkov, G. I., *J. Geophys. Res.*, **83**, 3721–3728 (1978).
25. Slavin, J. A. *et al.*, *J. Geophys. Res.*, **85**, 7625–7641 (1980).
26. Smirnov, V. N., Vaisberg, O. L., and Intriligator, D. S., *J. Geophys. Res.*, **85**, 7651–7654 (1980).
27. Slavin, J. A., and Holzer, R. E., *J. Geophys. Res.*, **86**, 11401–11418 (1981).
28. Zhang, T. L., Luhmann, J. G., and Russell, C. T., *J. Geophys. Res.*, **96**, 11145–11153 (1991).
29. Suess, S. T., *Rev. Geophys.*, **28**, 97–115 (1990).

The Effect of Self-consistent Stochastic Preacceleration of Pickup Ions on the Composition of Anomalous Cosmic Rays.

Frank C. Jones*, Matthew G. Baring† and Donald C. Ellison**

*NASA Goddard Space Flight Center
†Rice University
**North Carolina State University

Abstract. We have previously calculated the spectrum of anomalous cosmic rays (ACR) by employing the supposed spectrum of interstellar pickup ions as the seed population for a Monte Carlo model of the solar wind termination shock. This pickup ion spectrum was extrapolated from measurements some distance from the shock under the assumption that adiabatic loss was the only energy change process acting prior to reaching the shock. Our results while reasonable in many respects were underabundant in He^+ and O^+ ions relative to H^+ as determined by observation. le Roux and Ptuskin have shown that stochastic preacceleration is more effective for He^+ and O^+ ions than for H^+ ions thereby redressing this underabundance. We have employed the results of le Roux and Ptuskin as input to our previous model and will show to what extent the relative abundances are improved thereby.

INTRODUCTION

In a previous work (Ellison, Jones & Baring [1]) we employed our Monte Carlo shock acceleration model to explore the question of whether or not the Anomalous Cosmic Rays could be produced from the interstellar pickup ions by the solar wind termination shock using reasonable shock parameters. In this study we employed the values for pickup ions of Cummings,& Stone [2] and employed a variety of parameters for solar wind speed, density, diffusion parameters etc. The pick up ion density spectrum that we employed at the shock is shown in Figure(1).

As can be seen the phase space density cuts off at an ion velocity in the wind frame equal to the speed of the solar wind, the speed with which they are picked up. The low energy tail is from adiabatic deceleration of ions injected deeper in the wind but note that there are no ions with velocities higher than their injection velocity. Employing this injection spectrum we were able to obtain an ACR spectrum as shown Figure (2).

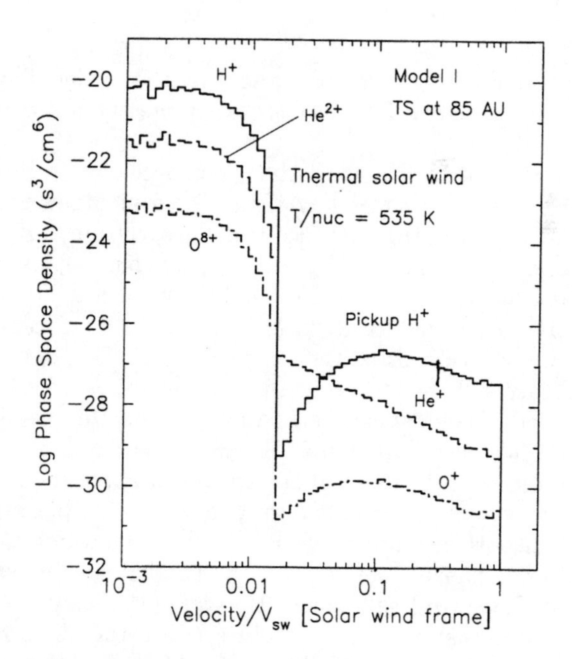

FIGURE 1. Pick up ion phase space density employed in Ellison, Jones & Baring [1].

Although most models could be made to fit the H^+ data, as can be seen by comparing to Voyager data (Cummings, Christian & Stone [3], Cummings,& Stone [2]) taken at an average radial location of 57 AU and the straight line representing an extrapolation of the H^+ data to the termination shock (Cummings,& Stone [2]) the flux of He^+ and O^+ are under

CP598, *Solar and Galactic Composition*, edited by R. F. Wimmer-Schweingruber

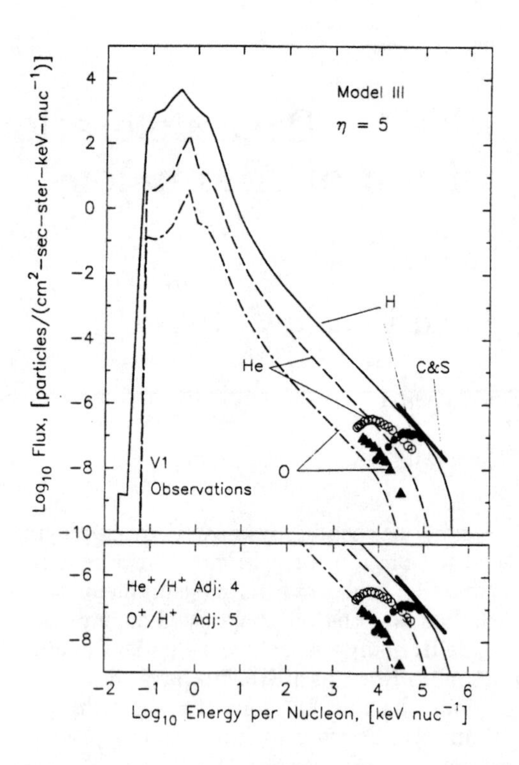

FIGURE 2. Spectrum of ACR derived in Ellison, Jones & Baring [1].

produced by factors of 4 and 5 by the model. This result held true for all values of parameters that were tried.

Since ions that start the shock acceleration process at a higher energy than others maintain this advantage in the final spectrum the spectrum of He^+ and O^+ would be enhanced if a way could be found to selectively heat the He^+ and O^+ while leaving the H^+ unchanged. le Roux & Ptuskin [4] have found such a mechanism. They calculated the effect of preheating of the pickup ions by turbulence in the solar wind. They demonstrated that O^+ and He^+ would be fully heated but that H^+ would be under abundant in the heated tail due to absorption of the resonant waves by the background solar wind plasma. There is just too much H^+ in the solar wind and in the pickup ions for it to be heated as effectivly as the much less bundant He^+ and O^+. They speculated that this effect could produce the observed under abundance of H^+ in the ACR. Such an effect was also found by Chalov, Fahr & Izmodenov [5] but they required preheating by interplanetary shocks to boost the He^+ in the heated tail while in a similar work Giacalone *et al.* [6] studied the heating of only the H^+ ions. This current work has been carried further by Zank *et al.* [7].

We have taken the new pickup ion densities found by le Roux & Ptuskin [4] (shown in Figure (3)) and used them in a new simulation of one of the models we previously investigated. The density of oxygen was multiplied by 0.16 to compensate for the fact that these authors used a value of the interstellar density that was larger than the one used by us.

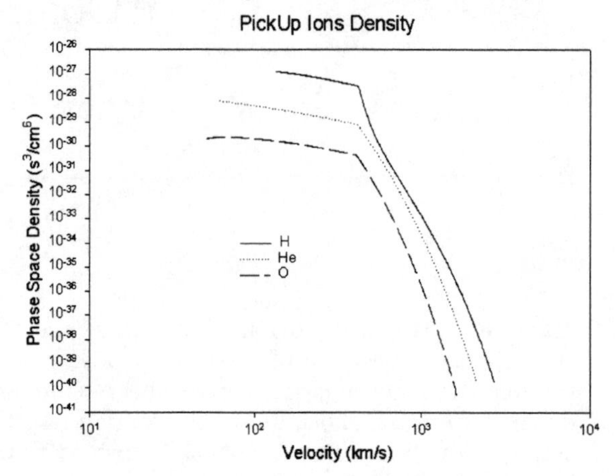

FIGURE 3. Phase space density of pick up ions with pre heated tail

As can be seen from this graph H^+ is seriously underabundant in the heated tail compared to He^+ and O^+. However, as can be seen from Figure (4), although adding the heated component to the shock process moves the ACR density in the right direction (enhancing He^+ and O^+ with essentially no effect on H^+) the effect does not seem to be nearly strong enough to explain the data.

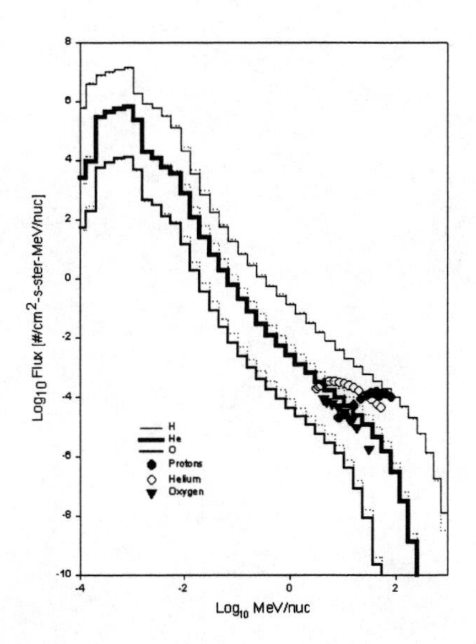

FIGURE 4. ACR spectrum showing the effect of pre-heating. The dotted lines show the slight increase in the flux of Helium and Oxygen and none for Hydrogen due to the selective effect of the heating

CONCLUSION

Although the selective heating effect of le Roux and Ptuskin moves things in the right direction it is insufficient, by itself, to explain the under abundance of H^+ in the ACR produced by our shock model. Clearly some other method of enhancing the He^+ and O^+ must be found or the estimates of interstellar abundances must be revised.

REFERENCES

1. Ellison, D. C., Jones, F. C., & Baring, M. G. 1999, ApJ, 512, 403
2. Cummings, A. C., & Stone, E. C. 1996, Space Sci. Rev., 78, 117
3. Christian, E. R., Cummings, A. C., & Stone, E. C. 1995, ApJ, 446, L105
4. le Roux, J. A., & Ptuskin, V. S. 1998, JGR, 103, 4799
5. Chalov, S. V., Fahr, H. J. & Izmodenov, V. 1997, A&A 320, 659
6. Giacalone J, *et al.* 1997, ApJ 486, 471
7. Zank, G. P. *et al.* 2001, ApJ 556, 494

Non-Shock Diffusive Acceleration in Regions of Solar-Wind Compression

J. Giacalone*, J.R. Jokipii* and J. Kóta*

*Dept. of Planetary Sciences, University of Arizona, Tucson

Abstract. We investigate the acceleration of charged particles in regions of solar-wind compression in the absence of shocks. The mechanism we describe is essentially the same as diffusive shock acceleration, except that here we show that a gradual compression of the plasma, rather than a shock, can efficiently accelerate the particles. Our method is to integrate the trajectories of an ensemble of test particles moving in synthesized electromagnetic fields, which are similar to what is currently known about corotating interaction regions. We show that compression regions at 1 AU, with widths ~ 0.03 AU can accelerate particles up to ~ 10 MeV.

INTRODUCTION

It is widely believed that energetic nuclei in corotating particle events observed near Earth's orbit are the result of backstreaming particles accelerated at the forward and reverse shocks bounding a corotating interaction region (CIR) at distances > 2 AU [1, 2]. The shock-accelerated particles propagate back to the observer at 1 AU [3]. Recent observations of energetic nuclei associated with CIRs do not seem to fit this picture [4]. Instead, they suggest that energetic particles are accelerated within the transition region from slow to fast solar wind, which is a more gradual change than the near instantaneous jump across a shock. The particle acceleration is observed *in situ*.

It is the purpose of the present work to investigate the acceleration of charged particles in regions of gradual compression, where the particle mean-free path, $\lambda_{\|}$, is large compared to the width of the compression, Δ_c. This may be the case for low-energy particles encountering a CIR near 1 AU where the forward and reverse shocks have not yet formed. If particles diffuse through the velocity gradient associated with the CIR formation, they will be efficiently accelerated due to the fact that this gradient is very large. Consequently, considerable acceleration can occur before the shocks form (< 2 AU) and this can lead to a peak in the intensity within the CIR. This might explain recent observations by Mason [4].

ANALYTICAL CONSIDERATIONS

In order for acceleration to occur at a compression region, we require that the diffusive length scale Δ_d be larger than the compression region width Δ_c. This gives,

$$\Delta_c < \Delta_d = \kappa_{rr}/U \tag{1}$$

where κ_{rr} is the diffusion coefficient along the propagation direction of the compression region (assumed to be radial) and U is the flow speed. It is readily demonstrated that in the limit $\Delta_c \ll \Delta_d$ all of the results of diffusive shock acceleration are recovered[5].

We consider the acceleration of interstellar pickup ions by a solar wind compression region which corotates with the sun. It is known that there will be a forward and reverse shock at distances beyond ~ 2 AU. Near 1 AU, however, there is a smooth transition from the fast to slow solar wind which occurs over a scale which is much larger than the gyroradius of the pickup ions. We further assume that the gradient associated with the plasma compression is aligned with the radial direction. Using these considerations, it is straightforward to show that Eq. (1) leads to $\Delta_c < (w/U)\lambda_{\|}$, where w is the particle speed. If, however, $\lambda_{\|}$ is too large, the particles will not scatter downstream (the same side as the sun) of the compression region and will instead be mirrored in the strong magnetic field near the sun. This will not lead to any significant acceleration. Consequently, we require both that the mean-free path is larger than the width of the compression, but smaller than about 1 AU (for particles with a speed equal to the flow speed). Pickup ions probably satisfy these criteria and, therefore, we expect them to be efficiently accelerated at solar

CP598, *Solar and Galactic Composition*, edited by R. F. Wimmer-Schweingruber

wind compression regions. The mechanism is similar to diffusive shock acceleration in that the particles are accelerated as they scatter off of converging scattering centers.

The acceleration can occur for any particles, provided that their mean-free path is large enough to sample the velocity gradient and small enough to be scattered downstream of the compression region. The highest energy attainable, therefore, can be estimated by setting the mean-free path to the heliocentric distance of the compression region. A compilation of observations of particle mean-free paths by Palmer[6] has shown that even 1-10 MeV particles can have mean-free paths which are smaller than, or are comparable to 1 AU. Consequently, we conclude that pickup ions can be accelerated by corotating compression regions (at distances smaller than where the forward and reverse shocks form) up to energies of about 1-10 MeV.

However, for the case of particle acceleration in the inner heliosphere where the particle mean-free paths may be comparable to the size of the system ($\sim$ AU), significant anisotropies are expected and we cannot use the diffusive approximation. Consequently, a numerical treatment is necessary.

NUMERICAL MODEL

In order to quantify the acceleration of charged particles by gradual compressions in the solar wind, we first construct a simple model of a corotating interaction region which has not yet formed the forward and reverse shocks. From this model we determine the electromagnetic fields using magnetohydrodynamics. We then integrate the trajectories of an ensemble of test particles in the model fields from which we obtain the distribution function by binning along the particle trajectories.

We consider a simple model in which the solar wind flow is radial in a frame fixed with respect to the Sun with a speed U, and in steady state in a frame which corotates with the sun. In this frame, the continuity equation yields

$$-\Omega_{\odot}\frac{\partial}{\partial\phi}(r^2\rho)+\frac{\partial}{\partial r}(r^2\rho U)=0 \tag{2}$$

Our method is to determine the density, ρ, for a given functional form of the flow speed U. We do this by solving Eq. (2). We assume the following form for the flow speed:

$$U(r,\phi)=U_s+\frac{1}{2}(U_f+U_s)\tanh\left(\frac{\phi_c-\phi-r\Omega_{\odot}/W}{\Delta\phi_c}\right)$$
$$-\frac{1}{2}(U_f-U_s)\tanh\left(\frac{\phi_{rf}-\phi-r\Omega_{\odot}/W}{\Delta\phi_{rf}}\right) \tag{3}$$

where W is a constant speed and $\Omega_{\odot}$ is the solar rotation frequency. U_s and U_f are the slow and fast solar-wind speeds, respectively.

This form for U contains both a compression region, with a azimuthal width $\Delta\phi_c$, and a rarefaction region with an azimuthal width $\Delta\phi_{rf}$. The compression region does not form into shocks in this simplified model. The constants ϕ_c and ϕ_{rf} specify the location the compression and rarefaction regions. It is readily shown that the thickness of the compression region along a given radius vector is given by $\Delta_c = W\Delta\phi_c/\Omega_{\odot}$. The speed W can be thought of as the speed at which these disturbances move radially outward in the inertial (not rotating) frame of reference. If W is slower than the slow solar-wind speed, then the disturbance will be a "reverse" compression (similar to a reverse shock); whereas for W faster than the fast solar-wind speed, the disturbance will be a "forward" compression.

A solution of Eq. (2) using Eq. (3) is given by

$$\rho(r,\phi)=\rho_s\frac{U_s-W}{U(r,\phi)-W}\left(\frac{r_{\odot}}{r}\right)^2 \tag{4}$$

where $r_{\odot}$ is the solar radius and ρ_s is the density in the slow solar wind at $r=r_{\odot}$. To avoid the singularity in Eq. (4), we consider cases where W is everywhere smaller, or larger, than U.

A divergence-free magnetic field is given by

$$\mathbf{B}(r,\theta,\phi)=B_r\hat{e}_r+B_{\phi}\hat{e}_{\phi} \tag{5}$$

where

$$B_r(r,\theta,\phi)\propto B_r(r_{\odot},\theta)\rho(r,\phi)U(r,\phi)$$
$$B_{\phi}(r,\theta,\phi)\propto r\rho(r,\phi)\Omega_{\odot}\sin\theta \tag{6}$$

Figure 1 shows magnetic field lines for the magnetic field given by Eqs. (5) and (6). The parameters are shown in the caption. The compression region is evident. Note that because we have chosen W to be slower than the slow solar-wind speed, there are no magnetic field lines originating in the slow wind which intersect the compression region. Hence, this is a reverse compression region.

We follow test particles (helium ions) in the magnetic field given by Eq. (5). The orbits are integrated numerically by solving the Lorentz force acting on each particle. The particle orbits are computed in the corotating frame of reference. Each particle is followed until it crosses an outer boundary which is placed at 3 AU. Particle distributions are computed by binning along the trajectory at a constant time interval. This method gives the steady-state distribution as a function of r, ϕ, and momentum $\mathbf{p}$.

Scattering is introduced in a phenomenological manner. The scattering conserves energy and is isotropic in a

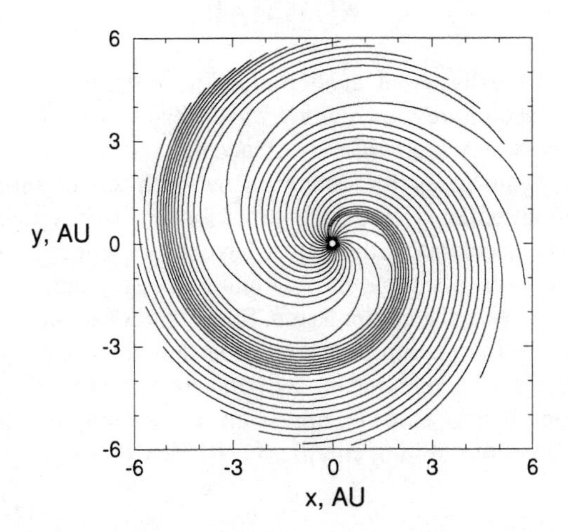

FIGURE 1. Magnetic field lines projected onto the solar equatorial plane for $\theta = 80°$. The parameters are: W=300 km/s, $\Delta\phi_c = 1°$ ($\Delta_c = 0.012$ AU), $\Delta\phi_{rf} = 25°$, U_f=800 km/s, U_s=400 km/s, $\phi_c = 90°$, and $\phi_{rf} = 20°$.

frame moving with the plasma. A scattering time is chosen from an exponential distribution with a given mean. The mean scattering time is related to the parallel mean-free path, $\lambda_{\parallel}$. We use two forms for $\lambda_{\parallel}$: $\lambda_{\parallel} \propto w^{2/3}$ (quasi-linear theory), and $\lambda_{\parallel}$=constant. We note, in passing, that cross-field diffusion is included here in the form of classical scattering (which is also often referred to as "hard-sphere" or "billiard-ball" scattering).

We consider two type of sources in this study. The first are ionized interstellar neutrals, or pickup ions. These are singly-ionized particles which we release from rest in the inertial frame of reference (the non rotating frame in which the solar wind flows radially outward). The spatial distribution is determined from the analytic model of Vasyliunas and Siscoe[7] using a characteristic ionization distance of 0.7 AU and an interstellar density of 0.01cm^{-3} which are representative of interstellar helium. For simplicity, we assume that the interstellar flow speed and temperature are both zero. The second type of source that we consider is solar-wind alpha particles. These particles are doubly ionized. A kappa distribution is assumed to determine their initial speed ($\kappa = 3$). Their density at 1 AU is assumed to be 0.5cm^{-3} and falls off as r^{-2}. Their temperature is determined from the adiabatic law (at 1 AU it is $2\text{x}10^5$K).

RESULTS AND DISCUSSION

Shown in Figure 2 are particle fluxes (lower frame) and plasma flow speed (top frame) as a function of the variable $\phi/\Omega_{\odot}$, which has units of time. The results shown are for the case of an interstellar pickup-ion source. An observer at rest with respect to the rotating sun would see this structure every solar rotation period. The energies for each of the curves shown are in the caption. The simulation parameters are the same as that shown in Figure 1 except that here we used $\Delta\phi_c = 2°$ which gives $\Delta_c = 0.025$ AU. Also for this case we used $\lambda_{\parallel} = 0.056(w/U_s)^{2/3}$ AU. Note that at all energies the particles have a mean-free path which is larger than the compression width.

Figure 2 shows that the particle fluxes rise with the passage of the compression region and that the fluxes drop off at different rates depending on their energy. These are both consistent with the observations described by Mason[4].

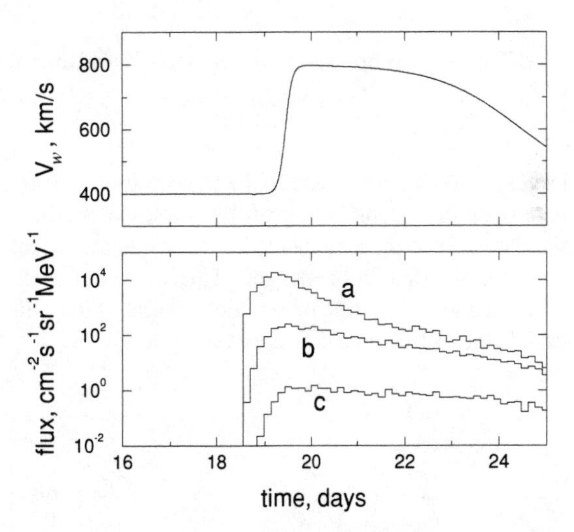

FIGURE 2. (top frame) flow speed as a function of time ($\phi/\Omega_{\odot}$. (bottom frame) energetic particle fluxes as a function of time for the case of a interstellar pickup-ion source. The energies are: (a) 50-250 keV (b) 250-500 keV and (c) 0.5-5 MeV.

Shown in Figure 3 is the energy spectrum at 1 AU integrated over all ϕ for the case of an interstellar pickup-ion source. The turnover at high energies is due to the fact that those particles have mean-free paths which are comparable to the distance from the Earth to the Sun. For this case, particles with a speed $75U_s$ (an energy of $\sim$ 5 MeV) have a mean-free path of 1 AU. This is somewhat higher than where the turnover in energy occurs. The discrepancy is due to the fact the inside 1 AU the mean-free path is significantly reduced by the stronger magnetic field near the sun.

Shown in Figure 4 are the energy spectrum observed at 1 AU for the case of a constant mean-free path for two input source distributions: a solar-wind source which has a high-energy tail (kappa distribution), and a pickup-ion source. Note that the abundance of solar-wind He^{++} and interstellar He^{+} are about equal at high energies. At

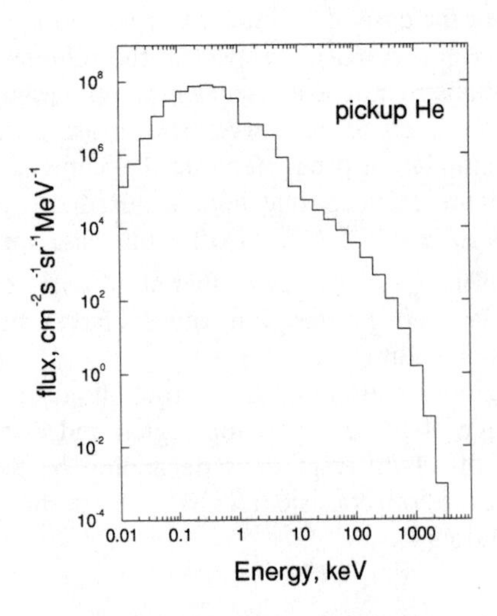

FIGURE 3. Energy spectrum of accelerated interstellar pickup helium at 1 AU.

larger heliocentric distances the pickup ions will dominate since their density falls off less rapidly than the solar wind density. Note also that the energy spectra do not exhibit a turnover at high energies. This is due to the choice of a constant mean-free path which is smaller than the radial distance from the compression region.

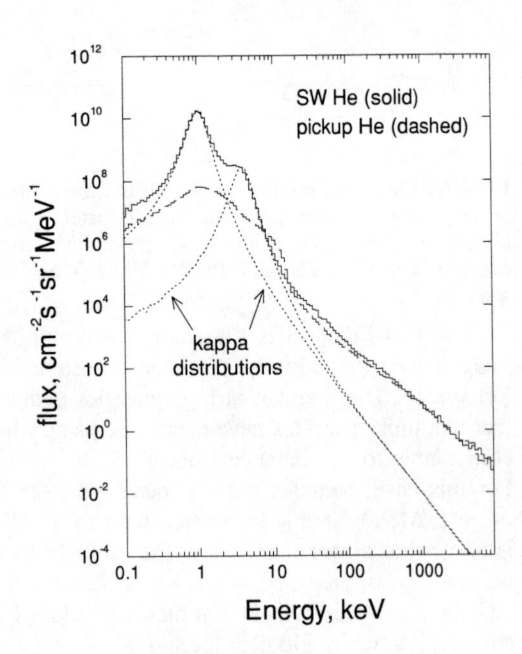

FIGURE 4. Energy spectra for solar-wind alpha particles (solid line) and interstellar pickup helium (dashed line) at 1 AU. The dotted curves shown are κ-distributions for the fast and slow winds which were used as the source for the solar-wind alpha particles. A constant mean-free path (0.056 AU) was considered for both species.

SUMMARY

We have discussed an alternate view of particle acceleration at compression regions, rather than shocks. This process is very much like shock acceleration in that the energy gain arises from scattering between converging scattering centers. However, in order for low-energy particles to be accelerated, they must have a long enough mean-free path to sample the gradual velocity gradient associated with the compression. Spacecraft observations indicate that pickup ions, and other low-energy particles, have much longer mean-free paths than one would expect from theory. Thus, compression regions may accelerate pickup ions to high energies (1-10 MeV).

ACKNOWLEDGMENTS

This work was supported by NASA under grants NAG5-2251 and NAG5-6620.

REFERENCES

1. Barnes, C. W., and J. A. Simpson, Evidence for interplanetary acceleration of nucleons in corotating interaction regions, Astrophys. J. 210 (1976) L91.
2. McDonald, F. B., B. J. Teegarden, J. H. Trainor, and T. T. von Rosenvinge, The interplanetary acceleration of energetic nucleons, Astrophys. J. 203 (1975) L149.
3. Fisk, L. A., and M. A. Lee, Shock acceleration of energetic particles in corotation interaction regions in the solar wind, Astrophys. J., 237, (1980) 620.
4. Mason, G. M., Composition and energy spectra of ions accelerated in corotating interaction regions, in Acceleration and Transport of Energetic Particles Observed in the Heliosphere , edited by R. A. Mewaldt, et al., American Institute of Physics, New York, (2000).
5. Drury, L., An introduction to the theory of diffusive shock acceleration of energetic particles in tenuous plasma, Rep. Prog. Phys., 46, (1983), 973.
6. Palmer, I. D., Transport coefficients of low-energy cosmic rays in interplanetary space, Rev. Geophys. 20 (1982) 335.
7. Vasyliunas, V. M., and G. L. Siscoe, On the flux and the energy spectrum of interstellar ions in the solar system, J. Geophys.Res., 81, (1976), 1247.

Time Variations in Elemental Abundances in Solar Energetic Particle Events

T. T. von Rosenvinge*, C. M. S. Cohen†, E. R. Christian*, A. C. Cummings†, R. A. Leske†, R. A. Mewaldt†, P. L. Slocum**, O. C. St. Cyr*, E. C. Stone† and M. E. Wiedenbeck**

*NASA/Goddard Space Flight Center, Greenbelt, MD 20771 USA
†California Institute of Technology, Pasadena, CA 91125 USA
**Jet Propulsion Laboratory, Pasadena, CA 91109 USA

Abstract. The Solar Isotope Spectrometer (SIS) on-board the Advanced Composition Explorer has a large collection power and high telemetry rate, making it possible to study elemental abundances in large solar energetic particle (SEP) events as a function of time. Results have now been obtained for more than 25 such events. Understanding the causes of these variations is key to obtaining reliable solar elemental abundances and to understanding solar acceleration processes. Such variations have been previously attributed to two models: (1) a mixture of an initial impulsive phase having enhanced heavy element abundances with a longer gradual phase with coronal abundances and (2) rigidity dependent escape from CME-driven shocks through plasma waves generated by wave-particle interactions. In this second model the injected abundances are assumed to be coronal. Both these models can be expected to depend upon solar longitude since impulsive events are associated with flares at longitudes well-connected magnetically to the observer, and shock properties and connection of the observer to the shock are also longitude dependent. We present results on temporal variations from event to event and within events and show that they appear to have a longitude dependence. We show that the events which have been well-explained by model (2) tend to be near central meridian or the west limb. In addition, we show that there are events with little time variation and heavy element enhancements similar to those of impulsive events. These events seem to be better explained by model (1) with only an impulsive phase.

INTRODUCTION

One of the goals of measurements of solar energetic particles (SEPs) has been to measure solar abundances of elements which are not easily determined by other methods. Cohen, *et al.* (1, this conference) present such results from the Solar Isotope Spectrometer (SIS)(2) onboard the Advanced Composition Explorer (ACE), averaged over the principal SEP events observed since the launch of ACE in August, 1997. The present paper illustrates the variations from event to event and within events that are integrated over by such averaging. These variations are substantial and illustrate the difficulties of this approach to obtaining solar abundances. Cohen, *et al.* (3), von Rosenvinge, *et al.* (4), and Boberg and Tylka (5) have also discussed variations in abundances measured by SIS, but for smaller numbers of events.

Event-to-event variations in elemental abundances have been known for a long time (e.g., 6,7,8,9; see also the review by Reames (10)). The causes of these variations, however, have been relatively slow in being understood. An important step was the work of Breneman and Stone (8), who showed that the elemental abundances in specific solar events could be ordered as power laws in Q_X/M_X , where Q_X and M_X are the mean ionic charge state and the mean elemental mass for element X. The corresponding power laws had exponents ranging from roughly -2.5 to +2.5, corresponding to both heavy element enhancements and heavy element depletions. No explanation for this behavior was provided other than to presume that particle rigidity was the essential parameter governing both particle acceleration and escape from the sun. While this approach was purely phenomenological, it did provide a means for estimating Q/M dependent biases between the observed abundances and solar abundances.

Reames, *et al.* (11) and Cane, *et al.* (12) were the first to report on an organization of abundance data by the longitude of the source region. They suggested that solar event elemental abundances as a function of lon-

CP598, *Solar and Galactic Composition,* edited by R. F. Wimmer-Schweingruber

gitude could be understood in terms of two particle sources: 1) particles originating out of flare-heated material and 2) particles accelerated at coronal and interplanetary shocks. Fe-rich, high Q-state, flare-heated material would only have access to the observer when the observer was magnetically well-connected to the flare site. Reames (10) has subsequently rejected this interpretation, instead attributing the longitude dependence of Fe enhancements to the fact that the observer is well-connected magnetically to the nose of a shock (where the shock is presumably the strongest) when the shock originates at a western location. The longitude dependence of SEP abundances has been largely ignored since the time of Cane, *et al.* (12) but will be revisited in the present paper. We note that the Cane, *et al.* (12) explanation is consistent with the correlation between Fe/O and $< Q_{Fe} >$ reported at low energies by Möbius, *et al.* (13) and at high energies by Leske, *et al.* (14, this conference).

Recently instruments have been built which have much greater collection power than was available to Cane, *et al.* (12) and which can therefore examine time variations within many more events and in greater detail (2,15). In principle this gives greater opportunities for understanding the causes of variations. Recent work by Reames, Tylka and Ng has explained some unexpected temporal behavior in solar event composition in terms of shock acceleration and the build-up of plasma waves, predominantly due to proton wave-particle interactions (e.g., 16,17,18; see also Reames (19), this conference). The theory of Ng, *et al.* (18), however, is incomplete in that it posits a spectrum which is continuously injected at a moving shock and accounts for how particles are released from the shock and travel to the observer. This release depends upon the temporal build-up of plasma waves in the vicinity of the shock. In other words, the Ng, *et al.* (18) model does not follow the acceleration of individual particles from low to high energies. Thus, for example, it does not readily explain why events such as the 20 April 1998 event are Fe-rich at the energies ($\sim 3-10$ MeV/nucleon) measured by the Low Energy Matrix Telescopes (LEMT) on-board the Wind spacecraft (16) and Fe-poor at SIS energies ($\sim 12-60$ MeV/nucleon) (4). Nonetheless, a qualitative explanation can be provided in the framework of the Ng, *et al.* (18) model as discussed by von Rosenvinge, *et al.* (4).

The Ng, *et al.* (18) model assumes that particles are injected into the shock with coronal abundances. This model can, however, create an enhancement in Fe/O above the coronal value at low energies for some observers for at least a portion of the event. In particular, it has been able to inject particles with coronal abundances and produce a peak in the modeled Fe/O time intensity-curve at a few MeV/nucleon as reported by Tylka, et al. (16) in the 1998 April 20 event. This peak was enhanced by a factor of 4 over the coronal value. A constant high value of Fe/O, however, would be difficult to achieve in this model.

The model of Ellison and Ramaty (20) produces a Q/M dependent energy spectrum by accounting for first order Fermi acceleration and rigidity dependent escape at a shock, but the predicted spectral shape, unlike the model of Ng, *et al.* (18), applies at the shock and not to the remote observer. Furthermore there is no time dependence in the Ellison and Ramaty (20) model. This model has fit differential intensity spectra quite well and has been useful in deriving Q/M ratios for different elements (21). This is possible since the model spectrum consists of a power law in momentum modified by an exponential factor, $\exp(-E/E_{0X})$, produced by the finite spatial extent of the shock. Here E is kinetic energy/nucleon and $E_{0X} = E_{0H}(Q_X/M_X)$, where X denotes element X and E_{0H} is the appropriate E_{0X} for hydrogen. The exponential factor gives rise to characteristic spectral knees which can, for example, account for depletions of Fe relative to O above $\sim E_{0Fe}$ since in general Q_{Fe}/M_{Fe} is less than Q_O/M_O. Tylka, *et al.* (21), by dividing event data into time intervals of 8 hours, have used the evolution of spectral shapes to infer charge states as a function of time during two events. However, it is hard to see how this model could account for the Fe/O ratio increasing with energy as was observed for the 6 November 1997 event (16,4). It is also not clear how it could account for the previously mentioned correlation between Fe/O and $< Q_{Fe} >$ or for enhancements above coronal abundances.

OBSERVATIONS

The Solar Isotope Spectrometer (SIS) on-board the Advanced Composition Explorer (ACE) has a large geometry factor ($\sim$ cm^2-sr), enabling abundances in solar energetic particle (SEP) events to be observed on a time scale of hours or less. We have identified 27 such events, which are listed in Table 1. The columns labeled 'Start UT' and 'End UT' give the start and end Universal Times of the time interval for which energetic particle intensities were studied for each event. Table 1 includes approximate times and solar longitudes for 'associated' solar flares. The identification of such flares is not always straight forward. Indeed, in models where acceleration in gradual events is due to acceleration by shocks driven by Coronal Mass Ejections (CMEs), 'associated' solar flares are not even considered to be relevant. Nonetheless, we have attempted to find such flares and their start times and longitudes for each particle event using H_α, X-ray, and Type II and Type III radio data taken mainly from NOAA's Solar Geophysical Data Reports (`http://julius.ngdc.noaa.gov:8000/Welcome_SGD.html`). Radio data was also

Table 1. SIS particle events.

#	Start UT	End UT	Flare UT	Longitude	Comments
1	1997 Nov 4 06:00	1997 Nov 6 12:00	Nov 5 05:52	W33	X2.1, 2B
2	1997 Nov 6 12:00	1997 Nov 10 00:00	Nov 6 11:49	W63	X9.4, 2B
3	1998 Apr 20 12:00	1998 Apr 27 00:00	Apr 20 09:38	W90	M1.4, no H_α
4	1998 May 2 12:00	1998 May 5 00:00	May 2 13:31	W15	X1.1, 3B
5	1998 May 6 08:00	1998 May 8 00:00	May 6 07:58	W65	X2.7, 1N
6	1998 May 9 04:50	1998 May 11 12:00	May 9 03:04	W90	M7.7, no H_α
7	1998 Aug 25 00:00	1998 Sep 1 00:00	Aug 24 21:50	E09	X1.0, 3B
8	1998 Sep 24 12:00	1998 Sep 25 12:00	Sep 23 07:13	E09	M7.1, 3B, strong ESP spike after Sep 24 12:00 UT
9	1998 Sep 30 12:00	1998 Oct 5 00:00	Sep 30 13:50	W78	M2.8, 2N
10	1998 Oct 19 00:00	1998 Oct 20 18:00	Oct 18 21:15	W90	no X-rays, no H_α, Type III
11	1998 Nov 14 06:00	1998 Nov 18 12:00	Nov 14 05:05	W90	weak X-rays, no H_α
12	1999 Jan 21 00:00	1999 Jan 27 12:00	Jan 20 19:06	W90	M5.2, no H_α
13	1999 Feb 16 00:00	1999 Feb 19 00:00	Feb 16 02:49	W14	M3.2, SF
14	1999 Apr 24 12:00	1999 Apr 26 12:00	Apr 24 12:36	W90	no X-rays, no H_α, CME
15	1999 May 4 00:00	1999 May 9 12:00	May 3 05:50	E32	M4.4, 2N
16	1999 May 9 18:00	1999 May 11 00:00	May 9 18:07	W90	M7.6, no H_α
17	1999 Jun 2 00:00	1999 Jun 4 08:00	Jun 1 19:30	W90	no X-rays, no H_α, CME
18	1999 Jun 4 08:00	1999 Jun 8 00:00	Jun 4 07:00	W69	M3.9, 2B
19	2000 Apr 4 12:00	2000 Apr 9 00:00	Apr 4 15:12	W66	C9.7, 2F
20	2000 Jun 7 00:00	2000 Jun 10 00:00	Jun 6 13:24	E12	X1.1, 3B
21	2000 Jun 10 12:00	2000 Jun 13 00:00	Jun 10 16:40	W38	M5.2, 3B
22	2000 Jul 14 00:00	2000 Jul 19 00:00	Jul 14 10:03	W07	X5.7, 3B
23	2000 Sep 12 12:00	2000 Sep 16 00:00	Sep 12 13:31	W05	
24	2000 Oct 16 00:00	2000 Oct 20 00:00	Oct 16 07:00	W90	M2.5, no H_α
25	2000 Oct 25 12:00	2000 Oct 29 00:00	Oct 25 09:30	W90	no X-rays, no H_α
26	2000 Nov 9 00:00	2000 Nov 15 00:00	Nov 8 23:28	W77	M7.4, 3F
			Nov 9 16:13	E10	M1.0, SF
27	2000 Nov 24 00:00	2000 Nov 29 00:00	Nov 24 05:02	W05	X2.0
			Nov 24 15:13	W07	X2.3
			Nov 24 21:59	W14	X1.8

taken from the WAVES investigation on the Wind spacecraft (`http://lep694.gsfc.nasa.gov/waves/waves.html`). CME data was provided from the LASCO instrument on the SoHO spacecraft. For the simplest cases, a hard X-ray event starts essentially at the same time as a Type III burst, followed by H_α and high-frequency Type II emission. Utilizing the Wind/WAVES radio data, the Type III data can be generally tracked down to $\sim 20-40$ kHz, the local plasma frequency at 1 A.U., indicating magnetic connection between the particle acceleration site all the way out to the observer at 1 A.U. The event solar longitude is then obtained from the H_α data. More difficult cases have no H_α event, and even no X-ray event. These events are generally from beyond the west limb of the sun, as can be verified by, for example, LASCO CME observations. While one can perhaps obtain a better location by extrapolating observations of active regions prior to their rotating out of view, we have simply entered the location in Table 1 as *W90*. In some cases there are multiple flare events and these are indicated in Table 1. Figures 1, 2, and 3 show event-averaged abundances for different events as observed by SIS in the energy interval 12-60 MeV/nucleon. The energy interval which SIS covers for each element varies with the atomic number Z. 12-60 MeV/nucleon is the broadest energy interval for which SIS can observe all elements from C to Fe. Figures 1, 2, and 3 show SIS abundances versus Z as ratios with respect to the average abundances for gradual events (open circles) and for impulsive events (closed circles) as reported by Reames (22). The average gradual event values are thought to represent average coronal values (22). Figures 1, 2, and 3 show data for sample events which are, respectively, depleted, unchanged, and enhanced in heavy elements relative to the average abundances in gradual events. These figures show that heavy element enhancements/depletions vary considerably from event to event but that each event generally shows a trend with respect to increasing Z.

The upper three panels of Figure 1 show large depletions of heavy elements corresponding to events east of central meridian and, in the rare case of the 20 April 1998 event, at $\sim W90$. The bottom panel corresponds to 2 widely separated events, one at W77 and one at E10, as shown by the corresponding Type III bursts which both reach 1 A.U. Figure 2 shows three events near central

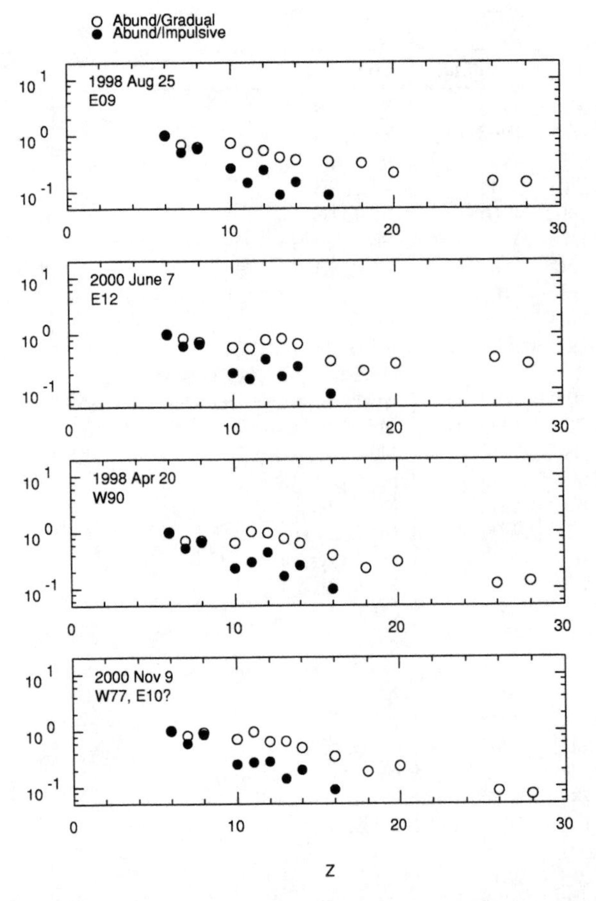

FIGURE 1. Shows the ratios of element abundances observed by SIS to those of average gradual events (open circles) and to those of average impulsive events (closed circles). Each panel shows these ratios normalized to C and plotted versus Z for a separate SIS particle event. These events were chosen because they show heavy-element depletions relative to average gradual event abundances.

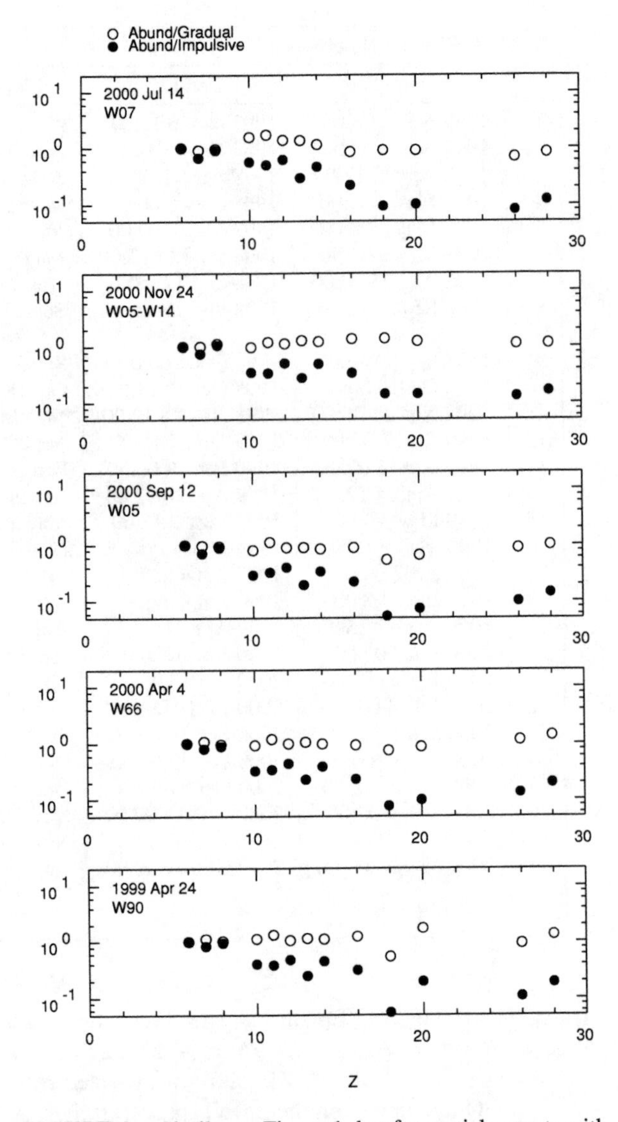

FIGURE 2. Similar to Figure 1, but for particle events with abundances similar to average gradual event abundances.

meridian, one at W66 and one from beyond the west limb. Except for event 19 at W66 these events follow a similar pattern to the events of Figure 1. Figure 3 shows that events with impulsive-like composition can come from all longitudes $\gtrsim W15$.

Figure 4 shows the enhancement of Fe/C over the average gradual event Fe/C ratio for multiple events versus the solar longitude of the associated flare event (i.e. the y-axis values are the same as for the open circles at Z=26 in Figures 1, 2, and 3). The ACE-SIS event number from Table 1 is printed for each event in place of a data point. This figure shows that all events occurring between $W15$ and $W85$ have Fe enhancements greater than one. All events with enhancements ~ 1 or less are either to the east of $W15$ or, in the case of the unusual 20 April 1998 event, at $\sim W90$. Events with large Fe enhancements also occur at or beyond the west limb, but not near central meridian.

Figure 5 shows, for each separate event, the ratio of the maximum Fe/O ratio to the minimum Fe/O ratio versus the solar longitude of the associated flare event. The ratios are determined at 14 MeV/nucleon. This then gives a measure of the variability of Fe/O during each event. We see that, with the exception of event 19, there is very little variability within events between $\sim W15$ and $\sim W80$. Events with relatively large variability are either close to central meridian or towards the west limb. Events with low variability, however, can occur at all longitudes.

Figure 6 shows time-intensity profiles (upper panel) and ratios (lower panel) at 14 MeV/nucleon normalized to average gradual event values for representative elements in the event of 2 May 1998. This normalization

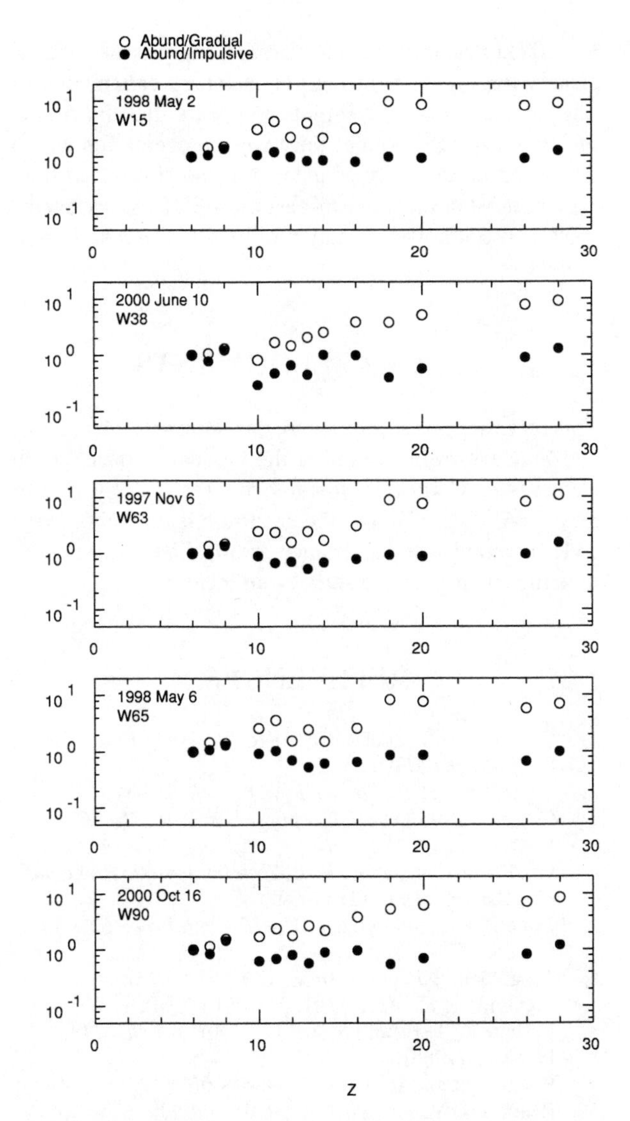

FIGURE 3. Similar to Figure 1, but for particle events with abundances similar to average impulsive event abundances.

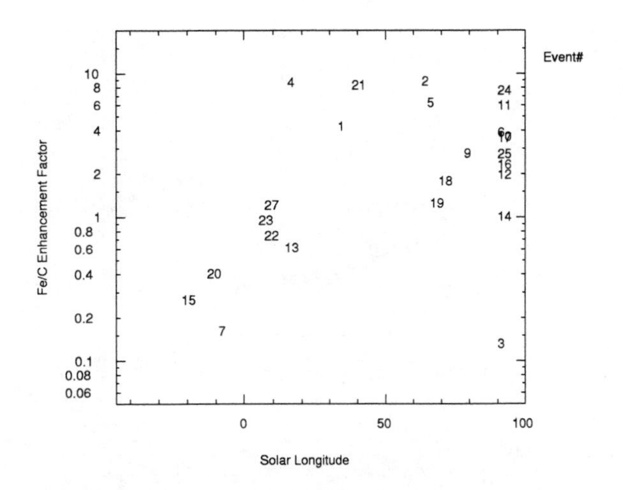

FIGURE 4. Shows the enhancement of Fe/C over the average gradual event abundance reported by Reames (1995) for multiple events versus the solar longitude of the associated flare event.

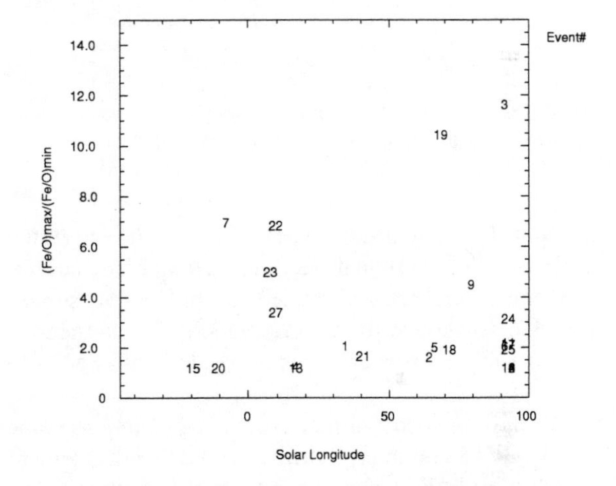

FIGURE 5. Shows the ratio of the maximum Fe/O ratio during each event to the minimum Fe/O ratio versus the solar longitude of the associated flare event.

means that ratios which are the same as the corresponding event averages given by Reames (22) are plotted with a value of 1.000 and enhancements above (or depletions below) the average values are immediately apparent. There was a corresponding flare at $W15$ at 1998 May 2 13:31 UT. Figure 6 shows that the normalized Fe/O ratio is quite constant throughout the event with a value of approximately 6. It also shows that the normalized Ca/O ratio is similarly constant with a similar value. The ratios He/O and Si/O show only slight enhancements. Note also the two interplanetary shocks shown in Figure 6 and separated by only 9 hours. The presence of these two shocks potentially complicates the interpretation of this event. On the other hand, as shown in Figures 4 and 5, well-connected events with relatively constant Fe/O during each event and large enhancements of Fe are fairly common. Three other events with relatively constant Fe/O (6 November 1997, $W33$, 14 November 1998, $> W90$, and 6 May 1998, $W65$) are shown in von Rosenvinge, *et al.* (4) and von Rosenvinge, *et al.* (23).

DISCUSSION

It is clear from the data presented here that the observed event-to-event average abundances depend on the solar longitude of associated solar flare events. It appears that the temporal variability of abundances within individual events may also depend upon solar longitude. At this point there is no accepted model which we know of that

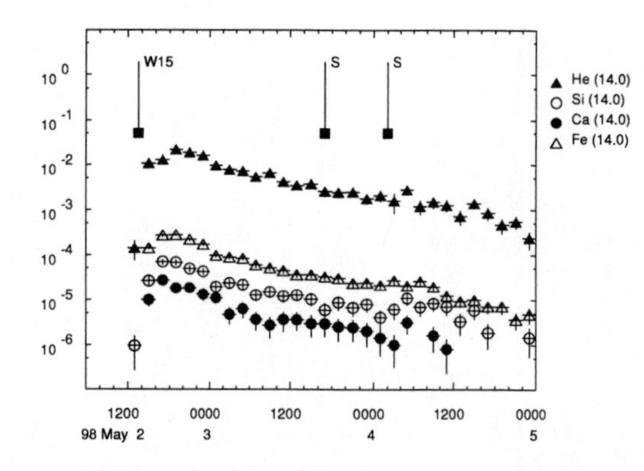

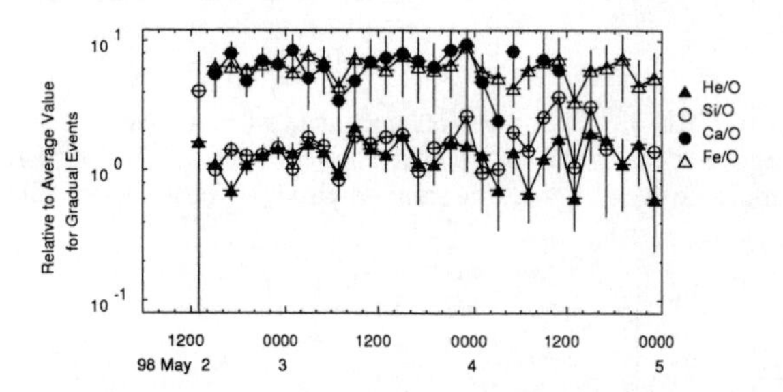

FIGURE 6. Example of an event where Fe/O is essentially constant throughout the event at an enhanced level.

accounts for these dependences, although the suggestion of Cane, *et al.* (21) that there can be a major flare accelerated component cannot be ruled out. In fact, this suggestion is consistent with the observation that the mean Fe charge state in large events is correlated with Fe/O (13, 14).

We have also shown that events originating from beyond $\sim W15$ commonly show large Fe/O enhancements which vary little during the course of the events. Such enhancements cannot be explained within the context of the Ellison and Ramaty (20) model and injection of relative abundances at low energies corresponding to coronal abundances. It is possible that injections from multiple impulsive events can fill the inner solar system with enhanced heavy elements which are then accelerated by CME related shocks (Mason, *et al.*, 24). However, events with large Fe/O enhancements have heavy element composition essentially identical to that of impulsive events, so this process would have to completely dominate the acceleration of normal coronal/solar wind material. And why, then, would there be no Fe-enhanced events east of $\sim W15$?

It is not clear either how the Ng, *et al.* (18) model can account for events with large, constant Fe/O enhancements. The Ellison and Ramaty (20) model, however, does appear to have successfully accounted for spectral shapes for elements from H to Fe for the event of 20 April 1998 (at the west limb) and the event of 25 August 1998 (at $E09$) (Tylka, *et al.*, 21). Similar analysis of spectral shapes for events with near constant Fe enhancements has yet to be done. It similarly appears that the model of Ng, *et al.* (18) applies very well to events for which the observer is poorly connected to the associated flare region, most notably events east of $\sim W15$. Many events west of $W15$, however, may require a flare-related component.

ACKNOWLEDGMENTS

This research was supported by the National Aeronautics and Space Administration at the Goddard Space Flight Center, the California Institute of Technology (under grant NAG5-6912), and the Jet Propulsion Laboratory. TvR acknowledges assistance from Hilary Cane with determining associated solar event locations.

REFERENCES

1. Cohen, C.M.S., et al., *ACE/SoHO Workshop Proc.*, Berne, this volume (2001).
2. Stone, E.C., et al., *Sp. Sci. Rev.*, **86**, 357 (1998).
3. Cohen, C.M.S., et al., *Geophys. Res. Lett.*, **26**, 2697 (1999).
4. von Rosenvinge, et al., *Proc. 26th Int. Cosmic Ray Conf. (Salt Lake City)*, **6**, 131 (1999).
5. Boberg, P.R. and Tylka, A.J., *AIP Conf. Proc.*, **528**, 115 (2000).
6. Teegarden, B.J., et al., *Ap. J.*, **180**, 571 (1973).
7. McGuire, et al., *Ap. J.*, **301**, 938-961 (1986).
8. Breneman, H.H. and Stone, E.C., *Ap. J. Lett.*, **299**, L57-L61 (1985).
9. Reames, et al., *Ap. J. Supp.*, **90**, 649-667 (1994).
10. Reames, D.V., *Sp. Sci. Rev.*, **90**, 413 (1999).
11. Reames, D.V., et al., *Ap. J.*, **357**, 259 (1990).
12. Cane, H.V., et al., *Ap. J.*, **373**, 675 (1991).
13. Möbius, et al., *AIP Conf. Proc.*, **528**, 131-134 (2000).
14. Leske, R.A., et al., *ACE/SoHO Workshop Proc.*, Berne, this volume, 2001.
15. von Rosenvinge, T.T., et al., *Sp. Sci. Rev.*, **71**, 155-206 (1995).
16. Tylka, A.J., et al., *Geophys. Res. Lett.*, **26**, 3585 (1999).
17. Reames, D.V., et al., *Ap. J. Lett.*, **531**, L83 (2000).
18. Ng, C.K., et al., *Geophys. Res. Lett.*, **26**, 2145 (1999).
19. Reames, D.V., *ACE/SoHO Workshop Proc.*, Berne, this volume, 2001.
20. Ellison, D., and Ramaty, R., *Ap. J.*, **298**, 400-408 (1985).
21. Tylka, A.J., et al., *AIP Conf. Proc.*, **528**, 147 (2000).
22. Reames, D.V., *Adv. Space Res.*, **15**, 41 (1995).
23. von Rosenvinge, T.T., et al., *AIP Conf. Proc.*, **528**, 111 (2000).
24. Mason, G.M., et al., *Ap. J. Lett.*, **525**, L133 (1999).

^{3}He-enrichments in Solar Energetic Particle Events: SOHO/COSTEP Observations

V. Bothmer[1], H. Sierks[1], E. Böhm[2], and H. Kunow[2]

[1]*Max-Planck-Institut für Aeronomie, D-37191 Katlenburg-Lindau, Germany*
[2]*Institut für Experimentelle und Angewandte Physik, Universität Kiel, D-24118 Kiel, Germany*

Abstract. We present first results based on a systematic survey of 4-41 MeV/N ^{3}He/^{4}He isotope abundances with ratios >0.01 detected by the COmprehensive SupraThermal and Energetic Particle analyzer (COSTEP) onboard the SOHO (SOlar and Heliospheric Observatory) spacecraft. During 53 out of 148 identified days with a ^{3}He/^{4}He ratio ≥0.01, the ratio was in the range 0.1-1.0. For days with sufficiently high detector count rates, the atomic mass plots could be resolved up to a time resolution of 1 hour. These days were most suitable for comparisons with in situ solar wind plasma and magnetic field measurements and SOHO's optical white-light and extreme ultraviolet (EUV) observations of the Sun. Here we present a brief overview of a ^{3}He/^{4}He-rich particle event detected on October 30, 2000 that was associated with the passage of a fast CME.

1. Introduction

The SOHO/COSTEP instrument measures solar energetic particles (SEPs) at MeV energies in the interplanetary medium. The solid state detectors are capable to detect ^{3}He/^{4}He-enrichments at these energies (Müller-Mellin et al., 1995). Usually, the ^{3}He/^{4}He-ratio in the solar wind is at the order of 10^{-4}, but occasionally ratios up to about values of ~1 or even above have been observed in SEP events (e.g., Mason et al., 1999). The origin of these isotope abundances has commonly been attributed to impulsive solar flares and wave-particle interaction mechanisms (Temerin and Roth, 1992). However, fully satisfying physical explanations are still lacking. Here we present first results from our start of a systematic survey of the He-measurements taken by COSTEP since launch in 1995 until the end of 2000.

2. Data

For this study we have analyzed SOHO/COSTEP measurements of 4.3-40.9 MeV/N helium particles as well as proton and electron measurements at energies below 10 MeV. These data were compared with magnetic field and plasma data from the Advanced Composition Explorer (ACE), SOHO/LASCO (Large Angle Spectroscopic COronagraph) observations of CMEs and X-ray flare measurements taken by the GOES satellite.

3. Identification of ^{3}He/^{4}He-rich events

Figure 1 shows an example of a ^{3}He/^{4}He-rich event as identified from the COSTEP mass separation plot for Oct. 30, 2000. The vertical axis in Figure 1 provides information about the detector count rates, the horizontal axis is labelled such that ^{4}He corresponds to a value of 0. The two largest peaks in the count rates at –0.6 and 0 correspond to proton (-0.6) and ^{4}He particles. The presence of a major contribution of ^{3}He isotopes at about –0.05 is very distinguished. The ratio of ^{3}He/^{4}He was 0.7 (see Table 1, DoY 304 in 2000). All identified ratios ≥0.01 are listed in Table 1.

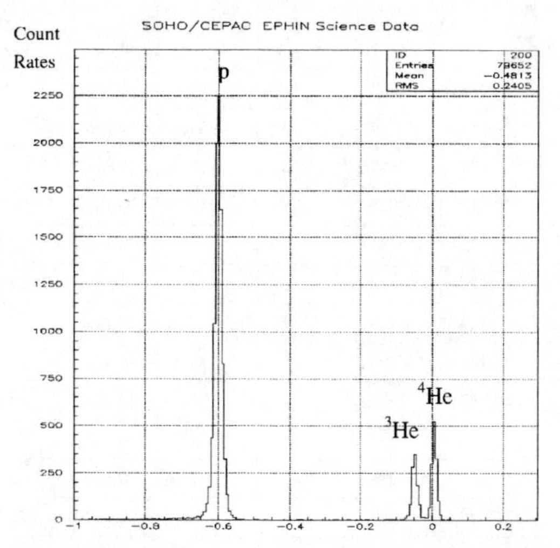

Figure 1. COSTEP mass plot for 30 October 2000. The peaks at –0.6, -0.05 and 0 correspond to H (p), ^{3}He and ^{4}He.

Correspondence to: H. Sierks (sierks@linmpi.mpg.de)

CP598, *Solar and Galactic Composition,* edited by R. F. Wimmer-Schweingruber

Table 1. $^3He/^4He$-rich (>0.01) days identified in SOHO/COSTEP data (4.3-40.9 MeV/N) until the end of 2000. Notes: p, ^3He and ^4He peaks are given in total counts (parallel detector segments) based on maximums of daily mass plots (as all mass plots are binned to the same histogram, this number can be used for relative comparison of ^3He and ^4He (and to protons)); the mass plots do not take into account special instrument configurations like ring segment offs; bold values denote $^3He/^4He$-ratios ≥0.1.

Year	DoY	p peak	^{3}He peak	^{4}He peak	^{4}He/p	^{3}He/^{4}He
1995						
1996	**121**	**14**	**5**	**20**	**1.429**	**0.250**
	191	645	6	211	0.327	0.028
	334	7962	9	537	0.067	0.017
	335	8148	12	803	0.099	0.015
1997	**063**	**17**	**6**	**16**	**0.941**	**0.375**
	092	1634	4	89	0.054	0.045
	223	**83**	**10**	**57**	**0.687**	**0.175**
	261	337	8	84	0.249	0.095
	262	**190**	**13**	**51**	**0.268**	**0.255**
	264	**1032**	**7**	**66**	**0.064**	**0.106**
	308	4940	37	3305	0.669	0.011
*	310	7517	49	2920	0.388	0.017
*	312	6986	62	5438	0.778	0.011
	318	4142	8	346	0.084	0.023
	332	**321**	**16**	**75**	**0.234**	**0.213**
	333	**157**	**10**	**35**	**0.223**	**0.286**
1998	095	3007	8	601	0.200	0.013
	097	3908	9	698	0.179	0.013
	113	6679	82	5314	0.796	0.015
	114	3617	35	3071	0.849	0.011
	115	3028	19	1624	0.536	0.012
*	122	2293	11	574	0.250	0.019
*	123	4368	27	1670	0.382	0.016
*	124	4130	23	1837	0.445	0.013
	126	6796	102	3486	0.513	0.029
	127	7621	33	2482	0.326	0.013
	128	2832	7	189	0.067	0.037
	132	1906	7	237	0.124	0.030
	148	9097	13	559	0.061	0.023
	149	**5209**	**48**	**383**	**0.074**	**0.125**
	169	9456	15	1258	0.133	0.012
	309	**207**	**9**	**22**	**0.106**	**0.409**
*	310	16424	28	2089	0.127	0.013
*	311	15831	66	2008	0.127	0.033
*	312	13171	178	3178	0.241	0.056
	313	**7436**	**56**	**300**	**0.040**	**0.187**
	314	**2907**	**6**	**58**	**0.020**	**0.103**
	326	5525	5	201	0.036	0.025
	327	4695	7	239	0.051	0.029
	344	**119**	**5**	**18**	**0.151**	**0.278**
	352	**650**	**5**	**24**	**0.037**	**0.208**
1999	070	657	7	38	0.058	0.184
	080	**127**	**8**	**20**	**0.157**	**0.400**
	081	**42**	**11**	**7**	**0.167**	**1.571**
*	128	14360	7	561	0.039	0.012
	129	7167	9	197	0.027	0.046
	130	6785	11	536	0.079	0.021
	132	**2523**	**12**	**65**	**0.026**	**0.185**
*	155	7574	40	3201	0.423	0.012
*	158	8584	18	1640	0.191	0.011
	164	2960	9	102	0.034	0.088
	169	**1521**	**45**	**168**	**0.110**	**0.268**
	170	**1807**	**36**	**139**	**0.077**	**0.259**
	171	1374	6	70	0.051	0.086
	174	6191	19	431	0.070	0.044
	175	4766	9	218	0.046	0.041
	176	10332	5	354	0.034	0.014
	181	**1019**	**8**	**54**	**0.053**	**0.148**
	182	1469	8	81	0.055	0.099
	183	**910**	**5**	**33**	**0.036**	**0.152**
	188	**399**	**6**	**53**	**0.133**	**0.113**
	193	**139**	**4**	**32**	**0.230**	**0.125**
	194	**1179**	**30**	**117**	**0.099**	**0.256**
	195	**421**	**8**	**34**	**0.081**	**0.235**
	218	**702**	**4**	**35**	**0.050**	**0.114**
	219	**648**	**7**	**10**	**0.015**	**0.700**
	226	**409**	**14**	**62**	**0.152**	**0.226**
	227	**174**	**6**	**28**	**0.161**	**0.214**
	236	**116**	**6**	**8**	**0.069**	**0.750**
	258	6771	6	185	0.027	0.032
	264	1488	5	103	0.069	0.049
	289	1621	5	131	0.081	0.038
	290	2409	5	128	0.053	0.039
	305	33	5	7	**0.212**	**0.714**

	329	**381**	**7**	**41**	**0.108**	**0.171**
	330	**270**	**5**	**39**	**0.144**	**0.128**
	337	**249**	**6**	**5**	**0.020**	**1.200**
	344	14042	8	144	0.010	0.056
	358	**1915**	**31**	**85**	**0.044**	**0.365**
	359	**1379**	**18**	**46**	**0.033**	**0.391**
	360	**424**	**8**	**15**	**0.035**	**0.533**
	361	**748**	**5**	**46**	**0.061**	**0.109**
	363	636	9	385	0.605	0.023
	365	9155	12	797	0.087	0.015
2000	002	10496	8	653	0.062	0.012
	018	1149	15	546	0.475	0.027
	019	5244	14	147	0.028	0.095
	030	11939	16	1450	0.121	0.011
	062	**639**	**4**	**30**	**0.047**	**0.133**
	065	**455**	**4**	**19**	**0.042**	**0.211**
	107	**95**	**4**	**36**	**0.379**	**0.111**
	118	**2856**	**7**	**23**	**0.008**	**0.304**
	122	884	17	1420	1.606	0.012
	125	404	7	122	0.302	0.057
	126	1681	10	294	0.175	0.034
	127	5393	6	292	0.054	0.021
	138	6469	16	1376	0.213	0.012
	144	7516	15	386	0.051	0.039
	145	**2003**	**137**	**638**	**0.319**	**0.215**
	146	953	6	106	0.111	0.057
	163	12458	20	1824	0.146	0.011
	165	13761	4	156	0.011	0.026
	168	4985	17	1134	0.227	0.015
	170	8404	21	1284	0.153	0.016
	171	**4101**	**53**	**299**	**0.073**	**0.177**
	172	**1045**	**10**	**59**	**0.056**	**0.169**
	174	4469	7	133	0.030	0.053
	175	9801	13	687	0.070	0.019
	176	4925	11	518	0.105	0.021
	177	9630	27	2218	0.230	0.012
	180	6454	6	230	0.036	0.026
	181	9515	9	476	0.050	0.019
	193	3690	8	187	0.051	0.043
	194	5672	10	534	0.094	0.019
	202	4767	28	2619	0.549	0.011
	223	4437	5	109	0.025	0.046
	224	12709	15	1115	0.088	0.013
	225	6280	18	1164	0.185	0.015
	235	**206**	**25**	**28**	**0.136**	**0.893**
	236	**75**	**19**	**22**	**0.293**	**0.864**
	259	13141	40	3699	0.281	0.011
	261	15400	51	3934	0.255	0.013
	262	8397	6	386	0.046	0.016
	263	4389	5	337	0.077	0.015
	264	7349	9	573	0.078	0.016
	265	6188	8	176	0.028	0.045
	271	**682**	**15**	**38**	**0.056**	**0.395**
	272	**788**	**3**	**16**	**0.020**	**0.188**
	284	2587	3	47	0.018	0.064
	285	8465	6	117	0.014	0.051
	286	8977	5	145	0.016	0.034
	287	5544	14	341	0.062	0.041
	288	**1494**	**25**	**49**	**0.033**	**0.510**
	289	**639**	**4**	**16**	**0.025**	**0.250**
	292	12006	29	2389	0.199	0.012
	293	10654	10	901	0.085	0.011
	300	16020	24	1504	0.094	0.016
	301	18067	13	812	0.045	0.016
	303	**4018**	**31**	**71**	**0.018**	**0.437**
	304	**2247**	**350**	**523**	**0.233**	**0.669**
	305	6761	42	1720	0.254	0.024
	316	3778	80	2318	0.614	0.035
	317	2016	20	1820	0.903	0.011
	318	1378	13	980	0.711	0.013
	319	1665	11	1024	0.615	0.011
	320	1634	12	845	0.517	0.014
	335	3778	10	936	0.248	0.011
	363	**518**	**11**	**52**	**0.100**	**0.212**

* SEP ^{3}He Enhancements reported by Mason et al., 1999.

3. Fine-scale characteristics

The ^{3}He/^{4}He-rich event on October 30, 2000 had sufficiently high count rates (compare with Table 1) that made it possible to investigate the fine-scale characteristics of the event by using a time resolution up to 1 hour as shown in Figure 2. In this event the ^{3}He increase occurred during transition from October 29 to 30 (DoYs 303-304) with the highest ^{3}He levels staying until the end of October 30 (DoY 304). After that time the ^{3}He level dropped in contrast to the further rising proton and helium values.

4. Comparison with solar wind measurements

Figure 3 shows a comparison with ACE solar wind plasma and magnetic field data (courtesy CDAW, D. J. McComas).

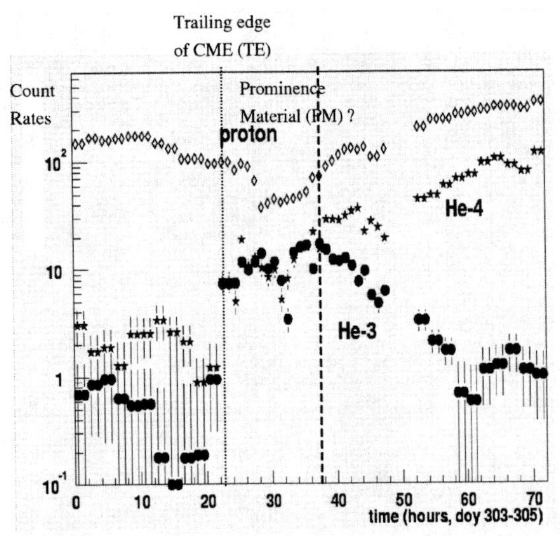

Figure 2. Hourly values of proton, ^{4}He and ^{3}He count rates measured for the period October 29 to 31 (doys 303-305) in 2000 by SOHO/COSTEP.

ACE detected a shock passage followed by a flux rope type CME (magnetic cloud). There might have been a reverse shock following the CME, but this needs further study.

Figure 3 shows somehow surprisingly, that the ^{3}He/^{4}He-enrichment measured at MeV energies corresponds to a solar wind He/p-enrichment following the trailing edge of the typical magnetic cloud signatures. The plasma and magnetic field signatures appear to be similar to the classical January 10/11, 1997 event that has been described by Burlaga et al. (1998). It has to be clarified by further analyses whether this association is by chance or might be of importance for the understanding of the origin and interplanetary evolution of such events.

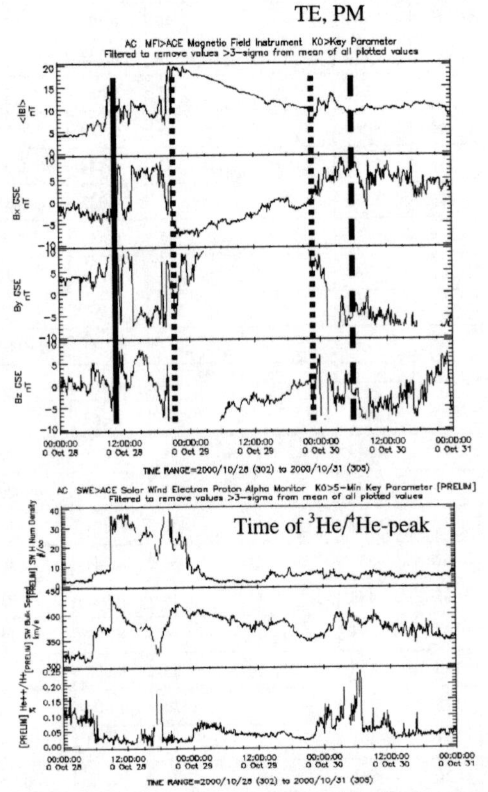

Figure 3. Solar wind plasma and magnetic field data measured by ACE from October 28 to October 31, 2000. Top to bottom: Magnetic field strength, cartesian components of the field, proton number density, bulk speed and He/p-ratio. The solid line denotes the shock ahead of the CME (dotted lines) followed by a region that might be prominence material (between second dashed and solid lines).

5. Associations with solar activity

The comparison with GOES X-ray flare observations showed no direct correspondence to the onset of the ^{3}He/^{4}He enrichment seen in the energetic particles. The in situ detected CME in the solar wind can be traced back to a halo CME on October 25 observed by the SOHO/ LASCO coronagraphs. The event is typical in the sense that we did not find a good correlation so far of large ($\geq$0.1) ^{3}He/^{4}He-enrichments at MeV energies with solar flares.

6. Summary

Through a systematic survey of He-measurements at MeV energies provided by the SOHO/COSTEP instrument we identified ^{3}He/^{4}He isotope abundances with ratios $\geq$0.01 from the beginning of the mission until the end of 2000. The events listed in Table 1 can serve as a valuable tool for further investigations.

So far the start of the analysis of large ($\geq$0.1) ^{3}He/^{4}He-enrichments showed not a direct correspondence to solar flares, but rather to in situ passages of CMEs. There is some indication for a possible association with large He/p-enhancements in the transient solar wind plasma at times of the CME passages, maybe indicative of special conditions of the solar source regions were the CMEs had been released. The solar source and in situ conditions of large ^{3}He/^{4}He-enrichments at MeV energies need further studies to clarify their origins in more detail.

Acknowledgements. The SOHO/COSTEP project is supported under grant No. 50 OC 9602 by the German Bundesminister für Bildung und Forschung (BMBF) through the Deutsches Zentrum für Luft- und Raumfahrt (DLR).

References

Burlaga, L., Fitzenreiter, R., Lepping, R., Ogilvie, K., et al., J. Geophys. Res., 103, 277, 1998.

Mason, G., Mazur, J., and Dwyer, J., Astrophys. J., 525, L133, 1999.

Müller-Mellin, R., Kunow, H., Fleißner, V., Pehlke, E., et al., Sol. Phys., 162, 483, 1995.

Temerin, M., and Roth, I., Astrophys. J., 391, L105, 1992.

Propagation of Impulsive Solar Energetic Particle Events

G.C. Ho*, G.M. Mason†, R.E. Gold*, J.R. Dwyer** and J.E. Mazur‡

*The Johns Hopkins University Applied Physics Laboratory, Laurel, MD 20723-6099
†Department of Physics, University of Maryland, College Park, MD 20742
**Department of Physics and Space Sciences, Florida Institute of Technology, FL 32901
‡Aerospace Corporation, El Segundo, CA 90245

Abstract. Impulsive solar energetic particle (SEP) events are associated with impulsive X-ray flares, energetic electrons, and enhanced heavy ion abundances. Using instrument on ACE, we examine the onset and ion's path-length for an impulsive SEP in May 2000. By assuming zero degree pitch-angle scattering, we fit the minimum ion's path-length and compare that with the nominal Parker spiral length. We found a small difference between the two path-lengths, with the fitted path-length 10% longer. We believe the difference can be due to ions traveling along the magnetic field with a finite pitch angle distribution, and/or from paths differing from the nominal Parker spiral configuration. The event is also found to have a close association with solar metric Type III radio bursts.

INTRODUCTION

Solar Energetic Particle (SEP) events have been generally classified into two major classes: gradual and impulsive. Gradual SEP events have been attributed to a shock which is driven by a coronal mass ejection as it moves out from the Sun. Impulsive SEP events on the other hand are thought to be accelerated directly by a solar flare (see review by [1]). While there are numerous studies investigating the propagation and transport of gradual SEP events, only a relatively small number of impulsive SEP events have been thoroughly investigated. Since an impulsive SEP event is observed only when there is good magnetic connection to the flare region back at the Sun, it is believed that by studying this class of SEP event we can gain insight into the SEP acceleration and transport mechanisms. However, the short duration and small fluence of impulsive SEP events makes it difficult to study its transport except for a few large fluence events in the past [2, 3].

In this paper, we present a technique for estimating the onset and path-length of energetic ions in impulsive SEP events utilizing high sensitivity instruments on the Advanced Composition Explorer (ACE) spacecraft.

OBSERVATIONS

Data from the Ultra-Low Energy Isotopic Spectrometer (ULEIS) [5] on the ACE spacecraft are presented in this paper. ACE was launched in August 1997 and is now in a halo orbit 200 Re upstream of the Earth [4]. The ULEIS instrument is one of ten particle and field instruments on ACE. ULEIS, in particular, measures both the elemental and isotopic composition of ~50 keV/nucleon to 10 MeV/nucleon ions using a redundant Time-Of-Flight system with a 50 cm flight path [5].

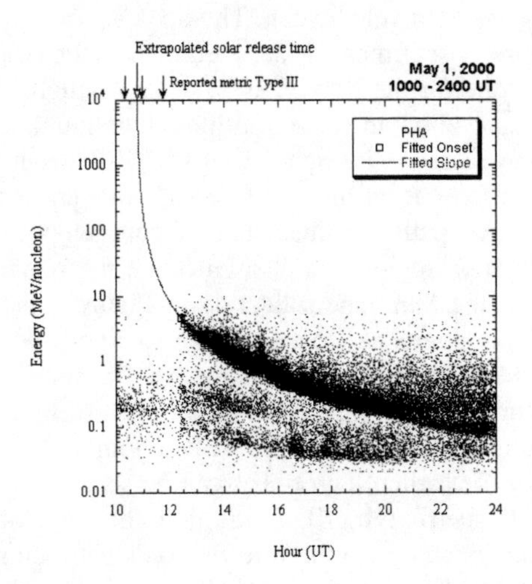

FIGURE 1. An impulsive SEP event observed on ACE on May 1, 2000. The plot shows ULEIS pulse height data in energy (MeV/nucleon) versus time format. Each data point represents one ion event (H-Fe) that has been measured in arrival time, mass and energy. Clear velocity dispersion is apparent in this figure. The open squares are the fitted onset times for each velocity bin and the solid curve is a $1/v$ dependence fit to the onset time. The extrapolated solar release for these ions is shown with an open arrow at the top of the figure.

CP598, *Solar and Galactic Composition,* edited by R. F. Wimmer-Schweingruber

Figure 1 shows an impulsive SEP event that was observed by ULEIS on May 1, 2000. Each data point in the figure represents an individual ion (H-Fe) whose mass and energy were determined by pulse-height-analysis (PHA) in the instrument, and whose time of detection of each ion was recorded. In the figure, higher speed solar energetic ions arrive at ACE before those at lower speed, leading to a characteristic curved pattern when plotted in energy versus time format. Some typical features of impulsive SEP events can be seen in this figure. The May 1, 2000 impulsive event lasted for less than 24 hours at the ULEIS energy range, in comparison with large SEP events that can last for more than 10 days. This event is also ^{3}He-enhanced with a (0.4 - 2.0 MeV/nucleon) ^{3}He/^{4}He ratio of 0.59±0.06 [6].

We developed an automated procedure to fit the velocity dispersion pattern in Figure 1 in order to estimate the event onset time and the ion pathlength for a range of energies. The routine first logarithmically bins the ion velocity into 50, 200 or 300 bins from 16.7 keV/nucleon to infinite velocity according to the event's ion intensity. The routine then divides the time axis into 90-second bins and examines the distribution of PHA events as a function of time for every velocity bin. The routine then fits a half-Gaussian distribution to the observed PHA for that velocity bin, thus estimating the maximum and width of the distribution. The routine finds the time two standard deviations prior to the maximum of the distribution, and defines that time as the onset for that velocity bin. Hence a set of onset points will be located as the routine scans each velocity bin. The open squares in Figure 1 are the onset times that are identified as the two-sigma onset time for the May 1, 2000 event. A more detailed discussion about the onset routine can be found in [6].

If we assume ions in impulsive SEP events travel "scatter-free" from the Sun to 1 AU, we can fit a minimum travel path to estimate the ion path-length from the acceleration region and also calculate the release time back at the Sun. The solid line in Figure 1 is a fit to the onset points as located by the routine. The extrapolated solar release time (open arrow) is shown on top of figure along with reported metric type III radio bursts (solid arrows) that were within one hour of the release time. An excellent agreement can be seen in this event with the metric type III. In addition, the slope ofthe fitted curve, which is related to the ion's path-length as it propagates from the Sun to 1 AU, gives us a path-length of 1.200±0.004AU in this event (the uncertainty of the path-length is due to fitting of the onset points only). Using a nominal Parker spiral field line with the observed solar wind speed during the onset of this event, the deduced path-length is 1.11±0.02 AU. Reames and Stone [7] examined trajectories that track the kilometric type III radio burst, that are associated with impulsive SEP events, as they move outward along the interplanetary magnetic field. They found that the trajectory could depart considerably from the ideal Parker spiral, with trajectory variations in and out the ecliptic. Hence, the measured in-situ solar wind speed can only gives an estimate of the path that energetic ions take in transit from the Sun to 1 AU. On the other hand, the difference between the measured path-length and the Parker spiral length can simply be due to ions traveling along the magnetic field with a finite pitch angle distribution. For example, a 30° mean pitch angle would account for 12.5% difference between the path-length as ions travel from the Sun to 1 AU [6].

DISCUSSION AND CONCLUSIONS

We have demonstrated a simple technique to estimate the onset for impulsive SEP events at 1 AU. With the estimated onset time at each velocity, we can fit a $1/v$ dependence curve to extrapolate the solar release time of ions from impulsive SEP events. In addition, the slope of the fitted $1/v$ line is also related to the ion's path-length while in transit from the Sun to 1 AU. Using this technique with data from high sensitivity instruments on ACE, we can study the onset and propagation of these impulsive SEP events in greater detail than ever before.

ACKNOWLEDGMENTS

The work is supported under NASA contract NAS5-97271, task order 009 at JHU/APL and grant PC251429 at the University of Maryland.

REFERENCES

1. Reames, D.V., *Space Sci. Rev.* **90**, 413 (1999).
2. Zwickl, R.D., E.C. Roelof, R.E. Gold, S.M. Krimigis, and T.P. Armstrong, *Astrophys. J.* **225**, 281 (1978).
3. Mason, G.M., C.K. Ng, B. Kleckler, G. Green, *Astrophys. J.* **339**, 529 (1989).
4. Stone, E.C., et al., *Space Sci. Rev.* **86**, 1 (1998).
5. Mason, G.M., et al., *Space Sci. Rev.* **86**, 409 (1998).
6. Ho, G.C., E.C. Roelof, D. Lario, R.E. Gold, G.M. Mason, J.R. Dwyer, J.E. Mazur, *Maunscript in preparation*, (2001).
7. Reames, D.V., and R.G. Stone, *Astrophys. J.* **308**, 902 (1986).

Applications and Constraints on Cosmochemistry

Applications of Abundance Data and Requirements for Cosmochemical Modeling

H. Busemann[1], W. R. Binns[2], C. Chiappini[3], G. Gloeckler[4], P. Hoppe[5], D. Kirilova[6], R. A. Leske[7], O. K. Manuel[8], R. A. Mewaldt[7], E. Möbius[9], R. Wieler[10], R. C. Wiens[11], R. F. Wimmer-Schweingruber[1] and N. E. Yanasak[7]

[1]*Physikalisches Institut, University of Bern, Sidlerstr. 5, 3012 Bern, Switzerland*
[2]*McDonnell Center for the Space Sciences, Washington University, St. Louis, MO, USA*
[3]*Observatory of Trieste, Trieste, Italy*
[4]*Department of Physics, University of Maryland, College Park, USA*
[5]*Cosmochemistry Department, Max-Planck-Institute for Chemistry, Mainz, Germany*
[6]*Institute of Astronomy, Bulgarian Academy of Sciences, Sofia, Bulgaria*
[7]*Space Radiation Laboratory, California Institute of Technology, Pasadena, CA, USA*
[8]*Department of Chemistry, University of Missouri, MO, USA*
[9]*Department of Physics, University of New Hampshire, Durham, NH, USA*
[10]*Institute for Isotope Geology and Mineral Resources, ETH Zürich, Zürich, Switzerland*
[11]*Space and Atmospheric Sciences, MS D466, Los Alamos National Laboratory, Los Alamos, NM 87545, USA*

Abstract. Understanding the evolution of the universe from Big Bang to its present state requires an understanding of the evolution of the abundances of the elements and isotopes in galaxies, stars, the interstellar medium, the Sun and the heliosphere, planets and meteorites. Processes that change the state of the universe include Big Bang nucleosynthesis, star formation and stellar nucleosynthesis, galactic chemical evolution, propagation of cosmic rays, spallation, ionization and particle transport of interstellar material, formation of the solar system, solar wind emission and its fractionation (FIP/FIT effect), mixing processes in stellar interiors, condensation of material and subsequent geochemical fractionation. Here, we attempt to compile some major issues in cosmochemistry that can be addressed with a better knowledge of the respective element or isotope abundances. Present and future missions such as Genesis, Stardust, Interstellar Pathfinder, and Interstellar Probe, improvements of remote sensing instrumentation and experiments on extraterrestrial material such as meteorites, presolar grains, and lunar or returned planetary or cometary samples will result in an improved database of elemental and isotopic abundances. This includes the primordial abundances of D, ^{3}He, ^{4}He, and ^{7}Li, abundances of the heavier elements in stars and galaxies, the composition of the interstellar medium, solar wind and comets as well as the (highly) volatile elements in the solar system such as helium, nitrogen, oxygen or xenon.

INTRODUCTION

This manuscript is the résumé of a working group at the joint SOHO-ACE workshop held in Bern, Switzerland, in March 2001. The goal of this working group was "to determine the importance of various elements from the point of view of discerning different models that address questions such as solar-system formation, stellar and Big Bang nucleosynthesis, and chemical evolution of the Galaxy". We were asked to identify elements, which may serve as good references or indicators for the key physical processes involved. Certainly, we cannot provide a complete overview of all questions in the various disciplines relevant to this goal. Therefore, we will follow the expertise of the working group members and discuss some of the most crucial questions in cosmochemistry that can and should be answered in the near future through improvements in instrumentation, observational techniques, theory, and with new dedicated missions.

CP598, *Solar and Galactic Composition*, edited by R. F. Wimmer-Schweingruber

We will start with issues concerning the formation and composition of the solar system, and then address the composition of the (local) interstellar medium (L)ISM and of galactic cosmic rays (GCR). While the solar system represents a sample of galactic matter from 4.5 billion years ago, the LISM is a current and local sample of our Galaxy. Both samples provide benchmarks for models of galactic chemical evolution (GCE) from the Big Bang until today. GCRs represent another sample of the current Galaxy from a wide range of distances that shows some additional characteristics of high-energy interactions. With this paper, we aim to contribute to the interdisciplinary discussions between planetary scientists, solar physicists, cosmochemists, and astrophysicists.

SOLAR SYSTEM ABUNDANCES

Highly Volatile Elements

Absorption-line spectra of the solar photosphere and laboratory-based analyses of the most primitive meteorites, the CI (Ivuna-type) carbonaceous chondrites, yield solar-system abundances of the elements (1, 2). For most elements, the agreement of these data sets is ~10% or better (Figure 1). However, meteorites do not represent solar abundances of the light or most volatile elements H, Li, Be, C, N, O, and the noble gases. The reasons are nucleosynthesis processes in the Sun's interior and the incomplete condensation during formation of the first solid matter in the solar system, respectively (2). Among the light elements, only the meteoritic abundance of boron agrees with the value recently determined in the photosphere (3). Therefore, even relatively imprecise measurements (compared to the precision usually obtained from meteorites) of all the mentioned elements in the Sun and the solar wind provide cosmochemically important information. The solar system isotopic composition of these elements is relatively poorly known. The light elements in meteorites are particularly subject to isotopic fractionation from originally solar system composition due to their volatility, the large relative mass differences of their isotopes and their chemical reactivity (4). It might well be that the isotopic composition of the noble gases in meteorites are not representative of the solar system at all (5), because the meteorite parent bodies or precursor planetesimals might never have incorporated these gases, in contrast to the much heavier planets which could have gravitationally captured gas from the nebula (6 and references therein, see also for alternative trapping mechanisms). The Sun and the gaseous giant planets Jupiter and Saturn, which formed only relatively small cores and hence remain largely undifferentiated, might represent isotopically undisturbed solar system composition or nearly so, although Jupiter appears to have a more evolved atmosphere, possibly of cometary origin (7).

The protosolar He abundance as well as its value in the present-day convective zone can be precisely determined by solar modeling and helioseismology (e.g., 8). However, for the solar Ne, Ar, Kr, and Xe abundances, we must rely on extrasolar sources, analyses of implanted solar wind (SW) in lunar soils, and the systematics of s-process nucleosynthesis, which leads to rather large uncertainties of 15-25% (1, 2). The abundance of these elements is important to assess the fractionation in the upper solar atmosphere according to first ionization potential or first ionization time (the so-called FIP/FIT effect, the relative enrichment of elements with FIP below ~10 eV in the low speed solar wind relative to photospheric abundances and high-FIP element abundances, e.g., 9 and references therein), compositional differences between the solar wind and solar energetic particles (SEP), temporal variability of the solar wind, or possible fractionation upon trapping in lunar soil (e. g., 10, 11). Most solar wind noble gas isotopic ratios as derived from measurements of implanted solar wind in lunar material have stated precisions of 1% or better (12 and references therein) but better values are needed especially for the less abundant light Kr and Xe isotopes (see below). The Apollo Solar Wind Composition experiment (13) and space missions such as Ulysses (14) and SOHO (15, 16) also provided isotopic ratios for He-Ar in the solar wind. Higher precision data from future missions, e.g., Genesis, are required, however, to test whether the values derived from lunar samples may be affected by isotopic fractionation upon or after trapping.

Solar elemental abundances of C, N, and O are believed to be known to within 8-12%, comparable with the estimated accuracy for most elements in CI chondrites (1, 2). Their isotopic ratios, however, are not sufficiently well known. We will discuss below how a more precise solar oxygen isotopic composition is of high importance with respect to studies of the homogeneity of the solar nebula as well as for an improved understanding of the GCE. Even more controversial is the solar $^{15}N/^{14}N$ ratio, as we will discuss below.

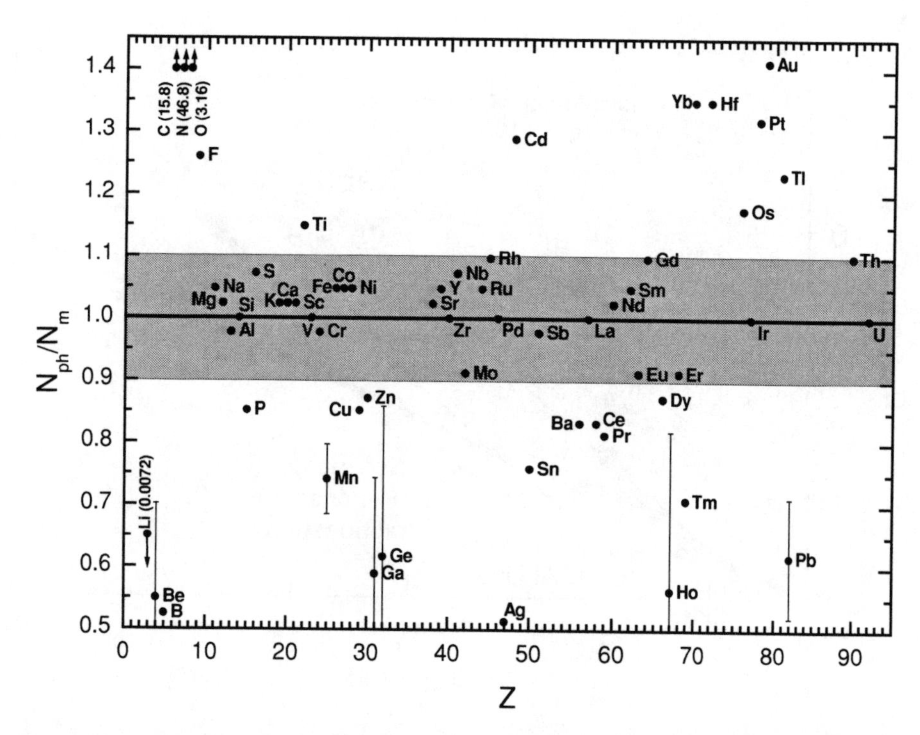

FIGURE 1. Comparison of solar photospheric (N_{ph}) and meteoritic "CI" (N_m) abundances (2). For most of the heavy elements, the abundances agree within 10%. Error bars are shown if they do not overlap with the range N_{ph}/N_m = 0.9-1.1. Only Mn and Pb show significant deviations not covered by the uncertainties. Figure after (2) with kind permission from Springer-Verlag, Heidelberg.

The isotopic signature of Li and B in the solar system, which is influenced by nuclear reactions, both in the upper solar atmosphere and in early condensed material, is discussed in detail elsewhere (e.g., 17, 18 and references therein).

Precise data on the abundances of the volatile elements in the solar wind will be essential to improve the knowledge of fractionation mechanisms that mask the solar source composition. Measurements of isotopic abundances of other, more refractory, elements in the solar wind such as Mg (15) will help to address this issue. Many of the future efforts, however, focus on the determination of the elemental and isotopic abundances of N, O, and the noble gases. Future missions to comets (19) will obtain precise abundances of the volatile elements that should fill some of the gaps left by meteorites, because comets condensed at large distances to the Sun at low temperatures and thus contain more pristine material than that found in meteorites (20). In the following, we will discuss some of the volatile elements and the issues they address in more detail.

The Genesis Mission and Solar Oxygen

The Genesis mission, launched on August 8, 2001, will collect solar wind particles for about two years in a variety of targets that will be analyzed on Earth. This will significantly improve the precision of solar wind isotope measurements for many elements. The highest priority objective is a precise determination of the oxygen isotopic composition in the solar wind as discussed below. This is a good starting point to discuss the importance of accurate solar abundance values for an improved understanding of solar system formation and current solar processes. These include nebular mixing issues needed to understand the differences among planetary compositions and planetary atmospheres. A further high priority objective is the measurement of the isotopic composition of nitrogen and the noble gases in the solar wind. These issues will be discussed in detail in the following paragraphs.

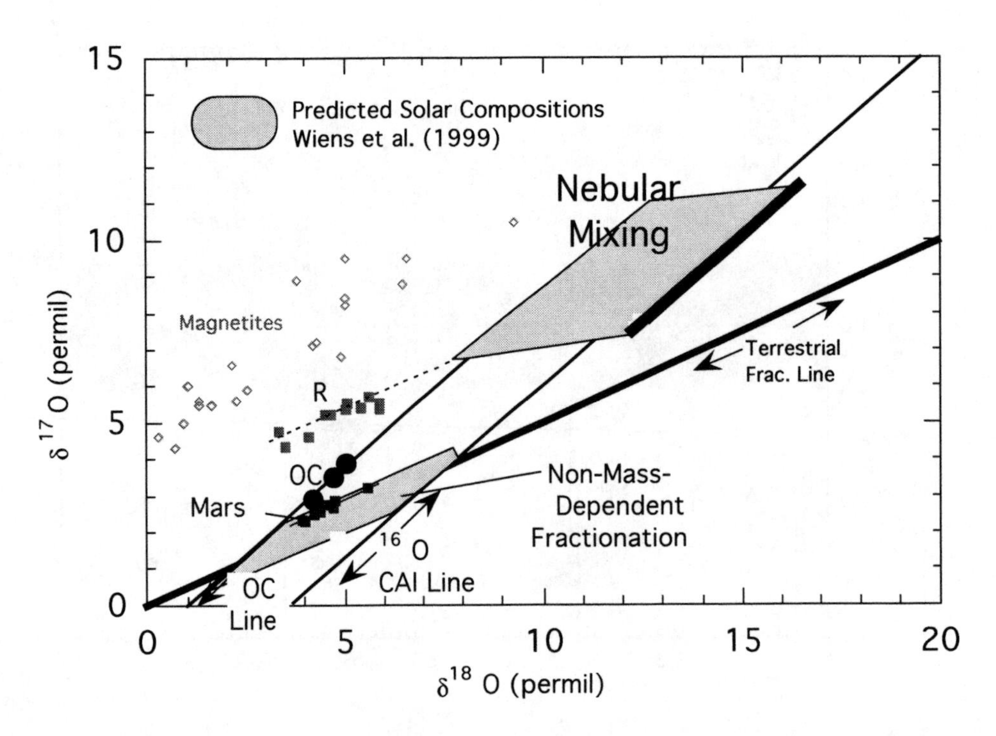

FIGURE 2. Solar oxygen isotopic compositions predicted by the nebular mixing model (24) and by the non-mass-dependent fractionation model (29). X and Y-axes give ^{18}O and ^{17}O enrichments relative to standard mean ocean water in ‰. Present solar composition uncertainties are on the order of ±200‰ for $^{18}O/^{16}O$, while the solar $^{17}O/^{16}O$ ratio is completely unconstrained (25-27). With the Genesis mission, it is hoped to measure solar wind oxygen isotopes to ±1‰, which is sufficient to distinguish between the models. OC = ordinary chondrites; R = R (Renazzo-type) chondrites; CAI = calcium-aluminum-rich inclusions. "Magnetites" refers to individual magnetite grains in unequilibrated ordinary chondrites.

The Genesis mission results will also help to determine nuclear processes operating on the surface of the Sun and in solar system precursor materials, and solar-specific processes operating either early or later in the solar system history. An example of the latter is a comparison of heavy and light elements relative to CI chondrites indicative of solar gravitational settling (21). Examples of the former are Li, Be, B isotopic and elemental abundances, which will help to understand the history of the solar convection zone (e.g., 22). Examples of objectives addressing nuclear processes are, e.g., the solar F abundance as a measure of integrated spallation production through time and comparisons of Kr and Xe abundances to meteoritic abundances of neighboring elements in the periodic table as a measure of solid-gas fractionation during formation of the Sun (23).

The three oxygen isotopes show clear differences between various solar system bodies of a few permil up to several percent (Figure 2). So far, we have measurements for Earth, Moon, Mars, Vesta, (via meteorites), ordinary and carbonaceous chondrites, and miscellaneous differentiated meteorites (24). Measurements of the outer planet or Hermean oxygen compositions, however, are missing. Oxygen compositions of the Sun (e.g., 25, 26, and 27, this volume) and comets (28) are known only at the ~10-20% level, and do not include ^{17}O.

At present, we do not understand the oxygen isotopic variations. One theory (e. g., 29) implies that *non-mass-dependent fractionation* acted on an initially homogeneous hot solar nebula reservoir to produce ^{16}O enrichments in the initially crystallizing refractory materials. The other theory (e.g., 24 and references therein), suggests that *nebular mixing* between an ^{16}O-enriched solid composition and a ^{16}O-poor nebular gas produced the variety of compositions seen now. In its simple form, the nebular mixing model predicts for the Sun a significantly more ^{16}O-depleted composition than the non-mass-dependent model, for which a composition essentially identical to the Earth, Moon, and Mars is inferred. A measurement of ±0.1% in $^{17}O/^{16}O$ and $^{18}O/^{16}O$ can distinguish between these (Figure 2, 30). This is a reduction by a factor of more than 100 from the current measurement uncertainties for solar oxygen (25-27). Within the confines of the nebular mixing model, there are also ways in which $^{18}O/^{16}O$ in the Sun could be significantly larger than the simplistic prediction. If so, a measurement at the ~1% level will suffice to distinguish between these models, but further precision will give insight into nebular processes such as possible fractionation between nebular CO and H_2O gas during meteorite formation or solid/gas enrichment in the meteorite-forming region (30).

Nitrogen in the Sun and the Solar System

Along with H, C, O, and the noble gases, nitrogen belongs to the incompletely condensed elements in meteorites whose elemental abundance and isotopic composition cannot therefore be deduced from CI chondrite data. Unlike for the noble gases, it has also proven very difficult to deduce the isotopic composition of solar nitrogen from analyses of samples that were exposed to the solar wind. The currently used solar system abundance compilations (e. g., 1, 2) therefore adopt the $^{15}N/^{14}N$ ratio in the terrestrial atmosphere (3.68×10^{-3}) as a solar system standard, although differences between solar and terrestrial noble gases illustrate that this is not necessarily correct.

A crucial but ill-understood observation is that the nitrogen isotopic composition trapped in lunar soils varies by up to 35% in different bulk samples and in different extraction steps of an individual soil (31, 32). Classically, these observations have been interpreted to indicate a secular increase of the $^{15}N/^{14}N$ ratio in the solar wind from about 2.9×10^{-3} several Gyrs ago to perhaps up to 4.1×10^{-3} for the recent solar wind (33). This interpretation implies that essentially all trapped nitrogen in lunar soil samples is from the solar wind. There are two major problems with the hypothesis of a secular change of the solar wind nitrogen composition. First, there is no generally accepted process for providing such a fractionation (34). For instance, isotopic fractionation in the solar wind has been shown to be small (at most a few %/amu) based on measurements of the isotopes of the refractory elements Mg and Si (15, 35, and 36, this volume, and references therein). Postulating an increased solar activity in the past can probably not explain the isotopic behavior either. Measurements of the Mg and Si isotopic compositions in coronal mass ejections (CME), the most dramatic manifestation of increased solar activity, show no fractionation (35). Second, the abundance ratio $N/^{36}Ar$ in lunar samples is about an order of magnitude higher than the respective solar wind ratio. The latter observation may indicate preferential loss of solar wind-implanted ^{36}Ar relative to N, e.g., by diffusion, but analyses of single lunar dust grains indicate that this is probably not the case, implying that nitrogen in the lunar regolith has a predominantly non-solar source (37). In summary, the nitrogen isotopic composition as deduced from lunar regolith samples is highly controversial. A value of $(3.82\pm0.02) \times 10^{-3}$ has been proposed from analyses of two relatively recently irradiated samples (38), whereas ion-probe measurements on single grains thought to contain relatively little non-solar nitrogen yielded a value of $<2.79 \times 10^{-3}$ (39), and multi-step analyses of single grains by incremental heating support such an isotopically light composition (40). These workers attribute the variable $^{15}N/^{14}N$ ratios in lunar samples to a variable admixture of a non-solar component possibly of meteoritic origin (41, this volume).

Comparing the various lunar estimates with the direct isotopic analysis of solar wind nitrogen by the CELIAS/MTOF instrument on SOHO (42) and a value for the Jovian atmosphere obtained by the Galileo probe (43) does not help to resolve the conflict. The SOHO value of $5.0^{+1.9}_{-1.1} \times 10^{-4}$ is compatible within its large uncertainty with the value by Kim et al. (38), whereas the Galileo probe value of $(2.3\pm0.3) \times 10^{-3}$ (43) and the value obtained with ISO-SWS ($1.9^{+0.9}_{-1.0}$) x 10^{-3}, 44) are consistent with the isotopically light value advocated by Hashizume and coworkers (39, Figure 3). The latter comparison assumes that the Jupiter value is representative for the (proto) solar composition. Owen et al. (43) argue that this is the case. In HCN in interstellar clouds, ^{15}N is enriched by up to 30% (45), but N in Jupiter is derived from protosolar N_2. The $^{15}N/^{14}N$ ratio in HCN in comet Hale-Bopp's coma is indeed about 30% higher than the Galileo value (Figure 3, see discussion in 43).

Given this situation, it becomes obvious why a precise determination of the nitrogen isotopic composition in the solar wind is one of the two highest-priority goals of the upcoming Genesis mission. Hopefully, this will not only allow us to come closer to a solution of the conundrum of the origin of nitrogen in the lunar regolith, but also yield important clues as to the relation between giant planets and the Sun. A well defined nitrogen isotopic composition in the Sun is also needed for an improved understanding of the highly variable composition of this element in the various meteorite classes, which show $^{15}N/^{14}N$ ratios in the range $(3.2\text{-}9.5) \times 10^{-3}$ (46, Figure 3). It remains unclear to what extent this variability in meteorites reflects a large isotopic heterogeneity in the early solar system or fractionation processes that might have occurred during condensation or later in the meteorite parent bodies.

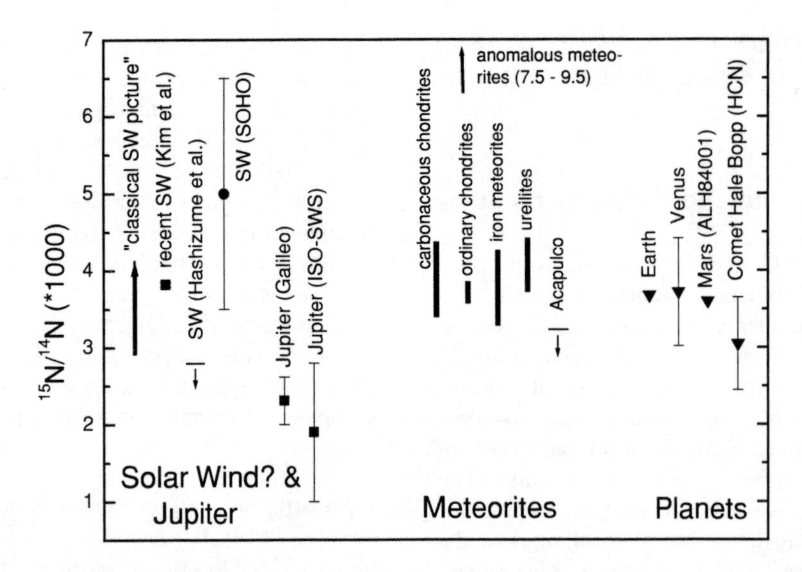

FIGURE 3. Various published $^{15}N/^{14}N$ ratios for the solar wind and Jupiter's atmosphere compared with the nitrogen composition of meteorites, planets, and comet Hale-Bopp. The "classical SW picture" postulates a secular increase of $^{15}N/^{14}N$ in the solar wind over the past several Gyrs (33). Data sources for SOHO, Galileo probe, and ISO-SWS values see text. Venus, Mars, Hale-Bopp, and meteorite values are taken from references (43, 46).

(Proto) Solar and Cometary Xenon Isotopic Composition

The chemically inert noble gases are important tracers to understand physical processes that formed the planets and their atmospheres. Distinct models discuss the noble gases in the terrestrial planets (e. g., 47, 48, 49, 50). Their atmospheres are strongly depleted in the light elements relative to the heavier elements and solar composition. The surprisingly uniform trapped primordial noble gas component in meteorites yields a similar element composition (51). However, distinct isotopic signatures in the atmospheres of Earth and Mars (they are similar to each other and "isotopically heavier" than meteoritic noble gases) and "missing Xe" in the atmospheres exclude a common precursor of meteoritic and atmospheric noble gases (49). As we will see below, the Xe isotopic composition of these atmospheres cannot be explained by simple fractionation of solar noble gases, although the interior of the Earth (6, 52, 53) and possibly Mars (54, 55) contain solar-like noble gases. Xenon, with nine stable isotopes, is very suitable to discuss many of the processes that led to the present-day compositions of the atmospheres such as fission and decay, impact-induced and hydrodynamic loss, as well as additions of cometary and meteoritic gas. We will discuss two models for the origin of terrestrial and Martian atmospheric Xe in the following:

(i) Pepin's model suggests *fractionation of a hypothetical primordial component dubbed U-Xe* (56). This fractionation is the result of the preferential loss of the light species relative to a heavier one upon an early hydrodynamic escape of a dense primordial atmosphere and can explain the "planetary" element pattern. If one fractionates solar Xe in a way that the isotopic ratios $^{124/126/128}Xe/^{130}Xe$ are in agreement with the terrestrial atmosphere, the heavier isotopes $^{134/136}Xe$ are already overabundant relative to the observed patterns (Figure 4a). This would thus not allow for contributions of $^{134/136}Xe$ from fission of ^{238}U and ^{244}Pu in the Earth's interior. To account for this fission Xe, the primordial U–Xe needs to have lower abundances of $^{134/136}Xe$ than the Sun (56).

Interestingly, fits on achondrite and chondrite Xe isotopic data seem to point to a common endmember component very similar to this independently deduced component. The ubiquitous and relatively uniform meteoritic Xe ("Xe-Q", 51) can be obtained by fractionation of U-Xe and addition of ^{134}Xe and ^{136}Xe ("Xe-H"), whereas solar Xe is unfractionated U-Xe plus significant amounts of Xe-H (Figure 4b). U-Xe, although a potential precursor of Xe in the terrestrial atmosphere, has not unambiguously been observed (57). It is unclear where and how the fractionation of U-Xe into Xe-Q in the different meteorite classes occurred and why the Sun does not contain the putative primitive U-Xe.

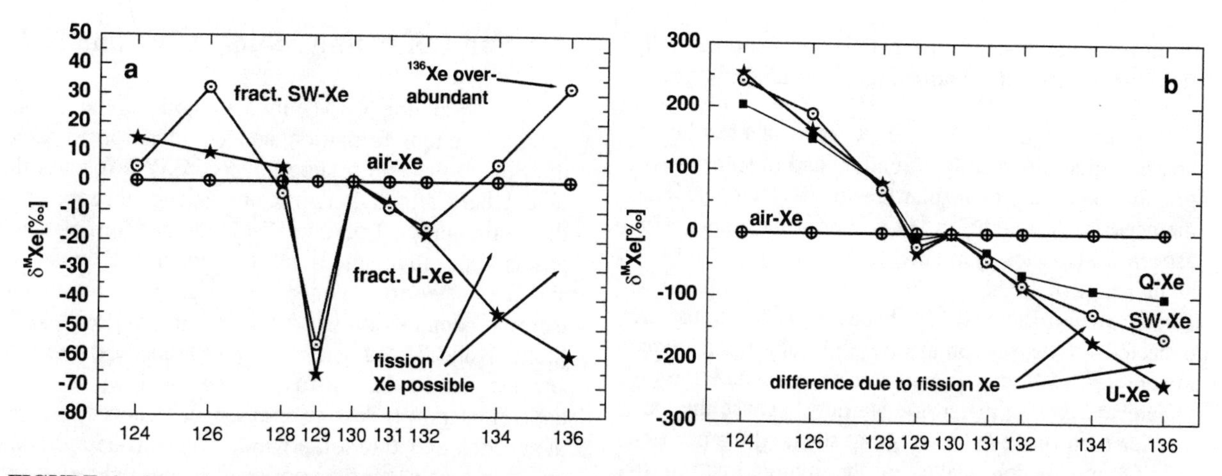

FIGURE 4. Isotopic composition of solar wind and U-Xe, plotted as deviation from terrestrial atmospheric Xe in ‰. Pepin suggested that the terrestrial atmosphere Xe could be deduced by fractionation of a solar-type Xe component (56). **a.** Fractionated solar wind already yields an excess of ^{134}Xe and ^{136}Xe, if the light isotopes of both components roughly correspond. Addition of fission Xe from the interior would increase the discrepancy. **b.** U-Xe, which is solar wind Xe depleted in the heavy isotopes, accounts for contributions of fission Xe from the interior. Fractionated and supplemented by heavy Xe, it also reproduces the common trapped Xe-Q component in meteorites. Data given in (6, 51, and 56).

(ii) If hydrodynamic escape is the reason for the similarly fractionated composition of both the Martian and terrestrial atmosphere, then this similarity must be coincidental, because of the different masses of these planets. Otherwise, atmospheric Xe on Mars and Earth may have their root in a *common fractionated source* (58). A cometary origin has been proposed by Owen and coworkers (48). They suggested that Ar, Kr, and Xe in the atmospheres of the terrestrial planets are a mixture of an internal component and a contribution from impacting comets that must have formed at a temperature of about 50K.

One important step to clarify this issue is to measure the Xe isotopes in comets and the outer planets. Most indicative are the ratios ^{134}Xe/^{132}Xe and ^{136}Xe/^{132}Xe. Uncertainties of ~1% are needed, however, to decide whether comets and the gaseous planets contain meteoritic or solar wind Xe. Ratios within 2% would resolve U-Xe and SW-Xe. For comparison, the Galileo probe measured the respective Xe ratios in Jupiter's atmosphere with uncertainties of ~10% (59, 60). Measurements of the solar wind Xe as trapped by the Genesis mission (see above) with expected uncertainties <1% will allow refining the models. At present the ratios ^{124}Xe/^{132}Xe and ^{126}Xe/^{132}Xe in the solar wind are known to within 2-4%. So far, only lunar soil has yielded solar wind Xe isotopic ratios (61, 62). Xe data from Genesis will facilitate a comparison of ancient with present solar wind, and check for possible fractionation processes during trapping on the moon.

Presolar Grains and Solar System Si- and O-Isotopic Ratios

Meteorites contain refractory grains of stardust believed to originate from stellar outflows and supernova ejecta. Large isotopic anomalies, e.g., of Ne and Xe, led to the identification of several types of presolar grains and provide important clues on stellar nucleosynthesis processes and galactic evolution (63, 64 and references therein, see also 65, this volume). Presolar SiC is the major carrier of Si among meteoritic stardust. Because its average ^{29}Si/^{28}Si and ^{30}Si/^{28}Si ratios are a few percent higher than the solar system ratios (66), other yet unidentified presolar Si-bearing mineral types with isotopically light Si may have contributed to the Si budget of the solar system. Potential candidates are silicates that are believed to be a major circumstellar condensate (67). The search for presolar silicates is complicated by the fact that the separation of presolar minerals from meteorites relies on the use of chemicals that destroy silicates. The successful identification of presolar silicates requires the use of non-destructive separation techniques, the in situ search in meteoritic thin sections, or the analysis of cometary samples which will become available from the STARDUST mission (19). The latter samples are of particular interest as comets represent the most primitive solar system matter, thus being a potentially rich source of yet unidentified presolar mineral types.

Similarly, the average $^{17}O/^{16}O$ ratio of presolar O-bearing minerals, mainly corundum and to some extent graphite, is higher than solar (see 63, 64), emphasizing the need for a search for ^{16}O-rich presolar grains. This aspect is also closely related to the puzzle of the low abundance of oxide grains from supernovae (SN) among meteoritic stardust because such grains are expected to be predominantly ^{16}O-rich. It has been argued that SN oxide grains have remained largely undetected among meteoritic stardust because they are smaller than those from red giant stars, which are rich in ^{17}O and which make up the major fraction of presolar oxide grains (68). A systematic search for smaller oxide grains as well as a search for yet unidentified mineral types may allow finding an answer on this question.

The $^{53}Cr/^{52}Cr$ Ratio as Indicator for the Formation Region of Planetary Bodies?

Several short-lived (half-lives of ~10^6 yrs) now extinct radionuclides have been incorporated into or produced in situ in early solar system material. The decay leads to enrichments of daughter nuclides and thus allows the relative chronology of early solar system processes, if the initial daughter/reference nuclide ratio is known. The short-lived ^{53}Mn decays to ^{53}Cr. The $^{53}Cr/^{52}Cr$ ratio of planetary samples (Earth & Moon, meteorites from Mars ("SNCs" - shergottites, nakhlites, chassignites) and Vesta ("HEDs" - howardites, eucrites, diogenites) apparently correlates with heliocentric distance of the place of formation (69, see also 65, this volume). If this is generally true, then the initial ^{53}Mn abundance could be used to constrain the place of formation of solar system samples of unknown origin. However, the radial heterogeneity of the initial ^{53}Mn in the solar nebula has been called into question (70). New data on chondrules in unequilibrated meteorites suggest volatility-controlled variation of Mn and Cr in the nebula instead. In this respect, it would be of particular interest to measure $^{53}Cr/^{52}Cr$ in other inner solar system objects, e.g., the Sun (or solar wind) or Mercury. The required accuracy is about 10 ppm, which cannot be achieved by space missions planned (e.g., to Mercury) in the foreseeable future, but would require sample return missions.

A Sun Consisting Mainly of Fe and Ni?

One co-author (O. M.) favors a non-standard model of solar system formation and composition in which the Sun formed on a supernova core. We discuss this model here although it has not gained acceptance in the community. The empirical comparison of noble gas isotope abundances in solar-wind-rich lunar soil and the meteorite Allende (see Figure 4b for the Xe isotopic composition in solar wind and meteorites) led Manuel and Hwaung (71 and references therein) to suggest intra-solar diffusion that follows a mass-dependent power law to explain differences between meteorites and the solar wind. This diffusion should result in a solar surface enriched in the light elements H and He, as observed, whereas the most abundant elements in the solar interior are Fe, Ni, O, Si, S, Mg, and Ca. This solar composition then resembles that found in bulk meteorites.

The enrichment of the lighter (mass m_L) relative to the heavier (m_H) isotope must also be discernible in the solar wind isotope composition of other elements. The power law describes the fractionation (f) with $f=(m_H/m_L)^{4.5}$ and thus predicts, e.g., a Mg isotope fractionation of ~20%/amu. The Mg concentration in lunar soils is much too high to detect solar wind Mg, but in situ measurements of the isotopic composition of the refractory elements Mg and Si in the solar wind show that they are fractionated by at most 2%/amu in the solar wind relative to terrestrial or meteoritic values (15, 35, 36). In addition, the Al/Mg ratio in the solar wind is indistinguishable from the solar system value obtained from meteorites (72), in contrast to the model by (71) that predicts a Mg enhancement of 70%. Furthermore, this model will have to prove its capability to match all observational constraints from helioseismology, solar neutrino flux observations, and the average solar density of ~1.41 g/cm^3.

THE INTERSTELLAR MEDIUM

The composition of the interstellar medium provides an important benchmark in the discussion of several cosmological and cosmochemical issues. It helps to decide whether the Sun's composition is unusual for this part of the Galaxy (73, this volume), and serves as reference composition for current galactic material and thus is essential for GCE models (see below). The composition of the dust and the gas in the ISM also serves as a baseline for the source of GCRs, thought to be mainly accelerated interstellar material (see below), and finally provides useful information for understanding Big Bang nucleosynthesis.

Optical and radio spectroscopy as well as direct measurements of infalling particles in the heliosphere provide elemental and isotopic abundances in the local interstellar medium (LISM). In the following, we will briefly review the state of satellite-based particle measurements, such as the detection of pickup ions and anomalous cosmic rays within the heliosphere. We will then discuss the prospects of future heliospheric missions to address these goals, culminating in an interstellar medium probe to beyond the heliosphere.

Measurements of the Interstellar Medium by Pickup Ions

Most pickup ions in the solar wind are particles originating from the interstellar medium that enter the heliosphere as neutral atoms. Subsequently, they become singly ionized and acquire energies in the range 0 to 4 E_{SW}, i.e. roughly 0-10 keV/nuc (The only exception is ^{3}He which is doubly ionized). Therefore, pickup ions provide an important source to our knowledge of both the composition of the Local Interstellar Cloud (LIC) as well as of the filtration processes occurring in the heliospheric interface region.

Element abundances of H, He, N, O, and Ne and the ^{3}He/^{4}He isotopic ratio have been determined (74, 75). Other heavy elements with low first ionization potential (FIP) are already mostly ionized in the interstellar medium and therefore do not enter the heliosphere. Pickup ions of low FIP elements, such as C, have been found, but they apparently mainly originate from the "inner source" (most likely solar-wind-loaded interplanetary dust in the solar system) and can be easily distinguished by their velocity distribution (76, 77).

Current knowledge of the elemental composition in the solar system and the LISM suggests only a slight, if any, overabundance of N and about an equal abundance of O and Ne in the LISM compared to the solar system (78, this volume). The uncertainties for these elements in both reservoirs are still very large. The uncertainty of elemental abundances in the LISM using pickup ions is at present ~25% or more. An accuracy of a few percent is necessary to better constrain models of GCE (see below). Therefore, significant improvements in the accuracy of abundances are urgently needed. Improvements need to be made in two ways. First, the counting statistics have to be improved. Because the density of all interstellar neutrals, except He, increases significantly between 1 and 3 to 10 AU, dedicated instruments with high resolution and large collection power on a spacecraft in a 1 by 3 AU orbit are required to increase the accuracy of measurements to less than one percent. An "Interstellar Pathfinder" mission with such performance characteristics has recently been proposed (79). Second, measurements of, and models describing the physical state of the ISM and the characteristics of the interface region between the heliosphere and the LIC should be improved. This is necessary in order to increase the accuracy of estimates of the degree of ionization of the LIC gas and the fraction of elements bound in interstellar dust, both of which are required to obtain the elemental composition of the LIC from pickup ions. The density of interstellar neutral H, N, and O in the heliosphere is systematically reduced from the corresponding densities in the LIC by filtration in the heliospheric interface region (e.g., 80, 81). Using pickup ion observations, absorption measurements from nearby stars and a simple model of the interface region, constraints on the amount of filtration of H, N, and O and on the ionization fractions of H and He were obtained (78, this volume, 82). More detailed probing of the filtration can be expected from direct observation of the interstellar neutral gas flow pattern in the inner heliosphere (e.g., 83). Nevertheless, the composition of low-FIP elements will remain inaccessible to in situ observations inside the heliosphere. To remove this ultimate restriction will require an "Interstellar Probe" mission into the neighboring interstellar medium (84).

Largely unaffected by filtration at the heliospheric boundary is the determination of isotopic composition of the LISM through in situ observations in the heliosphere. Especially important is the He isotopic composition, because it represents the present-day ^{3}He abundance in the LISM. In combination with the protosolar ^{3}He abundance, deduced from meteorites (85) and/or the Jovian atmosphere (59), it allows one to trace the evolution of the Galaxy over the past 4.56 Gyrs (see below). A ^{3}He/^{4}He ratio of $2.48^{+0.68}_{-0.62} \times 10^{-4}$ has been determined from pickup ions observed with SWICS on Ulysses (86). This ratio is, within the large uncertainties, comparable with $1.7^{+0.50}_{-0.42} \times 10^{-4}$ measured in foils directly exposed to the inflowing neutral interstellar gas onboard the space station MIR (87, this volume). Both ^{3}He/^{4}He ratios are known to within ~25%. Desirable is an accuracy of <5%, because this ratio yields, in combination with the present-day solar wind value, measured during several missions, the very high priority protosolar D/H ratio, which sets tight constraints on the baryon density of the universe. This example illustrates that more, and more precise, measurements of the LISM are extremely important. The recent re-determination of the $^3He/^4He$ ratio in the solar wind with SWICS on Ulysses, e.g., has already provided an averaged $^3He/^4He = (3.75\pm0.27) \times 10^{-4}$ in the outer convection zone (88).

Except for the $^3He/^4He$ ratio, all other isotopic ratios in the LISM measured with pickup ions are essentially unconstrained. The $^{20}Ne/^{22}Ne$ ratio is especially important, as it allows a comparison with the ratio obtained from anomalous cosmic rays (see below) and may help to understand the discrepancy between values obtained in meteorites, solar wind, and energetic particles. An "Interstellar Pathfinder"-like mission could provide the mass resolution and collecting power to achieve these goals (79).

Anomalous Cosmic Rays as Probe of the Interstellar Medium

The elements H, He, C, N, O, Ne, and Ar have been clearly detected in "anomalous cosmic rays" (ACR), which are called so, because their composition is neither solar nor similar to that in galactic cosmic rays (89). After ionization and pick up by the solar wind in the inner heliosphere, the formerly interstellar atoms are convected to the outer heliosphere, where they are accelerated at the termination shock to energies ~1-100 MeV/nuc (89). Possibly, Na, Mg, S, and Si have also been detected in ACRs (90, 91). However, their origin is not clear. All four elements have a low FIP <10.4 eV and should have a low neutral abundance in the interstellar medium (92).

Isotopically resolved measurements (especially at energies below 10 MeV/nuc) are most important. For example, the upper limit on the ACR $^{15}N/^{14}N$ ratio is about 10 times larger than that adopted for the Sun (93, Figure 5a), since it is limited by GCR background. Thus, a future measurement of N isotopes with 50% uncertainty at energies <10 MeV/nuc, where the GCR background is less, would be desirable. The $^{18}O/^{16}O$ ratio however needs to be measured to within some 10%, as the ratios deduced for the Sun and the ACR agree within a factor of two (93, Figure 5b). The $^{20}Ne/^{22}Ne$ isotopic ratio in ACRs has been found to match solar wind or meteoritic composition, with the former being more likely (Figure 5c, options "c" and "b", respectively).

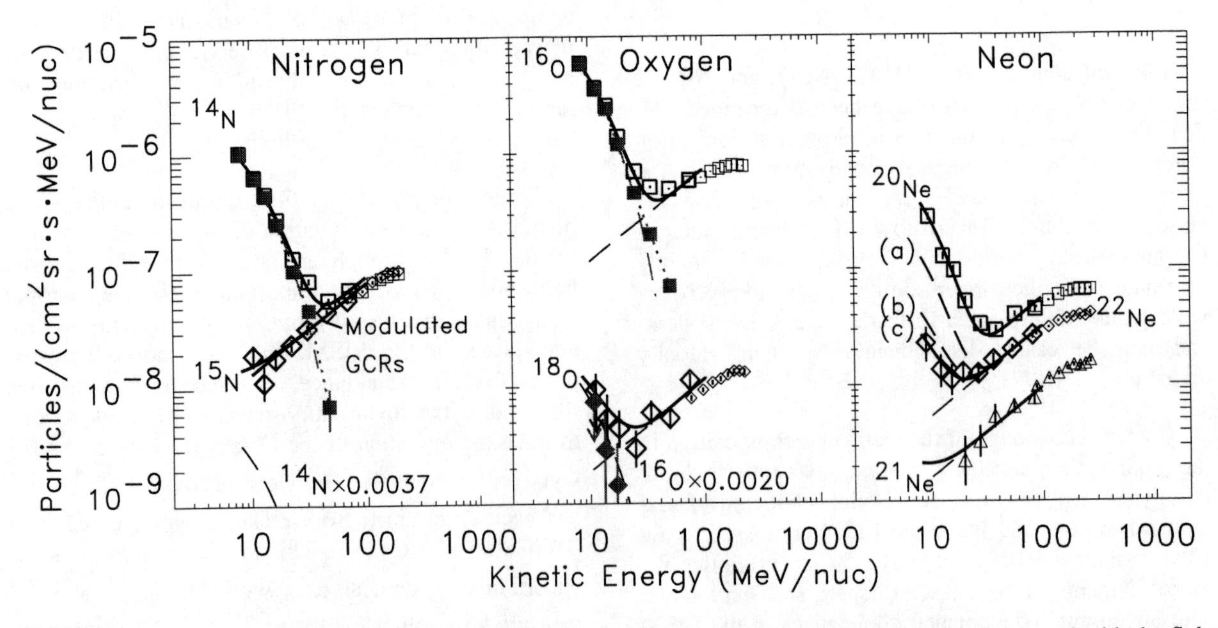

FIGURE 5. Isotopic composition of N, O, and Ne in the anomalous component of cosmic rays as measured with the Solar Isotope Spectrometer (SIS) on the Advanced Composition Explorer (ACE) (93). Open symbols are measured data points. The low energy excesses are due to anomalous cosmic rays. **a)** and **b)** Dashed lines are expected ^{15}N and ^{18}O abundances, respectively, using ^{14}N and ^{16}O abundances and assuming solar system composition. The figure indicates that any ^{15}N in the ACRs is not detectable at these energies above the galactic cosmic ray (GCR) background. After subtracting this background, the remaining ACR ^{18}O (filled symbols) is consistent (within a factor of ~2) with the $^{18}O/^{16}O$ isotopic ratio in the solar system. Figure c) shows similar calculations for ^{22}Ne assuming the ACR $^{22}Ne/^{20}Ne$ ratio to be similar to (a) the GCR source, (b) meteorites and (c) solar wind. The data indicate an ACR Ne isotopic composition similar to the solar wind.

Almost certainly, the Ne composition of the ACRs does not resemble that of GCRs (93, Figure 5c, option "a"), indicating that an additional source for ^{22}Ne other than just the ISM is required for GCRs. Better statistics could be reached with larger instruments and measurements in the outer heliosphere. Anomalous cosmic rays are also trapped and concentrated by a factor of 100 in the Earth's radiation belt. Measurements here could also be valuable (94, 95).

Interstellar Probe Mission

Figure 6 (84) illustrates the relative distribution of matter in the LISM in the three principal reservoirs, interstellar dust, neutral atoms, and plasma (data taken from 92). The fraction of material that is in the plasma state does not enter the heliosphere. Likewise, the majority of small interstellar grains are excluded from entry by the heliospheric magnetic field. Therefore, the most significant progress in sampling the properties of the interstellar medium would be gained by an interstellar probe mission (84, 96) that would provide direct access to these components of the LISM. The mission would allow an improved comparison of LISM and solar system composition, which should finally answer the question how closely the solar composition resembles that of the neighborhood in our Galaxy.

Furthermore, this mission could provide the measurement of cosmic rays outside of the heliosphere, inside which they experience energy loss. This would allow us to (possibly) determine the cosmic-ray spectrum in the ISM below 200 MeV/nuc. Cosmic-ray induced spallation produces the light elements Li, Be, and B by three channels: a) GCR p and He interacting with ISM C and O; b) GCR C, N, and O interacting with ISM p and He; c) GCR He interacting with ISM He (Li only). Approximately, all of the Li, Be, and B produced via channel (a) by GCRs with energies of 200 to 2000 MeV/nuc is at thermal energies or will quickly thermalize. GCR p and He spectra at these energies are measured *within* the solar system. This channel contributes most of the spallogenic Li, Be, and B. However, channel (b) contributes as well, although not as much as (a). Li, Be, and B produced by channel (b) are almost as energetic as their GCR parents. Effectively all of the Li, Be, and B produced above 200 MeV/nuc will escape before thermalizing. Therefore, channel (b) will only contribute if the GCR parents have energies of less than 200 MeV/nuc. However, such a GCR spectrum cannot be measured inside the heliosphere because of solar modulation effects. The interstellar probe would thus help us to narrow down the GCR spallogenic contribution to Li, Be, and B galactic abundances at thermal energies.

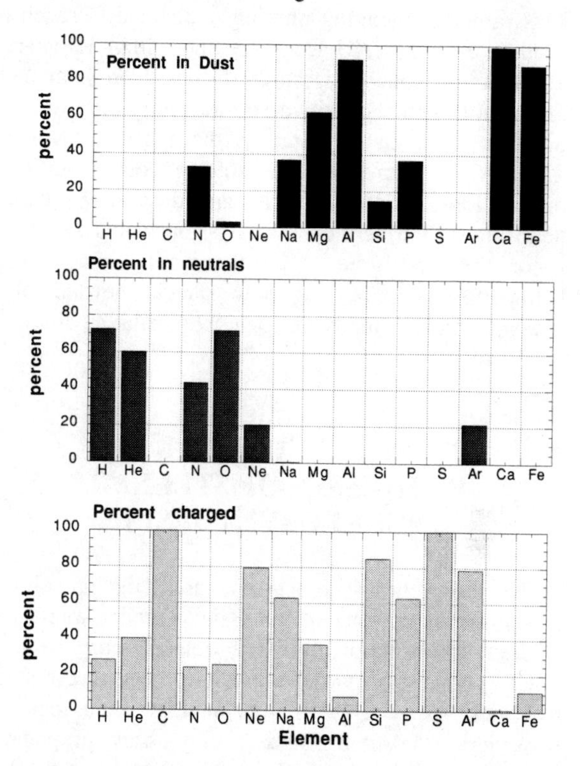

FIGURE 6. The distribution of the major elements in matter in the local interstellar medium as deduced theoretically (92). Figure reprinted from (84) with kind permission from Elsevier Science.

In addition to the investigations mentioned above, an Interstellar Probe mission would measure the magnetic field and plasma properties of the interstellar medium, investigate the interstellar spectrum of cosmic rays below ~300 MeV/nuc that are excluded from entering the heliosphere, and determine the source of the 2-3 kHz radio emission that apparently originates near the nose of the heliosphere. During its passage through the outer solar system, it could survey the density of small (1 to 100 km) Kuiper Belt objects from 30 to ~200 AU and thereby investigate the radial extent of the primordial solar nebula. This mission could also carry a small infrared telescope that would map the dust density in the outer solar system and measure the cosmic infrared background radiation in a wavelength region (5 to 100 μm) that is inaccessible to Earth-based telescopes because of obscuration by the zodiacal light (see 84 and 97 for additional information).

In order to sample the material effectively throughout its journey, an "Interstellar Probe" must contain plasma, neutral gas, dust, and cosmic ray instruments with sufficient mass and charge resolution. This rather challenging mission is planned to reach a distance of >200 AU from the Sun within 15 years. The mission concept is currently based on solar sail propulsion with a gravity assist trajectory close to the Sun (84, 96). On its journey, the spacecraft would measure the elemental and isotopic composition of pickup ions, ACRs, neutrals, and dust in the outer heliosphere. Furthermore, the mission would examine - for the first time - the termination shock, the heliopause, possibly a bow shock beyond the heliosphere, and finally the local interstellar medium.

GALACTIC COSMIC RAYS - WITNESSES OF STELLAR NUCLEOSYNTHESIS

Nucleosynthesis processes and the resulting compositions of stars in the Galaxy cannot solely be studied spectroscopically. The galactic cosmic rays (GCRs) consist of particles that have been accelerated by shock waves in the vicinity of stars to energies up to several GeV/nuc. Subsequently, they propagate through the Galaxy and allow us to directly probe the average composition of the nucleosynthetic products of several stellar sources. A thorough understanding of this cosmic-ray propagation is important to identify how the cosmic-ray abundances are altered during transport and to determine the initial averaged composition of the cosmic-ray source. Here, we will discuss models that describe the physical propagation process and data needed to delimit the distinct possibilities. Another way to directly sample stellar compositions and decipher processes that lead to the production of nuclei in stars is the analysis of presolar or circumstellar grains that formed in the outflow of certain stars and survived the formation of the solar system unprocessed in primitive meteorites. This topic has been reviewed recently (63, 64) and is discussed by Hoppe in these proceedings (65, this volume). Therefore, we only discuss the importance of presolar grains for solar system abundances and galactic chemical evolution (see below).

The Propagation of Galactic Cosmic Rays

Isotopic and elemental abundance measurements of the galactic cosmic rays using instruments aboard balloons and spacecraft such as ACE, HEAO-3 (e.g., 98), Ulysses (e.g., 99), ISEE-3 (e.g., 100), IMP-7/8 (e.g., 101) and others, have provided insight into the nature of cosmic-ray propagation of particles with energies E_{ISM} in the range of 200 MeV/nuc to 50 GeV/nuc. The seed nuclei for galactic cosmic rays (GCRs) are synthesized in massive stars, supernovae, and possibly Wolf-Rayet (WR) stars. The GCR source nuclei experience a time delay of more than 10^5 yrs between their synthesis and their initial acceleration. This age was determined by observing the essentially complete decay of the e-capture isotope ^{59}Ni (102), which will decay only if the GCR source material remains at low energies where its electrons can remain attached for a period significantly longer than the ^{59}Ni half-life. Stable species such as the isotopes of boron, ^{19}F, and ^{45}Sc are produced predominantly via fragmentation during propagation (e.g., 103), and the abundances of these species indicate an average path length of 1-10 g/cm^2 in the energy range given above (e.g., 98). However, uncertainties in modeling GCR propagation remain that can be overcome by new measurements.

Is GCR propagation described accurately by a "leaky-box" model or are more realistic diffusion models required? The leaky-box model assumes that all GCRs freely diffuse within a confinement region (or "box") containing an approximately homogeneous interstellar medium (ISM) matter density, with a finite probability for escape at the boundary. However, more realistic diffusion models suggest that one cannot treat the ISM as homogeneous, and that the average value of the ISM density probed by GCRs depends on the size of the propagation volume. One can test this dependence by measuring abundances of GCR species that are produced only via fragmentation during propagation and decay via β-emission. The abundances of these GCR "clocks" will depend on the time of cosmic-ray confinement in the Galaxy and the rate of fragmentation in the ISM (and in turn the ISM density in the propagation volume probed by each species). The surviving fractional abundances of these secondary β-decay radionuclides (e.g., ^{10}Be, ^{26}Al, ^{36}Cl, ^{54}Mn) measured at E_{ISM} = 200-500 MeV/nuc imply a propagation time τ_{esc} of 15 Myr and an average ISM hydrogen density ρ_{ISM} of 0.34 H atoms/cm^3 (104). All of the individual τ_{esc} and ρ_{ISM} values from measurements of these species are consistent within measured uncertainties with a unique value of ρ_{ISM} and τ_{esc} as assumed by the leaky-box model. However, the decay time of the GCR clock species will experience relativistic time dilatation at higher energies, so the propagation volume is correspondingly larger.

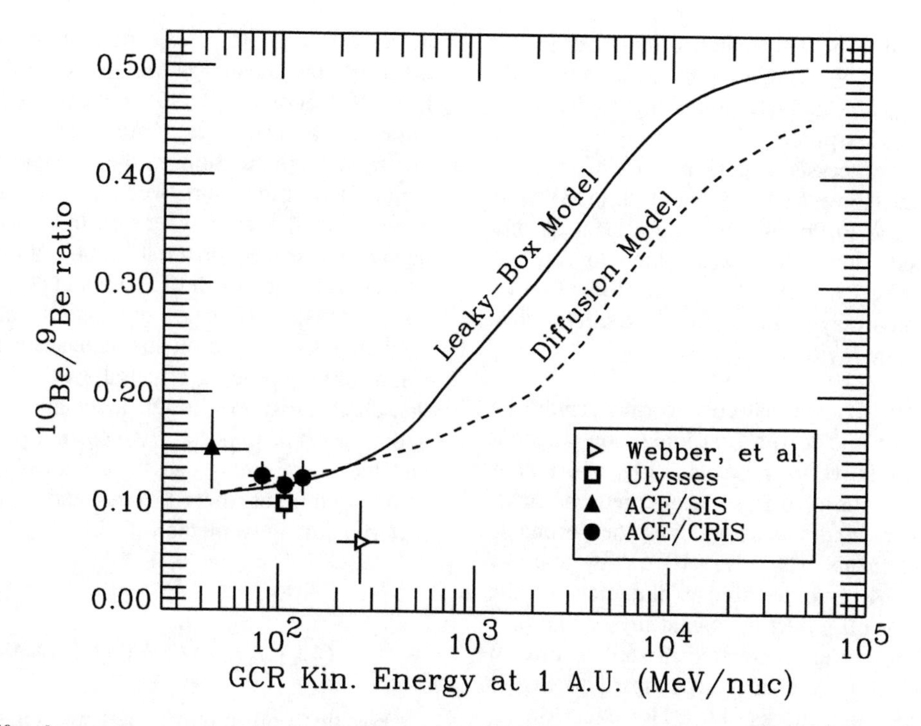

FIGURE 7. $^{10}Be/^{9}Be$ ratios as measured in the GCR at lower energies (104, 105, 108). Data points obtained for GCRs at energies in the range 1-10 GeV/nuc with sufficiently small uncertainties could distinguish between the models that describe the propagation of GCRs.

The inhomogeneity predicted in diffusion models leads to a decrease in the average ISM density ρ_{ISM} with increasing propagation volume. Thus, abundance measurements at both low and high energies will help distinguish between the two models (Figure 7).

Figure 7 shows low-energy GCR data measured by the ACE/CRIS and ACE/SIS instruments (104) and Ulysses (105) for the $^{10}Be/^{9}Be$ ratio. Also shown are $^{10}Be/^{9}Be$ ratio curves calculated from a leaky-box model (106) for a density of ρ_{ISM} =0.34 H atoms/cm^3 and a diffusion model (107) for a galactic halo size of z(h) = 2 kpc. Measurements of $^{10}Be/^{9}Be$ at higher energies have been performed by the balloon experiments of Webber and Kish (108), SMILI (109), and ISOMAX (110, 111). The recent ISOMAX results, which extend from ~0.3 to 2 GeV/nuc, give values of ~0.2 at 0.65 GeV/nuc and ~0.3 at ~1.5 GeV/nuc, consistently above the diffusion model in Figure 7. However, given the uncertainties, it is not yet possible to distinguish between the leaky-box model and a diffusion model, possibly characterized by a somewhat different halo size and/or diffusion coefficient. Measurements of the $^{10}Be/^{9}Be$ ratio (see figure 7) at two higher energies between 1-10 GeV/nuc to 11% uncertainty (systematic and statistical) or four energies to 15% uncertainty could distinguish between the competing models discussed above to within 3σ, or could motivate a new physical resolution to this problem. In addition, high-energy measurements of the $^{26}Al/^{27}Al$ ratio to a similar accuracy could identify whether a density variation probed by a particular species is the same for other clocks.

The Abundances of Cosmic Ray Nuclei beyond the Iron Peak

In nature, nuclei with $31 \leq Z \leq 92$ are extremely rare. Their abundances amount in total to only ~3x10^{-4} of the Fe abundance. In cosmic rays, nuclei with $Z \geq 30$ are of particular interest because they can address two important issues, the nature of the material that is accelerated to be cosmic rays and the age of the material since it was originally synthesized.

Most of the nuclei with $Z \geq 30$ are produced by neutron-capture processes that can be modeled in terms of two extremes – a slow-process (s-process) that permits β-decay before additional neutron captures, and a rapid process (r-process), resulting in multiple neutron captures before β-decay. Previous space experiments (see, e.g., 112 and references therein) observed an underabundance of nuclei in the "Pb-group" ($81 \leq Z \leq 83$) relative to those in the "Pt-group" ($74 \leq Z \leq 80$), and interpreted this as due to an overabundance of r-process nuclei in cosmic rays relative to the r-process/s-process mix of material in the solar system. However, Westphal et al. (113), who confirmed that the Pt/Pb ratio is enhanced in cosmic rays, suggested that this was due to atomic fractionation effects associated with the formation of dust grains, which may be the source of the refractory elements in cosmic rays (114).

It is well known that cosmic rays contain an overabundance of nuclei with first ionization potential (FIP) <10 eV, relative to those with FIP >10 eV. A similar pattern is observed in solar energetic particles and in the solar wind, indicating that the corona is enhanced in elements with FIP <10 eV by about a factor of ~3 to 4. One possible explanation for the occurrence of the FIP effect in cosmic rays is that they may originate as nuclei accelerated in stellar flares of Sun-like stars, which are then accelerated to higher energies by supernova shocks (115). However, Meyer, Drury, and Ellison (114) argued that the apparent correlation with FIP is coincidental, and that it is actually the volatility of the elements that is important. They suggested that cosmic rays originate from interstellar dust grains that are accelerated to energies of ~0.1 to 1 MeV/nuc. Ions sputtered from these grains (mostly refractory species) can then be efficiently accelerated to cosmic ray energies along with an appropriate mixture of interstellar gas (mostly volatile elements).

Although the atomic properties FIP and volatility are highly correlated, there are a number of elements in with $30 \leq Z \leq 83$ (e.g., ^{31}Ga, ^{32}Ge, ^{34}Se, ^{37}Rb, ^{47}Ag, ^{50}Sn, ^{55}Cs and ^{82}Pb) that can distinguish between these two models (114), if one could measure the abundances of individual cosmic rays species with sufficient accuracy. The proposed *Heavy Nuclei Explorer* (HNX) mission, now under study by NASA, will measure the abundances of cosmic rays with $10 \leq Z \leq 96$ with ~10 times the collecting power of previous experiments, and with individual element resolution (116). HNX consists of the ENTICE instrument, which measures nuclei with $10 \leq Z \leq 83$, and the ECCO instrument, which uses glass track-etch detectors to measure nuclei with Z >70.

In addition to answering the question as to whether cosmic rays originate as grain-destruction products, HNX would also collect anywhere from ~100 to ~300 actinide nuclei (those with $90 \leq Z \leq 96$), sufficient to measure the age of cosmic ray nuclei since their nucleosynthesis. If cosmic rays are accelerated from the ISM, the mean age will be several Gyrs, and at least 100 actinides would be expected in a 2-year mission. If cosmic rays are enhanced in r-process nuclei, a large fraction of these nuclei must be very young, permitting short-lived elements such as ^{96}Cm to survive. In the case where cosmic rays originate in newly synthesized material, ~300 actinides would be expected. In the model of Higdon et al. (117), in which cosmic rays originate in superbubbles, there is predicted to be a dramatic enhancement in freshly-synthesized r-process material, with an age of $\sim 10^7$ yrs, characteristic of an OB association. In addition to addressing the important issues of FIP vs. volatility, and the cosmic rays age since nucleosynthesis, HNX would determine the r-process and s-process mix of material among elements with $30 \leq Z \leq 83$.

GALACTIC ABUNDANCES

For this contribution, we have chosen situations, where measured solar or meteoritic and ISM abundance ratios can be used to set tight constraints both to chemical evolution and stellar evolution models.

Galactic Chemical Evolution and Presolar Grains

It has been argued by Clayton (118) that presolar dust found in meteorites preserves a memory of the evolution of the abundances of the chemical elements in the Galaxy with time, the galactic chemical evolution (GCE). In fact, the distribution of Si-isotopic ratios measured in presolar SiC grains separated from primitive meteorites (66) cannot be reconciled with evolutionary models of the parent stars (AGB stars, or more specifically carbon stars) but may reflect the GCE of the Si isotopes, both in time and space (119, 120, see also 65, this volume). As the stardust in meteorites formed at different times and locations in our Galaxy, its isotopic compositions can be used to test models of GCE. In general, the isotopic compositions of stardust are determined by the starting compositions of the parent stars (reflecting the GCE) and the nucleosynthesis and evolution of these stars during their lifetime. The GCE is best recognized if the effects of stellar nucleosynthesis and evolution on the isotopic compositions at the parent star's surface are small compared to the spread of isotopic compositions at the times the parent stars formed.

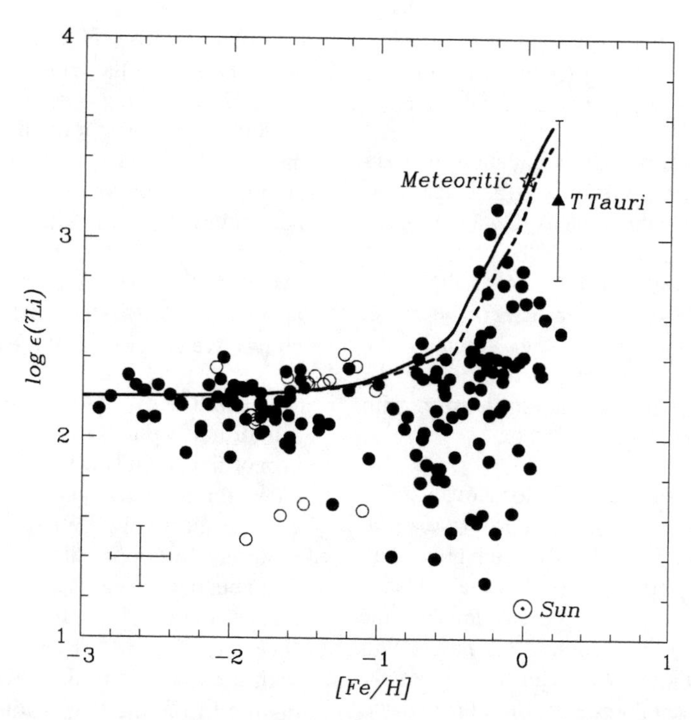

FIGURE 8. Log $\varepsilon(^7Li)$ (≡ "A(Li)" ≡ $\log(N_{7Li}/N_H)+12.0$) vs. [Fe/H] (≡ $\log(N_{Fe}/N_H)-\log(N_{Fe}/N_H)_\odot$) theoretical trends predicted by models of Romano et al. (124, figure reprinted with kind permission from EDP Sciences) including all astrophysical sites for ^{7}Li production (Data from 121, filled circles, and 149, open circles). Dashed line: Type II SNe + AGB stars + low-mass giants + novae. Continuous line: Type II SNe + AGB stars + low-mass giants + novae + GCRs (their best model).

Besides Si and Ti, elements of interest include Mg in spinel (and silicates) and Ca and Ti in hibonite and other Ca- or Ti-bearing minerals from red giant and AGB stars. As the expected isotopic effects due to GCE are on the order of 10% (cf. Si in SiC), isotopic compositions of individual presolar grains should be measured with accuracy on the order of a few %.

Contemporary interstellar dust will be collected by the STARDUST mission and brought to Earth in the year 2006 (19). A comparison of its isotopic compositions with those of presolar grains from meteorites that formed more than 4.6 Gyrs ago will allow to obtain additional information on the GCE of certain elements.

Evolution of Lithium in the Milky Way

To explain the ^{7}Li evolution in the Galaxy, several sites of production and destruction of this isotope have to be considered, namely Big Bang nucleosynthesis (BBN), spallation reactions between GCRs and the ISM, and stellar nucleosynthesis (121). In particular, the measurement of the ^{7}Li abundance in distinct stellar objects of different populations and its temporal variation as predicted by chemical evolution models lead to an estimate of the primordial abundance of ^{7}Li. This abundance as well as those of D, ^{3}He, and ^{4}He, produced during the Big Bang, are of fundamental importance to probe the consistency of the BBN theory and provide a valuable constraint on the baryon density of the universe (122).

The ^{7}Li abundances as observed in the solar neighborhood are shown in Figure 8. The upper envelope of the data, as plotted in the log $\varepsilon(^7Li)$ vs. [Fe/H] diagram, is generally believed to reflect the ^{7}Li enrichment history of the LISM in the solar neighborhood. Therefore, it can be used to constrain models of galactic chemical evolution. The envelope is characterized by i) a large plateau at low metallicity, the so-called Spite plateau (123) and ii) a steep rise afterwards (124 and references therein) due to the ^{7}Li production by stellar nucleosynthesis and spallation of cosmic rays.

Data points below the envelope indicate that Li was depleted due to mixing of the surface layers of the respective stars with their interiors (convection). The efficiency of this mixing depends essentially on the stellar temperature (125). The solar photosphere contains less than 1% of the Li present in the ISM at the time of formation of the Sun (Figure 8), which is well represented by Li in meteorites (1, cf. Figure 1).

Generally, ^{7}Li in warm halo population II stars with very low metallicity (Spite plateau) is considered to represent its primordial abundance (122, 126). Some depletion of ^{7}Li in pop II stars might be possible, but the low dispersion of the lithium data at the Spite plateau suggests that this depletion cannot be large. One way to address this issue is to compare the primordial ^{7}Li abundance, as estimated by using the deuterium-inferred value of the baryon density, with the Spite plateau value (127).

Adopting up-to-date stellar yields for Novae, AGB stars, Type II SNe and Carbon stars, Romano et al. (124) conclude that: i) Type II SNe and novae are necessary in order to reproduce the ^{7}Li abundance evolution represented by the upper envelope of the observational data; ii) when adopting the new stellar ^{7}Li for AGB stars, those stars can no longer be considered as a significant source of ^{7}Li in the Galaxy; iii) Novae (and probably low-mass giant stars) restoring their processed material on long timescales, are among the best candidates for reproducing the late rise from the Spite plateau.

The D, ^{3}He, and ^{4}He Galactic Evolution and their Abundance Gradient

Chemical evolution models are useful to derive both the primordial abundances of D, ^{3}He, and ^{4}He as well as their evolution. In the light of recent observations, Chiappini et al. developed a new model for the Galaxy, the so-called two-infall model (128, 129, this volume). Predictions of this model for the chemical evolution of D, ^{3}He, and ^{4}He in the solar vicinity and for their distribution along the galactic disk are described in the following.

The model assumes two main gas infall episodes for the formation of the halo (and part of the thick disk) and thin disk, respectively (see 128 for references). The timescale for the formation of the thin disk is much longer than that of the halo, implying that the infalling gas forming the thin disk not only comes from the halo but mainly from the intergalactic medium. The formation time of the thin disk is assumed to be a function of galactocentric distance, leading to an "inside-out" picture for the Galaxy disk buildup, where the inner part accreted much faster than the outer regions (130). Figure 9 shows the predictions of the two-infall model for the evolution of ^{3}He and D in the solar vicinity. The figure indicates a solution for a long-standing problem in chemical evolution, namely the overestimation of ^{3}He by the models compared to the values observed in the Sun and the interstellar medium. The solution (129, this volume) requires allowing for "extra-mixing" in low-mass stars ($M < 2.5\ M_{\odot}$,131). This "non-standard" extra-mixing occurs in RGB stars between the bottom of the convective envelope and the H-burning shell (131).

Chemical evolution models can also constrain the primordial value of the deuterium abundance. The primordial abundances by mass of D and ^{3}He were taken for the calculations to be 4.4×10^{-5} and 2.0×10^{-5}, respectively. While this D primordial value represents an upper limit (see Figure 9a), the adopted ^{3}He abundance must be a lower limit (Figure 9b). The present model (129, this volume) suggests a value of $(D/H)_{prim} < 3 \times 10^{-5}$ (by number). This is in agreement with the low primordial D/H value of $(3.0 \pm 0.4) \times 10^{-5}$ deduced from the Lyα feature in spectra of four highly redshifted ($z > 3$) low metallicity quasar absorption systems (QAS) (132, 133 and references therein). Other measurements indicating a significantly higher D/H may be explained with H contamination (132).

The primordial D abundance is considered to be the best baryometer, because it strongly depends on the baryon density. Observed abundances of D are lower limits on D_{prim} and thus upper limits of the baryon density, because D is only destroyed in post-BBN astrophysical processes. This ratio thus probes the consistency of BBN and constrains D chemical evolution. Vice versa, the low primordial D value from QAS may serve as a test for chemical evolution modeling. Observations of the LISM (134) and the solar system (135) represent tight constraints to the evolution of the D primordial abundance (Figure 9a; see 136 for possible D abundance variations in the LISM). Models that can reproduce the bulk of the observational data predict only a modest D destruction (in our case a factor <1.6; see 137).

The predicted radial abundance gradient for D (see 138) is positive and steep, due to the faster evolution of the inner disk regions compared with the outer parts (which are still in the process of formation and thus having an almost primordial composition). Of the various elements, D is probably most sensitive to radial variations in the timescale of disk formation. Thus, precise D abundance measurements in regions outside the solar vicinity are needed.

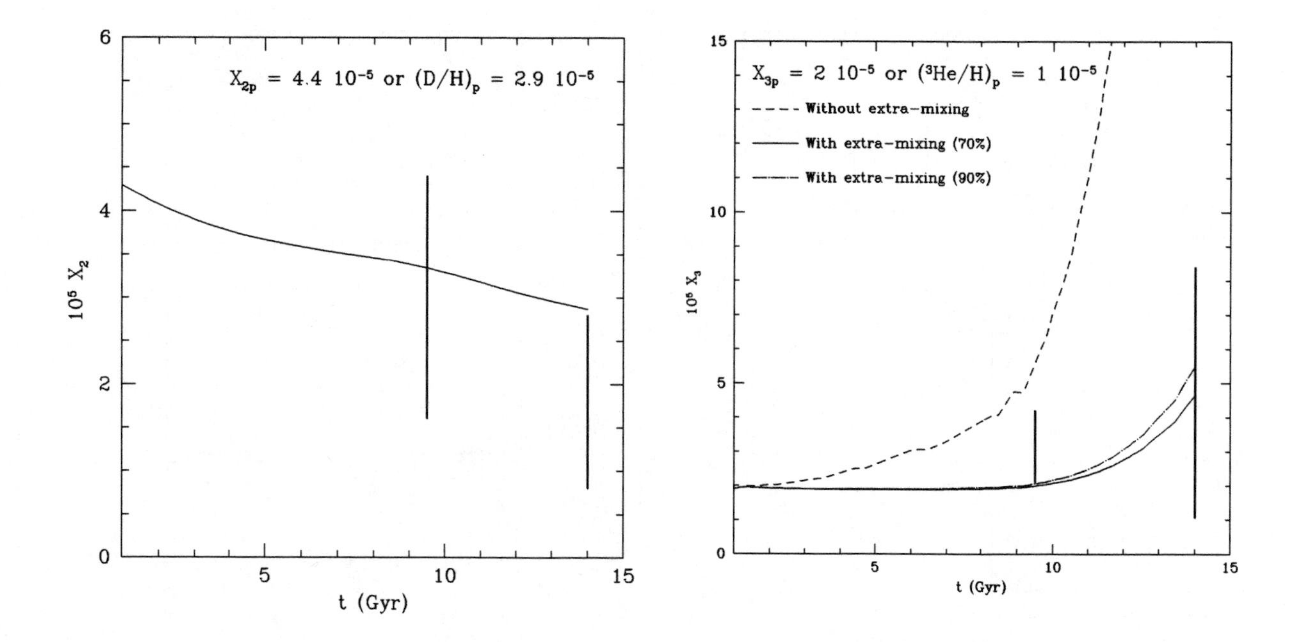

FIGURE 9. a) D and b) ^{3}He evolution (by mass) as predicted in 138, reprinted with kind permission from the Astronomical Society of the Pacific. The bars at 9.5 Gyr (4.5 Gyrs ago) and 14 Gyr (age of the Galaxy) represent the value of solar (2σ, 135) and ISM (2σ, 134) abundances, respectively.

For the ^{3}He abundance gradient (138), the assumption that a high fraction of low-mass stars suffers extra mixing leads to a flat gradient along the disk, which is in quite good agreement with the observed gradient measured in HI regions (139). Models without extra-mixing predict too much ^{3}He in the last Gyrs of the evolution of the Galaxy (Figure 9b). In this case, the ^{3}He abundance gradient would be sensitive to the adopted timescales of disk formation. Without extra-mixing and subsequent ^{3}He destruction in low-mass stars, we would predict that in the inner regions (older in the inside-out scenario) the contribution of these stars for the ^{3}He enrichment of the ISM would be more important than in the outer regions, and therefore a negative gradient would have been formed towards the outer regions of the disk. More data on the ^{3}He abundance at different galactocentric distances are thus very important to better constrain the chemical evolution models. These data as well constrain models of low-mass stellar nucleosynthesis.

Finally, adopting the primordial ^{4}He abundance suggested in (140), the model (138) yields a value for the galactic enrichment of He relative to the heavy elements $\Delta Y/\Delta Z \cong 2$ and a better agreement with the solar ^{4}He abundance (see summary of results and figures in 138). The ^{4}He gradient with galactocentric distance is rather flat ($\cong$ -0.003 dex/kpc over the 4-14 kpc galactocentric range) in agreement with the results on disk planetary nebulae (141). These results should be used with caution, because at present, the primordial ^{4}He abundance, Y_p, inferred from observations still suffers from systematic errors (Y_p = 0.228±0.005 vs. 0.244±0.002, 142, 143). Future measurements of helium with a decreased systematic error will allow using the ^{4}He abundance as a precision test for BBN, standard cosmology, and chemical evolution modeling. At present, a theoretically calculated Y_p value is usually used to constrain physics beyond the standard model (144, this volume, 145 and references therein).

It is interesting to notice that chemical evolution models can be used to constraint models of big bang nucleosynthesis (e.g., 144) as far as the primordial abundances of the light elements are inferred from observational data, accounting for their corresponding chemical and stellar evolution. On the other hand, standard BBN is a powerful tool for constraining models of chemical evolution, non-standard cosmological models and physics beyond the standard model, because the four light element abundances are predicted based on only one parameter - the baryon density. Hence, new observational data of the light elements D, ^{3}He, ^{4}He, and ^{7}Li are essential.

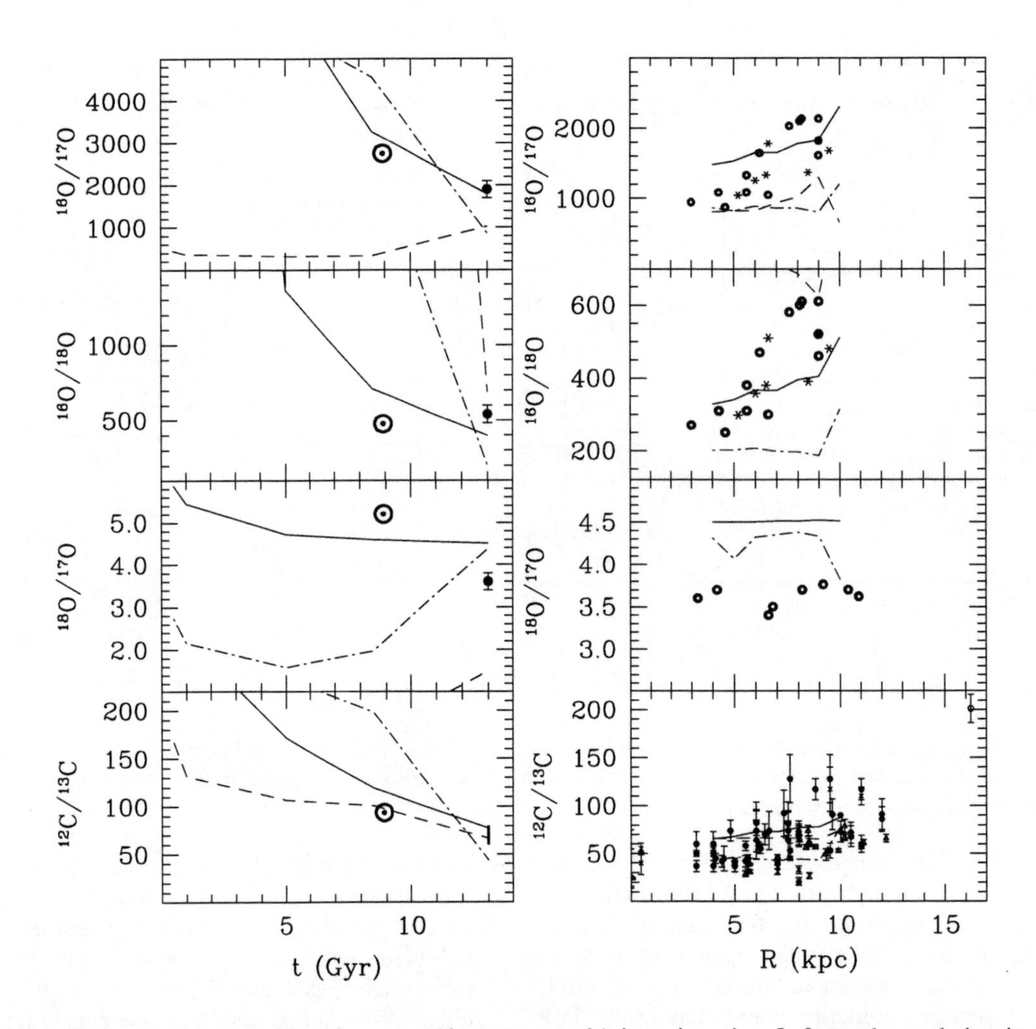

FIGURE 10. Carbon (bottom panels) and oxygen (three top panels) isotopic ratios. Left panels: evolution in the solar neighborhood. Right panels: current distributions with galactocentric distance. The solar symbol represents the solar ratio derived from Anders & Grevesse (1); all the other data are from radio observations of molecular clouds (146 and references therein). Models and figure are from Tosi (146, reprinted with kind permission from Kluwer Academic Publishers).

Abundance Ratios as a Function of Metallicity

Abundance ratios of a primary element (its production is independent of the initial heavy element abundance of the star) over a secondary one (its precursor element is produced in previous generation stars and its abundance is proportional to the initial heavy element abundance) are expected to decrease with time or metallicity (125). This can be useful to understand the origin of different elements and to give us information on the timescale of formation of a certain system such as the solar vicinity. Moreover, the abundance ratio of two primary elements that are restored to the interstellar medium by stars in different mass ranges would show almost the same behavior discussed above. One example is the $^{16}O/^{12}C$ ratio. Both elements are primary but since ^{12}C is mainly restored into the interstellar medium by intermediate mass stars (and hence on large timescales compared to the ^{16}O enrichment that comes mainly from massive stars), this ratio decreases as a function of metallicity.

The isotopic ratios of C, N, and O are of particular interest (Figure 10). As discussed recently by Tosi (146), chemical evolution models have difficulties in explaining the behavior of some of the isotopic ratios. The solar and ISM values for $^{13}C/^{12}C$, $^{17}O/^{16}O$, $^{18}O/^{16}O$ and $^{15}N/^{14}N$, as well as their radial profiles in the Galaxy, are very important as they can be used to constrain the mechanisms of disk formation and particular stellar evolution models.

For example, the steep rise of the $^{13}C/^{12}C$ with metallicity -or time- can be explained by novae contributions to ^{13}C or by the extra-mixing mechanism discussed above, which consumes ^{3}He and produces ^{13}C in low-mass stars (147, 148). The oxygen isotopes need further examination because the models (146) predict an increase of the $^{18}O/^{17}O$ ratio from the solar to the local ISM value and a corresponding decrease of $^{16}O/^{18}O$, contrary to what is observed (see 146 for references). The predictions involving ^{18}O are not in agreement with the observations, whereas chemical evolution models can well explain the observed behavior of $^{12}C/^{13}C$ and $^{16}O/^{17}O$. The discrepancies in ^{18}O seem to be related to stellar evolution computations, because the observed values for ^{18}O (at least that in the Sun) are, for this purpose, sufficiently well known. The increase of some of the ratios mentioned above as a function of galactocentric distance (Figure 10) could be understood based on the "inside-out" picture for galactic disk formation (129, this volume). At present, the inner parts of the Galaxy are more evolved and thus present a larger abundance of secondary elements compared with the outer regions.

ACKNOWLEDGMENTS

We thank the organizing committee of the SOHO/ACE Workshop held at the University of Bern, Switzerland, for bringing us together. This resulted in fruitful cross-disciplinary discussions, which we hope to be at least partly reflected in this contribution.

REFERENCES

1. Anders, E., and Grevesse, N., *Geochim. Cosmochim. Acta* **53**, 197-214 (1989).
2. Palme, H., and Beer, H., "Abundances of the elements in the solar system" in *Landolt Börnstein Vol. VI/3a, New Series, Astronomy and Astrophysics*, edited by H. H. Voigt, Berlin: Springer, 1993, pp. 196-221.
3. Cunha K., and Smith, V. V., *Astrophys. J.* **512**, 1006-1013 (1999).
4. Begemann, F., *Rep. Progr. Phys.* **43**, 1309-1356 (1980).
5. Huss, G. R., and Alexander Jr., E. C., *Proc. 7th Lunar Planet. Sci. Conf., J. Geophys. Res.* **92**, E710-E716 (1987).
6. Porcelli, D., and Pepin, R. O., "Rare gas constraints on early earth history" in *Origin of the Earth and Moon*, edited by R. M. Canup and K. Righter, Tucson: University of Arizona Press, 2000, pp. 435-458.
7. Owen, T., Mahaffy, P., Niemann, H. B., Atreya S., Donahue, T., Bar-Nun, A., and de Pater, I., *Nature* **402**, 269-270 (1999).
8. Christensen-Dalsgaard, J., *Space Sci. Rev.* **85**, 19-36 (1998).
9. Bochsler, P., *Rev. Geophys.* **38**, 247-266 (2000).
10. Feldman, U., *Space Sci. Rev.* **85**, 227-240 (1998).
11. Wieler, R., *Space Sci. Rev.* **85**, 303-314 (1998).
12. Pepin, R. O., Becker, R. H., and Rider, P. E., *Geochim. Cosmochim. Acta* **59**, 4997-5022 (1995).
13. Geiss, J., Bühler, F., Cerutti, H., Eberhardt, P., and Filleux, C., *Apollo 16 Prelim. Sci. Rep. NASA SP-315*, 14.1-14.10 (1972).
14. Bodmer, R., and Bochsler, P., *Astron. Astrophys.* **337**, 921-927 (1998).
15. Kallenbach, R., Ipavich, F. M., Kucharek, H., Bochsler, P., Galvin, A. B., Geiss, J., Gliem, F., Gloeckler, G., Grünwaldt, H., Hefti, S., Hilchenbach, M., and Hovestadt, D. *Space Sci. Rev.* **85**, 357-370 (1998).
16. Weygand, J. M., Ipavich, F. M., Wurz, P., Paquette, J. A., and Bochsler, P., *Geochim. Cosmochim. Acta*, in press (2001).
17. Chaussidon, M., and Robert, F., *Nature* **402**, 270-273 (1999).
18. McKeegan, K. D., Chaussidon, M., and Robert, F., *Science* **289**, 1334-1337 (2000).
19. Huntress Jr., W. T., *Space Sci. Rev.* **90**, 329-340 (1999).
20. Altwegg, K., Balsiger, H., and Geiss, J., *Space Sci. Rev.* **90**, 3-18 (1999).
21. Feldman, U., and Laming, J. M., *Physica Scripta* **61**, 222-252 (2000).
22. Boothroyd, A. I., Sackmann, I.-J., and Fowler, W. A., *Astrophys. J.* **377**, 318-329 (1991).
23. Wiens, R. C., Burnett, D. S., Neugebauer, M., and Pepin, R. O., *Proc. 22th Lunar Planet. Sci. Conf.*, 153-159 (1992).

24. Clayton, R. N., *Annual Rev. Earth Planet. Sci.* **21**, 115-149 (1993).

25. Collier, M. R., Hamilton, D. C., Gloeckler, G., Ho, G., Bochsler, P., Bodmer, R., and Sheldon, R., *J. Geophys. Res.* **103**, 7-13 (1998).

26. Wimmer-Schweingruber, R. F., Bochsler, P., and Gloeckler, G., *Geophys. Res. Lett.* **28**, 2763-2766 (2001).

27. Leske, R. A., Mewaldt, R. A., Cohen, C. M. S., Christian, E. R., Cummings, A. C., Slocum, P. L., Stone, E. C., von Rosenvinge, T. T., and Wiedenbeck, M. E., *AIP Conf. Proc.* **this volume** (2001).

28. Eberhardt, P., Reber, M., Krankowsky, D., and Hodges, R. R., *Astron. Astrophys.* **302**, 301-316 (1995).

29. Thiemens, M. R., and Heidenreich, J. E., *Science* **219**, 1073-1075 (1983).

30. Wiens, R. C., Huss, G. R., and Burnett, D. S., *Meteorit. & Planet. Sci. 34*, 99-107 (1999).

31. Kerridge, J. F., *Science* **188**, 162-164 (1975).

32. Clayton, R. N., and Thiemens, M. R., *Proc. Conf. Ancient Sun*, 463-473 (1980).

33. Kerridge, J. F., *Rev. Geophys.* **31**, 423-437 (1993).

34. Geiss, J. and Bochsler P., *Geochim.Cosmochim. Acta* **46**, 529-548 (1982).

35. Wimmer-Schweingruber, R. F., Bochsler, P., and Kern, O., *J. Geophys. Res.* **103**, 20621-20630 (1998).

36. Kallenbach, R., *AIP Conf. Proc.* **this volume** (2001).

37. Wieler, R., Humbert, F., and Marty, B., *Earth Planet. Sci. Lett.* **167**, 47-60 (1999).

38. Kim, J. S., Kim, Y., Marti, K., and Kerridge, J. F., *Nature* **375**, 383-385 (1995).

39. Hashizume, K., Chaussidon, M., Marty, B., and Robert, F., *Science* **290**, 1142-1145 (2000).

40. Hashizume, K., Marty, B., and Wieler, R., *Lunar Planet. Sci. Conf.* **XXX**, #1567 CD-ROM (1999).

41. Hashizume, K., Marty, B., Chaussidon, M., and Robert, F., *AIP Conf. Proc.* **this volume** (2001).

42. Kallenbach, R., Geiss, J., Ipavich, F. M., Gloeckler, G., Bochsler, P., Gliem, F., Hefti, S., Hilchenbach, M., and Hovestadt, D., *Astrophys. J.* **507**, L85-L88 (1998).

43. Owen, T., Mahaffy P. R., Niemann, H. B., Atreya, S., and Wong M., *Astrophys. J. Lett.* **553**, L77-L79 (2001).

44. Fouchet, T., Lellouch, E., Bézard, B., Encrenaz, T., Drossart, P., Feuchtgruber, H., and de Graauw, T., *Icarus* **143**, 222-243 (2000).

45. Terzieva, R., and Herbst, E., MNRAS **317**, 563-568 (2000).

46. Kerridge, J. F., *AIP Conf. Proc.* **341**, 167-174 (1995).

47. Pepin, R. O., *Icarus* **92**, 2-79 (1991).

48. Owen, T., Bar-Nun, A. , and Kleinfeld, I., *Nature* **358**, 43-46 (1992).

49. Zahnle, K., "Planetary noble gases" in *Protostars and planets III*, edited by E. H. Levy and J. I. Lunine, Tucson: The University of Arizona Press, 1993, pp. 1305-1338.

50. Ozima, M., Wieler, R., Marty B., and Podosek F. A., *Geochim. Cosmochim. Acta* **62**, 301-314 (1998).

51. Busemann, H., Baur, H., and Wieler, R., *Meteorit. Planet. Sci.* **35**, 949-973 (2000).

52. Farley, K. A., and Neroda, E., *Ann. Rev. Earth Sci.* **26**, 189-218 (1998).

53. Caffee, M. W., Hudson, G. B., Velsko, C., Huss, G. R., Alexander Jr., E. C., Chivas, A. R., *Science* **285**, 2115-2118 (1999).

54. Swindle, T. D., *AIP Conf. Proc.* **341**, 175-185 (1995).

55. Marti, K., and Mathew, K. J., *Proc. Indian Acad. Sci. (Earth Planet. Sci.)* **107**, 425-431 (1998).

56. Pepin, R. O., *Space Sci. Rev.* **92**, 371-395 (2000).

57. Busemann, H., and Eugster, O., *Meteorit. Planet. Sci.* **35**, A37 (2000).

58. Zahnle, K., Pollack, J. B., and Kasting, J. K., *Geochim. Cosmochim. Acta* **54**, 2577-2586 (1990).

59. Mahaffy, P. R., Niemann, H. B., Alpert, A., Atreya, S. K., Demick, J., Donahue, T. M., Harpold, D. N., and Owen, T. C., *J. Geophys. Res.* **105**, 15061-15071 (2000).

60. Mahaffy, P. R., pers. comm. to R. Wieler (2001).

61. Eberhardt, P., Geiss, J., Graf, H., Grögler, N., Mendia, M. D., Mörgeli, M., Schwaller, H., Stettler, A., Krähenbühl, U., and v. Gunten, H. R., *Proc. 3rd Lunar Sci. Conf.*, 1821-1856 (1972).

62. Wieler, R., and Baur, H., *Meteoritics* **29**, 570-580 (1994).

63. Zinner, E., *Annu. Rev. Earth Planet. Sci.* **26**, 147-188 (1998).

64. Hoppe, P., and Zinner, E., *J. Geophys. Res.* **105**, 10371-10385 (2000).

65. Hoppe, P., *AIP Conf. Proc.* **this volume** (2001).

66. Hoppe, P., and Ott, U., *AIP Conf. Proc.* **402**, 27-58 (1997).

67. Lequeux, J., *J. Geophys. Res.* **105**, 10249-10255 (2000).

68. Nittler, L. R., Alexander, C. M. O'D., and Wang, J., *Nature* **393**, 222 (1998).

69. Shukolyukov, A., and Lugmair, G. W., *Space Sci. Rev.* **92**, 225-236 (2000).

70. Nyquist, L., Lindstrom, D., Mittlefehldt, D., Shih, C.-Y., Wiesmann, H., Wentworth, S., and Martinez, R., *Meteorit. & Planet. Sci.* **36**, 911-938 (2001).

71. Manuel, O. K., and Hwaung, G., *Meteoritics* **18**, 209-222 (1983).

72. Bochsler, P., Ipavich, F. M., Paquette, J. A., Weygand, J. M., and Wurz, P., *J. Geophys. Res.* **105**, 12659-12666 (2000).

73. Holweger, H., *AIP Conf. Proc.* **this volume** (2001).

74. Gloeckler G., and Geiss, J., *Space Sci. Rev.* **86**, 127-159 (1998).

75. Kallenbach, R., Geiss, J., Gloeckler, G., and von Steiger, R., *Astrophys. Space Sci.* **274**, 97-114 (2000).

76. Geiss, J., Gloeckler, G., Fisk, L. A., and von Steiger, R., *J. Geophys. Res.* **100**, 23373 - 23377 (1995).

77. Gloeckler, G., Fisk, L. A., Geiss, J., Schwadron, N. A., and Zurbuchen, T. H., J. Geophys. Res. **105**, 7459-7463 (2000).

78. Gloeckler, G., and Geiss, J., *AIP Conf. Proc.* **this volume** (2001).

79. Gloeckler, G., Fisk, L. A., Zurbuchen, T. H., Möbius, E., Funsten, H. O., Witte, M., and Roelof, E. C. *EOS, Trans. Am. Geophys. Union* **80**, S237 (1999).

80. Baranov, V. B., *Space Sci. Rev.* **52**, 89-120 (1990).

81. Fahr, H. J., *Space Sci. Rev.* **78**, 199-212 (1996).

82. Gloeckler, G., Fisk, L. A., and J. Geiss, *Nature* **386**, 374-377 (1997).

83. Möbius, E., et al., *Adv. Space Res.* **in press** (2001).

84. Mewaldt, R. A., Liewer, P. C., and the Interstellar Probe Science and Technology Definition Team, *Adv. Space Res.* **in press** (2001).

85. Busemann, H., Baur, H., Wieler, R., *Lunar Planet. Sci. Conf.* **XXXII**, #1598 CD-ROM (2000).

86. Gloeckler G., and Geiss, J., *Space Sci. Rev.* **84** 275-284 (1998).

87. Salerno, E., Bühler, F., Bochsler, P., Busemann, H., Eugster, O., Zastenker, G. N., Agafonov, Y. N., and Eismont, N. A., *AIP Conf. Proc.* **this volume** (2001).

88. Gloeckler, G., and Geiss, J., in *The Light Elements and Their Evolution, IAU Symp.*, edited by L. da Silva, M. Spite, and J. R. de Medeiros, 2000, Vol. 198, pp. 224-233.

89. Klecker, B., *Adv. Space Res.* **23**, 521-530 (1999).

90. Reames, D. V., *Astrophys. J.* **518**, 473-479 (1999).

91. Cummings, A. C., Stone, E. C., and Steenberg, C. D., *Proc. 26th Int. Cosmic Ray Conf.* **7**, 531-534 (1999).

92. Slavin, J. D., and Frisch, P. C., *Astrophy. J.*, **in press** (2001).

93. Leske, R. A., *AIP Conf. Proc.* **516**, 274-282 (2000).

94. Biswas, S., Space Sci. Rev. 75, 423-451 (1996).

95. Mazur, J. E., Mason, G. M., Blake, J. B., Klecker, B., Leske, R. A., Looper, M. D., and Mewaldt, R. A., J. Geophys. Res. 105, 21015-21023 (2000).

96. Möbius, E., Gloeckler, G., Fisk, L. A., and Mewaldt, R. A. "To the boundaries of the heliosphere and beyond" in *The Outer Heliosphere: Beyond the Planets*; edited by K. Scherer, H. Fichtner, E. Marsch, Katlenburg-Lindau, Copernicus Gesellschaft e.V., 2000, pp. 357-393.

97. Liewer, P. C., Mewaldt, R. A., Ayon, J. A., and Wallace, R. A., *AIP Conf. Proc.* **504**, 911 (2000).

98. Engelmann, J. J., Ferrando, P., Soutoul, A., Goret, P., Juliusson, E., Koch-Miramond, L., Lund, N., Masse, P., Peters, B., Petrou, N., and Rasmussen, I. L., *Astron. Astrophys.* **233**, 96-111 (1990).

99. DuVernois, M. A., Simpson, J. A., and Thayer, M. R., *Astron. Astrophys.* **316**, 555-563 (1996).

100. Leske, R. A., *Astrophys. J.* **405**, 567-583 (1993).

101. Garcia-Munoz, M., and Simpson, J. A., *Proc. 16th Int. Cosmic Ray Conf.* **1**, 270-275 (1979).

102. Wiedenbeck, M. E., Binns, W. R., Christian, E. R., Cummings, A. C., Dougherty, B. L., Hink, P. L., Klarmann, J., Leske, R. A., Lijowski, M., Mewaldt, R. A., Stone, E. C., Thayer, M. R., Von Rosenvinge, T. T., and Yanasak, N. E. , *Astrophys. J. Lett.*, 523, L61-L64 (1999).

103. Meneguzzi, M., Audouze, J., and Reeves, H. *Astron. Astrophys.* **15**, 337-359 (1971).

104. Yanasak, N. E., Wiedenbeck, M. E., Mewaldt, R. A., Davis, A. J., Cummings, A. C., George, J. S., Leske, R. A., Stone, E. C., Christian, E. R., von Rosenvinge, T. T., Binns, W. R., Hink, P. L., and Israel, M. H., submitted to *Astrophys. J.* (2001).

105. Connell, J. J., *Astrophys. J. Lett.* **501**, L59-L62, (1998).

106. Yanasak, N. E., Wiedenbeck, M. E., Binns, W. R., Christian, E. R., Cummings, A. C., Davis, A. J., George, J. S., Hink, P. L., Israel, M. H., Leske, R. A., Lijowski, M., Mewaldt, R. A., Stone, E. C., von Rosenvinge, T. T., *Adv. Space Res.* **27**, 727-736 (2001).

107. Strong, A. W., and Moskalenko, I. V., *Adv. Space Res.* **27**, 717-726 (2001).

108. Webber, W. R., and Kish, J., *Proc. 16th Int. Cosmic Ray Conf.* **1**, 389-394 (1979).

109. Ahlen, S. P., Greene, N. R., Loomba, D., Mitchel, J. W., Bower, C. R., Heinz, R. M., Mufson, S. L., Musser, J., Pitts, J. J., Spiczak, G. M., Clem, J., Guzik, T. G., Lijowski, M., Wefel, J. P., McKee, S., Nutter, S., Tomasch, A., Beatty, J. J., Ficenec, D. J., and Tobias, S., *Astrophys. J.* **534**, 757-769 (2000).

110. Hams, T., Barbier, L. M., Bremerich, M., Christian, E. R., de Nolfo, G. A., Geier, S., Goebel, H., Gupta, S. K., Hof, M., Menn, W., Mewaldt, R. A., Mitchell, J. W., Schindler, S. M., Simon, M. and Streitmatter, R. E., to be published in *Proc. 27th Int. Cosmic Ray Conf.* (2001).

111. de Nolfo, G. A., Barbier, L. M., Bremerich, M., Christian, E. R., Davis, A. J., Geier, S., Goebel, H., Gupta, S. K., Hams, T., Hof, M., Menn, W., Mewaldt, R. A., Mitchell, J. W., Schindler, S. M., Simon, M. and Streitmatter, R. E., to be published in *Proc. 27th Int. Cosmic Ray Conf.* (2001).

112. Binns, W. R., Garrard, T. L., Gibner, P. S., Israel, M. H., Kertzman, M. P., Klarmann, J., Newport, B. J., Sotne, E. C., and Waddington, C. J., *Astrophys. J.* **346**, 997-1009 (1989).

113. Westphal, A. J., Price, P. B., Weaver, B. A., and Afanasiev, V. G., *Nature* **396**, 50-52 (1998).

114. Meyer, J.-P., Drury L. O'C., and Ellison, D. C., *Space Sci. Rev.* **86**, 179-201 (1988).

115. Meyer, J.-P., *Astrophys. J. Suppl. Ser.* **57**, 173-204 (1985).

116. Binns, W.R., Adams, J.H., Barbier, L.M., Christian, E.R., Craig, N. , Cummings, A.C., Cummings, J.R., Doke, T., Hasebe, N., Hayashi, T., Israel, M.H., Lee, D., Leske, R.A., Mark, D., Mewaldt, R.A., Mitchell, J.W., Ogura, K., Schindler, S.M., Stone, E.C., Tarlé, G., Tawara, H., Waddington, C.J., Westphal, A.J., Wiedenbeck, M.E., Yasuda, N., to be published in *Proc. 27th Int. Cosmic Ray Conf.* (2001).

117. Higdon, J. C., Lingenfelter, R. E., and Ramaty, R., *Astrophys. J* .**509**, L33-L36 (1988).

118. Clayton, D. D., *Astrophys. J.* **334**, 191-195 (1988).

119. Timmes, F. X., and Clayton, D. D., *Astrophys. J.* **472**, 723-741 (1996).

120. Lugaro, M., Zinner, E., Gallino, R., and Amari, S., *Astrophys. J.* **527**, 369-394 (1999).

121. Romano D., Matteucci, F., Molaro, P., and Bonifacio, P., *Astron. Astrophys.* **352**, 117-128 (1999).

122. Bonifacio, P., and Molaro, P., *MNRAS* **285**, 847-861 (1997).

123. Spite, F., and Spite, M., *Astron. Astrophys.* **115**, 357-366 (1982).

124. Romano, D., Matteucci, F., Ventura, P., D'Antona, F., *Astron. Astrophys.* **374**, 646-655 (2001).

125. Pagel, B. E. J., *Nucleosynthesis and chemical evolution of galaxies*, Cambridge: Cambridge University Press, 1997.

126. Ryan, S. G., Beers T. C., Olive K. A., Fields, B. D., and Norris, J. E. *Astrophys. J.* **530**, L57-L60 (2000).

127. Burles, S., Nollett, K. M., and Turner, M. S., *Astrophys. J.* **552**, L1-L5 (2001).

128. Chiappini, C., Matteucci, F., and Gratton, R., *Astrophys. J.* **477**, 765-780 (1997).

129. Chiappini, C., and Matteucci, F., *AIP Conf. Proc.* **this volume** (2001).

130. Chiappini, C., Matteucci, F., and Romano, D., *Astrophys. J.* **554**, 1044-1058 (2001).

131. Charbonnel, C., and do Nascimento Jr., J. D., *Astron. Astrophys.* **336**, 915-919 (1998).

132. Tytler, D.,O'Meara, J. M., Suzuki, N., and Lubin, D., *Physica Scripta* **T85**, 12-31 (2000).

133. O'Meara, J. M., Tytler, D., Kirkman, D., Suzuki, N., Prochaska, J. X., Lubin, D., and Wolfe, A. M., *Astrophys. J.* **552**, 718-730 (2001).

134. Linsky, J. L., *Space Sci. Rev.* **84**, 285-296 (1998).

135. Geiss, J., and Gloeckler, G., *Space Sci. Rev.* **84**, 239-250 (1998).

136. Sonneborn, G., Tripp, G. M., Ferlet, R., Jenkins, E. B., Sofia, U. J., Vidal-Madjar, A., Wozniak, R., *Astrophys. J.* **545** 277-289 (2000).

137. Tosi, M., Steigman, G., Matteucci, F., and Chiappini, C., *Astrophys. J.* **498**, 226-235 (1998).

138. Chiappini, C., and Matteucci, F., in *The Light Elements and Their Evolution, IAU Symp.*, edited by L. da Silva, M. Spite, and J. R. de Medeiros (2000), Vol. 198, pp. 540-546.

139. Bania, T. M., Rood, R. T., and Balser, D. S., in *The Light Elements and Their Evolution, IAU Symp.*, edited by L. da Silva, M. Spite, and J. R. de Medeiros, 2000, Vol. 198, pp. 214-223.

140. Viegas, S. M., Gruenwald, R., and Steigman, G., *Astrophys. J.* **531**, 813-819 (2000).

141. Maciel, W. J., in *The Chemical Evolution of the Milky Way*, edited by F. Matteucci and F. Giovanelli, Dordrecht: Kluwer, 2001, pp. 81-92.

142. Pagel, B. E. J., Simonson, E. A., Terlevich, R. J., and Edmunds, M. G., *MNRAS* **255**, 325-345 (1992).

143. Izotov, Y., and Thuan, T. X., *Astrophys. J.* **500**, 188-216 (1998).

144. Kirilova, D., *AIP Conf. Proc.* **this volume** (2001).

145. Kirilova D., and Chizhov M., *Nucl. Phys. B (Proc. Suppl.)* **100**, 360-362 (2001).

146. Tosi, M., "The chemical evolution of D and ^{3}He in the Galaxy in connection with CNO elements" in *The Chemical Evolution of the Milky Way*, edited by F. Matteucci and F. Giovanelli, Dordrecht: Kluwer, 2001, pp. 505-515.

147. Charbonnel, C., Brown, J. A., and Wallerstein, G., *Astron. Astrophys.*. **332**, 204-214 (1998).

148. Boothroyd, A. I., and Sackmann, I.-J., *Astrophys. J.* **510**, 232-250 (1999).

149. Ryan, S. G., Kajino, T., Beers, T. C., Suzuki, T. K., Romano, D., Matteucchi, F., Rosolankova K., *Astrophys. J.* **549**, 55-71 (2001).

Sun, solar wind, meteorites and interstellar medium: What are the compositional relations?

Peter Bochsler*, R. F. Wimmer-Schweingruber* and Peter Wurz*

*Physikalisches Institut, University of Bern, Sidlerstrasse 5, CH-3012 Bern, Switzerland

Abstract. Atomic properties of elements determine their chemical behavior, their ionization properties and their interaction with radiation in the solar atmosphere. Whereas the chemical properties influence the condensation process in the primordial solar nebula and, hence, meteoritic abundances, ionization properties seem to provide the most important ordering parameters for producing coronal -, solar wind -, and solar energetic particle abundances from the solar reservoir. Finally, since atomic properties also shape the interaction of solar matter with radiation, understanding these properties determines largely the experimental reliability of photospheric chemical abundances. Isotopic abundances must be derived from *nuclear* properties, which are almost insensitive to atomic processes. The solar spectrum is the result of *atomic* processes in the solar atmosphere. The derivation of isotopic abundances from the solar spectrum is impossible for most species, conversely, the insensitivity to chemical processes makes isotopes the first choice to trace the nucleosynthetic history and the degree of mixing of galactic matter from different astrophysical sources prior to formation of the solar system. The solar wind provides a representative sample of solar isotopes and - to some degree - also a rather trustworthy representation of elements with similar atomic properties, especially volatiles, which are difficult to derive from meteoritic abundances and from optical observations of the solar spectrum.

INTRODUCTION

ACE and SOHO, together with several other space missions have greatly enhanced our current knowledge of the chemical composition of the galaxy and of the Sun. Cosmic rays and heliospheric neutrals and pick-up ions provide information on the present-day local interstellar medium whereas the Sun represents a benchmark for the chemical evolution of the interstellar medium as it existed 4.6 Gy. ago at about 10 kpc from the galactic center. The purpose of this paper is to relate the compositional features between Sun, the solar wind, meteorites and the galactic interstellar medium, and to review mechanisms which could produce compositional variations. The solar wind has the potential of providing the most reliable information on the isotopic composition of the Sun. Heliospheric pick-up ions and anomalous cosmic-ray particles give direct evidence of the composition of the contemporary local interstellar medium surrounding the solar system.

The four topics, Sun, solar wind, meteorites and interstellar medium have a common relation originating in the primordial solar nebula which formed out of the local interstellar medium 4.6 Gy ago. The picture by Shu et al. [1] illustrates the action of the primordial solar wind, redistributing disk material between the accreting central body, the newborn Sun, and the planetary disk, thereby

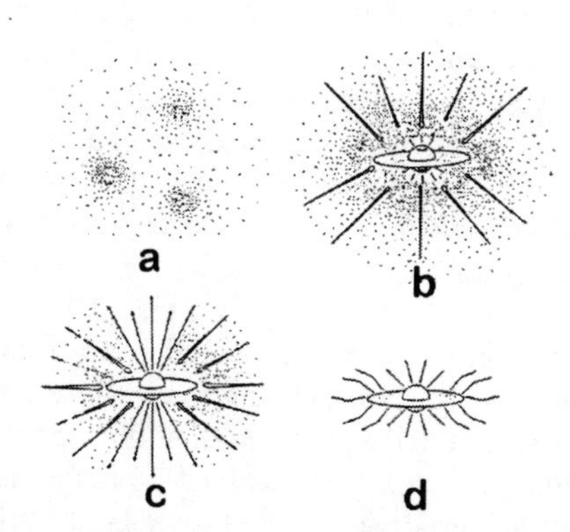

FIGURE 1. Four phases of the evolution of the protosun and protosolar disk according to Shu, Adams and Lizano [1]. Originating from a dense molecular cloud (a) the rapidly rotating Sun emerges (b) together with a circumsolar disk. Differential rotation of the Sun produces strong magnetic fields, which generate a vigorous polar outflow (c) while accretion continues in the equatorial plane. The outflow cone widens and (d) finally, the young, active Sun remains with a thin disk. Reprinted with permission from the Annual Review of Astronomy & Astrophysics, Volume 25 ©1987 by Annual Reviews www.AnnualReviews.org

CP598, *Solar and Galactic Composition,* edited by R. F. Wimmer-Schweingruber

fractionating gas and dust from which planetary bodies (including meteoritic parent bodies and comets) began to form. Accretion of the Sun from the protosolar nebula leads to its rapid spin-up. The increasing internal differential rotation of the Sun induces strong polar magnetic fields and a strong polar wind which prohibits further accretion in polar regions while infall of matter at low solar latitudes continues. A fraction of this infalling material, however, does not reach the Sun. Before it reaches the Alfvén point it encounters outflowing solar matter with which it is ejected into higher solar latitudes and falls back to the disk at some remote place. Hence, a fraction of the solar disk is recycled through this process of continuous infall and ejection. As the protosolar system evolves, the wind which originates above the polar regions, increases its aperture cone; magnetic field lines open just as the spokes of an opening umbrella. With each cycle in this turnover process, some material is lost to space and some matter drops to the solar surface. Matter falling back onto the planetary disk will be chemically fractionated; the degree of fractionation will depend on the cycling frequency, i.e., on the cycling time and on the travel distance. The fractionating process is most efficient for frequently cycled matter residing in the innermost parts of the solar system, whereas matter carried over greater distances remains largely unaltered. Hence, mass fractionation in the gas phase is expected to slightly enhance heavier species in the innermost parts of the solar system, whereas the outermost regions and the Sun itself retain their original composition. Experimental evidence about this process is scarce, and from theoretical models it is difficult to estimate even an order of magnitude of its efficiency. In view of the fact that large amounts of disk material have to be recycled within a rather short time, big effects are unlikely to occur.

Carbonaceous meteorites of the class CI are generally assumed to reflect solar composition most closely [2]. Condensation of refractory and moderately volatile elements is considered to be a non-equilibrium process with no isotopic fractionation involved. Volatile elements in planetary materials show generally a large variability in isotopic composition. Part of this variability is ascribed to isotopic fractionation, part must be due to variable mixing of materials from different sources, e.g. grain-gas mixture. In view of the conclusions from the above introductory section, it appears quite reasonable, that meteorites and cometary bodies, which condense farthest out in the solar system, and the Sun, which retained a representative sample of the primordial solar nebula can provide the samples least isotopically fractionated in the solar system. Unfortunately, direct measurements of the solar isotopic composition are impossible, except for a few elements, which form molecules at high temperatures.

FRACTIONATION IN THE PRIMORDIAL SOLAR NEBULA

From the scenario outlined in the introduction some speculations about the large-scale chemical and isotopic evolution of the solar system can be developed. Whereas it appears possible to produce gradients of chemical composition by circulation and redistribution of matter within the accretion disk, isotope effects are probably tiny, i.e., it is no surprise that refractory elements show almost no signs of isotopic fractionation throughout the solar system. In order to noticeably fractionate matter of the solar disk, a significant part of it has to be circulated within the lifetime of the disk. Consider a disk containing one percent of the solar mass. Circulating such an amount of mass during several 10 million years requires typical mass fluences of 10^{13} kg/s. With a typical flow speed of 100 km/s this process will consume only about the equivalent of one permill of the solar luminosity. We further assume that the X-point [1] of the protosolar wind is located at about 100 $R_\odot$. To maintain mass flux and speed as given above, a maximum density on the order of 10^{12} particles per m^3 at the X-point is required, much higher than densities observed at a similar location in the contemporary solar wind. As a consequence, collision frequencies between the dominant species (hydrogen) and minor species are exceedingly large and isotopic effects are small in such an intense wind. A weak mass fractionation in such a bipolar wind can occur through the difference of forces experienced by test particles of different masses. Heavier particles have a tendency to stay behind lighter particles in a drag exerted by a wind of light field particles. Below the critical point, some heavy particles will fall back to the Sun, while their lighter counterparts will be blown out to the solar disk. The effects are tiny and can be estimated from a simple consideration of momentum transfer in a series of elastic collisions from the coronal base at heliocentric distance R to the critical point at r_o with the following simplified expression:

$$f_{ij} = \frac{w_i}{w_j} = \frac{\left[1 - \left(\frac{M_i - m}{M_i + m}\right)^{N(r_o/R)(1-R/r)}\right]}{\left[1 - \left(\frac{M_j - m}{M_j + m}\right)^{N(r_o/R)(1-R/r)}\right]} \quad (1)$$

In this equation M_i and M_j indicate the masses of two different test particles, m is the mass of the field particles, and $N = n_o \sigma r_o$ a typical collision number on the way of the particles to the critical point, which is of order 3000.

Such a mechanism might naturally lead to the fractionation gradients, e.g. for the $^{52}Cr/^{53}Cr$ isotopic abundance ratio as discussed in the article by Hoppe [3] (see [4] for the original Cr data). Matter launched by a bipolar wind into the disk will generally be depleted in heavier

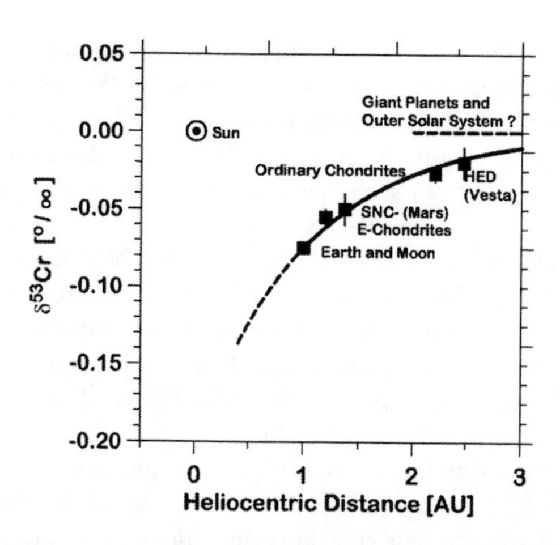

FIGURE 2. Chromium isotope systematics due to fractionated redistribution of matter in the primordial solar nebula

isotopes, whereas the Sun remains at its original composition due to the enormous size of the reservoir. Note, that at this stage the entire Sun is chemically and isotopically homogenized, since the rapidly rotating star is thoroughly mixed through a convective zone which reaches deep into the stellar core. Due to the circulation with the X-wind, the disk tends to undergo a tiny enrichment with light isotopes. The innermost parts of the protoplanetary disk will obtain the strongest enrichment of light isotopes due to the most thorough recycling with short characteristic transfer times, the outer parts will be much less frequently recycled; hence, isotopic effects will level out and the isotopic composition of matter will attain solar values in regions where recycling becomes negligible. Note, however, that these effects should not exceed small fractions of permills, even in the innermost parts of the planetary disk, due to the large densities and the strong coupling of the heavier elements to the bipolar outflow. Furthermore, this mechanism can only work as long as also refractory elements such as chromium are evaporated at the moment of the closest approach to the Sun. Furthermore, although recycling and sorting of dust grains might work in a similar manner in the protosolar disk, it seems highly unlikely that the complicated isotopic pattern of a partly rock-forming and partly volatile element such as oxygen could be explained with this simple idea.

ISOTOPIC FRACTIONATION IN THE CONTEMPORARY SOLAR WIND

It has been emphasized above that recycling of matter through the protoplanetary disk will hardly affect the bulk composition of the Sun. Hence, the Sun is generally considered to represent the most faithful sample of the protosolar nebula with respect to elemental and isotopic composition. Studying the isotopic composition of the solar wind is most interesting in this context, because the solar wind provides the least biased, accessible sample of isotopic solar composition. Nevertheless, it is important to note that the outer convective zone of the Sun has probably undergone some alteration in its elemental and isotopic composition due to the effect of gravitational settling across the boundary between the radiative core and the convective outer zone [5]. The effect amounts to typically a fraction of a percent per atomic mass unit [6].

Apart from fractionation in the source of the solar wind, i.e., matter from the solar atmosphere, which is in constant exchange with matter from the outer convective zone, some isotopic fractionation is also expected in the solar wind feeding and acceleration process. It has been known for a long time that the solar wind undergoes strong variations in its elemental composition, notably its He/H elemental ratio. The issue on the origin of these variations has not been conclusively settled, however, many theories which make an attempt to explain these variations involve the effect of inefficient Coulomb drag in the solar wind acceleration domain (e.g. [7]). The almost permanent depletion of helium relative to hydrogen in the solar wind seems to be a natural consequence of the weak Coulomb drag in the emanating proton flow, because the species $^4He^{++}$ has an exceptionally low drag factor. The three-fluid-models of Bürgi [8], making use of the superradial expansion in magnetic structures, which feed the low-speed solar wind, and which are located at the acceleration site, give a plausible explanation for the sometimes strong depletion of helium near the interplanetary current-sheet as demonstrated by Borrini et al. [9]. Wimmer-Schweingruber [10] found a significant depletion of helium relative to oxygen near sector-boundaries, underlining the strong discrimination of helium due to its weak Coulomb-coupling to hydrogen in coronal regions of strongly diverging magnetic fields. Similarly, Noci *et al.* [11] argued that the sometimes strong enrichment of oxygen relative to hydrogen as observed with SOHO/UVCS at the footpoints of streamers is the consequence of the superradial divergence of flow tubes above those streamers. That collisions play an important role in slow solar wind regimes, overruling the action of waves on particle distributions, has amply been demonstrated under many circumstances [12, 13], most recently by Hefti et al. [14]. In a similar manner Coulomb-collisions could also determine isotopic fractionation effects in the solar wind. Following these lines, Bodmer and Bochsler [15], have investigated the effects of flow geometry in the solar corona and the strength of Coulomb friction on isotopic and elemental fractionation. In their model they found that a significant depletion of helium is always accompanied by a solar wind depletion of the heavier isotopes of the order of a percent per atomic mass unit. According to the predictions of the model of Bodmer and Bochsler, the expected fractionation effect is weaker in the high-speed, coronal-hole associated solar wind. Waves play a much more important role driving minor species in this types of regimes. However, also in this case some Q/M-fractionation will occur. Preliminary estimates indicate that such effects will generally be of the order of a fraction of a percent per mass unit [6]. In fact, the first investigations of the isotopic composition of the refractory element magnesium, which shows only tiny variations in its isotopic composition among different planetary samples, revealed no difference between solar wind and terrestrial Mg [16]. Kallenbach and coworkers [17, 18] have carefully analyzed possible trends of the isotopic ratios $^{24}Mg/^{26}Mg$ and $^{24}Mg/^{25}Mg$ with the solar wind speed and the He/H abundance ratio. They found that the fastest solar wind streams contained magnesium which was depleted by typically 1 percent per mass unit and that there was a general trend of decreasing mass fractionation with increasing wind speed and increasing He/H ratio. In both cases the theoretical predictions are roughly consistent with the observations. However it should be noted that the experimental uncertainties are still substantial. A careful re-analysis of the data collected during five years of operations and a recalibration of the spare sensor of MTOF/CELIAS with terrestrial Mg is pending.

Apart from the fact that helium ions have the least favorable Coulomb drag factors for incorporation into the solar wind, helium is also the element with the highest first ionization potential ("FIP"), and the longest ionization times. The chromosphere looks rather homogeneous on a global scale; chromospheric properties under coronal holes do not differ from those under active regions. The latter are, however, exposed to higher EUV radiation intensities and to radiation spectra which differ from those under the open magnetic field regions. This, in turn, may have some influence on the solar wind feeding process in different chromospheric regimes. If indeed the systematic helium depletion in the solar wind has (partly) to be attributed to the radiation field and not to the inefficient Coulomb drag, the isotopic effects predicted to occur in correlation with the He/H abundance ratio, as discussed above, should rather be taken as upper limits.

ELEMENTAL FRACTIONATION IN THE SOLAR WIND

The elemental composition of the solar wind varies and differs at times significantly from the photospheric composition. It is well known that elements with ionization potentials above 10eV (e.g. He, Ne, Ar, O) are generally depleted compared to the low-FIP elements. Taking oxygen as a reference, this depletion amounts to typically a factor of 3 to 5 in low-speed solar wind whereas it is typically less than a factor of 2 in coronal-hole-associated high speed wind. The same phenomenon is observed in solar energetic particle abundances (see article by Reames [19]) and sometimes much more so in optical observations of active regions. No generally accepted explanation for this effect has been found yet. It is also not understood why the optical observations of Mg/O or Ne/O abundance ratios provide values which vary sometimes by factors of 10 [20], while the particle observations never have shown a clear case with a FIP-fractionation factor of an order of magnitude. We suspect that the solar wind is rather representative for the coronal base whereas optical observations must suffer from strong biases due to large differences in intensity of the different radiating elements of solar surface structures. Even if rather variable abundance ratios exist in different domains of the chromosphere, we believe that the solar wind is fed from much larger reservoirs, to be visualized as a tenuous haze spread above the turbulent dynamic and structured chromosphere.

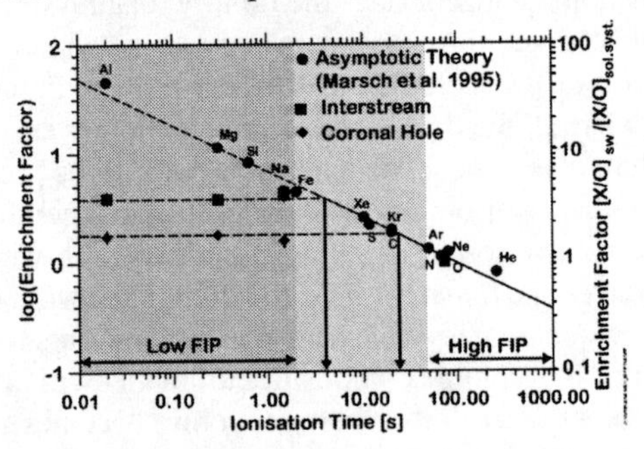

FIGURE 3. Abundances of low-FIP elements in the solar wind. The full line illustrates the result of the simple chromospheric diffusion model of Marsch et al.(1995) [21] which works for high- and intermediate-FIP elements. The preliminary investigations of Na and Al abundances of Ipavich et al. [22] and Bochsler et al. [23] give no evidence of fractionation among low-FIP elements.

Concerning the reliability of the witness samples of the solar atmosphere provided by the solar wind, the important question is whether the feeding process discriminates individually among the low-FIP elements or whether all low-FIP elements are incorporated without discrimination among each other into the solar wind. Of course, it would be most beneficial for abundance studies if the solar wind left low-FIP elements undiscriminated at all times. The next question then is whether the high-FIP elements left behind remain largely undiscriminated among themselves in the solar wind as well. The presently available experimental evidence is limited to results of a few preliminary studies: Aluminum and sodium abundances, two elements which have rather reliably determined photospheric and meteoritic abundances have been measured and compared with the abundance of magnesium in two selected periods of different solar wind regimes [23, 22]. The preliminary result looks promising: The analysis showed no fractionation among the three low-FIP elements Na, Al, and Mg in both types of solar wind although the regular fractionation factors between high/low-FIP elements have verified.

CONTEMPORARY INTERSTELLAR MATTER IN THE INNER HELIOSPHERE

Although the umbilical cord, which connected the interstellar medium and the solar system, has been cut long ago, the heliosphere is still pervaded by contemporary interstellar matter in the form of neutral atomic, molecular and dust particles. The interaction of solar radiation and the solar wind with gas and dust ionizes the gas and evaporates grains in the innermost parts of the heliosphere. Newly ionized interstellar particles are continuously swept to the boundaries of the heliosphere by means of the outward-moving interplanetary magnetic field, i.e., by the so-called pick-up process. Although particles of interstellar origin contribute the far largest fraction of diffuse matter within the boundaries of the heliosphere, their contribution to the solar and planetary reservoirs is minor. They can be distinguished from solar particles due to their different charge state distributions, their isotopic pattern and from their somewhat different energy distributions. Some indications for their presence in the record of the lunar soil and the possible relevance of this record for past encounters of the solar system with dense molecular clouds have been discussed [24, 25]. Pick-up ions from interstellar gas have been identified as the source of the anomalous cosmic rays [26]. Hence, anomalous cosmic rays provide another indirect source of information of the composition of the interstellar medium surrounding the heliosphere.

The inference of elemental abundances in the interstellar gas depends on the so-called "filtration factors". The properties of the interface between the heliosphere and the local interstellar cloud are determined by the interaction of the temporally variable solar wind with the magnetized, structured, partially condensed, and partially ionized interstellar medium. The probability (filtration factor) of a given species to enter the inner heliosphere through this interface depends crucially on the physical properties of this interface. As long as the state of ionization in the local interstellar cloud and the orientation and strength of the magnetic field in the local cloud are not well defined, a straightforward derivation of elemental filtration factors is problematic. Fahr [27], and Geiss and Witte [28] give a detailed account of the difficulties related to the inference of the abundance of volatile species in the local interstellar medium. Obviously, these difficulties become even more significant as soon as moderately volatile and refractory elements are included in such considerations. On the other hand, inferences about *isotopic* abundances of volatile elements are quite straightforward.

The $^3He/^4He$-isotopic ratio in the local interstellar cloud has been inferred from pick-up ion observations with Ulysses/SWICS [29]. More recently, this ratio has also been successfully been determined directly with the foil collection technique by means of the COLLISA experiment on the Mir space station [30]. Isotopic abundances of other elements can also be inferred in a rather straightforward manner from the Anomalous Component of cosmic rays (see, e.g. [31]).

CONCLUSIONS

Matter involved in almost any astrophysical process will undergo some alteration of its elemental and isotopic composition. This leads to the rich variety of compositional features observed in the galaxy, in the solar system, in meteorites, down to the smallest scales including the tiniest grains or macromolecules that only comprise hundred atoms or less. Cosmochemical and isotopic investigations always deal with a double-faced nature: On one hand one finds interest in studying the physical processes which lead to the compositional variety in order to learn about the history of an object, on the other hand one considers compositional signatures as fixed and uses them as tracers to track the flow and distribution of matter from cosmological scales into galaxies, and stars. Both approaches have been equally successful as has amply been demonstrated in the course of this workshop.

ACKNOWLEDGEMENTS

The authors gratefully acknowledge helpful comments by James Whitby. This work was supported by the Swiss National Science Foundation.

REFERENCES

1. Shu, F. H., Adams, F. C., and Lizano, S., *Ann. Rev. Astron. Astrophys.*, **25**, 23–81 (1987).
2. Anders, E., and Grevesse, N., *Geochim. Cosmochim. Acta*, **53**, 197–214 (1989).
3. Hoppe, P., *this volume* (2001).
4. Shukolyukov, A., and Lugmair, G., *Space Sci. Rev.*, **92**, 225–236 (2000).

5. Turcotte, S., Richer, J., Michaud, G., Iglesias, C., and Rogers, F., *Astrophys. J.*, **504**, 539–558 (1998).
6. Bochsler, P., *Rev. Geophys.*, **38**, 247–266 (2000).
7. Geiss, J., Hirt, P., and Leutwyler, H., *Solar Physics*, **12**, 458–483 (1970).
8. Bürgi, A., "Dynamics of alpha particles in coronal streamer type geometries", in *Solar Wind Seven. Proceedings of the 3rd COSPAR Colloquium held in Goslar, Germany 1991*, edited by E. Marsch and R. Schwenn, 1992, pp. 333–336.
9. Borrini, G., Gosling, J. T., Bame, S. J., Feldman, W. C., and Wilcox, J. M., *J. Geophys. Res.*, **86**, 4565–4573 (1981).
10. Wimmer-Schweingruber, R. F., *PhD. Thesis, University of Bern, Switzerland* (1994).
11. Noci, G., Kohl, J. L., Antonucci, E., Tondello, G., Huber, M. C., Fineschi, S., Gardner, L. D., Korendyke, C., Nicholosi, P., Romoli, M., Spadaro, D., Maccari, L., Raymond, J. C., Siegmund, O., Benna, C., Ciaravella, A., Giordano, S., Michels, J., Modigliani, A., Naletto, G., Panasyuk, A., Pernechele, C., Poletto, G., Smith, P. L., and Strachan, L., *In: The Corona and Solar Wind Near Minimum Activity, Proc. Fifth SOHO Workshop ESA SP-404*, pp. 75–84 (1997).
12. Marsch, E., Mühlhäuser, K.-H., Rosenbauer, H., Schwenn, R., and Neubauer, F. M., *J. Geophys. Res.*, **87**, 35–51 (1982).
13. Bochsler, P., Geiss, J., and Joos, R., *J. Geophys. Res.*, **90**, 10779–10789 (1985).
14. Hefti, S., Grünwaldt, H., Ipavich, F. M., Bochsler, P., Hovestadt, D., Aellig, M. R., Hilchenbach, M., Kallenbach, R., Galvin, A. B., Geiss, J., Gliem, F., Gloeckler, G., Klecker, B., Marsch, E., Möbius, E., Neugebauer, M., and Wurz, P., *J. Geophys. Res.*, **103**, 29697–29704 (1998).
15. Bodmer, R., and Bochsler, P., *J. Geophys. Res.*, **105**, 47–60 (2000).
16. Bochsler, P., Gonin, M., Sheldon, R. B., Zurbuchen, T., Gloeckler, G., Hamilton, D. C., Collier, M. R., and Hovestadt, D., "Abundance of solar wind magnesium isotopes determined with WIND/MASS", in *Solar Wind Eight. Proceedings of the Eighth International Solar Wind Conference*, edited by D. Winterhalter, J. Gosling, S. Habbal, W. Kurth, and M. Neugebauer, American Institute of Physics, Woodbury, N.Y., 1996, vol. 382, pp. 199–202.
17. Kallenbach, R., Ipavich, F. M., Kucharek, H., Bochsler, P., Galvin, A. B., Geiss, J., Gliem, F., Gloeckler, G., Grünwaldt, H., Hefti, S., Hilchenbach, M., and Hovestadt, D., *Space Sci. Rev.*, **85**, 357–370 (1998).
18. Kucharek, H., Ipavich, F. M., Kallenbach, R., Bochsler, P., Hovestadt, D., Grünwaldt, H., Hilchenbach, M., Axford, W. I., Balsiger, H., Bürgi, A., Coplan, M. A., Galvin, A. B., Geiss, J., Gliem, F., Gloeckler, G., Hsieh, K., Klecker, B., Lee, M. A., Livi, S., Managadze, G. G., Marsch, E., Möbius, E., Neugebauer, M., Reiche, K.-U., Scholer, M., Verigin, M. I., Wilken, B., and Wurz, P., *J. Geophys. Res.*, **103**, 26805–26812 (1998).
19. Reames, D. V., *this volume* (2001).
20. Feldman, U., *Physica Scripta*, **46**, 202–220 (1992).
21. Marsch, E., von Steiger, R., and Bochsler, P., *Astron. Astrophys.*, **301**, 261–276 (1995).
22. Ipavich, F. M., Bochsler, P., Lasley, S. E., Paquette, J. A., and Wurz, P., *EOS*, **80**, 256 (1999).
23. Bochsler, P., Ipavich, F. M., Paquette, J. A., Weygand, J. M., and Wurz, P., *J. Geophys. Res.*, **105**, 12659–12666 (2000).
24. Wimmer-Schweingruber, R. F., and Bochsler, P., *In: Acceleration and transport of energetic particles observed in the heliosphere. ACE-2000 Symposium (R.A. Mewaldt, J.R. Jokipii, M.A. Lee, E. Möbius and Th. H. Zurbuchen eds.) AIP Conference Proceedings 528*, pp. 270–273 (2000).
25. Wimmer-Schweingruber, R. F., and Bochsler, P., *this volume* (2001).
26. Fisk, L. A., Kozlowski, B., and Ramaty, R., *Astrophys. J.*, **190**, L35–L38 (1974).
27. Fahr, H. J., *Space Sci. Rev.*, **78**, 199–212 (1996).
28. Geiss, J., and Witte, M., *Space Sci. Rev.*, **78**, 229–238 (1996).
29. Gloeckler, G., *Space Sci. Rev.*, **78**, 335–346 (1996).
30. Salerno, E., Bühler, F., Bochsler, P., Busemann, H., Eugster, O., Zastenker, G. N., Agafonov, Y. N., and Eismont, N. A., *this volume* (2001).
31. Leske, R. A., Mewaldt, R. A., Christian, E. R., Cohen, C. M., Cummings, A. C., Slocum, P., Stone, E. C., von Rosenvinge, T. T., and Wiedenbeck, M. E., *In: Acceleration and transport of energetic particles observed in the heliosphere. ACE-2000 Symposium (R.A. Mewaldt, J.R. Jokipii, M.A. Lee, E. Möbius and Th. H. Zurbuchen eds.) AIP Conference Proceedings 528*, pp. 293–300 (2000).

Solar Krypton and Xenon in gas-rich Meteorites: New Insights into a Unique Archive of Solar Wind

Veronika S. Heber, Heinrich Baur and Rainer Wieler

Institute for Isotope Geology and Mineral Resources, ETH Zürich, CH-8092 Zürich, Switzerland, heber@erdw.ethz.ch

Abstract. We present elemental and isotopic ratios of Ar, Kr, and Xe implanted by the solar wind in different gas-rich regolithic meteorites. The noble gases were released by closed system stepwise etching with HF, which allows us to separate the solar from the primordial noble gases, at least in the first steps. The $^{36}Ar/^{84}Kr$ ratios of Noblesville and Fayetteville lie in the range of unfractionated bulk solar values. The rather constant $^{36}Ar/^{84}Kr$ ratio in Pesyanoe is about 40-50% lower than bulk solar values and in the range of solar wind values measured in lunar samples. In contrast, the $^{84}Kr/^{132}Xe$ ratio of all measured meteorites as well as all lunar samples is several times lower than the inferred bulk solar values. This Xe enhancement in the solar wind appears to be a result of the "FIP effect" or related processes in the solar wind source region, which leads to an enrichment of elements with a low first ionisation potential. The meteorite data presented here show that the Xe and Kr abundances in the solar wind in the past were more variable than indicated by lunar samples alone. The "FIP-effect" may have existed since 4 Ga ago or even earlier.

INTRODUCTION

Kr and Xe in the sun are difficult to analyse. So far the only direct information is from solar wind (SW) implanted in the recent or ancient past in dust on the surfaces of the moon and asteroids. Analyses in artificial targets or in situ measurements by spacecraft have not been possible yet, due to the low Kr- and Xe-concentrations. Photospheric abundance determinations of noble gases by spectroscopy are not feasible at all.

Relative Ar, Kr, and Xe abundances in the solar wind have been deduced from lunar samples and are variable with time on a 100–1000Ma scale (for review see [1]). These variations indicate a time-dependent element fractionation of the heavy noble gases in the solar wind source region. Besides lunar samples, "gas-rich" meteorites from surface dust on asteroids also present an archive of ancient solar wind. It is the goal of this work to enlarge the very scarce data base on solar Kr and Xe in meteorites [2,3,4], in order to compare their relative abundances with those in lunar samples.

Solar noble gases in lunar samples have been analysed by different noble gas release techniques, in particular by stepwise combustion and pyrolysis [5,6,7] and closed system stepwise etching (CSSE) [8,9,10]. CSSE on mineral separates and single-grain total fusion analyses showed that Ar, Kr, and Xe from the solar wind are retained unfractionated in lunar soils [9, 11]. In samples containing solar wind of variable ages, Xe, and to a lesser extent Kr, are enhanced relative to Ar and inferred solar abundances [12]. It is well known that elements with a first ionisation potential (FIP) <10eV are enriched in the solar wind [13], and the Xe enrichment factor of ~4 [9] recorded in lunar samples is similar to that of low FIP elements in the "slow" solar wind [13], although Xe has a FIP of 12.4eV.

Solar noble gas-rich meteorites are an attractive complement to lunar samples mainly for two reasons. First, it is often assumed that many of them acquired their solar wind earlier than most or all available lunar samples, perhaps more than 4 Ga ago. Second, asteroidal regoliths have much lower solar noble gas concentrations compared to lunar samples, due to larger distances from the sun and shorter residence times of grains near the parent body surface. Therefore noble gas saturation or other alteration effects should be minor. On the other hand, the analysis of the solar component of the heavy noble gases in meteorites is not easy. The major problem is that - unlike lunar samples - almost all meteorites contain primordial Kr and Xe and our main task here will be to find a way to separate the solar component from the primordial noble

CP598, *Solar and Galactic Composition,* edited by R. F. Wimmer-Schweingruber

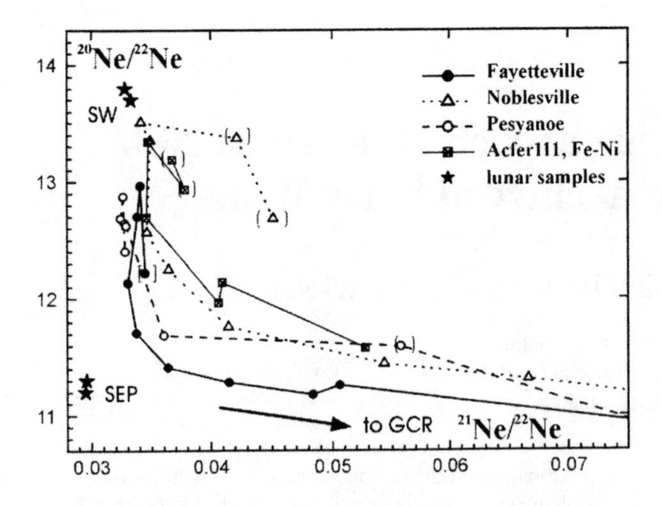

FIGURE 1. Ne-data of 4 solar-gas-rich meteorites. The gases were released by CSSE. 1σ error bars are smaller than symbol size, except for data in round and square brackets, which have errors of 1-4% and 10%, respectively. Solar wind (SW) and solar energetic particle (SEP) components obtained from lunar samples [25] are given for comparison. The arrow points to the galactic cosmic ray spallation component (GCR). The first steps of each meteorite have a composition comparable to, but somewhat lower than, pure SW-Ne. With continuous etching, the trapped Ne composition tends towards SEP but increasing "cosmogenic" contributions lead to the deviation towards GCR.

gases. These two components can hardly be separated by stepwise heating. A promising approach is CSSE by hydrofluoric acid, because the carbonaceous phases containing most of the primordial components should be much more resistant to HF than the solar-gas-bearing silicates. In addition, CSSE avoids noble gas fractionation due to thermal diffusion because the gas is released by destroying the carrier phase at room-temperature. Ideally, stepwise dissolution will provide a high-resolution depth profile of elemental and isotopic composition of implanted gases, which is crucial to resolve the different noble gas components.

SAMPLES AND EXPERIMENTAL PROCEDURE

Samples: We selected three H-chondritic regolithic breccias (Fayetteville, Noblesville, Acfer111) and the achondritic breccia Pesyanoe. The latter appeared to contain almost no primordial Xe [2]. Fayetteville and Noblesville are very solar-gas-rich. Acfer111 was selected because it contains unfractionated light solar noble gases [4]. The dark, solar-wind bearing matrix of all samples was gently crushed, sieved and washed in ethanol. Bulk silicate grain-size separates (without Fe-Ni-particles) in the size range of 64-125μm (approx. 400mg) were prepared for all samples. In addition an Fe-Ni separate of Acfer111 was also analysed. Prior to etching all bulk samples were preheated in vacuo for 10-20h at 100°C to remove adsorbed air. The Acfer111 Fe-Ni sample was extensively preheated (25d at 150°C).

Experimental procedure: Noble gases were released by CSSE in our Au-Pt-line directly connected to a mass-spectrometer [14]. All samples were etched by the vapour phase of concentrated hydrofluoric acid (37%). Only for Fayetteville the acid was distilled onto the sample towards the end of the run. HF background was monitored via mass 19, but no correction on the ^{20}Ne signal was ever required. Etch parameters for the individual steps are listed in [15]. For mass-spectrometric techniques and data reduction see [16]. Isotopic ratios of Kr and Xe are calibrated relative to the atmospheric composition given by [5]. All etch steps were corrected for the procedural blank of the extraction line (in 10^{-10} ccSTP: ^{4}He: 4; ^{20}Ne: 0.03; ^{40}Ar: 5; ^{84}Kr: 0.0007; ^{132}Xe: 0.0002). Additionally, acid blanks were measured at the beginning of each etch run (in 10^{-10}ccSTP: ^{4}He: 8; ^{20}Ne: 0.09; ^{40}Ar: 9; ^{84}Kr: 0.003; ^{132}Xe: 0.0009). They lie slightly above the procedural blank, but most probably decrease during the etch run. For most steps the blank correction corresponds to less than 1% of the sample gas, except for the least gas-rich steps, mainly the initial step of each run. Therefore no additional acid blank correction was made. Since etching was very short at low acid temperatures (10min, –20°C) for the initial step of each etch run, resulting in gas amounts hardly above blank levels for the heavy noble gases, we ignore these steps in the following. Xe in Pesyanoe was only measured in the four most gas-rich steps.

RESULTS AND DISCUSSION

The complete He, Ne, Ar, Kr, and Xe data are given in [15]. Some etch runs, e.g. Noblesville, were stopped before sizeable amounts of cosmogenic gases appeared. We therefore use the ^{36}Ar concentration of the bulk aliquot to normalise gas amounts released in these etch runs (Fig. 2 and 3).

1 Light noble gases

The Ne-data are shown in Fig. 1. All samples display the characteristic pattern for implanted solar Ne. The first steps have a composition comparable to that of SW-Ne, although $^{20}Ne/^{22}Ne$ ratios very close to the

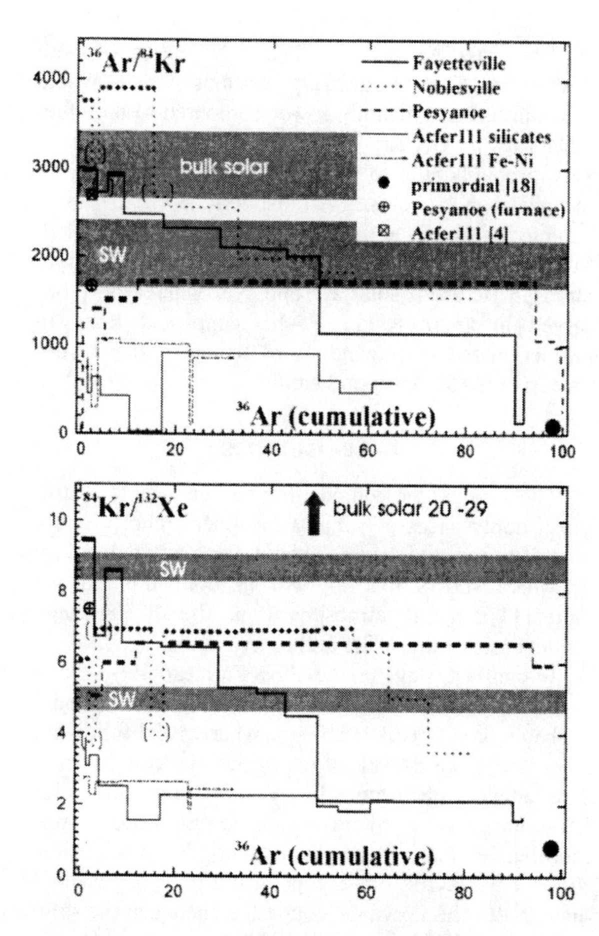

FIGURE 2. The ^{36}Ar/^{84}Kr ratios of the 4 meteoritic samples, analysed by CSSE. Gas concentrations are normalised to ^{36}Ar in the aliquot analysed by total fusion. 1σ errors are on the order of 5%. SW data obtained on lunar samples [9] and meteorites [4] as well as the inferred bulk solar values [12,17] are given for comparison. For Pesyanoe also our total extraction value (furnace) is shown.
The steps from which the solar wind compositions are deduced are highlighted. The first steps of both chondrites Fayetteville and Noblesville have high SW ^{36}Ar/^{84}Kr ratios, in the range of the inferred solar values. With progressive etching the ratios decrease due to increasing contribution of primordial noble gases. The achondrite Pesyanoe shows a rather constant ^{36}Ar/^{84}Kr ratio over the entire etch run in the range of SW values from lunar samples; only at the end small amounts of primordial gases arise. The ^{36}Ar/^{84}Kr ratio of both Acfer111 samples are compromised by adsorbed air.

FIGURE 3. Same as Fig. 2, but for ^{84}Kr/^{132}Xe. Like ^{36}Ar/^{84}Kr the ^{84}Kr/^{132}Xe ratios of Noblesville and Fayetteville have high values in the first etch steps, in the range of lunar-data, and decreasing values with continuing etching, again due to increasing contributions of primordial noble gases. Pesyanoe shows rather constant ratios. Adsorbed air compromised also the ^{84}Kr/^{132}Xe ratio of both Acfer111 fractions, leading to almost constant ratios around 2.7 throughout the run.

SW-value are observed only for Noblesville (13.51) and Acfer111 Fe-Ni (13.34). With progressive etching, the Ne composition becomes heavier and tends toward the SEP point. Increasing relative contributions of "cosmogenic" Ne, produced by galactic cosmic ray interactions, lead to a deviation of the data points of the last steps of each run towards the GCR point. He and Ar isotopes show a similar behaviour as Ne (not shown). The scatter of data points for some runs (e.g. the first two steps of Noblesville) is probably due to the different noble gas retentivities of various minerals and their different susceptibilities to etching by HF. The small sample amounts prevented preparation of monomineralic separates. In summary, the pattern of the light noble gases is well understood and will guide our interpretation of the Ar, Kr, and Xe results.

2 Element ratios of Ar, Kr, Xe

Figs. 2 and 3 show the evolution of the ^{36}Ar/^{84}Kr and ^{84}Kr/^{132}Xe ratios of all four meteorites. Solar wind compositions as deduced from lunar samples [9] and meteorites [2,4] as well as the inferred bulk solar values [12,17] are given for comparison (Table 1).
The etching was extremely gentle in the first steps, in order to obtain pure solar wind. Fayetteville and Noblesville show similar trends. Both meteorites start with ^{36}Ar/^{84}Kr ratios higher than the respective lunar data. For Fayetteville these first steps plot in the range of the unfractionated solar composition, considering both solar abundance estimates by [12] (3320) and [17] (2660). The first steps of Noblesville plot even somewhat above both solar abundance values. From these observations and the Ne results we conclude that we released pure SW-Ar and -Kr in the first few steps. Later, the ^{36}Ar/^{84}Kr ratio decreases steadily, very probably due to an increasing contribution of primordial noble gases, here represented by phase Q [18]. The near constancy of ^{36}Ar/^{84}Kr in the first few steps indicates that the very gentle initial etching with HF was successful to separate the solar component from primordial contributions, whereas later a reservoir of primordial noble gases must have been tapped, although phase Q is nominally resistant to HF attack. The ^{84}Kr/^{132}Xe ratios of both meteorites show a similar trend as ^{36}Ar/^{84}Kr. The first steps display values in the range defined by lunar samples, followed by a decrease, very probably again due to increasing contributions of primordial noble gases. In contrast to ^{36}Ar/^{84}Kr, the lunar samples as well as now the first steps of Fayetteville and Noblesville display ^{84}Kr/^{132}Xe ratios distinctly lower than the inferred bulk

TABLE 1. Heavy noble gas element ratios in the sun and solar wind.

source	$^{36}Ar/^{84}Kr$	$^{84}Kr/^{132}Xe$	data source/ procedure
solar abundances [12]	3320	20.6	Ar=SW; Kr,Xe nucleosynthesis calculations
solar abundances [17]	2660	29.3	Ar [12]; Kr,Xe nucleosynthesis calculations
"Young" Solar wind [9]	1870	8.7*1	lunar soils, CSSE
"Old" Solar wind [9]	2255*1	4.9*1	lunar soils, CSSE
Pesyanoe [2]	2270	7.5	bulk, oven total extraction
Acfer111 [4]	2690*2	-	FeNi, CSSE
this work			
Noblesville	3880	13.5±3*3 (6.88)*4	bulk, CSSE
Fayetteville	2940	9.0	bulk, CSSE
Pesyanoe	1620	6.45	bulk, CSSE
Pesyanoe	1660	7.5	bulk, oven total extraction

*1 for Kr/Xe ("young" SW) average value from 71501+67601; for Ar/Kr, Kr/Xe ("old" SW) average value from 79035+14301

*2 $^{36}Ar/^{84}Kr$ of uncorrected data, average of first 7 steps

*3 air corrected value

*4 measured value

solar values.

Unlike the chondritic samples the achondrite Pesyanoe may be expected to reveal constant heavy noble gas element ratios during the entire etch run, because of the putative absence of primordial gases [2, 3]. This is indeed true for the gas-rich steps. All the four major steps display relatively constant $^{36}Ar/^{84}Kr$ and $^{84}Kr/^{132}Xe$ ratios (although 85% of the gas was released in one step). The last four steps, with very low gas amounts, however, show strongly decreasing $^{36}Ar/^{84}Kr$ values (no $^{84}Kr/^{132}Xe$ data for these steps are available), indicating that also Pesyanoe contains minor amounts of primordial noble gases, which can be separated by the CSSE technique. The mean $^{36}Ar/^{84}Kr$ value of the etch run (1660) agrees well with the value of two total fusion analyses of aliquot samples (Table 1). However all these values are consistently lower than the $^{36}Ar/^{84}Kr$ ratio of 2270 reported by [2]. On the other hand the mean $^{84}Kr/^{132}Xe$ ratio of our etch run is ~14% lower than both the total fusion value and the value of [2], which might be caused by small amounts of atmospheric Xe adsorbed on the grains of the etch sample.

The $^{36}Ar/^{84}Kr$ and the $^{84}Kr/^{132}Xe$ ratios of both fractions of Acfer111 are very constant over the whole etch run but significantly lower compared to the other meteorites ($^{36}Ar/^{84}Kr$: ~1000; $^{84}Kr/^{132}Xe$: 2-3). This is very probably due to a severe contamination by atmospheric Kr and Xe. Atmospheric Kr and Xe is commonly observed in desert meteorite finds [19]. Surprisingly our etch runs were not able to separate atmospheric from solar Kr and Xe, whereas [4] observed in an Acfer111 Fe-Ni sample at least an unfractionated solar wind $^{36}Ar/^{84}Kr$ ratio after a very extensive (~180 days) preheating.

3 Xe- isotopes

The Xe isotopic composition can be used to distinguish noble gases of solar, atmospheric and possibly also of primordial origin. This is demonstrated in Fig. 4, which shows that the Xe in both fractions of Acfer111 is mainly atmospheric, as already suggested on the basis of the Ar/Kr and Kr/Xe ratios.

In contrast, Fayetteville shows a clear SW-Xe isotopic composition in the first etch steps 2 and 4 (step 3 probably had a blank problem), whereas the following steps tend towards SEP or the primordial composition. The latter is much more likely, as the element ratios have indeed revealed increasing contributions of primordial Xe in later steps. Atmospheric Xe is negligible in the Fayetteville run except in the gas poor steps 3 and 10-12. The Pesyanoe data fall in between the solar wind and SEP points, although with a considerable scatter. However, small amounts of atmospheric Xe cannot be excluded, which would explain the somewhat low $^{84}Kr/^{132}Xe$ ratios compared to the bulk samples (see above). Primordial Xe is negligible in Pesyanoe, as has been shown above. All data points of Noblesville plot close to each other in Fig. 4 as well as in other Xe-three-isotope diagrams (not shown). To explain the Xe-isotope data of Noblesville, two scenarios or a mixture thereof are conceivable: i) pure SEP-Xe or primordial Xe in all steps; ii) a rather constant mixture of SW-Xe and atmospheric Xe throughout the etch run. In analogy to Fayetteville and in view of the Ne isotope data we exclude a pure SEP-Xe as well as a pure primordial Xe composition throughout the run, but in particular for the first etch steps. At least for these steps we interpret the Xe-isotope pattern of Noblesville to indicate an ~50:50 mixture between SW-Xe and atmospheric Xe. This allows us to recalculate the SW $^{84}Kr/^{132}Xe$ ratio of Noblesville to 13.5±3, assuming that atmospheric Xe contributed 50±10 % of the total ^{132}Xe. Unfortunately, we cannot directly estimate a potential atmospheric Kr contamination in the same way, because the isotopic composition of SW-Kr and atmospheric Kr are rather close to each other [7,8]. However, "irreversible" adsorption of

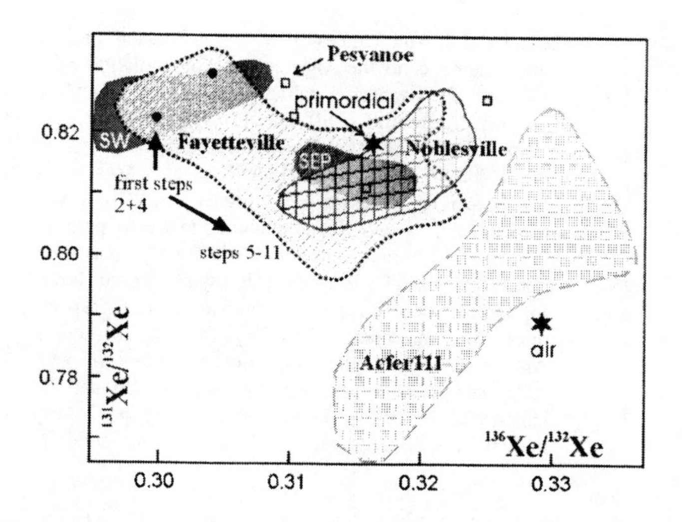

FIGURE 4. Xe-isotope composition of the four measured meteorites. The hatched fields encompass the data of the etch steps including their 1σ errors. SW and SEP composition obtained from lunar samples (grey fields) [8], atmospheric Xe [5] as well as primordial Xe [18] are shown for comparison. All steps of the Acfer111 Fe-Ni run (as well as all Acfer111 bulk silicate steps, not shown) plot near the atmospheric composition. The first steps of Fayetteville (black dots) have SW composition, whereas the following steps tend towards primordial Xe. The Pesyanoe data (squares) fall in between SW and SEP, although with a considerable scatter (1σ errors of 2% not shown). The Xe isotope composition of at least the first steps of Noblesville result from a mixture of atmospheric and SW-Xe.

atmospheric Kr has been reported to be roughly an order of magnitude less efficient than for Xe [20]. If so, the $^{84}Kr/^{132}Xe$ ratio of adsorbed air would be about 2.8, indeed in very good agreement with the observed ratios for both fractions of the desert find Acfer111 (Fig. 3). If this would also apply to Noblesville, the air-corrected $^{84}Kr/^{132}Xe$ ratio for the solar wind component would be 10.7±2. Nevertheless, atmospheric Kr probably did not contribute substantially to the total Kr inventory of Noblesville. This is indicated by the $^{36}Ar/^{84}Kr$ ratio. For a 50% atmospheric Xe contribution and a $^{84}Kr/^{132}Xe$ ratio of the atmospheric component of 2.8, the corrected $^{36}Ar/^{84}Kr$ ratio in Noblesville would be ~30% higher than the highest of the two bulk solar values. This seems unreasonable and we therefore do not correct for a potential atmospheric Kr contribution. Our preferred $^{84}Kr/^{132}Xe$ ratio of 13.5±3 indicates that solar Kr and Xe in Noblesville are less fractionated relative to the other meteorites studied here as well as to lunar samples.

4 Fractionation of solar wind noble gases; the FIP effect

In this section we compare the element ratios of the heavy noble gases in the solar wind as deduced from Fayetteville, Noblesville and Pesyanoe with the respective values reported earlier for lunar samples. Because He and Ne in bulk samples of solar noble-gas-rich meteorites usually show a lesser degree of element fractionation than lunar ilmenites, we can well assume that the Ar/Kr and Kr/Xe ratios deduced above for the solar component in the meteorites also represent the true abundance ratios in the solar wind at the time each meteorite was irradiated. Furthermore this is supported by the constant Ar/Kr and Kr/Xe ratios in etch runs on lunar ilmenite and plagioclase [9] as well as meteoritic enstatite (Pesyanoe, this work) implying that Ar, Kr, and Xe are not lost by diffusion. This gives us confidence that also in the chondritic bulk samples the true relative abundances of Ar, Kr, and Xe in the solar wind are recorded at least in the first 10-20% of gas released by etching.

In general the solar wind Ar/Kr and Kr/Xe ratios vary considerably in the three studied meteorites. Ar/Kr in Fayetteville and Noblesville is consistent with an unfractionated bulk solar composition, within the uncertainties of the latter, which is reflected by the variable estimates in the solar abundance compilations [12,17]. This is in contrast to all studied lunar samples as well as Pesyanoe, which show an enrichment of Kr relative to Ar of a factor of 1.4–2.0. On the other hand the Kr/Xe ratio in all three meteorites as well as in lunar samples is several times lower than estimates of the bulk solar value. Thus, all available data on the composition of the ancient solar wind show a Xe enrichment relative to Ar and bulk solar values, whereas some but not all of these samples also display a lesser enrichment of Kr (Fig. 5). In Fayetteville, the Xe enhancement relative to Ar is somewhat lower than that of two relatively recently irradiated (past ~100Ma) lunar samples. Both Kr and Xe enrichments of Pesyanoe compare better with two lunar samples irradiated one to several Ga ago. In contrast, the Noblesville data show almost no fractionation for Kr but also a smaller fractionation for Xe than all other samples.

It is well known that the abundances of elements with a low first ionisation potential (FIP) are enhanced in the solar wind by about a factor of four in the slow solar wind and almost a factor of two in the fast solar wind from coronal holes [13]. The enhancement of Xe (and Kr) in the solar wind also appears to be related to this "FIP effect", although both Kr and Xe have first ionisation potentials larger than the 10eV threshold. Such a behaviour is indeed predicted by models where the ionisation time in the chromosphere is the parameter that actually governs the "FIP-effect" [21, 22,23].

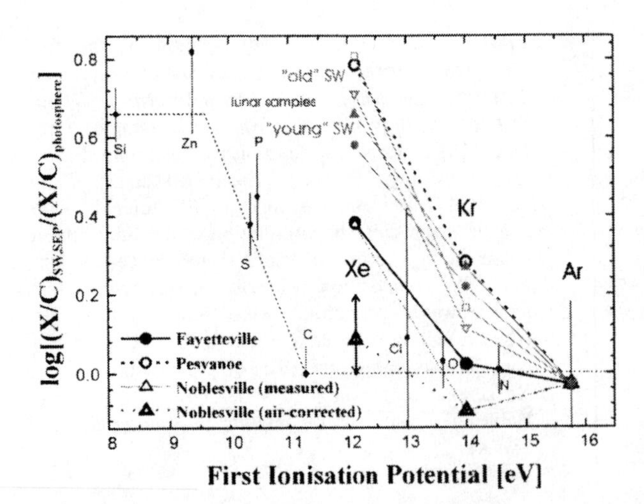

FIGURE 5. Abundances of Ar, Kr, and Xe and various other elements in the solar corpuscular radiation relative to "photospheric" values [12] versus their first ionisation potential (FIP) (see [13,9]). The Ar/Kr/Xe ratios are normalised to the inferred bulk solar values by Anders & Grevesse [12]. Ar/Xe_{sol} and Ar/Kr_{sol} by Palme & Beer [17] are ~15% higher and 20% lower, respectively. Kr and Xe are anchored to the Ar point of [12]. Lunar samples containing "old" (grey, open symbols) and "young" SW (grey, filled symbols) are shown for comparison. The arrow on the Noblesville-Xe-point marks the maximal uncertainty for the correction. Ar/Kr and Ar/Xe of the solar wind component of meteorites considerately expand the range of Xe and Kr enhancement factors in the solar wind previously known form lunar samples (see text for further details).

Wieler et al. [11] speculated that variable proportions of low- and high-speed solar winds in different epochs might explain part of the observed differences in the enhancement factors, although trends for Kr and Xe are sometimes opposite to each other, which would require additional processes at work. The lunar samples indicate a generally larger fractionation of Ar/Xe in the more distant past [1] as had already been suggested by [24] from bulk sample data. Unfortunately, for the meteorites here no reliable information on their time of solar wind irradiation (antiquity) is available, although it is often argued that gas-rich meteorites trapped their solar wind very early, perhaps more than 4 Ga ago. In Fig. 5 none of the meteorite data patterns match a lunar pattern well enough to allow us to tentatively assign an antiquity for a meteorite based on such a comparison of Ar-Xe abundance data. Instead, the meteorite data show, that the Xe and Kr abundances in the solar wind in the past have been even more variable than was known from the lunar samples. If at least some of the meteorites studied here were irradiated very early, the "FIP effect" was already active in the early sun.

REFERENCES

1. Wieler R., *Space Sci. Rev.* **85**, 303-314 (1998).
2. Marti K., *Science* **166**, 1263-1265 (1969).
3. Kim J. S. and Marti K., "Solar-type xenon: isotopic abundances in Pesyanoe" in *Proc. Lunar Planet. Sci.*, Houston, 1992, pp. 145-151.
4. Pedroni A. and Begemann F., *Meteoritics* **29**, 632-642 (1994).
5. Basford J. R., Dragon J. C., Pepin R. O., Coscio M. R., and Murthy V. R., "Krypton and Xenon in lunar fines" in *Proc.Lunar Sci.Conf. 4th*, Houston, 1973, pp.1915-1955.
6. Becker R. H. and Pepin R. O., *Geochim. Cosmochim. Acta* **53**, 1135-1146 (1989).
7. Pepin R. O., Becker R. H., and Rider P. E., *Geochim. Cosmochim. Acta* **59**(23), 4997-5022 (1995).
8. Wieler R. and Baur H., *Meteoritics* **29**, 570-580 (1994).
9. Wieler R. and Baur H., *Astrophys. J.* **453**(2 Part 1), 987-997 (1995).
10. Rider P. E., Pepin R. O., and Becker R. H., *Geochim. Cosmochim. Acta* **59**(23), 4983-4996 (1995).
11. Wieler R., Kehm K., Meshik A. P., and Hohenberg C. M., *Nature* **384**, 46-49 (1996).
12. Anders E. and Grevesse N., *Geochim. Cosmochim. Acta* **53**, 197-214 (1989).
13. von Steiger R., Geiss J., and Gloeckler G., "Composition of the solar wind", in *Cosmic winds and the heliosphere* edited by J. R. Jokipii, C. P. Sonett, and M. S. Giampapa, University of Arizona Press, Tucson, 1997, pp. 581-616.
14. Signer P., Baur H., and Wieler R., "Closed system stepped etching: an alternative to stepped heating" in *Alfred O. Nier Symposium on Inorganic Mass Spectrometry*, Durango, Colorado, 1993, pp. 181-202.
15. Heber V. S. (2001) PhD Thesis, ETH Zürich, in prep.
16. Graf T., Signer P., Wieler R., Herpers U., Sarafin R., Vogt S., Fieni C., Pellas P., Bonani G., Suter M., and Wölfli W., *Geochim. Cosmochim. Acta* **54**, 2511-2520 (1990).
17. Palme H. and Beer H., "Abundances of the elements in the solar system", in *Landolt-Börnstein, Group VI: Astronomy and Astrophysics*, Vol. 3(a) edited by H. H. Voigt, Springer Verlag, Berlin, 1993, pp. 196-221.
18. Busemann H., Baur H., and Wieler R., *Meteoritics & Planet. Sci.* **35**, 949-973 (2000).
19. Loeken T., Scherer P., Weber H. W., and Schultz L., *Chem. Erde* **52**, 249-259 (1992).
20. Niedermann S. and Eugster O., *Geochim. Cosmochim. Acta* **56**(1), 493-509 (1992).
21. Geiss J., Gloeckler G., and von Steiger R., *Phil. Trans. R. Soc. London* **A349** (1690), 213-226 (1994).
22. Marsch E., von Steiger R., and Bochsler P., *Astronomy & Astrophysics* **301**(1), 261-276 (1995).
23. Peter H., *Space Sci. Rev.* **85**(1-2), 253-260 (1998).
24. Kerridge J. F., "Secular variations in composition of the solar wind: Evidence and causes" in *Proc. Conf. Ancient Sun* 1980, pp. 475-489.
25. Benkert J.-P., Baur H., Signer P., and Wieler R., *J. Geophys. Res. (Planets)* **98**(E7), 13147-13162 (1993).

A New Look at Neon-C and SEP-Neon

R. A. Mewaldt*, R. C. Ogliore*, G. Gloeckler+ and G. M. Mason+

*California Institute of Technology, Pasadena, CA 91125 USA
+University of Maryland, College Park, MD 20742 USA

Abstract: Studies of the isotopic composition of neon in lunar soils, meteorites, and interplanetary dust particles have revealed several distinct components. In addition to implanted solar wind, which has a $^{20}Ne/^{22}Ne$-abundance ratio of 13.7, there is an additional component with $^{20}Ne/^{22}Ne \approx 11.2$, originally attributed to higher-energy solar energetic particles. Using data from the Advanced Composition Explorer, we have measured the fluence of solar wind, suprathermal particles, solar energetic particles and cosmic rays from ~0.3 keV/nucleon to ~300 MeV/nucleon over an extended time period. We use these measured spectra to simulate the present-day depth distribution of Ne isotopes implanted in the lunar soil. We find that the suprathermal tail of the solar wind, extending from a few keV/nucleon to several MeV/nucleon with a power law spectrum, can produce $^{20}Ne/^{22}Ne$ abundance ratios in the lunar soil that are similar to the measured composition, although there remain significant questions about the extent to which the present-day intensity of suprathermal ions is sufficient to explain the lunar observations.

INTRODUCTION

Solar-wind noble gases implanted in the lunar soil preserve a record of solar wind activity over the history of the solar system. Beginning with the return of the first Apollo samples from the moon, this record has revealed evidence for several distinct isotopic components implanted in lunar materials (e.g., Pepin et al. [1]). In addition to implanted solar wind and at least two primordial components, there is evidence for a component that contains excesses of the heavy isotopes of He, Ne, Ar, Kr, and Xe when compared to the composition of the present-day solar wind (see review by Wieler [2]). There is also evidence for an additional component of nitrogen, depleted in ^{15}N that may be related [3]. The best measurements of implanted noble gases have been obtained by step-wise etching of small grains of lunar and meteoritic material [2]. This process gradually releases gas from deeper and deeper sites that is then analyzed by mass spectroscopy techniques. A complication is that the depth from which gas is released during a given step is not well defined.

The noble gas components enriched in heavy isotopes were at first labeled He-C, Neon-C, etc [4]. They were later ascribed to solar energetic particles [5, 6], because they are implanted somewhat deeper in lunar samples than solar wind, and identified as the "SEP" component. However, it appears that the amount of implanted gas associated with this component is ~10% to 40% of the solar wind [7, 2], orders of magnitude greater than expected from the present-day frequency and size of solar-particle events associated with flares and coronal mass ejections.

A number of possible origins for the SEP component have been considered (see e.g., [2, 8, 9, 10]). Recently, Wimmer-Schweingruber and Bochsler [11, 12] suggested that this component was implanted as the Sun passed through dense interstellar clouds during its journey through the Galaxy. In their scenario the SEP component represents a sample of interstellar matter, the isotopic anomalies result from galactic evolution effects since the birth of the solar system, and the lunar soil provides a "travel diary" of the voyage of the solar system through the Galaxy.

In this paper we consider a solar/heliospheric origin for the SEP component. Using data from several instruments on the Advanced Composition Explorer (ACE), Mewaldt et al. [13] have measured the present-day fluences of energetic He, O, and Fe from ~0.3 keV/nucleon to ~300 MeV/nucleon, ranging from solar wind to cosmic ray energies. This period included high-speed streams, impulsive and gradual solar particle events, events associated with corotating interaction regions (CIRs) as well as solar-minimum and solar-maximum fluxes of anomalous and galactic cosmic rays. In the "suprathermal" energy range from ~10 keV/nucleon to ~10 MeV/nucleon, all species were found to have the same power-law spectral shape [13]. Using measured energy spectra of solar wind and suprathermal particles, we have simulated the depth distribution of Ne isotopes implanted in lunar soil. We compare these simulations with lunar sample data in an effort to shed light on the origin of the mysterious SEP component. In this discussion we will use the term SEP to refer to this unknown component without assuming that it is due to solar energetic particles.

CP598, *Solar and Galactic Composition*, edited by R. F. Wimmer-Schweingruber

FLUENCE MEASUREMENTS

The fluence measurements that are the basis for our simulations were obtained from four instruments on ACE during a 33-month period that starts with solar-minimum conditions in 1997 and ends with solar-maximum conditions in 2000 [13]. Energetic particle fluences above 40 keV/nucleon were integrated over the period from 1997:280 to 2000:184. The solar wind data were summed over an 11-month period from 1/99 to 11/99, and then multiplied by a factor of 3 to correspond to the 33-month period covered by the higher-energy instruments. Because solar wind is less variable than higher-energy solar/interplanetary particles, the 1999 period should provide a reasonable representation of the longer period.

The spectra in Fig. 1 have a common shape. The peak at ~0.8 keV/nucleon corresponds to the slow-speed solar wind with ~400 km/sec velocity. The contribution of occasional higher-speed streams reaching ~1000 km/sec is also apparent. Based on these spectra, it appears that the solar wind distribution extends to ~8 keV/nucleon, at which point there is a change in slope. Beyond this, a long, suprathermal tail extends with a power-law slope of –2 out to several MeV/nucleon. Near ~10 MeV/nucleon all three spectra exhibit a gradual "knee" and briefly steepen. Above ~50 MeV/nucleon, galactic cosmic rays (GCRs) make the dominant contribution. In the integral energy spectrum for oxygen in Figure 2 it is clear that the integrated fluence above ~7 keV/nucleon amounts to only a small fraction (~3 x 10^{-5}) of the solar-wind fluence.

In the energy range above ~10 keV/nucleon, measurements with SWICS/ACE show that suprathermal tails on the solar wind are continuously present, even in the absence of interplanetary shocks [14]. Although, the origin of these tails is presently a subject of investigation ([14, 15]), they do not appear to be due to large, individual solar particle events such as the July 14, 2000 event, to impulsive solar flares, or to CIR events [13]. A more or less continuous process appears to be the source of most of these particles.

Three-year fluence spectra of eight elements and ^{3}He from the ULEIS and SIS instruments on ACE all show the same E^{-2} power-law shape from ~0.04 MeV/nucleon to ~10 MeV/nucleon [13], suggesting that this behavior is characteristic of this extended suprathermal energy range. Studies of individual solar-particle events with ~1 and >10 MeV/nucleon indicate that the solar particle ^{20}Ne/^{22}Ne ratio is, on average, somewhat less than the solar wind value ([16, 17, 18, 19]). However, measurements of this ratio for the 3-year fluence period are not yet available.

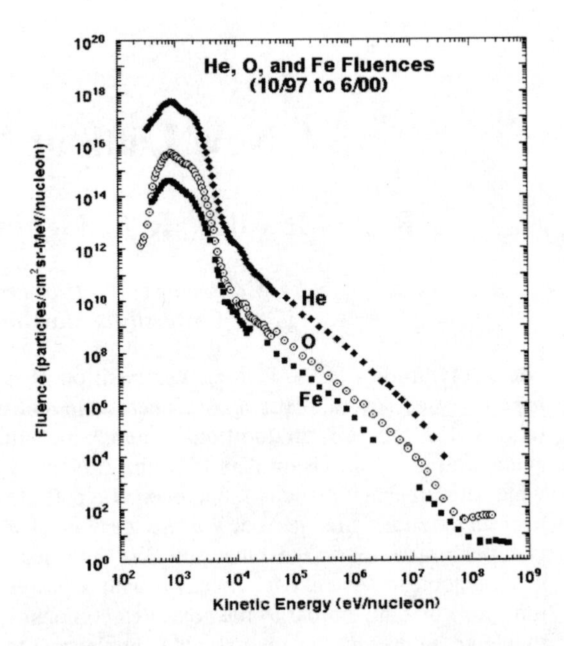

Figure 1: Fluences of He, O, and Fe nuclei measured during the period from 9/1997 to 6/2000 by the SWICS, ULEIS, SIS, and CRIS instruments on ACE. For additional information see Mewaldt et al. [13].

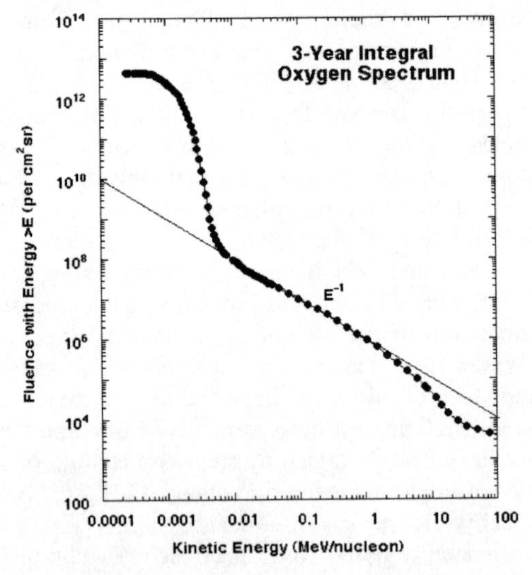

Figure 2: Integral spectrum of oxygen nuclei corresponding to the fluence spectrum in Figure 1.

SIMULATED DEPTH DISTRIBUTIONS

In order to simulate the depth distributions of energetic ions in the lunar soil we used the TRIM program [20], which performs 3-D Monte-Carlo calculations of individual ion trajectories in a variety of materials. By repeating the calculation, a distribution of penetration depths is obtained. Figure 3 shows the mean penetration depth of ^{22}Ne and ^{20}Ne as a function of incident kinetic energy. At a given energy/nucleon, ^{22}Ne has ~10% greater range than ^{20}Ne.

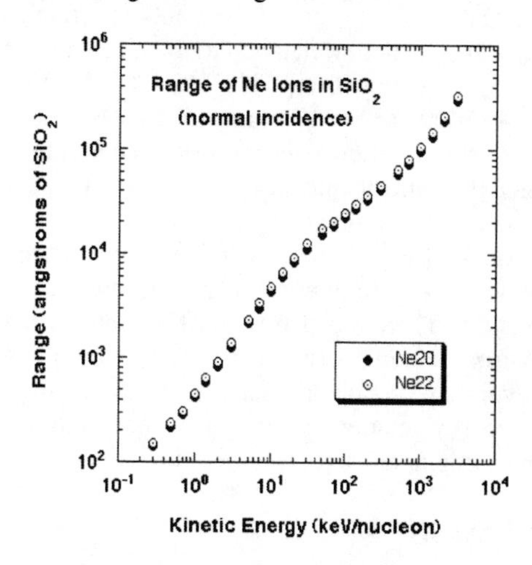

Figure 3: Range of ^{20}Ne and ^{22}Ne in SiO_2 at normal incidence calculated by the SRIM 2000 version of TRIM.

In order to simulate the spectrum and behavior of interplanetary ions the oxygen spectrum in Figure 1 was divided into ten energy-bins per decade with a logarithmic spacing factor of $10^{0.1}$ = 1.256. For example, ions with energies from E_1 keV/nucleon to $E_2 = 1.256E_1$ keV/nucleon were represented by a single energy $\langle E \rangle = (E_1E_2)^{1/2}$ keV/nucleon. A total of 42 bins were used to represent the observed spectrum from ~0.2 keV/nucleon to 3 MeV/nucleon.

The solar wind (which travels more or less as a beam) can only impact the Sun-facing side of the Moon, while particles with several times solar wind speed are assumed to have an isotropic distribution with access to all sides of the Moon. The solar wind (SW) and suprathermal tail distributions were therefore treated separately to allow for their different patterns of incidence and possibly different isotopic composition. Within each energy bin from 0.2 to 12 keV/nucleon, Monte Carlo trajectories were calculated with TRIM for ^{20}Ne and ^{22}Ne ions (a total of 2000 ions for each energy) distributed in eight angular intervals.

The distribution of incidence angles for the solar wind reflects the changing angle of incidence during the Moon's orbit about the Earth. The spectrum of energetic particles from 5 keV/nucleon to 3 MeV/nucleon (the suprathermal, or SEP component) was assumed to have a power-law spectrum with a slope of –2. Trajectories were calculated for 2000 ^{20}Ne and ^{22}Ne ions in each energy bin, in this case distributed between 16 angles to represent an isotropic flux. The depth distributions of ^{20}Ne and ^{22}Ne were accumulated in an array of bins from 1 angstrom to 3 x 10^5 angstroms, again with 10 logarithmically-spaced bins per decade (see Figure 4).

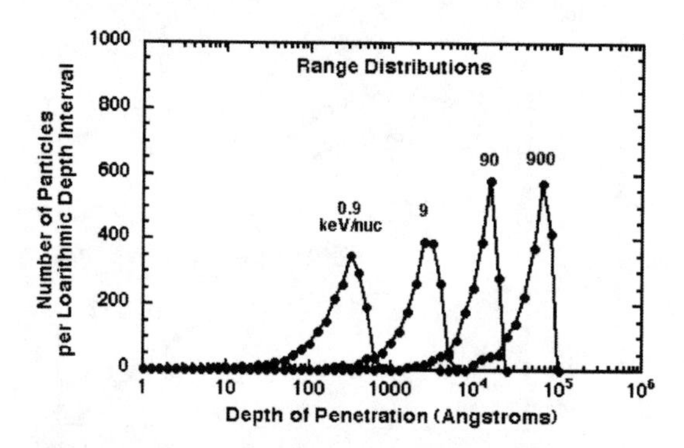

Figure 4: Depth distributions of ^{20}Ne isotopes at four energies (in keV/nucleon), taking into account the angular distributions of the incident ions (see text).

The information from these 96 runs can be combined to simulate arbitrary energy spectra and neon isotope compositions by adjusting the weighting of the various runs. For this study we assumed that the beam was pure solar wind for energies <5 keV/nucleon and pure suprathermal component for >12 keV/nucleon with a smoothly evolving mix at intermediate energies. The present-day solar-wind abundance ratio of ^{20}Ne/^{22}Ne = 13.7 [21] was assumed, independent of energy. The integrated interplanetary SEP/SW ratio is ~3 x 10^{-5} (see Figure 2).

Assuming SEP ions can access the lunar surface facing away from the Sun, the SEP/SW ratio in the implanted fluxes is ~10^{-4}. Figure 5 shows the depth distributions that result, including the integral distributions. The resulting ^{20}Ne/^{22}Ne ratio is shown in Figure 6, while Figure 7 shows ratios integrated over depth. We have not included the "cosmogenic" component, which includes roughly equal amounts of the three neon isotopes. This component, which results from cosmic-ray nuclear reactions with lunar material, is distributed uniformly throughout the samples.

Several features in the simulations are related to the input spectrum and the range-energy relation. Over a restricted interval the range-energy relation for Ne isotopes can be represented by $R(E,M) = a(M/20)E^b$, where a is a constant, M is the mass in amu, and $b \approx 1$ for $0.5 \leq E \leq 30$ keV/nucleon. Assuming a power-law spectrum ($dJ/dE = kE^s$) with index s, it can be shown that the relative abundance of stopping ^{20}Ne and ^{22}Ne, $(^{20}Ne/^{22}Ne)_d$, is related to the interplanetary ratio, $(^{20}Ne/^{22}Ne)_i$ by

$$(^{20}Ne/^{22}Ne)_d = (22/20)^{(s+1)/b}\,(^{20}Ne/^{22}Ne)_i. \quad (1)$$

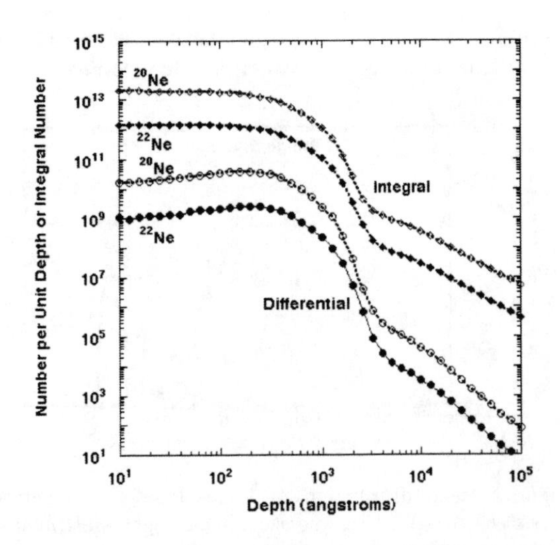

Figure 5: Simulated depth distributions of ^{22}Ne and ^{20}Ne based on the observed oxygen energy spectrum.

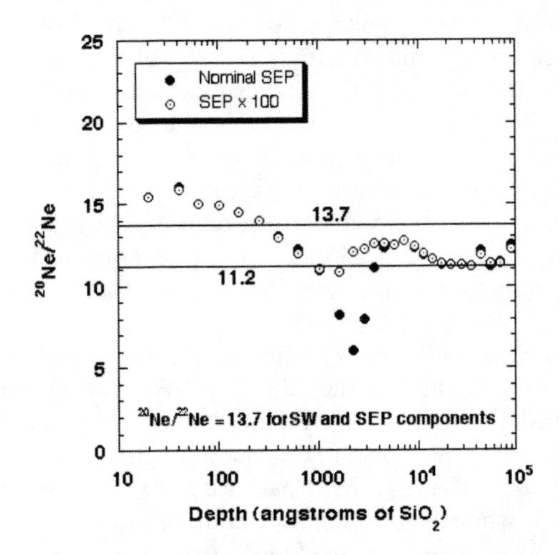

Figure 6: Ratio of ^{20}Ne/^{22}Ne vs. depth for spectra in Figure 5 (solid points). For the open points the SEP (or suprathermal component was multiplied by x100. Below ~2000 angstroms two adjacent bins were combined. Outside the 1000-5000 angstrom range, the open and closed points are essentially identical. The SW and SEP values are indicated.

The mean range of ~1 keV/nucleon solar wind is ~400 angstroms in SiO_2. At depths <400 angstroms the ^{20}Ne/^{22}Ne abundance ratio in the implanted ions is greater than the solar wind ratio of 13.7 because ^{22}Ne penetrates somewhat deeper on average. Thus, for $b \approx 1$ and $s > 0$ we expect $(^{22}Ne/^{20}Ne)_d$ >13.6 from equation (1). For depths from ~400 to ~3000 angstroms the solar wind spectral slope becomes more and more negative and the ^{20}Ne/^{22}Ne ratio at a given depth decreases, reaching a minimum at ~3000 angstroms at a point corresponding to the steepest part of the solar wind energy spectrum.

Particles at depths from ~4000 to ~12000 angstroms are derived from the suprathermal spectrum from ~8 to ~25 keV/nucleon, and we expect $(^{20}Ne/^{22}Ne)_d \approx 12.2$ for $s = -2$ and $b = 1$. From ~30 to ~300 keV/nucleon the range-energy relation flattens to $b \approx 0.5$ to 0.6 (see Figure 3), and we expect $(^{20}Ne/^{22}Ne)_d \approx 11.3$ for $s = -2$ from ~15000 to 50000 angstroms. Beyond ~300 keV/nucleon, the range-energy slope b again increases, causing the ratio to increase somewhat. The behavior predicted by the simulations differs somewhat from the predictions of Eq. (1) because of the range of angles of incidence and limitations of the power-law assumptions for b and s. There are also minor fluctuations that may result from beating between the energy and depth bins. However, we conclude that the spectral slope and range-energy relation govern the concentration of Ne isotopes implanted at a given depth.

DISCUSSION

The simulations in Figures 6 and 7 illustrate that even with a constant isotope ratio at all energies, one expects significant variations in ^{20}Ne/^{22}Ne as a function of depth as a result of changes in slope of the input spectrum and the range-energy relation. At a given depth the simulated ^{20}Ne/^{22}Ne ratio (Figure 6) ranges from ~16 down to ~6, and the interplanetary abundance ratio of ~13.7 is achieved only briefly. It is not clear to what extent these predicted features are reflected in the lunar sample data. In typical 3-isotope plots of step-wise etching data [7] the measured ratios range from ~13 to 14 down to ~11 for very low ^{21}Ne abundance. At greater depths, where cosmogenic neon dominates, the ^{20}Ne/^{22}Ne and ^{21}Ne/^{22}Ne ratios gradually approach ~1.

There are several reasons why the simulations in Figures 6 and 7 might not correspond directly with real measurements: (a) it is unlikely that samples have a uniformly flat surface oriented normal to the Sun and they have probably experienced several orientations; (b) there could be diffusion of Ne in the samples; (c) it appears that step-wise etching experiments release gas from a broad depth range (>1000 angstroms) at a time, and (d) the etch rate is not uniform over the surface (R. Wieler, personal communication). All of these effects tend to smooth out the measured gas-release pattern.

The plot in Figure 6, with depth on a log scale, emphasizes variations at small depths. If imagined on a linear scale, which corresponds more with the gas release process, it is more apparent that the expected ^{20}Ne/^{22}Ne ratio is similar to that of the SEP component over a large depth interval, from ~10^4 to ~10^5 angstroms (1 to 10 microns; see also Figure 7). The SEP component has been identified at depths up to ~35 microns in the lunar soil [7].

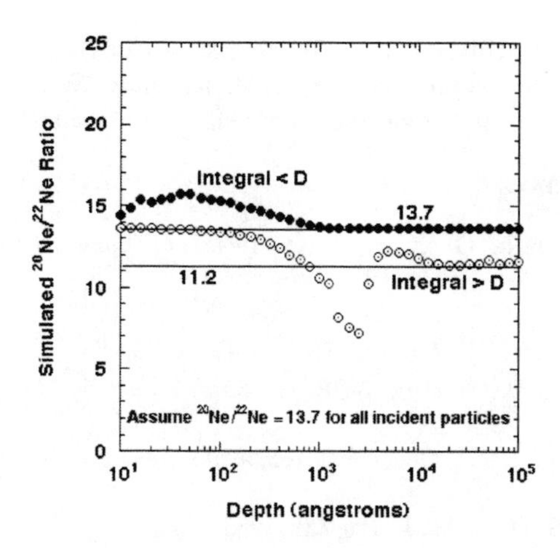

Figure 7: Ratios of $^{20}Ne/^{22}Ne$ in the integrated concentrations of Ne at depths <D and >D. The SW and SEP values are indicated.

If no material is lost from a sample and the first etch step removes at least 1000 angstroms of material, a ratio of 13.6 is expected (the small difference with the assumed SW value of 13.7 is because TRIM finds that slightly more ^{20}Ne than ^{22}Ne is backscattered from the lunar surface without being implanted). Beyond the peak in the solar wind spectrum (~400 angstroms) the expected ratio falls below 12 when ~10% of the implanted gas remains. At depths greater than a few thousand angstroms the ratios in the simulation approach the measured SEP ratio of ~11.2, but by this point only $\sim 10^{-5}$ of the implanted gas remains. This latter value is far below the estimate that SEP neon represents ~10% of the implanted Ne.

There are several possible ways to increase the SEP abundance in lunar samples. Wieler [2] points out that only ~10% of solar-wind Ne may be retained on impact. In addition, there may be weathering effects on the lunar surface that continually remove the outer layers of a grain. These two effects might increase the SEP/SW ratio by factor of ~10 to 100 [2]. There is support for these possibilities in the fact that none of the measurements in any of the five samples in Wieler et al. [7] give values as large as that in the contemporary solar wind (=13.7), suggesting that ^{20}Ne-rich material near the grain surface has been lost, and that the gas released in these measurements was implanted at a depth beyond ~500 angstroms (see Figure 6). In contrast, studies of ilmenite samples (which retain noble gases more effectively), do measure the solar wind composition in the outermost grain layers [22]. Becker and Wieler (see [2]) have conducted simulations that include surface sputtering which indicate that the true SW ratio is obtained at the grain surface once a steady state is achieved.

The depth resolution of the lunar measurements should also be considered. Our simulation (Figure 6) gives a $^{20}Ne/^{22}Ne$ ratio of <12.5 for all depths >600 angstroms, corresponding to ~20% of the total implanted neon. If the etch steps typically integrate over several thousand angstroms at a time and there is a substantial variation in the depth to which a given etch step penetrates, then it is possible to achieve $^{20}Ne/^{22}Ne$ abundance ratios significantly less that the solar wind value in ~10% of the released gas without increasing the intensity of the SEP component. However, it is not clear that this possibility would explain the many lunar-sample measurements to date.

It is interesting that none of the measurements that we are aware of find evidence for the region from ~1000 to 3000 angstroms where $^{20}Ne/^{22}Ne \approx 6 - 10$ (Figures 6 and 7). While the expected dip in the $^{22}Ne/^{20}Ne$ ratio at 1000-3000 angstroms might be smoothed out by uneven etch steps that sample a range of depths and by weathering effects, as discussed above, it seems unlikely that this would explain the lack of evidence for this dip in the many samples studied.

Our simulations show that it is possible to mask the effects of the steep negative slope of the SW spectrum by increasing the intensity of the SEP component by a factor of ~100 (see Figure 6). Thus, the lack of evidence for measurements with $^{20}Ne/^{22}Ne$ = 6 to 10 supports the possibility that the SEP/SW ratio was much greater when these samples were exposed. A somewhat smaller increase in SEP/SW might be adequate (and/or the expected $^{20}Ne/^{22}Ne$ ratio >600 angstroms further smoothed) if weathering (sputtering) effects are included. Further work is needed to see if these combined effects can explain the available data.

It is possible that the fluence of suprathermal ions was greater in the past than it is today. This might be due to a general increase in solar activity, or there could be other causes. If the suprathermal tail on the solar wind is due to statistical-acceleration by interplanetary turbulence [15], an increase in the level of turbulence would presumably lead to a corresponding increase in the interplanetary fluxes. Or if most of the particles >10 keV/nucleon are CIR-accelerated particles, an increase in the frequency or speed of high-speed streams would cause increased interplanetary particle intensities.

While it is difficult to explain an SEP/SW ratio of 0.1 or more, as inferred from lunar samples, our simulation results show that it is possible to mimic a separate, more penetrating, isotopically-heavy component with a combination of energy-spectrum and range-energy effects. Although some of the effects of the difference in range between ^{22}Ne and ^{20}Ne have been recognized previously, the possibility that they might help explain the SEP component has generally been discounted [9]. Our results show that reduced $^{20}Ne/^{22}Ne$ ratios arise quite naturally over an extended range of depths. For example, for 30 to 300 keV/nuc ions, Eq. (1) with $b \approx 0.5$ gives an SEP $^{20}Ne/^{22}Ne$ ratio of 11.3 with the following combinations of source ratio and spectral index: $^{20}Ne/^{22}Ne \approx 11.3$ with $dJ/dE \sim E^{-1}$, $^{20}Ne/^{22}Ne \approx 13.7$ with $dJ/dE \sim E^{-2}$, or $^{20}Ne/^{22}Ne \approx 16.5$ with $dJ/dE \sim E^{-3}$. While any of these is possible, Occam's Razor suggests assuming the SW ratio of 13.7 unless there is good reason not to. Changing the suprathermal isotope ratio might achieve a ratio close to the SEP ratio over a broader depth range, but this is outside the scope of this paper.The spectral and range-energy effects discussed here will also produce correlated enhancements of the heavy isotopes of He, Ar, Kr, and Xe, as observed in lunar and meteoritic samples [2], and they can potentially also cause apparent enhancements of Kr and Xe with respect to lighter noble gases.

SUMMARY

In this paper we have made an initial attempt to simulate the depth distribution of Ne isotopes implanted in lunar samples using available range-energy routines and newly-measured spectra of solar wind and heliospheric particles. We find that the falling energy spectra, combined with the longer range of ^{22}Ne, lead to an apparent enrichment in ^{22}Ne with respect to ^{20}Ne for all depths (in SiO_2) beyond ~400 angstroms. In particular, at depths of $\sim 10^4$ to 10^5 angstroms a $^{22}Ne/^{20}Ne = 13.7$ source ratio transforms to ~11.3, similar to the SEP component in lunar samples. The mechanism also causes apparent enrichments of heavy He, Ar, Kr, and Xe isotopes.

Although it is not yet clear to what extent these simulations are consistent with existing lunar sample data, these first results emphasize that interpretations of the composition of implanted ions require knowledge of their interplanetary energy spectrum. As for why the SEP component apparently comprises ~10-40% of all implanted ions, we find that the expected $^{22}Ne/^{20}Ne$ ratio at depths beyond ~600 angstroms (~20% of all implanted ions) is closer to the SEP value than to the (input) SW value. Improved agreement with the lunar data is achieved if the flux of suprathermal ions was ~100 times greater in the past. Further work is needed to explore the degree to which present-day solar wind and suprathermal spectra can help interpret available lunar data.

Acknowledgments: This work was supported by NASA at Caltech (under grant NAS5-6912) and at the University of Maryland. We thank D. Burnett and M. Wiedenbeck for help with TRIM simulations and R, Wimmer-Schweingruber for discussions of SEP neon. We greatly appreciate the assistance of R. Wieler in interpreting noble gas measurements in lunar samples.

References

1. Pepin, R. O. et al., *Proc. Apollo 11 Lunar Science Conference*, 1435, 1970.
2. Wieler, R., *Space. Science Reviews* 85, 303, 1998.
3. Kerridge, J. F., *Rev. Geophys.* 31, 423, 1993.
4. Black, D. C.and Pepin, R. O., *Earth Planet Sci. Lett.* 6, 395-405, 1969.
5. Black, D. C., *Geochim. Cosmochim. Acta* 36, 347-375, 1972.
6. Black, D. C., Ap. J., 266, 889, 1983,
7. Wieler, R., Baur, H., and Signer, P, *Geochim. Cosmochim. Acta*, 50, 1997-2017, 1986.
8. Geiss, J., and Bochsler, P., in *The Sun in Time*, C. P. Sonnett et al., eds, Univ. Arizona Press, Tucson 1991.
9. Wieler, R., Lunar and Planetary Science XXCVII, Lunar and Planetary Institute, Houston, p. 1551, 1997.
9. Wimmer-Schweingruber, R. F., *Lunar Soils: A Long-Term Archive for the Galactic Environmant of the Solar System*?, Habilitation Thesis University of Bern, 2000.
11. Wimmer-Schweingruber, R. F., and P. Bochsler, in *Acceleration and Transport of Energetic Particles Observed in the Heliosphere* (Mewaldt et al., eds.), AIP #528, p. 270, 2000.
12. Wimmer-Schweingruber, R, submitted to SOHO-ACE Workshop Proc. (AIP), 2001.
13. Mewaldt, R. A., et al., submitted to SOHO-ACE Workshop Proc. (AIP), 2001.
14. Gloeckler, G., Fisk, L. A., Zurbuchen, T. H., and Schwadron, N. A., in AIP Conf. Proc. 528, *Acceleration and Transport of Energetic Particles Observed in the Heliosphere*, ed. R. A. Mewaldt et al., (New York: AIP), 221, 2000.
15. Fisk, L. A., Gloeckler, G., Zurbuchen, T. H., and Schwadron, N. A., in AIP Conf. Proc. 528, *Acceleration and Transport of Energetic Particles Observed in the Heliosphere*, ed. R. A. Mewaldt et al., (New York: AIP), 229, 2000.
16. Leske, R. A. et al., submitted to SOHO-ACE Workshop Proc. (AIP), 2001.
17. Mason et al. *Ap. J.* 425, 843, 1994.
18. Dwyer, J. R., et al., *Proc. 26th Internat. Cosmic Ray Conf.* (Salt Lake City), 6, 147, 1999.
19, Slocum, P. S., et al., submitted to SOHO-ACE Workshop Proc. (AIP), 2001.
20. Ziegler, J. F., Biersack, J. P., and Littmark, U., *The Stopping and Range of Ions in Solids*, Vol. 1., Pergammon Press, New York, New York, 1985.
21. Geiss, J., et al. in *Apollo-16 Preliminary Science Report*, NASA SP-315, 231, p. 14-1, 1972.
22. Benkert J.P, Baur, H., Signer, P., and Wieler, R., *JGR* 98, 13147-13162, 1993.

Lunar Soils: A Long-Term Archive for the Galactic Environment of the Heliosphere?

Robert F. Wimmer-Schweingruber* and Peter Bochsler*

*Physikalisches Institut, Universität Bern, Sidlerstrasse 5, CH-3012 Bern, Switzerland

Abstract.
Solar wind implanted in surface layers ($\lesssim 0.03\mu$m) of lunar soil grains has often been analyzed to infer the history of the solar wind. In somewhat deeper layers, and thus presumably at higher implantation energies, a mysterious population, dubbed "SEP" for "solar energetic particle", accounts for the majority of the implanted gas - several orders of magnitude more than expected from the present-day flux of solar energetic particles. In addition, its elemental and isotopic composition is distinct from that of the solar system. While the heavy Ne isotopes are enriched relative to ^{20}Ne, ^{15}N is depleted relative to ^{14}N - a behavior that is hard to explain with acceleration of solar material. N is overabundant with respect to the noble gases (especially Ar). Here we show that interstellar pick-up ions (PUIs) which are ionized and accelerated in the heliosphere and subsequently implanted in lunar regolith grains can account for the properties of the "SEP" population. This implies that lunar soils preserve samples of the galactic environment of the solar system and may eventually be used as an archive for solar system "climate".

INTRODUCTION

Lunar soils contain a record of implanted noble gases since the formation of the lunar regolith (soil) more than $3 \cdot 10^9$ years ago. Figure 1 shows the ^{20}Ne/^{22}Ne isotope ratio vs. the cumulative release of ^{20}Ne for the Apollo 17 lunar soil sample 71501. The dominant contribution in the outermost layers of the grains which outgas first is of solar wind composition, hence this population is termed solar wind (SW). In later extraction steps, a population is found with a different composition than that of the the solar wind. It appears that this component has been implanted at energies far exceeding typical solar wind energies, but still below 100 keV/nuc [1], hence this population has been called "SEP" (for Solar Energetic Particles) [2]. Despite its apparent similarity with the composition of actual SEPs previously determined by *in-situ* observations with various spacecraft-borne instruments [3], this population cannot originate from SEPs. It exceeds by several orders of magnitude the amount expected from present-day fluxes of solar energetic particles. The "SEP" to solar wind abundance ratio measured in lunar soils (10% - 40%) disagrees with present-day expectations, even when allowing for the fact that solar wind neon may be trapped with an efficiency of only about 10%. Enhanced solar activity in the past is probably unable to explain the amount of the "SEP" component because the energy for acceleration needs to come

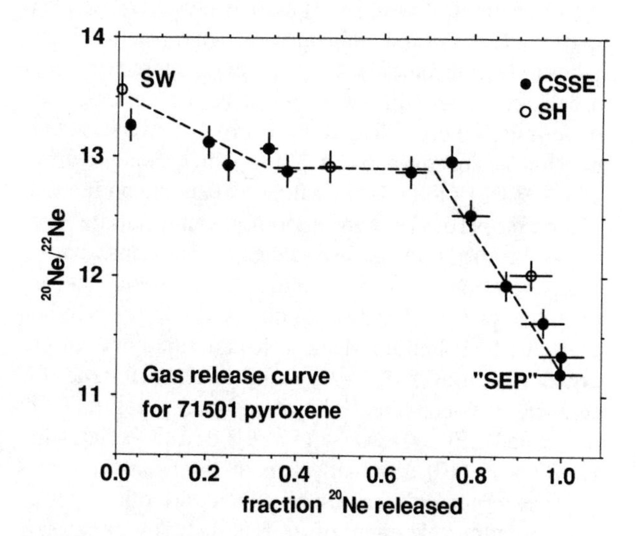

FIGURE 1. ^{20}Ne/^{22}Ne isotopic ratio versus the cumulative fraction of released neon for lunar soil 71501 pyroxene. Two different extraction techniques were applied, closed system stepped etching (CSSE, full symbols), and stepped heating (SH, open symbols). SW denotes solar wind isotopic composition, "SEP" that of the "SEP" component. Data are from Wieler *et al.* [2].

from the solar magnetic field which has not varied that much in the past [4, 5]. Here, we argue that this controversial "SEP" component is not solar, but originates

CP598, *Solar and Galactic Composition*, edited by R. F. Wimmer-Schweingruber

predominantly in interstellar pick-up ions (PUIs). We propose that this gas component should be more adequately termed "HEP" for "Heliospheric Energetic Particles". We will show that this new interpretation explains the compositional signatures of the HEP component, the amount of implanted HEP gas, as well as the implantation energy range.

INTERSTELLAR MATTER IN THE HELIOSPHERE

Interstellar neutral gas entering the inner heliosphere undergoes ionization mainly by charge exchange with solar wind ions or by ionization with solar ultraviolet. Once a particle is ionized, it experiences Lorentz forces due to the relative motion of the interplanetary magnetic field which is carried outwards from the sun with the solar wind plasma. The newly created ion is "picked up" and, together with the solar wind particles, it is gradually swept to the boundaries of the heliosphere. The pick-up process results in particle distributions which are extremely non-thermal with speeds ranging from about $-v_{sw}$ to $+v_{sw}$ in the solar wind reference frame. These velocity distributions are unstable to the generation of hydromagnetic waves [6, 7] which decay by generating turbulence in the ambient solar wind plasma. In regions of high turbulence, PUIs are much more efficiently accelerated than solar wind particles, because of their wide initial energy distribution. Strong power-law tails develop at speeds above v_{sw} [8] which contain up to $10\% - 20\%$ of the total PUI flux. Depending on the ionization properties of a given atom, it can penetrate more or less deeply into the heliosphere. Helium has particularly low ionization rates and, as a consequence, interstellar pick-up (HEP-) helium is similarly abundant as (true) SEP helium at the relevant suprathermal energies - even at 1 AU - resulting in characteristic differences between the He/O abundance ratios in CIR-associated SEPs (He/O $\sim$ 157 [9]) and the solar wind (He/O $\sim$ 75 [10]). Similarly, interstellar neon penetrates more deeply into the heliosphere than most other species (except helium) because of its relatively low ionization rates. Neon and helium are mainly ionized by solar ultraviolet [11], and the present-day flux of interstellar neon PUIs reaches its maximum near the orbit of the Earth. There it amounts to about 2.5×10^{-5} of the solar wind neon flux, depending somewhat on solar activity. Neutral hydrogen penetrates about as far as the orbit of Jupiter. Then it is partly ionized by ionization, but mainly by resonant charge exchange with solar wind protons [11]. The charge-exchange cross section of nitrogen is currently not known [11]. Nitrogen is probably ionized around 2-3 AU. Interstellar H, He, N, O, and Ne PUIs have been observed [12, 13, 14] at the orbit of the Earth or along the trajectory of the Ulysses spacecraft. The isotopic composition of interstellar pick-up helium, ^{4}He/^{3}He $\sim$ 4400 [13], or $\sim$ 5800 [15], differs significantly from the solar wind ^{4}He/^{3}He isotopic abundance ratio which amounts to typically 2300 [16, 17].

From in-situ investigations we know that interstellar He pick-up ions account for a large fraction of the suprathermal particle population [8, 18], ranging from 50% during solar quiet times to $\sim$ 10% during disturbed times[19]. Using the best currently known ionization rates for He and Ne [11] and the formalism of Vasyliunas and Siscoe [20] one can easily compute that the average ratio of interstellar Ne/He is depleted by about a factor $\sim$ 10 relative to its solar value. Hence we may expect that interstellar Ne accounts for between 1 - 5 % of the suprathermal flux of Ne. Indeed, Möbius et al.'[21] report that 8% of the suprathermal Ne associated with CIRs at 1 AU (and hence connected to a shock location beyond 1 AU) is singly charged.

CONTEMPORARY LOCAL INTERSTELLAR MEDIUM [LISM]

The density of the galactic interstellar medium varies over many orders of magnitude, from $\sim 0.01\text{cm}^{-3}$ to more than 10^6cm^{-3} in dense molecular clouds with a galactic average of about $\bar{n}_{\text{H gal}} \approx 1\ \text{cm}^{-3}$ [22]. Presently, the Sun and the heliosphere are located near the edge [23] of the so-called local interstellar cloud which has a density of $0.05 < n_H < 0.25\text{cm}^{-3}$ [24, 25, 26, 27, 28], a comparatively tenuous environment [29]. For this work we will adopt a value of $n_H = 0.1\text{cm}^{-3}$. The elemental composition of the local interstellar cloud is roughly comparable to the solar composition [30] although we would expect the "metals" to be enriched due to the chemical evolution of the ISM. The isotopic composition of the ISM is inferred from radio observations of interstellar molecules. The solar ^{14}N/^{15}N isotopic abundance ratio (200 ± 55) [31] is apparently somewhat lower than the terrestrial value ($\sim$ 272). The corresponding ratio in the ISM at the galactocentric distance of the Sun is 450 ± 22 [32]. It is generally assumed that the LISM has been subject to further nucleosynthetic processing since the solar system decoupled from galactic matter at the time of its formation. Thus, it appears that the abundance of ^{14}N increases with time with respect to ^{15}N. ^{14}N is the more 'secondary' isotope of the two (i. e. the species which is produced in stars and recycled by stellar winds and stellar explosions into the ISM after some 'primary' nuclei such as ^{12}C, ^{15}N, or ^{16}O have been produced in earlier generations of stars).

The isotopic composition of interstellar neon is

presently unknown. Since ^{20}Ne is certainly a primary nucleus, it is generally believed that the ISM neon has been decreasing its ^{20}Ne/^{22}Ne isotopic abundance ratio since the formation of the Sun. However, presently-available information from anomalous cosmic rays originating in accelerated PUIs that have been further accelerated at the heliospheric termination shock [33], gives no evidence for such a decrease within the present experimental uncertainties [34] .

We have summarized the isotopic and elemental signatures for other elements expected from galactic chemical evolution and the corresponding values measured in lunar soils in Table 1. The inferred HEP abundances are generally consistent with the behavior expected from galactic nucleosynthetic evolution. To our knowledge, no non-terrestrial D has been identified in lunar soils so far due to contamination issues. Because H and D PUIs reach their maximum production rate at several AU, well beyond the orbit of the Earth and the Moon, the HEP component should account for a considerably smaller fraction of implanted gas in lunar soils than in the case of He and Ne. Possibly, some asteroidal regoliths might have conserved a signature of interstellar HEP D. Note that in HEP N the heavier isotope is depleted with respect to the lighter one - a behavior that is hard to explain as due to accelerated solar wind or to to deeper implantation of the heavier isotope in lunar soil grains.

PAST AND FUTURE OF THE HELIOSPHERE

We now proceed to show that the amount and implantation energy of the HEP component follow naturally from the interpretation presented in this work. In its history, the solar system has orbited the galactic center about 18 times, assuming present-day orbital elements. Along its galactic orbit, the Sun and the heliosphere must have encountered interstellar clouds of varying density. In fact, the observation that the velocity dispersion of similar types of stars increases with their age is generally explained as the result of episodic encounters with dense molecular clouds [41], as is the possible outward galactic migration of the solar system to its present galactocentric distance suggested by Wielen *et al.* [42]. The observed size and density distributions of such clouds follow power laws, with the denser clouds being smaller than their more tenuous counterparts [e. g. 43].

Obviously, in a first approximation, the flux of interstellar PUIs will be proportional to the neutral density in the local environment, as long as their progenitor particles are atomic. With increasing density, molecular species will grow in importance, and, consequently, the radial distribution of the PUI flux will be altered because of the different ionization properties of molecules. This may explain why nitrogen, which forms very stable molecules and bounds strongly to interstellar dust particles, is too abundant in implanted gases when compared with e. g. Ar which, as a noble gas, does not form molecules.

ACCELERATION TO SUPRATHERMAL ENERGIES

Finding an estimate for the flux, Φ, of suprathermal, accelerated PUIs which are of prime importance for our interpretation of the HEP population of implanted gases in lunar soils, is somewhat more involved. In the following, we assume that it scales as a power law, $\Phi \sim n_\infty^{1+S}$. The case where $S = 0$ describes the situation in which the flux is strictly proportional to the interstellar neutral density and in which the ISM density has no influence on the acceleration process. The severe mass loading of the solar wind however, will lead to strongly enhanced turbulence, as is observed in the vicinity of comets [44, 45, 46, 47].

In-situ observations of accelerated PUIs strongly support the view that these particles are efficiently accelerated in turbulent regions of the heliosphere by a mechanism such as transit-time damping [48] of magnetic field fluctuations [8, 49] or possibly by other means. This type of acceleration can be viewed as a diffusion process in momentum space. The associated diffusion tensor, D_{pp}, increases with the square of the normalized power in the fluctuations of the magnetic field, $(\delta B/B)^2$. It is to be expected that the flux of accelerated PUIs, Φ, is proportional to the product of the number of PUIs available for acceleration and the efficiency of the injection or acceleration process. The number of PUIs available for acceleration is obviously proportional to the neutral density at heliocentric distance r, $n_{\mathrm{H}}(r)$. This in turn is proportional to the neutral number density in the interstellar medium, n_{H}. The efficiency of the injection and acceleration mechanism is inversely proportional to the acceleration time. This, in turn, is inversely proportional to D_{pp}. As we have already mentioned, D_{pp} is proportional to the amount of turbulence, $(\delta B/B)^2$, and this has been shown to be proportional to $n_{\mathrm{H}}(r)$ and hence to n_{H} [7]. Admittedly, the acceleration process of pick-up ions in turbulent regions in the heliosphere is not fully understood. Nevertheless, a power-law dependence of the acceleration efficiency on density is not without merits. Density inhomogeneities in the interstellar medium would certainly also contribute to a compression and subsequent enhancement of turbulence, for example. Moreover, in a dense interstellar environment, the fast magnetosonic speed in the heliosphere is enhanced due to the large pressure contribution of interstellar pick-up ions, result-

TABLE 1. Summary of solar and expected and observed behavior of the HEP component in lunar soils. * best estimate and/or expected temporal evolution. n.d.: not detectable, dec.: decreasing, inc.: increasing † assumed terrestrial ‡ J.M. Weygand, pers. comm., 2000 nrse: neutron rich species enriched

Signature	Solar	Reference	HEP*	Observed HEP	Reference
D/H	no deuterium	-	$\lesssim 2-3\times10^{-5}$ (dec.)	n.d.	[35, 36]
$^3He/^4He$	$\sim 4.3\times10^{-4}$	[16]	$< 4\times10^{-4}$ (inc.)	$(2.17\pm0.05)\times10^{-4}$	[37]
$^{12}C/^{13}C$	$\sim 89^\dagger$	-	< 89 (dec.)	-	-
$^{14}N/^{15}N$	200 ± 55	[31]	> 200 (inc.)	~ 300	[38]
$^{20}Ne/^{22}Ne$	13.7 ± 0.3	[16]	< 13.7 (dec.)	11.3 ± 0.3	[2]
$^{36}Ar/^{38}Ar$	$5.6\pm0.6^\ddagger$	[39]	< 5.6 (dec.)	~ 4.9	[37]
$^{84}Kr/^{86}Kr$	3.296 ± 0.013	[40]	< 3.3 (dec.)	3.151 ± 0.015	[40]
Xe isotopes	solar wind		heavier (nrse)	heavier (nrse)	[40]

ing in stronger interplanetary shocks and formation of corotating shocks nearer to one AU. Thus we assume that the flux of accelerated, suprathermal PUIs is proportional to a power of the neutral particle density, $\Phi \sim n_H^{1+S}$, where we believe that $S \approx 1$.

Therefore, during times when the heliosphere encounters dense clouds, the flux of suprathermal PUIs is considerably higher than today. Simplistically inserting the average neutral hydrogen density in the galaxy, $\bar{n}_{H\ gal} \approx 1\text{cm}^{-3}$, corresponding to the passage through a moderately dense cloud complex, we arrive at a flux 100 times greater than today.

In order to find the average flux to which lunar soils have been exposed during the history of the solar system, we need to weight the expected flux for a given density of the LIC with the occurrence rate for clouds of that density. In a previous paper [50], we considered a large volume, L_0^3, with $L_0 \sim 10^3$pc which contains one cloud of that dimension and a density n_0, two clouds with twice the density, and half its linear dimension, etc. down to a scale $L_1 \sim 0.1$pc (typically considered to be the size of the smallest and densest clouds). This reflects the mass-conserving, stationary state of fragmenting and collapsing clouds. The resulting mass distribution of the clouds, $n_m \sim m^{-2}$, lies close to observed values, $n_m \sim m^{-1.8}$[51]. Assuming that the average density of this model cloud conglomerate reflects $\bar{n}_{H\ gal} \approx 1\ \text{cm}^{-3}$, one can derive $n_0 \approx 0.05\ \text{cm}^{-3}$. Next, we need to account for the fact that only a fraction, η_{max}, of the available pick-up ions can be accelerated. η_{max} can be expressed as a size, L_η below which clouds contribute at most η_{max} of their density to suprathermal particles. Following our previous approach [50, 4] this can be evaluated and yields

$$\langle\Phi\rangle = (3-D)\frac{\varphi_{today}}{1-\left(\frac{L_1}{L_0}\right)^{3-D}}\frac{\rho_0}{\rho_{today}} * \left[\frac{\eta_0}{3-D-R(1+S)}\left(\frac{\rho_0}{\rho_{today}}\right)^S * \left(1-\alpha^{3-D-R(1+S)}\right) + \frac{\eta_{max}}{3-D-R}\left(\alpha^{3-D-R}-\left(\frac{L_1}{L_0}\right)^{3-D-R}\right)\right],$$

where D is the power-law index of the distribution of the number of clouds in a given size interval, R the power-law index for the density distribution in a given size interval. We have plotted the resulting enhancement factor of the long-term average compared to the present-day suprathermal pick-up flux $\langle\Phi\rangle/(\eta_0\varphi_{today})$ for various values of D in Figure 2. The solid curve in Figure 2 is for $D = 2.00$, the dashed and dotted curves for other values of D are arranged symmetrically and the values of D are indicated. We do not cover the entire range of values that D might have. Probable values for D are $1.2 < D < 2.7$ with the more extreme ones being less probable. For the curves in Figure 2, $L_0/L_1 = 10^4$, $\eta_{max} = 0.1$, $R = 1$, $n_{today} = 0.1\text{cm}^{-3}$ and todays ac-

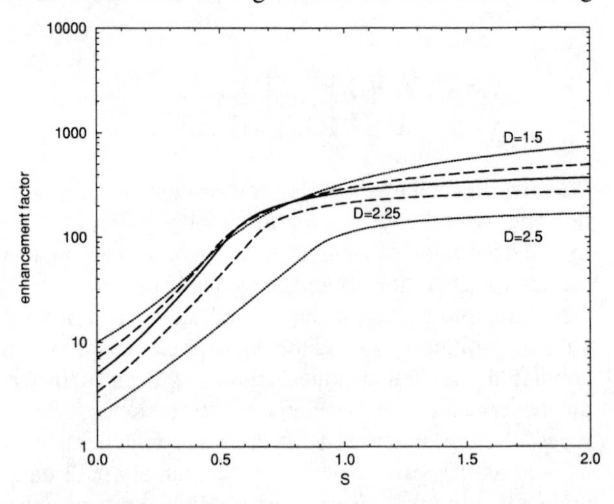

FIGURE 2. Enhancement factor $\Phi/(\eta_0\varphi_{today})$ vs. S for $D =$ 1.5, 1.75, 2.00, 2.25, and 2.50. The solid curve is for $D = 2.00$, the dashed and dotted curves for other values of D are arranged symmetrically and the values of D are indicated. For all curves, $L_1 = 10^{-1}$, $\eta_{max} = 0.1$, $\eta_0 = 10^{-3}$, $R = 1$, $\rho_{today} = 0.1\ \text{cm}^{-3}$.

celeration efficiency $\eta_0 = 10^{-3}$. The enhancement factor $\langle\Phi\rangle / (\eta_0 \varphi_{today})$ amounts to about 300 for $S = 1$ and $D = 2$. The long-term average of the flux of accelerated, i. e. suprathermal, pick-up ions is between 100 and 1000 times higher than today. The curves begin to flatten at $S \lesssim 1$ because then only the more tenuous clouds contribute in the non-linear fashion envisioned in our previous paper [50]. The denser clouds can only contribute a fixed maximum fraction (η_{max}) of their density. Therefore, the growth with S slows down.

CONCLUSIONS

We have shown that the non-solar composition of the HEP component measured in lunar soils may be explained by its interstellar origin. The accelerated PUIs have energies which lie in the appropriate range to explain the implantation depth of the HEP component. As we have already mentioned, it accounts for 10% - 40% of the total amount of implanted gas in lunar soils.

Mewaldt et al. [52] have simulated the isotopic composition of Ne implanted in lunar soils (assumed to be pure SiO_2) by applying the TRIM package [53] to the measured interplanetary spectrum of particles[54]. Assuming a solar value of 13.7 for the impinging solar and suprathermal $^{20}Ne/^{22}Ne$ isotope abundance ratio they find that beyond a depth of several 1000 Å this ratio lies at 11.3 in the implanted gas, essentially confirming work by Wieler et al. (mentioned in [1]), and Bochsler and Wimmer[55] that was not formally published. The reason for this depth-dependent variation in composition lies in the different ranges of the different isotopes in matter. The Ne isotopic ratio of the deeply implanted particles can be achieved with different combinations of energy spectral index and isotope ratio in the suprathermal particles [52]. The amount of the HEP component remains unexplained in their scenario, they find that an increase by a factor ~ 100 in the suprathermal tails is needed.

Considering that interstellar neon contributes several percent to the flux of suprathermal particles, our enhancement factor of ~ 300 shows that accelerated interstellar pick-up ions contributed a substantial fraction to the total flux of suprathermal particles in the past.

This interpretation has implications beyond the mere explanation of a badly understood gas component in the lunar soil. Since their formation more than $3 \cdot 10^9$ years ago, lunar soils have been preserving samples of the LISM through which the solar system has migrated. Deciphering this $> 3 \cdot 10^9$ year archive will not be easy. The moon is constantly being bombarded by micro-meteorites and occasionally by larger meteorites. This results in the process known as "lunar gardening", a constant tilling of the lunar soil to greatly varying depths. Encounters with dense interstellar clouds are only of short duration ($\sim 10^4 - 10^5$ y) and may possibly serve as markers for lunar soil dating, much as volcanic ashes in terrestrial sediments aid in understanding climate archives. Obviously, improvements in the dating of lunar soils are important. As we believe we have shown in this work, the lunar regolith may not only serve to understand the ancient solar wind (and thus processes active in the early history of the Sun), but also as an archive for heliospheric climate, and as a "travel diary" for the voyage of the solar system through the galaxy.

ACKNOWLEDGMENTS

We wish to thank Rainer Wieler, Otto Eugster, André Maeder, Fritz Bühler, and Kurt Marti for helpful discussions. This work was supported by the Schweizerischer Nationalfonds.

REFERENCES

1. Wieler, R., *Space Sci. Rev.*, **85**, 303 – 314 (1998).
2. Wieler, R., Baur, H., and Signer, P., *Geochim. et Cosmoschim. Acta*, **50**, 1997 – 2017 (1986).
3. Mewaldt, R. A., Spalding, J. D., and Stone, E. C., *Astrophys. J.*, **280**, 892 – 901 (1984).
4. Wimmer-Schweingruber, R. F., Lunar soils: A long-tern archive for the galactic environment of the solar system?, Habilitation Thesis (2000), Physikalisches Institut, Universität Bern, Switzerland.
5. Wimmer-Schweingruber, R. F., and Bochsler, P., *Adv. Space Sci.* (2001), in press.
6. Wu, C. S., and Davidson, R. C., *J. Geophys. Res.*, **77**, 5399 – 5406 (1972).
7. Lee, M. A., and Ip, W.-H., *J. Geophys. Res.*, **92**, 11041 – 11052 (1987).
8. Gloeckler, G., Geiss, J., Roelof, E. C., Fisk, L. A., Ipavich, F. M., Ogilvie, K. W., Lanzerotti, L. J., von Steiger, R., and Wilken, B., *J. Geophys. Res.*, **99**, 17637 – 17643 (1994).
9. Reames, D. V., *Adv. Space Sci.*, **15**, (7)41 – (7)51 (1995).
10. Bochsler, P., Geiss, J., and Kunz, S., *Sol. Phys.*, **103**, 177 – 201 (1986).
11. Ruciński, D., Cummings, A. C., Gloeckler, G., Lazarus, A. J., Möbius, E., and Witte, M., *Space Sci. Rev.*, **78**, 73 – 84 (1996).
12. Möbius, E., Hovestadt, D., Klecker, B., Scholer, M., Gloeckler, G., and Ipavich, F. M., *Nature*, **318**, 426 – 429 (1985).
13. Gloeckler, G., *Space Sci. Rev.*, **78**, 335 – 346 (1996).
14. Geiss, J., and Witte, M., *Space Sci. Rev.*, **78**, 229 – 238 (1996).
15. Salerno, E., Bühler, F., Bochsler, P., Busemann, H., Eugster, O., Zastenker, G. N., Aganof, Y. N., and Eismont, N. A., "Direct measurement of $^3He/^4He$ in the LISM with the COLLISA experiment", in *Solar and Galactic Composition*, edited by R. F. Wimmer-Schweingruber,

AIP conference proceedings, Woodbury, NY, 2001, this volume.
16. Geiss, J., Bühler, F., Cerutti, H., Eberhardt, P., and Filleux, C., "Solar wind composition experiment", in *Apollo 16 Prelim. Sci. Rep.*, 1972, pp. 14.1 – 14.10, NASA SP-315.
17. Bodmer, R., and Bochsler, P., *Astron. Astrophys.*, **337**, 921 – 927 (1998).
18. Hilchenbach, M., Grünwaldt, H., Kallenbach, R., Klecker, B., Kucharek, H., Ipavich, F. M., and Galvin, A. B., "Observations of Suprathermal Helium at 1 AU: Carge States in CIRs", in *Solar Wind Nine*, edited by S. R. Habbal, R. Esser, J. V. Hollweg, and P. A. Isenberg, American Institute of Physics, 1999, pp. 605 – 608.
19. Gloeckler, G., Fisk, L. A., Zurbuchen, T. H., and Schwadron, N. A., "Sources, Injection and Acceleration of Heliospheric Ion Populations", in *Acceleration and Transport of Energetic Particles Observed in the Heliosphere*, edited by R. A. Mewaldt, J. R. Jokipii, M. A. Lee, E. Moebius, and T. H. Zurbuchen., AIP conference proceedings, Woodbury, New York, 2000, pp. 221 – 228.
20. Vasyliunas, V. M., and Siscoe, G. L., *J. Geophys. Res.*, **81**, 1247 – 1252 (1976).
21. Möbius, E., Morris, D., Popecki, M. A., Klecker, B., Kistler, L. M., and Galvin, A. B., *Geophys. Res. Lett.* (2001), submitted.
22. Dyson, J. E., and Williams, D. A., *The Physics of the Interstellar Medium*, Institute of Physics Publishing, Bristol, 1997.
23. Linsky, J. L., Redfield, S., Wood, B. E., and Piskunov, N., *Astrophys. J.*, **528**, 756 – 766 (2000).
24. Suess, S. T., *Rev. Geophys.*, **28**, 97 – 115 (1990).
25. Frisch, P. C., *Science*, **265**, 1423 – 1427 (1994).
26. Davidsen, A. F., *Science*, **259**, 327 – 334 (1993).
27. Lallement, R., Bertaux, J. L., and Clarke, J. T., *Science*, **260**, 1095 – 1098 (1993).
28. Wang, C., Richardson, J. D., and Gosling, J. T., *Geophys. Res. Lett.*, **27**, 2429 – 2432 (2000).
29. Frisch, P. C., and York, D. G., "Interstellar Clouds Near the Sun", in *The Galaxy and the Solar System*, edited by R. Smoluchowski, J. N. Bahcall, and M. S. Matthews, 1986, pp. 83 – 100, tucson, Univ. of Ariz. Press.
30. Gloeckler, G., and Geiss, J., "Composition of the local interstellar cloud from observations of interstellar pickup ions", in *Solar and Galactic Composition*, edited by R. F. Wimmer-Schweingruber, AIP conference proceedings, Woodbury, NY, 2001, this volume.
31. Kallenbach, R., Geiss, J., Ipavich, F. M., Gloeckler, G., Bochsler, P., Gliem, F., Hefti, S., Hilchenbach, M., and Hovestadt, D., *Astrophys. J.*, **507**, L185 – L188 (1998).
32. Wielen, R., and Wilson, T. L., *Astron. Astrophys.*, **326**, 139–142 (1997).
33. Fisk, L. A., Kozlovsky, B., and Ramaty, R., *Astrophys. J.*, **190**, L35 – L37 (1974).
34. Leske, R. A., Mewaldt, R. A., Cummings, A. C., Stone, E. C., and von Rosenvinge, T. T., *Space Sci. Rev.*, **78**, 149 – 154 (1996).
35. Epstein, S., and Taylor Jr., H. P., "The concentration and isotopic composition of hydrogen, carbon and silicon in Apollo 11 lunar rocks and minerals", in *Proceedings of the Apollo 11 Lunar Science Conference*, Pergamon, New York, 1970, pp. 1085 – 1096.
36. Merlivat, L., Lelu, M., Nief, G., and Roth, E., "Deuterium, hydrogen, and water content of lunar material", in *Proceedings of the Fifth Lunar Conference*, 1974, pp. 1885 – 1895.
37. Benkert, J.-P., Baur, H., Signer, P., and Wieler, R., *J. Geophys. Res.*, **98**, 13147 – 13162 (1993).
38. Mathew, K., Kerridge, J., and Marti, K., *Geophys. Res. Lett.*, **25**, 4293 – 4296 (1998).
39. Weygand, J. M., Ipavich, F., Wurz, P., Paquette, J., and Bochsler, P., *ESA SP*, **446**, 22 – 25 (1999).
40. Wieler, R., and Baur, H., *Meteoritics*, **29**, 570 – 580 (1994).
41. Binney, J., and Tremaine, S., *Galactic Dynamics*, Princeton University Press, Princeton, New Jersey, 1987.
42. Wielen, R., Fuchs, B., and Dettbarn, C., *Astron. Astrophys.*, **314**, 438 – 447 (1996).
43. Henriksen, R. N., "Turbulence and Magnetic Fields in Molecular Clouds", in *IAU Symposium 147: Fragmentation of Molecular Clouds and Star Formation*, edited by E. Falgarone, F. Boulanger, and G. Duvert, Kluwer, Dordrecht, Netherlands, 1991, pp. 83 – 92.
44. Biermann, L., Brosowski, B., and Schmidt, H. U., *Sol. Phys.*, **1**, 254 – 284 (1967).
45. Tsurutani, B. T., and Smith, E. J., *Geophys. Res. Lett.*, **13**, 259 – 262 (1986).
46. Tsurutani, B. T., and Smith, E. J., *Geophys. Res. Lett.*, **13**, 263 – 266 (1986).
47. Smith, E. J., Tsurutani, B. T., Slavin, J. A., Jones, D. E., Siscoe, G. L., and Mendis, D. A., *Science*, **232**, 382 – 385 (1986).
48. Fisk, L. A., *J. Geophys. Res.*, **81**, 4633 – 4640 (1976).
49. Schwadron, N. A., Fisk, L. A., and Gloeckler, G., *Geophys. Res. Lett.*, **23**, 2871 – 2874 (1996).
50. Wimmer-Schweingruber, R. F., and Bochsler, P., "Is there a record of interstellar pick-up ions in the lunar regolith?", in *Acceleration and Transport of Energetic Particles Observed in the Heliosphere*, edited by R. A. Mewaldt, J. R. Jokipii, M. A. Lee, E. Moebius, and T. H. Zurbuchen., AIP conference proceedings, Woodbury, New York, 2000, pp. 270 – 273.
51. Stutzki, J., Genzel, R., Graf, U., Harris, A. I., Sternberg, A., and Güsten, R., "UV penetrated clumpy molecular cloud cores", in *IAU Symposium 147: Fragmentation of Molecular Clouds and Star Formation*, edited by E. Falgarone, F. Boulanger, and G. Duvert, Kluwer, Dordrecht, Netherlands, 1991, pp. 235 – 244.
52. Mewaldt, R. A., Ogliore, R. C., Gloeckler, G., and m. Mason, G., "A new look at neon-C and SEP-neon", in *Solar and Galactic Composition*, edited by R. F. Wimmer-Schweingruber, AIP conference proceedings, Woodbury, NY, 2001, this volume.
53. Ziegler, J. F., Biersack, J. P., and Littmark, U., *The Stopping and Range of Ions in Solids*, Pergamon, New York, 1992.
54. Mewaldt, R. A., Mason, G. M., Gloeckler, G., Christian, E. R., Cohen, C. M. S., Davis, A. C. C. A. J., Dwyer, J. R., Gold, R. E., Krimigis, S. M., Leske, R. A., Mazur, J. E., Stone, E. C., von Rosenvinge, T. T., Wiedenbeck, M. E., and h. Zurbuchen, T., "Long-Term fluences of energetic particles in the heliosphere", in *Solar and Galactic Composition*, edited by R. F. Wimmer-Schweingruber, AIP conference proceedings, Woodbury, NY, 2001, this volume.
55. Bochsler, P., and Wimmer-Schweingruber, R. F., *EOS, Trans. AGU*, **78**, F534 (1997), abstract.

Primordially produced helium-4 in the presence of neutrino oscillations

D.P. Kirilova

Institute of Astronomy, Bulgarian Academy of Sciences, blvd. Tsarigradsko Shosse 72, Sofia, Bulgaria

Abstract. The production of helium-4 during the cosmological nucleosynthesis in the presence of active–sterile neutrino oscillations, $\nu_e \leftrightarrow \nu_s$, efficient after decoupling of electron neutrino, is analyzed. All known oscillation effects on primordial nucleosynthesis, namely: increase of the effective degrees of freedom during nucleosynthesis, neutrino spectrum distortion, depletion of electron neutrino number density and generation of neutrino-antineutrino asymmetry, are precisely taken into account.

Primordially produced ^{4}He abundance is calculated, in a self-consistent study of the kinetics of the nucleons and the oscillating neutrinos, for the full range of parameters of the oscillation model with small mass differences: $\delta m^2 \leq 10^{-7}$ eV^2. A considerable relative increase of helium-4, up to 14% for non-resonant oscillations and up to 32% for resonant ones is registered for a certain interval of oscillations parameters values. Combined iso-helium contours $\delta Y_p/Y_p = 3\%, 5\%, 7\%$ for resonant and non-resonant oscillations are presented. Cosmological constraints on oscillation parameters and on the sterile solar neutrino solutions are discussed.

INTRODUCTION

In this work we investigate a modification of the standard Big Bang Nucleosynthesis with electron–sterile neutrino oscillations $\nu_e \leftrightarrow \nu_s$. The positive indications for oscillations, obtained at the greatest neutrino experiments (SuperKamiokande, Soudan 2, LSND, etc.) and the recent SNO result [1], turned the subject of neutrino oscillations into one of the hottest points of astrophysics and neutrino physics. The solar neutrino problem, atmospheric neutrino anomaly and the positive results of LSND experiments can be naturally resolved by the phenomenon of neutrino oscillations, implying nonzero neutrino mass and mixing. Massive neutrinos may also play the role of the hot dark matter component, needed for a successful large-scale structure formation in the Universe.

On the other hand, Big Bang Nucleosynthesis (BBN) is often used to probe the physical conditions of the early Universe. Requirements for an agreement between theoretically predicted and inferred from observational data primordial abundances of light elements D, ^{3}He, ^{4}He, ^{7}Li, restricts physics beyond the standard models. Neutrino oscillations present an indication of physics beyond the standard electroweak model. Hence, it is appropriate to study precisely the influence of neutrino oscillations on BBN and constrain neutrino oscillations parameters.

In this work we analyze ^{4}He primordial production taking into account all known effects of $\nu_e \leftrightarrow \nu_s$ oscillations on the primordial synthesis of ^{4}He. [2]

Although disfavored by the recent combined analyses of neutrino oscillations experimental data as a preferred oscillation channels for solving the solar and atmospheric neutrino problems, sterile neutrinos are worth considering because they are inevitable for the explanation of all the oscillation experiments data including LSND results, they are considered as a possible minor additional channels in both the solar and the atmospheric oscillation cases, and besides they may play essential role in structure formation and as a dark matter candidate in case they are massive.

We discuss the case when neutrino oscillations become effective after the electron neutrino decoupling from the plasma (i.e for $\delta m^2 \leq 10^{-7}$eV^2). The primor-

[1] The electron neutrino flux measured by the charged current reaction rate by SNO is 3.3σ apart from the SuperKamiokande precision value of the flux from elastic scattering reaction rate. This is considered an indication of non-electron flavor component in the solar neutrino flux, and may be interpreted in terms of neutrino oscillations.

[2] The effect of flavor neutrino oscillations to BBN is negligibly small because the temperatures and hence the densities of the neutrinos with different flavors are almost equal. On the contrary, the effects of active-to-sterile neutrinos may be considerable, because the sterile neutrinos may differ considerably (by temperature and density) from the active ones.

CP598, *Solar and Galactic Composition,* edited by R. F. Wimmer-Schweingruber

dial production of ^{4}He was calculated in the non-resonant and resonant oscillation cases. In both cases strong over-production of ^{4}He was found possible - up to 14% and 32%, correspondingly. Updated iso-helium contours are presented and the cosmological constraints on the oscillation parameters are discussed.

The oscillation effects on BBN and the required kinetic approach for their description are discussed in the next section. The results on ^{4}He primordially produced abundance in the presence of oscillations and the cosmological constraints on oscillation parameters are discussed in the last section.

COSMOLOGICAL NUCLEOSYNTHESIS WITH NEUTRINO OSCILLATIONS

Standard Cosmological Nucleosynthesis

According to the standard BBN, during the early hot and dense epoch of the Universe, light elements D, ^{3}He, ^{4}He, ^{7}Li were synthesized successfully. ^{4}He was the most abundantly produced. Only negligible amounts of D, ^{3}He and ^{7}Li were formed. Because of the low density, growing Coulomb barriers and stability gaps at A=5, A=8, the formation of larger nuclei was postponed until the formation of stars several billions of years later. The most reliable and abundant data are now available for ^{4}He, therefore it is the traditionally used element for the analysis of the oscillations effect on nucleosynthesis.

The contemporary values for the mass fraction of ^{4}He, Y_p, inferred from observational data, are 0.238–0.245 (the systematic errors are supposed to be around 0.007) [1].

^{4}He is a result of a complex network of nuclear reactions, proceeding after the neutron-to-protons freezing. It essentially depends on the freezing ratio $(n/p)_f$. The reactions :

$$\begin{aligned} \nu_e + n &\leftrightarrow p + e^- \\ e^+ + n &\leftrightarrow p + \tilde{\nu}_e. \end{aligned} \qquad (1)$$

maintained the equilibrium of nucleons at high temperature ($T > 1$ MeV). Their freeze-out occurred when in the process of Universe cooling the rates of these weak processes, Γ_w, became comparable and less than the expansion rate $H(t)$:

$$\Gamma_w \sim G_F^2 E_\nu^2 N_\nu \leq H(t) \sim \sqrt{g_{eff}}\, T^2$$

Then nucleon number densities departed from their thermal equilibrium values, and decreased slowly due to the weak interactions of eq.(1) and the neutrons decays that proceeded until the effective synthesis of D began.

The produced ^{4}He is a strong function of the number of the effective degrees of freedom at BBN epoch, g_{eff}, and the neutron mean lifetime τ_n, parametrizing the weak interactions strength. Besides, ^{4}He is a logarithmic function of the baryon-to-photon ratio η, due to the nuclear reactions dependence on nucleon densities, i.e. $Y_p(g_{eff}, \tau_n, \eta)$. Also it depends on the electron neutrino number density and spectrum, and on the neutrino-antineutrino asymmetry, which enter through Γ_w. In the standard BBN model the number of neutrino flavors equal to three, zero lepton asymmetry and equilibrium neutrino number densities and spectrum distribution is assumed:

$$n_{\nu_e}(E) = (1 + \exp(E/T))^{-1}.$$

Almost all neutrons, present at the beginning of nuclear reactions, are sucked into ^{4}He. So, the primordially produced mass fraction of ^{4}He can be approximated by $Y_p \sim 2(n/p)_f/(1+n/p)_f exp(-t/\tau_n)$.

Primordially produced ^{4}He abundance Y_p, is calculated with great precision within the standard BBN model [2], the theoretical uncertainty is less than 0.1% ($|\delta Y_p| < 0.0002$) within a wide range of η. The best baryometer now is deuterium. The predicted Y_p at the best fit value of η, obtained from deuterium measurements, $\eta = 5 \times 10^{-10}$, is $Y_p = 0.2462$. The recent CMB anisotropy data is also in remarkable agreement with the baryon density determined from deuterium measurements and BBN.

So, the predicted primordial ^{4}He abundance Y_p is in accordance with the observational data and is consistent with other light element abundances.

Effects of neutrino oscillations on nucleosynthesis

Cosmological nucleosynthesis with neutrino oscillations was studied in numerous publications [3]. In case neutrino oscillations are present in the Universe primordial plasma, they may lead to changes in the Big Bang Nucleosynthesis, depending on the oscillation channels and the way they proceed.

The basic idea of *oscillations* is that mass eigenstates ν_i are distinct from the flavor eigenstates ν_f:

$$\nu_i = U_{if}\, \nu_f \quad (f = e, \mu, \tau).$$

Then in the simple two-neutrino oscillation case, the probability to find at a distance l a given neutrino type in an initially homogeneous neutrino beam of the same type is:

$$P_{ff} = 1 - \sin^2 2\vartheta \sin^2\left(\frac{\delta m^2}{4E} l\right),$$

where δm^2 (the neutrino squared mass difference) and ϑ (the oscillations mixing angle) are the oscillation parameters. E is the neutrino energy.

The medium distinguishes between different neutrino types due to their different interactions with fermions present in the hot plasma of BBN epoch. This leads to different potentials for different neutrino types. In the adiabatic case the effect of the medium can be described by introducing matter oscillation parameters that are expressed through the vacuum ones and through the characteristics of the medium. The matter mixing angle is then

$$\sin^2\vartheta_m = \sin^2\vartheta/[\sin^2\vartheta + (Q \mp L - \cos 2\vartheta)^2],$$

where $Q = -bE^2T^4/(\delta m^2 M_W^2)$, $L = -aET^3L^\alpha/(\delta m^2)$, L^α is expressed through the fermion asymmetries of the plasma, a and b are positive constants different for the different neutrino types, $-L$ corresponds to the neutrino and $+L$ to the antineutrino case.

Although in general the medium suppresses oscillations by decreasing their amplitude, there also exists a possibility of enhanced oscillations transfer, in case a resonant condition between the parameters of the medium and the oscillations parameters holds:

$$Q \mp L = \cos 2\vartheta. \tag{2}$$

Then the mixing in matter becomes maximal, independently of the value of the vacuum mixing angle.

At BBN epoch with the cooling of the Universe, an interesting interplay between the two terms is observed. At high temperatures when $|Q| > |L|$, $\delta m^2 > 0$ corresponds to a non-resonant case, while $\delta m^2 < 0$ corresponds to a resonant case, and the resonance holds in both neutrino and antineutrino sectors. At low temperatures, when $|Q| < |L|$, the resonance is possible either for neutrinos in the case $\delta m^2 > 0$ or for antineutrinos in the case $\delta m^2 < 0$.

Oscillations are capable to shift neutrino number densities and spectrum from their equilibrium values. Besides, oscillations may change neutrino-antineutrino asymmetry and excite additional neutrino types. Thus, *the presence of neutrino oscillations invalidates the main BBN assumptions about three neutrino flavors, zero lepton asymmetry and equilibrium neutrino number densities and energy distribution.*

Through these effects oscillations affect the expansion rate and the weak interaction rates. Shifting particle densities and energy spectrum of the electron neutrinos from their equilibrium values, oscillations directly influence the kinetics of nucleons during the weak freeze-out.

Hence, neutrino oscillations may effect primordial nucleosynthesis by

- exciting additional degrees of freedom

Active-sterile oscillations may keep sterile neutrinos in thermal equilibrium [4] or bring them into equilibrium in case they have already decoupled. This leads to faster Universe expansion $H(t) \sim g_{eff}$, and to earlier n/p-freezing, $T_f \sim (g_{eff})^{1/6}$, at times when neutrons were more abundant [4, 5]:

$$n/p \sim \exp(-(m_n - m_p)/T_f)$$

This effect leads to ^{4}He overproduction. However, observational data on helium allows not more than one additional neutrino type, therefore forbids efficient production of sterile neutrinos due to oscillations.

- distorting the neutrino spectrum

Since oscillation rate is energy dependent $\Gamma \sim \delta m^2/E$ the low energy neutrinos start to oscillate first, and later the oscillations become noticeable for the more energetic neutrinos. Due to that, the neutrino spectrum may become strongly distorted. This effect was shown to be considerable for the active-sterile oscillations in vacuum [6] and in matter [7, 8]. The effect was proved important in the resonant oscillations case [9] and in the non-resonant one [10].

The neutrino spectrum distortion effect on ^{4}He primordial abundance has two aspects:

An average decrease of the energy of electron neutrinos leads to a decrease in Γ_w, and subsequently increases the freezing temperature and primordially produced ^{4}He.

On the other hand, due to the threshold for the reaction $\tilde{\nu}_e + p \to n + e^+$, when due to oscillations the energy of the greater part of the neutrinos becomes smaller than that threshold, the $(n/p)_f$-ratio decreases leading to a decrease of Y_p.

The total effect is an overproduction of ^{4}He primordial abundance.

- depleting the active neutrino number densities N_ν and $N_{\tilde{\nu}}$.

The effect was first studied for vacuum oscillations in ref. [6] and for matter oscillations in ref. [11]. It was precisely calculated with the account of spectrum spread in ref.[8]. Electron neutrino depletion slows down the weak rates, $\Gamma_w \sim N_\nu E_\nu^2$, and *leads to an earlier n/p-freezing and an overproduction of ^{4}He yield.* The evolution of electron neutrino depletion due to $\nu_e \leftrightarrow \nu_s$ oscillations was calculated and presented for different mass differences and for different mixing angles in ref. [10].

The net effect of spectrum distortion and neutrino depletion on the production of ^{4}He may be considerable (see Fig.2 from ref. [9]) and much stronger (several times larger) than the effect due to excitation of an additional neutrino type.

- neutrino-antineutrino asymmetry growth

The idea of neutrino-antineutrino asymmetry generation during the resonant transfer of neutrinos was first proposed in ref.[12]. Dynamically produced asymmetry exerts back effect to oscillating neutrino and may change its oscillation pattern [7, 13]. Thus it may effect indirectly BBN, even when its value is not sufficiently high to have a direct kinetic effect on the synthesis of light elements. For the case of small mass differences it was proven that even very small asymmetries $L << 0.01$ considerably influence nucleosynthesis through oscillations, and therefore asymmetry effect on nucleosynthesis should be accounted for during asymmetry's full evolution.

Dynamically produced asymmetry suppresses oscillations at small mixing angles, leading to less overproduction of ^{4}*He.* The effect of the oscillations generated asymmetry on ^{4}He was analyzed for hundreds of $\delta m^2 - \vartheta$ combinations in ref. [9]. In the resonant case the asymmetry effect on BBN was numerically analyzed and shown to be considerable – up to about a 10% relative decrease in ^{4}He compared with the case without asymmetry account.

The required kinetic approach

It is impossible to describe analytically, without some radical approximations, the non-equilibrium picture of active–sterile neutrino oscillations, producing non-equilibrium neutrino number densities, distorting neutrino spectrum and generating neutrino-antineutrino asymmetry.

We have provided self-consistent analysis of the evolution of the nucleons and the oscillating neutrinos in the high temperature Universe. Exact kinetic equations for the nucleons and for the neutrino density matrix *in momentum space* [8] were used. This allowed to describe precisely the spectrum distortion, neutrino depletion and neutrino asymmetry and its back effect *at each neutrino momentum.*

The equation for the neutron number densities in momentum space n_n reads:

$$\begin{aligned}
(\partial n_n/\partial t) &= H p_n\,(\partial n_n/\partial p_n) + \\
&+ \int \mathrm{d}\Omega(e^-,p,\nu)|\mathcal{A}(e^- p \to \nu n)|^2\,[n_{e^-} n_p(1-\rho_{LL}) - \\
&- n_n \rho_{LL}(1-n_{e^-})] \\
&- \int \mathrm{d}\Omega(e^+,p,\tilde{\nu})|\mathcal{A}(e^+ n \to p\tilde{\nu})|^2\,[n_{e^+} n_n(1-\bar{\rho}_{LL}) - \\
&- n_p \bar{\rho}_{LL}(1-n_{e^+})]\,. \qquad (3)
\end{aligned}$$

where $\mathrm{d}\Omega(i,j,k)$ is a phase space factor and $\mathcal{A}$ is the amplitude of the corresponding process, ρ_{LL} and $\bar{\rho}_{LL}$ at each integration step of eq. (3) are taken from the simultaneously performed integration of the set of equations for neutrino density matrix.

The equation provides a simultaneous account of the different competing processes, namely: neutrino oscillations (entering through ρ_{LL} and $\bar{\rho}_{LL}$), Hubble expansion (first term) and weak interaction processes (next terms).

The numerical analysis was performed for the temperature interval [2 MeV,0.3 MeV].

HELIUM-4 OVERPRODUCTION DUE TO $\nu_E \leftrightarrow \nu_S$ NEUTRINO OSCILLATIONS

Main results

We have calculated precisely the n/p-freezing, which is essential for the production of ^{4}He, till temperature 0.3 MeV. The analysis was provided for the non-resonant case in ref. [10] and for the resonant case in ref. [9]. Hundreds of $\delta m^2 - \vartheta$ combinations were explored. The neutron decay was accounted adiabatically till the beginning of nuclear reactions at about 0.09 MeV.

The overproduction of the primordial ^{4}He, $\delta Y_p = Y_p^{osc} - Y_p$ in the presence of $\nu_e \leftrightarrow \nu_s$ oscillations was calculated for the full set of oscillations parameters of the model: all mixing angles ϑ and mass differences $\delta m^2 \leq 10^{-7}$ eV^2.

The neutron-to-nucleons freezing ratio $X_n^f = (n/(n+p))_f$ as a function of neutrino mass differences for different mixings is shown in Fig.1 for $\delta m^2 > 0$. In the non-resonance case the oscillations effects become very small (less than 1%) for small mixings: as small as $sin^2 2\vartheta = 0.1$ for $\delta m^2 = 10^{-7}$ eV^2, and for small mass differences: $\delta m^2 < 10^{-10}$ eV^2 at maximal mixing. For very small mass differences $\delta m^2 \leq 10^{-11}$ eV^2, or at very small mixing angles $\sin^2 2\vartheta \leq 10^{-3}$, the effect on nucleosynthesis is negligible.

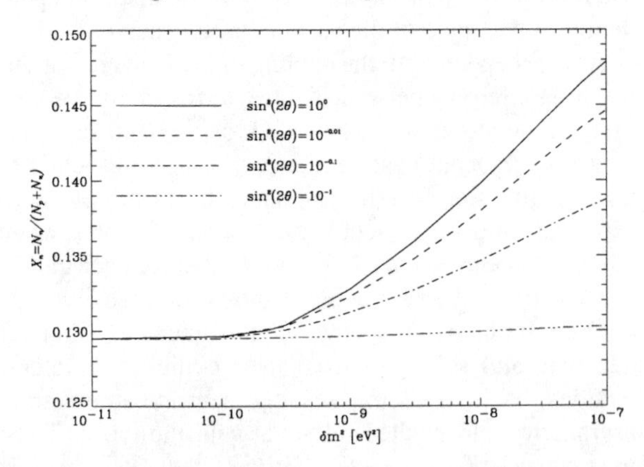

FIGURE 1. Neutron number density relative to nucleons as a function of the mass differences at different mixing angles.

The effect of oscillations in the non-resonant case is maximal at maximal mixing and greatest mass differences. In Fig. 2 (the lower curve) the maximal relative increase in the primordial ^{4}He as a function of neu-

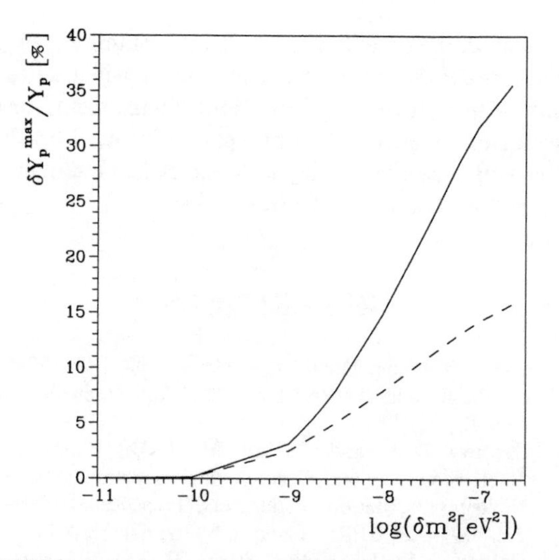

FIGURE 2. Maximum primordial helium-4 abundance for the resonant (upper curve) and the non-resonant oscillation case (lower curve), as a function of the neutrino mass differences. The non-resonant case is calculated at maximum mixing, while in the resonant case the helium abundance is calculated at the resonant mixing angle for the corresponding mass difference.

trino mass differences at maximal mixing: $\delta Y_p^{max}/Y_p = (Y_{osc}^{max} - Y_p)/Y_p = f(\delta m^2)_{|\theta=\pi/4}$ is presented.

In the resonant case for a given δm^2 there exists some resonant mixing angle, at which the oscillations are enhanced by the medium, and hence, the overproduction of ^{4}He is greater than that corresponding to the vacuum maximal mixing angle. This behavior of the helium production on the mixing angle is illustrated on the r.h.s. of Fig.3. The figure presents a combined plot (for the resonant and the non-resonant oscillation case) of δY_p dependence on the mixing angle for $\delta m^2 = 10^{-7}$ eV^2 and $\delta m^2 = 10^{-8}$ eV^2.

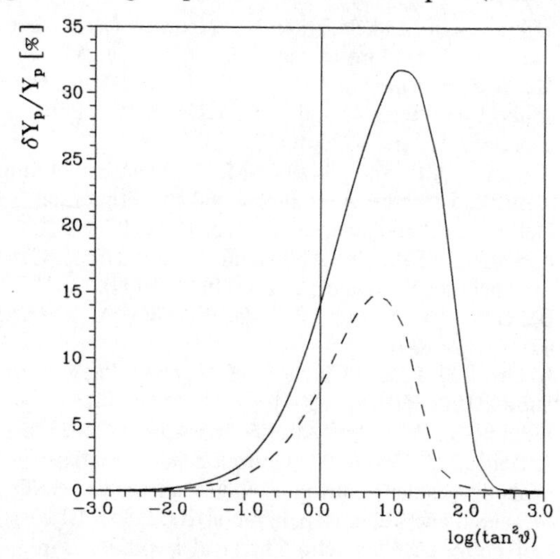

FIGURE 3. The dependence of the relative increase of primordial helium on the mixing angle for the resonant (r.h.s.) and non-resonant (l.h.s.) oscillation case. The upper curve corresponds to $\delta m^2 = 10^{-7}$ eV^2, the lower one to $\delta m^2 = 10^{-8}$ eV^2.

The upper curve in Fig.2 shows the maximal relative increase $\delta Y_p^{max}/Y_p$ in the resonant oscillations case as a function of mass differences, i.e. each maximum ^{4}He value corresponds to the resonant mixing angle for the concrete mass difference: $Y_p^{max}(\delta m^2, \vartheta_{\delta m}^{res})$. As can be seen from Figs.2 and 3, a considerable overproduction can be achieved: in the resonant case up to 32% and in the non-resonant one – up to 14%.

The kinetic effect of $\nu_e \leftrightarrow \nu_s$ neutrino oscillations comprises a major portion of the total effect - i.e. Y_p overproduction is mainly a result of electron neutrino depletion and spectrum distortion due to neutrino oscillations, which effect can be larger than the one corresponding to an additional degree of freedom.

Cosmological constraints on oscillation parameters

Observational data on primordial ^{4}He abundance limit the allowed oscillation parameters. Having in mind the large systematic uncertainty in the observational values for Y_p, we have calculated several iso-helium contours. The combined iso-helium contours for the non-resonant and the resonant case of electron–sterile oscillation parameters are shown in Fig. 4 corresponding to different values of relative increase of helium-4, namely $\delta Y_p = (Y_{osc} - Y_p)/Y_p = 3\%, 5\%, 7\%$.

Assuming the conventional observational bound on $\delta Y_p/Y_p = 3\%$, the cosmologically excluded region is situated above the 3% contour. The analytical fits to the

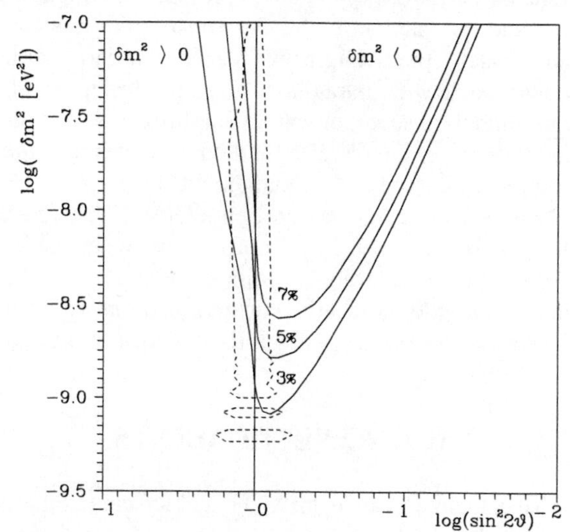

FIGURE 4. The iso-helium contours corresponding to 3%, 5% and 7% overproduction of primordial helium abundance. LOW sterile solar solution is given by the closed dashed curves.

exact constraints on oscillation parameters are:

$$\delta m^2 (\sin^2 2\vartheta)^4 \leq 1.5 \times 10^{-9} \mathrm{eV}^2 \qquad \delta m^2 > 0,$$
$$|\delta m^2| < 8.2 \times 10^{-10} \mathrm{eV}^2 \qquad \delta m^2 < 0, \text{ large } \vartheta.$$

LOW electron–sterile solar solution, obtained from the analysis of the 1258 days SuperKamiokande experimental data [14], is shown in the figure by the closed dashed curves.

These constraints on $\nu_e \leftrightarrow \nu_s$ neutrino oscillations exclude the active–sterile LOW solution to the solar neutrino puzzle [15] in addition to the LMA solution, excluded in the pioneer works. Even in the case of very high primordial helium-4 $\delta Y_p/Y_p = 7\%$, sterile LOW solution is still partially excluded.

This result is in agreement with the recent global analyses [16],[14], [17] of the neutrino data from SuperKamiokande, SNO, GALLEX+GNO, SAGE and Chlorine experiments, which does not favor $\nu_e \leftrightarrow \nu_s$ LOW solution.

CONCLUSIONS

We have analyzed the primordial production of ^{4}He in the presence of $\nu_e \leftrightarrow \nu_s$ oscillations with small mass differences. Enormous overproduction of ^{4}He (up to 32%) can be obtained for some interval of oscillation parameters values. Iso-helium contours are calculated for $\delta Y_p/Y_p = 3\%, 5\%, 7\%$. The results are used to constrain neutrino oscillation parameters. The $\delta Y_p/Y_p < 3\%$ limit excludes almost completely the LOW electron-sterile solution to the solar neutrino problem, in addition to the excluded sterile LMA solution in previous investigations This analysis can be useful for constraining nonstandard models, predicting active-sterile neutrino oscillations (models with extra-dimensions, producing oscillations, models of supernova bursts employing oscillations, etc.) It can be of interest also for models of chemical evolution, as discussed by C. Chiappini [18].

New observational data on light element abundances D, ^{3}He, ^{4}He, ^{7}Li will help improve our understanding of the conditions during the primordial synthesis of elements, provide more strict constraints on new physics and better our knowledge about the chemical evolution.

ACKNOWLEDGMENTS

I wish to thank C. Chiappini, G. Gloeckler and E. Salerno for useful discussions at the Workshop. I appreciate the comments and suggestions of R. Wimmer and the referees concerning the paper and I am also grateful to M. Chizhov for the overall help during its preparation.

I am obliged to the Organizing Committee of the SOHO-ACE Workshop, hosted by the Physikalisches Institut of the Universitat Bern, Bern, Switzerland, for the financial support of my participation in the Workshop. This work was finalized at the Abdus Salam International Centre for Theoretical Physics, Trieste.

REFERENCES

1. Izotov, Y. I., and Thuan, T. X., *Ap. J.*, **500**, 188 (1998).
2. Lopez, R., and Turner, M. S., *Phys. Rev. D*, **59**, 103502 (1999).
3. Kirilova, D. P., and Chizhov, M. V., Big Bang Nucleosynthesis and Cosmological Constraints on Neutrino Oscillations Parameters, Tech. Rep. CERN-TH/2001-020, CERN, Geneva, Switzerland (2001).
4. Dolgov, A. D., *Sov. J. Nucl. Phys.*, **33**, 700 (1981).
5. Fargion, D., and Shepkin, M., *Phys. Lett. B*, **146**, 46 (1984).
6. Kirilova, D. P., Neutrino Oscillations and Primordial Nucleosynthesis, Tech. Rep. JINR E2-88-301, Joint Institute for Nuclear Research, Dubna, Russia (1988).
7. Kirilova, D., and Chizhov, M., "Nonequilibrium Neutrino Oscillations and Primordial Helium Production", in *17 International Conference on Neutrino Physics and Astrophysics, NEUTRINO 96*, edited by K. Enqvist, K. Huitu, and J. Maalampi, World Scientific, Helsinki, 1996, pp. 478–484.
8. Kirilova, D. P., and Chizhov, M. V., *Phys. Lett. B*, **393**, 375 (1997).
9. Kirilova, D. P., and Chizhov, M. V., *Nucl. Phys. B*, **591**, 457–468 (2000).
10. Kirilova, D. P., and Chizhov, M. V., *Phys. Rev. D*, **58**, 073004 (1998).
11. Barbieri, R., and Dolgov, A., *Phys. Lett. B*, **237**, 440 (1990).
12. Mikheyev, S., and Smirnov, A., "Neutrino Oscillations in Matter with Varying Density", in *VI Moriond Meeting on Massive Neutrinos in Particle Physics and Astrophyisics*, edited by O. Fackler and J. Tran Thanh Van, Editions Frontiers, Tignes, 1986, pp. 355–372.
13. Kirilova, D. P., and Chizhov, M. V., Neutrino-Mixing Generated Lepton Asymmetry and the Primordial Helium-4 Abundance, Tech. Rep. ICTP IC/99/112; hep-ph/9908525, The Abdus Salam International Centre for Theoretical Physics, Trieste, Italy (1999).
14. Bahcall, J. N., Krastev, P. I., and Smirnov, A. Y., *JHEP*, **0105**, 015 (2001).
15. Kirilova, D. P., and Chizhov, M. V., *Nucl. Phys. B Proc. Suppl.*, **100**, 360–362 (2001).
16. Fukuda, S., *Phys.Rev.Lett.*, **86**, 5656–5660 (2001).
17. J. Bahcall, C. P.-G., M. Gonzalez-Garsia, Global analysis of solar neutrino oscillations including SNO CC measurement, Tech. Rep. hep-ph/0106258 (2001).
18. Chiappini, C., "Galactic Chemical Evolution", in *Joint SOHO-ACE Workshop on Solar and Galactic Composition*, AIP Conference Proceedings, Bern, 2001.

The Electron Temperature and ^{44}Ti Decay Rate in Cassiopeia A

J. Martin Laming

Naval Research Laboratory, Code 7674L, Washington DC 20375, USA

Abstract. The effects of plasma elemental composition and ionization state on the effective decay rate of ^{44}Ti are investigated. We essentially follow the methods of the first authors to treat this topic, Mochizuki et al., but use more realistic plasma models, including radiative cooling, to compute the evolution of the charge state distribution behind the reverse shock. For uniform density ejecta (i.e. no clumps or bubbles) we find a negligible change to the decay rate of ^{44}Ti. We discuss the effects of non-uniform ejecta. We also briefly consider the effects on these calculations of collisionless electron heating associated with weak secondary shocks propagating throughout the Cas A shell as a result of foward or reverse shock encounters with density inhomogeneities, recently suggested as an explanation for the hard X-ray tail seen in BeppoSAX and RXTE/OSSE spectra.

INTRODUCTION

Radioactive nuclei in the galaxy are mainly produced in supernova explosions. Among the most important of these is ^{44}Ti, which after ^{56}Ni and ^{56}Co is the main energy source for the ejecta. Its abundance is also sensitive to details of the explosion. The observation of γ rays from decay products of ^{44}Ti (i.e. ^{44}Sc to which ^{44}Ti decays with a lifetime of 85.4 ± 0.9 years [1, 2] and ^{44}Ca, to which ^{44}Sc decays with a lifetime of a few hours) in the Cassiopeia A supernova remnant by the COMPTEL instrument on the Compton Gamma Ray Observatory [3, 4] has sparked a re-examination of some of these ideas. Cassiopeia A is the youngest known supernova remnant in the galaxy, with a likely explosion date of 1680 A.D., and so is a good target to search for emission from the decay products of ^{44}Ti, since its lifetime is a significant fraction of the age of the remnant. The flux in the 1157 keV line of ^{44}Ca of $4.8 \pm 0.9 \times 10^{-5}$ photons $s^{-1}cm^{-2}$ implies an initial mass of ^{44}Ti of 2.6×10^{-4} $M_{\odot}$[3, 4]. The fluxes in the 67.9 and 78.4 keV lines of ^{44}Sc observed by OSSE and BeppoSAX are consistent with lower values of the initial ^{44} Ti mass synthesized in the explosion [5]. Neglecting the contribution of the continuum (a questionable assumption) the observed flux is $2.9 \pm 1.0 \times 10^{-5}$ photons s^{-1} cm^{-2}, marginally consistent with the ^{44}Ca flux.

The estimates of the ^{44}Ti mass based on the ^{44}Ca flux are at the high end of the range suggested by theory. Thielemann et al. [6] predict a ^{44}Ti mass of $1.7 \times 10^{-4}M_{\odot}$ for a $20M_{\odot}$ progenitor, whereas other workers [7, 8, 9] predict less than $10^{-4}M_{\odot}$. Since ^{44}Ti is produced only in the α rich freeze out (i.e. Si burning at low density, so that reactions involving α particles have negligible rates [10]), Nagataki et al. [11] speculated that the yield of ^{44}Ti could be increased if the explosion were axisymmetric rather than spherically symmetric, due to the existence for example of rotation or magnetic fields. The reason for this is that in the reduced symmetry, α rich freeze out occurs at higher entropy in the polar regions and hence produces more ^{44}Ti relative to ^{56}Ni. An aspherical explosion is also quite consistent with the morphology observed today in the remnant [12].

An equally ingenious solution was that of [13] who suggested that ^{44}Ti, if sufficiently highly ionized behind the reverse shock (see below), would have its decay rate to ^{44}Sc reduced. This occurs because ^{44}Ti decays mainly by capture of a K-shell electron, and if ionized to the hydrogenic or bare charge state, i.e. if the K shell electrons are removed, the decay rate is reduced. In this paper we revisit the work of [13] using some more recent ideas about the electron heating in the Cas A ejecta to investigate the effect on the ^{44}Ti decay rate and the inferred mass of ^{44}Ti.

THE IONIZATION STATE OF THE CASSIOPEIA A EJECTA

Following a supernova explosion, a spherical shock wave moves out through the medium surrounding the progenitor (a presupernova stellar wind in the case of Cas A). As this shock sweeps up more and more mass from the sur-

CP598, *Solar and Galactic Composition,* edited by R. F. Wimmer-Schweingruber
2001 American Institute of Physics 0-7354-0042-3

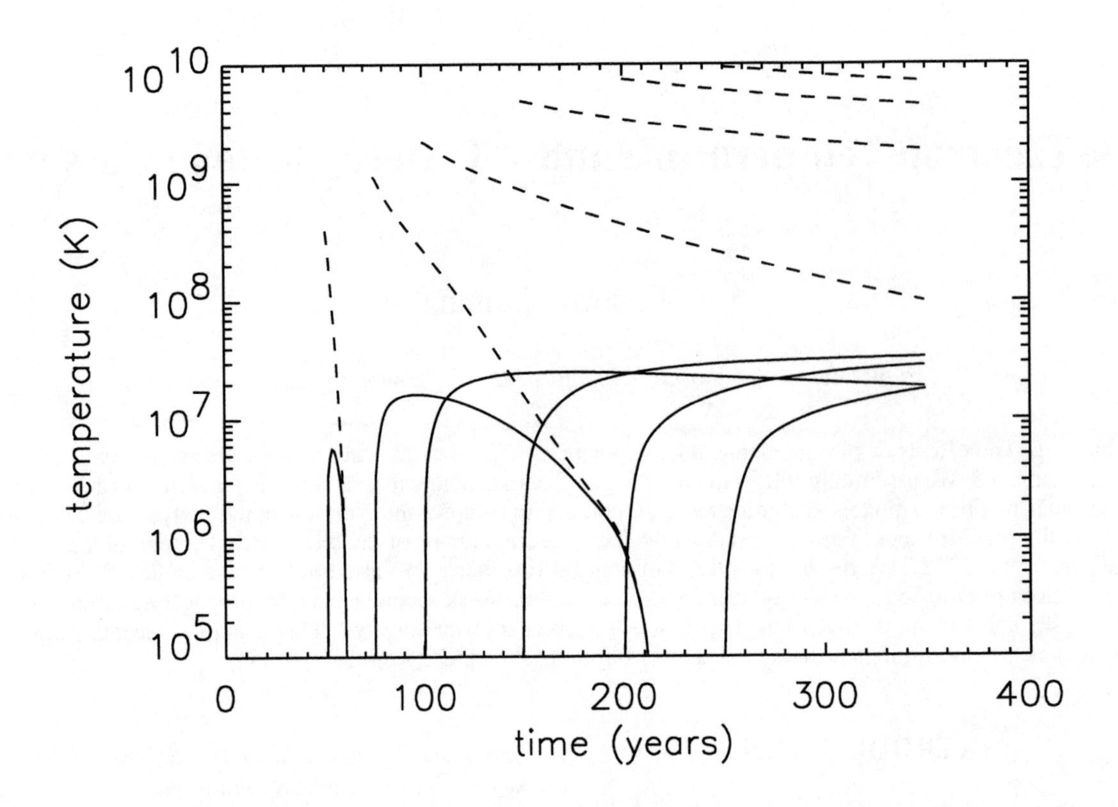

FIGURE 1. Electron (solid lines) and ion (dashed lines) temperatures in pure Fe ejecta passing through the reverse shock at 50, 75, 100, 150, 200, and 250 years after the initial explosion. The ejecta shocked at 50 years undergoes radiative cooling back to temperature below 10^5 K within about 10 years, ejecta shocked at 75 years takes 130 years to cool.

rounding medium, it slows down. This causes a reverse shock to develop, as freely expanding ejecta runs into more slowly expanding shocked ambient plasma. This reverse shock moves inwards in a Lagrangain coordinate system expanding with the plasma, and is responsible for heating the stellar ejecta up to X-ray emitting temperatures. In investigations of the element abundances produced by supernova explosions, the reverse shock is where we focus our attention.

The ionization state of the Cas A ejecta following passage of the reverse shock is computed using the formalism described in [14]. Assuming a total ejecta mass of $4M\odot$, an explosion energy of 2×10^{51} ergs and a circumstellar medium density of 3 hydrogen atoms cm^{-3}, the reverse shock velocity as a function of elapsed time since the explosion is taken from the analytical expressions in [15]. Ionization and recombination rates are taken from [16] (using subroutines kindly supplied by Dr P. Mazzotta). Radiative cooling is taken from [17]. Further cooling comes from the adiabatic expansion of the ejecta. A well known problem is that the shocks in supernova remnants are collisionless. Conservation of energy, momentum, and particle number results in shock jump conditions that predict shocked particle temperatures proportional to their masses, i.e. $T_{ion}/T_{electron} = m_{ion}/m_{electron} \simeq 1836 \times A_{ion}$ where A_{ion} is the atomic mass. These temperatures will equilibrate by Coulomb collisions, but the possibility remains that collective effects may induce faster equilibration. For the time being we neglect this possibility.

Figure 1 shows the electron and ion temperatures in pure Fe ejecta, for plasma encountering the reverse shock at times 50, 75, 100, 150, 200, and 250 years after explosion. Ejecta shocked 100 or more years after explosion reaches electron temperatures of $2-3 \times 10^7$ K, where appreciable ionization of ^{44}Ti to the H-like or bare charge states may occur. Ejecta shocked after 75 years reaches an electron temperature of 2×10^7 K briefly, but cools by radiation and adiabatic expansion to temperatures an order of magnitude lower at the present day. Ejecta shocked even earlier than this at 50 years never reaches electron temperatures above 10^7 K, and undergoes a thermal instability, cooling catastrophically to temperatures below 10^5 K within 15 years or so. Mochizuki et al.

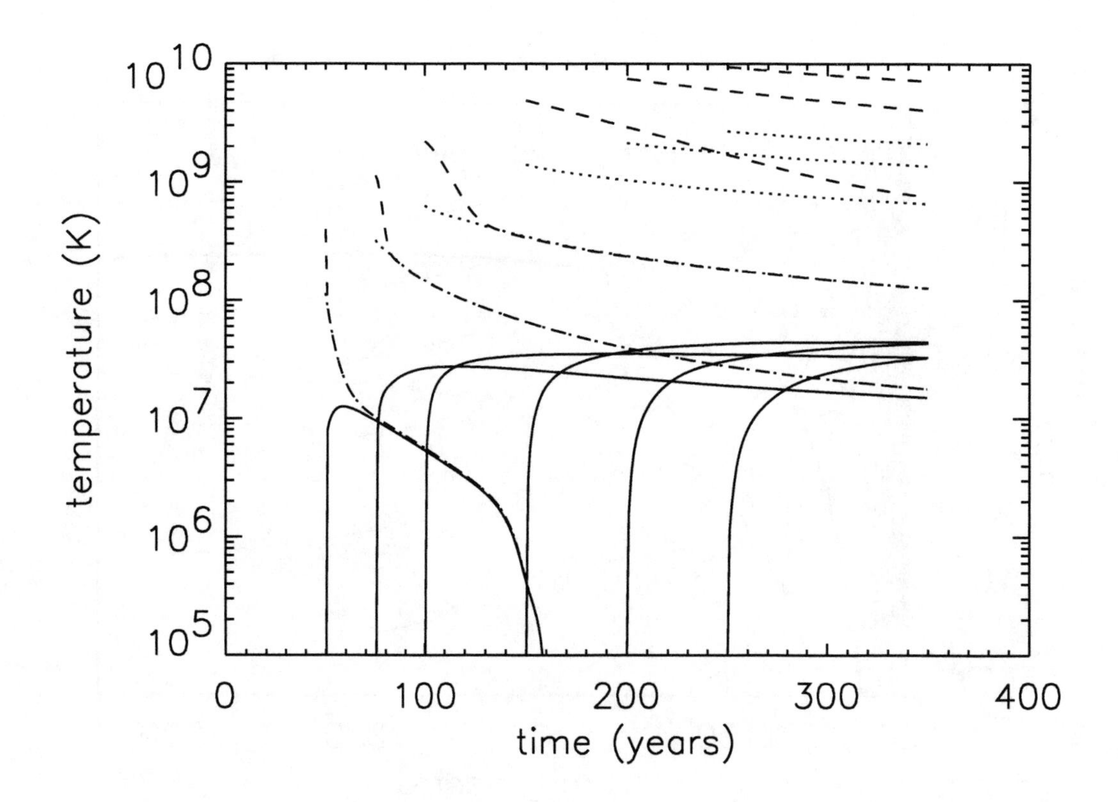

FIGURE 2. Electron (solid line), Fe ion (dashed line) and O ion (dotted line) temperatures in 90% O, 10% Fe (by mass) ejecta.

[13] only included Fe charge states from Ne-like to bare in their calculation (we use the full range of 27 charge states, starting everything off in the Fe^{+} state), and neglected to include radiation and so do not find the thermal instability. Figure 2 shows similar plots but this time of ejecta composed of 90% O and 10% Fe (by mass). Slightly lower electron temperatures are found in the plasma shocked later in the evolution of the remnant, but the reduced radiative cooling of this mixture gives higher electron temperatures in the ejecta shocked at 50-75 years. The temperatures here are very similar to those in pure O plotted in [14].

^{44}TI DECAY RATE AND RADIOACTIVITY

We now compute the ionization balance of trace amounts ^{44}Ti assumed to be expanding with the Fe ejecta, using the previously determined electron temperature and density profiles. ^{44}Ti decays by orbital electron capture, and the rate is proportional to the electron probability density at the nucleus. Thus only *s* states contribute to the decay rate, and because the electron probability density at the nucleus varies as n^{-3}, the $n = 1$ shell is the dominant contribution. In the following we assume that the $n = 1$ shell is the *only* contribution. Thus all charge states up to He-like will decay with the neutral decay rate, H-like will decay with half this rate and the bare charge state will not decay at all. Figure 3 shows the evolution of the Ti ionization balance following the reverse shock encounter 100 years after explosion. After a further 220 years, most of the Ti is in the He-like charge state (the dotted line), but the smaller fraction that has accumulated in the H-like (solid line) and bare (dashed line) charge states is sufficient to increase the amount of Ti by a factor of 1.025 over that determined from the neutral decay rate alone. The current emission in ^{44}Sc and ^{44}Ca is enhanced by a factor 1.009. Results for various reverse shock encounter times are given in Table 1 for pure Fe ejecta, and in Table 2 for mixed 90% O, 10% Fe (by mass) ejecta. Mochizuki et al. [13] find a maximum enhancement of ^{44}Ti radioactivity of about 2.5 320 years after explosion for ejecta at mass coordinate $q = 0.5 - 0.6$. We find our maximum enhancement at the same mass coordinate, but only a factor of about 1.03 in pure Fe ejecta. Mochizuki et al. take similar models to us, but in seek-

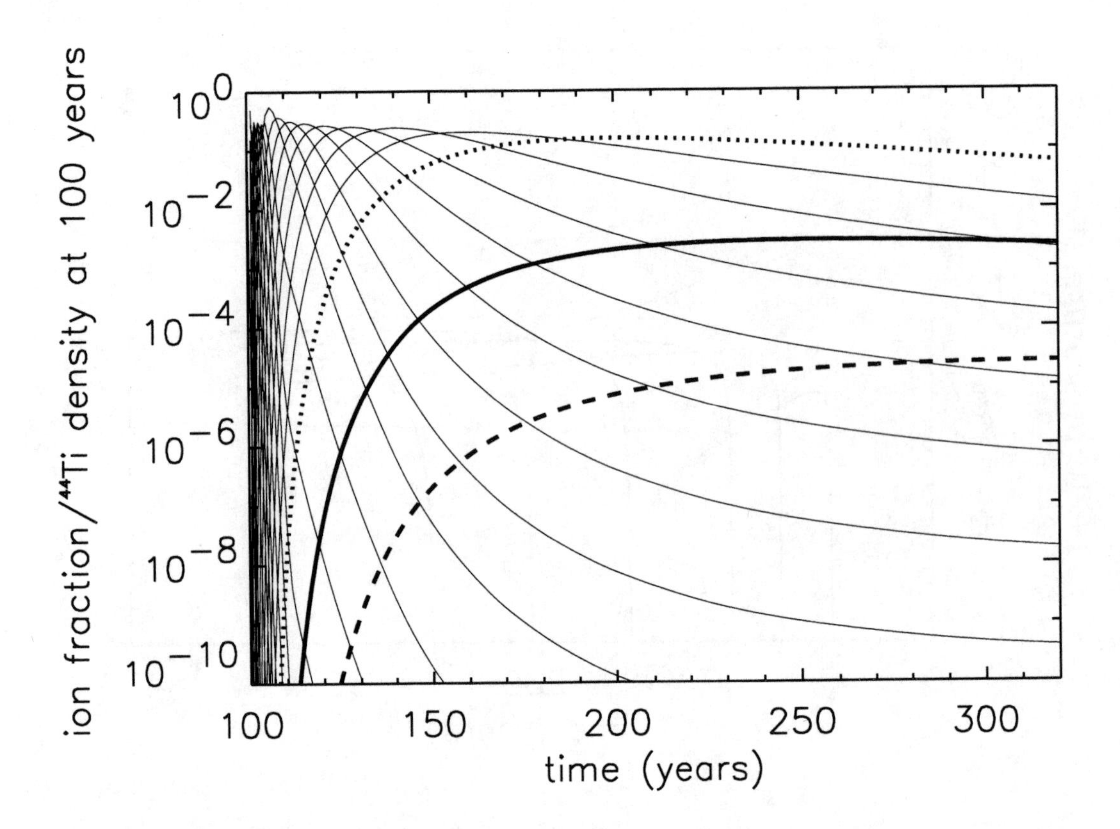

FIGURE 3. Ionization balance of ^{44}Ti in pure Fe ejecta following reverse shock at 100 years. The He-like charge state is shown as the thick dotted line, the H-like as the thick solid line and the bare charge state as the thick dashed line. The narrow solid lines represent all the lower stages of ionization. At the present day (320 years) the He-like charge state dominates the ionization balance.

TABLE 1. Enhancements in ^{44}Ti density and ^{44}Sc and ^{44}Ca emission due to suppression of the ^{44}Ti decay rate by plasma ionization. ^{44}Ti assumed embedded in pure Fe ejecta, for various reverse shock times after explosion.

time (years)	mass coordinate	^{44}Ti density	^{44}Sc, ^{44}Ca emission
50	0.72	1.0	1.0
75	0.63	1.0	1.0
100	0.55	1.025	1.009
150	0.43	1.001	1.0
200	0.34	1.0	1.0
250	0.28	1.0	1.0

TABLE 2. Same as Table 1, but giving ^{44}Ti enhancements etc in mixed 90% O, 10% Fe (by mass) ejecta.

time (years)	mass coordinate	^{44}Ti density	^{44}Sc, ^{44}Ca emission
50	0.72	1.0	1.0
75	0.63	1.11	1.04
100	0.55	1.09	0.995
150	0.43	1.01	1.002
200	0.34	1.0	1.0
250	0.28	1.0	1.0

ing to match a present day blast wave velocity of ~ 2000 km s^{-1} inferred in [18], adopted a circumstellar medium density much higher than ours. We took a lower circumstellar medium density to match the blast wave velocity of ~ 5000 km s^{-1} found by [19, 20]. The effect of this change is that the reverse shock in our model accelerates more slowly, generally producing lower temperatures in the shocked ejecta, with correspondingly lower degrees of ionization for the ^{44}Ti. Mochizuki et al. also assumed the Fe ejecta to expand in overdense (by a factor ~ 10) clumps, which reduces the reverse shock velocity and temperature somewhat, but increases the electron density and hence the ionization time, defined as the product of the electron density and the elapsed time since shock passage, assuming an electron temperature constant with time during this period.

FURTHER ELECTRON HEATING?

The fundamental reason for not obtaining the same suppresion of the ^{44}Ti decay rate as Mochizuki et al.[13] is that the electron temperatures in our probably more realistic model for Cas A do not reach high enough temperatures. Here we will briefly discuss two possibilities so far neglected.

First, [13] assumed the Fe to exist in clumps a factor of 10 more dense than the surrounding ejecta. In connection with Type Ia supernova, [21] argue that the Fe in such clumps is unlikely to have a radiogenic origin, and is probably ^{54}Fe. The reason is that the likely mechanisms to form clumps are hydrodynamic instabilities following Ni-Co-Fe bubble formation as a result of the extra pressure resulting from energy input mainly due to the Ni radioactivity in the first 10 days following explosion. Evidence for such bubbles is found in SN 1987A [22], and some modeling has already been performed [23]. The ^{44}Ti is likely to reside in these bubbles, being formed in the same regions as ^{56}Ni. If these ejecta bubbles persist into the remnant phase, then the reverse shock will accelerate into them due to their lower density. Higher temperatures in the shocked plasma will result, with correspondingly higher degrees of ionization and possibly an effect on the ^{44}Ti decay rate.

For the time being it is not clear where in Cas A this Fe ejecta exists. We expect it to be synthesized in the innermost regions of the progenitor, and so naively assuming spherical symmetry we should expect the Fe to pass through the reverse shock relatively late in the evolution of the remnant. However in the SE quadrant of the remnant, (the region most extensively studied in Chandra data to date) the Fe rich knots are remarkable for being *exterior* to those rich in Si and O [24, 25, 26]. These Fe rich knots also have high ionization times, implying high electron density and/or early encounter with the reverse shock. The high density also possibly suggests that this Fe is not associated with ^{44}Ti. However the Fe rich region gives mass ratios in the following ranges; Fe/O = 0.2 - 0.4, Si/O = 0.03 -0.06 [27]. Fe/O is consistent with or slightly higher than with solar system values, while Si/O is lower. Earlier X-ray observations, essentially spatially unresolved, of large regions of the remnant find Si/O, S/O, Ar/O broadly consistent with solar system values, while Fe/O is significantly lower [28, 29].

The second possibility is that the electrons are heated following shock passage by processes that are faster than the ion-electron Coulomb equilibration. Most discussion of this in the literature has focused on plasma wave excited by instabilities at the forward shock [see e.g. 30, 31, 32]. At the reverse shock most electrons are bound to heavy ions, and so cannot participate in collisionless heating [33]. However it is important to realize that most of the ejecta emission comes from plasma that passed through the reverse shock in the first 75 years or so after explosion. From Figures 1 and 2 it is apparent that the temperature of these ejecta at the present day is rather lower than the observed value of $\sim 4 \times 10^7$ K, spectacularly so for the ejecta shocked at around 50 years. This is due to the combined effects of radiation losses and adiabatic expansion. Additionally, this plasma has reached ion-electron temperature equilibrium, and so the inclusion of extra collisionless processes at the reverse shock itself, however unlikely, would not produce any change to this conclusion.

Recently it has been suggested that this ejecta is reheated by weak secondary shocks produced as the forward or reverse shocks run into density inhomogeneities producing reflected and transmitted shocks [34, 14]. The precise mechanism is a modified two stream instability arising as upstream ions are reflected from the shock front, move back through the preshock plasma and excite lower-hybrid waves as they go. These waves are electrostatic ion oscillations with wavevectors at angle nearly perpendicular to the magnetic field direction. Thus the wave phase velocity perpendicular to the magnetic field can resonate with the ions, and the wave phase velocity parallel to the field can resonate with the electrons, allowing fast energy transfer between the two [35, 36]. Such a mechanism is particularly appealing in Cas A. Arguing from the radio synchrotron luminosity, which is maximized when the cosmic ray electrons and magnetic field have similar energy densities, leads to the inference of magnetic field strengths of order 1 mG; significantly higher than usually assumed for supernova remnants. The electrons are accelerated into a characteristic distribution function along the magnetic field. Calculations of the bremsstrahlung spectrum emitted by such an accelerated electron distribution produced a remarkably good match to the BeppoSAX MECS and PDS spectra [see 34, for fuller description]. In fact the computed continuum spectrum for certain parameters (basically the Alfvén speed in the ejecta) can be consistent with the seemingly extreme assumption made in [5] of neglecting the continuum under the ^{44}Sc emission to derive a flux of ^{44}Sc decay photons consistent with the observation of the ^{44}Ca flux. This accelerated electron distribution equilibrates with the ambient plasma electrons by Coulomb collisions. The extra heating thus produced would increase the ionization of the plasma above that modeled in this paper, and could plausibly produce more ^{44}Ti in H-like or bare charge states, hence further suppressing its decay. However in order to avoid heating the ejecta of Cas A to temperatures significantly higher than those observed, such electron acceleration must have commenced relatively recently (i.e. within 50-100 years of the present day), if it is also responsible for the hard X-ray emission.

CONCLUSIONS

Cassiopeia A is an exciting physics laboratory for a wide variety of phenomena; particle heating and acceleration, magnetic field amplification, hydrodynamic instabilities and of course stellar nucleosynthesis. Initial ideas to study the element abundances produced in a core-collapse supernova can often end up with arguments based on observed element abundances to infer some other characteristic of the supernova or its remnant. Examples of this are provided in this paper by the plasma processes that may effect the decay rate of ^{44}Ti. We find significantly less change to the decay rate produced by the plasma ionization than previous work [13], but our model for the ejecta, while arguably more realistic, is still probably a long way from reality itself.

X-ray astronomers have only just begun to get to grips with the wealth of new data on Cassiopeia A and other supernova remnants, and we can expect many exciting new results as this effort matures. This, coupled with hard X-ray ray and γ-ray observations with the forthcoming INTEGRAL mission, might finally allow to unambiguously infer exactly how much ^{44}Ti was produced in the explosion of Cassiopeia A.

ACKNOWLEDGMENTS

I acknowledge illuminating discussions and correspondence with Una Hwang and Jacco Vink regarding the X-ray observations of Cas A. This work was supported by basic research funds of the Office of Naval Research.

REFERENCES

1. Ahmad, I., Bonino, G., Castagnoli, G. C., Fischer, S. M., Kutschera, W., and Paul, M., *Phys. Rev. Lett.*, **80**, 2550 (2000).
2. Görres, J., Meissner, J., Schatz, H., Stech, E., Tischauser, P., Wiescher, M., Bazin, D., Harkewicz, R., Hellstrom, M., Sherrill, B., Steiner, M., Boyd, R. N., Buchmann, L., Hartmann, D. H., and Hinnefeld, J. D., *Phys. Rev. Lett.*, **80**, 2554 (2000).
3. Iyudin, A. F., Diehl, R., Bloemen, H., Hermsen, W., Lichti, G. G., Morris, D., Ryan, J., Schönfelder, V., Steinle, H., Varendorff, M., de Vries, C., and Winkler, C., *Astron. Astrophys,*, **284**, L1 (1994).
4. Iyudin, A. F., Diehl, R., Lichti, G. G., and et al., *ESA*, **Sp-382**, 37 (1997).
5. Vink, J., Kaastra, J. S., Bleeker, J. A. M., and Bloemen, H., *Advances in Space Research*, **25**, 689 (2000).
6. Thielemann, F. K., Nomoto, K., and Hashimoto, M., *Astrophys. J.*, **460**, 408 (1996).
7. Woosley, S. E., and Weaver, T. A., *Astrophys. J.*, **101**, 181 (1995).
8. Woosley, S. E., Langer, N., and Weaver, T. A., *Astrophys. J.*, **448**, 315 (1995).
9. Timmes, F. X., Woosley, S. E., Hartmann, D. H., and Hoffman, R. D., *Astrophys. J.*, **464**, 332 (1996).
10. Arnett, D., *Supernovae and Nucleosynthesis*, Princeton University Press, 1996.
11. Nagataki, S., Hashimoto, M., Sato, K., Yamada, S., and Mochizuki, Y., *Astrophys. J.*, **492**, L45 (1998).
12. Fesen, R. A., *Astrophys. J. Supp.*, **133**, 161 (2001).
13. Mochizuki, Y., Takahashi, K., Janka, H.-T., Hillebrandt, W., and Diehl, R., *Astron. Astrophys.*, **346**, 831 (1999).
14. Laming, J. M., *Astrophys. J. submitted* (2001).
15. Truelove, J. K., and McKee, C. F., *Astrophys. J. Supp.*, **120**, 299 (1999).
16. Mazzotta, P., Mazzitelli, G., Colafrancesco, S., and Vittorio, N., *Astron. Astrophys. Supp.*, **133**, 403 (1998).
17. Summers, H. P., and McWhirter, R. W. P., *J. Phys. B*, **12**, 2387 (1979).
18. Borkowski, K. J., Szymkowiak, A. E., Blondin, J. M., and Sarazin, C. L., *Astrophys. J.*, **466**, 866 (1996).
19. Vink, J., Bloemen, H., Kaastra, J. S., and Bleeker, J. A. M., *Astron. Astrophys.*, **339**, 201 (1998).
20. Koralesky, B., Rudnick, L., Gotthelf, E. V., and Keohane, J. V., *Astrophys. J.*, **505**, L27 (1998).
21. Wang, C.-Y., and Chevalier, R. A., *Astrophys. J.*, **549**, 1119 (2001).
22. Li, H., McCray, R., and Sunyaev, R. A., *Astrophys. J.*, **419**, 824 (1993).
23. Borkowski, K. J., Blondin, J. M., Lyerly, W. J., and Reynolds, S. P., "Remnants of Core-Collapse SNe: Hydrodynamical Simulations with Fe-Ni Bubbles", in *Cosmic Explosions*, edited by S. S. Holt and W. W. Zhang, AIP Conference Proceedings 522, American Institute of Physics, New York, 2000, p. 173.
24. Hughes, J. P., Rakowski, C. E., Burrows, D. N., and Slane, P. O., *Astrophys. J.*, **528**, L109 (2000).
25. Hwang, U., Holt, S. S., and R., R. P., *Astrophys. J.*, **537**, L119 (2000).
26. Hwang, U., "X-ray Observations of Young Supernova Remnants", in *Young Supernova Remnants*, edited by S. S. Holt and U. Hwang, AIP Conference Proceedings in press, American Institute of Physics, New York, 2001.
27. Hwang, U., *private communication* (2001).
28. Vink, J., Kaastra, J. S., and Bleeker, J. A. M., *Astron. Astrophys.*, **307**, L41 (1996).
29. Favata, F., , Vink, J., del Fiume, D., Parmar, A. N., Santangelo, A., Mineo, T., Preite-Martinez, A., Kaastra, J. S., and Bleeker, J. A. M., *Astron. Astrophys.*, **324**, L49 (1997).
30. Cargill, P. J., and Papadopoulos, K., *Astrophys. J.*, **329**, L29 (1988).
31. Bykov, A. M., and Uvarov, Y. A., *JETP*, **88**, 465 (1999).
32. Dieckmann, M. E., McClements, K. G., Chapman, S. C., Dendy, R. O., and Drury, L. O., *Astron. Astrophys.*, **356**, 377 (2000).
33. Hamilton, A. J. S., and Sarazin, C. L., *Astrophys. J.*, **287**, 282 (1984).
34. Laming, J. M., *Astrophys. J.*, **546**, 1149 (2001).
35. Bingham, R., Dawson, J. M., Shapiro, V. D., Mendis, D. A., and Kellett, B. J., *Science*, **275**, 49 (1997).
36. Shapiro, V. D., Bingham, R., Dawson, J. M., Dobe, Z., Kellett, B. J., and Mendis, D. A., *J. Geophys. Res.*, **104**, 2537 (1999).

List of Participants

Lucia Abbo
Osserv. Astronom. di Torino
Via Osservatorio 20
10025 Torino
ITALY
abbo@to.astro.it

Frédéric Allegrini
Physikalisches Institut
Universität Bern
Sidlerstrasse 5
CH-3012 Bern
SWITZERLAND
frederic.allegrini@phim.unibe.ch

Kathrin Altwegg
Physikalisches Institut
Universität Bern
Sidlerstrasse 5
CH-3012 Bern
SWITZERLAND
kathrin.altwegg@phim.unibe.ch

Ester Antonucci
Osserv. Astron. di Torino
Via Osservatorio 20
10025 Torino
ITALY
antonucci@to.astro.it

Karin Bamert
Physikalisches Institut
Universität Bern
Sidlerstrasse 5
CH-3012 Bern
SWITZERLAND
karin.bamert@soho.unibe.ch

Robert Binns
Washington Univ.
Dept. of Physics
St. Louis
MO 63130
USA
wrb@howdy.wustl.edu

Peter Bochsler
Physikalisches Institut
Universität Bern
Sidlerstr. 5
CH-3012 Bern
SWITZERLAND
peter.bochsler@soho.unibe.ch

Volker Bothmer
MPI für Aeronomie
Max-Planck-Str. 2
D-37191 Katlenburg-Lindau
GERMANY
bothmer@kernphysik.uni-kiel.de

Frédéric Buclin
Physikalisches Institut
Universität Bern
Sidlerstrasse 5
CH-3012 Bern
SWITZERLAND
frederic.buclin@phim.unibe.ch

Fritz Bühler
Physikalisches Institut
Universität Bern
Sidlerstrasse 5
CH-3012 Bern
SWITZERLAND
fritz.buehler@phim.unibe.ch

Henner Busemann
Physikalisches Institut
Universität Bern
Sidlerstrasse 5
CH-3012 Bern
SWITZERLAND
henner.busemann@phim.unibe.ch

Marc Chaussidon
CRPG-CNRS
54501, Vandoeuvre-les-Nancy
FRANCE
Chocho@crpg.cnrs-nancy.fr

Cristina Chiappini
Osserv. Astron. di Trieste
Via G. B. Tiepolo, 11
I-34131 Trieste- TS
ITALY
chiappini@ts.astro.it

Eric Christian
NASA/GSFC/USRA
Code 661
Greenbelt, MD 20771
USA
erc@cosmicra.gsfc.nasa.gov

Christina Cohen
Caltech, MC 220-47
Pasadena, CA 91125
USA
cohen@srl.caltech.edu

Alan C. Cummings
Caltech, MC 220-47
Pasadena, CA 91125
USA
ace@srl.caltech.edu

Guilio Del Zanna
DAMTP
Univ. of Cambridge
Silver Street
CB3 9EW
UNITED KINGDOM
G.Del-Zanna@damtp.cam.ac.u

Mitsue Den
Communications Research Lab
3601 Isozaki
311-1202 Ibaraki Hitachinaka
JAPAN
den@crl.go.jp

Mihir Desai
Univ. of Maryland
Dept. of Physics
College Park, MD 20742-4111
USA
desai@umtof.umd.edu

Peter Eberhardt
Physikalisches Institut
Universität Bern
Sidlerstrasse 5
CH-3012 Bern
SWITZERLAND
peter.eberhardt@phim.unibe.ch

Bernard Fleck
ESA Space Sci. Dept.
NASA/GSFC, MC 682.3
Greenbelt, MD 20015
USA
bfleck@esa.nascom.nasa.gov

Jeff George
Caltech, MC- 220-47
Pasadena, CA 91125
USA
george@srl.caltech.edu

Silvio Giordano
Osserv. Astron. di Torino
Via Osservatorio 20
10025 Torino
ITALY
giordano@to.astro.it

George Gloeckler
Univ. of Maryland
Dept. of Physics
College Park
MD 20742-4111
USA
gloeckler@pop.umail.umd.edu

Peter Grieder
Physikalisches Institut
Universität Bern
Sidlerstrasse 5
CH-3012 Bern
SWITZERLAND
peter.grieder@phim. unibe.ch

Veronika Heber
Isotopengeol. &
Mineralogische Rohstoffe
ETH Zürich
CH-8092 Zürich
SWITZERLAND
heber@erdw.ethz.ch

George Ho
Johns Hopkins Univ., APL
11100 Johns Hopkins Rd.
Laurel, MD 20723
USA
george.ho@jhuapl.edu

Mirjam Hofer
ESA/ESTEC
Keperlaan 1
Postbus 299
2200 AG Noordwijk
THE NETHERLANDS
mhofer@so.estec.esa.nl

Markus Hohl
Physikalisches Institut
Universität Bern
Sidlerstr. 5
CH-3012 Bern
SWITZERLAND
markus.hohl@phim.unibe.ch

Hartmut Holweger
Institut f. Theo. Physik &
Astrophysik
Universität Kiel
D-24118 Kiel
GERMANY
holweger@astrophysik.uni-kiel.de

Peter Hoppe
MPI für Chemie
P.O. Box 3060
D-55020 Mainz
GERMANY
hoppe@mpch-mainz.mpg.de

Fred Ipavich
Univ. of Maryland
Dept. of Physics
College Park
MD 20742-4111
USA
ipavich@umtof.umd.edu

Karine Issautier
Physikalisches Institut
Universität Bern
Sidlerstrasse 5
CH-3012 Bern
SWITZERLAND
karine.issautier@soho.unibe.ch

Randy Jokipii
Univ. of Arizona
Dept. of Planetary Sciences
Bldg. 92
Tucson, AZ 85721
USA
jokipii@lpl.arizona.edu

Frank C. Jones
NASA/GSFC, Lab for High Energy Astrophysics
Greenbelt, MD 20771
USA
Frank.c.jones@gsfc.nasa.gov

Reinald Kallenbach
ISSI
Hallerstrasse 6
CH-3012 Bern
SWITZERLAND
rkallenbach@issi.unibe.ch

John Keller
NASA/GSFC, MC 691
Greenbelt, MD 20771
USA
john.w.keller.1@gsfc.nasa.gov

Hiroshi Kimura
Univ. of Münster
Institute of Planetology
D-48149 Münster
GERMANY
kimura@uni-muenster.de

Daniela Kirilova
Institute of Astronomy
Bulgarian Academy of Sciences
Blvd.Tsarigradsko Shosse 72
1000 Sofia
Sofia
BULGARIA
dani@libra.astro.bas.bg

Berndt Klecker
MPI f. extraterrestrische Physik
Postfach 1312
D-85741 Garching
GERMANY
bek@mpe.mpg.de

Yuan-Kuen Ko
NASA/GSFC
MC 682.3
Greenbelt, MD 20771
USA
kuen@uvcs6.nascom.nasa.gov

Leon Kocharov
Univ. of Turku
Dept. of Physics
Vesilinnantie 5
FIN-20014 Turku
FINLAND
kocharov@srl.utu.fi

Olga Kryakunova
Institute of Ionosphere
480020 Kazakhstan
Almaty
REPUBLIC OF KAZAKHSTAN
olga@ionos.alma-ata.su

Horst Kunow
IEAP
Universität Kiel
Leibnizstr. 11
D-24118 Kiel
GERMANY
kunow@physik.uni-kiel.de

Martin Laming
Naval Research Lab
MC 7674L
Washington, DC 20375
USA
jlaming@ssd5.nrl.navy.mil

Enrico Landi
Arcetri Astrophysical Obs.
Largo E. Fermi 5
50125 Florence
ITALY
enricol@arcetri.astro.it

Alan Lazarus
MIT, Room 37-687
77 Massachusetts Ave.
Cambridge, MA 02139
USA
ajl@space.mit.edu

Richard Leske
Caltech, MC 220-47
Pasadena, CA 91125
USA
ral@srl.caltech.edu

Yuri Litvinenko
Univ. of New Hampshire
Morse Hall
39 College Rd.
Durham, NH 03824-3525
USA
yuri.litvinenko@unh.edu

Oliver K. Manuel
Univ. of Missouri
Dept. of Chemistry
Rolla, MO 65401-0249
USA
oess@umr.edu

Glenn Mason
Univ. of Maryland
Dept. of Physics
College Park, MD 20742-4111
USA
glenn.mason@umail.umd.edu

Joe Mazur
Naval Research Lab
4555 Overlook Ave. S.
Washington, DC 20375
USA
Joseph.E.Mazur@aero.org

Richard Mewaldt
Caltech
MS 220-47 Downs Lab.
Pasadena, CA 91125
USA
rmewaldt@srl.caltech.edu

Eberhard Moebius
Univ. of New Hampshire
Morse Hall
39 College Rd.
Durham, NH 03824-3525
USA
eberhard.moebius@unh.edu

Keith Ogilvie
NASA/GSFC MC 692
Greenbelt, MD 20771
USA
keith.w.ogilvie.1@gsfc.nasa.gov

John Paquette
Univ. of Maryland
Dept. of Physics
College Park
MD 20742-4111
USA
paquette@umtof.umd.edu

Susanna Parenti
Arcetri Astrophysical Obs.
Largo E. Fermi 5
50125 Florence
ITALY
sparenti@arcetri.astro.it

Giannina Poletto
Arcetri Astrophysical Obs.
Largo E. Fermi 5
50125 Florence
ITALY
poletto@arcetri.astro.it

Mark Popecki
Univ. of New Hampshire
Morse Hall
39 College Rd.
Durham
NH 03824-3525
USA
mark.popecki@unh.edu

John Raymond
Center for Astrophysics
60 Garden Street
Cambridge, MA 02138
USA
jraymond@cfa.harvard.edu

Donald Reames
NASA/GSFC Code 661
Bldg. 2, Room 29
Greenbelt, MD 20771
USA
reames@milkyway.gsfc.nasa.gov

Alysha Reinard
AOSS Dept.
Space Research Bldg.
2455 Hayward Street
Ann Arbor, MI 48109
USA
alysha@umich.edu

Dan Reisenfeld
Los Alamos National Lab
NIS-1, Space & Atmospheric Sciences
Los Alamos, NM 87545
USA
dreisen@lanl.gov

Javier Rodriguez-Pacheco
Univ. of Alcala
Dept. of Physics
Apdo. 20
28871 Madrid Alcala de Henares
SPAIN
fsrodriguez@uah.es

Emma Salerno
Physikalisches Institut
Universität Bern
Sidlerstrasse 5
CH-3012 Bern
SWITZERLAND
emma.salerno@phim.unibe.ch

Juan Sequeiros-Ugarte
Univ. of Alcala
Dept. of Physics
Apdo. 20
28871 Madrid Alcala de Henares
SPAIN
juan.sequeiros@uah.es

Hironori Shimazu
Communications Research Lab
4-2-1 Nukukita
184-8795 Tokyo
Koganei
JAPAN
shimazu@crl.go.jp

Penny Slocum
Jet Propulsion Lab
MS 169 -327
4800 Oak Grove Ave.
Pasadena, CA 91106
USA
penny@heag1.jpl.nasa.gov

Ulysses Sofia
Whitman College
Dept. of Astronomy
Walla Walla, WA 99632
USA
sofiauj@whitman.edu

Daniele Spadoro
Osserv. Astrofisico di Catania
Cit. Universitat
Via S. Sofia 78
I-95125 Catania
ITALY
dsp@sunct.ct.astro.it

Steve Suess
NASA Marshall Space Flight Center
MS SD50
Huntsville, AL 35812
USA
steve.suess@msfc.nasa.gov

Luca Teriaca
Arcetri Astrophysical Obs.
Largo E. Fermi 5
50125 Florence
ITALY
lte@arcetri.astro.it

Friedrich Thielemann
Institut für Physik
Universität Basel
Klingelbergstrasse 82
CH-4056 Basel
SWITZERLAND
Fkt@quasar.physik.unibas.chh

Vladislav Timofeev
Institue of Cosmophysical Res. & Aeronomy
Lenin Ave. 31
Yakutsk 677891
RUSSIA
vetimofeev@ikfia.ysn.ru

Cecil Tranquille
ESA/ESTEC
Keperlaan 1
Postbus 299
2200 AG Noordwijk
THE NETHERLANDS
ctranqui@estec.esa.nl

Igor Veselovsky
Moscow State Aviation Technology Institute
Petrovka 27
103763 Moscow
RUSSIA
veselov@dec1.sinp.msu.ru

Jean-Claude Vial
Inst. d'Astrophysique Spatial I.A.S.
U. Paris-XI- CNRS, Bat. 121
91405 Orsay Cedex
FRANCE
vial@ias.fr

Tycho von Rosenvinge
NASA/GSFC MC 661
Greenbelt, MD 20771
USA
tycho@milkyway.gsfc.nasa.gov

Rudolf von Steiger
ISSI
Universität Bern
Hallerstrasse 6
CH-3012 Bern
SWITZERLAND
vsteiger@issi.unibe.ch

Peter Wenzel
ESA/ESTEC
Keperlaan 1
Postbus 299
2200 AG Noordwijk
THE NETHERLANDS
wenzel@estec.esa.nl

Mark Wiedenbeck
Caltech
JPL M/S 169-327
4800 Oak Grove Ave.
Pasadena, CA 91106
USA
mark.e.wiedenbeck@jpl.nasa.gov

Rainer Wieler
ETH Zürich
Isotopengeol. &
Mineralogische Rohstoffe
NO C61
ETH Zürich
CH-8092 Zürich
SWITZERLAND
wieler@erdw.ethz.ch

Roger C. Wiens
Los Alamos National Lab.
MS D466
Los Alamos, NM 87545
USA
rwiens@lanl.gov

Klaus Wilhelm
MPI für Aeronomie
Max-Planck-Str. 2
D-37191 Katlenburg-Lindau
GERMANY
Wilhelm@linmpi.mpg.de

Robert Wimmer-Schweingruber
Physikalisches Institut
Universität Bern
Sidlerstrasse 5
CH-3012 Bern
SWITZERLAND
robert.wimmer@phim.unibe.ch

Peter Wurz
Physikalisches Institut
Universität Bern
Sidlerstrasse 5
CH-3012 Bern
SWITZERLAND
peter.wurz@soho.unibe.ch

Nathan Yanasak
Caltech, MC 220-47
Pasadena, CA 91125
USA
yanasak@llif.srl.caltech.edu

Syed-Yahya Yusainee
Universiti Teknologi Mara
Lecture Bldg.
26400 Pahang Bandar Jengka
MALAYSIA
yusanis@tm.net.my

Luca Zangrilli
Dipartimento di Astronomia & Scienza dello Spazio
Largo E. Fermi 5
50125 Florence
ITALY
lz@arcetri.astro.it

Thomas Zurbuchen
Univ. of Michigan
2455 Hayward Street
Ann Arbor MI 48109-2143
USA
thomasz@umich.edu